# Phytophthora

## Its Biology, Taxonomy, Ecology, and Pathology

D. C. Erwin, S. Bartnicki-Garcia,
and P. H. Tsao, Editors

*Published by*
The American Phytopathological Society
St. Paul, Minnesota

The cover logo was designed for the *International Symposium on* Phytophthora: *Its Biology, Ecology and Pathology* by Mrs. Ildiko Bartnicki. It symbolizes two of the most distinctive morphological stages of the fungus.

Library of Congress Catalog Card Number: 82-73998
International Standard Book Number: 0-89054-050-0



Printed in the United States of America

The American Phytopathological Society
3340 Pilot Knob Road
St. Paul, Minnesota 55121, USA

# Contributors and Their Affiliations


| | |
|---|---|
| **Bartnicki-Garcia, Salomon** | Department of Plant Pathology, University of California, Riverside, California 92521, USA |
| **Brasier, Clive M.** | Pathology Branch, Forest Research Station, Alice Holt Lodge, Wrecclesham, Farnham, Surrey GU10 4LH, England, UK |
| **Broadbent, Patricia** | New South Wales Department of Agriculture, Biological and Chemical Research Institute, Private Mail Bag No. 10, Rydalmere, NSW 2116, Australia |
| **Canaday, Craig H.** | Department of Plant Pathology, The Ohio State University, Ohio Agricultural Research and Development Center, Wooster, Ohio 44691, USA |
| **Carlile, Michael J.** | Department of Pure and Applied Biology, Imperial College of Science and Technology, London SW7 2BB, England, UK |
| **Coffey, Michael D.** | Botany School, Trinity College, Dublin 2, Ireland. (Present address: Department of Plant Pathology, University of California, Riverside, California 92521, USA) |
| **Duniway, John M.** | Department of Plant Pathology, University of California, Davis, California 95616, USA |
| **Elliott, Charles G.** | Department of Botany, University of Glasgow, Glasgow G12 8QQ, Scotland, UK |
| **Elliott, Vern J.** | Department of Plant Pathology, The Pennsylvania State University, University Park, Pennsylvania 16802, USA |
| **Erjefält, Lennart** | Potato Department, Svalöf Seed Ltd. AB, Svalöv S-268 00, Sweden |
| **Erwin, Donald C.** | Department of Plant Pathology, University of California, Riverside, California 92521, USA |
| **Gallegly, Mannon E.** | Division of Plant and Soil Sciences, West Virginia University, Morgantown, West Virginia 26506, USA |
| **Gisi, Ulrich** | Institute of Botany, University of Basel, CH-4056 Basel, and SANDOZ Ltd., Basel, Switzerland |
| **Gregory, Philip H.** | Rothamsted Experimental Station, Harpenden, Hertfordshire AL5 2JQ, England |
| **Hemmes, Don E.** | Natural Sciences Division, University of Hawaii at Hilo, Hilo, Hawaii 96720, USA |
| **Hohl, Hans R.** | Institute of Plant Biology, University of Zürich, CH-8008 Zürich, Switzerland |
| **Kannwischer-Mitchell, Mary E.** | Department of Plant Pathology, University of Florida, Gainesville, Florida 32611, USA |
| **Keen, Noel T.** | Department of Plant Pathology, University of California, Riverside, California 92521, USA |
| **Kidney, Barbara A.** | Department of Plant Pathology, The Pennsylvania State University, University Park, Pennsylvania 16802, USA |
| **King, Eileen D.** | Department of Plant Pathology, The Pennsylvania State University, University Park, Pennsylvania 16802, USA |
| **MacKenzie, David R.** | Department of Plant Pathology, The Pennsylvania State University, University Park, Pennsylvania 16802, USA |

| | |
|---|---|
| **Malajczuk, Nicholas** | Division of Land Resources Management, CSIRO, Private Bag, Wembley, Western Australia 6014, Australia |
| **Mitchell, David J.** | Department of Plant Pathology, University of Florida, Gainesville, Florida 32611, USA |
| **Newhook, Frank J.** | Department of Botany, University of Auckland, Private Bag, Auckland, New Zealand |
| **Nilsson, Barbro A.** | Potato Department, Svalöf Seed Ltd. AB, Svalöv S-268 00, Sweden |
| **Ribeiro, Olaf K.** | Department of Plant Pathology, University of California, Riverside, California 92521, USA. (Present address: Peninsu-Lab, 23976 N.E. Newellhurst Court, Kingston, Washington 98346, USA) |
| **Royer, Matt H.** | Department of Plant Pathology, The Pennsylvania State University, University Park, Pennsylvania 16802, USA |
| **Schmitthenner, August F.** | Department of Plant Pathology, The Ohio State University, Ohio Agricultural Research and Development Center, Wooster, Ohio 44691, USA |
| **Schwinn, Franz J.** | Research and Development Department, Agricultural Division, CIBA-GEIGY Ltd., CH-4002 Basel, Switzerland |
| **Shaw, David S.** | School of Plant Biology, University College of North Wales, Bangor LL57 2UW, Wales, UK |
| **Shea, Sydney R.** | Forests Department, Research Station, Dwellingup, Western Australia 6213, Australia |
| **Stamps, D. Jean** | Commonwealth Mycological Institute, Ferry Lane, Kew, Surrey TW9 3AF, England, UK |
| **Theberge, Robert L.** | Department of Plant Pathology, The Pennsylvania State University, University Park, Pennsylvania 16802, USA |
| **Tsao, Peter H.** | Department of Plant Pathology, University of California, Riverside, California 92521, USA |
| **Umaerus, Magnhild** | Department of Plant and Forest Protection, Swedish University of Agricultural Sciences, P.O. Box 7044, Uppsala, S-750 07, Sweden |
| **Umaerus, Vilhelm** | Department of Plant and Forest Protection, Swedish University of Agricultural Sciences, P.O. Box 7044, Uppsala, S-750 07, Sweden |
| **Wang, Ming Chang** | Department of Plant Pathology, University of California, Riverside, California 92521, USA |
| **Waterhouse, Grace M.** | Commonwealth Mycological Institute, Ferry Lane, Kew, Surrey TW9 3AF, England, UK |
| **Weste, Gretna** | Botany School, University of Melbourne, Parkville, Victoria 3052, Australia |
| **Wilson, Una E.** | Botany School, University College, Dublin 2, Ireland |
| **Yoshikawa, Masaaki** | Laboratory of Plant Pathology, Faculty of Agriculture, Kyoto Prefectural University, Kyoto 606, Japan |
| **Zentmyer, George A.** | Department of Plant Pathology, University of California, Riverside, California 92521, USA |

# Preface

The publication of this book culminates a long endeavor. For several years, we felt the need to bring together the existing knowledge on *Phytophthora* and to explore and define the numerous questions that have confronted investigators of this important genus of plant-pathogenic fungi. The idea of organizing an international symposium at which scientists from many countries could meet, speak, and discuss these issues was conceived in 1979. The impending retirement in 1981 of our esteemed colleague, Professor George A. Zentmyer, a lifelong "phytophthorologist," provided the impetus to make this dream a reality. On April 1–4, 1981, more than 300 scientists from 24 nations attended the *International Symposium on* Phytophthora: *Its Biology, Ecology and Pathology* at the University of California, Riverside, where they exchanged views on *Phytophthora* and paid tribute to Professor Zentmyer, to whom the symposium was dedicated. Attendance by most of the major investigators of *Phytophthora* and the enthusiasm of the participants confirmed our initial belief that the symposium was necessary and timely.

This book, which was planned in conjunction with the symposium, is the first comprehensive account of the biology, taxonomy, ecology, and pathology of *Phytophthora*. It contains current research as well as in-depth reviews of almost all the major areas of knowledge related to *Phytophthora*. Although we strove to make the volume coherent and uniform within the limitations imposed by a multiauthor book, the depth of coverage, style, and terminology often differ from chapter to chapter. A number of phenomena or topics are covered more than once in different chapters, but we retained this duplication because we felt that each chapter should be reasonably self-contained. Use of different specific or subspecific epithets for the same fungus was sometimes unavoidable because of differences of opinion on classification or nomenclature. We worked with the authors to revise chapters whenever greater clarity of expression was desirable or gaps appeared to occur in the exposition; however, the opinions of each author were respected.

We wish to acknowledge the contributions of Tomás Rivera, Chancellor of the University of California at Riverside, and W. M. Dugger, former Dean of the College of Natural and Agricultural Sciences, who gave us substantial financial assistance and support at the critical initial planning phase, which made it possible to organize the symposium relatively quickly. Students, staff, and faculty in our Department of Plant Pathology provided help of every kind. The author and subject indexes were prepared by Julie Moore and Associates, Riverside, CA. A number of agricultural organizations and chemical companies, the National Science Foundation, the U.S. Department of Agriculture, and the U.S. Department of Energy contributed generous financial support. We also acknowledge a grant from the Foundation for Microbiology for preparation of this book. To all these groups and individuals we extend our sincere thanks and appreciation.

**Donald C. Erwin**
**Salomon Bartnicki-Garcia**
**Peter H. Tsao**

Riverside, California

# Contents

## Pathology of *Phytophthora*



## Conclusions and Prospects



## Indexes

# 1

# The World of *Phytophthora*

George A. Zentmyer

CONTENTS



## INTRODUCTION

The organization and presentation of this First International Symposium on *Phytophthora* demonstrate the increasing recognition of the importance and extraordinary nature of this large genus of plant-pathogenic fungi. Anton de Bary (Fig. 1) first coined the name *Phytophthora* ("plant destroyer") in 1876, when he described the potato late blight fungus, *Phytophthora infestans,* as the type species for the new genus. In the slightly more than a century since de Bary's description, interest in and research on the genus have expanded dramatically.

For several years, *P. infestans,* widely publicized as the cause of the very destructive and economically important potato blight disease, was the sole representative of the genus that was recognized in the plant pathology literature. Gradually, other species of *Phytophthora* were described, beginning with *P. cactorum* in 1886, *P. phaseoli* in 1889, and *P. nicotianae* in 1896 (Table 1). Many species (43, according to Waterhouse et al [Chapter 10]) are now known to cause a wide range of diseases on an ever increasing number of plants, including major food crops, forest trees, subtropical and tropical fruit trees, berries, nut trees, many ornamental plants, and valuable tropical export crops.

Species of *Phytophthora* have a number of unusual features that distinguish them from many other pathogenic fungi. These features include:

1) Motile zoospores with a remarkable ability to synthesize a cyst cell wall within minutes of encystment (Bartnicki-Garcia and Wang, Chapter 9). Complete differentiation of zoospores occurs within the sporangium before they are expelled into the vesicle or released through the apex of the sporangium (Gisi, Chapter 8). Zoospores have a tinsel and a whiplash flagellum (Carlile, Chapter 7). These unique infective units provide tremendous and explosive disease potential.
2) "Not one but several sporangia successively formed at the end of each branch of the tree-like sporangiophore," as de Bary noted to distinguish *Phytophthora* from *Peronospora* (Tucker, 1931). Another feature of *Phytophthora* is the caducity or noncaducity of the sporangia in different species (Waterhouse, 1974).
3) Cell walls distinctly different from those of most fungi in that they are made up primarily of glucans ($\beta$-1,3 and $\beta$-1,6 linkages) plus some cellulose, instead of chitin (Bartnicki-Garcia and Wang, Chapter 9).
4) Both heterothallic and homothallic species. A single oospore is normally formed within each oogonium. Antheridia are usually single and may be paragynous or amphigynous.
5) A primarily diploid life cycle with meiosis occurring in the gametangia.
6) Certain unique physiologic responses. Some species contain a unique form of storage polysaccharide, the mycolaminarans. Also, species of *Phytophthora* respond differently to sterols than do most of the higher fungi. *Phytophthora* spp. are unable to synthesize sterols and,

**Fig. 1.** Anton de Bary.

although they do not require sterols for vegetative growth, they must have exogenously supplied sterols for sporulation. Inasmuch as the presence of sterols is a requisite for sensitivity to polyene antibiotics, the tolerance of *Phytophthora* to these antibiotics, another unique feature, is obviously related to the lack of sterols.

Thus, *Phytophthora* is truly a remarkable genus of plant-pathogenic fungi. Its unique morphological, genetic, and physiologic features, combined with the wide variety of diseases caused on a tremendous number of plants, make *Phytophthora* one of the most fascinating subjects for investigation. The complicated life cycle, with very distinct and strikingly different spore forms ranging from motile zoospores to thick-walled oospores, makes control of diseases caused by *Phytophthora* difficult and challenging.

Outstanding progress has been made in recent years in understanding and controlling the destructive diseases induced by *Phytophthora* and in solving the mysteries involved in the growth and development of these pathogens. This progress was described in detail in the papers presented at the *Phytophthora* Symposium, which form the basis for the chapters in this volume. Detailed studies are provided in the areas of biology, taxonomy, ecology, and pathology of the genus *Phytophthora.*

Direct studies of the biology of the pathogen (Chapters 2–9) give basic and detailed insights into *Phytophthora.* These studies include descriptions of the cytology of the vegetative and reproductive stages and investigations of the biochemistry of morphogenesis, including the role of $\beta$-glucans. Similar studies expand our understanding of nutritional relationships among species and of physiologic processes and chemicals involved in sexual and asexual sporulation. Other studies address the roles of various physical factors, such as water, oxygen, and other gases, in developmental processes. The interactions of host and pathogen, including chemotaxis and other tactic processes, and the mechanism for invasion and infection of the host provide fascinating research topics. Also significant are the contributions of cytogenetics and genetics, which have established that species of *Phytophthora,* along with other Oomycetes, are diploid in the vegetative state, with meiosis occurring in the gametangia.

Taxonomic studies (Chapters 10–13) have also provided much information on variability within and among species and on new criteria for classification, including protein and enzyme patterns, serology, and numerical taxonomy. Although these new criteria help us to understand taxonomic relationships among species, morphology still provides the primary basis for identification.

Progress in ecological studies of *Phytophthora* has been equally remarkable, as demonstrated by Chapters 14–19. Significant and highly useful data are being obtained on the roles of physical and chemical factors in disease development, on the activity of various microbial agents in antagonism to *Phytophthora,* and on the devious means by which these pathogens develop, survive, and infect plants. Substantial progress is being made on isolation and quantification of *Phytophthora* in soil.

Pathological investigations (Chapters 20–26) include research on epidemiology and on the complex interactions involved in the chemical and physical bases for disease susceptibility and resistance. They also describe promising new developments in the areas of cultural, biological, and chemical control, especially the control of root diseases by new approaches and new chemicals.

TABLE 1. Dates of description of species of *Phytophthora*[a]

| Date | Species |
|---|---|
| 1876 | *P. infestans* (Montagne) de Bary |
| 1886 | *P. cactorum* (Lebert & Cohn) Schroeter |
| 1889 | *P. phaseoli* Thaxter |
| 1896 | *P. nicotianae* van Breda de Haan |
| 1900 | *P. colocasiae* Raciborski |
| 1909 | *P. syringae* (Klebahn) Klebahn |
| 1913 | *P. arecae* (Coleman) Pethybridge |
| | *P. erythroseptica* Pethybridge |
| | *P. parasitica* Dastur |
| 1918 | *P. meadii* McRae |
| 1919 | *P. cryptogea* Pethybridge and Lafferty |
| | *P. lepironiae* Sawada |
| | *P. palmivora* (Butler) Butler |
| 1922 | *P. capsici* Leonian |
| | *P. cinnamomi* Rands |
| 1923 | *P. mexicana* Hotson & Hartge |
| 1925 | *P. citrophthora* (Smith & Smith) Leonian |
| | *P. hibernalis* Carne |
| 1927 | *P. boehmeriae* Sawada |
| | *P. cambivora* (Petri) Buisman |
| | *P. citricola* Sawada |
| | *P. gonapodyides* (Petersen) Buisman |
| | *P. richardiae* Buisman |
| 1929 | *P. heveae* Thompson |
| 1931 | *P. drechsleri* Tucker |
| | *P. megasperma* Drechsler |
| | *P. porri* Foister |
| 1935 | *P. cyperi* (Ideta) Ito |
| 1940 | *P. fragariae* Hickman |
| | *P. macrospora* (Saccardo) Ito & Tanaka |
| | *P. verrucosa* Alcock & Foister |
| 1942 | *P. lateralis* Tucker & Milbrath |
| 1947 | *P. quininea* Crandall |
| 1949 | *P. inflata* Caroselli & Tucker |
| 1952 | *P. primulae* Tomlinson |
| 1953 | *P. cyperi-bulbosi* Seethalakashmi & Ramakrishnan |
| 1957 | *P. ilicis* Buddenhagen |
| | *P. vignae* Purss |
| 1969 | *P. botryosa* Chee |
| | *P. vesicula* Anastasiou & Churchland |
| 1971 | *P. iranica* Ershad |
| 1974 | *P. japonica* Waterhouse |
| 1976 | *P. melonis* Katsura |
| 1979 | *P. katsurae* Ko & Chang |

[a] From Waterhouse (1970), Newhook et al (1978), and Ko and Chang (1979).

**Fig. 2.** Young avocado tree affected with Phytophthora root rot.

## TWO SPECIES OF *PHYTOPHTHORA*

This chapter emphasizes some highlights of our research on two species of *Phytophthora, P. cinnamomi* and *P. palmivora,* both of which are extensively distributed geographically and on many hosts. These species are used to illustrate some of the unique aspects of recent research on the genus. Both species are primarily pathogens of plants in mild temperate, subtropical, and tropical regions, and both cause severe losses on hundreds of species of plants.

These two fungi differ significantly in many ways. First, *P. cinnamomi* has nearly 1,000 hosts (Zentmyer, 1980), compared with several hundred hosts of *P. palmivora* (Chee, 1969). Only a few of these hosts, such as papaya and cinchona, are common to both species.

Second, *P. cinnamomi* is primarily a soilborne pathogen attacking roots; *P. palmivora* is an airborne and water-borne pathogen of fruit and foliage. These distinctive pathological characteristics provide an opportunity for further differentiation between the two fungi, in terms of host invasion, epidemiology, and control.

Third, *P. palmivora* produces sporangia readily and abundantly on many agar media, but *P. cinnamomi* does not. Sporangium production by *P. cinnamomi* is much more difficult in the laboratory because sporangia are produced only in liquid culture or water, and only when stimulated by some substance(s) from a nonsterile soil extract or after young mycelia are washed thoroughly and incubated in salt solution. In addition, in *P. palmivora,* sporangia are caducous with short stalks (a later section describes the different species in the *P. palmivora* complex); in *P. cinnamomi,* sporangia are persistent. Also, the typical sympodium in *P. palmivora* differs significantly from the random production of sporangia on separate sporangiophores in *P. cinnamomi.*

Fourth, in *P. palmivora,* both mating types are widely distributed throughout the tropical regions of the world. The A2 mating type of *P. cinnamomi* is widely distributed, but the A1 has an extremely limited distribution.

My interest and involvement in research on *Phytophthora* began in 1944 on my appointment to a position for research on diseases of avocado and other subtropical and tropical fruit trees at the Citrus Experiment Station of the University of California. It was soon apparent that the primary disease problem facing the avocado industry in California was Phytophthora root rot, then generally termed "decline" and attributed primarily to excess soil moisture. Just 3 yr before, however, Vincent Wager, a plant pathologist from South Africa on a study leave in California, had isolated and identified *P. cinnamomi* from avocado trees affected with "decline" (Wager, 1942). Subsequent research by Wager indicated that this species of *Phytophthora* was usually associated with the disease and apparently played an important role. Our subsequent studies at Riverside, including those by W. T. Horne, established the causal role and the importance of *P. cinnamomi* in avocado "decline," which we subsequently named avocado root rot or Phytophthora root rot of avocado.

### *Phytophthora cinnamomi*

My research at that time was concentrated on *P. cinnamomi,* primarily in relation to the destructive root rot of avocado in California (Fig. 2), but also in relation to some other hosts of the fungus in California, including many ornamental plants (heather, camellia, azalea, rhododendron, *Myrtus*), conifers of many types, and macadamia trees. Thus, I continued a tradition of *Phytophthora* research begun by H. S. Fawcett and L. J. Klotz at Riverside in the 1920s in their pioneering investigations of species that attack citrus.

The principal host plant in much of my research over the past 37 yr has been the avocado, that remarkable subtropical evergreen fruit tree, *Persea americana.* The avocado is a native tree in subtropical and tropical America; its nutritious fruit has been a favored food of the natives of those areas for centuries. Aztec pictographs depict the avocado tree, and the Aztec name for the tree, *ahuacaquahuital,* is the basis for the Spanish name *aguacate* and for our common English name.

The international aspects of these studies of *Phytophthora* began in the early 1950s, when a search began for resistant species of *Persea* to serve as rootstocks to control Phytophthora root rot of avocado. About 80 species of *Persea* are native to Latin America (Kopp, 1966). To begin the collecting project, I visited most of the major herbarium collections in botanical gardens in the United States (New York Botanical Garden, Chicago Natural History Museum, Missouri Botanical Garden, and Gray Herbarium, for example) and noted locations where various species of *Persea* had been collected in countries throughout Latin America. The fascinating project of collecting seed and/or budwood of many of the native species of *Persea* began in 1951 with a trip to Mexico. This search has continued for more than 30 yr, in nearly every country in Central and South America and the Caribbean and, with the cooperation of many scientists and native inhabitants of many countries, has resulted in the collection of thousands of seeds and budwood samples of some 20 species of *Persea* and numerous and varied clones of the avocado itself.

This collecting project also provided an opportunity to investigate other species of *Phytophthora* in Latin America, including *P. palmivora* on cacao (*Theobroma cacao*), papaya, and black pepper; *P. capsici* on pepper; *P. cinnamomi* on pineapple, cinchona, macadamia, and avocado; and *P. nicotianae* var. *parasitica* on citrus and various flower crops. Investigation of *Phytophthora* in other parts of the world included 1) studies of *P. cinnamomi* on avocado, macadamia, and eucalyptus in Australia (particularly the devastating jarrah dieback disease of *Eucalyptus marginata*), and on avocado in Hawaii, Fiji, the Trust Territory of the Pacific Islands, and in South Africa, where the fungus also attacks various native species of *Protea*, and 2) studies of *P. palmivora* on cacao in West Africa, Brazil, Venezuela, Colombia, Costa Rica, Fiji, New Guinea, and Sri Lanka. (Research on *P. palmivora* is covered in more detail in the next section.)

Studies of *P. cinnamomi* have emphasized variation, host range, genetics, mating type, chemotaxis, saprophytism, indigenous home of the fungus, basis for pathogenicity, response to fungicides and fumigants, resistance, sporulation, and environmental relationships. In this chapter, a few of these aspects are noted. They are covered in more detail in a monograph on *P. cinnamomi* (Zentmyer, 1980).

#### Host Range

Since it was first described by Rands (1922) on cinnamon trees (*Cinnamomum burmannii* Blume) in Sumatra, *P. cinnamomi* has been described from 67 countries as the cause of many diseases on nearly 1,000 host plants. It is now recognized as one of the most ubiquitous and destructive plant pathogens, the most widely distributed species of *Phytophthora,* and the species with the largest host range. What a striking and fascinating contrast between *P.*

*cinnamomi* and *P. infestans* or *P. fragariae,* two species with a more specialized mode of parasitism and an extremely limited host range! There must be some sharp differences in physiology and enzyme relationships between the species.

*P. cinnamomi* causes great devastation in forests, as exemplified by the spectacular epidemic on eucalyptus ("jarrah dieback") and on more than 90 other indigenous plants in Western Australia and by root and crown rots and cankers on a variety of food and ornamental crops, on pineapple in plantations all over the world, on a variety of nursery plants (causing great monetary losses in this industry), and on chestnut, shortleaf pine, Monterey pine, and many other hosts in addition to the avocado (Zentmyer, 1980).

**Origin of the Species**

Speculation has been rife on the origin of this species. Rands (1922), in his description of *P. cinnamomi* from Sumatra, commented, "The causal fungus is believed to have come into the plantings along with its native host which is indigenous to this region." This provides the first suggestion of a possible Asian origin of this pathogen.

Where do we find environmental conditions favorable for a fungus such as *P. cinnamomi*? This fungus grows best in mild temperate or subtropical regions. It does not grow or survive at low (below 6° C) or at high (above 32–35° C) soil temperatures. Optimal temperatures are in the range of 21–27° C. The pathogen does not survive well or spread under conditions of low soil moisture but develops rapidly when soil moisture is abundant. The moist subtropics seem to provide the best possibilities; this includes many areas in the Americas from Mexico to Peru and Brazil and possibly in the higher elevations in the tropical part of that large region. Other likely areas are northeastern Australia and the higher elevations of the area from Papua, New Guinea, north through the Malaysian archipelago, and possibly farther north and east toward Taiwan.

Crandall and Gravatt (1967), who studied the chestnut root disease in the southern United States, suggested that *P. cinnamomi* probably had an "Asian" origin. They further commented that the pathogen could have been carried with plant material through the Pacific Islands, and similarly to Africa, Europe, and various ports in the Americas, by early Spanish, Portuguese, French, and English explorers. They concluded that *P. cinnamomi* must have been imported into the United States before 1780, probably through ports at Savannah or Mobile.

There are some indications that the center of origin is from northeastern Australia and New Guinea north into the original "home" of the pathogen in Sumatra. Ko et al (1977) also recently presented information suggesting the origin of *P. cinnamomi* in Taiwan; they isolated both mating types from native vegetation in several parts of that country.

One can also look at susceptible hosts, by analogy to a speculation (discussed in more detail in a later section) made by Waterhouse on the origin of *P. palmivora* (*personal communication*). A number of native plants in southeastern Asia and Australia are hosts, including cinnamon, *Nothofagus, Grevillea, Macadamia, Macrozamia, Eucalyptus, Endiandra, Brachenridgea, Cryptocarya,* and *Cryptomeria.* Also, several native hosts of *P. cinnamomi* occur in Central and South America, including *Persea, Cinchona, Carica papaya, Ananas comosus,* and *Jacaranda.* Some of these plants could have been primary hosts in the native forest.

Our studies over the past 30 yr have included attempts to determine whether the fungus is indigenous to southern California and also to the tropical and subtropical areas of Latin America where the avocado and other species of the genus *Persea* are native (Zentmyer, 1977). All of the sampling data indicate that it is unlikely that *P. cinnamomi* is indigenous to those areas. In the Latin American samples, for example, roots were cultured from 373 trees (avocado and other species of *Persea*) in 17 countries. In no case was the pathogen recovered from any native trees of *Persea* growing in undisturbed areas (rain forest, tropical swamp, montane forest, etc.) remote from cultivation, even though some samples were taken from areas of very high rainfall and in soils containing considerable clay. No avocado trees or other trees of *Persea* spp. with root rot were found in any of the native undisturbed soils. More detailed surveys should be made in Latin America, involving additional samples and trapping tests with susceptible hosts that might detect low populations of *P. cinnamomi.* Some of the area would certainly fulfill many of the climatic and floristic requirements of the fungus. Similarly, in southern California, *P. cinnamomi* was not recovered from any of more than 400 samples collected in native brushland above cultivated areas. In addition to culturing roots from these samples, we trapped the soils for the presence of *Phytophthora,* using avocado or *Persea indica* seedlings.

The data indicate that *P. cinnamomi* is not indigenous to the soils of southern California. Furthermore, the climate of southern California, with long, hot, dry periods from April to November, was not conducive for such a fungus to survive and thrive before the advent of irrigated agriculture. It seems much more likely that the fungus was imported into California from tropical America, Hawaii, or Australia in the late 19th century or early 20th century, with plant material and accompanying soil, before restrictions were placed on movement of such material.

The curious and limited distribution of the A1 mating type of *P. cinnamomi,* in contrast to the worldwide distribution of the A2 mating type (Zentmyer, 1976), adds an additional dimension to the search for an origin of this species. To date, the A1 mating type has been reported from only six countries: Australia, Malagasy Republic (Madagascar), Papua New Guinea, South Africa, Taiwan, and the United States. The host range of the A1 type is also extremely limited. *P. cinnamomi* is thus quite different from *P. infestans,* in which the A2 type is restricted to Mexico, and from *P. palmivora,* in which the two mating types are both widely distributed in many countries.

Another potentially highly significant development concerning the origin of *P. cinnamomi* is the recent discovery by von Broembsen (*personal communication*) of the A1 type in native vegetation and as zoospores in mountain streams draining remote areas of the Cape Province in South Africa. The Cape Province contains many susceptible native hosts of *P. cinnamomi,* especially in the Proteaceae; infection has been found on these in the field.

**Sexual Reproduction**

The sexual stage of *P. cinnamomi* can be produced by interspecific and intraspecific crosses of the normally heterothallic species, i.e., pairing A1 and A2 mating types, and by initiation of sexuality by selfing in the presence of several stimuli.

Roots of the avocado contain a chemical or chemicals extractable with water or organic solvents that will initiate production of large numbers of oospores by the A2 mating type of *P. cinnamomi* (Fig. 3) and of *P. drechsleri* and *P. capsici.* The substance(s) does not affect the A1 mating type of *P. cinnamomi* (Zentmyer, 1979). Because avocado root extracts contain oleic acid, and because synthetic oleic acid initiates oospore production in the A2 but not the A1 mating type of *P.*

*cinnamomi* (Zaki et al, 1982), oleic acid appears to be the active factor. Additional research is needed on this phenomenon.

Stimulation of oospore production by *Trichoderma viride* in A2 mating types of *P. cinnamomi* and several other species of *Phytophthora* has also been reported by Brasier (1971) and by Reeves and Jackson (1972). The basis for this stimulation is not known, but there is some indication that a toxic volatile chemical produced by *Trichoderma* produces morphological changes that may be related to oospore formation. A metabolite of *T. viride* has been identified in culture (Moss et al, 1975), but its role in sexual reproduction is questionable. Reeves and Jackson (1974) also showed that oospores can be initiated in *P. cinnamomi* by mechanically wounding mycelium of the A2 mating type. Recently, Ko (1978) reported initiation of sexual reproduction in *P. cinnamomi* and several other species of *Phytophthora* by a hormonelike substance from an opposite mating type that passes through polycarbonate membranes. In this system, oospore production occurs in A1 as well as A2 mating types; the phenomenon therefore is basically quite different from the other systems described above.

Thus, the A2 mating type of *P. cinnamomi* can also function as a homothallic fungus when the proper stimulus is provided. This may be a nutritional response, a response to staling products (as indicated by the production of oospores in old cultures), a response to a more specific chemical (as in the case of the avocado root extract and the *Trichoderma* stimulus), or a response to a "hormone" in some cases, as indicated by Ko's research.

**Variability**

Some of our recent research (Zentmyer et al, 1976) shows that *P. cinnamomi* is much more variable than was apparent in the early stages of my research on this pathogen 30 or 35 yr ago. We have found considerable intraspecific variability in colony morphology, pathogenicity, response to antibiotics and fungicides, response to temperature, morphology (Fig. 4) and abundance of oospores and sporangia, responses to nutritional substances (such as carbon and nitrogen sources and sterols), abundance of chlamydospores, production of pigment on tyrosine media, and other physiologic responses. As one example of differences in pathogenicity, isolates from camellia are pathogenic to both camellia and avocado, whereas isolates from avocado cause disease on avocado but not on camellia (Zentmyer and Guillemet, 1981).

## *Phytophthora palmivora*

My interest in *P. palmivora* dates from my travels in Central America in search of avocado rootstock material resistant to *P. cinnamomi*. In the early 1960s in Costa Rica, I had the opportunity to observe and isolate *P. palmivora* from cacao with black pod disease in the germ plasm planting at the then International Institute of Agricultural Sciences in Turrialba. The first cultures from cacao in 1964 formed the basis for the present extensive collection at Riverside of cultures of "*P. palmivora*" (over 400) from many countries, which complements the equally large *P. cinnamomi* collection. The *Phytophthora* culture collection at the University of California, Riverside, now numbers nearly 1,700 cultures representing 26 species of the genus.

**Origin of the Species**

Waterhouse (*personal communication*) provided some interesting speculation on the possible origin of *P. palmivora*, a very important species of *Phytophthora* in the tropics. Noting that most hosts of this pathogen are native to either Asia or Central and South America, she suggested those areas as prime suspects for centers of origin of *P. palmivora*. Waterhouse further proposed that the Central and South American region is the more likely center because of the native occurrence of a number of woody plants very susceptible to the pathogen, including cacao (*Theobroma cacao*), rubber (*Hevea brasiliensis*), papaya (*Carica papaya*), avocado (*Persea americana*), and other subtropical and tropical plants.

Gregory introduces an interesting concept into the speculation on the origin of *P. palmivora* when he states (Chapter 20) that "the pathogen, *P. palmivora*, seems to have migrated with its host from the American tropics to new plantations overseas. The cocoa tree also encountered additional *Phytophthora* spp. in its travels." Still, we have no

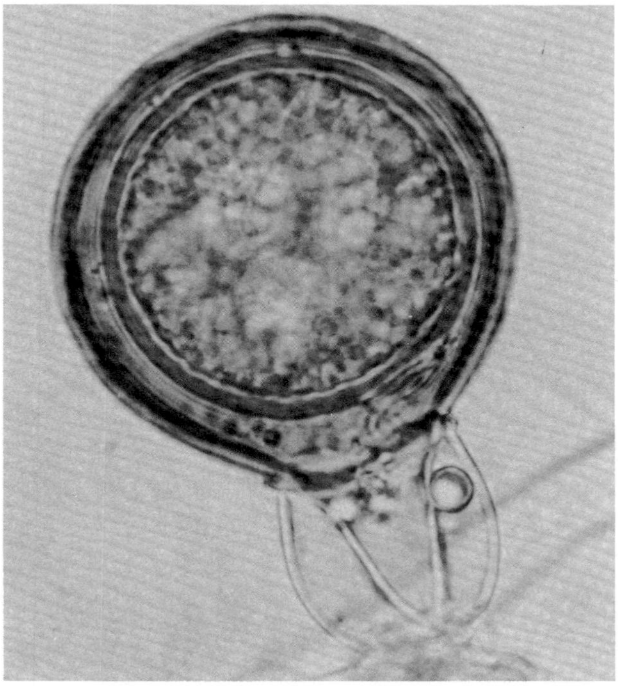

**Fig. 3.** Oospore of *Phytophthora cinnamomi* produced by A2 mating type isolate in avocado root extract.

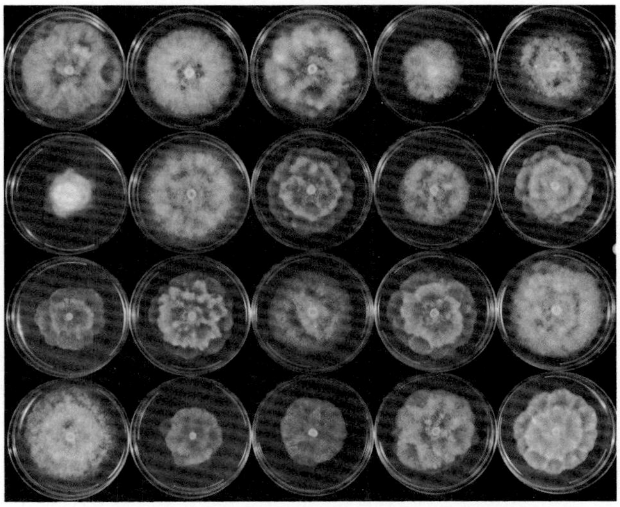

**Fig. 4.** Variable phenotype of cultures of *Phytophthora cinnamomi*, grown for 6 days on potato-dextrose agar at 25°C.

definitive evidence yet on the existence of *P. palmivora* as an indigenous species on native hosts in Latin America. Further explorations in that huge and diverse region, however, are certainly in order. The genus *Theobroma* has an extremely wide distribution in Central and South America, and further examination of these hosts in remote parts of their areas of natural distribution, as well as isolation from roots and soil in those areas, should be undertaken.

*P. palmivora* has at least partially become adapted to a nonaquatic environment; it causes diseases of the aerial portions of many of its hosts, in contrast to species such as *P. cinnamomi,* which are primarily root- and soil-inhabiting. A pathogen of leaves, stems, and fruit should be spread more rapidly as spores by wind or windblown rain than a strictly soilborne pathogen would be. Theoretically, this feature should have resulted in more rapid and widespread development from a given center of origin, but evidence of this type of development is not available and is difficult to obtain.

**Species that Cause Black Pod**

Any discussion of "*P. palmivora*" must be qualified by the changes in the past few years in classification of the fungi that can cause black pod and canker of cacao. Our research, and that in several other laboratories, has recently demonstrated that the *Phytophthora* isolates originally designated by Butler and others as *P. palmivora* were really a complex of several species, now designated as *P. palmivora, P. megakarya, P. capsici, P. citrophthora,* and possibly *P. megasperma.* These species may be distinguished by the deciduous or persistent nature of the sporangia and—in the deciduous species (*P. palmivora, P. megakarya,* and *P. capsici*)—by the length of the sporangial pedicel. There are other differences in the morphology of sporangia, in chlamydospore production, and, in some of the species at least, in chromosome number and size. This provides a striking contrast to the situation that existed for many years, in which one variable species, *P. palmivora,* was regarded as the sole cause of black pod.

This clarification in identity of "*P. palmivora*" isolates from cacao dates from the Cocoa *Phytophthora* Workshop held at the Rothamsted Experimental Station in England in 1976. As a result of discussions and information presented at that workshop, and of additional recent research, the taxonomic situation is now as follows (Griffin, 1977; Zentmyer et al, 1977): Typical *P. palmivora* (temporarily designated morphological form [MF] 1 at the workshop) has short, thick sporangial stalks less than 5 μm long. This is the most common species pathogenic to cacao worldwide. *P. megakarya,* temporarily designated MF 3 at the workshop, has thin sporangial stalks ranging usually from 5 to 15 μm in length (Brasier and Griffin, 1979). This species has limited distribution, but it is the primary black pod pathogen in Nigeria. Sansome et al (1975) and Brasier and Griffin (1979) reported that *P. palmivora* and *P. megakarya* differ in size and number of chromosomes; *P. palmivora* has 9–10 small chromosomes, and *P. megakarya* has five large chromosomes.

The type temporarily designated MF 4 has long sporangial stalks, up to 250 μm long, and sporangia with length to breadth ratios often over 2.0. This species of *Phytophthora* is closest to *P. capsici* (Kaosiri, 1978; Kaosiri et al, 1978; Zentmyer et al, 1981), and we have designated it as *P. capsici.* This species is found in Central and South America and in West Africa (but only in Cameroon); it is the principal black pod pathogen in Brazil. This strange distribution, quite widespread in Latin America but in only one country in West Africa, indicates that this species must have been carried from Latin America, probably from Brazil, to Cameroon in cacao material, possibly early in the development of the cocoa industry in West Africa.

Recently, a fourth species, *P. citrophthora,* was found to be a cause of black pod and canker of cacao in Brazil (Kellam and Zentmyer, 1981a). This species has persistent sporangia, rather than the caducous sporangia of the above three species, and also forms chlamydospores.

A fifth species, *P. megasperma,* has been described by L. deReyes (*personal communication*) as the cause of "watery pod rot" of cacao in Venezuela; this is a homothallic species, with persistent sporangia.

Differences in the epidemiology of black pod disease that have been observed in different parts of the world may be due to the presence or absence of these different species of *Phytophthora.* Use of resistant cultivars and other control measures must be adapted to the life cycle and physiology of the particular pathogenic species that predominates in a particular area.

In our own research, we have used gel electrophoresis to characterize protein patterns as an aid in distinguishing the several species now recognized as pathogens of cacao (Idosu and Zentmyer, 1978; Kaosiri and Zentmyer, 1980).

**Soil Phase**

Another important aspect of research on the "*P. palmivora*" complex is the study of the occurrence, survival, pathogenesis, and persistence of these fungi in soil. Gregory comments on this in Chapter 20. Turner and Asomaning (1962) showed that *P. palmivora* could attack cacao roots. Similarly, our research at Riverside (Zentmyer et al, 1968) demonstrated that under ideal conditions for pathogenesis, in hydroponic culture, a typical *P. palmivora* isolate from Costa Rica could invade cacao roots and cause considerable rotting.

Studies in Nigeria showed that *P. megakarya* is well adapted to existence in soil, although no data were given on the means of survival in soil (Gregory and Maddison, 1981). Kellam and Zentmyer (1981b) found oospores of the *P. capsici* type in old diseased pods in the field in Brazil. In some of our field studies in Costa Rica, we observed chlamydospores of *P. palmivora* in decaying pod tissue in the soil under cacao trees; these spores could certainly be important in the persistence of the fungus in the soil. One of the great mysteries, for both *P. palmivora* and *P. cinnamomi,* is the nature of the surviving and perennating structure(s) in soil. What are the primary structures responsible for soil infectivity: sporangia, chlamydospores, oospores, or zoospores? Some aspects of the survival of propagules of *P. palmivora, P. capsici,* and *P. citrophthora* in soil are being studied by M. K. Kellam in my laboratory.

Despite increasing recognition that the "*P. palmivora*" complex can exist in the soil, further information is needed on the survival of the different species in soil and on their possible roles as pathogens of cacao roots. Because the different species vary in their ability to produce the different spore stages, the means of survival and of dissemination in soil could vary considerably. Further studies in this area are urgently needed.

Other aspects of research on "*P. palmivora*" will be affected by the changes in classification of the cacao pathogens. These include studies of the pathogenicity of various isolates from different countries to the variety of cacao clones, studies of nutrition of the pathogens, and studies of interrelationships between host nutrition and disease production.

## FUTURE RESEARCH

As to the future of the "World of *Phytophthora,*" there are many exciting topics for further study, in addition to those summarized in this volume—new approaches and new concepts. Much more research, with new and different orientation, is needed on control of diseases induced by species of *Phytophthora.* For example, with Phytophthora root rot of avocado, promising avenues for research include the use of combination treatments (integrated control), such as combining resistant rootstocks with some of the new effective organic chemicals and with biological control (including use of suppressive soils, amendments, and antagonists). Other potentially profitable areas to explore in depth are hypovirulence, protoplast culture, and utilization of optimal ecological and climatic conditions to favor the host and reduce the activity of the pathogen. With all of the recent emphasis on molecular genetics, fields such as recombinant DNA, plasmids, gene splicing, and genetic engineering may well play a significant role in the future of research on understanding and controlling fungi such as *Phytophthora.*

## LITERATURE CITED


Brasier, C. M. 1971. Induction of sexual reproduction in single $A^2$ isolates of *Phytophthora* species by *Trichoderma viride.* Nature (London) New Biol. 231:283.

Brasier, C. M., and Griffin, M. J. 1979. Taxonomy of '*Phytophthora palmivora*' on cocoa. Trans. Br. Mycol. Soc. 72:111-143.

Chee, K. H. 1969. Hosts of *Phytophthora palmivora.* Rev. Appl. Mycol. 48:337-344.

Crandall, B. S., and Gravatt, G. F. 1967. The distribution of *Phytophthora cinnamomi.* Ceiba 13:43-53.

de Bary, A. 1876. Researches into the nature of the potato fungus, *Phytophthora infestans.* J. R. Agric. Soc. Engl. 12:239-269.

Gregory, P. H., and Maddison, A. C., eds. 1981. Epidemiology of *Phytophthora* on cocoa in Nigeria. Phytopathol. Pap. 25. Commonw. Mycol. Inst., Kew, Surrey, England. 188 pp.

Griffin, M. J. 1977. Cocoa *Phytophthora* Workshop. Rothamsted Experimental Station, England, 24–26 May, 1976. PANS 23:107-110.

Idosu, G. O., and Zentmyer, G. A. 1978. *Phytophthora palmivora:* A comparative study of "typical" and "atypical" isolates from cacao (*Theobroma cacao*). Mycologia 70:1101-1112.

Kaosiri, T. 1978. Morphological, taxonomic, and cytological studies of *Phytophthora palmivora.* Ph.D. dissertation, Univ. Calif., Riverside. 148 pp.

Kaosiri, T., and Zentmyer, G. A. 1980. Protein, esterase, and peroxidase patterns in the *Phytophthora palmivora* complex from cacao. Mycologia 72:988-1000.

Kaosiri, T., Zentmyer, G. A., and Erwin, D. C. 1978. Stalk length as a taxonomic criterion for *Phytophthora palmivora* isolates from cacao. Can. J. Bot. 56:1730-1738.

Kellam, M. K., and Zentmyer, G. A. 1981a. Isolation of *Phytophthora citrophthora* from cocoa in Brazil. (Abstr.) Phytopathology 71:230.

Kellam, M. K., and Zentmyer, G. A. 1981b. Natural occurrence of oospores of "*Phytophthora palmivora* MF 4" (*P. capsici*) in cocoa pods in Brazil. (Abstr.) Phytopathology 71:885.

Ko, W. H. 1978. Heterothallic *Phytophthora:* Evidence for hormonal regulation of sexual reproduction. J. Gen. Microbiol. 107:15-18.

Ko, W. H., and Chang, H. S. 1979. *Phytophthora katsurae*, a new name for *P. castaneae.* Mycologia 71:840-844.

Ko, W. H., Chang, H. S., and Su, H. J. 1977. Distribution of *Phytophthora cinnamomi* mating types in Taiwan. (Abstr.) Proc. Am. Phytopathol. Soc. 4:98.

Kopp, L. E. 1966. A taxonomic revision of the genus *Persea* in the Western Hemisphere (Persea-Lauraceae). Mem. N.Y. Bot. Gard. 14:1-120.

Moss, M. O., Jackson, R. M., and Rogers, D. 1975. The characterization of 6-(pent-1-enyl)-α-pyrone from *Trichoderma viride.* Phytochemistry 14:2706-2708.

Newhook, F. J., Waterhouse, G. M., and Stamps, D. J. 1978. Tabular key to the species of *Phytophthora* de Bary. Mycol. Pap. 143. Commonw. Mycol. Inst., Kew, Surrey, England. 20 pp.

Rands, R. D. 1922. Streepkanker van kaneel, veroorzaakt door *Phytophthora cinnamomi* n. sp. Meded. Inst. Plantenziekt. No. 54. 53 pp.

Reeves, R. J., and Jackson, R. M. 1972. Induction of *Phytophthora cinnamomi* oospores in soil by *Trichoderma viride.* Trans. Br. Mycol. Soc. 59:156-159.

Reeves, R. J., and Jackson, R. M. 1974. Stimulation of sexual reproduction in *Phytophthora* by damage. J. Gen. Microbiol. 84:303-310.

Sansome, E. R., Brasier, C. M., and Griffin, M. J. 1975. Chromosome size differences in *Phytophthora palmivora,* a pathogen of cocoa. Nature (London) 255:704-705.

Tucker, C. M. 1931. Taxonomy of the genus *Phytophthora* de Bary. Univ. Mo. Agric. Exp. Stn. Bull. 153. 208 pp.

Turner, P. D., and Asomaning, E. J. A. 1962. Root infection of *Theobroma cacao* by *Phytophthora palmivora.* Trop. Agric. 39:339-343.

Wager, V. A. 1942. *Phytophthora cinnamomi* and wet soil in relation to the dying-back of avocado trees. Hilgardia 14:519-532.

Waterhouse, G. M. 1970. The genus *Phytophthora* de Bary. Mycol. Pap. 122. 2nd ed. Commonw. Mycol. Inst., Kew, Surrey, England. 104 pp.

Waterhouse, G. M. 1974. Other *Phytophthora* species recorded on cacao. Pages 71–79 in: *Phytophthora* Disease of Cocoa. P. H. Gregory, ed. Longman, London. 348 pp.

Zaki, A. I., Zentmyer, G. A., Sims, J. J., and Keen, N. T. 1982. Stimulation of sexual reproduction in the A2 mating type of *Phytophthora cinnamomi* by oleic acid and lipids from avocado roots. Phytopathology. (In press)

Zentmyer, G. A. 1976. Distribution of the $A^1$ mating type of *Phytophthora cinnamomi.* Phytopathology 66:701-703.

Zentmyer, G. A. 1977. Origin of *Phytophthora cinnamomi.* Evidence that it is not an indigenous fungus in the Americas. Phytopathology 67:1373-1377.

Zentmyer, G. A. 1979. Stimulation of sexual reproduction in the $A^2$ mating type of *Phytophthora cinnamomi* by a substance in avocado roots. Phytopathology 69:1129-1131.

Zentmyer, G. A. 1980. *Phytophthora cinnamomi* and the diseases it causes. Monogr. 10. Am. Phytopathol. Soc., St. Paul, MN. 96 pp.

Zentmyer, G. A., and Guillemet, F. B. 1981. Evidence for strains of *Phytophthora cinnamomi.* Plant Dis. 65:475-477.

Zentmyer, G. A., Mircetich, S. M., and Mitchell, D. M. 1968. Tests for resistance of cacao to *Phytophthora palmivora.* Plant Dis. Rep. 52:790-791.

Zentmyer, G. A., Leary, J. V., Klure, L. J., and Grantham, G. L. 1976. Variability in growth of *Phytophthora cinnamomi* in relation to temperature. Phytopathology 66:982-986.

Zentmyer, G. A., Kaosiri, T., and Idosu, G. 1977. Taxonomic variants in the *Phytophthora palmivora* complex. Trans. Br. Mycol. Soc. 69:329-332.

Zentmyer, G. A., Kaosiri, T., Idosu, G. O., and Kellam, M. K. 1981. Morphological forms of *Phytophthora palmivora.* Pages 291-295 in: Proc. Int. Cocoa Res. Conf. 7th, Douala, Cameroon, Nov. 4-12, 1979. J. de Lafforest, ed. Transla-Inter Ltd., London.

# 2

# Cytology of *Phytophthora*

Don E. Hemmes

CONTENTS



## INTRODUCTION

In this chapter on the cytology of *Phytophthora*, I would like to summarize and illustrate the ultrastructural cytology of spore development and germination in *Phytophthora*, placing emphasis on wall morphogenesis and the role of selected organelles in wall deposition. This information has been gathered by a variety of modern ultrastructural research techniques, including interference contrast microscopy, transmission electron microscopy of thin sections, scanning electron microscopy, and freeze fracture. Studies that have incorporated histochemical staining and cellular fractionation into the analysis of cellular organelles have been especially useful in correlating structural components to their function. Before beginning, I think one should pause to admire the marvelous, exacting cytologic work accomplished by early investigators of *Phytophthora* who were able to describe nuclear behavior and cytoplasmic changes occurring in dense spores with only the aid of light microscopy (Blackwell, 1943; Murphy, 1918; Pethybridge, 1913). This groundwork gives us a basis for our ultrastructural investigations into the cell biology of *Phytophthora*.

*Phytophthora* has proven to be attractive for cytologic studies of morphogenesis and wall formation because of the relative ease with which large numbers of synchronously developing propagules can be produced. In some species, sporangial development can be induced by exposure to light (Aragaki and Hine, 1963), zoospores can be triggered to encyst by agitation with a Vortex mixer (Tokunaga and Bartnicki-Garcia, 1971a), and gametangial interaction can be studied within hours after combining minced mycelia of the proper mating types (Hemmes and Bartnicki-Garcia, 1975). With these advantages of rapid and synchronous development, and with many studies of related fungi to make comparisons and extend our knowledge (Bartnicki-Garcia and Hemmes, 1976; Heath, 1976), considerable progress has been made during the past 10 yr in understanding the cytologic events accompanying morphogenesis and differentiation in *Phytophthora*.

## CYTOLOGIC COMPOSITION AND ORGANELLES

The cytoplasm of *Phytophthora* undergoes drastic changes in composition as the mycelium and spores develop and age. Hyphal tips and young, developing spores of *Phytophthora* contain "functional" cytoplasm consisting of a large proportion of typical eucaryote organelles, including nuclei, mitochondria, dictyosomes, rough endoplasmic reticulum, unattached ribosomes, microbodies, and an assortment of vesicles. In aged hyphae and sporangia and in fully developed chlamydospores and oospores, there is a dramatic change to a "reserve" cytoplasm dominated by lipid bodies and vacuoles. Nuclei are sequestered in small pockets of cytoplasm between these reserve materials.

### Endomembrane System

A functional relationship between cisternae of endoplasmic reticulum and/or the nuclear envelope, dictyosomes, and the production of secretory vesicles, as elucidated by Morré et al (1971), is obvious in *Phytophthora*. Dictyosomes are invariably found in close association with a cisterna of endoplasmic reticulum or the nuclear envelope (Fig. 1). Small vesicles appear to bud from the endoplasmic reticulum or nuclear envelope and fuse to form the first cisterna of the forming face of the dictyosome (Fig. 8). Cisternae at the distal side of the dictyosomes inflate to form secretory vesicles of varying types. In some cases, the contents of alternate cisternae within a dictyosome differ in density, suggesting that more than one secretory product may be packaged within a single dictyosome (Fig. 1) (Hemmes and Ribeiro, 1977).

### Mitochondria

Mitochondria are abundant in the "functional" cytoplasm of *Phytophthora* (Figs. 2, 4, 22, 25, 34, and 40). Similar in morphology to those found in animal and plant cells, these organelles typically measure $0.5 \times 1.0$ $\mu$m and are surrounded by two unit membranes, the innermost having numerous fingerlike cristae. Longer mitochondria, up to 10 $\mu$m long, are often seen in hyphae. Calcium-binding substrates have been

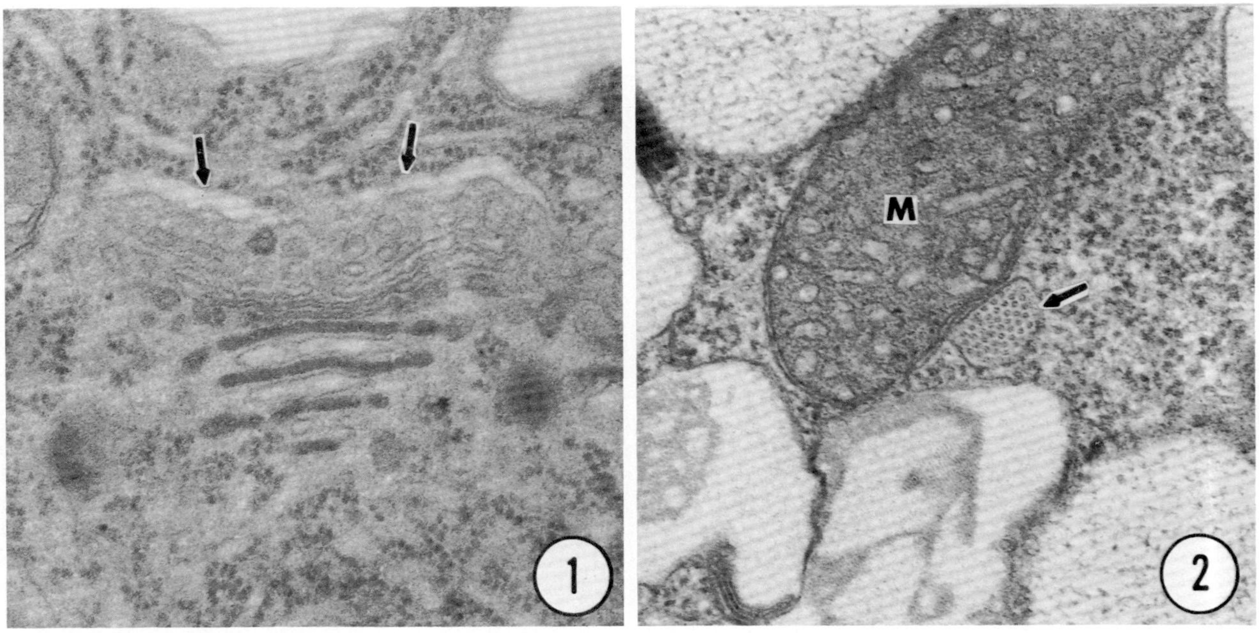


**Figs. 1 and 2. 1,** Dictyosome in oogonium of *Phytophthora megasperma* f. sp. *glycinea* in which the contents of alternate cisternae differ in density. Cisterna of endoplasmic reticulum associated with dictyosome (arrows). (×30,000) (Reprinted, by permission, from Hemmes and Ribeiro, 1977) **2,** A cisterna of endoplasmic reticulum (arrow) containing microtubles as commonly seen in sporangia of *P. palmivora*. M = Mitochondrion. (×40,000)

localized in the space between the outer and inner membranes of mitochondria of *P. palmivora* (Fig. 24) (Hemmes and Pinto da Silva, 1980).

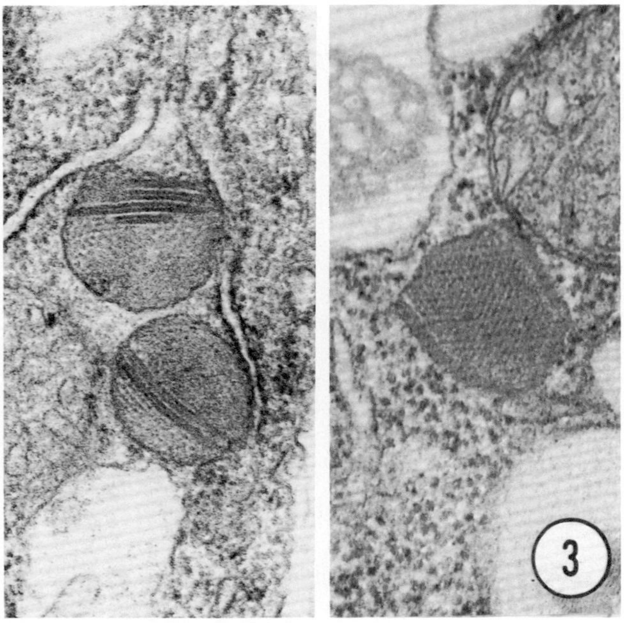


**Fig. 3.** The typical bullet-shaped microbody of *Phytophthora*, as seen in a sporangium of *P. palmivora*, contains an array of fibrils and a marginal plate (right). Occasionally, a series of plates traverse the interior of these organelles, as in the two microbodies in a hypha of *P. megasperma* f. sp. *medicaginis* (left). (×50,000)

**Microtubules**

Classical cytoplasmic microtubules with a cross-sectional diameter of 25 nm are rarely observed in hyphae or spores of *Phytophthora*, although they are found in the mitotic spindle and the complex of "rootlets" anchoring flagella in zoospores. On rare occasions, I have observed single microtubules traversing the cytoplasm in hyphae, but their involvement in morphogenesis or directed growth remains unclear. However, within tubular cisternae of rough endoplasmic reticulum, bundles of 20–50 smaller microtubules, each microtubule approximately 10 nm in diameter, are commonly observed in sporangia and occasionally in other stages of the life cycle (Fig. 2). Although the fate of these microtubules has not been traced in *Phytophthora*, in *Pythium middletonii* they appear to be transferred through dictyosomes and dictyosome vesicles to the surface of flagella, where they appear as flimmer hairs or mastigonemes (Bracker et al, 1970). Large numbers of these cisternae with their microtubular components accumulate in the cytoplasm of aged sporangia of *P. palmivora* (Hemmes and Hohl, 1975).

**Microbodies**

Small bullet-shaped organelles 0.2–0.3 μm wide and 0.3–0.5 μm long (Figs. 3, 7, 22, and 25) have been found in sporangia and zoospores of *Phytophthora*; they have been called procentrioles (Hohl and Hamamoto, 1967), parastrasomes (Williams and Webster, 1970), and microbodylike organelles (Bimpong and Hickman, 1975; Hemmes and Hohl, 1969, 1971; Ho et al, 1968a; Reichle, 1969b). These cytoplasmic structures are bounded by a single membrane, 6.5–7.5 nm thick, and are filled with a finely granular matrix interspersed with an array of tiny fibrils approximately 7.5 nm in diameter (Fig. 3). A marginal plate 16 nm thick usually underlies the membrane in the U-shaped portion of the organelle; occasionally, a series of longitudinal plates traverse the interior of the body between the fibrils (Fig. 3).

Early investigators suggested that these organelles might represent procentrioles, because they were often found in pairs lying at right angles to each other in the sporangium (Hohl and Hamamoto, 1967; Williams and Webster, 1970). Some doubt was cast on this idea when large numbers of these inclusions were observed scattered under the plasma membrane of zoospores and when the internal ultrastructure did not resemble that of a centricle. More recently, histochemical and biochemical studies have demonstrated catalase activity associated with these organelles (Philippi et al, 1975), giving evidence that their chemical properties are similar to microbodies found in higher plants and animals.

Several subclasses of microbodies, including peroxisomes and glyoxisomes, have been characterized in plant and animal cells. Peroxisomes contain oxidases that oxidize amino acids and uric acid in animals and function in glycolate oxidation during photorespiration in plants. Catalase oxidizes hydrogen peroxide generated in these reactions to water and oxygen in a detoxification reaction. Glyoxisomes contain the enzymes for fatty acid oxidation and for the conversion of fats to carbohydrates through the glyoxylate cycle. Although the bullet-shaped microbodies do not resemble glyoxisomes described in plants, it is tempting to speculate that these microbodies, so prevalent at the periphery of zoospores (Fig. 22), might contain, in addition to catalase, the glyoxylate cycle enzymes, isocitrate lyase and malate synthetase, known to be present in zoospores of *Phytophthora* (Bimpong, 1975).

These bullet-shaped microbodies become randomly distributed in the cytoplasm after encystment (Bimpong and Hickman, 1975) and often appear without the marginal plate (Hemmes and Hohl, 1971). Other microbodylike organelles with concentric lamellae or paracrystalline inclusions are occasionally seen in other stages of the life cycle of *Phytophthora* (Hemmes and Hohl, 1971; Hemmes and Ribeiro, 1977).

### "Peripheral" Vesicles

Large, spherical vesicles 0.7–1.0 $\mu$m in diameter with finely fibrillar contents are commonly observed in sporangia and at the periphery of zoospores of *Phytophthora* (Fig. 22) and other oomycetes (Grove and Bracker, 1978; Hoch and Mitchell, 1972b; Lunney and Bland, 1976). These characteristic vesicles were named "peripheral" vesicles by Grove (1971) for *Pythium* and have been variously described in *Phytophthora* spp as "fibrillar" vacuoles (Hemmes and Hohl, 1971), vesicles with "granular to fibrous structure" (Reichle, 1969b), "granular" vesicles (Bimpong and Hickman, 1975), and "peripheral" vesicles (Hemmes and Pinto da Silva, 1980; Pinto da Silva and Nogueira, 1977; Sing and Bartnicki-Garcia, 1975a). These vesicles are preformed at some unknown site in the mycelium or in the cortical cytoplasm of germinating chlamydospores and oospores. For example, numerous newly formed peripheral vesicles can be seen in the cortical cytoplasm of a germinating oospore of *P. megasperma* f. sp. *medicaginis* (Fig. 40). The vesicles often abut on clusters of lipidlike bodies in the germinating oospore, but the exact mode of development of the peripheral vesicles is difficult to interpret in micrographs. The peripheral vesicles are seen in hyphae and flow into the sporangium during the sporangial expansion phase, where they become closely associated with cleavage vesicles (Fig. 12). Thus upon cleavage, peripheral vesicles are automatically situated at the zoospore periphery where they function in the encystment process. When zoospores encyst, the peripheral vesicles fuse with the plasma membrane to form bilayer membrane diaphragms, which then rupture to release the vesicle contents to the exterior of the zoospore (Hemmes and Hohl, 1971; Pinto da Silva and Nogueira, 1977). The released materials include glycoprotein with receptor sites for concanavalin A (Sing and Bartnicki-Garcia, 1975b) and possibly enzymes and/or precursors for microfibril synthesis. The glycoprotein appears to function in cyst wall formation by forming a matrix for wall neogenesis and also appears to be responsible for adhesion of zoospores to various substrates (Sing and Bartnicki-Garcia, 1975a).

### "Fingerprint" Vacuoles

"Fingerprint" vacuoles (referred to as vacuoles with crystalline contents by Ho et al [1968a] and Reichle [1969b], fingerprint vacuole by Hemmes and Hohl [1971], liposomes by Williams and Webster [1970], vesicles with striated configuration by King et al [1968], and crystalline vesicle by Bimpong and Hickman [1975]) are large, ovoid vacuoles 0.7–5.0 $\mu$m in diameter occurring in nearly all stages of the life cycle of *Phytophthora* (Figs. 5–7, 17, 22, and 25) and other oomycetes (Grove and Bracker, 1978; Hoch and Mitchell, 1972a; Lunney and Bland, 1976). In the "functional" cytoplasm of sporangia, zoospores, cysts, and developing oogonia, these vacuoles contain electron-dense material that exhibits a paracrystalline, lamellar array resembling a fingerprint pattern (Hemmes and Bartnicki-Garcia, 1975; Nogueira et al, 1977). This lamellated material condenses into solid, electron-opaque, spherical globules in sporangia 7–15 days old (Hemmes and Hohl, 1975; King et al, 1968) and "reserve" cytoplasm of chlamydospores (Hemmes and Wong, 1975) and oospores (Hemmes and Bartnicki-Garcia, 1975; Hemmes and Ribeiro, 1977). After this conversion, the vacuoles usually fuse, depositing the opaque globules into a large, central vacuole. The ooplast, the refractile sphere seen in oospores by light microscopy, appears to form by the coalescence of these dense globules (Figs. 38 and 39) (Hemmes and Bartnicki-Garcia, 1975).

Some early studies suggested that the densely staining lamellated configuration might represent lipid materials. The vacuoles gave positive reactions with Sudan and Nile blue dyes and became electron transparent when extracted with methanol-chloroform (Williams and Webster, 1970). However, the striated structure was not removed after extraction with several lipid solvents in *Aphanomyces euteiches* (Hoch and Mitchell, 1972a). More recently, "fingerprint" vacuoles have been isolated from zoospores of *P. palmivora* and shown to have a high content of the reserve polysaccharides mycolaminaran and mycolaminaran phosphate (Powell and Bracker, 1977). Furthermore, studies of the reserve polysaccharides of *Phytophthora* have shown a direct correlation between the presence of fingerprint vacuoles with a lamellate appearance and the occurrence of mycolaminaran phosphate (Wang and Bartnicki-Garcia, 1980).

### Nuclear Behavior

Interphase nuclei containing dense nucleoli and peripheral chromatin are commonly observed in thin sections of hyphae and developing sporangia, chlamydospores, and oogonia of *Phytophthora*. Grazing sections of these organelles illustrate the presence of numerous, evenly spaced pores over the surface

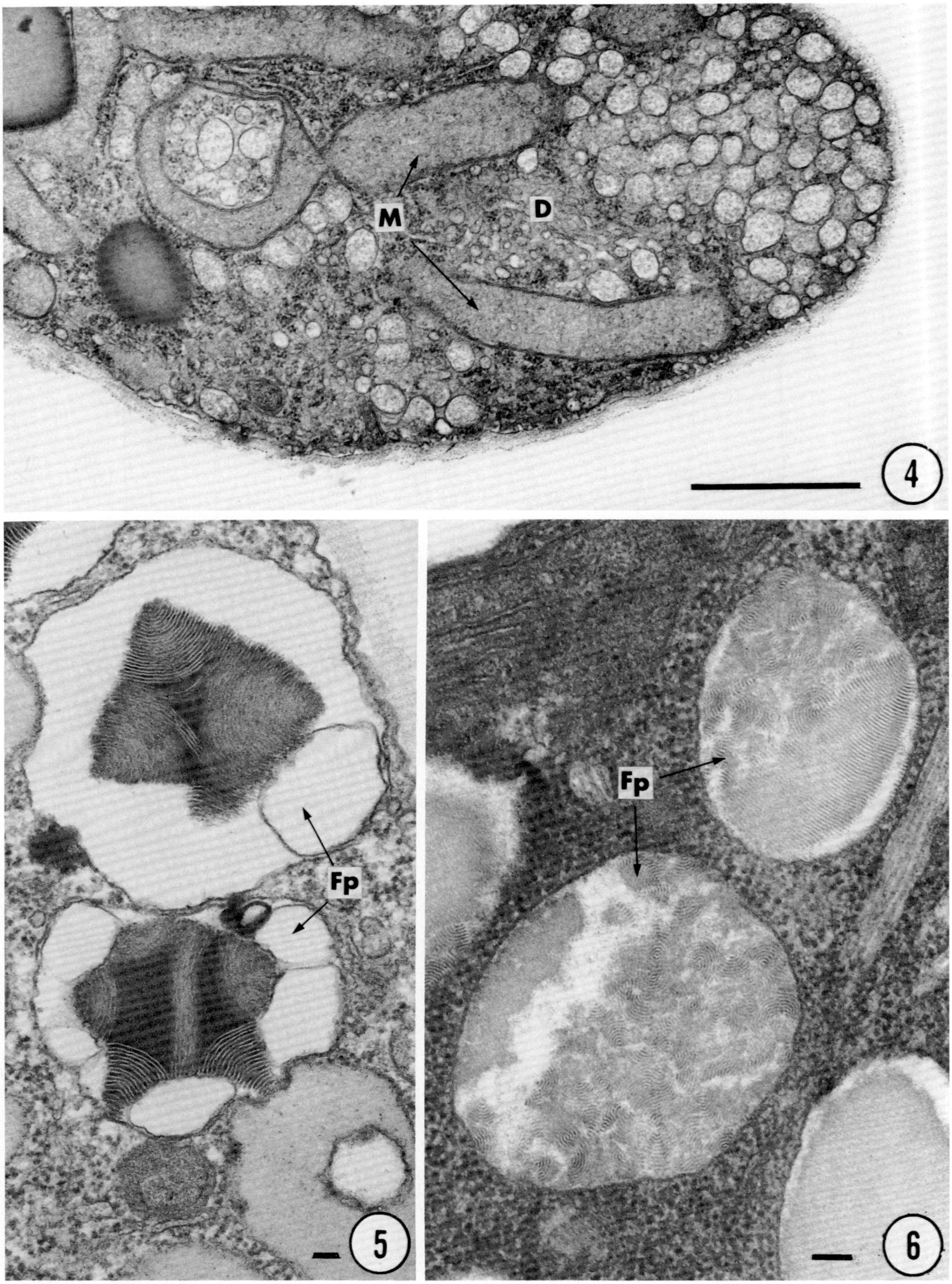


**Figs. 4–6. 4,** Hyphal tip of *Phytophthora megasperma* f. sp. *medicaginis*. Note the two types of secretory vesicles at the tip. D = Dictyosome, M = mitochondria. Bar represents 1 μm. (×29,000) **5,** Fingerprint vacuoles (Fp) in a recently encysted zoospore of *P. palmivora* fixed in calcium-containing glutaraldehyde. Bar represents 0.1 μm. (×45,000) **6,** Fingerprint vacuoles (Fp) in a zoospore of *P. palmivora*. Bar represents 0.1 μm. (×70,000)

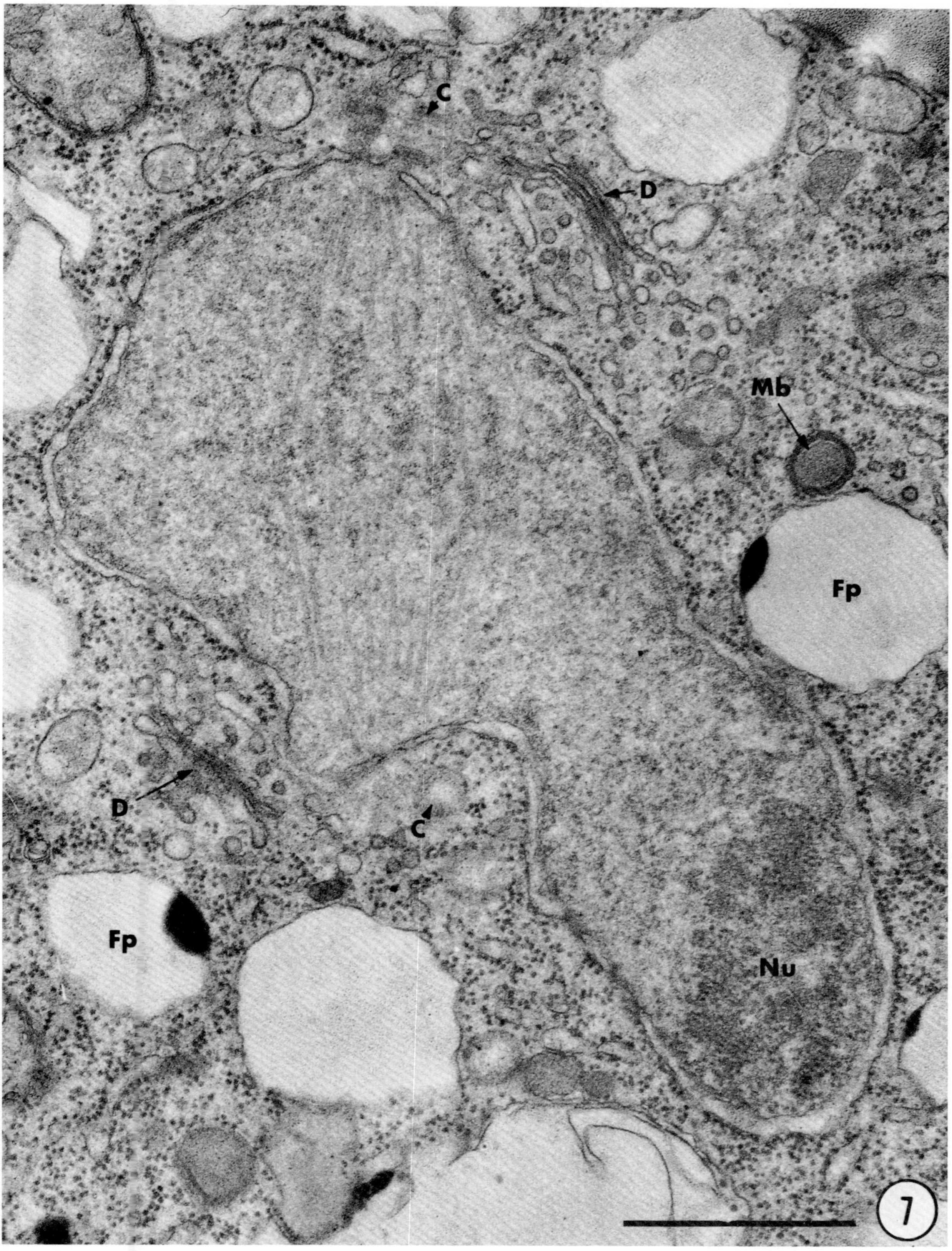


**Fig. 7.** Dividing nucleus of *Phytophthora palmivora*, showing the intranuclear spindle and persistent nucleolus (Nu). Note the nuclear pores and ribosomes attached to the cytosol face of the outer nuclear membrane. D = Dictyosomes, C = centrioles, Fp = fingerprint vacuoles with condensed contents, Mb = microbody. Bar represents 1 $\mu$m. (×40,000) (Reprinted, by permission, from Hemmes and Hohl, 1973)

of the nuclear envelope. The cytosol surface of the envelope has attached ribosomes and can occasionally be seen to be continuous with the endoplasmic reticulum.

Locating stages in the life cycle of *Phytophthora* where synchronously dividing mitotic or meiotic nuclei can be studied by electron microscopy is difficult. Fortunately, mitosis occurs fairly frequently and predictably in a stage of the life cycle of *Phytophthora*, that of aged sporangia of *P. palmivora*, where a round of mitosis occurs in nuclei just before secondary sporangial formation (Hemmes and Hohl, 1973).

Mitosis in *Phytophthora* is intranuclear, as the nuclear envelope is retained throughout the division process. The nucleolus also persists outside the spindle apparatus in an

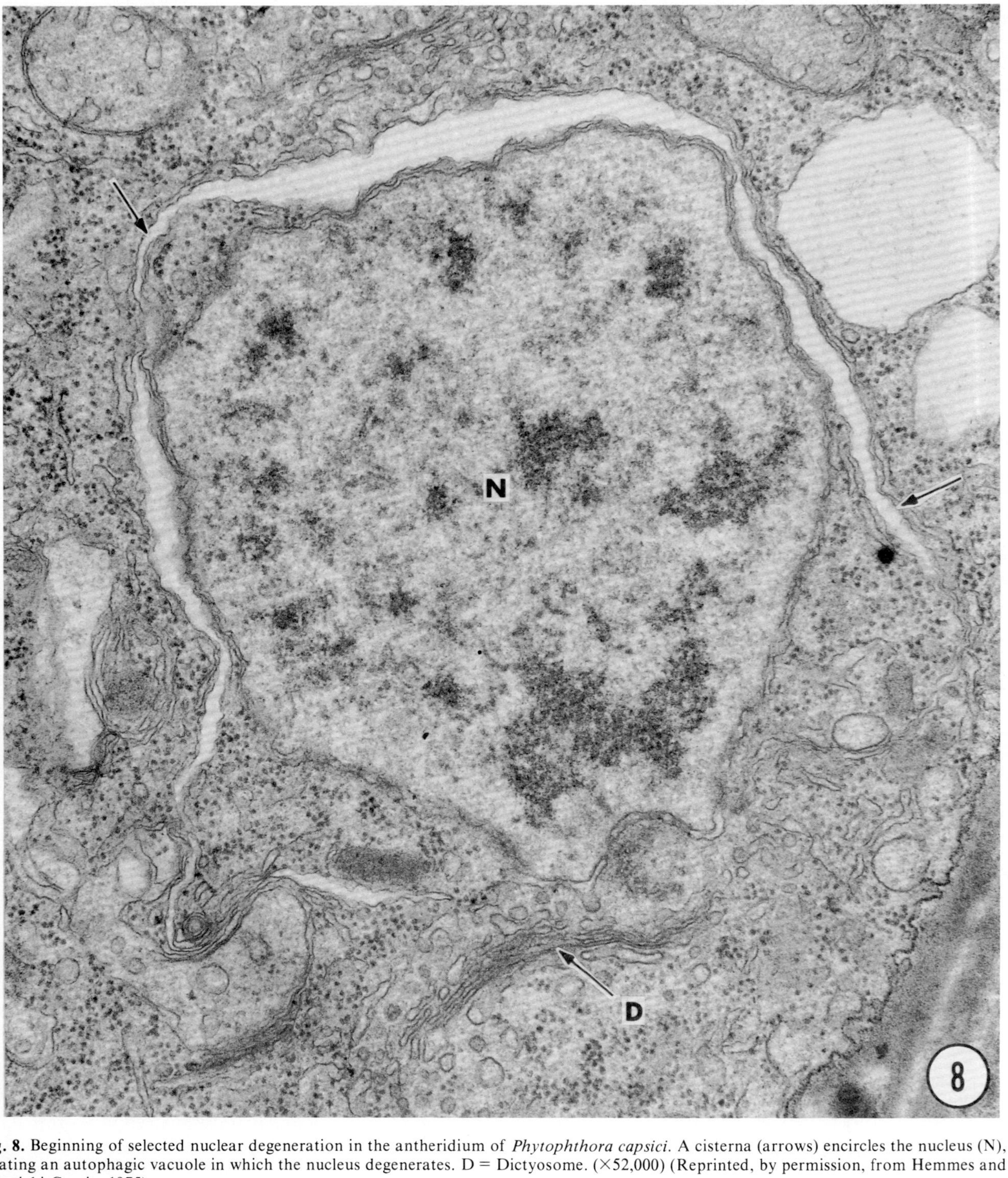


**Fig. 8.** Beginning of selected nuclear degeneration in the antheridium of *Phytophthora capsici*. A cisterna (arrows) encircles the nucleus (N), creating an autophagic vacuole in which the nucleus degenerates. D = Dictyosome. (×52,000) (Reprinted, by permission, from Hemmes and Bartnicki-Garcia, 1975)

outpocketing of the nuclear envelope (Fig. 7). The centrioles divide at an early stage in mitosis and migrate to the spindle poles. During anaphase and telophase, the nucleus elongates and pinches in the middle to separate the daughter nuclei.

The ultrastructure of meiotic division in *Phytophthora* has not been reported. Studies of gametangial interaction in *Phytophthora* thus far have concentrated on wall deposition and morphogenesis rather than on nuclear behavior. An extensive study of early gametangial interaction by thin sectioning and electron microscopy should be fruitful in locating synaptonemal complexes as in other oomycetous fungi (Beakes, 1980; Ellzey, 1974; Ellzey and Huizar, 1977).

Nuclei are commonly eliminated from sporangial and gametangial cytoplasm during phases of aging and cytoplasmic reduction by selective autolysis (Hemmes and Bartnicki-Garcia, 1975 Hemmes and Hohl, 1973). Cisternae and lysosomelike organelles of unknown origin encircle the targeted nuclei, forming an autophagic vacuole in which the nuclear envelope and nucleoplasm disintegrate (Fig. 8).

## DEVELOPMENTAL STAGES

### Mycelium

A gradual transition from a largely "functional" cytoplasm filled with nuclei, mitochondria, dictyosomes, and secretory vesicles to a "reserve" cytoplasm dominated by vacuoles and lipid bodies is apparent as one progresses from hyphal tips to subapical zones of hyphae of *Phytophthora* (see Grove et al [1970a] for a spectacular illustration of this gradation in hyphae of *Pythium ultimum*). Nuclei are sequestered in pockets of "functional" cytoplasm in older parts of the mycelium.

The cytoplasmic composition of mycelia of *Phytophthora* growing under differing culture conditions has not been systematically examined. Such a study might reveal subtle differences in cytoplasmic composition occurring early in mycelial growth that relate to the eventual spore types produced. For example, mycelia of *P. palmivora* growing on V-8 juice agar produce sporangia and no other spore type. The mycelium of this species under these culture conditions contains numerous peripheral vesicles that eventually flow into the sporangia and function during zoospore encystment. In contrast, mycelia of *P. capsici* growing in V-8 juice broth, which produce only oospores under these conditions, contain few, if any, peripheral vesicles (Hemmes and Bartnicki-Garcia, 1975).

#### Hyphal Tips

Hyphal tips of *Phytophthora* contain accumulations of secretory vesicles that arise from dictyosomes and/or endoplasmic reticulum in the subapical hyphal cytoplasm and migrate to the plasma membrane, where they fuse with the membrane and release their contents into the wall region (Fig. 4) (Hemmes and Hohl, 1969) (see also Grove et al [1970] and Grove and Bracker [1970] for reviews on the organization of hyphal tips in the oomycetes). Whether the hyphae are simply growing vegetatively in a culture medium (Hemmes and Hohl, 1969), functioning as gametangial initials or fertilization tubes (Hemmes and Bartnicki-Garcia, 1975), rupturing through spore walls in direct germination (Hemmes and Hohl, 1969; Hemmes and Wong, 1975), or penetrating host plant cell walls in an infection process (Fig. 26), they contain tip vesicles of two basic types: small, spherical vesicles 70–100 nm in diameter and larger, ovoid ones 200–300 nm in diameter. Few fusion profiles or vesicles of intermediate size are seen that would suggest that the smaller vesicles fuse to form the larger ones. In gametangial initials of *P. megasperma* f. sp. *glycinea* (Hemmes and Ribeiro, 1977) and *P. capsici* (Hemmes and Bartnicki-Garcia, 1975), the smaller vesicles have distinctly denser contents, whereas in germinating chlamydospores of *P. cinnamomi*, the larger vesicles have denser contents (Hemmes and Wong, 1975).

The membranes, and to a lesser extent the contents of both small and large tip vesicles, stain positively for carbohydrate and are thought to transport wall precursors and wall-synthesizing enzymes to the hyphal tip (Meyer et al, 1976). For continued growth of the hyphal tip, the apex must be kept in a pliable condition by a balance between wall-synthesizing and wall-lysing enzymes (Bartnicki-Garcia, 1973). The requisite enzymes, β-glucanase and uridine diphosphate glucose transferase, have been isolated from germinating cysts of *P. palmivora* (Meyer et al, 1976) and may be transported in the two types of vesicles observed.

### Sporangia

Sporangia of the various species of *Phytophthora* vary somewhat in shape, usually being ovoid-ellipsoid to pyriform (Fig. 9), and range in size from the very small sporangia of *P. botryosa* (14–16 μm wide × 26–31 μm long) to the large sporangia of *P. macrospora* (50–65 μm wide × 85–112 μm long (Ribeiro, 1978). Some structural variations such as pedicel length and size of the apical papilla also can be discerned by light microscopy and are useful diagnostic characters for species identification (Newhook et al, 1978; Ribeiro, 1978). Thin sections of sporangia of several different species of *Phytophthora* have been examined by electron microscopy (Chapman and Vujičić, 1965; Christen and Hohl, 1972; Ehrlich and Ehrlich, 1965; Elsner et al, 1970; Hemmes and Hohl, 1969; Hohl and Hamamoto, 1967; King et al, 1968; Williams and Webster, 1970). Sporangia, which are multinucleate and contain largely "functional" cytoplasm, develop flagella and a system of cleavage vesicles or cisternae

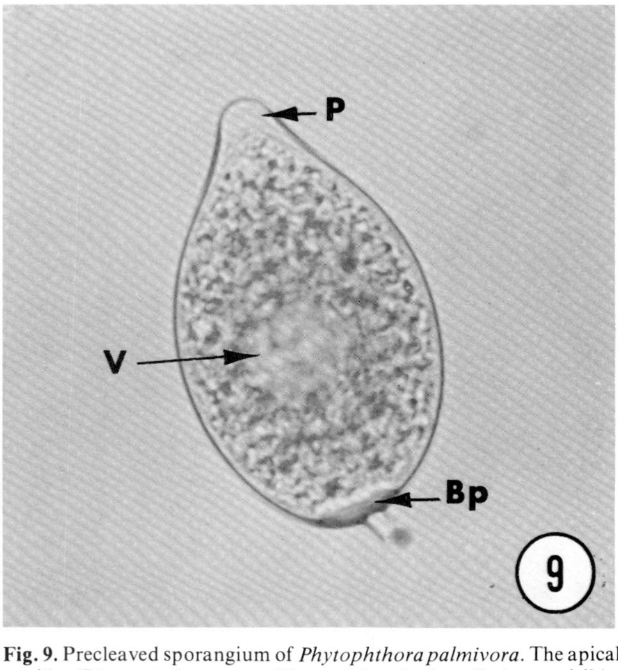


**Fig. 9.** Precleaved sporangium of *Phytophthora palmivora*. The apical papilla (P), central vacuole (V), and basal plug (Bp) are visible. (×1,200)

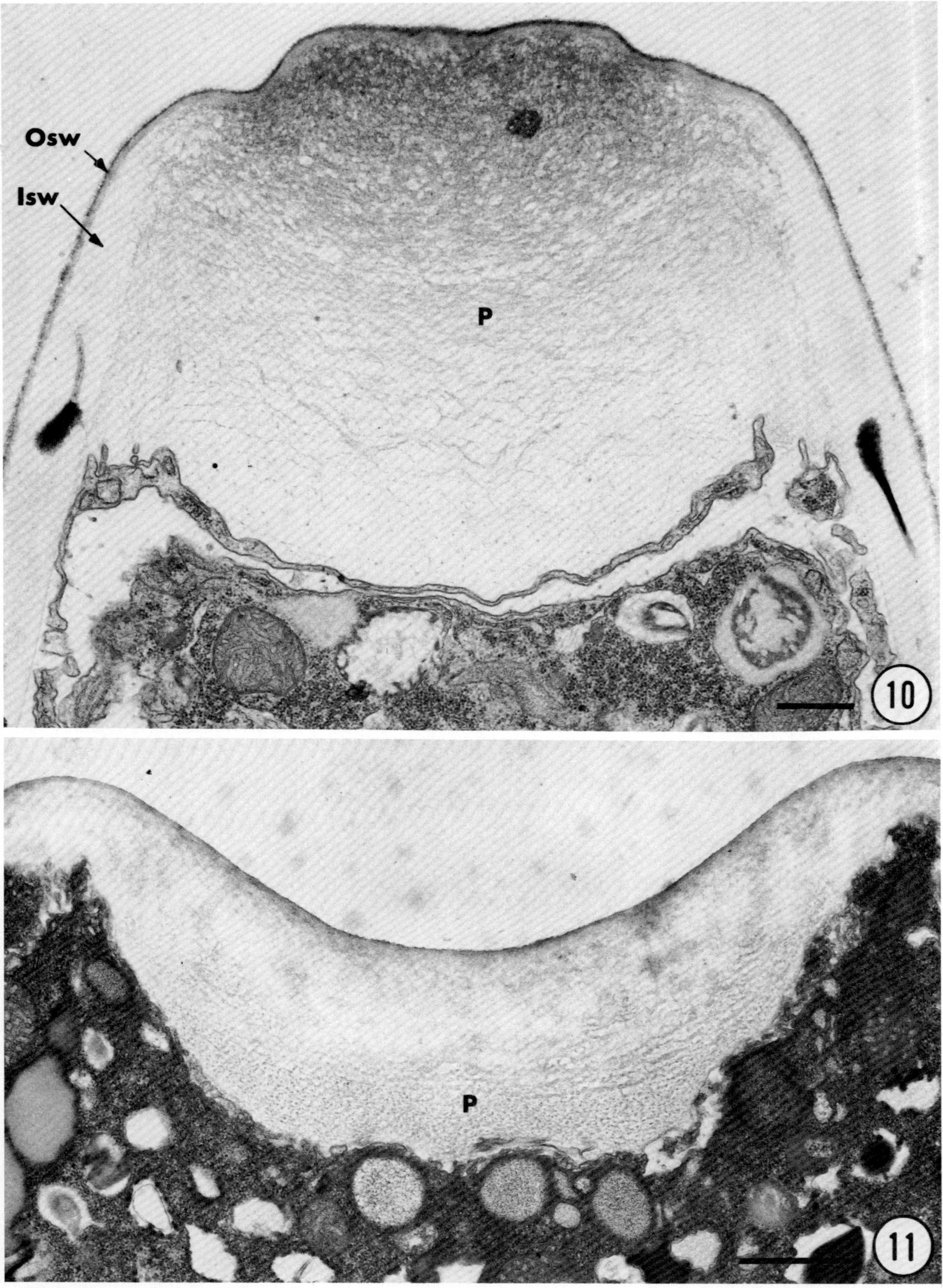


**Figs. 10 and 11.** Apical exit pores of *Phytophthora* sporangia. **10,** Papillate exit pore of *P. heveae*, showing extensive fibrillar plug material (P), which balloons into a discharge vesicle during zoospore release. Osw = Outer sporangial wall, Isw = inner sporangial wall. Bar represents 1 μm. (×13,500) **11,** Nonpapillate exit pore of *P. megasperma* var. *sojae*, showing the small amount of plug material (P) in this species. Bar represents 1 μm. (×18,000)

that fuse to cleave the cytoplasm into uninucleate zoospores (Fig. 12). Variations seen at the ultrastructural level include the thickness of the sporangial wall (from 0.5 μm in *P. megasperma* f. sp. *glycinea* to 1.5 μm in *P. palmivora*) and the thickness of the plug material in the apical exit pore (from 4–8 μm in *P. palmivora* and *P. heveae* [Fig. 10] to only 1 μm in *P. megasperma* f. sp. *glycinea* [Fig. 11]) and basal septum (from 4–6 μm in *P. palmivora* [Fig. 13] and *P. capsici* to only 1 μm in *P. megasperma* f. sp. *glycinea* [Fig. 14]).

**Sporangial Development**

A combination time-lapse and electron microscopy study of the development of sporangia of *P. palmivora* (Christen and Hohl, 1972) documented the expansion of hyphal tips into the ovoid-ellipsoidal shape of the sporangium and the flow of "functional" cytoplasm from the underlying mycelium into the expanding sporangium. The sporangial wall remains thin, approximately 0.1 μm thick, during the expansion phase. The cytoplasm flowing into the expanding sporangium contains nuclei, mitochondria, dictyosomes, microbodies, rough endoplasmic reticulum, unattached ribosomes, lipid bodies, fingerprint vacuoles, and peripheral vesicles. Although mitotic figures have been reported in expanding sporangia fixed for light microscopy (Laviola, 1974), these figures are rarely seen by electron microscopy. It appears that most nuclei are formed in the mycelium and flow into the sporangium with the cytoplasm.

Near the end of the expansion phase, a basal plug or septum and an apical exit pore containing wall-like plug material are differentiated. Also at this time, the sporangial wall may thicken considerably, starting at the basal end of the sporangium and thickening up to the exit pore. Clusters of vesicles approximately 0.25 μm in diameter can be seen budding from the cisternae of the dictyosomes and underlying the plasma membrane in regions where wall deposition is occurring. Although few fusion profiles are seen in these areas,

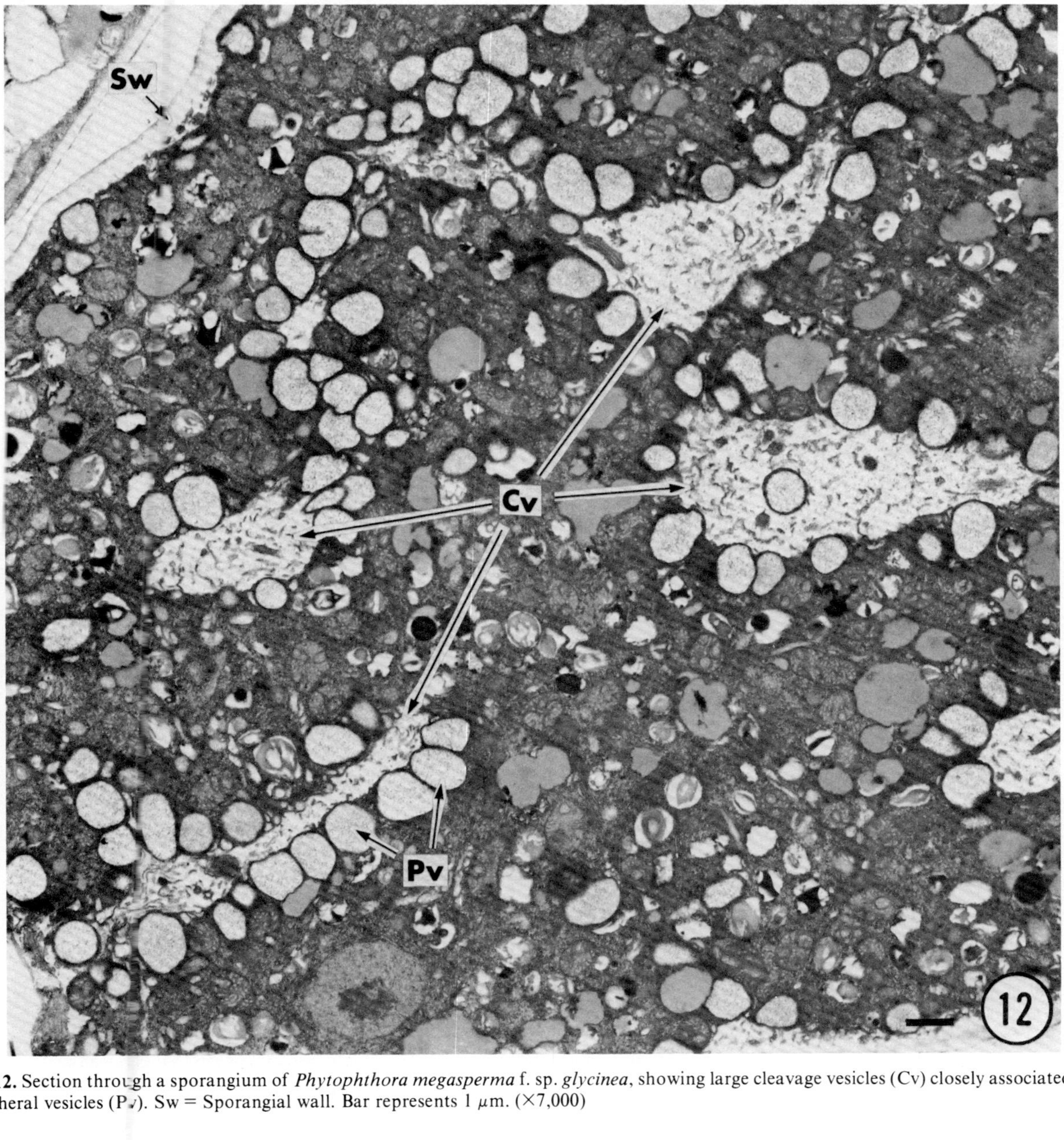


**Fig. 12.** Section through a sporangium of *Phytophthora megasperma* f. sp. *glycinea*, showing large cleavage vesicles (Cv) closely associated with peripheral vesicles (Pv). Sw = Sporangial wall. Bar represents 1 μm. (×7,000)

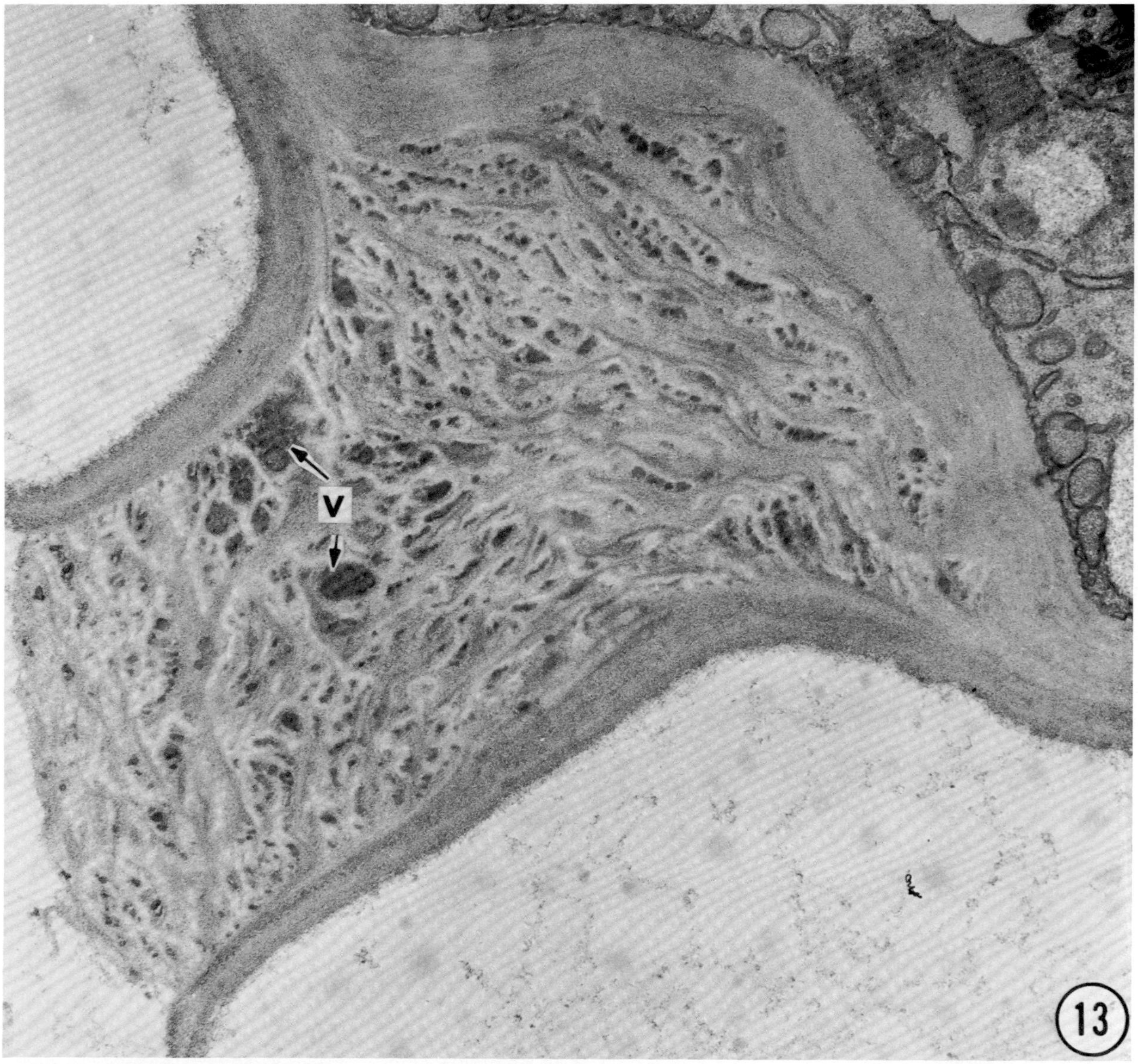


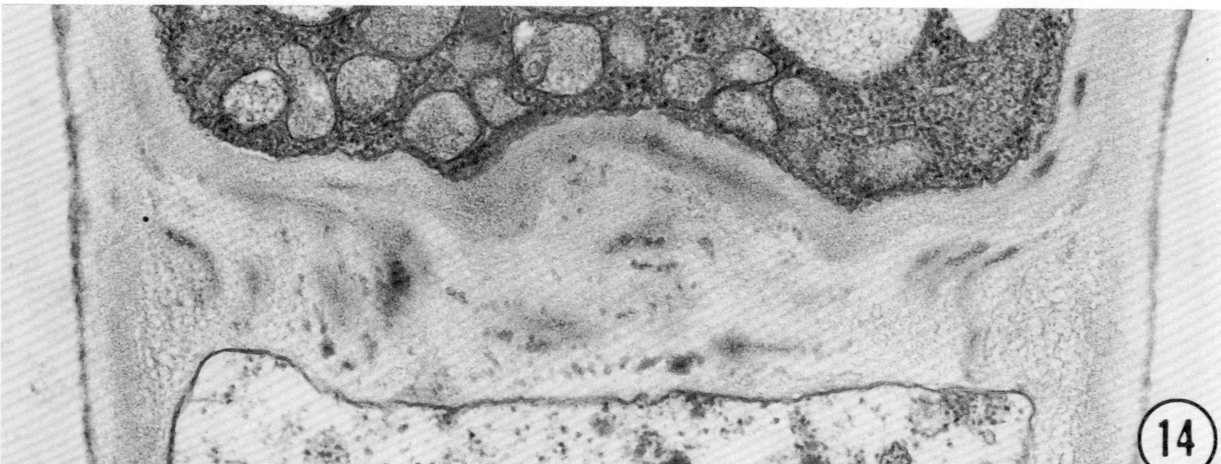


**Figs. 13 and 14. 13,** Basal sporangial plug of *Phytophthora palmivora* containing many entrapped vesicular elements (V). This is an example of the more extensive sporangial plugs observed in *Phytophthora*. Fixation was with glutaraldehyde followed by permanganate. (×28,000) (Reprinted, by permission, from Hemmes and Hohl, 1969) **14,** The relatively thin basal sporangial septum of *P. megasperma* f. sp. *glycinea*. (×27,000)

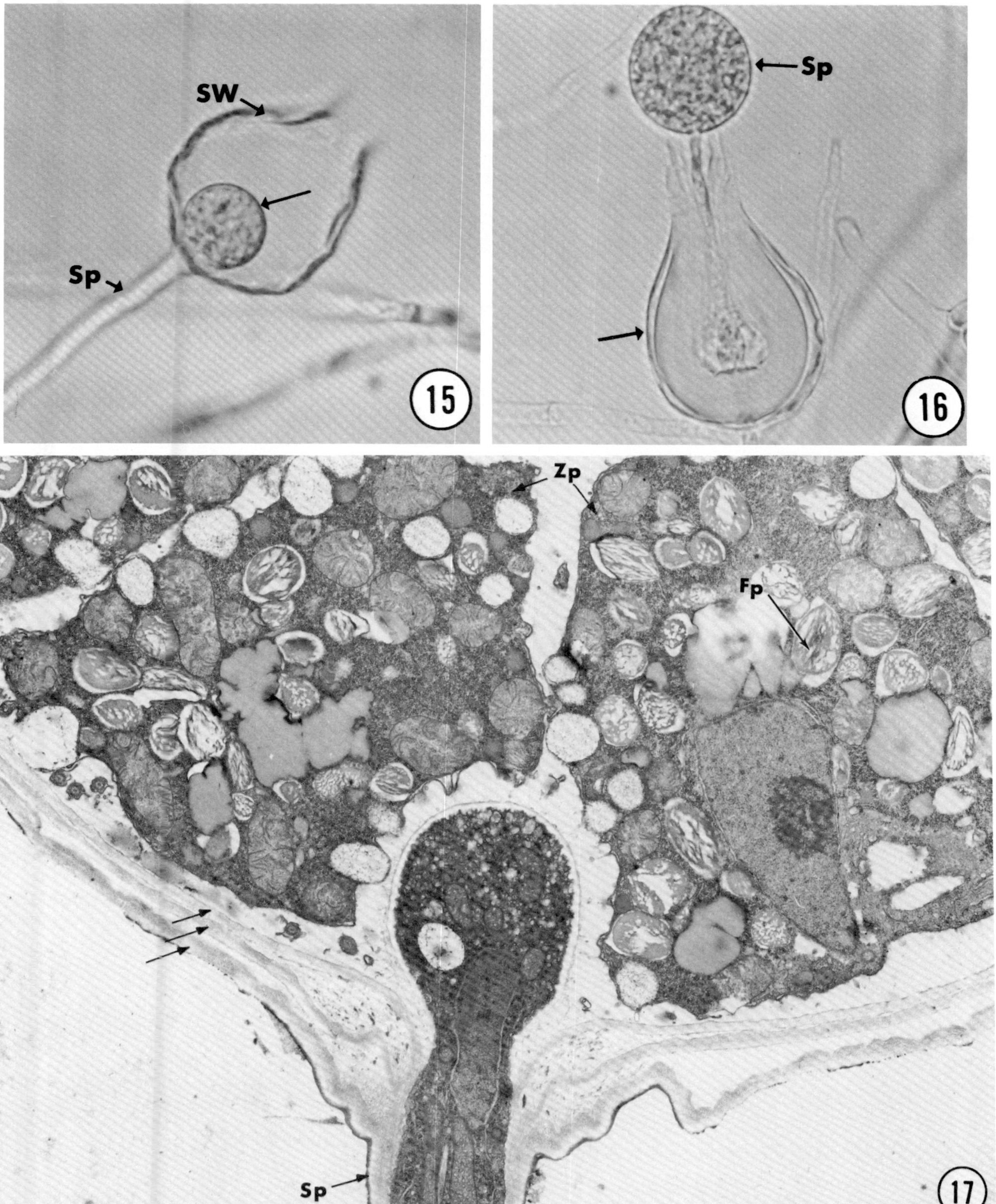


**Figs. 15–17.** "Nested" proliferation of sporangia in a high-temperature isolate of *Phytophthora megasperma.* **15,** A sporangial initial (arrow) is enlarging by cytoplasmic inflow from the sporangiophore (Sp) inside sporangial walls (SW) left by preceding sporangia. (×1,200) **16,** The primary sporangial chamber (arrow) is nearly occluded with sporangial walls left by the previous sporangia. A sporangiophore has extended through the narrow opening and is inflating to form another primary sporangium (Sp) in a repeat of the "nesting" phenomenon. (×1,200) **17,** Cytoplasm of an incipient sporangium begins to flow from the sporangiophore (Sp) into the preceding sporangium before its zoospores (Zp) are released. The three sporangial walls (arrows) surrounding the zoospores indicate that this will be the fourth sporangium to develop at this hyphal apex. Fp = Fingerprint vacuole. (×12,600) (Cultures of the high-temperature isolate of *P. megasperma* were provided by Olaf Ribeiro, University of California, Riverside)

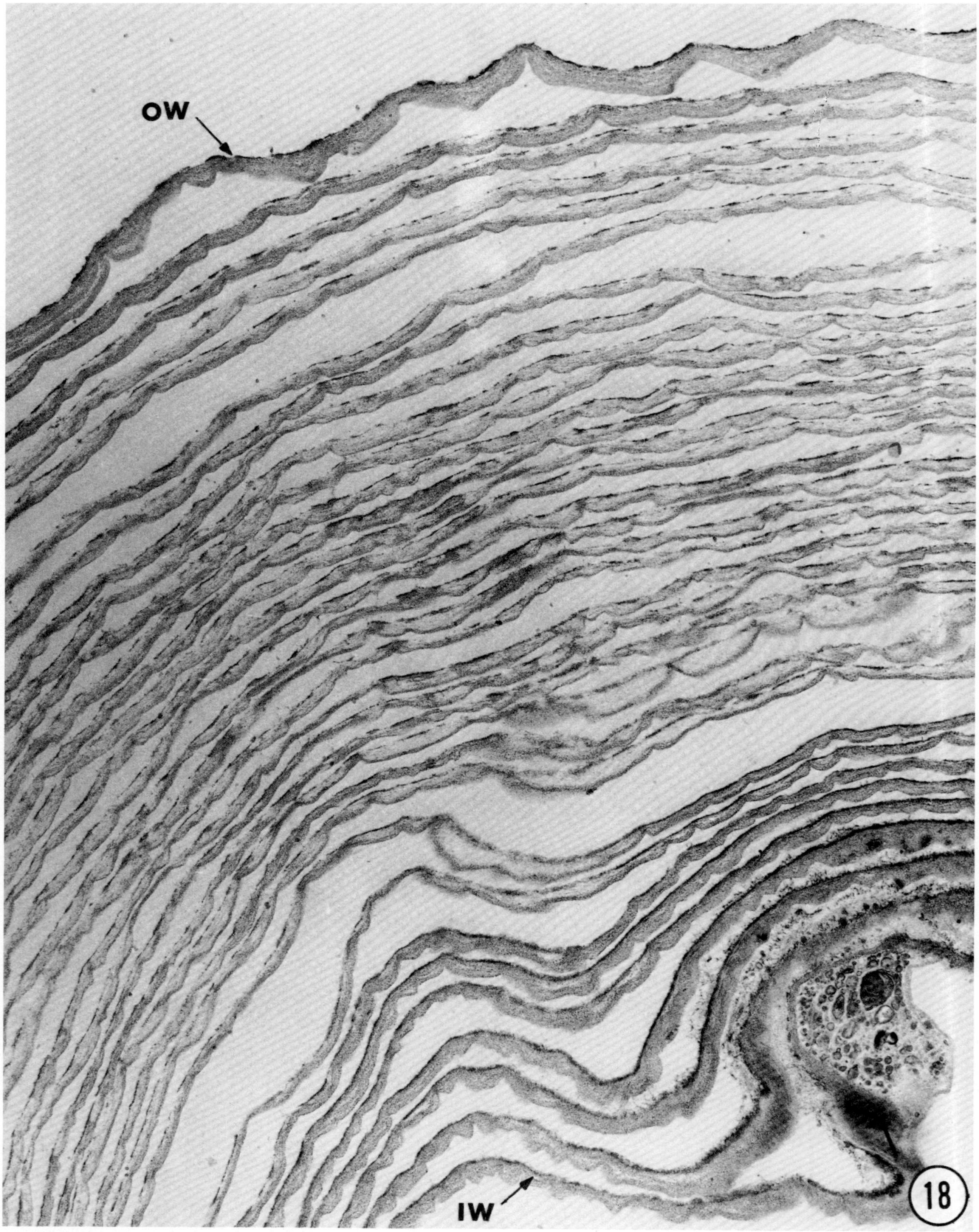


**Fig. 18.** Repeated sporangial development. Multiple sporangial walls laid down by a high-temperature isolate of *Phytophthora megasperma* after 1 wk of growth ("nested" sporangial proliferation). OW = Outermost sporangial wall, IW = innermost sporangial wall. (×14,400)

it is assumed that the dictyosome-derived vesicles fuse with the plasma membrane to deposit wall materials (Christen and Hohl, 1972). For example, vesicles are seen in the sporangiophore when irregular deposits of wall-like material are formed in apposition to the sporangiophore wall. These deposits grow inward, entrapping cytoplasmic materials and eventually forming an extensive plug 4–6 $\mu$m thick that completely occludes the sporangiophore and bulges into the base of the sporangium of *P. palmivora* (Fig. 13) and *P. capsici* (Christen and Hohl, 1972; Williams and Webster, 1970). Similar vesicles, along with smaller spherical ones 50–100 nm in diameter, accumulate at the sporangial apex when the exit pore plug material is deposited. Once these structures are formed and the sporangial wall is fully thickened, the activity of dictyosomes and the number of wall-destined vesicles decrease drastically.

The first sign of the development of flagella and a cleavage system is the reactivation of dictyosomes and the proliferation of clusters of vesicles and cisternae close to each sporangial nucleus. Some of these membranous elements disperse into the cytoplasm to function as cleavage vesicles (Hohl and Hamamoto, 1967). These vesicles range from 1 to 2 $\mu$m in diameter and appear empty except for a thin layer of amorphous material (the presumptive "fuzzy" coat of the zoospore [Nogueira et al, 1977]) coating the inner face of the vesicle membrane and are thus readily distinguished from wall-destined vesicles and peripheral vesicles. As mentioned previously, the cleavage vesicles form a close association with peripheral vesicles, already found in the cytoplasm, so that immediately upon cleavage the peripheral vesicles lie on the cytosol face of the zoospore plasma membrane (Fig. 12). Other vesicles within the aggregates accumulate around the basal bodies and fuse with one another to form the flagellar vacuole (Hemmes and Hohl, 1971; King and Butler, 1968). Flagellar axonemes then develop by outgrowth from the basal bodies into the enlarging vacuoles.

### Zoosporogenesis (Indirect Sporangial Germination)

The fusion of cleavage vesicles and release of zoospores occur spontaneously in distilled water in most species of *Phytophthora* and can be enhanced by chilling sporangia at about 10°C for 30 min before immersing them in distilled water. The cleavage vesicles line up close to the sporangial wall and in planes equidistant from neighboring nuclei and then fuse to cleave the cytoplasm into uninucleate zoospores (Hohl and Hamamoto, 1967; King et al, 1968). Zoospore cleavage begins within minutes after chilling, with zoospore release following shortly thereafter. In aged sporangia of *P. palmivora* (7–10 days old), cleavage is often aberrant so that binucleate and multinucleate zoospores are released (Hemmes and Hohl, 1975). Once cleavage has occurred, the apical exit pore plug material balloons outward to form a discharge "vesicle" varying in size, depending on the amount of plug material present (Gisi et al, 1979). The zoospores are then expelled into the discharge "vesicle" and break through its thin, plastic wall to escape.

### Repeated Sporangial Development

In some *Phytophthora* species, new sporangia are repeatedly produced inside one another in a process called "nested" proliferation (Ribeiro, 1978). While cleaved zoospores are still in the primary sporangium, cytoplasm of the sporangiophore dissolves its way through the basal septal plug and expands inside the primary sporangium to form a secondary sporangium (Figs. 15–17). After the secondary sporangium has fully expanded inside the empty primary sporangial wall, a new apical exit pore and basal septum are differentiated and the cytoplasm undergoes cleavage again. This developmental process can be repeated more than 30 times within 1 wk in cultures of a high-temperature isolate of *P. megasperma* (Ribeiro et al, 1978) incubated at 25°C in distilled water. As seen in Figure 18, a total of 33 thin concentric sporangial walls appear in thin sections of a sporangium of this isolate fixed after 1 wk of incubation (D. E. Hemmes, *unpublished*).

Other phytophthoras, such as *P. megasperma* f. sp. *glycinea*, proliferate sporangia in a variation of this process. Sporangia are repeatedly produced one inside the other, but in this case cytoplasm dissolves through the basal septum as a hyphal tip after the primary zoospores have been released and then elongates within the empty primary sporangial wall, extends out through the open apical pore, and inflates and differentiates into a secondary sporangium. This process is repeated so that long chains of sporangia develop, the terminal one always containing cytoplasm. (See Ribeiro, Chapter 4, Fig. 4B.)

### Sporangial Aging

Sporangia of *Phytophthora* have been shown to age on the mycelium (Hemmes and Hohl, 1975; King et al, 1968). After incubation of sporangia of *P. palmivora* for 7 days at 25°C, the flagellar vacuoles vesiculate, leaving the flagellar axonemes to disintegrate in the cytoplasm. Massive numbers of tubular cisternae of rough endoplasmic reticulum enclosing microtubules (presumptive flagellar hairs) accumulate, and the cytoplasm becomes highly vacuolated. Concomitant with flagellar degeneration, the sporangia lose their ability for cleavage and zoospore release. Within a few days, a thin hyphal tube is produced near the apical papillum that inflates to form a secondary sporangium. After a round of nuclear division and some nuclear degeneration (Hemmes and Hohl, 1973), all the cytoplasm flows from the primary sporangium into the secondary sporangium. New flagella, cleavage vesicles, an apical exit pore, and a basal septum are differentiated, and the aging process repeats itself. In *P. cinnamomi*, chains of 15–18 sporangia have been observed formed in this manner with cytoplasm found only in the terminal sporangium (D. E. Hemmes, *unpublished*). Each progressive sporangium is slightly smaller than the previous one, much as in the sporangial proliferation discussed previously. Long chains of sporangia may terminate in a uninucleate sporangium.

### Direct Sporangial Germination

Sporangia may emit germ tubes directly through the exit pore plug material (Fig. 19) and sporangial walls when incubated in a nutritive medium with high osmotic pressure, such as papaya juice or the commercially available oligosaccharide mixture, dextrin (Hemmes and Hohl, 1969). In direct sporangial germination, the flagella and cleavage system degenerate before germ tube formation (Hemmes and Hohl, 1972). A new germination wall is deposited underneath the sporangial wall and extends to form the hyphal wall as the germ tubes rupture through the sporangial wall to the exterior. Numerous dictyosome-derived vesicles underlie the regions of wall deposition.

## Zoospores and Cysts

### Zoospore Ultrastructure

The morphology and ultrastructure of *Phytophthora*

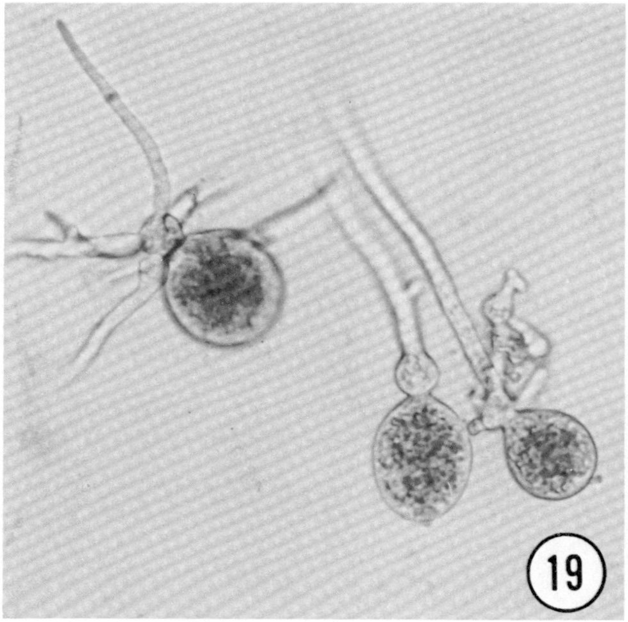


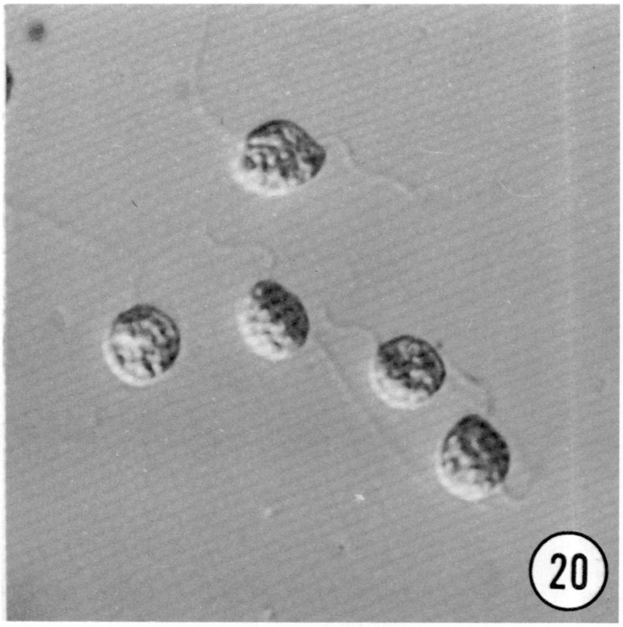


**Figs. 19 and 20. 19,** Directly germinating sporangia of *Phytophthora* in 25% papaya extract. (×300) **20,** Zoospores of *P. palmivora* as seen by Normarski interference optics. (×1,200)

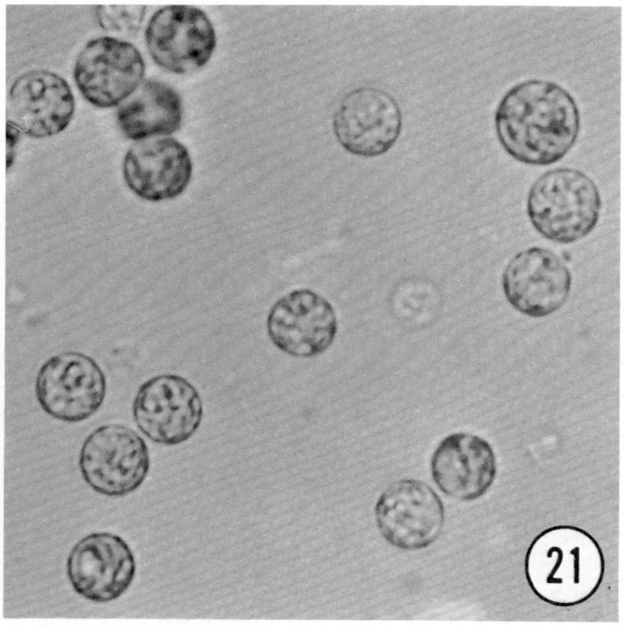


**Fig. 21.** Cysts of *Phytophthora palmivora* induced to encyst by agitation in a Vortex mixer. (×1,200)

zoospores have been thoroughly examined by interference contrast light microscopy, freeze fracture, and scanning and transmission electron microscopy (Bimpong and Hickman, 1975; Desjardins et al, 1973a; Ho et al, 1968a, 1968b; Nogueira et al, 1977; Reichle, 1969b). Zoospores are basically ovoid in shape (Fig. 20) and possess a groove that runs longitudinally along the zoospore from the bluntly pointed anterior end to the rounded posterior end of the cell. The groove is shallow at each end and deepest in the center where the two flagella, an anterior tinsel flagellum and a posterior whiplash flagellum, are inserted at a common point (see Carlile, Chapter 7, Fig. 1; Desjardins et al, 1973a; Ho et al, 1968b). A uniform "fuzzy coat" approximately 20–25 nm thick has been shown to cover the outer surface of the plasma membrane and surface of the flagella (Hemmes and Pinto da Silva, 1980; Nogueira et al, 1977). The role of this surface coat in the physiology of the zoospore or encystment process is not known.

The anteriorly directed tinsel flagellum has lateral hairs about 2 $\mu$m long projecting from the entire surface of the flagellum. These hairs are 30 nm wide over the first three-quarters of their length but then constrict into a narrow distal region only 13 nm wide (Desjardins et al, 1969; Reichle, 1969b). Much smaller lateral and tip "hairs" (500 nm × 15 nm) have also been visualized by shadow casting and negative staining on the whiplash flagellum of several species (see Carlile, Chapter 7, Fig. 3) (Desjardins et al, 1973b).

The ultrastructure of zoospores in thin section has been described for several species of *Phytophthora* (Fig. 22) (Bimpong and Hickman, 1975; Ho et al, 1968a; Reichle, 1969b). The centrally positioned pyriform nucleus has the constricted end oriented toward the groove region of the zoospore and the two basal bodies of the flagella. The basal bodies lie at an angle of approximately 90° to each other in a slightly raised portion of the groove and are anchored into the zoospores by a system of microtubules and microfibrils. A major "rootlet" consisting of eight fused microtubules extends from the basal bodies along the longitudinal axis of the zoospore to the end of the cell in *P. parasitica* (Reichle, 1969b). Microtubules arise at regular intervals on both sides of the "rootlet" to form a riblike frame just inside the plasma membrane.

A water expulsion apparatus is located just anterior to the nucleus and is partially surrounded by a Golgi apparatus consisting of elongated dictyosomes. Numerous small coated vesicles 50–100 nm in diameter surround the main expulsion vesicle and often have profiles continuous with the expulsion vesicle membranes (Figs. 22 and 23).

Peripheral vesicles and cisternae line the periphery of zoospores except in the groove region. Mitochondria and

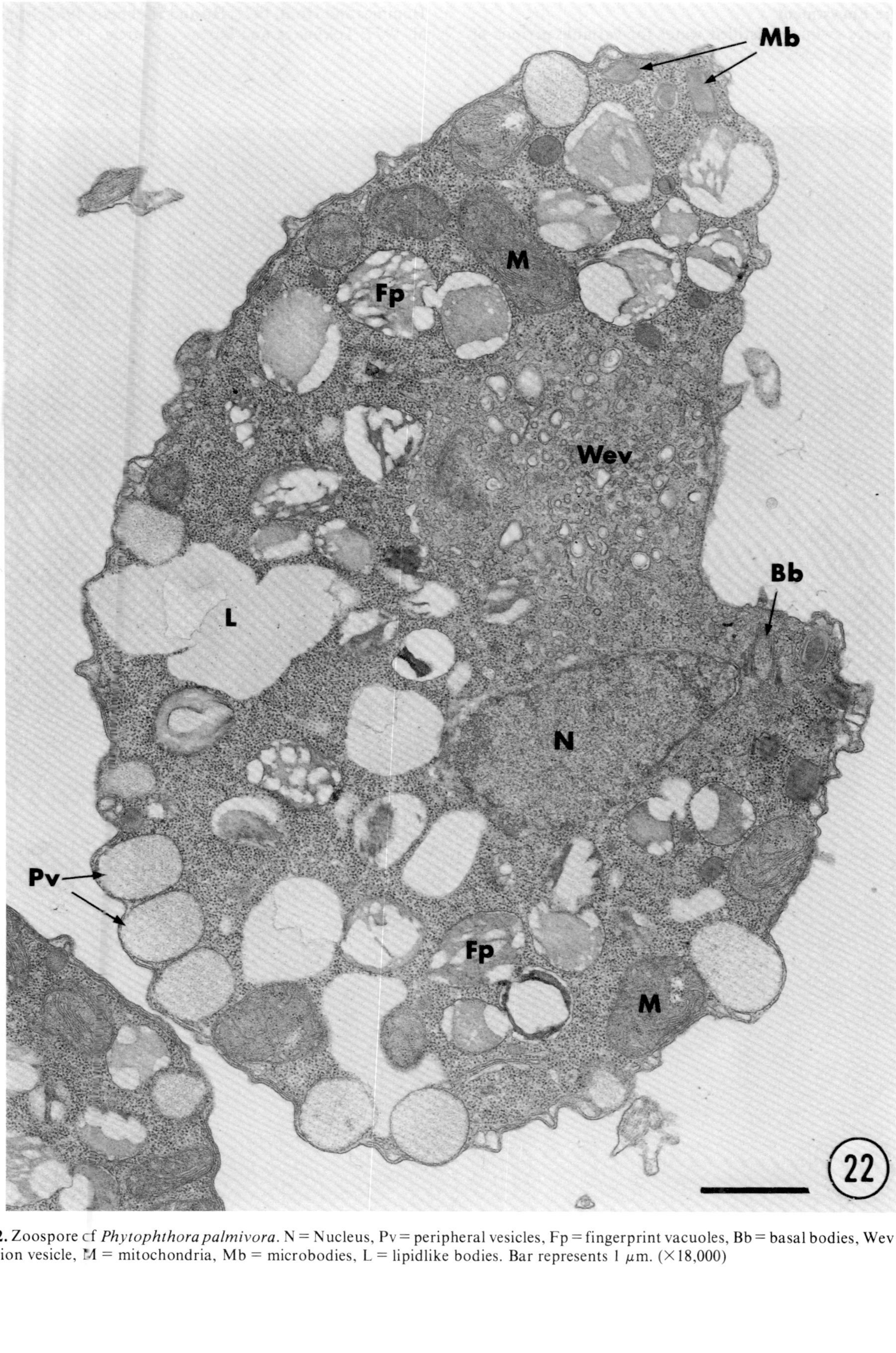


**Fig. 22.** Zoospore of *Phytophthora palmivora*. N = Nucleus, Pv = peripheral vesicles, Fp = fingerprint vacuoles, Bb = basal bodies, Wev = water expulsion vesicle, M = mitochondria, Mb = microbodies, L = lipidlike bodies. Bar represents 1 $\mu$m. ($\times$18,000)

the basal bodies and are shed. The discarded flagella can be seen in recently encysted populations of zoospores with the coiled axoneme inside the flagellar membrane (Fig. 26). Detachment of flagella during encystment appears to be the most common means of flagella disposal in *Phytophthora*, although retraction of flagella has been reported (Reichle, 1969a).

Within the zoospores, the peripheral vesicles fuse with the plasma membrane to form bilayer membrane diaphragms, which rupture to release glycoprotein and possibly other cell wall precursor materials (Pinto da Silva and Nogueira, 1977; Sing and Bartnicki-Garcia, 1975a, 1975b). Calcium-binding substrates, visualized along the peripheral cisternae and peripheral vesicles by fixing encysting zoospores in calcium-containing glutaraldehyde (Hemmes and Pinto da Silva, 1980), appear to be involved in triggering this membrane fusion and exocytosis in encysting zoospores (Fig. 24). As peripheral vesicles are released in incipient cysts, the cyst wall begins to form as a mixture of an amorphous substance (presumably adhesive material) and microfibrils (Sing and Bartnicki-Garcia, 1975a) (see also Bartnicki-Garcia and Wang, Chapter 9).

### Cysts and Cyst Germination

The cytoplasmic organization of cysts is different from that of zoospores. The cell is spherical (Figs. 21 and 25) and surrounded by a uniform, thin wall 0.1–0.15 μm thick (Bimpong and Hickman, 1975; Hemmes and Hohl, 1971). The polarized, unique distribution of organelles characteristic of the zoospore is no longer present. The nucleus has rounded and assumed a central position in the cell. The basal bodies are still found close to the nucleus, but the anchoring system of "rootlets" has disappeared. Peripheral vesicles, peripheral cisternae, and the water expulsion apparatus are also no longer seen. Mitochondria, dictyosomes, and microbodies are evenly distributed.

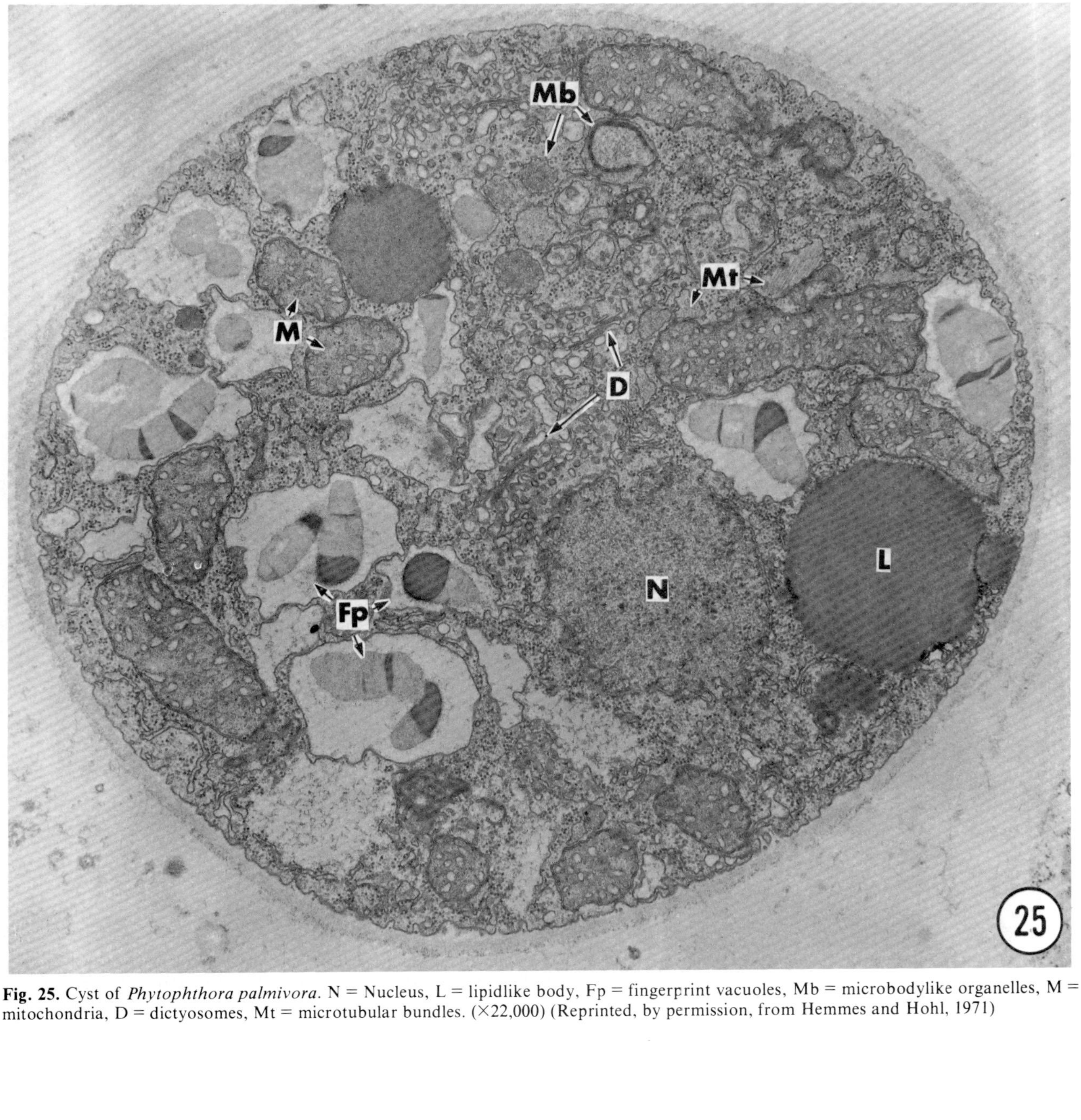


**Fig. 25.** Cyst of *Phytophthora palmivora*. N = Nucleus, L = lipidlike body, Fp = fingerprint vacuoles, Mb = microbodylike organelles, M = mitochondria, D = dictyosomes, Mt = microtubular bundles. (×22,000) (Reprinted, by permission, from Hemmes and Hohl, 1971)

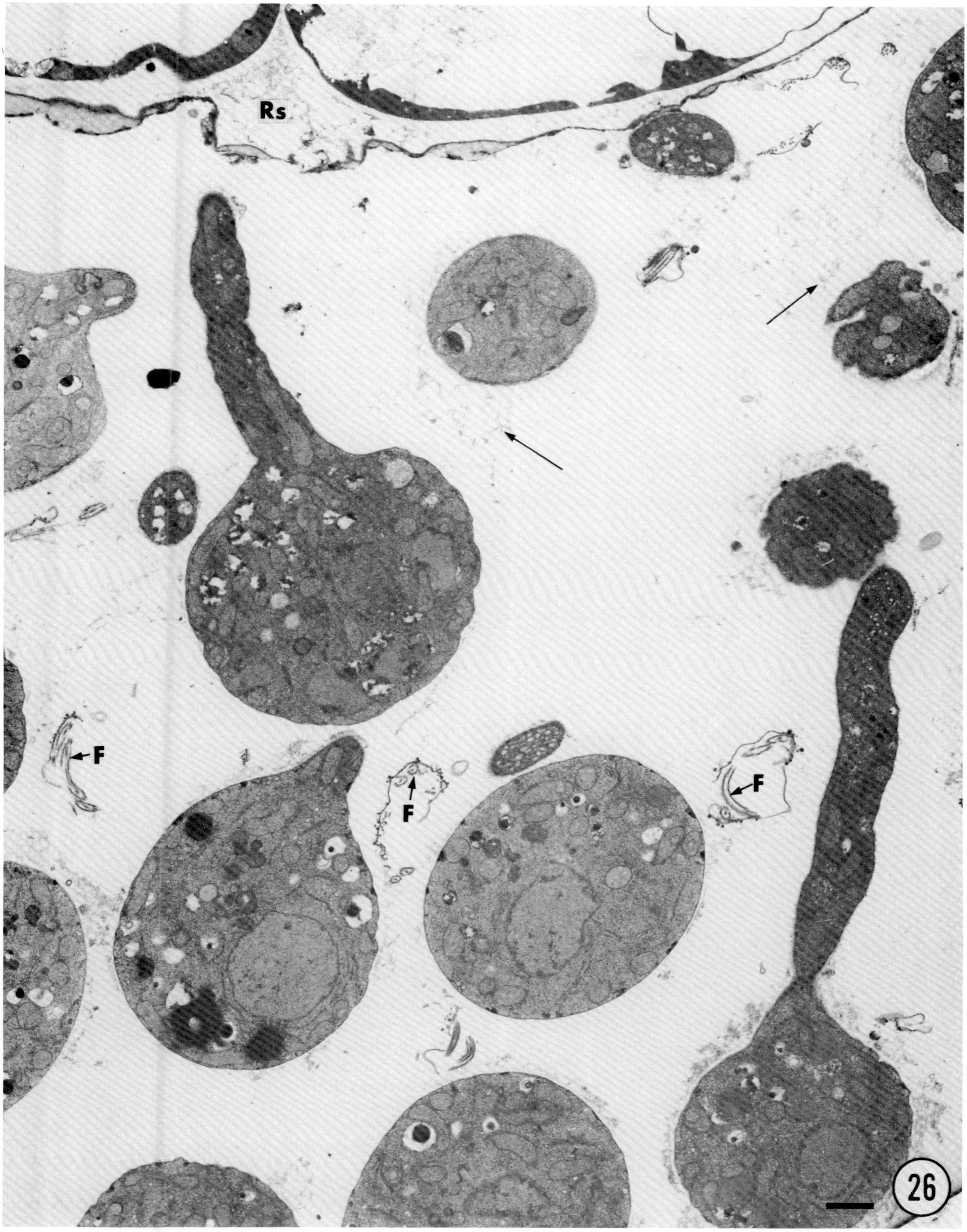


**Fig. 26.** Germinating cysts of *Phytophthora palmivora* that have encysted on the root surface (Rs) of a host papaya plant (*Carica papaya*). Note the shed flagella (F) and fibrous material (arrows), which may be adhesive material released from peripheral vesicles during encystment. Bar represents 1 μm. (×8,000)

Cysts, like sporangia, can germinate directly by producing a germ tube (Fig. 26) or indirectly by differentiating another zoospore and an exit pore (Hemmes and Hohl, 1971; Ho and Hickman, 1967). The incubation of cysts in a medium of organic substrates, or at a low pH (approximately 5.0), or in a solution of calcium salts usually stimulates cysts to germinate directly within 1 hr. Dictyosome-derived vesicles accumulate against the cyst wall in one area, and the wall bulges out to form a germ tube. Electron microscopy of isolated germinating cyst walls shows that the hyphal wall is conspicuously different from the cyst wall in texture. The cyst walls consist of a thin layer of interwoven, randomly oriented microfibrils, whereas the hyphal walls are microfibrillar on the internal surface but have an amorphous external coat (Hegnauer and Hohl, 1973; Tokunaga and Bartnicki-Garcia, 1971b) (see Bartnicki-Garcia and Wang, Chapter 9, Figs. 4 and 6).

If cysts are incubated in distilled water with no contaminating calcium from the original culture medium, cysts will not germinate directly but will differentiate zoospores and exit pores (Hemmes and Hohl, 1971). At present, how many times this repeated zoospore emergence can occur is not known.

## Chlamydospores

Chlamydospores of *Phytophthora* develop by spherical expansion of hyphal tips or by localized swelling of hyphal tubes to produce terminal or intercalary spores ranging from 8 to 15 μm in diameter. Chlamydospores may be produced singly, in chains, or in clusters interspersed with undifferentiated hyphal swellings. Most fully expanded chlamydospores are spherical, although some species produce irregular to oblong forms (Cother and Griffin, 1973). After the spore is fully expanded, the walls of chlamydospores thicken to varying degrees depending on the species and growth conditions (0.5 μm for *P. cinnamomi* [Hemmes and Wong, 1975]; 1.0–1.5 μm for *P. drechsleri* [Cother and Griffin, 1973]; less than 1 to more than 1.5 μm for *P. palmivora* [Kadooka and Ko, 1973]). The delimitation of these terminal and intercalary spores by septa is the usual criterion for distinguishing chlamydospores from hyphal swellings.

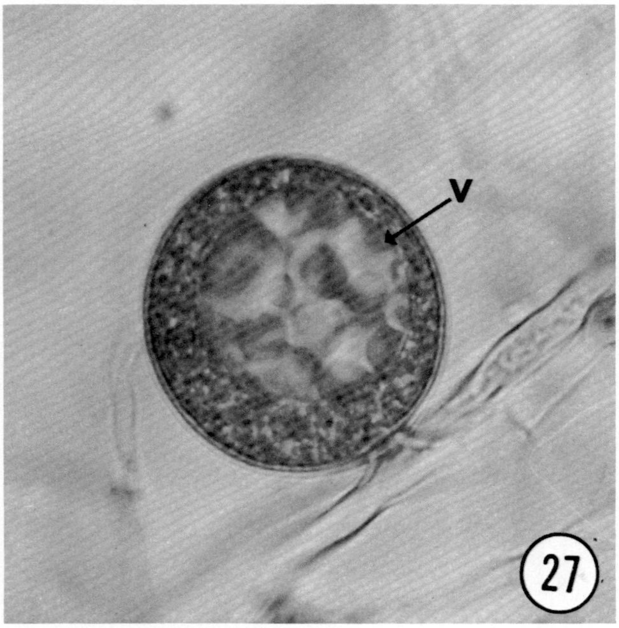


**Fig. 27.** Thin-walled chlamydospore of *Phytophthora palmivora*. Note the large central vacuole (V) with dense, spherical droplets. (×1,350)

Little internal organization of chlamydospores can be seen by light microscopy: a thickened wall, the delineating septa, and the presence of a central vacuole are the only features usually apparent (Fig. 27). Ultrastructural studies of the thin-walled chlamydospores of *P. cinnamomi* (Hemmes and Wong, 1975), produced on V-8 agar plates, and the "thin"- and "thick"-walled chlamydospores of *P. palmivora* (D. E. Hemmes, *unpublished*), produced in submerged, liquid cultures, reveal the basic cytoplasmic changes occurring during development and the cytoplasmic organization of fully expanded chlamydospores (Figs. 28 and 29).

The cytologic development and final cytoplasmic organization are similar in both thin- and thick-walled chlamydospores. Developing chlamydospores are filled with "functional" cytoplasm, which flows into the spore from the subtending mycelium (Hemmes and Wong, 1975). Vesicles, 0.25 μm in diameter, underlie the plasma membrane during the expansion phase, but they are fewer in number than those seen in oogonial expansion and their origin has yet to be traced. No micrographs of the wall-thickening phases have been obtained.

In later stages of expansion, there is a large increase in "reserve" cytoplasm, so that by the time chlamydospores are fully expanded, the proportion of "functional" cytoplasm is greatly reduced. The "reserve" cytoplasm is dominated by fingerprint vacuoles with dense, spherical inclusions. The vacuoles enlarge considerably during the expansion phase and thus squeeze the nuclei and "functional" cytoplasm into small pockets throughout the spore (Fig. 29). In older chlamydospores, some of the fingerprint vacuoles coalesce to form a large central vacuole.

In thin sections of chlamydospore walls, a demarcation line between the wall present during the expansion phase and the secondary wall laid down after expansion is often seen. The thin outer layer and thick inner layer of chlamydospores of *P. palmivora* seen in freeze fracture and replicas (Hegnauer and Hohl, 1978) probably correspond to these two wall layers.

## Gametangia and Oospores

### Gametangial Development

The cytology of gametangial interaction and oospore development in both paragynous and amphigynous species of *Phytophthora* has been studied by electron microscopy (Hemmes and Bartnicki-Garcia, 1975; Hemmes and Ribeiro, 1977; Vujičić, 1971). Although the early gametangial interaction is more complicated in amphigyny, both modes of development result in the firm attachment of an antheridium near the base of the oogonium (Figs. 30 and 33). The expansion of the oogonium, development of the fertilization tube, and maturation of the oospore are similar in both paragynous and amphigynous species.

The early stages in the development of the paragynous configuration are quite simple. The gametangial initials make contact before any morphological or cytologic differentiation can be observed by microscopy and form an adhesive contact zone that is manifested by the formation of a dense layer in the contact area between the gametangial walls (Hemmes and Ribeiro, 1977). Numerous vesicles can be observed on the antheridial side of the contact zone, suggesting that the adhesive material may be secreted by the antheridium. After contact, both initials begin an expansion phase. The oogonium expands considerably more than the antheridium, resulting in the final paragynous configuration—an oogonium with a much smaller antheridium cemented to its base near the oogonial stalk (Figs. 30 and 32).

The early stages in the development of the amphigynous configuration are more complex. Light microscopy studies have shown that the oogonial initial contacts a slightly swollen antheridial initial and then proceeds to penetrate it and to rupture through the other side (Figs. 31, 33, and 38) (Dastur, 1913; Pethybridge, 1913; Vujičić, 1971). Transverse and longitudinal sections of the amphigynous configuration at later stages viewed by transmission electron microscopy (Hemmes and Bartnicki-Garcia, 1975; Vujičić, 1971) as well as by scanning electron microscopy (Ho, 1979; Ho et al, 1977) substantiate this interpretation. As the oogonial initial penetrates the antheridial initial, the antheridial plasma membrane appears to remain intact, surrounding the oogonial initial until it reaches the other side of the antheridium. There the membrane probably fuses with the antheridial plasma membrane on that side of the antheridium. The wall of the penetrating oogonial hypha (the future oogonial stalk wall) consists of a central electron-dense layer sandwiched between electron-transparent layers. The presence of wall-destined

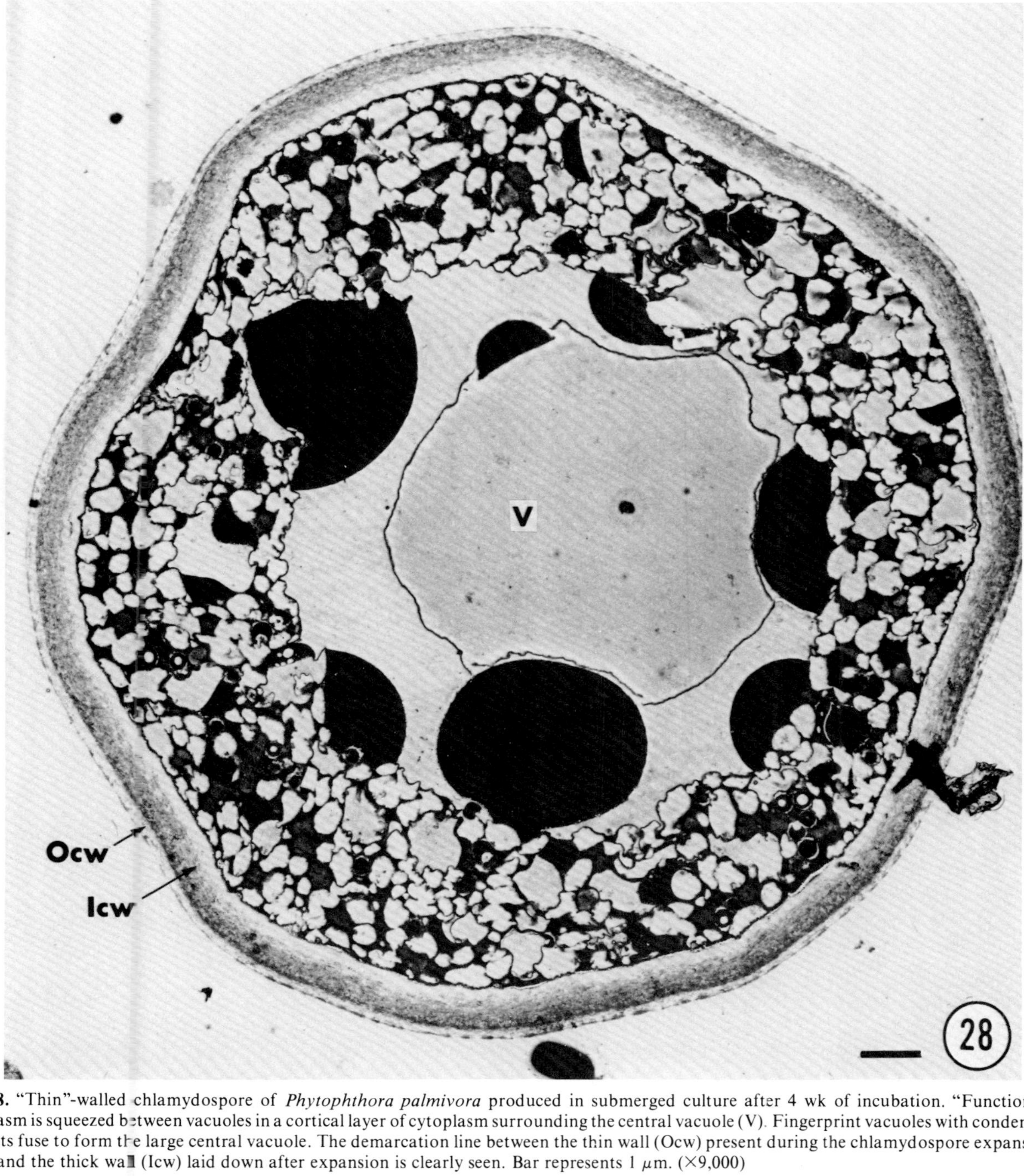


**Fig. 28.** "Thin"-walled chlamydospore of *Phytophthora palmivora* produced in submerged culture after 4 wk of incubation. "Functional" cytoplasm is squeezed between vacuoles in a cortical layer of cytoplasm surrounding the central vacuole (V). Fingerprint vacuoles with condensed contents fuse to form the large central vacuole. The demarcation line between the thin wall (Ocw) present during the chlamydospore expansion phase and the thick wall (Icw) laid down after expansion is clearly seen. Bar represents 1 $\mu$m. (×9,000)

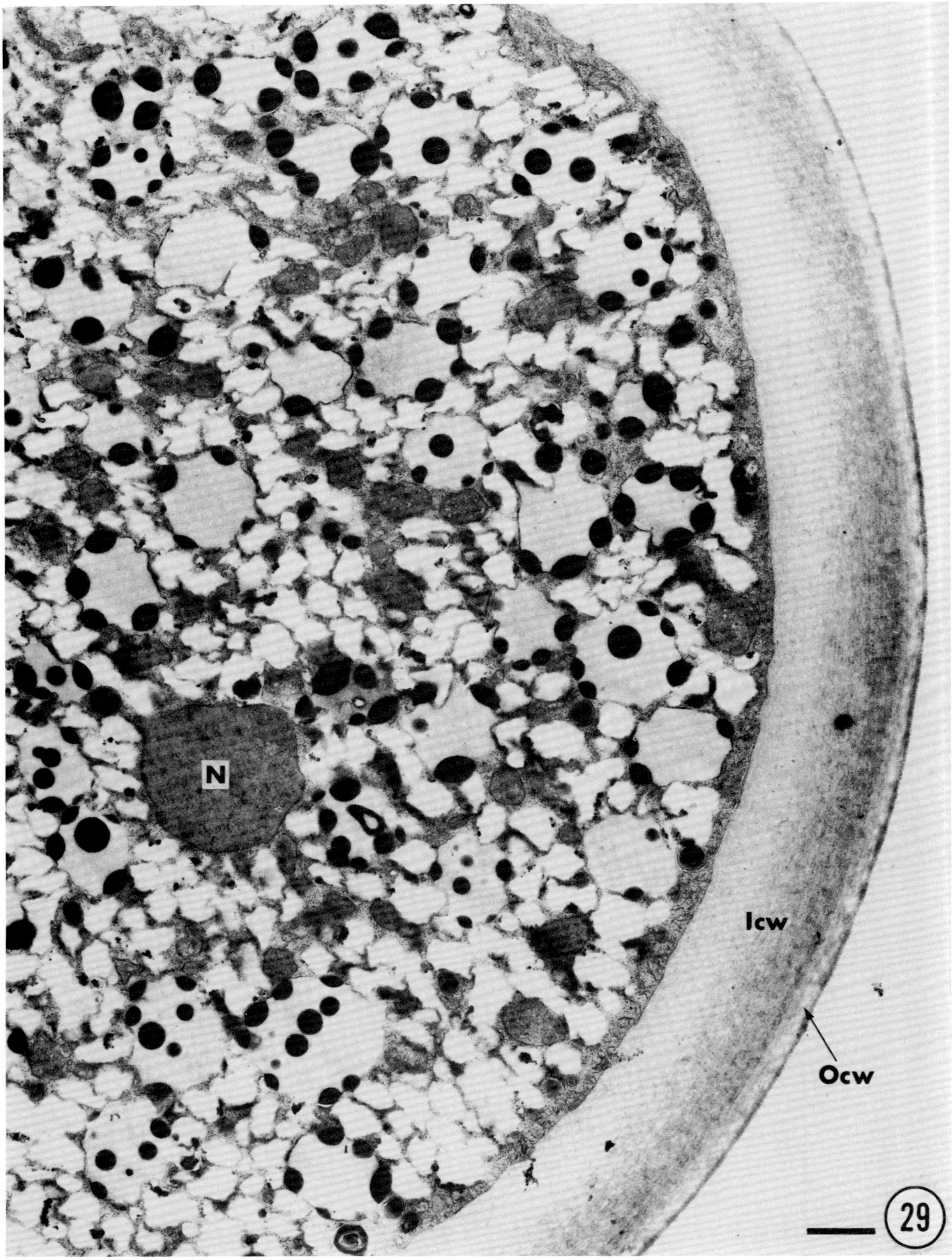


**Fig. 29.** "Thick"-walled chlamydospore of *Phytophthora palmivora* produced in papaya juice after 8 wk of incubation. The cytoplasm is dominated by vacuoles with dense, spherical inclusions (fingerprint vacuoles in which the lamellated inclusions have condensed). The demarcation between the thin wall present during the expansion phase of the chlamydospore (Ocw) and the secondary wall formed after expansion (Icw) is faintly visible. N = Nucleus. Bar represents 1 $\mu$m. ($\times$11,000)

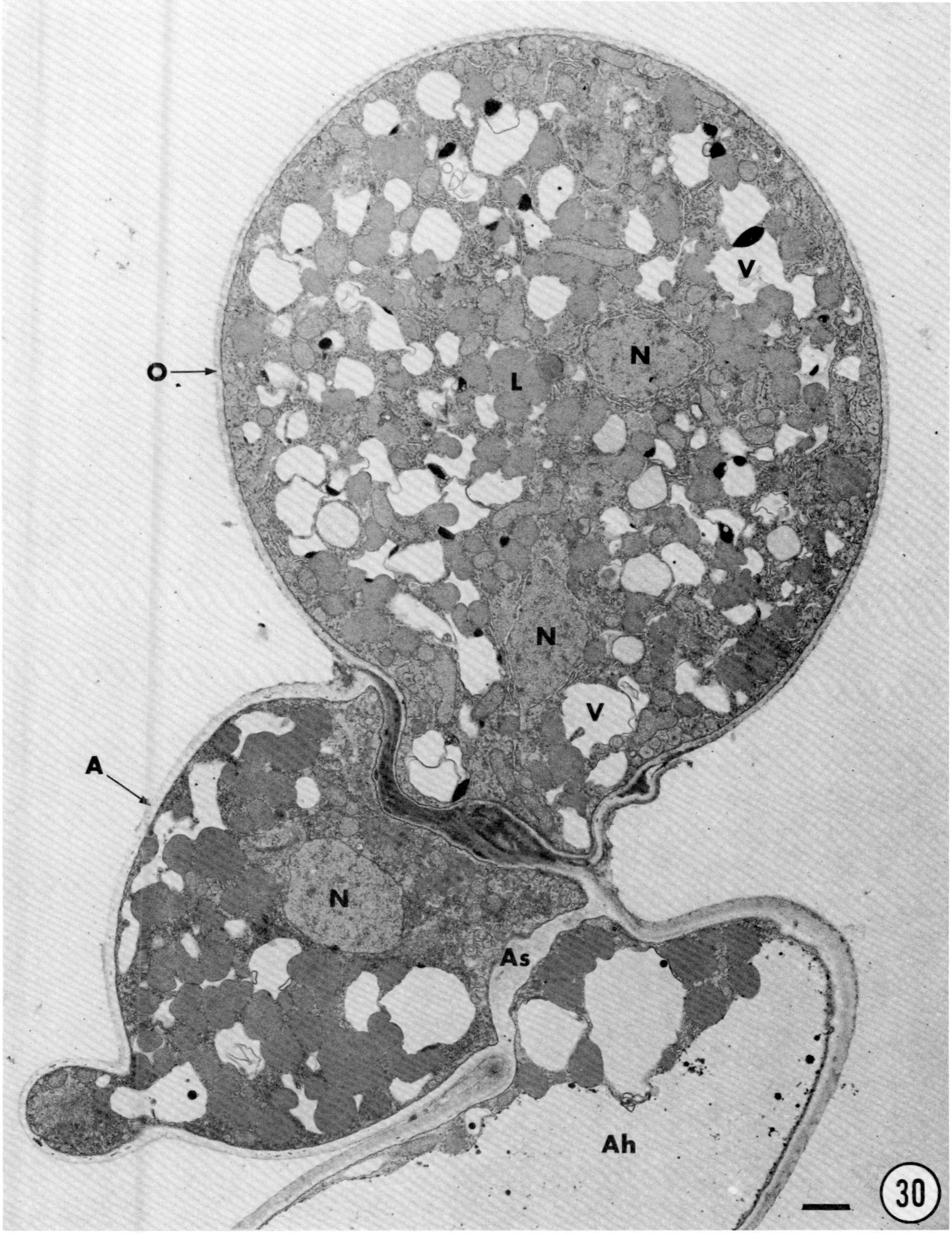


**Fig. 30.** Median longitudinal section of an oogonium and antheridium of *Phytophthora megasperma* var. *sojae* at the beginning of the oogonial expansion phase. The expanding oogonium (O) contains about half "functional" cytoplasm—i.e., nuclei (N), mitochondria, dictyosomes, and rough endoplasmic reticulum—and about half "reserve" cytoplasm—i.e., lipid bodies (L) and vacuoles (V). Note the concentration of vesicles within the antheridium (A) near the zone of attachment. As = Antheridial septum, Ah = antheridial hypha. Bar represents 1 $\mu$m. (×8,000) (Reprinted, by permission, from Hemmes and Ribeiro, 1977)

vesicles on both sides of this wall suggests that it is a hybrid wall jointly formed by the oogonial and antheridial initials.

As the oogonial initial ruptures through the other side of the antheridium, it leaves a frayed collar of wall material at the emergence site as evidence of the penetration (Hemmes and Bartnicki-Garcia, 1975; Ho, 1979). No mingling of the two gametangial cytoplasms occurs in the penetration event. Transverse sections through the antheridium show that the antheridial cytoplasm completely encircles the oogonial stalk with no evidence that the antheridial initial wrapped around the oogonial stalk to form this configuration. Even as penetration is occurring, the oogonium begins its expansion phase, rupturing through the antheridial initial and rapidly expanding, leaving the antheridium as a doughnut-shaped structure encircling the oogonial stalk (see Fig. 38 for a diagrammatic summary of the various stages in amphigyny).

During the oogonial expansion phase, the oogonial stalk remains open, and largely "functional" cytoplasm flows into the oogonium from the underlying mycelium. Vesicles of two distinct sizes bud from dictyosomal cisternae and underlie the entire oogonial plasma membrane during expansion: smaller, spherical ones 70–100 nm in diameter with dense contents and larger, ovoid ones 200–300 nm in diameter. In *P. megasperma* f. sp. *glycinea*, the two types of vesicles appear to form from alternate cisternae of each dictyosome (Fig. 1) (Hemmes and Ribeiro, 1977). During the expansion phase, the oogonial wall remains thin (about 200 nm), similar to the wall thickness of the hyphal initial.

Upon completion of the expansion phase, the oogonial stalk is plugged with wall-like material containing dense inclusions, and the entire oogonial wall thickens from 200 to 800 nm. In *P. capsici*, anywhere from two to 12 uneven and often confluent dense layers are laid down in the thickened wall (Hemmes and Bartnicki-Garcia, 1975), whereas the final oogonial wall of *P. megasperma* f. sp. *glycinea* consists of an outer, loosely organized layer 300–400 nm thick, a middle dense layer 100–200 nm thick, and an inner "fibrillar" layer 300–400 nm thick (Hemmes and Ribeiro, 1977). The two types of cytoplasmic vesicles observed underlying the plasma membrane during the expansion phase are also present during the thickening phase.

The cytoplasmic changes and behavior of oogonial nuclei during the development of the oosphere and fertilization tube have been studied in detail by light microscopy for *P. cactorum* (Blackwell, 1943) and probably represent the pattern of nuclear events common to *Phytophthora*. The supernumerary oogonial nuclei migrate to the periphery of the oogonium, where they are sequestered in the periplasm along with a portion of cytoplasm, thus leaving the presumptive fusion nucleus alone in the oogonial cytoplasm. The fertilization tube ruptures the oogonial wall and extends up to 10 $\mu$m into the oogonial cytoplasm to deposit the antheridial nucleus. Once the antheridial nucleus is deposited, a unit membrane is formed to delimit the periplasm and oosphere.

The presence of widely dispersed nuclei in the expanding oogonium and the presence of nuclei sequestered in the periplasm and delimited by an oosphere membrane have been confirmed by electron microscopy (Hemmes and Bartnicki-Garcia, 1975; Hemmes and Ribeiro, 1977), although the mechanism involved in nuclear migration and the cytoplasmic details of oosphere membrane formation have not been observed. Likewise, the fertilization tube has been observed by electron microscopy. The tube is filled with numerous vesicles of two sizes, 150–300 and 60–80 nm. In longer fertilization tubes, the larger vesicles are congregated at the tip of the tube, whereas the smaller, spherical vesicles fill the remainder of the tube (Fig. 34). A single nucleus is present in the fertilization tube, although the actual deposition of the nucleus into the oogonium has not been observed.

By the time the oosphere has formed, the composition of the cytoplasm has changed drastically from the largely "functional" cytoplasm of the oogonium to a "reserve"

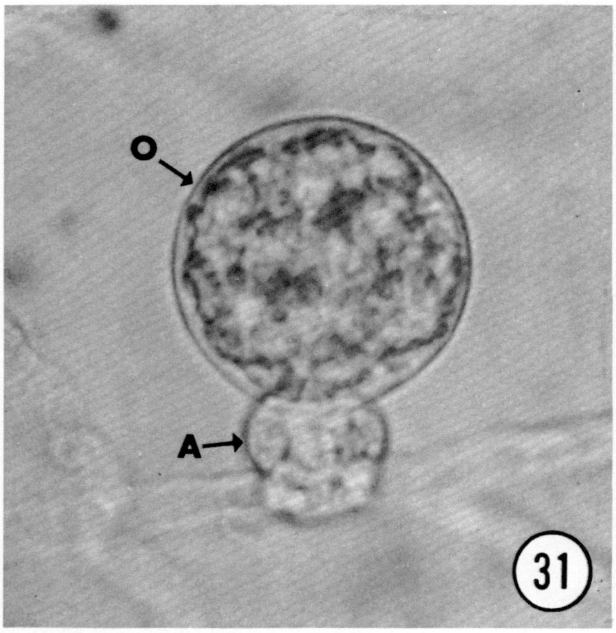


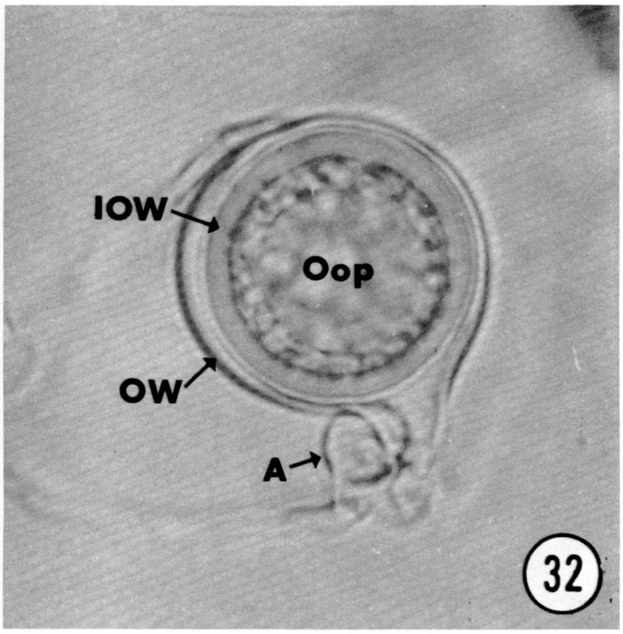


**Figs. 31 and 32. 31,** Developing gametangium of *Phytophthora parasitica*, showing the amphigynous antheridium. O = Oogonium, A = antheridium. (×1,200) **32,** Oospore of *P. citricola*, showing the paragynous antheridium. Note that the outline of the ooplast (Oop) can be discerned. OW = Oogonial wall, IOW = inner oospore wall, A = antheridium. (×1,200)

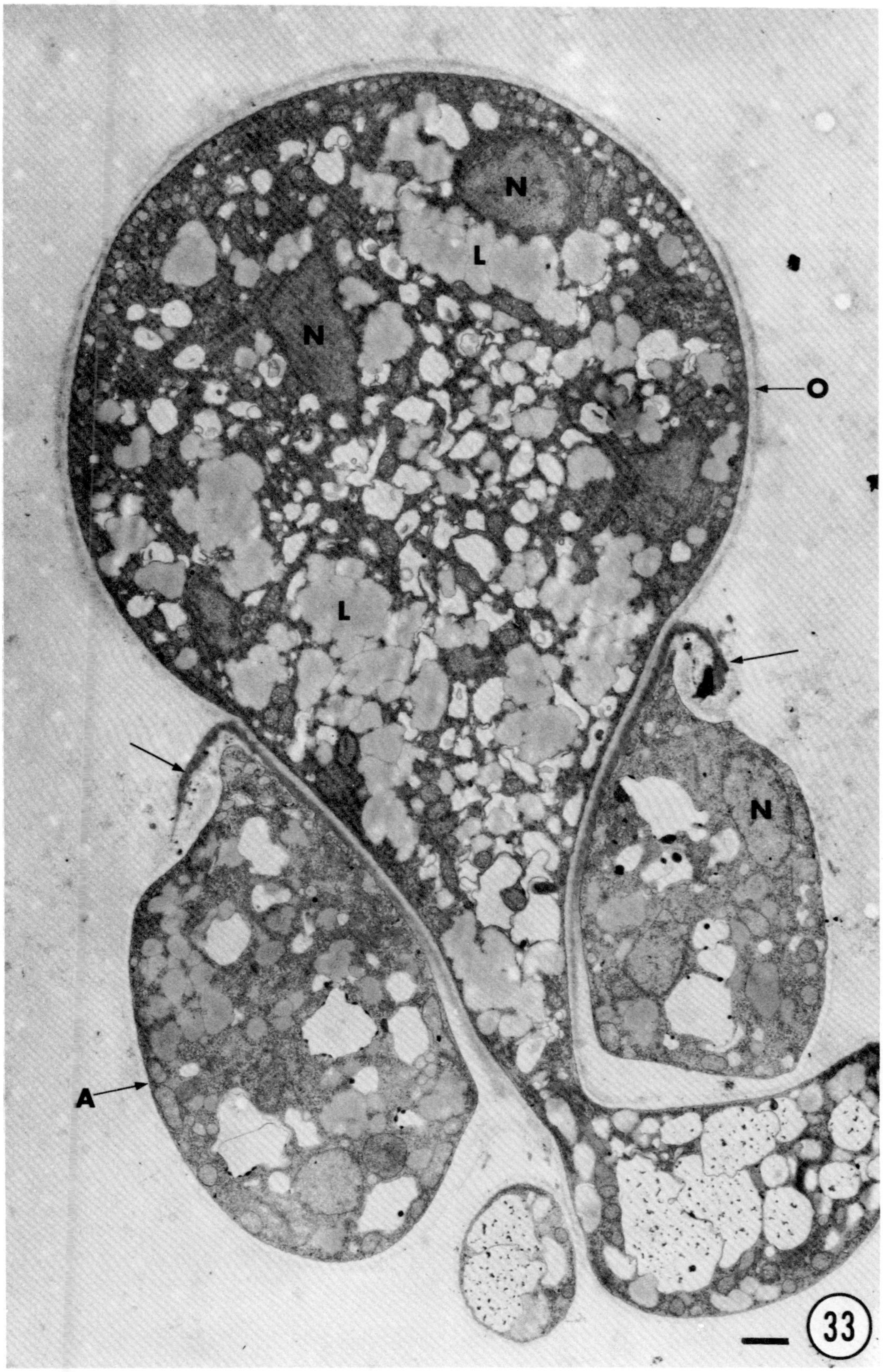


**Fig. 33.** Longitudinal section through a young expanding oogonium and antheridium of *Phytophthora capsici*. The ruptured oogonial stalk wall and antheridial wall have formed a curled collar (arrows). Note that the oogonial wall is of the approximate thickness of the hyphal wall and that numerous vesicles underlie the oogonial plasmalemma. N = Nuclei, L = lipidlike bodies, A = antheridium, O = oogonium. Bar represents 1 μm. (×7,500) (Reprinted by permission, from Hemmes and Bartnicki-Garcia, 1975)

cytoplasm dominated by roughly equal numbers of fingerprint vacuoles and lipid bodies. The fingerprint vacuoles have expanded from 1–2 μm in diameter in the oogonium to 3–5 μm in the oosphere, squeezing the remaining "functional" cytoplasm into pockets between the reserve vacuoles and lipid.

The final stages of oospore development include the development of the oospore wall and the rearrangement of the "reserve" cytoplasm into a central ooplast surrounded by a cortex of lipid bodies. Oospore wall formation begins in the periplasm, where dense, membrane-bound cisternae apparently condense to form a thin, electron-dense outer oospore wall as the periplasmic cytoplasm degenerates (Figs. 35 and 36). This outer oospore wall consists of one to five dense layers, each about 20 nm thick. The thick inner oospore wall (0.7–1.0 μm) is laid down inside this thin layer. The inner oospore wall is electron-transparent and does not exhibit any layering. No special concentration of organelles has been observed in the peripheral cytoplasm when the inner oospore wall is produced (Fig. 37). A "mature" oospore of *P. megasperma* f. sp. *medicaginis* within the oogonium is shown (Fig. 39) with an attached, empty paragynous antheridium and remnant of the fertilization tube.

**Oospore Germination**

The dissolution of the thick, inner oospore wall during oospore germination can be observed by light microscopy and

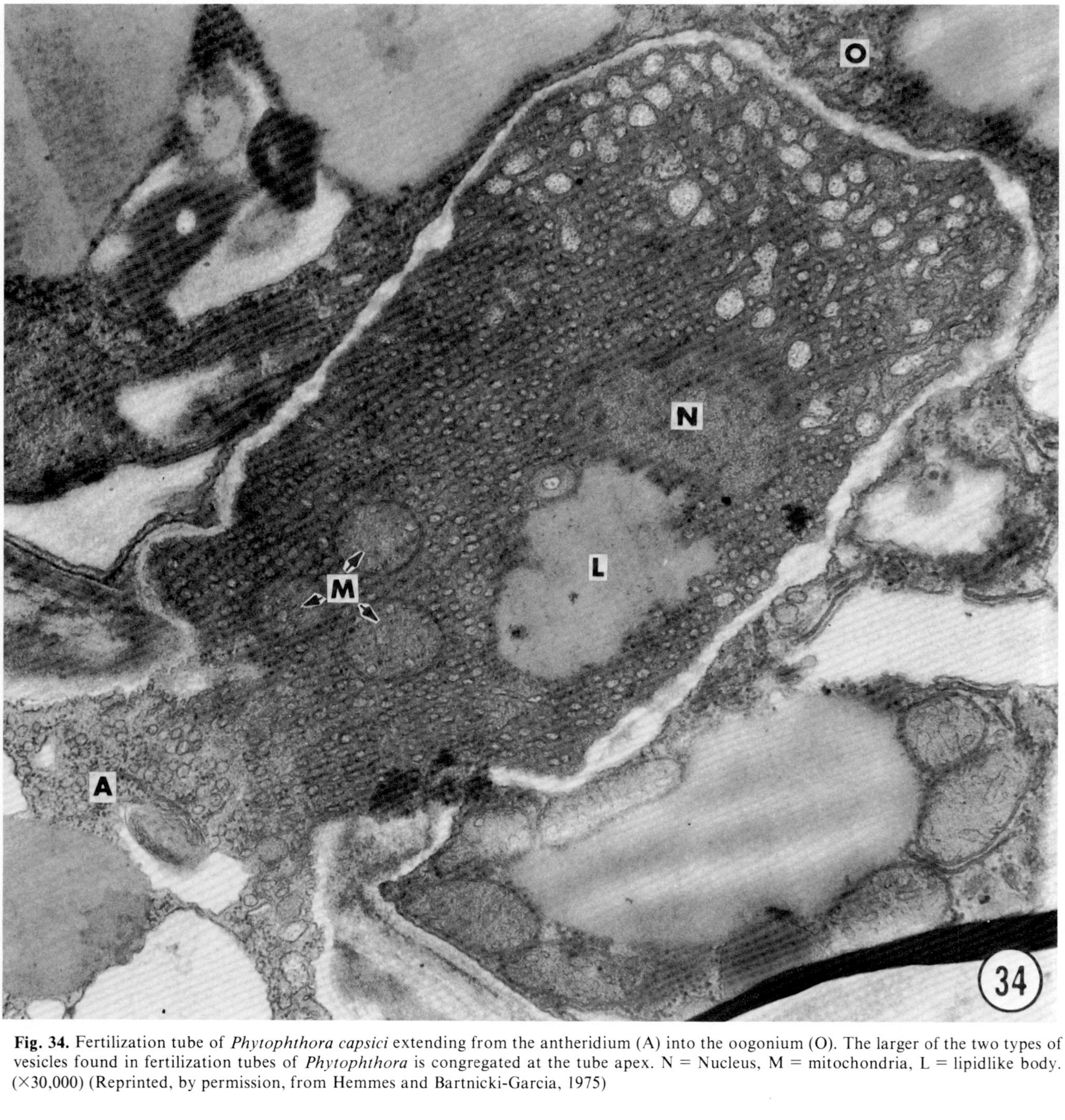


**Fig. 34.** Fertilization tube of *Phytophthora capsici* extending from the antheridium (A) into the oogonium (O). The larger of the two types of vesicles found in fertilization tubes of *Phytophthora* is congregated at the tube apex. N = Nucleus, M = mitochondria, L = lipidlike body. (×30,000) (Reprinted, by permission, from Hemmes and Bartnicki-Garcia, 1975)

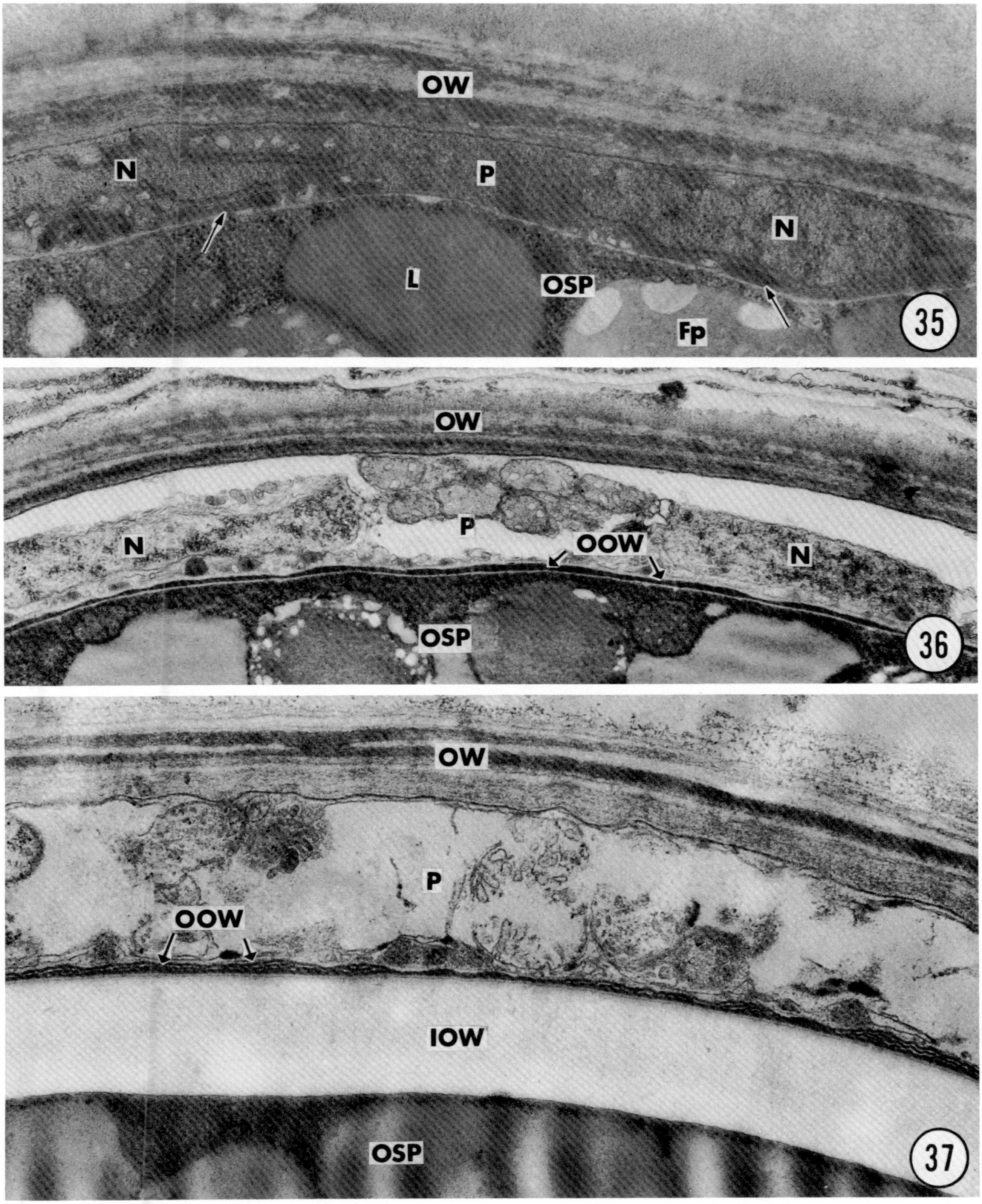


**Figs. 35–37.** Periplasm and oospore wall formation in *Phytophthora capsici*. **35,** Nuclei and mitochondria are sequestered in the periplasm. Note the dense organelles (arrows) at the inner side of the periplasm. (×27,000) **36,** Periplasm degeneration and outer oospore wall formation. (×18,300) **37,** Inner oospore wall formation. (×29,000) OW = Oogonial wall, P = periplasm, OOW = outer oospore wall, IOW = inner oospore wall, OSP = oosphere, N = nuclei, L = lipidlike body, Fp = fingerprint vacuole. (Reprinted, by permission, from Hemmes and Bartnicki-Garcia, 1975)

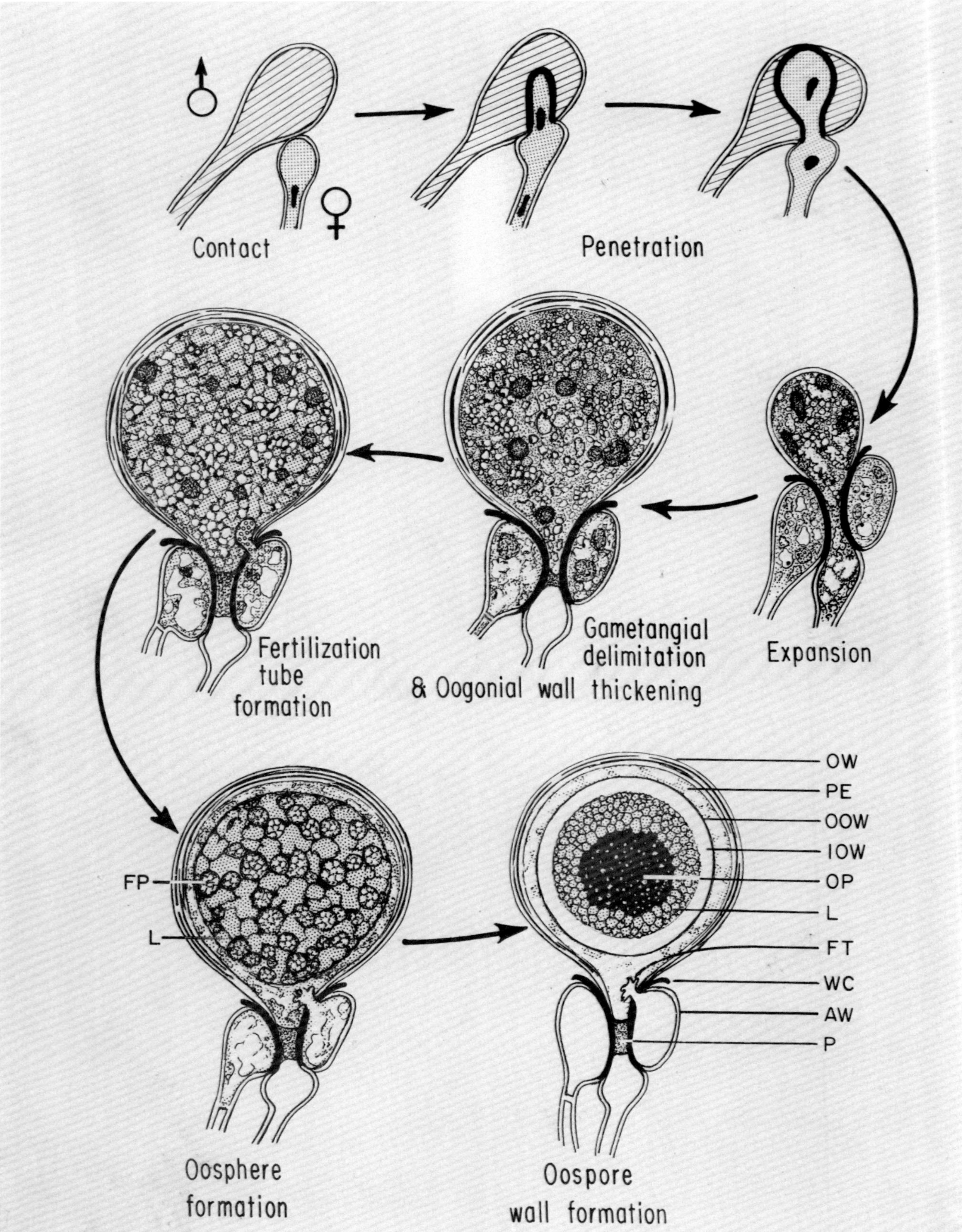


**Fig. 38.** Sexual morphogenesis in *Phytophthora capsici*: gametangial interplay and oospore development. AW = Antheridial wall, FP = fingerprint vacuole, FT = fertilization tube, IOW = inner oospore wall, L = lipidlike body, OOW = outer oospore wall, OP = ooplast, OW = oogonial wall, P = oogonial plug, PE = periplasm, WC = wall collar. (Reprinted, by permission, from Hemmes and Bartnicki-Garcia, 1975)

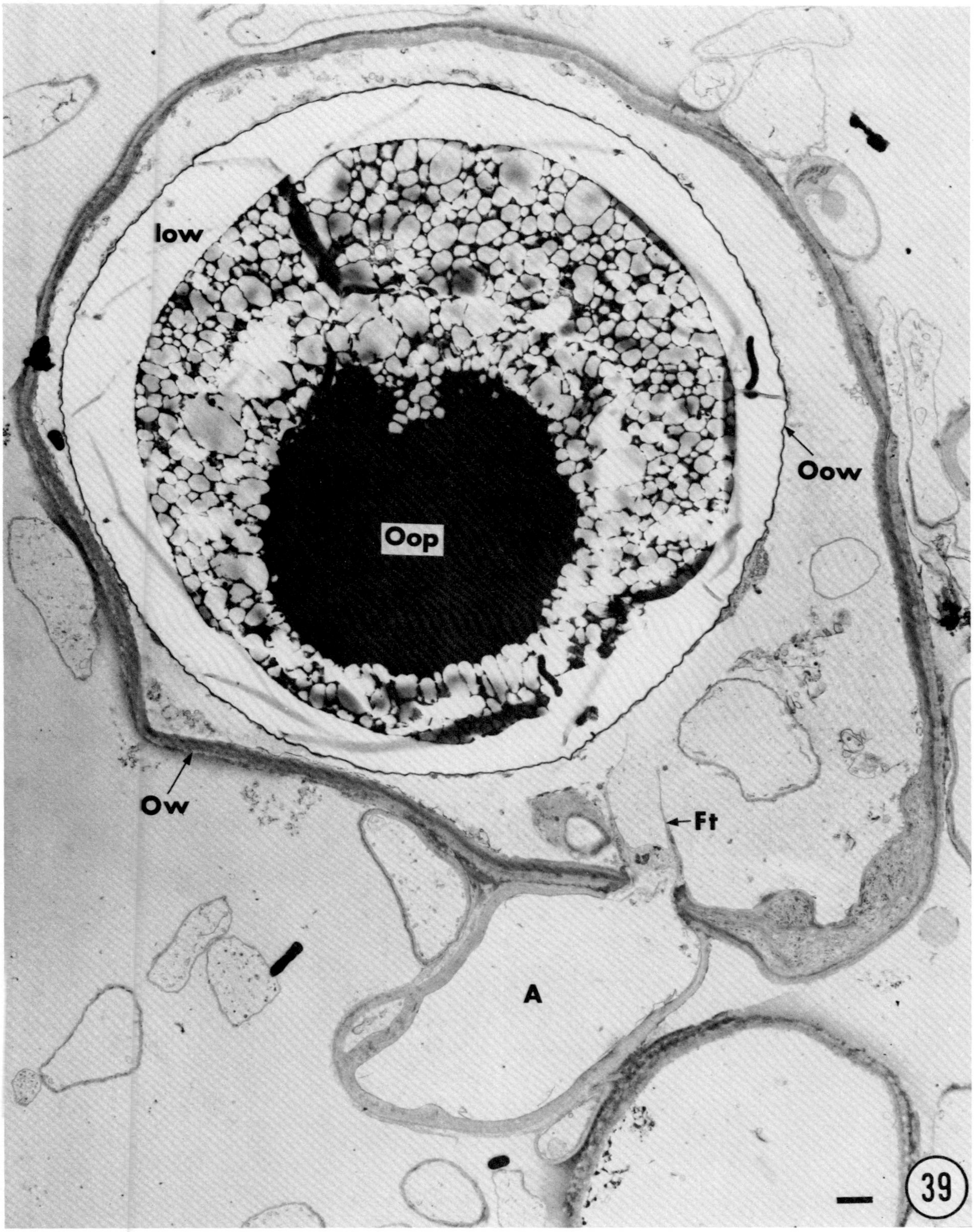


**Fig. 39.** Oospore of *Phytophthora megasperma* f. sp. *medicaginis* within the oogonium. The empty antheridium (A) and fertilization tube (Ft) can be seen attached to the oogonium. Oop = Ooplast, Iow = inner oospore wall, Oow = outer oospore wall, Ow = oogonial wall. Bar represents 1 μm. (×5,500)

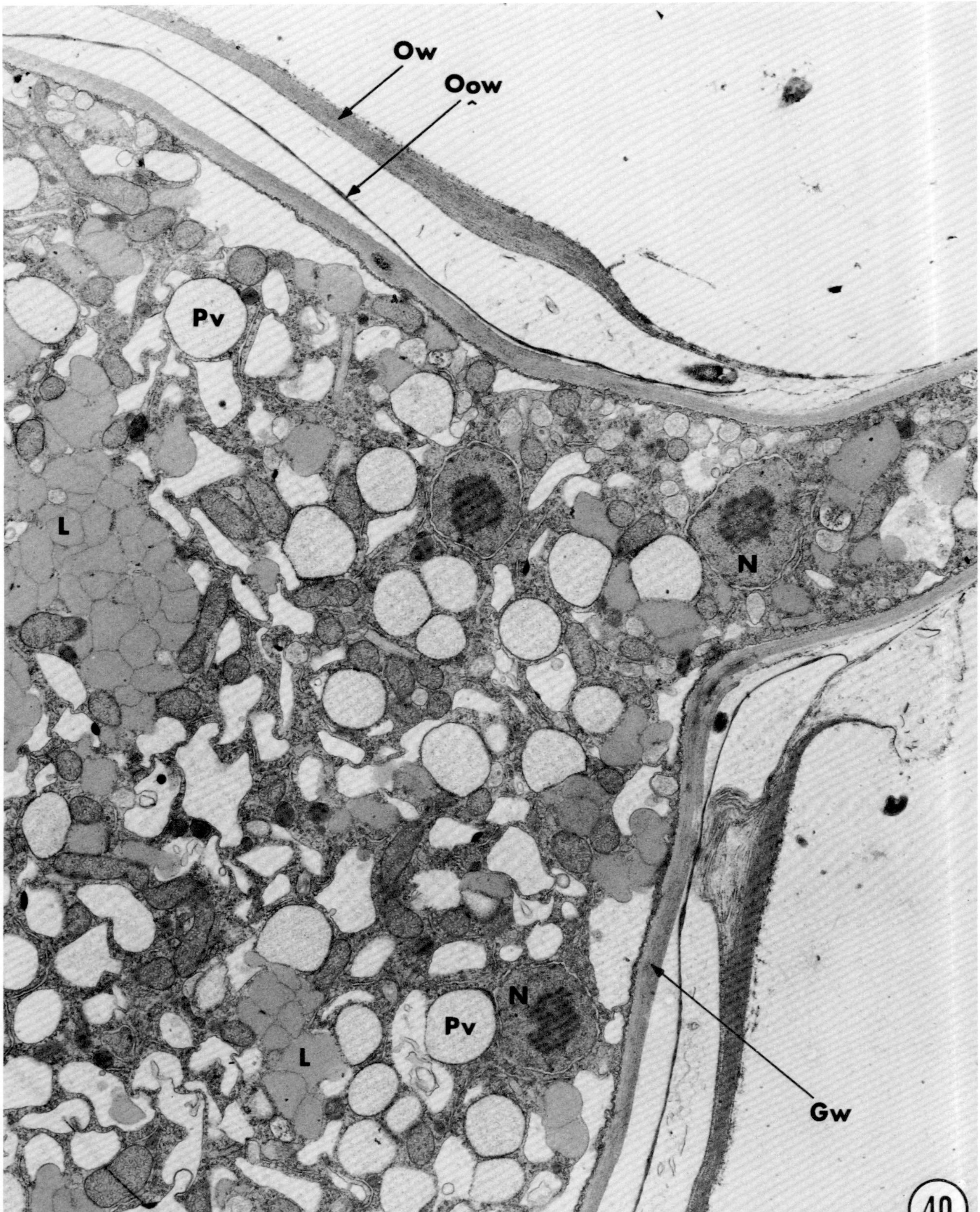


**Fig. 40.** Germinating oospore of *Phytophthora megasperma* f. sp. *medicaginis*, showing a germ tube rupturing through the outer oospore wall (Oow) and oogonial wall (Ow). The ooplast and inner oospore wall have been consumed in the germination process, but the lipidlike bodies (L), which formed a cortical layer in the oospore, flow with the newly synthesized "functional" cytoplasm into the germ tube. A germination wall (Gw) has been synthesized in the region previously occupied by the inner oospore wall and continues out of the germinating spore as the hyphal wall. N = Nuclei, Pv = peripheral vesicles. Bar represents 1 $\mu$m. (×9,000) (Germinating oospores were provided by H. Förster, University of California, Riverside)

has been confirmed by electron microscopy (Fig. 40) (D. E. Hemmes, *unpublished*). The ooplast is also consumed during germination, but some of the lipidlike bodies, which form a cortical layer in the oospore, flow with the newly synthesized "functional" cytoplasm into the germ tube. "Peripheral" vesicles commonly abut on lipid bodies in the newly synthesized cytoplasm, but their mode of development is not clear. A germination wall is synthesized in the region previously occupied by the inner oospore wall and continues out of the germinating spore as the hyphal wall. In most germinating oospores, a single germ tube is produced that ruptures the thin outer oospore wall and the oogonial wall near the oogonial stalk (Fig. 40) (D. E. Hemmes, *unpublished*).

## SUMMARY

The cytoplasmic organization of nearly all stages of the life cycle of *Phytophthora* has been described from thin sections viewed by transmission electron microscopy. The cytoplasm contains a complement of eucaryotic organelles similar in morphology to those of plants and animals, including nuclei, mitochondria, rough endoplasmic reticulum, and dictyosomes. In addition, four organelles have been described that are peculiar to *Phytophthora* and most other oomycetes: peripheral vesicles, which eventually locate at the periphery of zoospores, where they are released during encystment; fingerprint vacuoles, which contain a reserve polysaccharide, mycolaminaran; bullet-shaped microbodies with crystalline arrays of fibrils; and clusters of microtubules contained within cisternae of rough endoplasmic reticulum, which are secreted as flagellar hairs.

In areas of cytoplasm actively engaged in wall deposition, such as at hyphal tips or along the periphery of developing sporangia or oogonia, the endomembrane system is active in forming wall-destined vesicles. The contents of these vesicles are not fully known but probably include wall precursors as well as wall-synthesizing and wall-softening enzymes.

The most thoroughly studied example of wall deposition in *Phytophthora* is in the secretion of peripheral vesicles during zoospore encystment. Several investigations involving thin sectioning and freeze fracture have documented the fusion of the "peripheral" vesicles to the plasma membrane to form membrane bilayer diaphragms and the release of adhesive materials and wall-forming precursors to the exterior of the cell. The formation of the fibrillar architecture of the cyst wall has been followed by studying replicas of the cyst surface at various times during encystment (see Bartnicki-Garcia and Wang, Chapter 9).

The early development of sporangia, chlamydospores, and oogonia of *Phytophthora* appears to occur largely by cytoplasmic inflow from the underlying mycelium. The spore wall remains thin during the expansion phase, whereas considerable wall thickening occurs after expansion. The cytoplasm of developing chlamydospores and oospores, as well as the cytoplasm of aged hyphae and aged sporangia, undergoes a drastic change from a "functional" cytoplasm filled with synthesizing organelles such as nuclei, mitochondria, endoplasmic reticulum, and dictyosomes to a "reserve" cytoplasm dominated by lipidlike bodies and fingerprint vacuoles. Upon germination, there is a rapid conversion of cytoplasm back to the "functional" state.

## ACKNOWLEDGMENTS

I thank Eleanor Lippman, John Kitasako, Priscilla Vanus, and Richard Kunimoto for technical assistance, Olaf Ribeiro and Helga Förster for valuable advice and discussions, and La Vanda Warren and Gail Inouye for patiently typing this manuscript. Special appreciation is expressed to S. Bartnicki-Garcia for providing the research facilities for the greater part of this work. The research was supported in part by National Institutes of Health Minority Biomedical Support Grant No. RR08073.

## LITERATURE CITED

Aragaki, M., and Hine, R. B. 1963. Effect of radiation on sporangial production of *Phytophthora parasitica* on artificial media and detached papaya fruit. Phytopathology 53:854-856.

Bartnicki-Garcia, S. 1973. Fundamental aspects of hyphal morphogenesis. Pages 245-267 in: Microbial Differentiation. J. M. Ashworth and J. E. Smith, eds. Cambridge Univ. Press, Cambridge.

Bartnicki-Garcia, S., and Hemmes, D. E. 1976. Some aspects of the form and function of oomycete spores. Pages 593-641 in: The Fungal Spore: Form and Function. D. J. Weber and W. M. Hess, eds. John Wiley & Sons, New York. 895 pp.

Beakes, G. W. 1980. Electron microscopic study of oospore maturation and germination in an emasculate isolate of *Saprolegnia ferax*. 4. Nuclear cytology. Can. J. Bot. 58:228-240.

Bimpong, C. E. 1975. Changes in metabolic reserves and enzyme activities during zoospore motility and cyst germination in *Phytophthora palmivora*. Can. J. Bot. 53:1411-1416.

Bimpong, C. E., and Hickman, C. J. 1975. Ultrastructural and cytochemical studies of zoospores, cysts, and germinating cysts of *Phytophthora palmivora*. Can. J. Bot. 53:1310-1327.

Blackwell, E. 1943. The life history of *Phytophthora cactorum* (Leb. & Cohn) Schroet. Trans. Br. Mycol. Soc. 26:71-89.

Bracker, C. E., Heintz, C. E., and Grove, S. N. 1970. Structural and functional continuity among endomembrane organelles in fungi. Pages 103-104 in: Proc. Int. Congr. Electron Microsc., 7th. Soc. Fr. Microsc. Electron., Paris.

Chapman, J. A., and Vujičić, R. 1965. The fine structure of sporangia of *Phytophthora erythroseptica* Pethyb. J. Gen. Microbiol. 41:275-282.

Christen, J., and Hohl, H. R. 1972. Growth and ultrastructural differentiation of sporangia in *Phytophthora palmivora*. Can. J. Microbiol. 18:1959-1964.

Cother, E. J., and Griffin, D. M. 1973. Formation of chlamydospores by *Phytophthora drechsleri*. Trans. Br. Mycol. Soc. 61:379-402.

Dastur, J. F. 1913. On *Phytophthora parasitica* sp. nov. A disease of the castor oil plant. Mem. Dep. Agric. India Bot. Ser. 5:177-213.

Desjardins, P. R., Zentmyer, G. A., and Reynolds, D. 1969. Electron microscopic observations of the flagellar hairs of *Phytophthora palmivora* zoospores. Can. J. Bot. 47:1077-1079.

Desjardins, P. R., Wang, M. C., and Bartnicki-Garcia, S. 1973a. Electron microscopy of zoospores and cysts of *Phytophthora palmivora*: Morphology and surface texture. Arch. Mikrobiol. 88:61-70.

Desjardins, P. R., Zentmyer, G. A., Chen, D., DeWolfe, T. A., Klotz, L. J., and Reynolds, D. A. 1973b. Flagellar hairs on zoospores of *Phytophthora* species: Tip hairs on the whiplash flagellum. Experientia 29:240-241.

Ehrlich, M. A., and Ehrlich, H. G. 1965. Ultrastructure of *Phytophthora infestans*. Phytopathology 55:1056.

Ellzey, J. T. 1974. Ultrastructural observations of meiosis within antheridia of *Achlya ambisexualis*. Mycologia 66:32-47.

Ellzey, J. T., and Huizar, E. H. 1977. Synaptonemal complexes in antheridia of *Achlya ambisexualis* E87. Arch. Microbiol. 112:311-313.

Elsner, P. R., VanderMolen, G. E., Horton, J. C., and Bowen, C. C. 1970. Fine structure of *Phytophthora infestans* during sporangial differentiation and germination. Phytopathology 60:1765-1772.

Gisi, U., Hemmes, D. E., and Zentmyer, G. A. 1979. Origin and significance of the discharge vesicle in *Phytophthora*. Exp. Mycol. 3:321-339.

Grove, S. N. 1971. Protoplasmic correlates of hyphal tip initiation and development in fungi. Ph.D. dissertation, Purdue Univ., Lafayette, IN.

Grove, S. N., and Bracker, C. E. 1970. Protoplasmic organization of hyphal tips among fungi: Vesicles and Spitzenkorper. J. Bacteriol. 104:989-1009.

Grove, S. N., and Bracker, C. E. 1978. Protoplasmic changes during zoospore encystment and cyst germination in *Pythium aphanidermatum*. Exp. Mycol. 2:51-98.

Grove, S. N., Bracker, C. E., and Morré, D. J. 1970. An ultrastructural basis for hyphal tip growth in *Pythium ultimum*. Am. J. Bot. 57:245-266.

Heath, I. B. 1976. Ultrastructure of freshwater phycomycetes. Pages 603-650 in: Recent Advances in Aquatic Mycology. E. B. G. Jones, ed. John Wiley & Sons, New York. 749 pp.

Hegnauer, H., and Hohl, H. R. 1973. A structural comparison of cyst and germ tube wall in *Phytophthora palmivora*. Protoplasma 77:151-163.

Hegnauer, H., and Hohl, H. R. 1978. Cell wall architecture of sporangia, chlamydospores, oogonia, and oospores in *Phytophthora*. Exp. Mycol. 2:216-233.

Hemmes, D. E., and Bartnicki-Garcia, S. 1975. Electron microscopy of gametangial interaction and oospore development in *Phytophthora capsici*. Arch. Microbiol. 103:91-112.

Hemmes, D. E., and Hohl, H. R. 1969. Ultrastructural changes in directly germinating sporangia of *Phytophthora parasitica*. Am. J. Bot. 56:300-313.

Hemmes, D. E., and Hohl, H. R. 1971. Ultrastructural aspects of encystation and cyst-germination in *Phytophthora parasitica*. J. Cell Sci. 9:175-191.

Hemmes, D. E., and Hohl, H. R. 1972. Flagellum degeneration in the fungus *Phytophthora palmivora* (formerly *Phytophthora parasitica*). J. Gen. Microbiol. 73:345-351.

Hemmes, D. E., and Hohl, H. R. 1973. Mitosis and nuclear degeneration: Simultaneous events during secondary sporangia formation in *Phytophthora palmivora*. Can. J. Bot. 51:1673-1675.

Hemmes, D. E., and Hohl, H. R. 1975. Ultrastructural changes and a second mode of flagellar degeneration during ageing of *Phytophthora palmivora* sporangia. J. Cell Sci. 19:563-577.

Hemmes, D. E., and Pinto da Silva, P. 1980. Localization of secretion-related, calcium binding substrates in encysting zoospores of *Phytophthora palmivora*. Biol. Cell. 37:235-240.

Hemmes, D. E., and Ribeiro, O. K. 1977. Electron microscopy of early gametangial interaction in *Phytophthora megasperma* var. *sojae*. Can. J. Bot. 55:436-447.

Hemmes, D. E., and Wong, L. D. S. 1975. Ultrastructure of chlamydospores of *Phytophthora cinnamomi* during development and germination. Can. J. Bot. 53:2945-2957.

Ho, H. H. 1979. The nature of amphigyny in *Phytophthora*. Mycologia 71:1057-1064.

Ho, H. H., and Hickman, C. J. 1967. Asexual reproduction and behavior of zoospores of *Phytophthora megasperma* var. *sojae*. Can. J. Bot. 45:1963-1981.

Ho, H. H., Zachariah, K., and Hickman, C. J. 1968a. The ultrastructure of zoospores of *Phytophthora megasperma* var. *sojae*. Can. J. Bot. 46:37-41.

Ho, H. H., Hickman, C. J., and Teleford, R. W. 1968b. The morphology of zoospores of *Phytophthora megasperma* var. *sojae* and other phycomycetes. Can J. Bot. 46:88-89.

Ho, H. H., Zentmyer, G. A., and Erwin, D. C. 1977. Morphology of sex organs of *Phytophthora cambivora*. Mycologia 69:641-646.

Hoch, H. C., and Mitchell, J. E. 1972a. The ultrastructure of *Aphanomyces euteiches* during asexual spore formation. Phytopathology 62:149-160.

Hoch, H. C., and Mitchell, J. E. 1972b. The ultrastructure of zoospores of *Aphanomyces euteiches* and of their encystment and subsequent germination. Protoplasma 75:113-138.

Hohl, H. R., and Hamamoto, S. T. 1967. Ultrastructural changes during zoospore formation in *Phytophthora parasitica*. Am. J. Bot. 54:1131-1139.

Kadooka, J. Y., and Ko, W. H. 1973. Production of chlamydospores by *Phytophthora palmivora* in culture media. Phytopathology 63:559-562.

King, J. E., and Butler, R. D. 1968. Structure and development of flagella of *Phytophthora infestans*. Trans. Br. Mycol. Soc. 51:689-697.

King, J. E., Colhoun, J., and Butler, R. D. 1968. Changes in the ultrastructure of sporangia of *Phytophthora infestans* associated with indirect germination and ageing. Trans. Br. Mycol. Soc. 51:269-281.

Laviola, C. 1974. Cytological observations on the asexual structure of *Phytophthora infestans* (Mont.) DeBary. Ann. Fac. Agrar. Univ. Bari 20:1-12.

Lunney, C. Z., and Bland, C. E. 1976. Ultrastructural observations of mature and encysting zoospores of *Pythium proliferum* de Bary. Protoplasma 90:119-137.

Meyer, R., Parish, R. W., and Hohl, H. R. 1976. Hyphal tip growth in *Phytophthora*. Arch. Microbiol. 110:215-224.

Morré, D. J., Mollenhauer, H. H., and Bracker, C. E. 1971. Origin and continuity of Golgi apparatus. Pages 82–126 in: Results and Problems in Cell Differentiation. Vol. 2. Origin and Continuity of Cell Organelles. T. Reinert and H. Ursprung, eds. Springer-Verlag, Berlin.

Murphy, P. A. 1918. The morphology and cytology of the sexual organs of *Phytophthora erythroseptica* Pethyb. Ann. Bot. (London) 32:115-153.

Newhook, F. J., Waterhouse, G. M., and Stamps, D. J. 1978. Tabular key to the species of *Phytophthora* de Bary. Mycol. Pap. 143. Commonw. Mycol. Inst., Kew, Surrey, England. 20 pp.

Nogueira, M. L. B., Pinto da Silva, P., and Bartnicki-Garcia, S. 1977. Freeze-fracture and freeze-etching study of encystment of *Phytophthora palmivora* zoospores. J. Gen. Microbiol. 102:149-155.

Pethybridge, G. H. 1913. On the rotting of potato tubers by a new species of *Phytophthora* having a method of sexual reproduction hitherto undescribed. Sci. Proc. R. Dublin Soc. 13:529-567.

Philippi, M. L., Parish, R. W., and Hohl, H. R. 1975. Histochemical and biochemical evidence for the presence of microbodies in *Phytophthora palmivora*. Arch. Microbiol. 103:127-132.

Pinto da Silva, P., and Nogueira, M. L. 1977. Membrane fusion during secretion. A hypothesis based on electron microscope observation of *Phytophthora palmivora* zoospores during encystment. J. Cell Biol. 73:161-181.

Powell, M. J., and Bracker, C. E. 1977. Isolation of zoospore organelles from *Phytophthora palmivora*. Page 533 in: Proc. Int. Mycol. Congr., 2nd, Tampa, FL.

Reichle, R. E. 1969a. Retraction of flagella by *Phytophthora parasitica* var. *nicotiana* zoospores. Arch. Mikrobiol. 66:340-347.

Reichle, R. E. 1969b. Fine structure of *Phytophthora parasitica* zoospores. Mycologia 61:30-51.

Ribeiro, O. K. 1978. A Sourcebook of the Genus *Phytophthora*. J. Cramer, Lehre, Germany. 420 pp.

Ribeiro, O. K., Erwin, D. C., and Khan, R. A. 1978. A new high-temperature *Phytophthora* pathogenic to roots of alfalfa. Phytopathology 68:155-161.

Sing, V. O., and Bartnicki-Garcia, S. 1975a. Adhesion of *Phytophthora palmivora* zoospores: Electron microscopy of cell attachment and cyst wall fibril formation. J. Cell Sci. 18:123-132.

Sing, V. O., and Bartnicki-Garcia, S. 1975b. Adhesion of *Phytophthora palmivora* zoospores: Detection and ultrastructural visualization of concanavalin A receptor sites appearing during encystment. J. Cell Sci. 19:11-20.

Tokunaga, J., and Bartnicki-Garcia, S. 1971a. Cyst wall formation and endogenous carbohydrate utilization during synchronous encystment of *Phytophthora palmivora* zoospores. Arch. Mikrobiol. 79:283-292.

Tokunaga, J., and Bartnicki-Garcia, S. 1971b. Structure and differentiation of the cell wall of *Phytophthora palmivora*: Cysts, hyphae and sporangia. Arch. Mikrobiol. 79:293-310.

Vujičić, R. 1971. An ultrastructural study of sexual reproduction of *Phytophthora palmivora*. Trans. Br. Mycol. Soc. 57:525-530.

Wang, M. C., and Bartnicki-Garcia, S. 1980. Distribution of mycolaminarans and cell wall $\beta$-glucans in the life cycle of *Phytophthora*. Exp. Mycol. 4:269-280.

Williams, W. T., and Webster, R. K. 1970. Electron microscopy of the sporangium of *Phytophthora capsici*. Can. J. Bot. 48:221-227.

# 3

# Nutrition of *Phytophthora*

Hans R. Hohl

CONTENTS



## INTRODUCTION

Proper nutrition of the vegetative thallus represents a most basic aspect of the fungal life cycle. The food sources available have a profound influence on growth rate measured as linear extension or increase in cell mass. They also influence the chances of survival under various environmental conditions and form the basis for reproduction and even spore germination.

*Phytophthora* has interesting nutritional aspects because it encompasses species with a wide variety of nutritional demands, ranging from the very simple to the highly complex (Hohl, 1975). It also represents an intermediate form between the preferentially saprophytic genera and the obligate parasites of the Peronosporales.

The first attempts to grow species of *Phytophthora* in the laboratory must probably be credited to de Bary, who cultivated and observed *P. infestans* (de Bary, 1876) and *P. omnivora* (now *P. cactorum*) (de Bary, 1881) inoculated onto various plant fragments. More serious investigations, aimed at growing these fungi on complex substrates such as variously amended decoctions of, for example, fruits, tubers, or seeds, date back to the turn of this century (e.g., Clinton, 1908; Matruchot and Molliard, 1903; Pethybridge and Murphy, 1913). They followed the pioneering efforts by Brefeld (1881), who was probably the first to grow *P. infestans* under strictly saprophytic conditions (Brefeld, 1883). Despite an increasingly widespread use of media such as oatmeal or lima bean agar, work on defined media did not materially progress until the 1930s, when the vitamin requirements of these organisms were uncovered (Fries, 1938; Kögl and Fries, 1937; Leonian, 1935; Leonian and Lilly, 1938; Robbins, 1938; Ronsdorf, 1935).

This review will concentrate on the purely nutritional aspects of vegetative growth of *Phytophthora* species (Table 1), although other environmental factors such as temperature, light, availability of oxygen, and humidity may profoundly influence growth of a species. This is not intended to be a compendium of media for growing species of *Phytophthora*, since that has recently been accomplished by Ribeiro (1978). The need for particular ingredients and the evidence available to substantiate the need will be emphasized. Nevertheless, requirements for vegetative growth can differ substantially from those for reproduction, a question considered in later chapters of this volume.

## NUTRITION

Plants require 10 macroelements for growth—C, O, H, N, S, P, K, Ca, Mg, and Fe—and they are able to assimilate these elements from inorganic salts, water, and $CO_2$. Fungi, on the other hand, are heterotrophic and require an organic source of carbon. They might have additional demands for specific growth factors or for specific compounds containing certain food sources (Cantino and Turian, 1959).

### Vitamins

Leonian (1935, 1936) and Ronsdorf (1935) were among the first to study vitamin requirements in *Phytophthora*. Leonian described a substance from corn roots or canned pea, probably thiamin, that stimulated growth of *P. cactorum*. Ronsdorf tested the growth-promoting activity of "Bios" extracted from yeast and known to contain biotin, among other biologically active ingredients. She was able to show that addition of "Bios," from which thiamin (vitamin $B_1$) was isolated 2 yr later (Kögl and Fries, 1937), greatly improved the growth of *P. cactorum, P. citrophthora, P. erythroseptica,* and *P. parasitica* var. *nicotianae*. Furthermore, she demonstrated the interesting fact that species of the closely related genus *Pythium* differed greatly by not being affected by this vitamin, a situation later confirmed and extended by Ridings et al (1969). With purified thiamin at hand, a series of studies emerged that firmly established the need for thiamin among species of *Phytophthora* (Cameron, 1966; Leonian and Lilly, 1938; Robbins, 1938; Robbins and Kavanagh, 1942; Roncadori, 1965; Saksena and Bhargava, 1943).

Only small quantities are needed to satisfy the thiamin requirement. According to Erwin and Katznelson (1961), 12–25 $\mu$g/L suffices, although other authors recommend higher amounts, e.g., 200 $\mu$g/L (Payette and Perrault, 1944), 200–800 $\mu$g/L (Lopatecki and Newton, 1956), 100 $\mu$g/L (Sakai, 1961), and 50 $\mu$g/L (Hwang et al, 1965). According to Fries (1938), concentrations between 200 and 1,000 $\mu$g/L increase yields but not growth rates.

Other vitamins have been tested for growth-promoting activities, alone and in combination, but no good evidence for an additional vitamin essential for the growth of

TABLE 1. Nutritional studies of *Phytophthora* species (on defined media)

| Species[a] | Studies[b] |
|---|---|
| *arecae* | 4,5,31,41,50,52 |
| *bahamensis* | 1 |
| *boehmeriae* | 4,15,22,31,44,52 |
| *botryosa* | 31 |
| *cactorum* | 4,5,7,8,12,13,22,28,30,31,39–42,45, 47,49,50,52,53,58,61 |
| *cambivora* | 4,5,28,30,31,40,50,52 |
| *capsici* | 4,5,22,30,31,36,40,48,50,52,58 |
| *castaneae* | 31 |
| *cinnamomi* | 3,4–6,14,22,28,30,31,40,41,50–52 |
| *citricola* | 4,5,30,31,40,41,47,50,52,58 |
| *citrophthora* | 4,5,28,30,31,48,50,52,53,57 |
| *colocasiae* | 31,44,50 |
| *cryptogea* | 4,5,15,30,31,50,52,58 |
| *cyperi* | ··· |
| *cyperi-bulbosi* | ··· |
| *drechsleri* var. *drechsleri* | 4,5,15,31,50,52,62 |
| *drechsleri* var. *cajani* | ··· |
| *epistomium* | 2 |
| *erythroseptica* var. *erythroseptica* | 4,5,18,22,30,31,41,43,45,50,52,53 |
| *erythroseptica* var. *pisi* | ··· |
| *fragariae* | 4,5,11,17,22,28,30,31,46,50 |
| *gonapodyides* | 31 |
| *heveae* | 5,31,39–42,50,52 |
| *hibernalis* | 31,52 |
| *ilicis* | 4,5,40,52 |
| *infestans* | 5,9,10,16,18–22,26–32,34,35,37,38, 52,56,59 |
| *inflata* | ··· |
| *iranica* | ··· |
| *japonica* | ··· |
| *lateralis* | 4,5,22,30,31,50,52 |
| *lepironiae* | ··· |
| *macrospora* | ··· |
| *meadii* | 47,50 |
| *megakarya* | 31 |
| *megasperma* var. *megasperma* | 4,5,28,30,31,40,45,50,52 |
| *megasperma* var. *sojae* | 14,15,22,28,31,50,52 |
| *melonis* | ··· |
| *mexicana* | ··· |
| *mycoparasitica* | ··· |
| *nicotianae* | 3–5,15,18,22–25,28,30,31,44,45, 47,50–53,58,63 |
| *palmivora* | 4,5,22,28,30,31,33,40,44,48,50,52,60 |
| *phaseoli* | 52,55 |
| *porri* | 31 |
| *primulae* | ··· |
| *quininea* | 31,52 |
| *richardiae* | 31,47 |
| *spinosa* var. *lobata* | 1 |
| *spinosa* var. *spinosa* | 1,31 |
| *syringae* | 5,30,31,50,52 |
| *verrucosa* | ··· |
| *vesicula* | 1,30,31 |
| *vignae* | 31 |

[a] Nomenclature according to Waterhouse (1963) and Newhook et al (1978), of marine species according to Fell and Master (1975).

[b] 1 = Bahnweg, 1978; 2 = Bahnweg, 1980; 3 = Bartnicki-Garcia, 1966; 4 = Cameron, 1966; 5 = Cameron and Milbrath, 1965; 6 = Chee and Newhook, 1965a; 7 = Christie, 1958; 8 = Christie, 1959; 9 = Clarke, 1966; 10 = Cuppett and Lilly, 1973; 11 = Davies, 1959; 12 = Elliott, 1972; 13 = Elliott et al, 1966; 14 = Erwin, 1968; 15 = Erwin and Katznelson, 1961; 16 = Fehrmann, 1971; 17 = Fleetwood Walker, 1955; 18 = Fothergill and Child, 1964; 19 = French, 1953; 20 = [*continued in next column*]

TABLE 1—*Continued*

Gaertner, 1959; 21 = Hall, 1959; 22 = Hendrix, 1965; 23 = Hendrix and Apple, 1964; 24 = Hendrix and Guttman, 1970; 25 = Hendrix et al, 1969; 26 = Henniger, 1959; 27 = Henniger, 1963; 28 = Ho and Foster, 1972; 29 = Hodgson, 1958; 30 = Hohl, 1975; 31 = Hohl, Chapter 3, this volume; 32 = Hollomon, 1966; 33 = Huguenin and Boccas, 1971; 34 = Huang et al, 1980; 35 = Hwang et al, 1965; 36 = Katsura and Nishioka, 1958; 37 = Kusaka and Kitai, 1978; 38 = Langcake, 1974; 39 = Leal et al, 1964; 40 = Leal et al, 1967a,b; 41 = Leal et al, 1970; 42 = Leal et al, 1971; 43 = Leonian and Lilly, 1938; 44 = Lin and Liang, 1965; 45 = Lopatecki and Newton, 1956; 46 = Maas, 1972; 47 = Mehrotra, 1951; 48 = Mitchell and Zentmyer, 1971; 49 = Nes et al, 1979; 50 = Ribeiro et al, 1975; 51 = Ridings et al, 1969; 52 = Roncadori, 1965; 53 = Ronsdorf, 1935; 54 = Sakai, 1961; 55 = Saksena and Bhargava, 1943; 56 = Scheepens and Fehrmann, 1978; 57 = Schiffmann-Nadel and Cohen, 1972; 58 = Schwinn, 1959; 59 = Seidel, 1961; 60 = Singh, 1975; 61 = Sneh, 1972; 62 = Timmer et al, 1970; 63 = Wills, 1954.

*Phytophthora* has been disclosed (Cameron, 1966; Chee and Newhook, 1965a; Erwin and Katznelson, 1961; Fothergill and Child, 1964; French, 1953; Hodgson, 1958; Payette and Perrault, 1944). Ascorbic acid (vitamin C) is an apparent exception; it has been noted to promote growth of several species under some conditions (Bahnweg, 1978; Cuppett and Lilly, 1973; Hodgson, 1958; Hollomon, 1966; Singh, 1975). However, ascorbic acid must be applied in greatly elevated amounts, e.g., 160 mg/L (Hodgson, 1958) and 200 mg/L (Singh, 1975), and appears to act primarily as a reducing agent or even as a carbon source (Bahnweg, 1978). In all these instances, it can be replaced by some other reducing compounds.

### Carbon Sources

A series of sugars and additional compounds have been tested for their ability to support growth of *Phytophthora* spp. Table 2 summarizes some of the results obtained from studies using liquid culture techniques. By combining these results with some additional data (Chee and Newhook, 1965a; Erwin and Katznelson, 1961; Henniger, 1959; Ho and Foster, 1972; Hodgson, 1958; Katsura and Nishioka, 1958; Kusaka and Kitai, 1978; Lin and Liang, 1965; Lopatecki and Newton, 1956; Schiffmann-Nadel and Cohen, 1972; Schwinn, 1959; Volkonsky, 1934; Wills, 1954), a general picture emerges. The best and also the most widely used carbon sources are sucrose and glucose, followed by fructose, maltose, and starch. All others are less suitable and some, such as sorbose, do not appear to support growth at all. In general, hexoses are preferred.

Of the alcohols, only glycerol (a sugar alcohol) proved to be a good source in the few cases tested (Bahnweg, 1978; Hodgson, 1958). Other sugar alcohols (Bahnweg, 1978; Mehrotra, 1951), amino acids (Davies, 1959; Wills, 1954), organic acids (discussed in the next section), and several polysaccharides such as chitin, cellulose, or pectin (Clarke, 1966; Hall, 1959) appear to be very poor sources, despite the fact that the latter two compounds are major components of host cell walls penetrated by these fungi during parasitic development.

The amount of glucose or sucrose incorporated into media varies but is usually between 1 and 3% (w/v). Higher concentrations often lead to higher dry weight production, although the efficiency of sugar utilization decreases. Kusaka and Kitai (1978) recommend 6% soluble starch plus 6% sucrose for optimal growth of *P. infestans*. This fungus is even capable of growing on a 15% sucrose concentration (H. R.

Hohl, *unpublished*), indicating a tolerance for high osmotic pressures.

Lipids such as fats and fatty acids have also been reported to be good carbon sources for *Phytophthora* (Hendrix and Apple, 1964; Sneh, 1972). Evidence for a possible role of lipids other than providing carbon skeletons for energy production and growth will be provided in a following section.

### Organic Acids

A number of organic acids have been tested for their growth-promoting activity in several species. The most frequently tested compounds were malate, succinate, fumarate, tartrate, citrate, acetate, and oxalate, but glyoxylate, pyruvate, malonate, glutarate, adipate, pimelate, α-ketoglutarate, ascorbate, and oxaloacetate were also included in these studies. Test species included *P. vesicula, P. bahamensis, P. spinosa* (all marine species) (Bahnweg, 1978), *P. cinnamomi* (Chee and Newhook, 1965a), *P. fragariae* (Davies, 1959), *P. infestans* (Henniger, 1959; Hodgson, 1958; Huang et al, 1980; Hwang et al, 1965; Sakai, 1961; Seidel, 1961), *P. parasitica* (Hwang et al, 1965; Lin and Liang, 1965), *P. parasitica* var. *nicotianae* (Hwang et al, 1965; Lin and Liang, 1965), and *P. boehmeriae, P. colocasiae*, and *P. palmivora* (Lin and Liang, 1965).

Researchers generally agree that organic acids do not support growth of *Phytophthora* spp. when added as sole carbon sources. In 1934, Leonian (Leonian and Lilly, 1940) first noticed growth enhancement in *Phytophthora* when 0.2% succinate was added to a proteose peptone-dextrose medium. The observation that supplementing a carbohydrate-containing (e.g., glucose) medium with an organic acid has a beneficial effect on growth has been made several times since (Leal et al, 1971; Roncadori, 1965; Sakai, 1961; Seidel, 1961), particularly in the presence of ammonium as N source.

Addition of organic acids may also lead to growth reduction, depending on the basal composition of the medium (Henniger, 1959) and the strain used (French, 1953). Extensive investigations by Chinese researchers (Huang et al, 1980; Hwang et al, 1965; Lin and Liang, 1965) have shown that the detrimental effects of organic acids are overcome by the addition of sufficient amounts of calcium. According to these authors, in a medium containing, for instance, fumarate, $Ca^{2+}$, glucose, ammonium, and inorganic salts, even fastidious strains such as those of *P. infestans* are able to grow luxuriantly.

Hwang et al (1965) and Lin and Liang (1965) have suggested that organic acids may provide Krebs cycle intermediates. Furthermore, organic acids apparently serve as buffers to minimize the acidification associated with ammonium

TABLE 2. Effect of carbon sources on growth of *Phytophthora* species in suspension cultures, expressed as a percentage of the highest value (from different authors)

| | Species | | | | | | | | | | | | | | | |
|---|---|---|---|---|---|---|---|---|---|---|---|---|---|---|---|---|
| | *vesicula* isolates[a] | | | *spinosa* | *cactorum* study | | | | *parasitica* | *fragariae* study | | *infestans* study | | | | |
| Carbon source | Pacific | Miami | *baha-mensis*[a] | var. *spinosa*[a] | 1[b] | 2[c] | *citri-cola*[c] | *meadii*[c] | var. *nicotianae*[c] | 1[d] | 2[d] | 1[e] | 2[f] | 3[g] | 4[h] | *palmi-vora*[i] |
| Hexoses | | | | | | | | | | | | | | | | |
| D-fructose | 88 | 100 | 100 | 98 | 0 | 38 | 67 | 53 | 64 | 21 | 22 | 100 | 0 | 94 | 4 | 82 |
| D-galactose | 0 | 0 | 0 | 0 | 17 | 16 | 67 | 62 | 71 | 0 | 18 | 23 | 12 | 54 | ··· | 20 |
| D-glucose | 93 | 80 | 82 | 91 | 42 | 75 | 77 | 62 | 71 | 95 | 100 | 91 | 100 | 98 | 100 | 87 |
| L-mannose | 94 | 69 | 94 | 75 | ···[j] | 28 | 33 | 31 | 36 | 29 | 33 | 55 | ··· | ··· | ··· | 37 |
| L-rhamnose | 0 | 0 | 0 | 0 | ··· | 44 | 50 | 38 | 71 | ··· | ··· | 5 | ··· | ··· | ··· | 15 |
| L-sorbose | 0 | 0 | 0 | 0 | ··· | ··· | ··· | ··· | ··· | 0 | 0 | ··· | ··· | ··· | ··· | 0 |
| Pentoses | | | | | | | | | | | | | | | | |
| L-arabinose | 0 | 0 | 0 | 0 | 0 | 41 | 58 | 52 | 71 | ··· | ··· | 9 | ··· | ··· | 11 | 11 |
| D-ribose | 59 | 52 | 0 | 0 | ··· | ··· | ··· | ··· | ··· | ··· | ··· | ··· | ··· | ··· | ··· | ··· |
| D-xylose | 93 | 64 | 0 | 0 | 0 | 31 | 33 | 25 | 32 | 10 | 10 | 10 | ··· | ··· | 0 | 11 |
| Di- and tri-saccharides | | | | | | | | | | | | | | | | |
| cellobiose | 100 | 79 | 92 | 0 | ··· | ··· | ··· | ··· | ··· | ··· | ··· | 6 | ··· | ··· | ··· | ··· |
| lactose | 90 | 0 | 77 | 0 | ··· | 62 | 62 | 55 | 48 | 0 | 10 | 44 | 10 | 9 | ··· | 20 |
| maltose | 99 | 48 | 0 | 0 | 100 | 94 | 87 | 75 | 80 | 88 | 90 | 15 | 10 | 24 | ··· | 14 |
| trehalose | 0 | 0 | 0 | 0 | ··· | ··· | ··· | ··· | ··· | ··· | ··· | 10 | ··· | ··· | ··· | ··· |
| sucrose | 99 | 65 | 0 | 100 | 74 | 100 | 100 | 100 | 100 | 100 | 96 | 95 | 50 | 100 | 52 | 100 |
| raffinose | 79 | 0 | 0 | 45 | ··· | 47 | 10 | 42 | 18 | 38 | 49 | 28 | ··· | ··· | 11 | ··· |
| melibiose | 0 | 0 | 0 | 0 | ··· | ··· | ··· | ··· | ··· | ··· | ··· | ··· | ··· | ··· | ··· | ··· |
| Polysaccharides | | | | | | | | | | | | | | | | |
| starch | 84 | 0 | 0 | 0 | 0 | 94 | 90 | 62 | 96 | 17 | 22 | ··· | 0 | ··· | ··· | 16 |
| dextran | 67 | 0 | 0 | 0 | ··· | 47 | 55 | 42 | 36 | 67 | 75 | ··· | ··· | ··· | ··· | 21 |
| cellulose | 0 | 0 | 0 | 0 | ··· | ··· | ··· | ··· | ··· | ··· | ··· | ··· | ··· | ··· | ··· | ··· |
| inulin | 0 | 0 | 0 | 0 | 11 | 62 | 50 | 36 | 64 | ··· | ··· | ··· | 34 | ··· | ··· | ··· |
| chitin | 0 | 0 | 0 | 0 | ··· | ··· | ··· | ··· | ··· | ··· | ··· | ··· | ··· | ··· | ··· | ··· |

[a] Bahnweg (1978).
[b] Christie (1958).
[c] Mehrotra (1951).
[d] Davies (1959).
[e] Henniger (1963).
[f] Sakai (1961).
[g] Seidel (1961).
[h] Hall (1959).
[i] Singh (1975).
[j] Not determined.

assimilation. Thus, the growth inhibition frequently observed in the presence of ammonium ions (Leal et al, 1970; Lin and Liang, 1965; Saksena and Bhargava, 1943) is attributable to a considerable degree to their effect on the pH of the medium but might also represent a direct action of ammonia itself (Leal et al, 1970). Replacement of organic acids by inorganic buffers might produce a similar growth improvement, as has been shown by Morton and Macmillan (1954) for fungi other than *Phytophthora*.

## Nitrogen Sources

Although *Phytophthora* spp. are generally considered to prefer organic nitrogen sources, only very few species appear to have an absolute requirement for them. Among these are strains of *P. infestans, P. fragariae*, and *P. bahamensis*. The claim that *P. phaseoli* has a specific need for alanine (Saksena and Bhargava, 1943) appears questionable in the light of Roncadori's (1965) report that *P. phaseoli* grows well on ammonium. As will be shown in this section, most phytophthoras thrive on inorganic nitrogen sources.

Individual amino acids presented to the organism as sole nitrogen source stimulate growth to different degrees. L-amino acids are better than their D-isomers (Newton, 1956). From the results of 14 studies performed with different *Phytophthora* species on various basal media, a list can be compiled of amino acids able to serve as sole nitrogen sources (the number of positive ratings is given in brackets, but not all amino acids were tested in all 14 studies). Good nitrogen sources were L-asparagine (10), L-arginine (8), L-proline (8), L-alanine (8), L-histidine (7), L-serine (6), glycine (6), L-threonine (5), L-glutamic acid (5), L-aspartic acid (5), and L-glutamine (4). L-ornithine, L-cysteine, L-valine, L-phenylalanine, and L-tyrosine were mentioned in one to three cases, but all five were also listed at least once as not supporting growth. Poor nitrogen sources were L-cystine, L-hydroxyproline, L-isoleucine, L-leucine, L-lysine, L-methionine, and L-tryptophan. These 14 studies were made by Bahnweg (1978), Cameron and Milbrath (1965), Chee and Newhook (1965a), Christie (1958), Davies (1959), Erwin and Katznelson (1961), Henniger (1963), Hodgson (1958), Leal et al (1971), Leonian and Lilly (1938), Sakai (1961), Saksena and Bhargava (1943), Singh (1975), and Wills (1954).

TABLE 3. Amino acid mixtures recommended for growth of *Phytophthora infestans*

| | Values (mg/L) from study[a] | | | | | |
|---|---|---|---|---|---|---|
| Amino acid | 1 | 2 | 3a | 3b | 4 | 5 |
| Alanine | 100 | 100 | | 89 | | |
| Arginine | 100 | 200 | | | | |
| Asparagine | | | 130 | 1,000 | 500 | 240 |
| Aspartic acid | 100 | 400 | | | | |
| γ-aminobutyric acid | 200 | | | | | 102 |
| Cysteine | 200 | 200 | 60 | | 500 | |
| Glycine | | 200 | | | | 8 |
| Glutamine | | 400 | | | | 214 |
| Histidine | | | 190 | 190 | | 17 |
| Isoleucine | | | | | | 21 |
| Leucine | 200 | 100 | 65 | 65 | | 9 |
| Lysine | | | | | | 17 |
| Proline | | | | | 1,000 | |
| Serine | | | 50 | 50 | | 18 |
| Threonine | | | | | | 90 |
| Tryptophan | 80 | | | | | |
| Valine | | | | | | 30 |

[a] Studies: 1 = Henniger (1959), 2 = Henniger (1963), 3 (a and b) = Hohl (1975), 4 = Kusaka and Kitai (1978), 5 = Scheepens and Fehrmann (1978).

With the exception of strains of *P. infestans* and *P. fragariae*, the nitrogen requirement for most species can be met by the addition of a single amino acid, e.g., L-asparagine. *P. infestans* is the most widely studied and probably also the most complex species in its nitrogen requirement (Hohl, 1975). Although some of the strains grow reasonably well on asparagine alone or even on ammonium (Cuppett and Lilly, 1973; Hwang et al, 1965; Seidel, 1961), none grows well on nitrate, and some require more than one amino acid for optimal growth under the conditions tested. Apart from strain 211.46, which appears to have an absolute requirement for L-histidine (Hohl, 1975), several mixtures of amino acids are capable of sustaining good growth of these strains (Table 3). This indicates a high degree of substitution of one amino acid for another. *P. fragariae* represents a somewhat simpler case. It does grow fairly well in the presence of only one amino acid such as L-alanine, L-proline, or L-asparagine (Davies, 1959), provided $Ca^{2+}$ is present in the medium, but it grows considerably better in the presence of amino acid mixtures (e.g., casein hydrolysate [Maas, 1972]) and additional growth factors (Hohl, 1975).

As outlined in the previous section, ammonium can be assimilated well by the majority of species (Roncadori, 1965), and even by strains of *P. infestans* (Henniger, 1959; Lin and Liang, 1965; Sakai, 1961; Seidel, 1961), provided the medium is buffered, preferably with an organic acid.

Whereas amino acids and ammonium contain nitrogen in its reduced state, in nitrate the nitrogen occurs in an oxidized form. The ability to assimilate nitrate nitrogen is considered a primitive trait (Cantino and Turian, 1959; Robbins, 1937), yet most phytophthoras obviously grow well on nitrate, provided all other factors are optimal (Bahnweg, 1978; Christie, 1958; Erwin and Katznelson, 1961; Hohl, 1975; Leal et al, 1971; Lopatecki and Newton, 1956; Roncadori, 1965). For sexual reproduction, an organic nitrogen source may be necessary (Klemmer and Lenney, 1965; Ribeiro et al, 1975). Hendrix and Guttman (1970) have clearly documented that *P. parasitica* var. *nicotianae* has a calcium or sterol requirement for growth on a nitrate medium, and Hohl (1975) was able to stimulate growth of many species on nitrate by adding lecithin to the medium. The not uncommon practice of using ammonium nitrate as a nitrogen source might not be advisable, because ammonium has been reported to block assimilation of nitrate (Lin and Liang, 1965).

*P. fragariae* does not grow on nitrate (Davies, 1959). The same applies to *P. infestans* (French, 1953; Henniger, 1959; Hodgson, 1958; Hohl, 1975; Hwang et al, 1965; Sakai, 1955, 1961; Seidel, 1961), to *P. phaseoli* (Roncadori, 1965; Saksena and Bhargava, 1943), and to the marine species *P. bahamensis* (Bahnweg, 1978). These species have apparently lost the capability to assimilate oxidized nitrogen efficiently. According to the scheme of Robbins (1937), most *Phytophthora* species fit into group II (i.e., they assimilate nitrate efficiently). *P. infestans* (some strains), *P. epistomium*, and *P. phaseoli* belong to group III (i.e., they assimilate ammonium and organic N efficiently), whereas *P. fragariae, P. infestans* (many strains), *P. bahamensis, P. cambivora, P. colocasiae, P. megasperma* f. sp. *medicaginis* (some strains), *P. quininea*, and *P. richardiae* should be placed into group IV (i.e., they require organic nitrogen). This classification is still preliminary because of the limited number of strains tested.

A purine base proved advantageous for growth of two strains of *P. infestans*, but was an absolute requirement for one strain only (Hohl, 1975). The small amounts needed (50 mg/L) tend to indicate that the nucleotide was not primarily used as a nitrogen source but rather as a growth factor. Chee and Newhook (1965b) reported that two of the five strains of *P. cinnamomi* produce more oospores in the presence of hypoxanthine, and Clark et al (1978) consider hypoxanthine to be a key intermediate in the purine metabolism directly associated with spore formation and development in *P. infestans*.

### Sulfur and Phosphorus

Both sulfur and phosphorus are unquestionably essential elements for all organisms and are routinely incorporated into *Phytophthora* media, usually as inorganic sulfates and phosphates. Still, the question may be raised about the best form in which these two elements are assimilated by *Phytophthora*. Apart from a study by Mehrotra in 1950 (cited by Fothergill and Child [1964]), only one report deals with this question systematically (Fothergill and Child, 1964). These authors studied a wide range of inorganic and organic sulfur and phosphorus compounds (30 mg/L of S compounds, 300 mg/L of P) in an asparagine-glucose medium somewhat modified from that of Lopatecki and Newton (1956). Some of the results are summarized in Tables 4 and 5.

According to Fothergill and Child (1964), both sodium sulfate and sodium thiosulfate were very good sources of sulfur for all three test organisms. Sodium metabisulfite, sodium dithionate, sodium sulfite, sodium sulfide, cystine, and methionine were good sources for *P. erythroseptica* and *P. parasitica* but not for *P. infestans* (Table 4). In general then, the oxidation level of the sulfur atom had no consistent significant effect on mycelial growth. Yields from the best inorganic sulfur sources were comparable to those obtained from the sulfur-containing amino acids methionine and cystine. In contrast, the Saprolegniales do not grow on inorganic sulfur compounds (Volkonsky, 1933). Also, *P. infestans* has a narrower range for sulfur compounds than the other two species tested but is still capable of using inorganic as well as organic sulfur sources (Sakai, 1961).

As seen in Table 5, sodium dihydrogen orthophosphate, sodium metaphosphate, and lecithin were excellent sources of phosphorus. Whereas tetrasodium pyrophosphate, sodium $\beta$-glycerophosphate, and nucleic acid produced high yields for *P. erythroseptica* and *P. parasitica*, the corresponding yields for *P. infestans* were greatly reduced, again pointing to the more restricted nutritional range of *P. infestans*. As a whole, inorganic phosphorus sources appeared to surpass the organic compounds tested. Thus, *Phytophthora* is capable of assimilating effectively even the most oxidized forms of this element.

The remarkable effect of lecithin (Table 5) may be due not only to its phosphorus content but also to its content of unsaturated fatty acids, as will be shown in a following section.

### Potassium, Calcium, and Magnesium

$K^+$ and $Mg^{2+}$ are unquestionably universal requirements. Hodgson's study (1958) confirmed these requirements for a species of *Phytophthora* (*P. infestans*).

$Ca^{2+}$ has been identified in at least four cases as an essential element for *Phytophthora* spp. Davies (1959) demonstrated an absolute $Ca^{2+}$ requirement for *P. fragariae* independent of the carbon or nitrogen source present. This need was expressed even in the presence of $Mg^{2+}$, another divalent alkaline earth element. Strontium partly replaced calcium, but barium did not. A second case concerns *P. colocasiae*, which did not grow on a nitrate medium in the absence of $Ca^{2+}$. The third case is presented by Hendrix and Guttman (1970), who described an "absolute" requirement for $Ca^{2+}$ for *P. parasitica* var. *nicotianae* when the fungus was grown on nitrate in the absence of sterol. The sterol was able to replace $Ca^{2+}$. The combined action of sterol and $Ca^{2+}$ was strongly synergistic, i.e., greater than the sum of the individual effects. In the presence of an organic nitrogen source (asparagine), $Ca^{2+}$ promoted growth but was not essential (Hendrix and Guttman, 1970). A comparable result had been obtained previously by Erwin (1968). He found that $Ca^{2+}$ stimulated growth of *P. megasperma* and of *P. cinnamomi* by a factor of two to three; yet some growth occurred even in the apparent absence of $Ca^{2+}$ and despite the precautions taken to avoid accidental contamination of the medium (in particular the

TABLE 4. Utilization of inorganic and organic sulfur compounds by *Phytophthora erythroseptica* (PE), *P. parasitica* (PP), and *P. infestans* (PI)

| | PE | | | PP | | | PI | |
|---|---|---|---|---|---|---|---|---|
| Sulfur source | Yield[a] | Time[b] (days) | Final[c] pH value | Yield[a] | Time[b] (days) | Final[c] pH value | Yield[a] | Final[c] pH value |
| Sodium sulfate | 100 | 20 | 5.8 | 100 | 17 | 5.7 | 100 | 5.8 |
| Potassium persulfate | 75 | 20 | 5.8 | 100 | 17 | 5.5 | 24 | 3.9 |
| Sodium metabisulfite | 124 | 25 | 5.7 | 110 | 17 | 5.5 | 29 | 4.0 |
| Sodium dithionite | 100 | 25 | 5.1 | 90 | 17 | 5.3 | 33 | 4.2 |
| Sodium thiosulfate | 81 | 20 | 5.8 | 100 | 17 | 5.4 | 100 | 6.0 |
| Sodium sulfite | 124 | 25 | 5.8 | 132 | 13 | 5.3 | 81 | 5.8 |
| L-cysteine hydrochloride | 115 | 25 | 5.8 | 98 | 17 | 5.3 | 43 | 5.9 |
| Cystine | 86 | 25 | 5.7 | 93 | 17 | 5.6 | 95 | 5.1 |
| Thiourea | 27 | 20 | 5.9 | 28 | 17 | 5.4 | 57 | 7.3 |
| Sodium sulfide | 127 | 25 | 5.7 | 103 | 17 | 5.4 | 62 | 5.6 |
| Sulfanilic acid | 20 | 20 | 6.2 | 15 | 13 | 5.6 | 57 | 6.4 |
| Sulfamic acid | 26 | 20 | 5.8 | 14 | 13 | 5.8 | 43 | 4.1 |
| Sulfanilamide | 10 | 25 | 6.0 | 16 | 13 | 5.5 | 57 | 7.4 |
| Sulfosalicylic acid | 18 | 25 | 6.1 | 22 | 9 | 5.5 | 19 | 3.9 |
| Methionine | 110 | 25 | 5.8 | 104 | 17 | 5.4 | 86 | 5.7 |
| Thiazamide | 20 | 20 | 6.2 | 17 | 17 | 6.0 | 53 | 7.5 |
| Control (no sulfur) | 17 | 20 | 6.1 | 16 | 17 | 5.5 | 24 | 4.7 |

[a] Highest yield (milligrams of dry weight of mycelium per flask, average of five replications), expressed as percentage of yield from $Na_2SO_4 \cdot 10H_2O$, taken as 100.

[b] Incubation period; for PI, 16 days.

[c] Initial pH values: PE = 6.6, PP = 6.6, PI = 4.7.

*Source*: Fothergill and Child (1964).

asparagine) by $Ca^{2+}$. Here, too, $Sr^{2+}$ was able to replace $Ca^{2+}$.

The observation that $Ca^{2+}$ improves growth of *Phytophthora* spp. has been confirmed and extended to the following 12 species: *P. boehmeriae* (Erwin and Katznelson, 1961; Lin and Liang, 1965), *P. cactorum* (Elliott, 1972), *P. colocasiae* (Lin and Liang, 1965), *P. cinnamomi* (Erwin, 1968), *P. erythroseptica* (Fothergill and Child, 1964; Lopatecki and Newton, 1956), *P. drechsleri* (Erwin and Katznelson, 1961), *P. fragariae* (Davies, 1959), *P. infestans* (Fothergill and Child, 1964; Gaertner, 1959; Hwang et al, 1965; Kusaka and Kitai, 1978; Sakai, 1961), *P. megasperma* (Erwin and Katznelson, 1961, referred to there as *P. cryptogea* [Erwin, 1968]; Lopatecki and Newton, 1956), *P. palmivora* (Lin and Liang, 1965; Singh, 1975), *P. parasitica* (Erwin and Katznelson, 1961; Lin and Liang, 1965; Lopatecki and Newton, 1956), and *P. parasitica* var. *nicotianae* (Hendrix and Guttman, 1970; Hendrix et al, 1969; Lin and Liang, 1965). A few reports conflict; Lopatecki and Newton (1956) did not observe a growth-promoting effect of $Ca^{2+}$ for *P. cactorum*; neither did Chee and Newhook (1965a) for *P. cinnamomi* nor Hodgson (1958) for *P. infestans*. Chee and Newhook (1965a) used agar media, which are known to contain traces of metals and minerals.

In general, optimal concentrations of $Ca^{2+}$ range from 50 to 100 mg/L. Yet, Lin and Liang (1965) and Hwang et al (1965) indicated that the amount of calcium required for optimal growth varies with the amount of organic acid present in the medium, the demand for calcium increasing with rising concentrations of organic acids. Most likely, counteracting detrimental effects of organic acids is not the only function of calcium, but a better understanding of its role for *Phytophthora* must await a more accurate evaluation of the role this element plays in fungi generally.

### Iron and Other Trace Elements

Iron holds an intermediate position between the macroelements and the microelements, as it is usually required in the milligram range compared with the microgram range of the "normal" trace elements. Its possible significance for *Phytophthora* has been under scrutiny for many years. Even in 1938, Fries incorporated iron into his medium for *P. cactorum*. No one doubts that iron is essential for growth of *Phytophthora*; it is part of several essential enzymes such as the cytochromes. Yet, in no documented case has this essentiality been accurately determined experimentally for any of the species, most likely because of technical problems in ridding media, glassware, or inocula from contaminating iron compounds. The strong promotion of growth obtained by the addition of small amounts (0.1–1.0 mg/L) of iron to a medium speaks well for its crucial role for *Phytophthora* (Cuppett and Lilly, 1973; Fothergill and Child, 1964; Fries, 1938; Gaertner, 1959; Hendrix et al, 1969; Hodgson, 1958; Kusaka and Kitai, 1978; Lopatecki and Newton, 1956; Sakai, 1961; Singh, 1975).

Cuppett and Lilly (1973) emphasized the importance of the valency of iron for growth of *P. infestans*, with the ferrous ($Fe^{2+}$) being favored over the ferric ($Fe^{3+}$) form. Since ferrous iron would usually be oxidized to the ferric state within a few days, the addition of reducing agents might improve the culture media by reducing the iron compounds. This, in fact, was observed, as evidenced by shortened lag phases and improved growth rates.

As media become progressively more defined and chemical ingredients increase in purity, the question of the importance of trace elements for growth of *Phytophthora* becomes acute, particularly when one is working with media not solidified with agar. Here again, several trace elements must be essential because they are part of essential enzyme systems. However, the difficulties of performing trace element studies are even more pronounced than those of calcium or iron studies because of the minute quantities required. Lopatecki and Newton (1956) convincingly, though not stringently, demonstrated the need of *P. parasitica* for zinc, copper, and manganese. Zinc (1 mg/L), copper (50 μg/L), and manganese (50 μg/L) promoted growth of the fungus markedly when all three were added to the medium. Growth was reduced when one of these trace elements was omitted. Molybdenum did not have an effect in this study and was presumed to be present as a contaminant. Sakai (1961), on the other hand, was able to show that molybdenum added to a medium did improve growth of *P. infestans*, whereas zinc and manganese had only slight stimulatory effects. Hendrix et al (1969) enhanced growth by adding zinc in *P. parasitica* var. *nicotianae*. Most authors working with liquid culture media routinely include a microelement solution in their formulas with varying but often positive effects on growth.

TABLE 5. Utilization of phosphorus compounds by *Phytophthora erythroseptica* (PE), *P. parasitica* (PP), and *P. infestans* (PI)

| Phosphorus source | PE | | PP | | PI |
|---|---|---|---|---|---|
| | Yield[a] | Time[b] (days) | Yield[a] | Time[b] (days) | Yield[a] |
| Sodium dihydrogen orthophosphate | 100 | 25 | 100 | 17 | 100 |
| Sodium metaphosphate | 94 | 25 | 96 | 17 | 129 |
| Tetrasodium pyrophosphate | 101 | 25 | 97 | 13 | 4 |
| Sodium β-glycerophosphate | 100 | 20 | 87 | 13 | 7 |
| Sodium phosphite | 8 | 20 | 19 | 17 | 4 |
| Sodium hypophosphite | 11 | 20 | 14 | 13 | 13 |
| Nucleic acid | 97 | 20 | 116 | 17 | 49 |
| Lecithin | 119 | 20 | 77 | 13 | 81 |
| Casein | +[c] | | +[c] | | +[c] |
| Control (no phosphate) | 23 | 20 | 11 | 13 | 11 |

[a] Highest yield (milligrams of dry weight per flask, average of five replications), expressed as percentage of yield from $Na_2H_2PO_4$, taken as 100. All initially pH 4.7.
[b] Incubation period; for PI, 16 days.
[c] Good mycelial growth but weight not determinable.

*Source*: Fothergill and Child (1964).

### Reducing Agents

In 1942, Peterson (cited by Hodgson [1958]) noticed that ascorbic acid stimulated growth of *P. infestans*. Hodgson (1958) observed that two of his four strains of *P. infestans* would grow on a defined medium only if the reducing agents ascorbic acid or sodium thioglycollate in concentrations of 160 mg/L were added to the medium. Glutathione or cysteine-HCl, also reducing agents, had no effect in this experiment. The stimulating action of various reducing compounds was confirmed in several laboratories (Fehrmann, 1971; Gaertner, 1964; Henniger, 1963; Hollomon, 1966; Hwang et al, 1965). All of these reports refer to *P. infestans*, which apparently prefers a high redox potential in the medium, perhaps to prevent formation of toxic oxidation products or, as in the case of iron (Cuppett and Lilly, 1973), to prevent oxidation to the ferric state. Whether the situation is peculiar to *P. infestans* or involves other species is at present uncertain. Effects of ascorbic acid and of ferric iron in *P. palmivora* were minimal, according to Singh (1975).

## Lipids

Several groups of lipids—including fats, fatty acids, phospholipids like lecithin, and sterols—have strong growth-promoting properties. Sakai (1961) noticed that a fatty acid fraction from cow liver stimulated growth of *P. infestans* considerably. Hendrix and Apple (1964), Hendrix (1965), and Hendrix et al (1966) performed experiments to determine the effect of fats, oils, and hydrocarbons on the growth of *P. parasitica* var. *nicotianae*. They were able to show that 1) vegetable oils served as very good carbon sources, surpassing even glucose, 2) many of these compounds supported good growth in the presence of glucose, and 3) most of the oils supporting good growth in liquid media were composed largely of triglycerides containing unsaturated fatty acids, while those containing more saturated fatty acids were less active. Tweens containing unsaturated fatty acids (e.g., Tween 80) have similar growth-stimulating activities (Hendrix and Apple, 1964; Nes et al, 1979). As will be discussed later, part of these activities is attributable to contaminating sterols in these crude lipid fractions.

In a nutritional study on a variety of *Phytophthora* species, Hohl (1975) noticed the very marked growth-promoting activity of lecithin. Lecithin is a complex substance containing unsaturated fatty acids, such as oleic and linoleic acids, esterified with glycerol. Furthermore, the molecule contains an orthophosphate group and an amino alcohol, choline. Of these moieties, the fatty acids (in particular, linoleic acid) appear to harbor the major growth-stimulating activity (Table 6). Figure 1 illustrates the dose-response curves of lecithin for three species of *Phytophthora*. Concentrations as low as 10 mg/L have a marked effect on radial growth rate (Fig. 1), and dry-weight increases up to 10-fold have been observed with higher concentrations (Hohl, 1975, Table 6). This indicates that lecithin and linoleic acid probably do not act as carbon sources primarily but serve more specific purposes, e.g., in membrane stability or enzyme activation. The lecithin effect is very pronounced in a majority of *Phytophthora* species. However, neither lecithin nor any of the other lipids appear to be absolutely required for vegetative growth.

Because unsaturated fatty acids in fungi are formed by desaturation of saturated fatty acids (Weete, 1980) one is tempted to predict that most *Phytophthora* species would have low activities in fatty acid desaturases.

A number of sterols (and sterol esters) enhance vegetative growth of many *Phytophthora* species (Elliott, 1977; Hendrix, 1970) in addition to having a profound influence on reproduction. With these substances, too, the small amount needed to get optimal effects (10–20 mg/L) (Elliott, 1968; Hendrix, 1965) indicates that they do not act primarily as a carbon source but belong to the general group of growth factors and are probably involved in membrane function and fungal hormone biosynthesis (Elliott, 1977; Hendrix, 1970; Nes, 1974). Because *Phytophthora* species are capable of growing without sterols, a highly unusual situation shared only by bacteria (Nes, 1974), they appear to be able to keep cell leakage at a sufficiently low level and to maintain the operation of the plasma membrane in the absence of these compounds. Cell leakage may be reduced in the presence of calcium and/or fatty acids, which might explain their growth-promoting capacity, at least in part.

Not all sterols are equally efficient in enhancing growth, and they may act differently on vegetative growth and

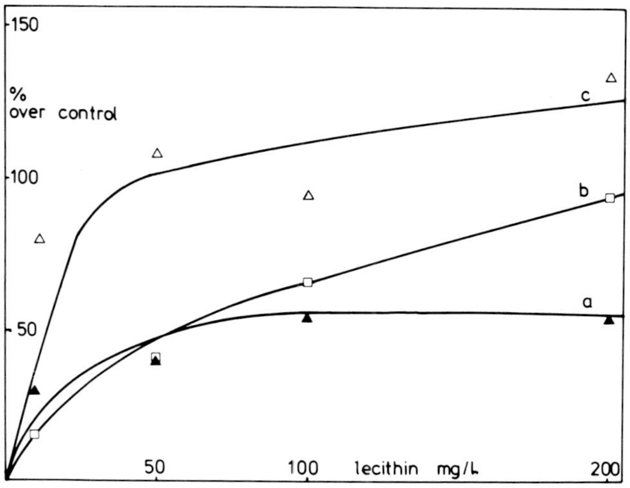


**Fig. 1.** Influence of lecithin on linear growth of *Phytophthora palmivora* P113 (a), *P. cactorum* 236 (b), and *P. cambivora* 77,374 (c). Growth on the medium without lecithin is set to 100%. (Modified from Hohl, 1975)

TABLE 6. Growth-promoting activities[a] of lipids (fatty acids, tristearin, and lecithins) in *Phytophthora* species

| | Fatty acids[b] | | | | | | Lecithins | | | |
|---|---|---|---|---|---|---|---|---|---|---|
| | | | | | | | | | Synthetic | |
| Strain | $C_{16:0}$ | $C_{18:0}$ | $C_{18:1}$ | $C_{18:2}$ | Tristearin[c] | Mixture[d] | Commercial grade[e] | A-grade[f] | dipalmitoyl[g] | dilineoyl[h] |
| *P. cactorum* S | 14 | 42 | 67 | 109 | 21 | 92 | 100 | 86 | 0 | 26 |
| *P. cactorum* 236 | 37 | 31 | 68 | 100 | 46 | 87 | 100 | ··· | ··· | 31 |
| *P. capsici* | 8 | 0 | 28 | 92 | 0 | 54 | 100 | 117 | 0 | 0 |
| *P. cinnamomi* | 6 | 0 | 0 | 31 | 0 | 22 | 100 | 100 | 6 | 149 |
| *P. citrophthora* | 0 | 7 | 0 | 65 | 0 | 0 | 100 | 195 | 30 | 125 |
| *P. palmivora* P 113 | 13 | 4 | 26 | 91 | 15 | 65 | 100 | 110 | 26 | 10 |
| Average | 13 | 14 | 32 | 81 | 14 | 53 | 100 | 172 | 12 | 68 |

[a] Values expressed as percentage of the linear growth obtained with the normally used (commercial grade) lecithin.
[b] Palmitic, stearic, oleic and linoleic acids, 99% pure, used at 50 mg/L.
[c] Over 90% pure, used at 50 mg/L.
[d] Fatty acids and tristearin, 25 mg/L each.
[e] Type II-E from egg yolk, 200 mg/L.
[f] Type III-E from egg yolk, 200 mg/L.
[g] $\beta$, $\gamma$-dipalmitoyl-DL-$\alpha$-lecithin, 200 mg/L.
[h] Dilineoyl-L-$\alpha$-lecithin 150 mg/L.

*Source*: Hohl (1975).

reproduction (Elliott, 1977; Hendrix, 1970; see also Elliott, Chapter 5). Langcake (1974) presented evidence that stigmasterol induces a much faster linear growth than cholesterol does but reduces sporangium production in *P. infestans* compared with cholesterol, whereas the increase in cell mass is more pronounced with cholesterol than with stigmasterol. This indicates that the mechanism of action of sterols on growth and reproduction and even on linear versus mass growth might be different (Nes, 1974).

### Nutritional Balance

The ultimate aims of nutritional studies are manifold. One might be interested in the minimal requirements of an organism, in how to get fastest growth or highest yields, or in how to achieve most efficient use of the nutrients available. To achieve each goal, different types, amounts, and relative proportions of nutrients might have to be formulated. Within the framework of some limiting factors—such as osmotic value of the medium (never determined accurately for *Phytophthora*); pH (usually between 4 and 7); and temperature, which is species-dependent and ranges from close to zero to 37° C (Ribeiro, 1978)—a wide range of possibilities exists for combining nutrients in absolute and relative amounts to attain the desired goal. Almost no systematic studies have been devoted to these problems, although some consideration in this direction is certainly implicit in many media formulations. Because components of a medium interact, such an endeavor would involve an extensive series of factorial analyses. A start in this direction was made by Fothergill and Child (1964) in their exploration of optimal mineral nutrition of *P. erythroseptica, P. infestans*, and *P. parasitica*. Negative interference of certain components, including glucose and several amino acids such as leucine, valine, and asparagine, has been noted by Leal et al (1966) and by Leal and Gomez-Miranda (1967). Inhibition of growth and sexual reproduction has been attributed to the accumulation of organic acids in the medium and the production of toxic levels of ammonia. With the extensive present knowledge of ingredients essential or favorable for *Phytophthora*, this field remains wide open for future studies.

## NUTRITION AND PATHOGENESIS

Alten and Orth (1940) studied the effect of various fertilizer treatments on the susceptibility of potatoes to *P. infestans*. Treatments that lowered the concentrations of total amino acids in the plant also reduced susceptibility, and even fluctuations of amino acid contents during plant growth were reflected in corresponding changes in susceptibility.

Grainger (1956, 1968) made similar studies but concentrated on the carbohydrate content of the host. Very young and old potato plants showed high susceptibility to the blight fungus and also high carbohydrate levels, whereas plants of intermediate age displayed both lower susceptibility and lower carbohydrate levels. However, Warren et al (1973) lowered carbohydrate levels experimentally without getting a corresponding drop in susceptibility. Furthermore, injections of sugar solutions increased susceptibility without affecting total carbohydrate levels correspondingly.

The theory that sterol availability in the host might represent a resistance factor was tested in potatoes affected by late blight. Yet, using susceptible and resistant cultivars, Langcake (1974) and Stössel and Hohl (1980) were unable to detect significant changes in sterol concentration or composition that might be considered important for disease development. Furthermore, the suggestion by Elliott and Knights (1969) that plant tissues containing a high ratio of sterol precursors (e.g., cycloartenol) to sterols might be more resistant than those with lower ratios could not be confirmed (Langcake, 1974; Stössel and Hohl, 1980). Actually, in both systems studied, this relation was inverse—maybe a point worth following up. Following a somewhat different line of investigation, Kaosiri and Hendrix (1972) found that sterols did not change the capacity of propagules of *Pythium* (predominantly chlamydospores) to survive in soils.

Other nutritional factors such as thiamin (Erwin and Katznelson, 1961; Rossetti and Bitancourt, 1952), reducing substances (Hollomon, 1966), or root exudates (Malajczuk and McComb, 1977) have been implicated as agents determining to some degree the success or failure of an attack by a soilborne pathogen. However, hardly any solid evidence is available to substantiate these interesting possibilities.

Similarly undecided is the issue of whether or not race specificity has some nutritional basis. Such a possibility in *P. infestans* was claimed by French (1953) and supported by Hodgson (1958) but has been disputed by Thurston et al (1959) and by Sakai (1961). Although nutritional variability is marked within this species, more recent investigations argue in favor of specific recognition substances as the major basis for race specificity (see Chapter 21).

In summary, the various leads indicating that the nutritional status of the host plant might be a factor in its resistance to disease have, in most cases, not been followed up with enough detail to make a definite assessment. Nevertheless, the possibility exists that nutritional factors do play a role in horizontal resistance.

## TAXONOMIC IMPLICATIONS

### Intraspecies Variation

Variations in nutritional requirements among isolates of a given species have been noted on several occasions but have not been studied or described in great detail (Cuppett and Lilly, 1973; Davies, 1959; French, 1953; Henniger, 1963; Hodgson, 1958; Sakai, 1961; Singh, 1975; Thurston et al, 1959). Hohl (1975) set up a system consisting of three media with and without added lecithin to get some overall view of intraspecies and interspecies variation.

Since then, the test system has been expanded to include four media, which are used in the presence or absence of lecithin (200 mg/L). The first three media[1] differ only in their nitrogen component: medium P-3 contains asparagine; medium P-4 contains $NaNO_3$; and medium P-5 contains $NH_4Cl$. The fourth medium[2], termed P-1, is more elaborate in its basic composition and contains an amino acid mixture as nitrogen source. The above media when supplemented with lecithin are referred to as P-3L, P-4L, P-5L, and P-1L, respectively.

Average radial (linear) growth rates were determined for every strain. For purposes of comparison, the responses were divided into five categories: category 1 encompasses values from 80 to 100%; category 2, 60–79%; category 3, 30–59%; category 4, 10–29%; and category 0, 0–9% of the highest value

[1] Basal medium for P-3, P-4, and P-5 contains, per liter: $KH_2PO_4$, 0.5 g; $K_2HPO_4$, 0.5 g; $MgSO_4 \cdot 7H_2O$, 0.5 g; $CaCl_2 \cdot 2H_2O$, 0.02 g; thiamin, 0.001 g; glucose, 20 g; and agar, 15 g. The following nitrogen sources were added: to medium P-3, L-asparagine, 1.0 g; to medium P-4, $NaNO_3$, 1.5 g; and to medium P-5, $NH_4Cl$, 1.0 g.

[2] Medium P-1 contains, per liter: $NaNO_3$, 3.0 g; $KH_2PO_4$, 1 g; $Na_2HPO_4 \cdot 2H_2O$, 0.35 g; $MgSO_4 \cdot 7H_2O$, 0.5 g; KCl, 0.5 g; $CaCl_2 \cdot 2H_2O$, 0.1 g; $FeSO_4 \cdot 7H_2O$, 0.01 g; $FeCl_3$, 0.001 g; $ZnCl_2$, 0.001 g; ethylenediamine tetraacetic acid, 0.01 g; thiamin·HCl, 0.0005 g; guanine, 0.05 g; asparagine·$H_2O$, 0.130 g; histidine·HCl, 0.190 g; serine, 0.05 g; leucine, 0.065 g; cysteine, 0.06 g; sucrose, 30 g; and agar, 15 g.

reached by the same strain on one of the eight media tested. In addition, the ratio of growth rate on P-4L to that on P-5L was determined. The effect of lecithin was tested on the four media and rated from 0 to 4−. The numerical score indicates the number of media in which lecithin caused a stimulation of at least 20% in growth rate. The plus sign indicates a strong response to lecithin, i.e., a growth rate increase of 50% or more in at least one of the test media.

Intraspecies variation was studied in more detail in *P. palmivora, P. infestans* and *P. megasperma* (Tables 7–9).

***P. palmivora***

Of the 19 strains tested (Table 7), five belong to the recently described *P. megakarya* (morphological form [MF] 3) (Brasier and Griffin, 1979). This species cannot clearly be separated on purely nutritional grounds from the other strains of this rather uniformly reacting "*P. palmivora* complex." MF 3 has a tendency for lower growth rates and a somewhat more pronounced reaction to lecithin, whereas MF 1 and MF 4 strains from cocoa behave rather similarly in most respects. Somewhat apart from these MF 4 strains are two of the three

TABLE 7. Nutritional characteristics of strains of *Phytophthora palmivora* and *P. megakarya* (morphological form [MF] 3)

| Strain | Host | MF | Growth (mm/day) | Relative growth rates[a] on medium[b] | | | | P-4L/ P-5L[c] | Lecithin effect[d] |
|---|---|---|---|---|---|---|---|---|---|
| | | | | P-4L | P-5L | P-3L | P-1L | | |
| P 113 | papaya | 1 | 5.6 | 1 | 2 | 1 | 1 | 1.4 | 4 |
| 106,326 | cocoa | 1 | 5.0 | 1 | 1 | 1 | 1 | 1.0 | 3 |
| 66 | cocoa | 1 | 4.8 | 1 | 2 | 1 | 1 | 1.5 | 4 |
| 69 | cocoa | 1 | 6.6 | 1 | 1 | 1 | 1 | 1.2 | 4 |
| 130 | cocoa | 1 | 6.1 | 1 | 1 | 1 | 1 | 1.1 | 4 |
| P 106 | cocoa | 3 | 2.3 | 2 | 3 | 1 | 1 | 1.2 | 4+ |
| P 182 | cocoa | 3 | 2.9 | 1 | 1 | 1 | 1 | 0.9 | 4 |
| P 307 | cocoa | 3 | 2.3 | 1 | 2 | 1 | 1 | 1.3 | 4+ |
| P 310 | cocoa | 3 | 3.0 | 2 | 3 | 1 | 1 | 1.3 | 4+ |
| 106,327 | cocoa | 3 | 5.7 | 1 | 1 | 1 | 1 | 1.2 | 4 |
| P 272 | cocoa | 4 | 6.0 | 1 | 2 | 1 | 1 | 1.2 | 4+ |
| P 273 | cocoa | 4 | 7.6 | 1 | 2 | 1 | 1 | 1.3 | 3 |
| 13A-3 | black pepper | 4 | 4.1 | 1 | 1 | 1 | 1 | 1.3 | 1 |
| 1393 | black pepper | ND[e] | 0.6 | 1 | 4 | 1 | 1 | 4.5 | 0 |
| 182,711 | black pepper | ND | 4.4 | 1 | 1 | 1 | 1 | 1.2 | 3 |
| 154.24 | coconut | ND | 5.8 | 2 | 1 | 1 | 1 | 0.7 | 1 |
| 7245 | ..[f] | ND | 4.7 | 1 | 1 | 1 | 2 | 1.2 | 2 |
| 1366 | cocoa | ND | 1.8 | 2 | 1 | 3 | 3 | 0.7 | 0 |
| 1394 | "orchid" | ND | 3.4 | 1 | 1 | 1 | 2 | 1.2 | 0 |

[a]1 = highest, 4 = lowest, 0 = essentially no growth.
[b]Lecithin-supplemented medium. Nitrogen components: P-4L, $NO_3^-$; P-5L, $NH_4^+$; P-3L, asparagine; P-1L (complex medium), an amino acid mixture.
[c]Ratio of growth rates on P-4L and P-5L.
[d]0 = weakest, 4+ = strongest growth promotion.
[e]Not determined; possibly 4 for strains 1393 and 182,711.
[f]Unknown.

TABLE 8. Intraspecies variation of nutritional parameters in *Phytophthora infestans*

| Strain | Host | Growth (mm/day) | Relative growth rates[a] on medium[b] | | | | P-4L/ P-5L[c] | Lecithin effect[d] |
|---|---|---|---|---|---|---|---|---|
| | | | P-4L | P-5L | P-3L | P-1L | | |
| B 25 | *Solanum tuberosum* | 2.9 | 0 | 0 | 0 | 1 | ND[e] | 1 |
| B 130 | *S. tuberosum* | 1.9 | 0 | 0 | 0 | 1 | ND | 2 |
| 211.46 | –.[f] | 1.9 | 0 | NM[g] | 0 | 1 | ND | 1 |
| 515 | *S. tuberosum* | 3.6 | 0 | 4 | 0 | 1 | ND | 2 |
| 223/33 | *S. tuberosum* | 4.1 | 0 | 4 | 3 | 1 | ND | 1 |
| 131,558 | *S. tuberosum* | 2.2 | 4 | 4 | 3 | 1 | ND | 2 |
| 131,559 | *S. lycopersici* | 1.9 | 4 | NM | 3 | 1 | ND | 3 |
| S | –.[f] | 3.6 | 0 | 3 | 4 | 1 | 0 | 2 |
| 78/1 | *S. tuberosum* | 2.2 | 4 | 2 | 1 | 1 | 0.4 | 3 |
| 36 | *S. lycopersici* | 2.4 | 4 | 2 | 1 | 2 | 0.3 | 1 |
| CG 96 | ..[f] | 2.4 | 4 | 2 | 1 | 2 | 0.4 | 1 |
| 116 | *S. tuberosum* | 1.6 | 4 | 4 | 1 | 3 | ND | 2 |
| 141,568 | *Mirabilis jalapa* | 3.6 | 3 | 1 | 2 | 2 | 0.4 | 3 |

[a]1 = highest, 4 = lowest, 0 = essentially no growth.
[b]Lecithin-supplemented medium. Nitrogen components: P-4L, $NO_3^-$; P-5L, $NH_4^+$; P-3L, asparagine; P-1L (complex medium), an amino acid mixture.
[c]Ratio of growth rates on P-4L and P-5L.
[d]0 = weakest, 4+ = strongest growth promotion.
[e]Not accurately determinable (very low growth).
[f]Unknown.
[g]Not measured.

strains from black pepper. Black pepper strains have strong affinities to MF 4 and also to *P. capsici* (Al-Hedaithy and Tsao, 1979; Kunimoto et al, 1976; Tsao and Tummakate, 1977; Tsao et al, 1981; Waterhouse, 1974; Zentmyer et al, 1977). More work is needed to show whether or not black pepper strains are nutritionally distinct from other MF 4 strains.

### *P. infestans*

As may be seen in Table 8, the nutritional patterns in *P. infestans* are characterized by a pronounced diversity. Although some isolates grew only on the complex P-1L medium, others were able to grow efficiently on asparagine as the only nitrogen source or even on ammonium. With one exception (a strain from an unusual host, *Mirabilis jalapa*), none of the strains showed any appreciable growth on nitrate. The species appears to have lost this "primitive" (Cantino and Turian, 1959) trait. The species is also characterized by its relatively low response to lecithin. The heterogeneity of *P. infestans* is remarkable and deserves more attention.

### *P. megasperma*

The 16 strains of *P. megasperma* (Table 9) fall into two clearly separable nutritional groups. Group 1 strains grow well on nitrate and, therefore, have high P-4L/P-5L ratios of 1.9–4.3. Group 2 strains grow much less well on nitrate, and the ratios are low (0.8–1.3). The first group encompasses strains that belong to *P. megasperma* f. sp. *glycinea* (Kuan and Erwin, 1980). The second encompasses the strains from alfalfa and other strains with small (below 45 μm) oogonia. The two strains of *P. megasperma* var. *megasperma* (Waterhouse, 1963) possessing large oogonia (over 45 μm) belong to group 2 also, but because only two strains have been tested so far, it is not possible to ascertain whether or not this is a general trend.

The occurrence of two nutritional groups within the small oogonial type of *P. megasperma* strains (*P. megasperma* var. *sojae* sensu Waterhouse [1963]) supports and extends the distinction of the two taxonomic entities based up to now primarily on pathogenicity tests (Hamm and Hansen, 1981; Kuan and Erwin, 1980). This is probably the first case in *Phytophthora* where a nutritional trait may be used for taxonomic purposes. However, the P-4L/P-5L ratio has a strictly operational definition and is not necessarily invariable with different basal media. Further work is needed to ascertain whether or not the *glycinea* strains form a valid new species.

TABLE 9. Differential growth rates of *Phytophthora megasperma* strains on two media

| Strain | Forma specialis[a] or variety[b] | Pathogenic on | Oogonia size (μm) | P-4L/P-5L[c] |
|---|---|---|---|---|
| 892 | *glycinea*[a] | soybean | <45 | 3.3 |
| R1 | *glycinea* | soybean | <45 | 3.0 |
| R6 | *glycinea* | soybean | <45 | 3.2 |
| 900 | *glycinea* | soybean | <45 | 1.9 |
| 382.61 | *glycinea* | soybean | <45 | 2.3 |
| 290 | *glycinea* | soybean | <45 | 4.3 |
| P174 | *glycinea* | soybean | 36[a] | 3.9 |
| P405 | *glycinea* | soybean | 39[a] | 2.7 |
| K18 | *glycinea* | soybean | ≤45[a] | 2.6 |
| P410 | *medicaginis*[a] | alfalfa | 40[a] | 0.9 |
| P1057 | *medicaginis* | alfalfa | 37[a] | 0.8 |
| K3 | *sojae*[b] | almond | 41[a] | 0.8 |
| P147 | *sojae* | sugarcane | 36[a] | 1.0 |
| K12 | *sojae* | grape | 44–48[a] | 0.8 |
| 33 | *megasperma*[b] | chestnut | >45 | 0.8 |
| 162,674 | *megasperma* | ...[d] | >45 | 1.3 |

[a] From Kuan and Erwin (1980).
[b] From Waterhouse (1963).
[c] Ratio of growth rates on P-4L (nitrogen as $NO_3^-$) and on P-5L (nitrogen as $NH_4^+$).
[d] Unknown.

## Interspecies Variation

To obtain an overall impression of the nutritional status of the entire genus, a series of strains belonging to 32 species were subjected to the previously described growth test on the eight test media. The results from the four most important media are shown in Table 10 (values for *P. palmivora* and *P. infestans* were given in Tables 7 and 8, respectively). Many species grow well on all test media listed in Table 10, i.e., they can use nitrate, ammonium, and organic nitrogen with similar efficiency. Notable exceptions are *P. colocasiae, P. epistomium, P. fragariae, P. infestans, P. megasperma, P. syringae, P. vesicula*, and *P. vignae*. Strains that grow well on nitrate usually also grow well on ammonium or organic nitrogen; strains that grow well on ammonium also grow well on organic nitrogen, as predicted by Robbins (1937). Four notable exceptions are *P. megasperma* f. sp. *glycinea, P. spinosa*, and *P. vesicula*, which prefer nitrate nitrogen above all the other sources tested, and *P. epistomium*, which grows best by a wide margin on ammonium nitrogen.

The nutritional demands of *Phytophthora* spp. are summarized in Table 11, where species and strains are listed according to the most "primitive" medium on which they are able to grow well relative to the other media, i.e., with a growth rate not less than 80% of that on the best medium. Growth on nitrate is considered a more primitive trait than growth on ammonium, on asparagine, or on the complex medium P-1L for reasons outlined by Robbins (1937) and discussed previously.

Table 11 shows that a few species grow well on a simple nitrate medium (P-4). Their number increases dramatically when lecithin is added (P-4L). Several species need a reduced form of nitrogen and have a marked preference for organic N sources. Finally, *P. fragariae*, most strains of *P. infestans*, and some strains of *P. megasperma* var. *megasperma* prefer the complex medium P-1L. Thus, the genus *Phytophthora* contains a pronounced gradient of nutritional complexity; the least demanding species such as *P. vesicula, P. drechsleri*, and *P. botryosa* are separated by several steps from the most demanding *P. infestans* and *P. fragariae*.

## SUMMARY

The discovery that thiamin (vitamin $B_1$) is essential for growth of *Phytophthora* set the stage for detailed nutritional investigations using defined media. So far, thiamin appears to be the only vitamin supplement the fungus requires.

Of the carbon sources, glucose, sucrose, and other glucose-containing saccharides are generally preferred, although trehalose, cellobiose, or cellulose are poorly utilized. Sugars may be replaced by fats; alcohols (except glycerol), organic acids, and amino acids are unsuitable carbon sources. However, organic acids may promote growth in media containing ammonium as nitrogen source, provided sufficient $Ca^{2+}$ is present. This effect is probably the result of the buffering capacity of these organic acids, which prevents the drastic drop of the pH in the medium associated with the uptake of the ammonium ions.

Of the nitrogen sources, nitrate, ammonium (buffered), or single amino acids such as asparagine are suitable for the majority of species. Some species such as *P. infestans* and *P. fragariae* do not grow on nitrate N but need N in its reduced form, and a few require an amino acid mixture for optimal growth.

A variety of organic and inorganic sulfur and phosphorus compounds may be utilized by *Phytophthora* to comparable

TABLE 10. Nutritional characteristics of *Phytophthora* species (for *P. infestans, P. megakarya,* and *P. palmivora,* see Tables 7 and 8)

| Species | Strain | Growth (mm/day) | Relative growth[a] on medium[b] | | | | P-4L/P-5L[c] | Lecithin effect[d] |
|---|---|---|---|---|---|---|---|---|
| | | | P-4L | P-5L | P-3L | P-1L | | |
| *arecae* | 305.62 | 6.7 | 1 | 1 | 1 | 1 | 1.0 | 4 |
| | 306.62 | 2.2 | 1 | 3 | 2 | 2 | 2.3 | 4+ |
| *boehmeriae* | 291.29 | 6.0 | 3 | 1 | 1 | 1 | 0.7 | 2 |
| | 357.52 | 3.3 | 1 | 2 | 1 | 1 | 1.2 | 4 |
| *botryosa* | 581.69 | 7.3 | 2 | 1 | 1 | 1 | 1.0 | 4+ |
| | 1434 | 4.2 | 1 | 1 | 1 | 1 | 1.3 | 1 |
| *cactorum* | S | 2.6 | 1 | 1 | 1 | 1 | 1.0 | 4+ |
| | 236 | 4.4 | 1 | 2 | 1 | 1 | 1.5 | 4+ |
| | 77 | 4.5 | 1 | 2 | 1 | 1 | 1.3 | 4 |
| *cambivora* | 159,577 | 3.6 | 2 | 3 | 3 | 1 | 1.8 | 0 |
| | 77,374 | 4.4 | 1 | 2 | 1 | 1 | 1.3 | 4+ |
| *capsici* | 128.23 | 3.8 | 1 | 1 | 1 | 1 | 1.1 | 4+ |
| | 178.26 | 4.7 | 1 | 1 | 1 | 1 | 1.1 | 2 |
| *castaneae* | 198,426 | 5.5 | 1 | 1 | 1 | 1 | 1.0 | 4 |
| *cinnamomi* | 36 | 8.0 | 1 | 2 | 1 | 1 | 1.1 | 4+ |
| | 77,375 | 7.6 | 1 | 1 | 1 | 1 | 1.0 | 4+ |
| | 144.22 | 6.8 | 1 | 1 | 1 | 1 | 1.0 | 3+ |
| *citricola* | 129,904 | 3.6 | 1 | 2 | 1 | 1 | 1.2 | 2 |
| | 211,638 | 5.4 | 1 | 1 | 1 | 1 | 0.9 | 3 |
| | 181.25 | 6.7 | 1 | 1 | 1 | 1 | 1.0 | 2 |
| | 273.32 | 5.0 | 1 | 1 | 1 | 1 | 1.0 | 4 |
| *citrophthora* | 32,034 | 3.6 | 1 | 2 | 1 | 1 | 1.2 | 3 |
| | 273.33 | 5.0 | 1 | 1 | 1 | 1 | 1.0 | 4 |
| *colocasiae* | 182.25 | 1.9 | 2 | 3 | 1 | 1 | 1.3 | 4+ |
| *cryptogea* | 45,168 | 5.8 | 1 | 2 | 1 | 1 | 1.3 | 2 |
| | 31 | 3.6 | 1 | 1 | 1 | 1 | 1.2 | 2 |
| | 32 | 6.5 | 1 | 2 | 1 | 1 | 1.3 | 2 |
| | 113.19 | 7.2 | 2 | 2 | 1 | 1 | 1.0 | 4 |
| *drechsleri* | 359.52 | 5.5 | 1 | 1 | 1 | 1 | 1.0 | 3 |
| | 291.35 | 3.6 | 2 | 2 | 1 | 1 | 1.0 | 3 |
| | 6503 | 3.6 | 1 | 2 | 1 | 1 | 1.2 | 1 |
| *epistomium* | F 55 | 0.3 | 4 | 1 | 4 | 3 | 0.2 | 4+ |
| *erythroseptica* | 34,684 | 5.4 | 2 | 1 | 1 | 1 | 0.8 | 3 |
| | 380.61 | 5.7 | 1 | 1 | 1 | 1 | 0.9 | 4 |
| *fragariae* | 131,557 | 1.8 | 0 | 2 | 2 | 1 | 0.0 | 3 |
| | 209.46 | 1.3 | 4 | 3 | 1 | 1 | 0.6 | 3 |
| *gonapodyides* | 554.67 | 2.7 | 1 | 2 | 1 | 1 | 1.3 | 4+ |
| *heveae* | 296.29 | 3.5 | 1 | 1 | 1 | 1 | 1.1 | 4+ |
| *hibernalis* | 270.31 | 6.0 | 1 | 1 | 1 | 1 | 1.0 | 3 |
| *lateralis* | 40,503 | 1.4 | 1 | 2 | 1 | 1 | 1.2 | 4+ |
| *megasperma* | 162,674 | 5.4 | 2 | 2 | 2 | 1 | 1.3 | 1 |
| | 33 | 4.7 | 2 | 1 | 2 | 1 | 0.8 | 0 |
| | P147 | 5.5 | 1 | 1 | 1 | 1 | 1.0 | 0 |
| | R1 | 0.6 | 1 | 3 | 3 | 4 | 3.0 | 0 |
| | R6 | 0.6 | 1 | 3 | 3 | 4 | 3.2 | 3 |
| *nicotianae* | 77,972 | 5.8 | 1 | 2 | 1 | 1 | 1.2 | 4 |
| | 310.62 | 3.0 | 2 | 1 | 2 | 2 | 0.8 | 4+ |
| | 109.17 | 3.5 | 1 | 1 | 1 | 1 | 0.9 | 4+ |
| | 35,087 | 4.8 | 2 | 1 | 1 | 1 | 0.9 | 4+ |
| *porri* | 366.59 | 3.5 | 1 | 2 | 1 | 1 | 1.1 | 3 |
| *quininea* | 406.48 | 4.7 | 3 | 2 | 1 | 1 | 0.6 | 4+ |
| *richardiae* | 240.30 | 4.0 | 2 | 3 | 1 | 1 | 1.8 | 4+ |
| *spinosa* var. *spinosa* | 112 | 0.3 | 1 | 3 | 3 | 4 | 3.0 | 4+ |
| *syringae* | 40,529 | 0.9 | 1 | 3 | 1 | 1 | 1.4 | 1 |
| | 166,373 | 0.9 | 3 | 3 | 1 | 3 | 0.7 | 3 |
| | 272.55 | 1.7 | 1 | 3 | 1 | 1 | 1.7 | 1 |
| *vesicula* | 139,645 | 2.2 | 1 | 3 | 1 | 2 | 1.6 | 0 |
| | 113 | 2.3 | 1 | 3 | 2 | 2 | 1.8 | 1 |
| *vignae* | 217,596 | 2.4 | 3 | 3 | 1 | 3 | 0.8 | 4+ |

[a] 1 = highest, 4 = lowest, 0 = essentially no growth.

[b] Lecithin-supplemented medium. Nitrogen components: P-4L, $NO_3^-$; P-5L, $NH_4^+$; P-3L, asparagine; P-1L (complex medium), an amino acid mixture.

[c] Ratio of growth rates on P-4L and P-5L.

[d] 0 = weakest, 4+ = strongest growth promotion.

degrees. Inorganic sources (sulfate and orthophosphate) are more commonly employed.

The need for potassium and magnesium is unquestionable, although the latter has not been established experimentally for *Phytophthora*. A definite requirement for calcium has been demonstrated in several instances where media containing nitrate as nitrogen source were used.

Of the trace elements, iron, zinc, manganese, and copper, in that order, clearly promote growth. For molybdenum, the evidence is less convincing; no attempts have been made to prove the essentiality of boron.

Growth of *P. infestans* is favored by reducing agents such as ascorbic acid, thioglycollate, glutathione, or cysteine.

Lipids such as fats or lecithins containing unsaturated fatty acids, particularly linoleic acid, promote growth of most species yet are not essential for growth. The same may be said for sterols, which, however, are essential for the reproductive phases of pythiaceous fungi.

The relationship between nutrition and pathogenesis is still unclear, and the possible role of nutrients available in the host as factors of horizontal resistance needs further exploration.

Intraspecies nutritional variation has been studied in only a few species. These differences are not pronounced within the "*P. palmivora* complex" but are strongly so in *P. infestans*. In *P. megasperma*, the two formae speciales *glycinea* and *medicaginis* may be distinguished by their differential growth rates on a nitrate and an ammonium nitrogen medium.

Interspecies variation is conspicuous in *Phytophthora*. Although some species grow well on a simple nitrate medium, others require additions of lipids, organic nitrogen, specific amino acids, and/or additional growth factors for optimal vegetative growth. The least demanding species, such as *P. botryosa* or *P. vesicula*, are clearly separated from the highly demanding *P. fragariae* and *P. infestans*.

TABLE 11. Summary of interspecies nutritional variations in *Phytophthora*

| Simplest medium[a] on which species grows adequately[b] | Species[c] |
|---|---|
| P-4 | *botryosa* (1), *drechsleri* (1), *megasperma* f. sp. *glycinea* (3), *palmivora* (4), *syringae* (1), *vesicula* (2) |
| P-4L | *arecae* (2), *boehmeriae* (1), *botryosa* (1), *cactorum* (3), *cambivora* (1), *capsici* (2), *castaneae* (1), *cinnamomi* (3), *citricola* (4), *citrophthora* (2), *cryptogea* (3), *drechsleri* (1), *erythroseptica* (1), *gonapodyides* (1), *heveae* (1), *hibernalis* (1), *lateralis* (1), *megakarya* (2), *megasperma* f. sp. *glycinea* (6), *megasperma* f. sp. *medicaginis* (1), *nicotianae* (2), *palmivora* (10), *porri* (1), *spinosa* var. *spinosa* (1), *syringae* (1) |
| P-5L | *boehmeriae* (1), *epistomium* (1), *erythroseptica* (1), *infestans* (1), *megasperma* f. sp. *medicaginis* (1), *megasperma* var. *megasperma* (1), *nicotianae* (1), *palmivora* (2) |
| P-3L | *colocasiae* (1), *cryptogea* (1), *drechsleri* (1), *fragariae* (1), *infestans* (4), *megakarya* (2), *megasperma* f. sp. *medicaginis* (2), *quininea* (1), *richardiae* (1), *syringae* (1), *vignae* (1) |
| P-1L | *cambivora* (1), *fragariae* (1), *infestans* (8), *megasperma* var. *megasperma* (1) |

[a] Media are listed in order of increasing complexity. Nitrogen component: P-4, $NO_3^-$; P-5, $NH_4^+$; P-3, asparagine; P-1 (complex medium), an amino acid mixture. L = Supplemented with lecithin.
[b] Not less than 80% of growth on best medium.
[c] Number in brackets indicates number of strains.

## ACKNOWLEDGMENTS

I thank Katja Iselin for her untiring help in performing the nutritional studies, some of which are reported here for the first time, and also Erika Rüegg for typing the manuscript. I am also thankful to D. J. Stamps, A. M. Tan, P. M. Smith, G. Bahnweg, K. Bohnen, C. M. Brasier, M. Coffey, D. C. Erwin, J. W. Fell, N. T. Keen, O. K. Ribeiro, F. J. Schwinn, D. S. Shaw, P. H. Tsao, and G. A. Zentmyer for providing cultures of *Phytophthora*. My studies on *Phytophthora* are supported by Swiss National Science Foundation grants 3.673-0.75 and 3.370-0.78.

## LITERATURE CITED

Al-Hedaithy, S. S. A., and Tsao, P. H. 1979. Sporangium pedicel length in *Phytophthora* species and the consideration of its uniformity in determining sporangium caducity. Trans. Br. Mycol. Soc. 72:1-13.

Alten, F., and Orth, H. 1940. Untersuchungen über den Aminosäurengehalt und die Anfälligkeit der Kartoffel gegen die Kraut- und Knollenfäule (*Phytophthora infestans* de By.). Phytopathol. Z. 13:243-271.

Bahnweg, G. 1978. Physiology of Peronosporales (*Phytophthora* spp. and *Pythium* sp.) associated with the decay of leaf litter in marine environment. Veröff. Inst. Meeresforsch. Bremerhaven 17:33-44.

Bahnweg, G. 1980. Phospholipid and steroid requirements of *Haliphthoros milfordensis*, a parasite of marine crustaceans, and *Phytophthora epistomium*, a facultative parasite of marine fungi. Bot. Marina 23:209-218.

Bartnicki-Garcia, S. 1966. Chemistry of hyphal walls of *Phytophthora*. J. Gen. Microbiol. 42:57-69.

Brasier, C. M., and Griffin, M. J. 1979. Taxonomy of '*Phytophthora palmivora*' on cocoa. Trans. Br. Mycol. Soc. 72:111-143.

Brefeld, O. 1881. Botanische Untersuchungen über Schimmelpilze. Untersuchungen aus dem Gesammtgebiet der Mykologie. IV. Heft. 1. Culturmethoden zur Untersuchung der Pilze. pp. 1-35. Verlag von Arthur Felix, Leipzig.

Brefeld, O. 1883. Botanische Untersuchungen über Hefenpilze. Untersuchungen aus dem Gesammtgebiet der Mykologie. V. Heft. Die Brandpilze I. 1. Die künstliche Kultur parasitischer Pilze. pp. 1-28. Verlag von Arthur Felix, Leipzig.

Cameron, H. R. 1966. Variability in the genus *Phytophthora*. II. Effects of vitamins on growth. Phytopathology 56:812-815.

Cameron, H. R., and Milbrath, G. M. 1965. Variability in the genus *Phytophthora*. I. Effects of nitrogen sources and pH on growth. Phytopathology 55:653-657.

Cantino, E. C., and Turian, G. F. 1959. Physiology and development of lower fungi (Phycomycetes). Annu. Rev. Microbiol. 13:97-124.

Chee, K.-H., and Newhook, F. J. 1965a. Nutritional studies with *Phytophthora cinnamomi* Rands. N.Z. J. Agric. Res. 8:523-529.

Chee, K.-H., and Newhook, F. J. 1965b. Variability in sexual reproduction of *Phytophthora cinnamomi* Rands. N.Z. J. Agric. Res. 8:947-950.

Christie, T. 1958. Nutritional studies of *Phytophthora cactorum* (Leb. and Cohn) Schroet. I. Utilisation of nitrogen and carbon compounds. N.Z. J. Sci. 1:83-90.

Christie, T. 1959. Nutritional studies of *Phytophthora cactorum* (Leb. and Cohn.) Schroet. II. Influence of pH of a culture medium on utilization of some nitrogen and carbon compounds. N.Z. J. Sci. 2:320-322.

Clark, M. C., Melanson, D. L., and Page, O. T. 1978. Purine metabolism and differential inhibition of spore germination in *Phytophthora infestans*. Can. J. Microbiol. 24:1032-1038.

Clarke, D. D. 1966. Factors affecting the development of single zoospore colonies of *Phytophthora infestans*. Trans. Br. Mycol. Soc. 49:177-184.

Clinton, G. P. 1908. Artificial cultures of *Phytophthora*, with special reference to oospores. Conn. Agric. Exp. Stn., Annu. Rep. 32:891-907.

Cuppett, V. M., and Lilly, V. G. 1973. Ferrous iron and the growth of twenty isolates of *Phytophthora infestans* in synthetic media. Mycologia 65:67-77.

Davies, M. E. 1959. The nutrition of *Phytophthora fragariae*. Trans. Br. Mycol. Soc. 42:193-200.

de Bary, A. 1876. Researches into the nature of the potato-fungus (*Phytophthora infestans*). William Clowes & Sons, London. pp. 1-33.

de Bary, A. 1881. Untersuchungen über die Peronosporeen und Saprolegnieen und die Grundlagen eines natürlichen Systems der Pilze. Pages 22-28 in: Beiträge zur Morphologie und Physiologie der Pilze. A. de Bary and M. Woronin, eds. IV. Reihe. 5. *Phytophthora omnivora*.

Elliott, C. G. 1968. Competition and synergism between cholesterol and cholestanol in oospore formation in *Phytophthora cactorum*. J. Gen. Microbiol. 51:137-143.

Elliott, C. G. 1972. Calcium chloride and growth and reproduction of *Phytophthora cactorum*. Trans. Br. Mycol. Soc. 58:169-172.

Elliott, C. G. 1977. Sterols in fungi: Their functions in growth and reproduction. Adv. Microb. Physiol. 15:121-173.

Elliott, C. G., and Knights, B. A. 1969. Interactions between steroids in the growth of *Phytophthora*. J. Sci. Food Agric. 20:406-408.

Elliott, C. G., Hendrie, M. R., and Knights, B. A. 1966. The sterol requirement of *Phytophthora cactorum*. J. Gen. Microbiol. 42:425-435.

Erwin, D. C. 1968. The effect of calcium on mycelial growth of *Phytophthora megasperma* and *P. cinnamomi*. Mycologia 60:1112-1116.

Erwin, D. C., and Katznelson, H. 1961. Studies on the nutrition of *Phytophthora cryptogea* Can. J. Microbiol. 7:15-25.

Fehrmann, H. 1971. Obligater Parasitismus phytopathogener Pilze. I. Vergleichende Wachstumsversuche mit *Phytophthora infestans* und verwandten Arten. Phytopathol. Z. 70:89-113.

Fell, J. W., and Master, I. M. 1975. Phycomycetes (*Phytophthora* spp. nov. and *Pythium* sp. nov.) associated with degrading mangrove (*Rhizophora mangle*) leaves. Can. J. Bot. 53:2908-2922.

Fleetwood Walker, P. M. 1955. Observations on the physiology of *Phytophthora fragariae*. Trans. Br. Mycol. Soc. 38:169.

Fothergill, P. G., and Child, J. H. 1964. Comparative studies of the mineral nutrition of three species of *Phytophthora*. J. Gen. Microbiol. 36:67-78.

French, A. M. 1953. Physiologic differences between two physiologic races of *Phytophthora infestans*. Phytopathology 43:513-516.

Fries, N. 1938. Über die Bedeutung von Wuchsstoffen für das Wachstum verschiedener Pilze. Symb. Bot. Ups. 3(2):1-188.

Gaertner, A. 1959. Versuche zur künstlichen Kultur von *Phytophthora infestans* de Bary. I. Bedarf und Wirkung einiger anorganischer Salze im Nährmedium. Arch. Mikrobiol. 32:261-269.

Gaertner, A. 1964. Versuche zur künstlichen Kultur von *Phytophthora infestans* de Bary. Z. Pflanzenkr. 71:143-148.

Grainger, J. 1956. Host nutrition and attack by fungal parasites. Phytopathology 46:445-456.

Grainger, J. 1968. $C_p/R_s$ and the disease potential of plants. Hortic. Res. 8:1-40.

Hall, A. M. 1959. The culture of *Phytophthora infestans* in artificial media. Trans. Br. Mycol. Soc. 42:15-26.

Hamm, P. B., and Hansen, E. M. 1981. Host specificity of *Phytophthora megasperma* from Douglas fir, soybean, and alfalfa. Phytopathology 71:65-68.

Hendrix, J. W. 1965. Influence of sterols on growth and reproduction of *Pythium* and *Phytophthora* spp. Phytopathology 55:790-797.

Hendrix, J. W. 1970. Sterols in growth and reproduction of fungi. Annu. Rev. Phytopathol. 8:111-130.

Hendrix, J. W., and Apple, J. L. 1964. Fats and fatty acid derivatives as growth stimulants and carbon sources for *Phytophthora parasitica* var. *nicotianae*. Phytopathology 54:987-994.

Hendrix, J. W., and Guttman, S.-M. 1970. Sterol or calcium requirement by *Phytophthora parasitica* var. *nicotianae* for growth on nitrate nitrogen. Mycologia 62:195-198.

Hendrix, J. W., Norman, C., and Apple, J. L. 1966. Chemical and physical factors influencing growth of *Phytophthora parasitica* var. *nicotianae* on vegetable oils. Physiol. Plant. 19:159-166.

Hendrix, J. W., Guttman, S.-M., and Wightman, D. L. 1969. Cation and sterol effects on growth of *Phytophthora parasitica* var. *nicotianae*. Phytopathology 59:1620-1624.

Henniger, H. 1959. Versuche zur Kultur verschiedener Rassen von *Phytophthora infestans* (Mont.) de By. auf künstlichen Nährböden. Phytopathol. Z. 34:285-306.

Henniger, H. 1963. Zur Kultur von *Phytophthora infestans* auf vollsynthetischen Nährsubstraten. Z. Allg. Mikrobiol. 3:126-135.

Ho, H. H., and Foster, B. 1972. Starch utilization by *Phytophthora* spp. Mycopathol. Mycol. Appl. 46:335-339.

Hodgson, W. A. 1958. Growth of four races of *Phytophthora infestans* (Mont.) de Bary in synthetic media. Can. J. Plant Sci. 38:145-154.

Hohl, H. R. 1975. Levels of nutritional complexity in *Phytophthora*: Lipids, nitrogen sources and growth factors. Phytopathol. Z. 84:18-33.

Hollomon, D. W. 1966. Reducing compounds and the growth of *Phytophthora infestans*. J. Gen. Microbiol. 45:315-324.

Huang, H., Xu, D., and Wang, C. 1980. The effect of organic acid and calcium on *Phytophthora infestans* (Mont.) de Bary. Acta Phytopathol. Sinica 1:95-102.

Huguenin, B., and Boccas, B. 1971. Rôle de quelques facteurs dans la formation et la germination des oospores chez le *Phytophthora palmivora* Butl. Ann. Phytopathol. 3:353-371.

Hwang, H., Hsu, T.-Y., and Lin, C.-K. 1965. Further studies on synthetic media for *Phytophthora infestans*. Acta Microbiol. Sinica 11:463-469. (In Chinese)

Kaosiri, T., and Hendrix, J. W. 1972. Sterols in relation to ability of *Pythium sylvaticum* to survive in soil. Can. J. Bot. 50:891-896.

Katsura, K., and Nishioka, M. 1958. Effect of carbon sources upon the mycelial growth and sporulation of the fungus, *Phytophthora capsici* Leonian. Sci. Rep. Saikyo Univ. Agric. 10:93-99.

Klemmer, H. W., and Lenney, J. F. 1965. Lipids stimulating sexual reproduction and growth in pythiaceous fungi. Phytopathology 55:320-323.

Kögl, F., and Fries, N. 1937. Über den Einfluss von Biotin, Aneurin und Meso-Inosit auf das Wachstum verschiedener Pilzarten. Z. Physiol. Chem. 249:93-110.

Kuan, T.-L., and Erwin, D. C. 1980. Formae speciales differentiation of *Phytophthora megasperma* isolates from soybean and alfalfa. Phytopathology 70:333-338.

Kunimoto, R. K., Aragaki, M., Hunter, J. E., and Ko, W. H. 1976. *Phytophthora capsici*, corrected name for the cause of Phytophthora blight of macadamia racemes. Phytopathology 66:546-548.

Kusaka, T., and Kitai, S. 1978. New synthetic medium for the cultivation of *Phytophthora infestans*. J. Takeda Res. Lab. 37:96-103. (In Japanese)

Langcake, P. 1974. Sterols in potato leaves and their effects on growth and sporulation of *Phytophthora infestans*. Trans. Br. Mycol. Soc. 63:573-586.

Leal, J. A., and Gomez-Miranda, B. 1967. Effect of amino acids and organic acids on the sexual reproduction of species of *Phytophthora* and *Pythium*. Trans. Br. Mycol. Soc. 50:77-84.

Leal, J. A., Friend, J., and Holliday, P. 1964. A factor controlling sexual reproduction in *Phytophthora*. Nature 203:545-546.

Leal, J. A., Gomez-Miranda, B., and Nicolas, G. 1966. Formation of toxic substances by fungi in the degradation of some amino acids. Can. J. Microbiol. 12:1073-1076.

Leal, J. A., Gallegly, M. E., and Lilly, V. G. 1967a. The relation of the carbon-nitrogen ratio in the basal medium to sexual reproduction in species of *Phytophthora*. Mycologia 59:953-964.

Leal, J. A., Lilly, V. G., and Gallegly, M. E. 1967b. The utilization of L-asparagine by two species of *Phytophthora*. Proc. W. Va. Acad. Sci. 39:95-101.

Leal, J. A., Lilly, V. G., and Gallegly, M. E. 1970. Some effects of ammonia on species of *Phytophthora*. Mycologia 62:1041-1056.

Leal, J. A., Gallegly, M. E., and Lilly, V. G. 1971. The value of 21 amino acids as nitrogen sources for *Phytophthora cactorum* and *P. heveae*. Can. J. Microbiol. 17:1319-1325.

Leonian, L. H. 1935. The effect of auxins upon *Phytophthora cactorum*. J. Agric. Res. 51:277-286.

Leonian, L. H. 1936. Control of sexual reproduction in *Phytophthora cactorum*. Am. J. Bot. 23:188-190.

Leonian, L. H., and Lilly, V. G. 1938. Studies on the nutrition of fungi. I. Thiamin, its constituents, and the source of nitrogen. Phytopathology 28:531-548.

Leonian, L. H., and Lilly, V. G. 1940. Studies on the nutrition of fungi. IV. Factors influencing the growth of some thiamin-requiring fungi. Am. J. Bot. 27:18-26.

Lin, C.-K., and Liang, P.-Y. 1965. Studies on nitrogen, calcium and organic acid requirements with reference to pH relations in the nutrition of some species of *Phytophthora*. Acta Microbiol. Sinica 11:470-479. (In Chinese)

Lopatecki, L. E., and Newton, W. 1956. The nutrition of *Phytophthora*. Can. J. Bot. 34:751-757.

Maas, J. L. 1972. Growth and reproduction in culture of ten *Phytophthora fragariae* races. Mycopathol. Mycol. Appl. 48:323-334.

Malajczuk, N., and McComb, A. J. 1977. Root exudates from *Eucalyptus calophylla* R. Br. and *Eucalyptus marginata* Donn. ex Sm. seedlings and their effect on *Phytophthora cinnamomi* Rands. Aust. J. Bot. 25:501-514.

Matruchot, L., and Molliard, M. 1903. Sur le *Phytophthora infestans*. Ann. Mycol. 1:540-543.

Mehrotra, B. S. 1951. Physiological studies of some phytophthoras. III. Carbon requirements. Lloydia 14:122-128.

Mitchell, D. J., and Zentmyer, G. A. 1971. Effects of oxygen and carbon dioxide tensions on growth of several species of *Phytophthora*. Phytopathology 61:787-791.

Morton, A. G., and Macmillan, A. 1954. The assimilation of nitrogen from ammonium salts and nitrate by fungi. J. Exp. Bot. 5:232-252.

Nes, W. D., Patterson, G. W., and Bean, G. A. 1979. The effect of steroids and their solubilizing agents on mycelial growth of *Phytophthora cactorum*. Lipids 14:458-462.

Nes, W. R. 1974. Role of sterols in membranes. Lipids 9:596-612.

Newhook, F. J., Waterhouse, G. M., and Stamps, D. J. 1978. Tabular key to the species of *Phytophthora* de Bary. Mycol. Pap. 143. Commonw. Mycol. Inst., Kew, Surrey, England. 20 pp.

Newton, W. 1956. The nonutilization of D-alanine and growth stimulation by L-alanine in the nutrition of *Phytophthora parasitica*. Can. J. Bot. 34:759-761.

Payette, A., and Perrault, C. 1944. Action de la thiamine sur le *Phytophthora infestans* (Mont.) de Bary. Can. J. Res. Sect. C 22:127-132.

Pethybridge, G. H., and Murphy, P. A. 1913. On pure cultures of *Phytophthora infestans* de Bary and the development of oospores. Proc. R. Dubl. Sci. Soc. 13:566-588.

Ribeiro, O. K. 1978. A Source Book of the Genus *Phytophthora*. J. Cramer, Vaduz, FL. 417 pp.

Ribeiro, O. K., Erwin, D. C., and Zentmyer, G. A. 1975. An improved synthetic medium for oospore production and germination of several *Phytophthora* species. Mycologia 67:1012-1019.

Ridings, W. H., Gallegly, M. E., and Lilly, V. G. 1969. Thiamine requirements helpful in distinguishing isolates of *Pythium* from those of *Phytophthora*. Phytopathology 59:737-742.

Robbins, W. J. 1937. The assimilation by plants of various forms of nitrogen. Am. J. Bot. 24:243-250.

Robbins, W. J. 1938. Thiamin and growth of species of *Phytophthora*. Bull. Torrey Bot. Club 65:267-276.

Robbins, W. J., and Kavanagh, V. 1942. Vitamin deficiencies of the filamentous fungi. Bot. Rev. 8:411-471.

Roncadori, R. W. 1965. A nutritional comparison of some species of *Phytophthora*. Phytopathology 55:595-599.

Ronsdorf, L. 1935. Vergleichende Untersuchungen über die Wirkung verschiedener Wuchsstoffe auf das Wachstum einiger Pilze. Arch. Mikrobiol. 6:309-325.

Rossetti, V., and Bitancourt, A. A. 1952. Thiamin and the growth substances for *Phytophthora* in the bark of citrus trees. Science 115:205-206.

Sakai, R. 1955. Physiological studies on the *Phytophthora infestans* (Mont.) de Bary. II. Nitrogen source of *Phytophthora infestans* (Mont.) de Bary. Ann. Phytopathol. Soc. Jpn. 19:141-145. (In Japanese)

Sakai, R. 1961. Studies on the nutritional physiology of *Phytophthora infestans* (Mont.) de Bary. Hokkaido Natl. Agric. Exp. Stn. Rep. 57:1-158.

Saksena, R. K., and Bhargava, K. S. 1943. Nitrogen requirements and vitamin deficiencies of *Phytophthora phaseoli* Thaxter. Proc. Indian Acad. Sci. Sect. B 18:45-51.

Scheepens, P. C., and Fehrmann, H. 1978. Cultivation of *Phytophthora infestans* on defined nutrient media. Phytopathol. Z. 93:126-136.

Schiffmann-Nadel, M., and Cohen, E. 1972. Effect of carbon sources upon the mycelial growth of *Phytophthora citrophthora*. Mycopathol. Mycol. Appl. 48:367-370.

Schwinn, F. J. 1959. Untersuchungen zur Systematik der Gattung *Phytophthora* de Bary. Arch. Mikrobiol. 33:223-252.

Seidel, H. 1961. Untersuchungen über den Nährstoffbedarf und die Toxinbildung des Pilzes *Phytophthora infestans* (Mont.) de Bary in vollsynthetischen Nährlösungen. Phytopathol. Z. 41:1-26.

Singh, N. 1975. Effect of nutrition on growth and sporangial production of two West African isolates of *Phytophthora palmivora* Butl. Ann. Bot. 39:129-136.

Sneh, B. 1972. An agar medium for the isolation and macroscopic recognition of *Phytophthora* spp. from soil on dilution plates. Can. J. Microbiol. 18:1389-1392.

Stössel, P., and Hohl, H. R. 1980. Sterols in *Phytophthora infestans* and their role in the parasitic interactions with *Solanum tuberosum*. Ber. Schweiz. Bot. Ges. 90:118-128.

Thurston, H. D., Wilde, P., and Suida, T. W. 1959. Growth of six races of *Phytophthora infestans* on artificial media. Phytopathology 49:450-451.

Timmer, L. W., Erwin, D. C., and McCormick, W. H. 1970. Glucose inhibition and inheritance of pigment-producing ability in *Phytophthora*. Mycologia 62:967-977.

Tsao, P. H., and Tummakate, A. 1977. The identity of a *Phytophthora* species from black pepper in Thailand. Mycologia 69:631-637.

Tsao, P. H., Alizadeh, A., and Tsao, P. W. 1981. Relationship between *Phytophthora capsici* and the black pepper isolates of '*P. palmivora*' morphological form (MF) 4. (Abstr.) Phytopathology 71:262.

Volkonsky, M. 1933. Sur les conditions de culture et le pouvoir de synthèse de *Saprolegnia* sp. Ann. Inst. Pasteur 50:703-730.

Volkonsky, M. 1934. Sur la nutrition de quelques champignons saprophytes et parasites. Ann. Inst. Pasteur 52:76-101.

Warren, R. C., King, J. E., and Colhoun, J. 1973. Reaction of potato plants to *Phytophthora infestans* in relation to their carbohydrate content. Trans. Br. Mycol. Soc. 61:95-105.

Waterhouse, G. M. 1963. Key to the species of *Phytophthora* de Bary. Mycol. Pap. 92:1-22.

Waterhouse, G. M. 1974. *Phytophthora palmivora* and some related species. Pages 51–70 in: *Phytophthora* Disease of Cocoa. P. H. Gregory, ed. Longman, London. 348 pp.

Weete, J. D. 1980. Lipid Biochemistry of Fungi and Other Organisms. Plenum Press, New York. 388 pp.

Wills, W. H. 1954. The utilization of carbon and nitrogen compounds by *Phytophthora parasitica* Dastur var. *nicotianae* (Breda de Haan) Tucker. J. Elisha Mitchell Sci. Soc. 70:231-235.

Zentmyer, G. A., Kaosiri, T., and Idosu, G. 1977. Taxonomic variants in the *Phytophthora palmivora* complex. Trans. Br. Mycol. Soc. 69:329-332.

# 4

# Physiology of Asexual Sporulation and Spore Germination in *Phytophthora*

Olaf K. Ribeiro

CONTENTS



## INTRODUCTION

Anton de Bary is perhaps best known to those of us working with the genus *Phytophthora* for his investigations into the causal agent of late blight of potato, which culminated in his description of the new genus *Phytophthora* in 1876. De Bary, however, is also credited with being the first to describe various aspects of fungus physiology in his book *Morphologie und Physiologie der Pilze,* published in 1866. Although this book gave impetus to the study of fungus physiology, very little if any work was published on the physiology of *Phytophthora* by de Bary and others investigating the genus.

Not until Leonian's remarkable investigations into the nutrition of *Phytophthora* (Leonian, 1925, 1927, 1934) was the physiology of this genus studied in any depth. These studies, together with several other investigations over the last four decades, have resulted in discoveries that have greatly enhanced our understanding of the physiology of *Phytophthora.* In this chapter, an attempt is made to bring together many of these contributions that have helped elucidate the factors involved in reproduction and spore germination in this important genus of plant pathogens.

## SPORANGIUM PRODUCTION

Sporangia in the genus *Phytophthora* form on solid as well as in liquid substrates. The abundance of sporangia either on solid or in liquid substrates varies with the species and within isolates of the same species (Brasier, 1967, 1969; Ribeiro et al, 1976b). Sporangia are usually produced at or near the air-substrate interface. Sporulation is a complex process involving water potential; nutrients; sterols; aeration; light; temperature; cations such as $Ca^{2+}$, $Fe^{3+}$, $Mg^{2+}$, and $K^{+}$; age of the culture; certain bacteria; root exudates; and soil extracts. The importance of each of these factors varies with the species and even within isolates of the same species. This will become more apparent in the following discussion of these factors and their roles in sporangium formation.

### Water Potential

Water potential is perhaps the most significant factor influencing sporangium production of all *Phytophthora* species. In general, relative humidity approaching 100% or matric potentials approaching zero are highly conducive to sporangium formation. For instance, Sneh and McIntosh (1974) found that numerous sporangia of *P. cactorum* formed at matric potentials of −0.1 and −0.3 bar (corresponding to soil water contents between near-saturation and field capacity), whereas few or no sporangia formed at −3.0 bars. Similarly, Duniway (1975a) reported that matric potentials between −0.025 and −0.3 bar were optimal for sporangium production of *P. drechsleri* in sterilized as well as unsterilized soils. Sporangia even formed at soil water potentials as low as −4 bars (Duniway, 1975b). In contrast, Pfender et al (1977) found that *P. megasperma* produced sporangia abundantly in flooded soil, with fewer sporangia forming at −0.05 and −0.1 bar matric potential. None were produced at −2.8 bars. Reeves (1975) investigated the effect of soil moisture regimes on the production of sporangia by *P. cinnamomi* and concluded that soil water suction pressure (matric potential) rather than absolute water content was the factor controlling production of sporangia in soil. Detailed accounts of the effect of water potential on sporangium formation and germination are given by Duniway (Chapter 14) and Gisi (Chapter 8) and in the recent review by Duniway (1979).

## Nutrition

Decoctions of lima bean, cornmeal, V-8 juice, or alfalfa in distilled water or a sucrose-asparagine medium stimulated production of sporangia (Schmitthenner, 1959, 1973). We have routinely used a dilution of four parts of distilled water to one part of V-8 juice for sporangium production in a number of *Phytophthora* species. In general, sporangium formation can be induced in most species by either 1) transferring a small portion of the mycelium (from the edge of an actively growing colony) to a natural broth such as V-8 juice or lima bean diluted with distilled water to provide a low nutrient level (a 1:4 dilution is usually sufficient), or 2) transferring actively growing cultures directly to sterile distilled water in petri dishes. Why sporangia form under these conditions will become apparent in the following discussion. It should be noted that some isolates of *Phytophthora* form sporangia readily on agar media and sparsely or not at all in liquid media. For instance, an isolate of *P. heveae* formed abundant sporangia on a carrot agar medium at temperatures above 25° C but failed to form any in Petri's mineral salts solution at any temperature between 15 and 30° C, whereas some isolates of *P. palmivora* formed sporangia abundantly in both liquid and solid media (Brasier, 1969).

Glucose is generally stimulatory to sporangium formation in concentrations up to 0.5 g/L. Above this concentration, sporangium formation is usually inhibited (Ayers and Zentmyer, 1971; Yamamoto and Tanino, 1961; Yoshikawa and Masago, 1977).

The stimulatory effect of various nitrogen sources depends on the isolate. Nitrogen sources that are stimulatory and/or inhibitory to sporangium production are given in Table 1. Brasier (1967) found that sporangium formation by *P. palmivora* increased when either asparagine or serine was added to the medium up to an optimal concentration of 2–3 g/L serine or 1.0 g/L asparagine; above this, sporangium production dropped off sharply. As would be expected, the carbon-nitrogen (C/N) ratio greatly influenced sporangium formation. However, Yoshikawa and Masago (1977) made the significant discovery that the inhibition of sporangium production by glucose, citric acid cycle intermediates, and amino acids was reversed by the addition of 3′, 5′ cyclic AMP. Repression of sporangium production by glucose at concentrations greater than 5 m*M* was not reversed by cyclic AMP.

The stimulatory effect of cations such as $Ca^{2+}$, $Mg^{2+}$, $Fe^{3+}$, and $K^{+}$ for sporangium production is well documented (Allen and Nandra, 1975; Halsall, 1977; Halsall and Forrester, 1977; Kennedy and Erwin, 1961). Halsall and Forrester (1977) demonstrated that $Ca^{2+}$ and $Fe^{3+}$ had a significant effect on sporangium production of *P. cambivora, P. cinnamomi, P. cryptogea,* and *P. drechsleri.* As the concentration of these cations was increased, sporangium production increased to a maximum and then declined. Halsall and Forrester (1977) then transposed the data to fit a mathematical model that allowed them to study the effects and possible interactions of each cation. It was found that each cation had a different optimal concentration for sporangium production of the four *Phytophthora* species examined. It was also found that there was no interaction between $Cu^{2+}$, $Ca^{2+}$, $Mg^{+}$, $K^{+}$, and $Fe^{3+}$ when these were added to the medium in various combinations (Halsall, 1977). The cation concentration required by different species is perhaps best illustrated by data obtained with $Ca^{2+}$. The optimal concentrations of $Ca^{2+}$ for sporangium production of *P. cambivora, P. cinnamomi, P. cryptogea,* and *P. drechsleri* were 0.23, 0.71, 0.68, and 0.12 m*M,* respectively (Halsall and Forrester, 1977), whereas for *P. nicotianae* var. *nicotianae, P. nicotianae* var. *parasitica,* and *P. palmivora,* sporulation increased as concentrations of $Ca^{2+}$ increased from 2 to 10 m*M* (Allen and Nandra, 1975). Other investigators (Gooding and Lucas, 1959; Schiffmann-Nadel and Cohen, 1966; Tucker, 1931; Yamamoto and Tanino, 1961) have increased the sporulation of many *Phytophthora* species by adding $KNO_3$ at low concentrations (0.01 *M*) to the sporulation medium.

Although the majority of *Phytophthora* species respond to one or more inorganic chemicals, *P. cinnamomi* is unique in requiring a precise balance of $Ca^{2+}$, $K^{+}$, $Mg^{2+}$, and $Fe^{2+}$ for sporangium formation (Chen and Zentmyer, 1970).

Sporangium formation is inhibited by very low concentrations of certain ions. For example, 0.2 mg/L $Cu^{2+}$ completely inhibited sporulation of *P. megasperma* (Kennedy and Erwin, 1961). This inhibition could be overcome by the addition of KEDTA (89 *μM*) to the medium. It is of interest to note that copper pipes can be the source of contamination of water with $Cu^{2+}$, and it is therefore imperative that double-distilled water be used for all experiments involving the asexual physiology of *Phytophthora.* Potassium and sodium chloride in low concentration are also known to inhibit sporulation (Zentmyer and Marshall, 1959).

TABLE 1. Effect of various chemicals on sporangium formation[a]

| Stimulatory | Inhibitory |
|---|---|
| Succinic acid | *cis*-Aconitate |
| Malic acid | Citric acid |
| L-Alanine | Fumaric acid |
| L-Aspartic acid | L-Aspartic acid |
| L-Glutamic acid | L-Glutamic acid |
| Glycine | L-Cysteine |
| Histidine | Leucine |
| Proline | Methionine |
| DL-Serine | Phenylalanine |
| Threonine | Valine |
| Lysine | |

[a]Species tested were *Phytophthora capsici, P. cryptogea, P. megasperma* f. sp. *glycinea, P. megasperma* f. sp. *medicaginis, P. heveae,* and *P. palmivora.*

*Sources*: Singh (1975), Yoshikawa et al (1976), O. K. Ribeiro (*unpublished*).

## Sterols

Sterols exert profound effects on the life cycle of *Phytophthora* (Figs. 1 and 2). In low concentrations, they increase growth, stimulate asexual reproduction, influence zoospore formation, and play a key role in the formation of sexual reproductive structures. The effect of sterols on sporulation in *Phytophthora* is discussed by Elliott (Chapter 5).

Hendrix (1964) was the first to demonstrate a sterol requirement for asexual sporulation of *Phytophthora* species. He later demonstrated that *P. cinnamomi, P. lateralis, P. capsici, P. palmivora,* and *P. parasitica* formed sporangia on media containing cholesterol at 20 mg/L. On media without sterol, cultures were essentially mycelial, with few or no sporangia (Hendrix, 1965). Several reports have since confirmed these findings (Alconero and Santiago, 1972; Brasier, 1967; Chee and Turner, 1965; Englander and Roth, 1980; Platonova et al, 1977; and Singh, 1975). In investigations on the relationship of sterol and light to sporangium formation, Hendrix (1967) found that functional sporangia were produced abundantly by *P. palmivora* and *P. capsici* in the light only when cholesterol at 20 mg/L was added to the medium. On sterol-free medium in the light, sporangia were produced, but they degenerated in a few days, indicating that sterols were necessary for sporangia to be functional. A few sporangia were produced in the dark on a medium containing sterol. No sporangia, however, were formed in the dark on media lacking sterols. With *P. lateralis,* Englander and Roth (1980) reported that sporangium production on a V-8 agar medium supplemented with β-sitosterol (10 μg/ml) and irradiated with light at 650 $\mu W/cm^2$ was approximately four times that obtained on sterol-free media incubated either in light or in the dark (Fig. 1).

Sterols also influence sporangium size. Brasier (1967) reported that the size of *P. palmivora* sporangia increased with increasing concentrations of β-sitosterol up to 20 mg/L.

## Aeration

Aeration significantly influences sporangia formation. Leonian (1925) noted early in his investigations that one of the limiting factors for sporangium formation was the availability of atmospheric oxygen. Gooding and Lucas (1959) found that *P. parasitica* var. *nicotianae* produced abundant sporangia only on moistened mycelial mats; submerged mats rarely produced sporangia, presumably because of a lack of oxygen. Duniway (1975a) reported that mycelium of *P. drechsleri* formed numerous sporangia on a soil surface at zero matric potential, whereas few sporangia formed in soil at a depth of 5 mm at matric potentials of 0 and −0.01 bar. He attributed this inhibition of sporangia formation in saturated soil to a lack of aeration. Mitchell and Zentmyer (1971) carefully monitored $O_2$ and $CO_2$ concentrations required for sporangium production of several *Phytophthora* species. They reported that sporulation was reduced by decreasing $O_2$ or increasing $CO_2$ concentrations from those in air. The degree of sporulation at different concentrations tested varied with the species. Abundant sporangia were formed in 5% $O_2$ by *P. nicotianae* var. *parasitica* and *P. capsici* (A1), while *P. palmivora, P. capsici* (A2), and *P. citrophthora* developed sporangia less abundantly. In contrast, sporangia production by *P. cactorum* was greatly reduced at 5% $O_2$. Because the concentrations of $CO_2$ and $O_2$ were not expressed in terms of partial pressures of these gases, it is difficult to compare these results with those obtained by other investigators (Duniway, 1975a; Gooding and Lucas, 1959; Leonian, 1925); the actual amounts of $O_2$ and $CO_2$ in the atmosphere depend on altitude, which was not specified.

## Light

Light enhances (Aragaki and Hine, 1963; Brasier, 1969; Englander and Roth, 1980; Harnish, 1965; Hendrix, 1967; Huguenin and Jacques, 1973; Lilly, 1966; Ribeiro et al, 1976a; Rocha and Machado, 1973; Schiffmann-Nadel and Cohen, 1966), inhibits (Cohen et al, 1975), or has no influence on

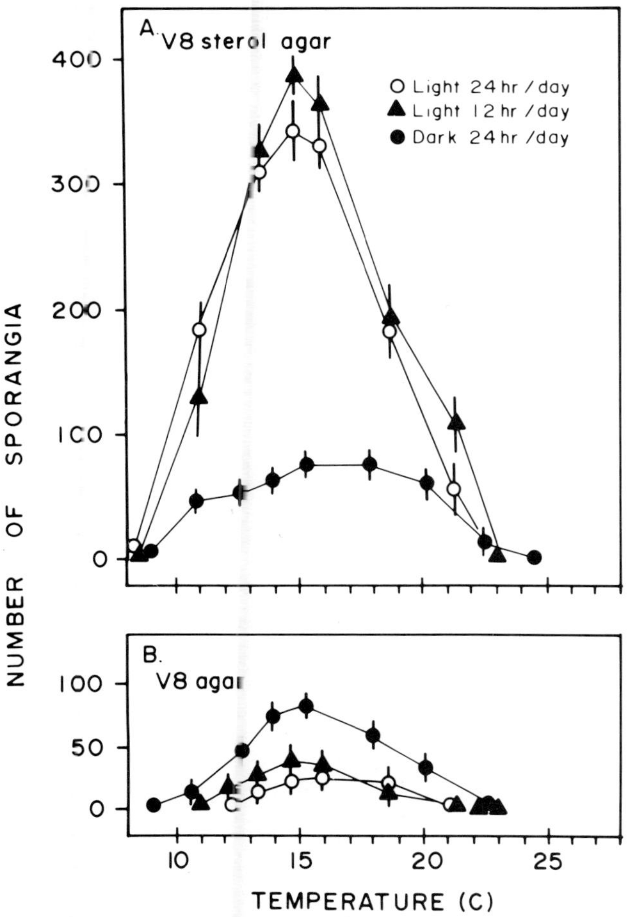


**Fig. 1.** Effect of sterols and light on sporangia production of *Phytophthora lateralis* on agar amended with 20 μg/ml β-sitosterol (A) and agar not amended with supplemental sterol (B). Vertical bars through data points designate confidence intervals ($P$ = 0.05). (Reprinted, by permission, from Englander and Roth, 1980)

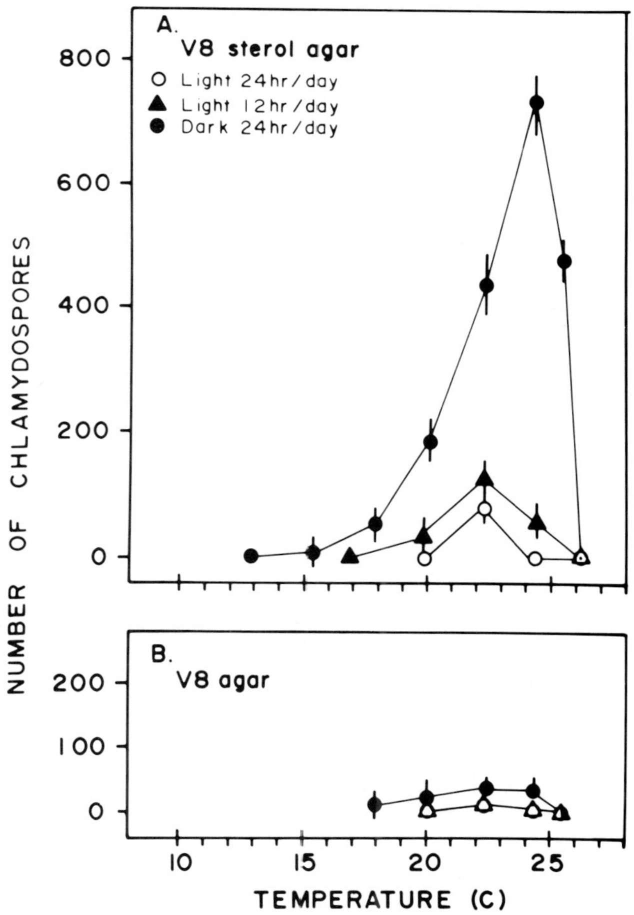


**Fig. 2.** Chlamydospore production by *Phytophthora lateralis* on agar amended with 20 μg/ml β-sitosterol (A) and agar not amended with supplemental sterol (B). Vertical bars through data points designate confidence intervals ($P$ = 0.05). (Reprinted, by permission, from Englander and Roth, 1980)

(Aragaki and Hine, 1963; Harnish, 1965; Waterhouse, 1931) sporangia production of *Phytophthora* species. Some species, e.g. *P. hibernalis* and *P. syringae*, sporulate better in the dark (Harnish, 1965). Investigations using monochromatic radiation have indicated that asexual sporulation of all species tested in axenic culture is significantly enhanced by wavelengths in the near-ultraviolet (UV) and blue regions of the spectrum (Englander and Roth, 1980; Huguenin and Jacques, 1973; Ribeiro et al, 1976b). On a natural substrate, sporulation by *P. infestans* was inhibited by radiation in the blue region of the spectrum (Cohen et al, 1975). Sporulation by *P. cinnamomi* was stimulated only by irradiation in the near-UV region up to 100 $\mu W/cm^2$ (Zentmyer and Ribeiro, 1977). Higher intensities were not significantly different from that obtained at 100 $\mu W/cm^2$; however, these sporangia failed to form zoospores.

Light also has a striking effect on the morphology of sporangia. Hendrix (1967) reported that sporangia of *P. palmivora* and *P. capsici*, formed in the light, differed considerably in size and shape from those produced in the dark. For instance, the length-to-breadth ratios of sporangia from cultures grown in the dark were found to be significantly lower than those reported for this species in taxonomic keys. Also, the shape of sporangia produced in dark-grown cultures varied to the point that *P. palmivora* could be confused with other closely related species (Hendrix, 1967). Similarly, when *P. cinnamomi* was irradiated with near-UV light, sporangia were greatly elongated (150–210 $\mu$m) compared to sporangia produced under standard light conditions (42–85 $\mu$m) (Zentmyer and Ribeiro, 1977).

The use of sporangium size for taxonomic purposes must therefore be treated with caution until uniform conditions for production of sporangia have been established.

### Temperature

Although sporangia are formed within a wide range of temperatures, the optimal temperature range is usually very narrow. Table 2 lists some representative temperatures for sporangium production of several *Phytophthora* species.

### Age of Culture

Juvenile mycelium, when depleted of nutrients, produced more sporangia than older mycelium (Ayers and Zentmyer, 1971; Eye et al, 1978; O. K. Ribeiro, *unpublished*). One-day-old mycelium of *P. megasperma* f. sp. *glycinea* produced four times as many sporangia as 5-day-old mycelium (Eye et al, 1978), whereas Ayers and Zentmyer (1971) found that sporangium formation by *P. cinnamomi* decreased with increased culture age under axenic conditions. Gooding and Lucas (1959) reported an overall increase in sporangium production by *P. parasitica* var. *nicotianae* as the culture aged.

TABLE 2. Temperature range for sporangium formation of some *Phytophthora* species

| Species | Range (°C) | Optimum (°C) | Reference |
|---|---|---|---|
| *P. infestans* | 3–26 | 18–22 | Crosier (1934) |
| *P. parasitica* | 15–30 | 20–25 | Wills (1954) |
| *P. parasitica* var. *nicotianae* | ··· | 20–25 | Gooding and Lucas (1959) |
| *P. capsici* | 21–33 | 24–27 | Divinagracia (1969) |
| *P. palmivora* | 15–30 | 25 | Aragaki and Hine (1963) |
| *P. palmivora* | 15–30 | 27–30 | Brasier (1969) |
| *P. lateralis* | 10–20 | 14–16 | Englander and Roth (1980) |

### pH

Sporangia may form within a wide pH range from 4 to 9 (Divinagracia, 1969; Gooding and Lucas, 1959; Kennedy and Erwin, 1961; O. K. Ribeiro, *unpublished*; Zentmyer and Marshall, 1959), although the optimal pH for sporangium formation varies with the species and among isolates of the same species. In reading the literature, it is often not possible to separate the salt components of the medium from the pH effects (e.g. Wills, 1954), thus making it difficult to interpret data in terms of a pH effect.

### Soil Extracts and Bacteria

It is well known that certain soil extracts stimulate sporangium production (Kennedy and Erwin, 1961; Marx and Bryan, 1969; Mehrlich, 1935; Zentmyer and Marshall, 1959). It was postulated that bacteria and other microorganisms in soil extracts either depleted the substrate of nutrients, thus favoring sporulation, or provided specific metabolites that stimulated sporulation (Ayers and Zentmyer, 1971). Ayers (1971) obtained an acetone extract from a soil pseudomonad that stimulated sporulation of *P. cinnamomi*. Acetone extracts of soil and of mixed populations of microorganisms from soil also enhanced sporangium production. Sporulation was induced by dilutions as high as $10^{-9}$ and was enhanced by the presence of $Ca^{2+}$ ions. Ayers (1971) characterized this stimulatory substance as a nonvolatile, polar, heat-stable compound soluble in water and several organic solvents.

Soil bacteria belonging to the genera *Pseudomonas* and *Chromobacterium* have been implicated in sporangium formation of *P. cinnamomi* (Chee and Newhook, 1966; Manning and Crossan, 1966; Marx and Haasis, 1965; Zentmyer, 1965). However, Broadbent and Baker (1974) could find no evidence that implicated these two genera in sporangium formation under their test conditions.

## GERMINATION OF SPORANGIA

Sporangia germinate either by formation of a germ tube(s) that eventually grows to form a mycelium (direct germination) or by differentiation of the cytoplasm within the sporangium into discrete zoospores that are then released through an exit pore (indirect germination). The sequence involved in the release of zoospores in *P. megasperma* f. sp. *medicaginis* is shown in Figure 3. After indirect germination, sporangium proliferation can occur in a nested or extended fashion (Fig. 4).

### Germination by Germ Tube

Direct germination depends on temperature and nutrition. Temperatures at and above the optimum for mycelial growth favor germ tube development by sporangia. For example, the optimal temperature for germ tube development of *P. infestans* sporangia is 24°C, whereas zoospore differentiation occurs at 12°C (Crosier, 1934).

Direct germination was enhanced in a medium consisting of 1% (w/v) peptone and 1% yeast extract or by the addition of 0.5% (w/v) of fructose, galactose, lactose, maltose, or sucrose to the medium (Clerk, 1972). Amino acids such as arginine, aspartic acid, glutamine, glycine, leucine, and tryptophan also enhanced direct germination and inhibited indirect germination (Clerk, 1972). A concentration of 10 m*M* $CaCl_2$ or 1–10 m*M* $MgSO_4$ also favored direct germination.

### Germination by Differentiation into Zoospores

In general, sporangia differentiate and release zoospores in the presence of free water and at temperatures below those favoring optimal growth of the fungus. Soil matric potentials between 0 and −1 mbar were reported to favor differentiation and release of zoospores (MacDonald and Duniway, 1978). This process was partially inhibited at −5mb and completely inhibited at −25 mb. Zoospore release was also shown to be affected by varying levels of osmotica. Although zoospores were released in solutions with water potentials as low as −9 bars, MacDonald and Duniway (1978) indicated that zoospore discharge in *P. cryptogea* and *P. megasperma* was influenced more by matric potential than by the total water content of the soil. Duniway (1975a) carefully monitored zoospore release from sporangia of *P. drechsleri* at several different matric potentials. He found that the release of zoospores was greatest at zero matric potential and then declined with decreasing moisture content. However, he also observed some zoospore release at matric potentials of −300 to −400 mb, indicating that saturated conditions were not absolutely necessary for germination. The influence of the matric and osmotic components of water on zoospore discharge are described in detail by Gisi (Chapter 8).

Anionic effects on sporangium germination were extensively investigated by Gisi et al (1977). Anions were found to enhance germination in the order sulfate, chloride, carbamate. The sulfate salts of $Na^+$, $K^+$, $Mg^{2+}$, and $Ca^{2+}$ stimulated sporangium differentiation more than the chloride forms of these cations. Differentiation and release of zoospores were delayed by increasing levels of NaCl (Gisi et al, 1977).

Light also has an effect on sporangium differentiation. Although sporangium production is significantly enhanced in the near-UV region, fewer of these sporangia released zoospores compared to those formed under daylight conditions (Ribeiro et al, 1976a).

Brasier (1967) found that sporangia of *P. palmivora* did not form zoospores in the absence of β-sitosterol, indicating that sterols are required for zoosporogenesis. Similar results were reported by Hendrix (1967) and Platonova et al (1977).

Adenine, hypoxanthine, and the riboside of dimethyl amino

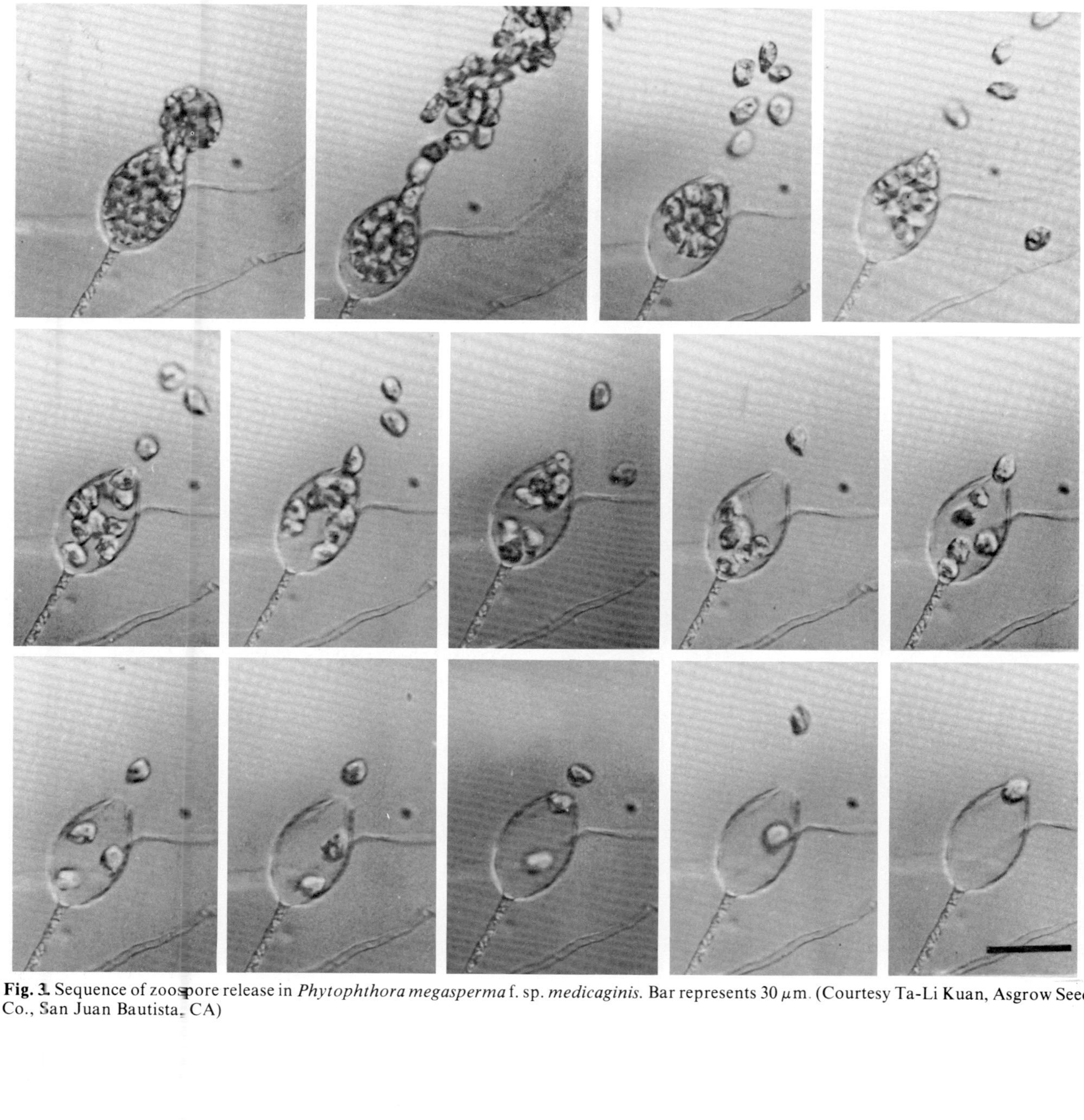

**Fig. 3.** Sequence of zoospore release in *Phytophthora megasperma* f. sp. *medicaginis*. Bar represents 30 μm. (Courtesy Ta-Li Kuan, Asgrow Seed Co., San Juan Bautista, CA)

purine were reported to stimulate indirect sporangium germination (Clark et al, 1978). Labeling studies in vivo and in vitro indicated that hypoxanthine was a key intermediate in sporangium development (Clark and Page, 1968). Varying concentrations of hypoxanthine significantly increased the rate of indirect germination both in vivo and in vitro. Enhanced germination obtained by the use of other purines was believed to reflect their conversion to hypoxanthine (Clark et al, 1978). Results obtained using antibiotics that inhibit protein synthesis indicate that protein synthesis associated with zoospore differentiation probably utilizes preformed RNA (Clark et al, 1978; Yoshikawa et al, 1976).

Indications that other biochemical pathways may influence zoospore differentiation were recently reported by Andreoli and Boistard (1977), who found that ornithine (0.03–0.04 mg/ml) resulted in zoospores with abnormal flagella, while higher concentrations resulted in release of undifferentiated cytoplasm.

TABLE 3. Influence of temperature on duration of zoospore motility

| Species | Temperature | | | Reference |
|---|---|---|---|---|
| | <10°C | 10–15°C | 16–25°C | |
| *Phytophthora infestans* | 22 hr | ··· | 30 min | Crosier (1934) |
| *P. citrophthora* | ··· | 24 hr | 1 hr | Klotz (1940) |
| *P. phaseoli* | 24 hr | 12 hr | 3–6 hr | Hyre and Cox (1953) |
| *P. fragariae* | 15 min | 5 hr | 15–30 min | Goode (1956) |

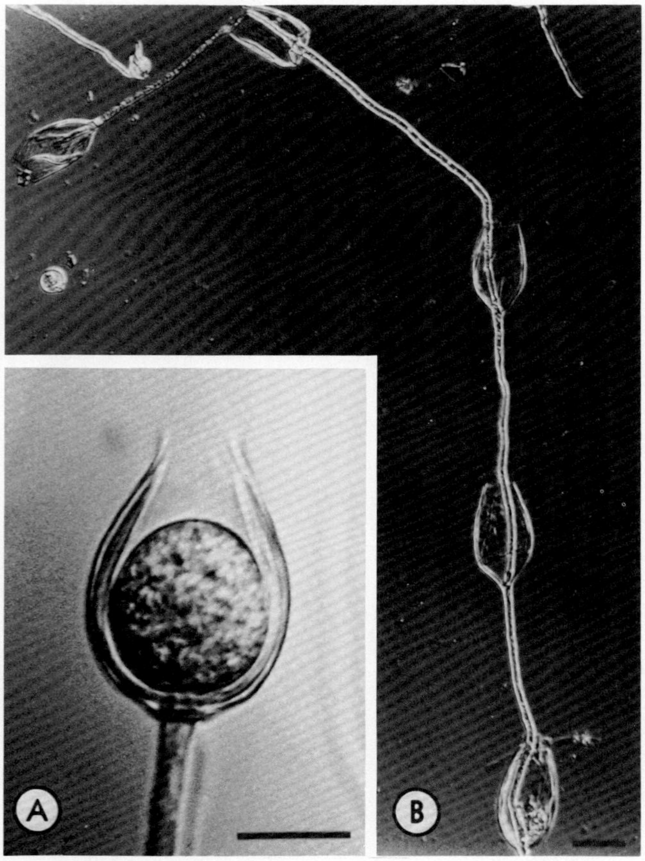


**Fig. 4.** Proliferating sporangia. **A,** Nested type (*Phytophthora megasperma* high-temperature isolate) (bar represents 35 μm). **B,** Extended proliferation (*P. megasperma* f. sp. *glycinea*) (bar represents 15 μm).

## Zoospore Motility

The duration of zoospore motility is influenced by the physical environment, particularly by temperature. In general, the lower the temperature, the greater the duration of motility. Representative times for the duration of motility of some *Phytophthora* species are given in Table 3. Zoospore motility is independent of exogenous nutrients; endogenous lipids appear to be the main energy source (Bimpong, 1975). The detection of measurable amounts of isocitrate ligase and malate synthetase suggests a functional glyoxylate cycle in motile zoospores (Bimpong, 1975). All the enzymes of the citric acid cycle except α-ketoglutarate dehydrogenase were also shown to be present. Although both cycles appear to be operative in motile zoospores, Bimpong (1975) suggested that the glyoxylate cycle was the major route for fatty acid utilization.

$^{14}$C-labeled D-glucose, acetate, L-serine, L-glutamate, and L-aspartate were readily metabolized by motile zoospores (Barash et al, 1965). However, these substances did not extend the duration of motility.

## Zoospore Germination

Following a period of motility lasting from a few minutes to several days depending on the physical environment, zoospores undergo a remarkable architectural rearrangement that results in the loss of the flagella, de novo synthesis of cell walls, and the emergence of a germ tube (Fig. 5) (see Hemmes, Chapter 2 and Bartnicki-Garcia and Wang, Chapter 9). The germination process is believed to be more dependent on exogenous nutrients that the motile state (Barash et al, 1965). Menyonga (1966) found that the germination of *P. parasitica* zoospores in vitro was enhanced by amino acids when compared to that obtained in distilled water, $NH_4NO_3$, or sugar solutions. Studies by Clark et al (1978) indicated that zoospore germination required de novo synthesis of mRNA.

Once the germ tube has emerged from the encysted zoospore, it is sensitive to antibiotics that inhibit active respiration and protein biosynthesis (Érsek, 1975; Yoshikawa et al, 1976).

# CHLAMYDOSPORE PRODUCTION

Chlamydospores are globose, thick- or thin-walled, asexual structures produced by some species of *Phytophthora.* The factors that influence chlamydospore production are sterols, the carbon-nitrogen ratio of the medium, water potential, aeration, light, and temperature.

### Sterols

Sterols enhance chlamydospore production in several *Phytophthora* species (Englander and Roth, 1980; Englander and Turbitt, 1979; Hendrix, 1970; Huguenin, 1974). *P. cinnamomi* formed three times as many chlamydospores in a V-8 juice agar medium supplemented with β-sitosterol (10–40 μg/ml), compared to unsupplemented medium (Englander and Turbitt, 1979). *P. lateralis* also produced more chlamydospores on sterol-supplemented medium than in unsupplemented medium (Fig. 2).

### Nutrition

Chlamydospores, unlike sporangia, can form readily in rich nutrient media, e.g. V-8 juice or papaya juice supplemented with 0.1% $CaCO_3$, clarified by centrifugation, and diluted with an equal volume of water (Chee, 1973; Kadooka and Ko, 1973).

Huguenin (1974) assessed the effects of temperature, pH, aeration, sterols, N-source, and C/N ratio of the medium on the formation of chlamydospores by an isolate of *P. palmivora* on a chemically defined medium. He concluded that the C/N ratio of the medium strongly influenced the formation of chlamydospores by *P. palmivora.* For instance, C/N ratios of 15/6 and 28/9 were optimal for chlamydospore formation, while C/N ratios of 10/1 or 50/6 resulted in significantly fewer chlamydospores.

Nitrogen sources such as $KNO_3$, $NH_4NO_3$, L-asparagine, urea, and casein hydrolysate stimulated chlamydospore production by *P. palmivora;* $(NH_4)_2SO_4$ inhibited formation of chlamydospores (Huguenin, 1974).

### Water Potential and Aeration

Tsao (1971) reported that chlamydospores of *P. parasitica* developed abundantly when cultures were submerged to a depth of about 50 mm in a liquid medium. Depletion of nutrients and low aeration were considered to be ideal conditions. Similarly, Huguenin (1974) found that production of chlamydospores by *P. palmivora* was favored by a lack of aeration. In contrast, Chee (1973) reported that abundant chlamydospores were produced only when mycelial mats of *P. palmivora* were barely wetted. A few chlamydospores were observed when mycelial mats were floated on the surface of the incubating solution, and no chlamydospores were formed when mats were submerged. No significant differences were noted when the mycelial mats were floated on either pea broth

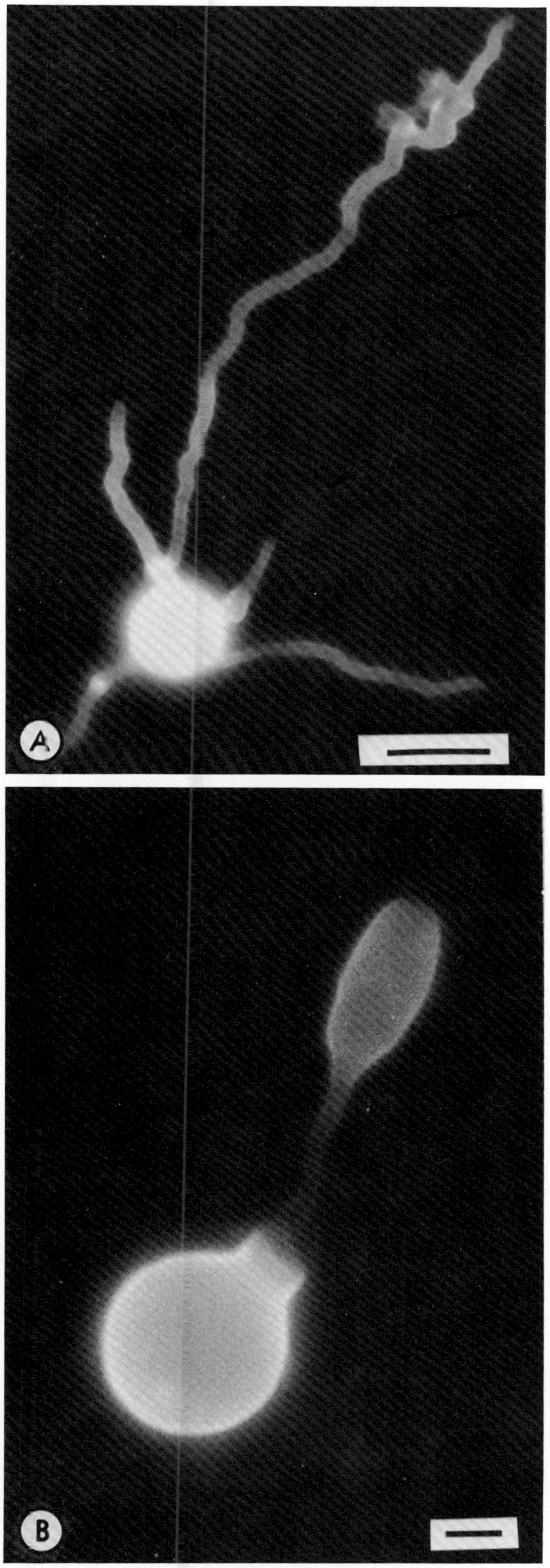


**Fig. 5.** Germination of a zoospore of *Phytophthora megasperma* f. sp. *medicaginis.* **A,** Multiple germ tubes (bar represents 10 $\mu$m). **B,** Germ tube terminating in a microsporangium (bar represents 5 $\mu$m). (Courtesy H. Förster, Universität Konstanz, West Germany)

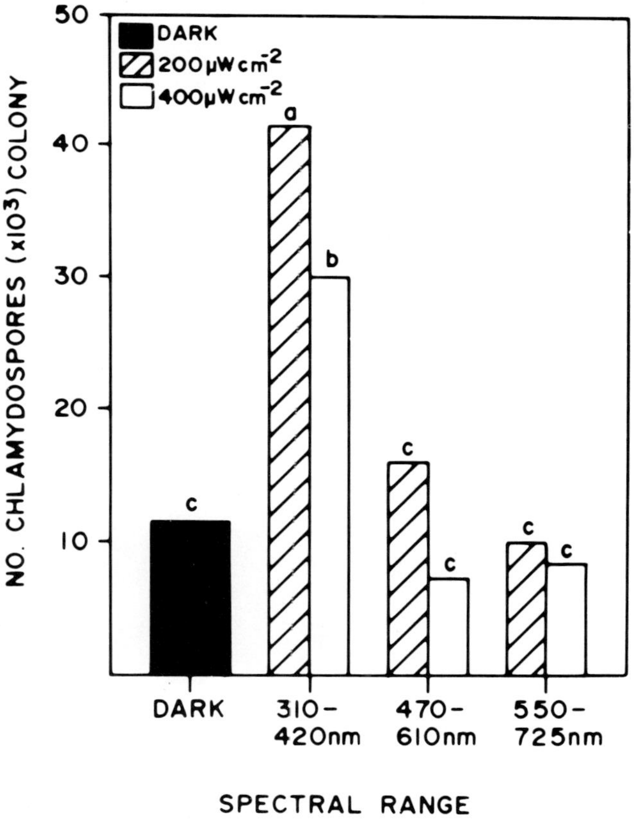


**Fig. 6.** Effect of different regions of the visible spectrum on chlamydospore production by *Phytophthora cinnamomi.* Bars with the same letters are not significantly different ($P = 0.05$). (Reprinted, by permission, from Englander and Turbitt, 1979)

at various dilutions (250 g of frozen whole garden peas in 1 L of distilled water) or V-8 broth (200 ml of V-8 juice plus 800 ml of distilled water). The type of medium and amount of aeration were also found to influence chlamydospore morphology; for example, Kadooka and Ko (1973) found that more than 70% of the chlamydospores produced by *P. palmivora* in the above-mentioned papaya juice medium were thick-walled, whereas only thin-walled chlamydospores were produced when Tsao's (1971) submerged culture technique was used.

The conflicting reports on chlamydospore formation in liquid media and the types of experiments that have been performed make it difficult to evaluate whether chlamydospore production was the result of an aeration effect, water potential, or both.

### Light

Chlamydospore production is generally more abundant in the light than in the dark. Chlamydospore production by *P. cinnamomi* was optimal when cultures were irradiated with 200 $\mu W/cm^2$ in the near-UV region on media amended with $\beta$-sitosterol (Englander and Turbitt, 1979). Above 400 $\mu W/cm^2$, chlamydospore production was inhibited. The influence of the spectral quality of light on the formation of chlamydospores is well illustrated by the data obtained with *P. cinnamomi*. Englander and Turbitt (1979) found no significant differences in chlamydospore production between cultures of *P. cinnamomi* irradiated with light of a spectral range between 470 and 725 nm and cultures incubated in the dark (Fig. 6), whereas chlamydospores formed abundantly in cultures irradiated in the near-UV region (310–420 nm). The magnitude of the response to light appears to depend on the species. In some species, e.g. *P. lateralis*, chlamydospores were suppressed in the light at intensities between 85 and 680 $\mu W/cm^2$ but were produced in the dark (Englander and Roth, 1980).

TABLE 4. Cardinal temperatures for chlamydospore production of some *Phytophthora* species

| Species | Minimum (°C) | Optimum (°C) | Maximum (°C) | Reference |
|---|---|---|---|---|
| *P. parasitica* | <12 | 15–18 | 27–30 | Tsao (1971) |
| *P. palmivora* | <15 | 32 | ? | Chee (1973) |
| *P. palmivora* | <10 | 18 | 33 | Huguenin (1974) |
| *P. cinnamomi* | 18 | 25 | 30.5 | Englander and Turbitt (1979) |
| *P. lateralis* | 12 | 24–25 | 26 | Englander and Roth (1980) |

### Temperature

Chlamydospores are produced over a wide temperature range. Representative temperatures for chlamydospore production of some *Phytophthora* species are given in Table 4.

### pH

Data concerning the effects of pH on the formation of chlamydospores by *Phytophthora* species are lacking. The data presently available concern only one *Phytophthora* species, *P. palmivora*. For an isolate of *P. palmivora* from rubber, Chee (1973) reported that the optimal pH for chlamydospore formation was 5.8, with only a few forming at pH 7.0. In contrast, an isolate of *P. palmivora* from citrus had an optimal pH of 6.3 for chlamydospore formation; none formed at pH 3.1 or above 6.9 (Huguenin, 1974).

## CHLAMYDOSPORE GERMINATION

Chlamydospores germinate by germ tubes that either continue growing to form mycelium or terminate in a sporangium (Fig. 7). Continued germ tube formation is enhanced in glucose- or asparagine-amended soils (Mircetich et al, 1968) and also at temperatures between 21 and 29°C (Chee, 1973). Formation of germ tubes that terminate in sporangia (Fig. 7A) is enhanced by low nutrient conditions, natural soils (Mircetich et al, 1968; Tsao, 1969), and temperatures between 32 and 35°C (Chee, 1973).

Chlamydospore germination requires exogenous nutrients (Tsao, 1969). The major factors influencing chlamydospore germination are water potential, the C/N ratio of the substrate, amino acids, and temperature.

### Water Potential

Sterne et al (1977) demonstrated that as much as 70% germination of *P. cinnamomi* chlamydospores could be obtained at matric potentials from 0 to −0.1 bar, whereas only

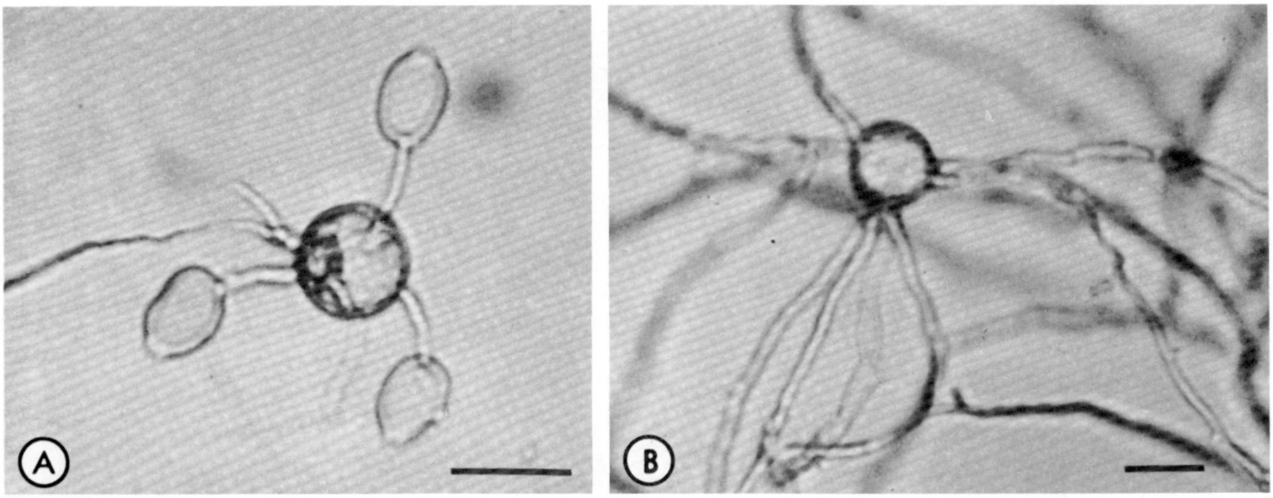


**Fig. 7.** Germination of chlamydospores of *Phytophthora parasitica*. **A,** Germ tubes terminating in sporangia (bar represents 30 $\mu m$). **B,** Multiple germ tube formation (bar represents 30 $\mu m$). (Reprinted, by permission, from Tsao, 1969)

40–50% germinated at −0.25 bar. Amending soil with glucose and asparagine at −0.25 bar resulted in germination equivalent to that obtained in untreated soil at 0 to −0.1 bar. On water agar amended with a mixture of sodium chloride, sodium sulfate, and potassium chloride, Cother and Griffin (1974) found that chlamydospore germination of *P. drechsleri* occurred at osmotic potentials as low as −97.3 bars. No germination occurred at −110 bars. The greatest lag in germination occurred at −97.3 bars; no germination was observed at this osmotic potential at 24 hr, whereas at 48 hr, 26.2% germination was recorded. The mechanism for chlamydospore germination at osmotic potentials lower than that for mycelial growth (−50 bars) has not been satisfactorily explained. It would be of interest to ascertain if continued growth of the germ tube to form mycelium is possible at osmotic potentials lower than −50 bars.

### Carbon-Nitrogen Ratio of the Substrate

Mircetich et al (1968) found that glucose, fructose, and sucrose did not induce germination of *P. cinnamomi* chlamydospores at concentrations as low as 2.5 m*M*, whereas low numbers (6%) germinated in demineralized water. Germination of chlamydospores was considerably enhanced, however, in a synthetic medium containing 15 g/L sucrose, 2 g/L asparagine, and $Fe^{2+}$, $Cu^{2+}$, $Mo^{6+}$, $Mn^{2+}$, and $Ca^{2+}$. Citric acid (0.025 *M*) also enhanced germination. The influence of organic and inorganic sources of nitrogen on the germination of *P. cinnamomi* chlamydospores has been investigated by Mircetich et al (1968). Asparagine (0.125 *M*) produced 83% germination, whereas $(NH_4)_2SO_4$, $NaNO_3$, and $KNO_3$ had no effect. Combinations of glucose and asparagine also enhanced germination (85%), whereas glucose + $(NH_4)_2SO_4$ and glucose + $NaNO_3$ were not stimulatory. All 11 amino acids tested stimulated chlamydospore germination (Mircetich et al, 1968). Similar results were also obtained with *P. parasitica* (Tsao, 1969; Tsao and Bricker, 1968). It was concluded that exogenous nutrients were required for chlamydospore germination under aseptic conditions. However, chlamydospore germination of *P. palmivora* does not require exogenous nutrients (Chee, 1973; Huguenin, 1974; Kadooka and Ko, 1973). The percentage germination of *P. palmivora* chlamydospores in water agar supplemented with glucose and asparagine was no different from that in water agar alone (Huguenin, 1974). Also, amino acids inhibited the germination of *P. palmivora* chlamydospores. This inhibition could be overcome by adding glucose to the medium.

Amending soil with glucose (0.05 *M*), asparagine (0.0125 *M*), or combinations of glucose and asparagine or glucose and $NaNO_3$ stimulated germination of *P. cinnamomi* chlamydospores (Mircetich and Zentmyer, 1969). In unamended natural soils, most of the chlamydospores that germinated formed a germ tube terminating in a germ sporangium, whereas those germinating in glucose- or asparagine-amended soils developed as many as 16 germ tubes per spore (Mircetich and Zentmyer, 1969). The enhancement of germination in natural soils by glucose or asparagine amendments was believed to be the result of the annulment of soil fungistasis by the increase in the level of nutrients in the soil. However, Tsao (1969) found that germination of *P. parasitica* chlamydospores in soil was not enhanced by the addition of glucose (0.4%); asparagine (0.4%) or $NH_4NO_3$ (0.4%) suppressed germination.

Thus, species of *Phytophthora* apparently respond differently to the presence of carbon and nitrogen in the soil and in artificial media.

### Temperature

Chlamydospores germinate over a wide temperature range; for example, the cardinal temperatures for germination of *P. cinnamomi* chlamydospores are minimum 9–12°C, optimum 20–30°C, and maximum 33–36°C (Mircetich et al, 1968).

### Age

Age appeared to have no influence on the percentage germination of *P. cinnamomi* (Mircetich et al, 1968) or *P. parasitica* (Holdaway and Tsao, 1972) chlamydospores. In contrast, Chee (1973) found that the highest chlamydospore germination was obtained from 1-wk-old cultures of *P. palmivora*. No germination was observed in 4-mo-old cultures.

### Effect of Wall Thickness

High germination (95%) of thin-walled chlamydospores occurred in V-8 juice agar, compared to only 5% germination for thick-walled chlamydospores of *P. palmivora* (Kadooka and Ko, 1973). In water agar, germination of thin-walled and thick-walled chlamydospores was between 90 and 95%, whereas in demineralized water, germination was 17% for thick-walled and 82% for thin-walled chlamydospores.

### pH

Mircetich et al (1968) found that chlamydospores of *P. cinnamomi* germinated over a wide pH range. Germination was optimal between pH 5.0 and 7.0, although good germination (over 50%) occurred at pH 3 and 9. No germination of *P. cinnamomi* chlamydospores occurred at pH 2.5. This, however, does not seem to hold true for other *Phytophthora* species. For instance, Chee (1973) reported no chlamydospore germination of *P. palmivora* at pH 4.0 or 8.0 (optimal pH was 5.8).

## OOSPORE GERMINATION

De Bary (1881) was the first to report the germination of oospores of a *Phytophthora* species. He found that some oospores of *P. cactorum* germinated after incubation in water for several months. His observations led him to believe that a chilling period was a prerequisite for oospore germination. Other early reports of oospore germination include those of Pethybridge (1914) with *P. erythroseptica*, Rosenbaum (1917) with *P. cactorum*, Hotson and Hartge (1923) with *P. mexicana*, and Blackwell and Waterhouse (1931) with *P. cactorum*. All these investigators reported that germination occurred only after oospores had been subjected to long periods of chilling either indoors or outdoors. Blackwell (1943) published precise instructions for successful germination of *P. cactorum* oospores: allow the oospore to mature for at least a month at 10–15°C under moist conditions; allow a further period of dormancy under moist conditions for 1–8 mo; refrigerate the oospores just above the freezing point for 1–2 wk; soak them in water (tap water of pH 8.0 or soil solutions of pH 7.0), changing the solution daily or at frequent intervals; and wait 1–3 wk (or longer) for germination. Despite these detailed instructions, it was found over the years that oospores of most *Phytophthora* species were difficult to germinate consistently and in sufficient numbers to conduct genetic or physiologic studies. It was subsequently demonstrated (see following discussions) that

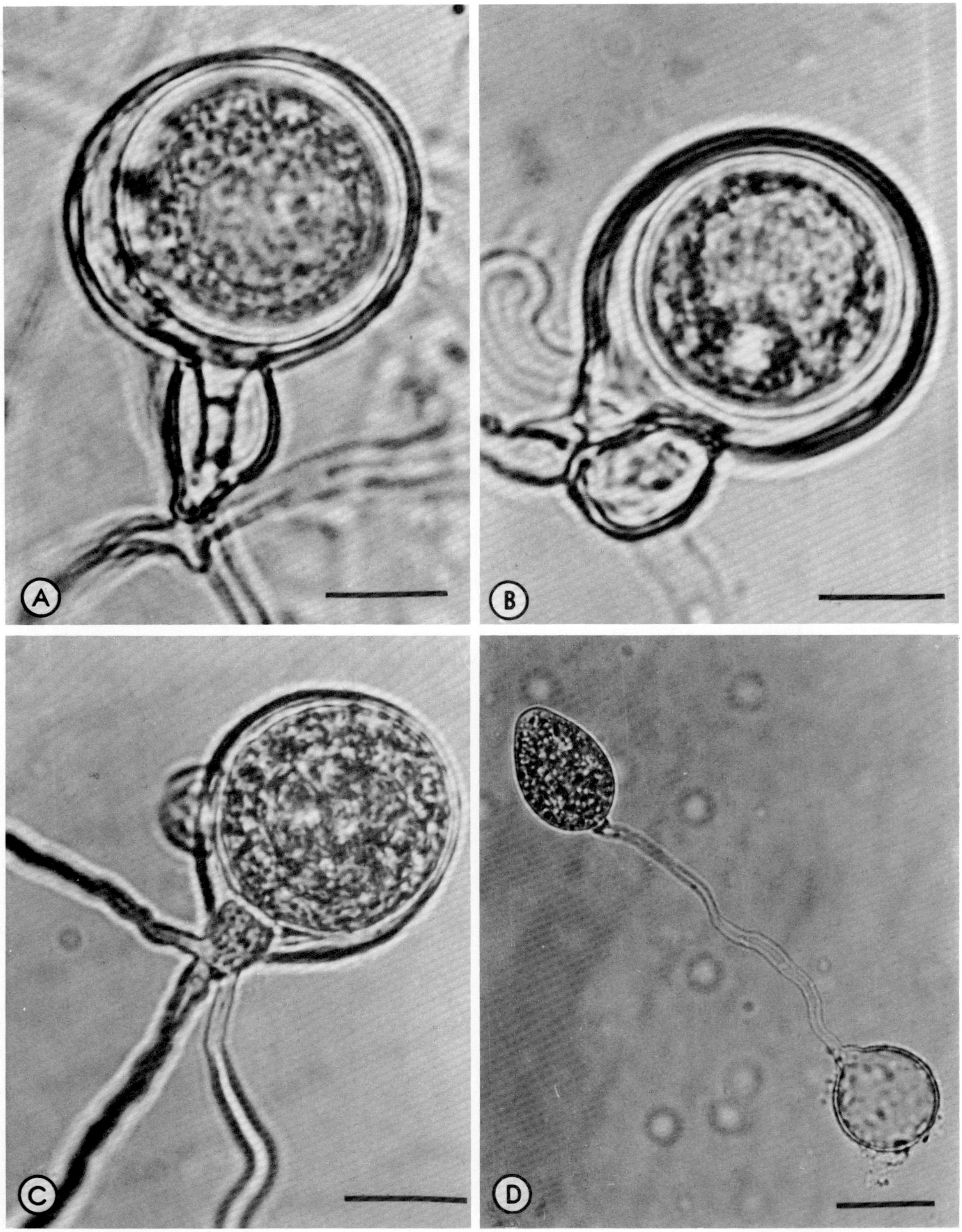


**Fig. 8.** Oospore germination of *Phytophthora megasperma* f. sp. *medicaginis*. **A,** Ungerminated oospore with amphigynous antheridium (bar represents 10 μm). **B,** Ungerminated oospore with paragynous antheridium (bar represents 10 μm). **C,** Germination by germ tube formation (bar represents 10 μm). **D,** Germination by a germ tube terminating in a sporangium (bar represents 32 μm). (Courtesy H. Förster, Universität Konstanz, West Germany)

other critical factors besides those mentioned by Blackwell affected oospore germination.

Oospores germinate either by single or multiple germ tubes or by germ tubes terminating in germ sporangia (Fig. 8C and D).

The following factors influence germination of oospores of *Phytophthora* species: light, temperature, nutrition, enzymes, culture growth media, and certain chemical, physical, and genetic factors.

### Light

A light requirement for germination of oospores has been reported by several investigators (Banihashemi and Mitchell, 1976; Berg and Gallegly, 1966; Erwin and McCormick, 1971; Henry and Stelfox, 1968; Huguenin and Boccas, 1971; Kaosiri et al, 1980; Klisiewicz, 1970; Leal and Gomez-Miranda, 1965; Lilly, 1966; Meyer, 1975; Polach and Webster, 1972; Ribeiro et al, 1975b; Romero and Erwin, 1969; Satour and Butler, 1968).

Shaw (1967) reported that only 3% of the oospores of *P. cactorum* germinated in the dark, whereas 38% germinated in the light. Similarly, Meyer (1975) found that 7-mo-old oospores of *P. cactorum* germinated better in the light (67%) than in the dark (32%). Banihashemi and Mitchell (1976) reported a relationship between the duration of irradiation and oospore germination of *P. cactorum*. They found that exposure to 108 hr of light at 2,594 lux resulted in 91.2% germination, whereas a 24-hr exposure to the same intensity followed by a dark period of 84 hr resulted in only 31% germination. The intensity of irradiation was also shown to influence oospore germination. At 104 lux for 92 hr, only 3.9% of the oospores of *P. cactorum* germinated, whereas 86.7% germination was obtained after exposure to 4,330 lux for the same length of time (Banihashemi and Mitchell, 1976). Above 4,330 lux, germination declined.

Germination of several *Phytophthora* species was enhanced by blue light (400–480 nm) (Banihashemi and Mitchell, 1976, 1980; Berg and Gallegly, 1966; Cardoso, 1971; Huguenin and Boccas, 1971; Ribeiro et al, 1975b, 1976a). Oospore germination was also enhanced by low levels of radiation (2 $\mu W/cm^2$) in the far-red region (Ribeiro et al, 1975b). There was also a difference in response to the blue or far-red wavelengths, depending on the *Phytophthora* species; e.g., germination of oospores of *P. megasperma* f. sp. *glycinea* increased when irradiated in both the blue (450 nm) and far-red (750 nm) regions of the spectrum, whereas *P. cinnamomi* was stimulated more in the blue than in the far-red wavelengths. These responses were elicited regardless of the light regime used (Ribeiro et al, 1975b).

### Oospore Maturity

In general, germination in many species increases with increasing age of the oospore up to a certain point (Banihashemi and Mitchell, 1976; Förster et al, 1981; Kaosiri et al, 1980; Klisiewicz, 1970; Meyer, 1975; Satour and Butler, 1968; Shaw, 1967).

Shaw (1967) observed no germination of 4-day-old oospores of *P. cactorum* but 67% germination with 25-day-old oospores. Satour and Butler (1968) reported 2.6% and 20% oospore germination of *P. capsici* from 14-day-old and 60-day-old cultures, respectively. Klisiewicz (1970) conducted extensive studies to determine the factors affecting germination of *P. drechsleri* oospores. He found that the maturity of the oospores was one of the factors that influenced oospore germination; e.g., less than 5% germination was obtained from 21-day-old oospores, 15% from 56-day-old oospores, and 51% from 102-day-old oospores.

Oospores of some species will germinate soon after formation under appropriate conditions of light, temperature, and nutrition (Duncan, 1977; Förster et al, 1981; Ribeiro et al, 1971). However, consistently high germination frequencies are found to be best with oospores from cultures at least 30 days old (Förster et al, 1981; Ribeiro et al, 1975b).

### Temperature

Oospores often require a period of pretreatment at a higher or lower temperature than ambient temperature before they will germinate. Pretreatment at 33°C for 2–8 days enhanced oospore germination of *P. drechsleri,* whereas pretreatments at 36°C inhibited germination (Klisiewicz, 1970). Similarly, Kaosiri et al (1980) in their studies with *P. palmivora* found that pretreatments at 12 and 30°C for 1 wk enhanced germination of *P. palmivora,* whereas pretreatment at 33°C suppressed germination. Duncan (1977) found that incubation of *P. fragariae* oospores for 24–48 hr at 25°C greatly increased the proportion of oospores that germinated, whereas incubation at 35°C proved to be inhibitory. In an extensive study involving several thousand individually monitored oospores, Long et al (1975) found that germination of oospores of several isolates of *P. megasperma* f. sp. *glycinea* was highest (over 50%) when the oospores were presoaked for 48 hr at 36°C before being plated on water agar. Presoaking oospores at 25°C for 48 hr or plating them directly on water agar resulted in significantly lower oospore germination (15–18%).

Exposures to low temperatures were reported to stimulate oospore germination of some *Phytophthora* species (Blackwell, 1943; Erwin and McCormick, 1971; Romero and Erwin, 1969). In contrast, Klisiewicz (1970) reported that chilling oospores of *P. drechsleri* at 3°C did not improve germination. Germination of *P. palmivora* oospores was inhibited after they were chilled at 5 or 9°C (Kaosiri et al, 1980).

Erwin and McCormick (1971) reported that the incubation temperature during oospore development was important for subsequent oospore germination. At 15°C, 13% germination of *P. megasperma* f. sp. *glycinea* oospores was obtained; at 24°C, germination was 65%; at 27°C, 78% germinated; and only 7% germination was recorded from oospores produced in cultures incubated at 30°C. Similarly, germination of oospores of *P. cactorum* at 16, 20, 24, and 28°C was 41, 88, 62, and 0%, respectively (Banihashemi and Mitchell, 1976).

### Nutrition

Sugars such as glucose, fructose, and sucrose generally inhibit oospore germination of several *Phytophthora* species (Banihashemi and Mitchell, 1976; Förster et al, 1982; Leal and Gomez-Miranda, 1965). However, fructose and mannitol were reported to be stimulatory for germination of *P. erythroseptica* oospores (Leal and Gomez-Miranda, 1965). The nitrogen source has also been found to influence oospore germination; e.g., 90% oospore germination was obtained with *P. heveae* in a synthetic medium supplemented with 5 g/L glucose and 2 g/L valine as the nitrogen source. Using cystine as the nitrogen source, 70% germination was recorded, whereas no germination occurred when citrulline was used as the nitrogen source (Leal and Gomez-Miranda, 1965). The effects of some carbon and nitrogen sources on oospore germination of *P. megasperma* f. sp. *medicaginis* are given in Table 5.

TABLE 5. Effect of some carbon and nitrogen sources on germination of oospores of *Phytophthora megasperma* f. sp. *medicaginis*

| Treatment | Relative germination rates[a] |
|---|---|
| Double-distilled water (control) | 100 |
| Glucose ($2.5 \times 10^{-4}$ *M*) | 13 |
| Asparagine ($7.5 \times 10^{-6}$ *M*) | 113 |
| Acetate ($10^{-5}$ *M*) | 144 |
| Acetate ($10^{-6}$ *M*) | 110 |
| Glucose ($2.5 \times 10^{-4}$ *M*) + asparagine | 70 |
| Glucose + acetate ($10^{-5}$ *M*) | 44 |
| Glucose + acetate ($10^{-6}$ *M*) | 107 |

[a] About 20% of the oospores germinated in the distilled water control.

*Source*: Förster et al (1982).

Cations such as $Ca^{2+}$ have also been reported to enhance oospore germination; for example, with *P. cactorum*, germination was 3% in demineralized distilled water and 50% when supplemented with 1 μg/ml of $CaCl_2$ (Banihashemi and Mitchell, 1976).

### Enzymes

Treatments with the snail gut enzyme helicase or glusulase (a mixture of β-glucuronidase and aryl sulfatase) at concentrations between 1 and 5% enhanced oospore germination of several *Phytophthora* species (Banihashemi and Mitchell, 1976; Förster et al, 1981; O. K. Ribeiro, *unpublished;* Salvatore et al, 1973). However, enzyme treatments do not appear to be an effective treatment for all *Phytophthora* species; for example, Long et al (1975) reported that treatments with helicase concentrations as low as 0.025% inhibited germination of *P. megasperma* f. sp. *glycinea* oospores. *P. capsici* oospores germinated well when treated with a 1% helicase solution (50% or better), compared to 1–8% for water-treated oospores. The germination of *P. cinnamomi*, on the other hand, was not enhanced by helicase treatments (O. K. Ribeiro, *unpublished*). Förster et al (1981) reported 45% germination of *P. megasperma* f. sp. *medicaginis* after treatment of oospores with 5% (w/v) glusulase for 12 hr. Prolonged enzymic treatment (more than 24 hr) reduced germination. Oospore germination was also stimulated with protease, lipase, and cellulase.

The increase in germination after passage of oospores through snails (Gregg, 1957; Salvatore et al, 1973; Shaw, 1967) is attributed to the action of β-glucuronidase and aryl sulfatase. Why these enzymes enhance oospore germination is not known. It should be noted that data concerning the effect of enzyme treatments for oospore germination are equivocal. Different species and even isolates of the same species appear to react differently to the same enzyme treatments. Helicase treatments in particular do not seem to be a panacea for the germination of oospores of most species tested thus far.

### Substrate

Klisiewicz (1970) reported that oospores of *P. drechsleri* produced on safflower stems had a germination rate 11–17% higher than that of oospores produced on agar media. Oospores of *P. capsici* produced in a V-8 juice medium that was buffered with $CaCO_3$ and clarified by centrifugation germinated less frequently than those produced in an unclarified V-8 juice medium (Satour and Butler, 1968). Oospores of *P. megasperma* f. sp. *medicaginis* germinated better in an alfalfa root extract (73–87%) than in double-distilled water (1–15%) (Förster et al, 1981). Extracts from older alfalfa plants (87 days old) had a stronger stimulatory effect on germination than extracts from younger plants (7–25 days old) (Förster et al, 1981). Although oospores often form in greater abundance in natural media, this is not necessarily correlated with higher percentage germination (Kaosiri et al, 1980; Ribeiro et al, 1975a), since oospores formed on a chemically defined medium germinated better than those formed on a V-8 juice agar medium (Leal and Gomez-Miranda, 1965; Ribeiro et al, 1975a).

### Other Factors

Smoot et al (1958) reported that oospores of *P. infestans* germinated best in an infusion of horse dung. Various physical treatments, such as alternate wetting and drying, aging 5–6 mo in agar culture or under moist soil at 12 and −20°C, and freezing and thawing (−20 to 20°C) for several months, had no effect. Likewise, pretreatments with chemicals such as ethyl formate, ethylene chlorohydrin, naphthalene acetic acid, oxalic acid, 2,4,5-trichlorophenoxyacetic acid, isopropylacetate, potassium acetate, salicylaldehyde, sodium citrate, sodium oleate, sodium thiocyanate, pyridine, D-xylose, furfural, D-glucose, hypoxanthine, and the mineral salts ammonium nitrate, anhydrous potassium carbonate, calcium carbonate, dipotassium phosphate, magnesium sulfate, potassium nitrate, monobasic sodium phosphate, and sodium sulfate in various combinations had no effect on oospore germination of *P. infestans* (Smoot et al, 1958). Chemicals that inhibit oospore germination of *P. megasperma* f. sp. *medicaginis* include glucose and sucrose (10 g/L); oleic acid (10 μg/ml); linoleic acid (100 μg/ml); palmitic acid (10 μg/ml); β-sitosterol (100 μg/ml); thiamin (0.1 μg/ml); ascorbic acid (0.1 μg/ml); pyridoxine (1 μg/ml); biotin (0.1 μg/ml); respiratory inhibitors such as KCN (200 μg/L); oligomycin (50 μg/ml), and antimycin A (10 g/ml); protein synthesis inhibitors such as cycloheximide (1 μg/ml) and chloramphenicol (10 μg/ml); and RNA synthesis inhibitors such as 5-fluorouracil (1 μg/ml), 8-azaguanine (10 μg/ml), and 6-azauracil (1 μg/ml) (H. Förster and O. K. Ribeiro, *unpublished*). The inhibition of glucose could be partly or completely overcome by the addition of asparagine ($7.5 \times 10^{-6}$ *M*) or acetate ($10^{-6}$ *M*) to the incubation medium (Table 5).

Oospore germination is apparently not greatly influenced by the pH of the incubation medium. Reports indicate that oospores of *P. capsici* germinate in a wide pH range from 4.5 to 7.8 (O. K. Ribeiro, *unpublished*; *P. drechsleri* oospores germinated between pH 4.5 and 9.5, with an optimum at pH 7.5 (Klisiewicz, 1970); and *P. palmivora* had an optimal pH of 5–6 for oospore germination (Huguenin and Boccas, 1971).

Oospores of some isolates of the same species germinate better than others (Ribeiro et al, 1971; Romero and Erwin, 1969). The reason for the genetic proneness of these isolates to produce oospores that germinate readily is not well understood.

## SUMMARY AND FUTURE DIRECTIONS

Klebs (1900) formulated four basic principles based on his physiologic studies of fungus reproduction. Stated briefly, these are: 1) the external environment determines when growth or reproduction will take place; 2) conditions for reproduction are generally less favorable for growth; 3) reproduction generally occurs over a narrower temperature range than growth; and 4) the degree of vegetative growth determines the caliber of reproduction. Our cumulative data on asexual reproduction of *Phytophthora* show that these principles are as valid today (with slight modifications) as when they were first proposed more than 80 yr ago. The major factors involved in asexual reproduction and spore germination in *Phytophthora* are summarized in Figure 9.

Some progress has been made toward understanding the basic mechanisms involved in asexual reproduction of *Phytophthora.* We now know that all species tested thus far and even isolates of the same species differ qualitatively and often quantitatively in their responses to chemical and physical stimuli, although the reasons for this are still not clear. In general, sporangium formation is strongly stimulated by light (315–475 nm) and by sterols. However, the basic unifying principle(s) for sporangium production still eludes us. We need to know more concerning the biochemical events preceding sporangium formation in order to predict its onset with greater precision.

In this context, we need to know the role of ion fluxes in sporangium development; what are the roles of cations such as $Ca^{2+}$, $K^{+}$, $Fe^{3+}$, $Mg^{2+}$, etc., in sporangium formation? Second, how critical is the role of light in asexual reproduction? Since asexual and sexual reproduction are affected by different regions of the spectrum, is more than one photoreceptor involved, as postulated by Ribeiro et al (1976a)? What is the nature of the photoreceptor? Third, what are the signals or triggers that initiate sporulation? Fourth, what are the biochemical pathways leading to sporangium, chlamydospore, and oospore production? What reactions or pathways do they have in common, and where do they diverge? Fifth, what role does cyclic AMP play in reproduction? Sixth, how does temperature affect sporulation? Are light, temperature, and

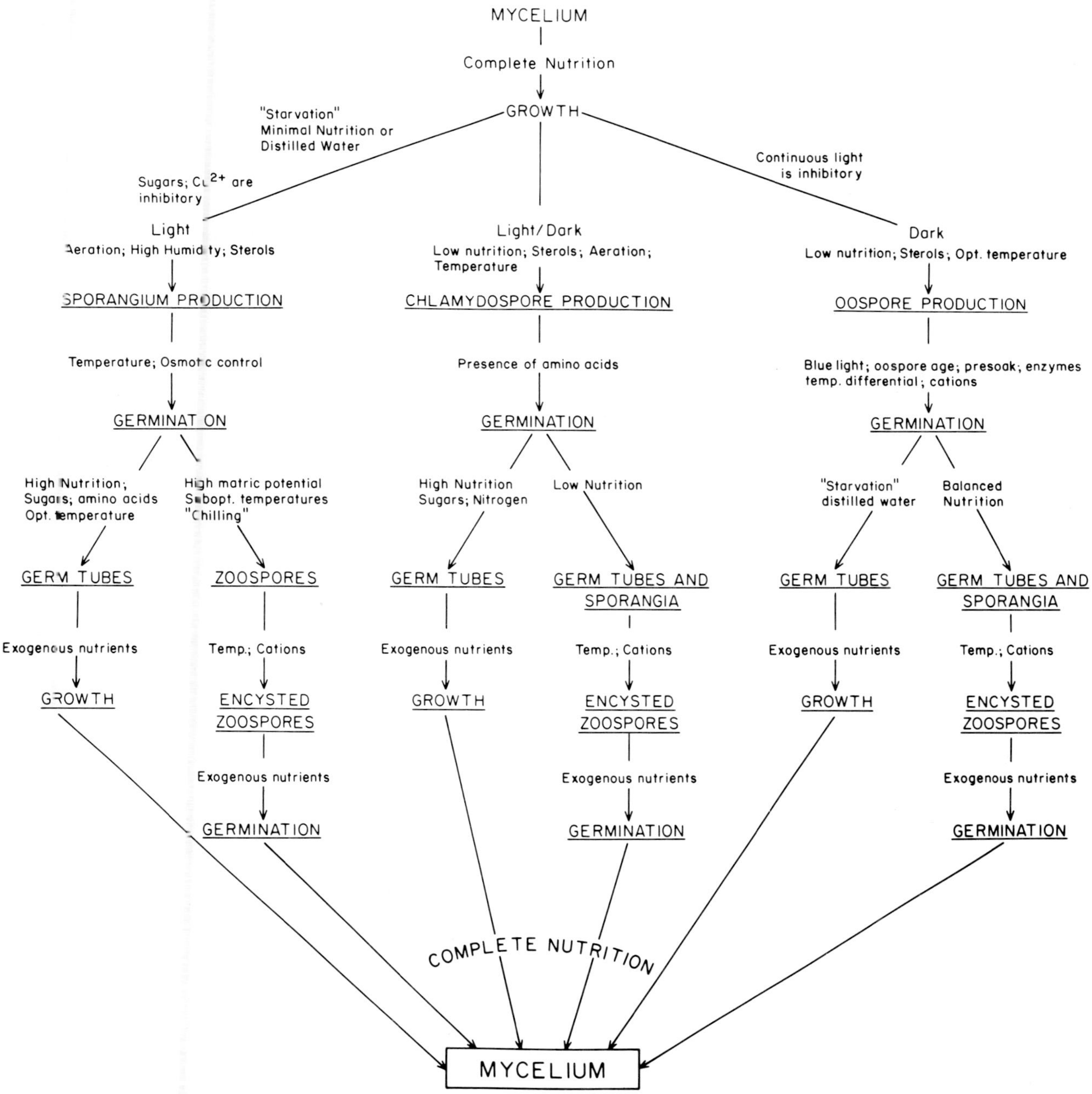


**Fig. 9.** Major factors influencing asexual and sexual reproduction and spore germination in *Phytophthora.*

sterols interrelated, or do they function at different levels of the developmental process? And seventh, how do sterols increase sporulation?

Several questions concerning sporangium germination also need to be answered. For instance, there is a paucity of data relating to cytoplasmic cleavage and subsequent zoospore formation. Also, how do zoospores devoid of walls survive without spontaneous lysis in environments where they are often subjected to large osmotic gradients? Are the signals or triggers that initiate encystment of a general or specific nature?

Although much progress has been made in producing and germinating oospores in the laboratory, here again we still do not know the nature of dormancy. We have a few clues from electron microscopy (see Hemmes, Chapter 2) and from biochemical studies by Leary et al (1974) indicating the lack of intact functional ribosomes in oospores. Also, Ribeiro et al (1976b) have shown that mature oospores lack cytochromes a and b. However, very little is known concerning macromolecular synthesis during germination such as that described for other dormant fungal spores (Sussman and Douthit, 1973). Why do oospores of different species react so differently to enzyme treatments? What are the functions of wall glucans? Beakes (1980) suggests that the mobilization of wall glucans may play a role in breaking dormancy. He also postulates that the enhancement of oospore germination by light and temperature may be caused by the induction of glucanase production.

The nature of dormancy of oospores raises many questions that have yet to be answered. Oospore dormancy is generally believed to be constitutive (Cochrane, 1974; Schmitthenner, 1968; Zentmyer and Erwin, 1970). This interpretation needs to be treated with some caution. We now know that oospores of some *Phytophthora* species, e.g. *P. capsici, P. infestans,* and *P. megasperma* f. sp. *medicaginis*, will germinate soon after formation, provided proper external conditions are present. Exogenous nutrients are required for the germination of oospores of most of the species investigated. Because we do not know the precise mechanism involved in the germination of *Phytophthora* oospores, we need to carefully define our terminology when discussing dormancy and germination of *Phytophthora* oospores. As pointed out by Blackwell (1943) and Cochrane (1974), we need to distinguish between the maturation process and dormancy. The often described constitutive dormancy of *Phytophthora* oospores might be simple immaturity. The nature of dormancy in *Phytophthora* oospores appears to be quite different from that in other fungal spores; perhaps it would be best not to refer to dormancy (according to current terminology: Cochrane, 1974; Sussman and Douthit, 1973) until more is known about the actual process of maturation and germination in *Phytophthora.*

Many of the investigations on oospore germination have been empirical; only recently have we begun to approach the problem more rationally with basic investigations on the biochemistry and ultrastructure of this process. Through these kinds of investigations, we will be able to understand the unique mechanisms that marshall seemingly isolated components in the cell into a careful and precise unit(s) capable of directing the synthesis of sporangium, chlamydospore, or oospore formation, development, and germination.

Our understanding of the physiology of asexual and sexual reproduction and germination of *Phytophthora* has made great strides since de Bary's formal recognition of the genus, and we can expect even greater advancements in the next decade, because we now have at our disposal more powerful and sophisticated techniques to help unravel many puzzling questions that were beyond reach as little as a decade ago.

## LITERATURE CITED


Alconero, R., and Santiago, A. 1972. Characteristics of asexual sporulation in *Phytophthora palmivora* and *P. parasitica* var. *nicotianae.* Phytopathology 62:993-997.

Allen, D. J., and Nandra, S. S. 1975. Effects of pH and calcium concentration on the sporulation of *Phytophthora* isolates from *Agave.* Plant Dis. Rep. 59:555-558.

Andreoli, C., and Boistard, P. 1977. Relationship between ornithine and sporangial differentiation in *Phytophthora nicotianae* var. *parasitica.* (Abstr.) Ann. Phytopathol. 9:90. (In French)

Aragaki, M., and Hine, R. B. 1963. Effect of radiation on sporangial production of *Phytophthora parasitica* on artificial media and detached papaya fruit. Phytopathology 53:854-856.

Ayers, W. A. 1971. Induction of sporangia in *Phytophthora cinnamomi* by a substance from bacteria and soil. Can. J. Microbiol. 17:1517-1523.

Ayers, W. A., and Zentmyer, G. A. 1971. Effect of soil solution and two soil pseudomonads on sporangium production by *Phytophthora cinnamomi.* Phytopathology 61:1188-1193.

Banihashemi, Z., and Mitchell, J. E. 1976. Factors affecting oospore germination in *Phytophthora cactorum,* the incitant of apple collar rot. Phytopathology 66:443-448.

Banihashemi, Z., and Mitchell, J. E. 1980. Photobiology of oospores of *Phytophthora cactorum* and *P. citricola.* (Abstr.) Proc. Am. Phytopathol. Soc. 14:88.

Barash, I., Klisiewicz, J. M., and Kosuge, T. 1965. Utilization of carbon compounds by zoospores of *Phytophthora drechsleri* and their effect on motility and germination. Phytopathology 55:1257-1261.

Beakes, G. W. 1980. Electron microscopic study of oospore maturation and germination in an emasculate isolate of *Saprolegnia ferax.* 2. Wall differentiation. Can. J. Bot. 58:195-208.

Berg, L. A., and Gallegly, M. E. 1966. Effect of light on oospore germination in species of *Phytophthora.* (Abstr.) Phytopathology 56:583.

Bimpong, C. E. 1975. Changes in metabolic reserves and enzyme activities during zoospore motility and cyst germination in *Phytophthora palmivora.* Can. J. Bot. 53:1411-1416.

Blackwell, E. 1943. Presidential address. On germinating the oospores of *Phytophthora cactorum.* Trans. Br. Mycol. Soc. 26:93-103.

Blackwell, E., and Waterhouse, G. 1931. Spores and spore germination in the genus *Phytophthora.* Trans. Br. Mycol. Soc. 15:294-310.

Brasier, C. M. 1967. Physiology of reproduction in *Phytophthora.* Ph.D. thesis. Univ. of Hull, England. 220 pp.

Brasier, C. M. 1969. The effect of light and temperature on reproduction in vitro in two tropical species of *Phytophthora.* Trans. Br. Mycol. Soc. 52:105-113.

Broadbent, P., and Baker, K. F. 1974. Association of bacteria with sporangium formation and breakdown of sporangia in *Phytophthora* species. Aust. J. Agric. Res. 25:139-145.

Cardoso, E. J. B. N. 1971. Influence of light on the oospore germination of several species of *Pythium* and *Phytophthora.* Ph.D. dissertation. Ohio State Univ., Columbus. 74 pp.

Chee, K. H. 1973. Production, germination and survival of chlamydospores of *Phytophthora palmivora* from *Hevea brasiliensis.* Trans. Br. Mycol. Soc. 61:21-26.

Chee, K. H., and Newhook, F. J. 1966. Relationship of microorganisms to sporulation of *Phytophthora cinnamomi* Rands. N.Z. J. Agric. Res. 9:32-43.

Chee, K. H., and Turner, N. A. 1965. A steroid factor in pea (*Pisum sativum* L.) influencing growth and sporulation of *Phytophthora cinnamomi* Rands. N.Z. J. Agric. Res. 8:104-108.

Chen, D. W., and Zentmyer, G. A. 1970. Production of sporangia by *Phytophthora cinnamomi* in axenic culture. Mycologia 62:397-402.

Clark, M. C., and Page, O. T. 1968. Accumulation of hypoxanthine in potato leaves and its effect on spore germination in *Phytophthora infestans.* Can. J. Bot. 46:267-270.

Clark, M. C., Melanson, D. L., and Page, O. T. 1978. Purine metabolism and differential inhibition of spore germination in *Phytophthora infestans.* Can. J. Microbiol. 24:1032-1038.

Clerk, G. C. 1972. Germination of sporangia of *Phytophthora palmivora* (Butl.) Butl. Ann. Bot. 36:801-807.

Cochrane, V. W. 1974. Dormancy in spores of fungi. Trans. Am.

Microsc. Soc. 93:559-609.

Cohen, Y., Eyal, H., and Sadon, T. 1975. Light-induced inhibition of sporangial formation of *Phytophthora infestans* on potato leaves. Can. J. Bot. 53:2680-2686.

Cother, E. J., and Griffin, D. M. 1974. Chlamydospore germination in *Phytophthora drechsleri.* Trans. Br. Mycol. Soc. 63:273-279.

Crosier, W. 1934. Studies in the biology of *Phytophthora infestans* (Mont.) de Bary. Cornell Univ. Agric. Exp. Stn. Mem. 155. 40 pp.

de Bary, A. 1876. Researches into the nature of the potato fungus, *Phytophthora infestans.* J. R. Agric. Soc. Engl. Ser. 2. 12:239-269.

de Bary, A. 1881. Zur Kenntniss der Peronosporeen. Bot. Zeitung Jahrgang 39:33-39.

Divinagracia, G. G. 1969. Sporangial and oospore formation by *Phytophthora capsici.* Philipp. Agric. 53:148-165.

Duncan, J. M. 1977. Germination in vitro of *Phytophthora fragariae* oospores from infected root tissue. Trans. Br. Mycol. Soc. 69:391-395.

Duniway, J. M. 1975a. Formation of sporangia by *Phytophthora drechsleri* in soil at high matric potentials. Can. J. Bot. 53:1270-1275.

Duniway, J. M. 1975b. Limiting influence of low water potential on the formation of sporangia by *Phytophthora drechsleri* in soil. Phytopathology 65:1089-1093.

Duniway, J. M. 1979. Water relations of water molds. Annu. Rev. Phytopathol. 17:431-460.

Englander, L., and Roth, L. F. 1980. Interaction of light and sterol on sporangium and chlamydospore production by *Phytophthora lateralis* Phytopathology 70:650-654.

Englander, L., and Turbitt, W. 1979. Increased chlamydospore production by *Phytophthora cinnamomi* using sterols and near-ultraviolet light. Phytopathology 69:813-817.

Érsek, T. 1975. The sensitivity of *Phytophthora infestans* to several antibiotics. Z. Pflanzenkr. Pflanzenschutz 82:614-617.

Erwin, D. C., and McCormick, W. H. 1971. Germination of oospores produced by *Phytophthora megasperma* var. *sojae.* Mycologia 63:972-977.

Eye, L. L., Sneh, B., and Lockwood, J. L. 1978. Factors affecting zoospore production by *Phytophthora megasperma* var. *sojae.* Phytopathology 68:1766-1768.

Förster, H., Ribeiro, O. K., and Erwin, D. C. 1981. Germination of oospores of *Phytophthora megasperma* f. sp. *medicaginis.* (Abstr.) Phytopathology 71:874.

Förster, H., Ribeiro, O. K., and Erwin, D. C. 1982. Factors affecting oospore germination of *Phytophthora megasperma* f. sp. *medicaginis.* Phytopathology. (In press)

Gisi, U., Oertli, J. J., and Schwinn, F. J. 1977. Wasser-und Salzbeziehungen der Sporangien von *Phytophthora cactorum* (Leb. et Cohn) Schroet. in vitro. Phytopathol. Z. 89:261-284.

Goode, P. M. 1956. Infection of strawberry roots by zoospores of *Phytophthora fragariae.* Trans. Br. Mycol. Soc. 39:367-377.

Gooding, G. V., and Lucas, G. B. 1959. Factors influencing sporangial formation and zoospore activity in *Phytophthora parasitica* var. *nicotianae.* Phytopathology 49:277-281.

Gregg, M. 1957. Germination of oospores of *Phytophthora erythroseptica.* Nature (London) 180:150.

Halsall, D. M. 1977. Effects of certain cations on the formation and infectivity of *Phytophthora* zoospores. 2. Effects of copper, boron, cobalt, manganese, molybdenum, and zinc ions. Can. J. Microbiol. 23:1002-1010.

Halsall, D. M., and Forrester, R. I. 1977. Effects of certain cations on the formation and infectivity of *Phytophthora* zoospores. 1. Effects of calcium, magnesium, potassium, and iron ions. Can. J. Microbiol. 23:994-1001.

Harnish, W. N. 1965. Effect of light on production of oospores and sporangia in species of *Phytophthora.* Mycologia 57:85-90.

Hendrix, J. W. 1964. Sterol induction of reproduction and stimulation of growth of *Pythium* and *Phytophthora.* Science 144:1028-1029.

Hendrix, J. W. 1965. Influence of sterols on growth and reproduction of *Pythium* and *Phytophthora* spp. Phytopathology 55:790-797.

Hendrix, J. W. 1967. Light-cholesterol relationships in morphogenesis of *Phytophthora palmivora* and *P. capsici* sporangia. Mycologia 59:1107-1111.

Hendrix, J. W. 1970. Sterols in growth and reproduction of fungi. Annu. Rev. Phytopathol. 8:111-130.

Henry, A. W., and Stelfox, D. 1968. Comparative behavior of the oospores and oogonia of *Phytophthora citricola* during germination on an artificial medium. Can. J. Bot. 46:1419-1421.

Holdaway, B. F., and Tsao, P. H. 1972. Longevity of chlamydospores of *Phytophthora parasitica* in water. Plant Dis. Rep. 56:971-973.

Hotson, J. W., and Hartge, L. 1923. A disease of tomato caused by *Phytophthora mexicana* sp. nov. Phytopathology 13:520-530.

Huguenin, B. 1974. Influence des conditions de culture sur la formation et la germination in vitro des chlamydospores du *Phytophthora palmivora.* Ann. Phytopathol. 6:425-440.

Huguenin, B., and Boccas, B. 1971. Rôle de quelques facteurs dans la formation et la germination des oospores chez le *Phytophthora palmivora* Butl. Ann. Phytopathol. 3:353-371.

Huguenin, B., and Jacques, R. 1973. Établissement du spectra d'action de la lumière sur la sporogenèse et l'oogenèse de souches du *Phytophthora palmivora* Butl. C. R. Hebd. Seances Acad. Sci. Ser. D 276:725-728.

Hyre, R. A., and Cox, R. S. 1953. Factors affecting viability and growth of *Phytophthora phaseoli.* Phytopathology 43:419-425.

Kadooka, J. Y., and Ko, W. H. 1973. Production of chlamydospores by *Phytophthora palmivora* in culture media. Phytopathology 63:559-562.

Kaosiri, T., Zentmyer, G. A., and Erwin, D. C. 1980. Oospore morphology and germination in the *Phytophthora palmivora* complex from cacao. Mycologia 72:888-907.

Kennedy, B. W., and Erwin, D. C. 1961. Some factors influencing sporangium formation of a *Phytophthora* species isolated from lucerne in certain salt solutions. Trans. Br. Mycol. Soc. 44:291-297.

Klebs, G. 1900. Zur Physiologie der Fortflanzung einiger Pilze. III. Allegemeine Betrachtungen. Jahrb. Wiss. Bot. 35:80-203.

Klisiewicz, J. M. 1970. Factors affecting production and germination of oospores of *Phytophthora drechsleri.* Phytopathology 60:1738-1742.

Klotz, L. J. 1940. Studies on *Phytophthora citrophthora.* (Abstr.) Phytopathology 30:14.

Leal, J. A., and Gomez-Miranda, B. 1965. The effect of light and darkness on the germination of the oospores of certain species of *Phytophthora* on some synthetic media. Trans. Br. Mycol. Soc. 48:491-494.

Leary, J. V., Roheim, J. R., and Zentmyer, G. A. 1974. Ribosome content of various spore forms of *Phytophthora* species. Phytopathology 64:404-408.

Leonian, L. H. 1925. Physiological studies on the genus *Phytophthora.* Am. J. Bot. 12:444-498.

Leonian, L. H. 1927. The effect of different hosts upon the sporangia of some *Phytophthora* species. Phytopathology 17:483-490.

Leonian, L. H. 1934. Identification of *Phytophthora* species. W. Va. Univ. Agric. Exp. Stn. Bull. 262. 35 pp.

Lilly, V. G. 1966. The effects of sterols and light on the production and germination of *Phytophthora* spores. Pages 259-271 in: The Fungus Spore. M. F. Madelin, ed. Butterworths, London.

Long, M., Keen, N. T., Ribeiro, O. K., Leary, J. V., Erwin, D. C., and Zentmyer, G. A. 1975. *Phytophthora megasperma* var. *sojae.* Development of wild-type strains for genetic research. Phytopathology 65:592-597.

MacDonald, J. D., and Duniway, J. M. 1978. Influence of the matric and osmotic components of water potential on zoospore discharge in *Phytophthora.* Phytopathology 68:751-757.

Manning, W. J., and Crossan, D. F. 1966. Effects of a particular soil bacterium on sporangial production in *Phytophthora cinnamomi* in liquid culture. Phytopathology 56:235-237.

Marx, D. H., and Bryan, W. C. 1969. Effect of soil bacteria on the mode of infection of pine roots by *Phytophthora cinnamomi.* Phytopathology 59:614-619.

Marx, D. H., and Haasis, F. A. 1965. Induction of aseptic sporangial formation in *Phytophthora cinnamomi* by metabolic diffusates of soil microorganisms. Nature (London) 206:673-674.

Mehrlich, F. P. 1935. Nonsterile soil leachate stimulating to zoosporangia production by *Phytophthora* sp. Phytopathology 25:432-435.

Menyonga, J. M. 1966. Nature of response of zoospores of *Phytophthora parasitica* to soil fungistasis. Ph.D. dissertation. Univ. Calif., Riverside. 106 pp.

Meyer, D. 1975. Zur Keimung der Oosporen von *Phytophthora*

*cactorum* (Leb. et Cohn) Schroet. Z. Pflanzenkr. Pflanzenschutz 82:168-172.

Mircetich, S. M., and Zentmyer, G. A. 1969. Effect of carbon and nitrogen compounds on germination of chlamydospores of *Phytophthora cinnamomi* in soil. Phytopathology 59:1732-1735.

Mircetich, S. M., Zentmyer, G. A., and Kendrick, J. B., Jr. 1968. Physiology of germination of chlamydospores of *Phytophthora cinnamomi*. Phytopathology 58:666-671.

Mitchell, D. J., and Zentmyer, G. A. 1971. Effects of oxygen and carbon dioxide tensions on sporangium and oospore formation by *Phytophthora* species. Phytopathology 61:807-812.

Pethybridge, G. H. 1914. Further observations on *Phytophthora erythroseptica* and on the disease produced by it in the potato plant. Sci. Proc. R. Dublin Soc. N.S. 14:179-198.

Pfender, W. F., Hine, R. B., and Stanghellini, M. E. 1977. Production of sporangia and release of zoospores by *Phytophthora megasperma* in soil. Phytopathology 67:657-663.

Platonova, T. A., Vasyukova, N. I., and Darydova, M. A. 1977. Effect of sterol deficiency on the sporulation of *Phytophthora infestans* (Mont.) De Bary. Prikl. Biokhim. Mikrobiol. 13:907-914. (In Russian)

Polach, F. J., and Webster, R. K. 1972. Identification of strains and inheritance of pathogenicity in *Phytophthora capsici*. Phytopathology 62:20-26.

Reeves, R. J. 1975. Behaviour of *Phytophthora cinnamomi* Rands in different soils and water regimes. Soil Biol. Biochem. 7:19-24.

Ribeiro, O. K., Gallegly, M. E., and Young, R. J. 1971. Oospore production, viability, and germination in relation to establishment of $F_2$ cultures of *Phytophthora infestans*. (Abstr.) Phytopathology 61:907-908.

Ribeiro, O. K., Erwin, D. C., and Zentmyer, G. A. 1975a. An improved synthetic medium for oospore production and germination of several *Phytophthora* species. Mycologia 67:1012-1019.

Ribeiro, O. K., Zentmyer, G. A., and Erwin, D. C. 1975b. Comparative effects of monochromatic radiation on the germination of oospores of three *Phytophthora* species. Phytopathology 65:904-907.

Ribeiro, O. K., Zentmyer, G. A., and Erwin, D. C. 1976a. The influence of qualitative and quantitative radiation on reproduction and spore germination of four *Phytophthora* species. Mycologia 68:1162-1173.

Ribeiro, O. K., Zentmyer, G. A., and Erwin, D. C. 1976b. Cytochromes in the mycelium and oospores of *Phytophthora capsici*. Phytopathology 66:172-174.

Rocha, H. M., and Machado, A. D. 1973. Influencia da luz, temperatura e umidade relativa na esporulacao do *Phytophthora* palmivora (Butl.) Butl. em frutos de cacau. Rev. Theobroma 3:22-25.

Romero, S., and Erwin, D. C. 1969. Variation in pathogenicity among single-oospore cultures of *Phytophthora infestans*. Phytopathology 59:1310-1317.

Rosenbaum, J. 1917. Studies of the genus *Phytophthora*. J. Agric. Res. 8:232-276.

Salvatore, M. A., Gray, F. A., and Hine, R. B. 1973. Enzymatically induced germination of oospores of *Phytophthora megasperma*. Phytopathology 63:1083-1084.

Satour, M. M., and Butler, E. E. 1968. Comparative morphological and physiological studies of the progenies from intraspecific matings of *Phytophthora capsici*. Phytopathology 58:183-192.

Schiffmann-Nadel, M., and Cohen, E. 1966. The effect of agar culture media and light on mycelial growth and sporangium formation of *Phytophthora citrophthora* (R. E. Sm. and E. M. Sm.) Leonian. Isr. J. Bot. 15:204-206.

Schmitthenner, A. F. 1959. The effect of media concentration on sporangia production in *Phytophthora*. (Abstr.) Phytopathology 49:550.

Schmitthenner, A. F. 1968. Significance of populations of major plant pathogens in soil: *Phytophthora* and *Pythium*. (Abstr.) Page 173 in: Proc. Int. Congr. Plant Pathol. 1st.

Schmitthenner, A. F. 1973. Isolation and identification methods for *Phytophthora* and *Pythium*. Proc. First Woody Ornamental Dis. Workshop 1:94-110.

Shaw, D. S. 1967. A method of obtaining single-oospore cultures of *Phytophthora cactorum* using live water snails. Phytopathology 57:454.

Singh, N. 1975. Effect of nutrition on growth and sporangial production of two West African isolates of *Phytophthora palmivora* Butl. Ann. Bot. 39:129-136.

Smoot, J. J., Gough, F. J., Lamey, H. A., Eichenmuller, J. J., and Gallegly, M. E. 1958. Production and germination of oospores of *Phytophthora infestans*. Phytopathology 48:165-171.

Sneh, B., and McIntosh, D. L. 1974. Studies on the behavior and survival of *Phytophthora cactorum* in soil. Can. J. Bot. 52:795-802.

Sterne, R. E., Zentmyer, G. A., and Kaufmann, M. R. 1977. The influence of matric potential, soil texture, and soil amendment on root disease caused by *Phytophthora cinnamomi*. Phytopathology 67:1495-1500.

Sussman, A. S., and Douthit, H. A. 1973. Dormancy in microbial spores. Annu. Rev. Plant Physiol. 24:311-352.

Tsao, P. H. 1969. Studies on the saprophytic behavior of *Phytophthora parasitica* in soil. Pages 1221-1230 in: Proc. Int. Citrus Symp. 1st, Vol. 3. H. D. Chapman, ed. Univ. Calif., Riverside.

Tsao, P. H. 1971. Chlamydospore formation in sporangium-free liquid cultures of *Phytophthora parasitica*. Phytopathology 61:1412-1413.

Tsao, P. H., and Bricker, J. L. 1968. Germination of chlamydospores of *Phytophthora parasitica* in soil. (Abstr.) Phytopathology 58:1070.

Tucker, C. M. 1931. Taxonomy of the genus *Phytophthora* de Bary. Mo. Agric. Exp. Stn. Res. Bull. 153. 208 pp.

Waterhouse, G. M. 1931. The production of conidia in the genus *Phytophthora*. Trans. Br. Mycol. Soc. 15:311-321.

Wills, W. H. 1954. Sporangium formation of *Phytophthora parasitica* Dastur var. *nicotianae* (Breda de Haan) Tucker. J. Elisha Mitchell Sci. Soc. 70:235-243.

Yamamoto, M., and Tanino, J. 1961. Physiological studies on the formation and germination of sporangia of *Phytophthora infestans* (Mont.) De Bary. Forsch. Geb. Pflanzenkr. (Shokubutsu Byogai Kenkyu) 7:7-22.

Yoshikawa, M., and Masago, H. 1977. Effect of cyclic AMP on catabolite-repressed zoosporangial formation in *Phytophthora capsici*. Can. J. Bot. 55:840-843.

Yoshikawa, M., Tanaka, H., Miyata, Y., and Masago, H. 1976. Evaluation by specific inhibitors of metabolic dependence of asexual differentiation processes in *Phytophthora capsici*. Sci. Rep. Kyoto Prefect. Univ. Agric. 28:57-62.

Zentmyer, G. A. 1965. Bacterial stimulation of sporangium production in *Phytophthora cinnamomi*. Science 150:1178-1179.

Zentmyer, G. A., and Erwin, D. C. 1970. Development and reproduction of *Phytophthora*. Phytopathology 60:1120-1127.

Zentmyer, G. A., and Marshall, L. A. 1959. Factors affecting sporangial production by *Phytophthora cinnamomi*. (Abstr.) Phytopathology 49:445.

Zentmyer, G. A., and Ribeiro, O. K. 1977. The effect of visible and near-visible radiation on sporangium production by *Phytophthora cinnamomi*. Phytopathology 67:91-95.

# 5

# Physiology of Sexual Reproduction in *Phytophthora*

Charles G. Elliott

## CONTENTS



## INTRODUCTION

Sexual reproduction in *Phytophthora,* resulting in the production of oospores, provides the means of long-term survival, as the oospores are resistant structures, often difficult to germinate. Genetic variation may also be generated.

To understand the physiology of sexual reproduction, we need to study the fungus in pure culture. In many investigations, complex media such as lima bean extract, carrot or oatmeal agar, or V-8 juice have been used, and much has been learned from them. However, for a more precise understanding of the biochemical processes involved, media of defined composition are required. Their use has shown up two features of particular interest: the importance of the nitrogen source and the need for an exogenous supply of sterol for reproduction to occur. Physical factors of the environment are also important.

## EFFECT OF TEMPERATURE AND LIGHT

Turner (1961) and Satchuthananthavale (1963) found that oospores of *P. palmivora* were formed at low temperatures (16 or 20° C) but not at higher temperatures (28° C), and the experiments of Brasier (1969) and Koffi-Dongo (cited by Huguenin and Boccas, 1971) confirmed that the optimal temperature for oospore formation in *P. palmivora* is lower than that for growth rate. This is also the case in *P. parasitica* (Honcur and Tsao, 1974) and *P. cinnamomi* (Zentmyer et al, 1979). On the other hand, the optimal temperatures for oospore formation and hyphal growth of *P. heveae* are the same (Brasier, 1969).

Harnish (1965) found that oospore production in several homothallic species occurred readily in the dark but was inhibited by continuous bright light (700 ft-c), though not in most cases by low-intensity light (3 ft-c). Similar results were obtained for heterothallic species by Mertz (cited by Lilly, 1966). Harnish (1965) and Brasier (1969) noted that whereas light inhibited oospore production, it stimulated sporangial production, and Harnish further noted that only blue light inhibited oospore formation and stimulated sporangial development; red light did neither.

Huguenin and Boccas (1971) studied the effect of light intensity and observed that, over a considerable range of intensities, the inhibitory effect of constant illumination on oosporogenesis was double that of 12 hr of light and 12 hr of darkness. Thus the Bunsen-Roscoe law applies, and this is prima facie evidence that the inhibition is due to a simple photochemical reaction.

The effect of different wavelengths was studied in more detail by Huguenin and Jacques (1973) and Ribeiro et al (1976). These studies established that red and far-red light stimulated oospore production above the level at which it occurred in the dark. In *P. palmivora,* the greatest stimulation occurred at 630 nm (Huguenin and Jacques, 1973). In *P. cinnamomi,* Ribeiro et al (1976) found stimulation even by daylight and blue light at low intensity.

One possible interpretation of these results is that in mated cultures of heterothallic species, the normal course of development leads to sexual reproduction but that the sequence of events can be interrupted by blue light of sufficient intensity, which diverts the fungus resources into sporangium formation (Huguenin and Jacques, 1973).

More recently, the effects of light and temperature on heterothallic species have been studied by Yu et al (1981), with the cultures of the compatible strains separated by a polycarbonate membrane. As will be shown below, in this system a substance (hormone) produced by one strain diffuses through the membrane and induces the other strain to produce sexual organs and oospores by selfing. In their experiments, Yu and co-workers used *P. colocasiae* A2 as the hormone producer to induce oospore production in *P. parasitica* A1. They found that when the cultures were paired for a short time (24 hr) and the inducing strain was then removed, oospores developed in the *P. parasitica* culture, provided the pairing took place in the dark; light had little effect on the maturation of oospores, but it greatly reduced effective induction. They also found that hormone production was optimal at 25° C and very poor at 30° C; however, oospores developed well if, after induction at 25° C, the induced culture was incubated at 30° C.

## NITROGEN METABOLISM AND REPRODUCTION

Leal and Gomez-Miranda (1967) investigated oospore formation in cultures of *P. heveae, P. cactorum,* and *Pythium butleri* grown with different nitrogen sources. With some, growth was poor, but growth and oospore production were said not to be correlated, although no figures were given. Reproduction was poor or did not occur with valine and leucine. This was attributed to the toxicity of their metabolites, isobutyrate and isovalerate, which could be formed from them by deamination and decarboxylation and which were found in culture filtrates (Leal et al, 1966). Aspartic and glutamic acids were good nitrogen sources; this was thought to be correlated with the beneficial effects of adding Krebs cycle acids to a medium with ammonium sulfate. (An exception was oxalacetate, which is the deamination product of aspartate.) Serine also was generally favorable to reproduction. Leal and Gomez-Miranda (1967) found that in most cases oospore formation declined with increasing concentration of the nitrogen source. Leal et al (1967) showed that a low C/N ratio (glucose, 2 g/L; asparagine, 1 g/L) was unfavorable to reproduction in many species, but oospores developed at high C/N ratios (e.g., glucose, 30 g/L; asparagine, 2 g/L). These results were attributed to ammonia toxicity (Leal et al, 1970). The detailed picture, however, varies from species to species (Leal et al, 1967).

The study of *Pythium* sp. PRL 2142 (*Pythium acanthicum*) by Child et al (1969) is more detailed than that on any phytophthora. This study gives figures for both mycelial growth and oospore production over a wide range of C/N ratios on three nitrogen sources; the ratio was altered both by varying the sugar concentration (nitrogen constant) and by varying the nitrogen concentration (sugar constant). The C/N ratio that gave optimal growth also gave maximal oospore production. Asparagine was a better nitrogen source for both growth and oospore production than alanine or arginine; leucine and valine gave poor growth and few oospores.

To understand the effect of the nitrogen source on reproduction, we apparently need much more precise information on the correlation between growth (properly measured, preferably as dry weight) and oospore production, to say nothing of knowledge of how the amino acid or other nitrogen source is metabolized.

TABLE 1. Activity of sterols in inducing oospore formation in *Phytophthora cactorum*

| Group[a] | Sterol | Structural difference from 5α-cholestan-3β-ol |
|---|---|---|
| 1 | Sitosterol | $\Delta^5$, 24α-ethyl |
| | Clionasterol | $\Delta^5$, 24β-ethyl |
| | Stigmasterol | $\Delta^{5,22}$, 24α-ethyl |
| | Poriferasterol | $\Delta^{5,22}$, 24β-ethyl |
| | Spinasterol[b] | $\Delta^{7,22}$, 24α-ethyl |
| | Fucosterol[c] | $\Delta^5$, (24E) 24-ethylidene |
| 2 | Chondrillasterol | $\Delta^{7,22}$, 24β-ethyl |
| | Campesterol | $\Delta^5$, 24α-methyl |
| | Stigmastanol | 24α-ethyl |
| 3 | $\Delta^5$-Ergostenol | $\Delta^5$, 24β-methyl |
| | Brassicasterol | $\Delta^{5,22}$, 24β-methyl |
| | $\Delta^7$-Ergostenol[b] | $\Delta^7$, 24β-methyl |
| | Ergosterol[b] | $\Delta^{5,7,22}$, 24β-methyl |
| | Cholesterol | $\Delta^5$ |
| | $\Delta^7$-Cholestenol | $\Delta^7$ |
| | 7-Dehydrocholesterol | $\Delta^{5,7}$ |
| | *trans*-22-Dehyrocholesterol | $\Delta^{5,22}$ |
| | Campestanol[d] | 24α-methyl |
| 4 | Cholestanol | ... |
| | Ergostanol | 24β-methyl |
| | 20-Isocholesterol | $\Delta^5$, 20S |
| | 20α-Hydroxycholesterol | $\Delta^5$, 20α-OH |
| | 29-Hydroxycholesterol | $\Delta^5$, 29-OH |
| | Saringosterol | $\Delta^5$, 24ξ-OH, 24-vinyl |
| | $\Delta^4$-Cholesten-3-one | $\Delta^4$, 3-oxo |
| | 7-Oxocholesterol | $\Delta^5$, 7-oxo |

[a] Sterols in group 1 have the highest activity; those in group 4 have low or no activity. Grouping is according to Nes et al (1980), with additions from Elliott (1972, 1979) and other information.
[b] These sterols produced fewer oospores than others of the group, according to Elliott (1972, 1979).
[c] Fucosterol was placed in a lower group by Nes et al (1980) but was found equal to sitosterol by Elliott (1972).
[d] Campestanol produced about half the number of oospores recorded for cholesterol (Nes et al, 1980).

## EFFECT OF GASEOUS ENVIRONMENT

Mitchell and Zentmyer (1971) found that more oospores were produced in an atmosphere of 1 or 5% oxygen in nitrogen than in air. Oospores could be produced in $CO_2$ concentrations of up to 15%, but the effect of $CO_2$ became more deleterious as oxygen concentration increased. The optimal conditions, they thought, were those to be expected within infected plant tissues.

## STEROL REQUIREMENTS

A number of species of *Phytophthora* (both homothallic and heterothallic) and of *Pythium* have been found to require addition of sterol to the medium for reproduction to occur. The fungi are unable to synthesize sterols, as shown by the absence of detectable sterol when the fungus is grown on a sterol-free medium, and by failure of incorporation of labeled mevalonate or acetate into a fraction with properties of sterols (see Hendrix [1970] for references). *P. cinnamomi* synthesizes squalene but does not convert it to squalene epoxide or cyclize it (Wood and Gottlieb, 1978a, 1978b). Furthermore, lanosterol and cycloartenol, when taken up by the mycelium, appear not to be converted to sterols by *P. cactorum* (Nes and Patterson, 1981), although Langcake (1975) claimed that some metabolic conversion of added lanosterol occurs in *P. infestans*. This synthetic deficiency is made good in nature by the host plant, and adaptation of fungus to host is shown in that the sterols most effective in inducing oospore formation are the common sterols of higher plants, especially sitosterol and stigmasterol.

The activity of various sterols has been studied mainly in *P. cactorum* (Elliott, 1972, 1979; Elliott et al, 1966; Nes et al, 1980) (Table 1). The principal features required for high oospore-inducing activity are a hydroxyl group at C-3, the absence of methyl groups at C-4 and C-14, and the presence of a double bond at C-5 and an alkyl substituent at C-24. (The numbering of carbon atoms is shown in Figure 1.) A 24-ethyl substituent gives greater activity than methyl. The orientation is also important; 24α-methyl sterols are more active than 24β-methyl, but among $\Delta^5$-sterols, the difference in the effects of 24α-ethyl and 24β-ethyl substituents is less. Nes et al (1980) found $\Delta^7$-sterols to be as active as $\Delta^5$-sterols, but they tested their compounds at 10 mg/L only; I found the activity of $\Delta^7$-sterols to fall off with decreasing concentration more rapidly than with $\Delta^5$-sterols (Elliott, 1979). *Trans* double bonds at C-22 have no effect on activity. Inversion of the stereochemistry at C-20 (20-isocholesterol) abolishes activity, as do hydroxyl groups at C-20, C-24, and C-29.

### Hypothesis That Sterols Are Precursors of Hormones Controlling Reproduction

But what is the function of sterols in reproduction? Before this question can be answered, it must be emphasized that many sterols promote vegetative growth as well as reproduction. Their effect on vegetative growth is assumed to be due to their ability to fit well into phospholipid membranes and affect membrane fluidity and permeability without undergoing chemical change themselves (Nes et al, 1979). The features of sterols that have high activity in inducing reproduction are in many respects also those that confer the ability to fit well into phospholipid membranes. Thus, sterol-induced reproduction could be merely the result of improved nutrition, except that there are cases of lack of agreement between effects on growth and reproduction. The principal examples of this lack of agreement concern 1) sitosterol and cholesterol, 2) sitosterol and stigmasterol, 3) cholesterol and cholestanol, and 4) certain triterpenoids.

Information is available on the comparative effects of cholesterol, sitosterol, stigmasterol, and cholestanol on model membrane systems, namely, lipid monolayers spread on water (the area occupied by the molecules is measured) and lipid bilayers in the form of liposomes (by measuring permeability and electron spin resonance with spin labels). These studies indicate that, in general, cholesterol, sitosterol, and cholestanol have similar effects, but stigmasterol and ergosterol are less effective (Butler and Smith, 1978; Butler et al, 1970; Demel et al, 1972a, 1972b; Ghosh and Tinoco, 1972). The effect of ergosterol as compared to cholesterol results from the difference in the side chain, not the B ring (Semer and Gelerinter, 1979). Cholesterol and sitosterol make liposomes more susceptible to filipin, but stigmasterol and ergosterol make them less so (Kleinschmidt et al, 1972). The effects of incorporating these sterols into the membranes of mycoplasmas on the growth of these organisms have also been studied, with generally similar conclusions (Lemcke and Burrows, 1980; Rottem et al, 1971; Smith, 1964; Smith and Lynn, 1958).

Thus, work with model membranes shows that the effects of cholesterol and sitosterol on the membranes' physical properties are similar, whereas the effects of stigmasterol and sitosterol differ. However, in *P. cactorum,* sitosterol promotes oospore formation much more effectively than cholesterol, and stigmasterol's activity is similar to that of sitosterol.

When we consider the effects of these sterols on vegetative growth in *Phytophthora* and ask whether this reflects the model membrane work, we are faced with the decision of how to measure growth. Nes et al (1979) measured the dry weight of cultures of *P. cactorum* and found no difference in the effects of sitosterol and cholesterol on growth rate. The results in Table 2, however, indicate that hyphal extension is more rapid with sitosterol than with cholesterol. Table 2 also shows an increased extension rate with sterols with a double bond at C-22 (stigmasterol and brassicasterol) compared to those with saturated side chains (sitosterol and $\Delta^5$-ergostenol). Langcake (1974) also recorded faster hyphal growth with stigmasterol than with sitosterol but no difference in dry weight. A double bond at C-22 of the sterol is known to affect the stretchability of the protoplast of yeast (Hossack and Rose, 1976), so perhaps the difference in extension growth of *Phytophthora* is a related phenomenon. However, these differences do not appear to be associated with a difference in reproduction. Why extension growth is faster with sitosterol than with cholesterol is not clear. Possibly an association between sterols and membrane proteins could be important (Nes and Heftmann, 1981).

Cholesterol and cholestanol have very different effects on reproduction in *P. cactorum.* With cholestanol, most of the oogonia abort at the stage when meiosis is about to begin (Elliott, 1968, 1972; Elliott and Sansome, 1977). Although Nes et al (1979) found growth rate to be lower with cholestanol, my data (Elliott, 1968 [dry weight]; Elliott et al, 1966 [colony diameter]) suggest little difference.

Finally, tetrahymanol and the triterpenoids $\alpha$-amyrin and $\beta$-amyrin stimulate the growth of *P. cactorum,* apparently being incorporated into membranes without undergoing any chemical change, but they do not induce oospore formation (Nes and Patterson, 1981) or do so only very poorly (Nes et al, 1980).

These results all point to the conclusion that the effect of sterols on reproduction is not so much due to the properties conferred on membranes by their presence but to some other activity, perhaps to their transformation to hormones. A compelling argument in favor of a hormonal role for sterols is that in another oomycete, *Achlya,* reproduction is controlled by steroid hormones, which are derived biosynthetically from sterols synthesized by the fungus. Because *Phytophthora* spp. are heterotrophic for sterols, the sterol that has to be supplied in the medium would be the precursor for any steroid hormones.

*Achlya* synthesizes fucosterol, from which the steroid hormones antheridiol and oogoniol are derived (McMorris, 1978) (Fig. 2). In *A. ambisexualis* and *A. bisexualis,* we find "strong males" and "strong females," which act as antheridial and oogonial parents, respectively, when mated. Antheridiol is produced by the female strain and taken up by the male strain, which responds by synthesizing rRNA and, later, proteins (including cellulase) and by producing antheridial branches and the hormone oogoniol. Oogoniol induces the differentiation of oogonia in the female plant. Continued

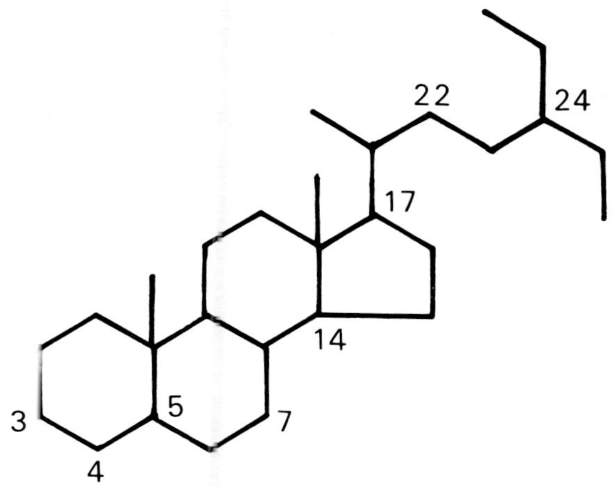


**Fig. 1.** Numbering of the carbon atoms of the sterol molecule, indicating the positions of importance affecting reproduction in *Phytophthora.*

TABLE 2. Mean rate of growth (mm/day) of hyphae of *Phytophthora cactorum* with various sterols[a]

| Sterol added | Experiment 1 | Experiment 2 | Experiment 3 |
|---|---|---|---|
| Cholesterol | 5.50 | 5.56 | 5.01 |
| $\Delta^5$-Ergostenol | 5.81 | 6.14 | 4.83 |
| Brassicasterol | 6.16 | 6.63 | 5.25 |
| Sitosterol | 6.23 | 5.83 | 5.80 |
| Stigmasterol | 6.36 | 7.77 | 6.20 |
| None | 4.94 | 5.37 | 4.55 |
| Least significant difference (5% level) | 0.154 | 0.204 | 0.270 |

[a] Colonies were grown on sucrose-asparagine medium supplemented with sterol at 10, 4, 1.6, and 0.64 mg/L (experiments 1 and 2) or 10, 3.3, and 1.1 mg/L (experiment 3). The effect of sterol concentration was very small, and the growth rates shown are means over all concentrations for 32 dishes in experiment 1, 20 in experiment 2, and 27 in experiment 3. (Data from C. G. Elliott, *unpublished*)

production of antheridiol by the female plant leads to directed growth of the antheridial branches to the oogonia and the delimitation of antheridia (Barksdale, 1967; McMorris, 1978). Another hormone, hormone D of Raper (1939), may be responsible for delimitation of oospheres in the oogonia.

Another important feature of reproduction in *Achlya* is the occurrence of "relative sexuality," by which we mean that in addition to strong males and strong females, there are intermediates that act as female to strong males and as male to strong females (Barksdale, 1960; Raper, 1940). Their behavior depends on their production of, and response to, antheridiol and oogoniol (Barksdale and Lasure, 1973, 1974).

Hormonal control of reproduction in other oomycetes is supported by strong evidence in *Dictyuchus* (Sherwood, 1966) and *Sapromyces* (Bishop, 1940) and by suggestive evidence in the heterothallic *Pythium sylvaticum*. In this last, interaction of male and female strains was observed before their hyphae touched (Papa et al, 1967); however, according to Pratt and Green (1971), the male and female hyphae of *Pythium sylvaticum* make contact before antheridial hyphae proliferate. Relative sexuality is also commonly observed in these fungi—in *Dictyuchus* (Couch, 1926), *Sapromyces* (Bishop, 1940), and *Pythium sylvaticum* (Pratt and Green, 1971).

## Metabolism of Sterols in Relation to Reproduction

### Cholesterol and Sitosterol

We have attempted to explain the different effects of cholesterol and sitosterol on reproduction by looking for differences in their metabolism (Elliott and Knights, 1981b). When *P. cactorum* is fed a mixture of the two, more of the sitosterol is esterified and more of it is converted to polar metabolites, which suggests that sitosterol may be a better substrate for the enzymes synthesizing the esters and polar metabolites. We have also found (B. A. Knights and C. G. Elliott, *unpublished*) quantitative differences in the amounts of various polar metabolites formed by these sterols, again suggesting differences in substrate efficiency. If the polar metabolites include the precursors of hormones, the greater production of polar metabolites from sitosterol could explain its greater activity in promoting reproduction.

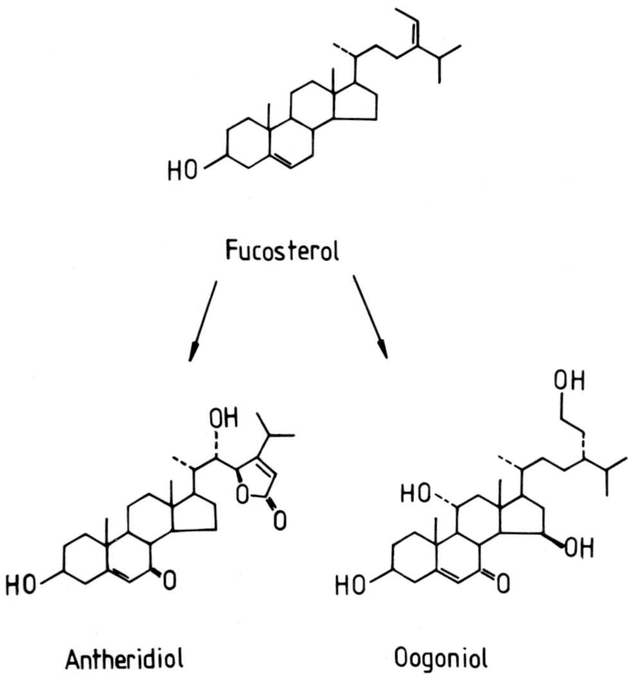


**Fig. 2.** Structure of fucosterol and the sexual hormones of *Achlya*, antheridiol and oogoniol, derived from it.

### Cholesterol and Cholestanol

The addition of a small amount of cholestanol to a medium with a low concentration of cholesterol can increase oospore production; adding more cholestanol is inhibitory (Elliott, 1968). This observation was interpreted as indicating two distinct functions for sterols (Elliott, 1968); i.e., 1) oogonium and antheridium formation, and 2) meiosis, oosphere, and oospore development. We suggested (Elliott and Sansome, 1977) that although both could function as precursors of hormones controlling the first stage, cholesterol could function as a precursor of the hormone(s) controlling the second stage, but cholestanol could not.

We have looked for differences in the metabolism of cholesterol and cholestanol (B. A. Knights and C. G. Elliott, *unpublished*). A much smaller proportion of cholestanol than of cholesterol is converted to polar material (Table 3). Figure 3 shows column chromatograms of the polar material formed from these two sterols and thin-layer chromatograms of the several zones from the cholesterol column. Seven compounds can be recognized in the cholesterol derivatives. The difference between cholesterol and cholestanol is in zones A and B. Zones C–E show the same three compounds with $R_f$ between 0.4 and 0.5 (acyl steryl glycosides). Cholestanol also gives rise to the compound in zone A with $R_f$ 0.35 (steryl glycoside), but the conspicuous compounds of zone B with $R_f$ 0.7–0.85 are absent from the cholestanol polar metabolites. One of these is possibly 7-oxocholesterol; its production would be greatly facilitated by the presence of the double bond at C-5. 7-Oxocholesterol as sole supplement of the medium does not promote oospore formation, but we wondered whether it might promote oospore formation when added to cultures grown with cholestanol. We added cholestanol to the medium at the beginning and the 7-oxocholesterol at the beginning (solid and liquid media) or after several days' growth (liquid media) but no oospores were produced. When cholesterol was added in the same way as the 7-oxocholesterol, oospores were produced. Despite the negative results of these experiments with 7-oxocholesterol and of similar ones with $\Delta^4$-cholesten-3-one, we feel strongly that the differences in metabolism of cholesterol and cholestanol must be associated with their different effects on reproduction.

Nes et al (1980) reported that numerous oospores form in cultures supplemented with stigmastanol, although fewer than in cultures with sitosterol. Evidently, the 24$\alpha$-ethyl group to some extent overcomes the disability conferred by lack of the double bond at C-5.

TABLE 3. Metabolism of cholesterol and cholestanol by *Phytophthora cactorum*[a]

| Supplement | Polar | Free sterol | Unmetabolized acetate | Long chain esters |
|---|---|---|---|---|
| Cholesteryl acetate | 4.3 | 56.3 | 4.2 | 35.2 |
| Cholestanyl acetate | 1.4 | 53.5 | 6.6 | 38.6 |

[a] Fungus was grown for 7 days on sterol-free medium; [$^{14}$C]cholesteryl acetate or [$^{14}$C]cholestanyl acetate was then added. Mycelia were harvested after 48 hr, dried, and extracted, first with acetone and then with chloroform-methanol (2:1). Combined extracts were analyzed by reverse-phase chromatography on lipophilic Sephadex, as described by Elliott and Knights (1981a). Data are percent of total radioactivity in the four zones. (Data from B. A. Knights and C. G. Elliott, *unpublished*)

**Reproduction with Δ⁷-Sterols: Effect of AY 9944**

Δ⁷-Sterols at low concentration promote development of fewer oospores than do the corresponding Δ⁵-sterols (Elliott, 1979). Δ⁷-Sterols are metabolized to Δ⁵-sterols; mycelium supplied with Δ⁷-sterols contains both Δ⁷-sterol and Δ⁵-sterol (Knights and Elliott, 1976). Because a double bond at C-5 apparently plays a critical role in the control of reproduction, the relative ineffectiveness of Δ⁷-sterols could well be the result of their inadequate conversion to Δ⁵-sterols. Especially at low sterol levels, not enough Δ⁵-sterols would be produced from the Δ⁷-compound, as compared to the amount in mycelium supplied directly with Δ -sterols. This idea was examined by treating mycelium supplied with Δ⁷-sterols with the compound AY 9944, which inhibits the saturation of the double bond at C-7 in rat tissues (Dvornik et al, 1966; Horlick, 1966). (In the alga *Chlorella ellipsoidea,* AY 9944 treatment reduces the amount of Δ⁵-sterol, but whether it inhibits Δ⁷-reductase is not clear, because the earlier steps in biosynthesis, Δ⁸→Δ⁷ isomerization and Δ¹⁴ reduction, are inhibited [Patterson et al, 1974].) Some degree of inhibition by AY 9944 of conversion of Δ⁷-ergostenol to Δ⁵-ergostenol in *P. cactorum* is shown in Table 4. Figure 4 shows the relation between concentration of AY 9944 and growth rate and oospore production for *P. cactorum* supplied with Δ⁷-ergostenol and Δ⁵-ergostenol. Growth rate is slower when Δ⁷-sterol is added to the medium and, with both sterols, is decreased by addition of AY 9944. Significantly, however, oospore production is reduced at much lower concentrations (fivefold) of AY 9944 with Δ⁷-ergostenol than with Δ⁵-sterol and at concentrations that have little effect on growth rate. However, the effects of AY 9944 on reproduction are not significantly different with Δ⁷-campestenol and Δ⁵-campestenol, perhaps because the activity of 24α-methyl sterols is much greater than that of 24β-methyl sterols. Child and Haskins (1976) found that AY 9944 inhibited the growth and reproduction of *Pythium acanthicum* growing on medium with cholesterol. They concluded that the inhibition was nonspecific, as indeed would be the case because the sterol supplement was not a Δ⁷-sterol.

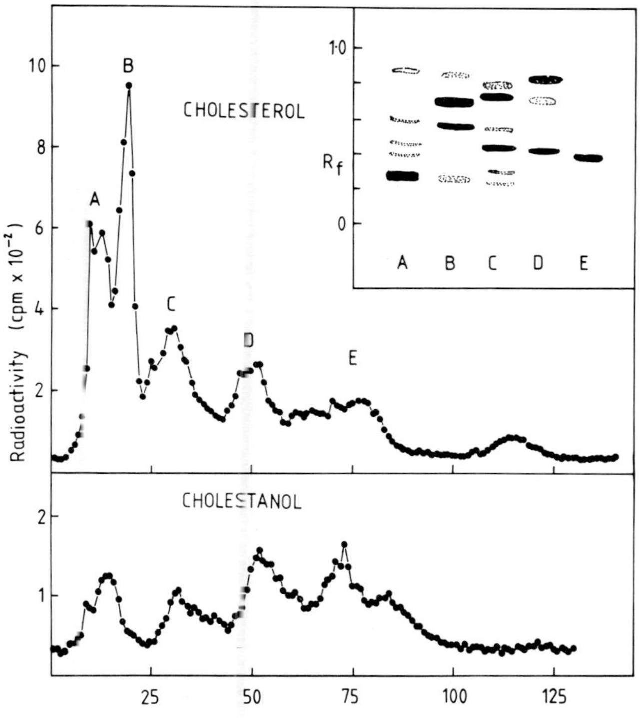


**Fig. 3.** Chromatograms of polar metabolites of cholesterol and cholestanol. Cultures of *Phytophthora cactorum* on sterol-free medium were supplemented at 7 days with [¹⁴C]cholesteryl acetate or [¹⁴C]cholestanyl acetate and were harvested 48 hr later. Mycelial extracts were chromatographed on lipophilic Sephadex with acetone-heptane-water (220:23:5 , as described by Elliott and Knights (1981a). The polar material obtained from this column was rechromatographed on lipophilic Sephadex with acetone-water (9:1) as mobile phase; the results are shown in the figure. The zones (A–E) were subjected to thin-layer chromatography on silica gel, developed with chloroform-methanol (9:1). Thin-layer chromatography of the cholesterol zones is shown in the inset. (Courtesy B. A. Knights and C. G. Elliott)

**Inhibition of Reproduction by Estradiol**

Hendrix and Guttman (1968, 1969) showed that estradiol inhibits cholesterol-induced oospore production in several *Pythium* species, although not in *Phytophthora cactorum.*

TABLE 4. Inhibition by AY 9944 of conversion of Δ⁷-ergostenol to Δ⁵-ergostenol in *Phytophthora cactorum*[a]

| Concentration of AY 9944 (mg/L) | Replication | Free sterol in extract (μg/mg of dry mycelium) Δ⁵ | Δ⁷ | Ratio |
|---|---|---|---|---|
| 100 | 1 | 0.08 | 0.35 | 0.24 |
| | 2 | 0.07 | 0.29 | 0.24 |
| 25 | 1 | 0.11 | 0.26 | 0.40 |
| | 2 | 0.12 | 0.37 | 0.32 |
| 6.25 | 1 | 0.16 | 0.39 | 0.41 |
| | 2 | 0.13 | 0.33 | 0.40 |
| 0 | 1 | 0.20 | 0.28 | 0.73 |
| | 2 | 0.13 | 0.35 | 0.39 |

[a]Fungus was grown for 6 days on sterol-free liquid medium; Δ⁷ ergostenol (10 mg/L) and AY 9944 at stated concentration were then added. Mycelia were harvested after 5 days' further incubation, dried, and extracted with acetone. Extracts were analyzed by gas-liquid chromatography with added stigmasterol as standard. (Data from C. G. Elliott, *unpublished*)

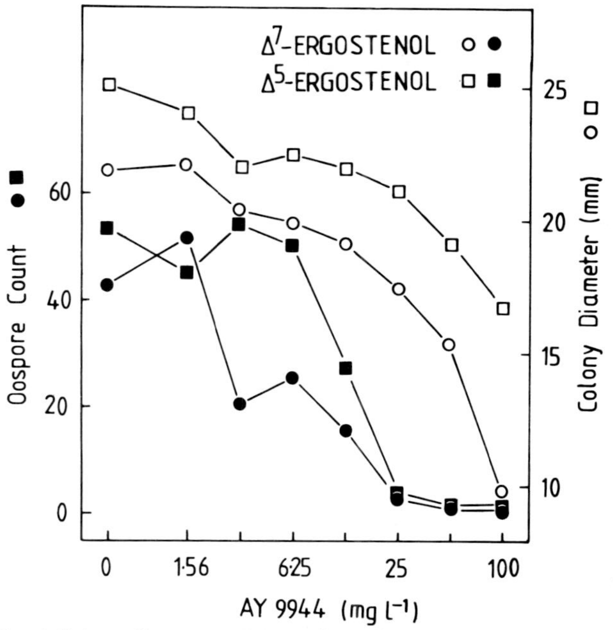


**Fig. 4.** Colony diameter at 7 days (O——O, □——□) and oospore production (●——●, ■——■) in *Phytophthora cactorum* with Δ⁷-ergostenol (O——O, ●——●) and Δ⁵-ergostenol (□——□, ■——■), as affected by AY 9944.

However, I have found that reproduction is inhibited in *P. cactorum* by estradiol and estrone and to a lesser extent by progesterone (Table 5). This result is not easy to interpret. However, although 17$\beta$-estradiol is inhibitory, 17$\alpha$-estradiol and 17$\beta$-estradiol-17-acetate are not (Hendrix and Guttman, 1972). The side chain of cholesterol has the 17$\beta$-configuration, so the effect may be related to improper fit to some binding site of a molecule with a hydroxyl group rather than a hydrocarbon side chain in this configuration. A similar situation is found in the lack of activity of 20$\alpha$-hydroxycholesterol in promoting reproduction (Nes et al, 1980). It would be interesting to know whether the latter compound also inhibits reproduction induced by cholesterol.

## HETEROTHALLIC SPECIES

Ashby (1922) was the first to find that isolates of *P. palmivora* that did not produce oospores in single culture could do so when paired with a complementary strain of the opposite mating type. This heterothallic behavior was subsequently found in a number of species. In *P. infestans,* the existence of two mating types was demonstrated by Smoot et al (1958) and Galindo and Gallegly (1960). The same two groups (called A1 and A2) were demonstrated in other species by Savage et al (1968) as a result of pairing isolates of heterothallic phytophthoras with two standard tester strains of *P. infestans;* the interaction is interspecific as well as intraspecific.

Galindo and Gallegly (1960) showed that hybridization occurred between strains of opposite mating type, one strain providing the antheridium, the other the oogonium, and genetic recombination has been demonstrated in the progeny of crosses (see Shaw, Chapter 6). Galindo and Gallegly also showed that the mated strains were bisexual; in a particular mating, the A1 strain sometimes produced the antheridium and the A2 the oogonium, and sometimes the reverse. However, they also showed the occurrence of relative sexuality: certain isolates were strong males, generally forming the antheridia, whereas others were strong females and still others were intermediate. Relative sexuality was also demonstrated in *P. palmivora* by Huguenin (1973). The situation thus resembles that in *Achlya,* except that superimposed on relative sexuality is an incompatibility system. There is a gradation from strong male to strong female *within each mating type.*

TABLE 5. Effect of various steroids on cholesterol-induced oospore production and mycelial growth of *Phytophthora cactorum*[a]

| Steroid | Concentration of steroid (mg/L) 0 | 1 | 2 | 4 | 8 |
|---|---|---|---|---|---|
| Oospore counts | | | | | |
| $\Delta^5$-Pregnenolone | 59.4 | 55.8 | 57.0 | 43.5 | 43.3 |
| Progesterone | 59.4 | 52.2 | 60.8 | 30.8 | 19.8 |
| Estradiol | 51.8 | 35.8 | 24.4 | 13.2 | 0.2 |
| Estrone | 51.8 | 43.7 | 39.3 | 10.2 | 0.9 |
| Colony diameter (mm) | | | | | |
| Estradiol | 30.4 | 26.4 | 23.8 | 23.2 | 21.8 |
| Estrone | 30.4 | 28.7 | 28.4 | 28.4 | 24.2 |

[a] All media contained cholesterol (10 mg/L), with added steroid at stated concentration. Oospore numbers are means of 20 counts (four in each of five petri dishes). Colony diameters are means of measurements made at 9 days. (Data from C. G. Elliott, *unpublished*)

However, in addition to hybridization, selfing of one or both partners in pairing of compatible strains can occur. This has been shown in interspecific matings by species-related differences in the morphology of oospores and hyphae (Stamps, 1953). Huguenin (1973) distinguished his strains of *P. palmivora* by labeling one of them with a fluorescent dye. Most convincingly, Sansome et al (1979) have demonstrated both crossing and selfing in pairings of *P. palmivora* (sensu restricto) and *P. megakarya,* by the differences in size and number of chromosomes of the strains, and in their nuclear morphology. Genetic evidence leading to the same conclusion has been given by Boccas (1981).

Thus, when compatible strains of opposite mating types interact, selfing of one or both partners and hybridization between them can occur; in the latter case, the frequency with which one forms the oogonium and the other the antheridium depends on their positions in the relative sexual scale. This mating behavior is well shown in the data of Huguenin (1973) for *P. palmivora* (Table 6). More data of this kind would be highly desirable.

The idea that the interacting strains induce each other to self was enunciated by Ashby in 1929 as follows:

> When strains of the two groups of *P. palmivora* are paired, the development of oospores after a few days is due to a biochemical stimulation of the one vegetation by the other, resulting in the development of both antheridia and oogonia on each mycelium or dominantly on one of them; this view does not exclude crossing but subordinates it to autogamy.

That a strain of one mating type can induce one of the opposite mating type to produce oospores by selfing has been demonstrated in the elegant experiments of Ko (1978), who placed the two strains on opposite sides of a polycarbonate membrane and observed oospore formation in each. The hyphae do not penetrate the membrane; a block of uninoculated medium remains sterile when placed on one side of a membrane with a culture on the other. The oospores, therefore, result from selfing. Something produced by one strain must diffuse through the membrane to induce reproduction in the other. No reproduction occurs when strains of the same mating type are paired across the polycarbonate membrane.

Ko has postulated a hormone $\alpha 1$, produced by strains of mating type A1, which induces reproduction in A2 strains, and a hormone $\alpha 2$, produced by A2 strains, which induces reproduction in A1 strains. Ko's results, and those of Yu and Chang (1980), show that the hormones can act interspecifically. Quantitative differences are found in the interactions of different strains, and A1s generally produce more spores than A2s. Young cultures are generally better producers of hormone, and older cultures are better responders (Ko, 1978). Differences in hormone production by a number of strains of

TABLE 6. Frequencies of selfing and crossing, and relative sexuality, in *Phytophthora palmivora*

| Cross A1 × A2 | Percent self-fertilized A1 | A2 | Percent crossed | A2 parent forming Oogonia | Antheridia |
|---|---|---|---|---|---|
| 570 × L | 6.7 | 33.2 | 60.1 | 71.5 | 28.5 |
| K × L | 1.3 | 28.6 | 70.0 | 36.8 | 63.2 |
| 26 × L | 1.0 | 33.0 | 66.0 | 3.1 | 96.9 |
| 570 × 36 | 0.9 | 48.5 | 50.7 | 73.0 | 27.0 |
| K × 36 | 0 | 33.3 | 66.7 | 100 | 0 |
| 26 × 36 | 2.7 | 59.0 | 38.3 | 62.0 | 38.0 |

*Source*: Huguenin (1973).

one mating type are shown by pairing each with a common strain of opposite mating type and observing the different numbers of oospores produced; similarly, differences in response of strains can be shown by pairing each with a common strain of opposite mating type as hormone producer (Ko and Kunimoto, 198 ).

Homothallic species, when paired with heterothallic strains across a polycarbonate membrane, can induce reproduction in the heterothallic strain (Ko, 1980; cf Brasier, 1972). The different kinds of strains that can be distinguished on the basis of production of $\alpha 1$ or $\alpha 2$ hormones and response to them are shown in Table 7 (Ko, 1980). Strains can be heterothallic because they lack receptors for the hormones they produce and so do not respond to them (types 1, 2, 4, 5, and 7) or because they do not produce hormones (types 3, 6, and 8). Homothallic strains respond to the hormones they produce, which may be $\alpha 1$, $\alpha 2$, or both. Four of the possible eight types of heterothallic strains have been found (Ko, 1980; Ko and Kunimoto, 1981). The homothallic species tested produced three kinds of response in the two mating type testers (spores produced in one, in the other, and in both), but which of the seven possible sexual types they belong to is uncertain (Ko, 1980).

Yu et al (1981) have shown that the responding strain need not be in contact with the hormone producer continuously. In their experiment, the hormone producer, *P. colocasiae* 1T, was removed from pairings with *P. parasitica* P991 after various times of contact; in 7 hr, enough hormone was transferred from the *P. colocasiae* culture to induce a small number of spores to form subsequently in *P. parasitica* (none was present when the cultures were separated). The numbers of oospores produced by *P. parasitica* increased with increasing time of contact with the homone producer up to 48 hr.

Thus, Ko's experiments show that a hormone produced by a strain of one mating type can induce reproduction in one of opposite mating type that is separated from it by a polycarbonate membrane. I have argued above from results with homothallic species that reproduction is controlled by steroid hormones. Presumably, these hormones are synthesized from sterols by enzymes whose production in homothallic species is normally derepressed. A plausible hypothesis is that in heterothallic species, the synthesis of such enzymes is normally repressed but that the $\alpha$ hormone brings about their derepression, and production of antheridia and oogonia in the induced strain then ensues. When hyphae of compatible strains can make contact, the $\alpha$ hormones produced by each mating type would similarly induce synthesis of hormones leading to antheridium and oogonium production, with the added possibility that the interacting antheridia and oogonia could now be from opposite strains. If this hypothesis, or something similar, is correct, we next need to know what determines the extent of selfing and crossing. (Relative sexuality determines the frequency with which each strain is the male or female partner when they cross.) Some light would be shed on this problem by comparing the interaction of strains when separated by a membrane and when able to make contact.

There are many reports of oospore formation in single cultures of species generally considered heterothallic (e.g., Brasier, 1972; Pethybridge and Murphy, 1913; Shepherd, 1978; Tsao et al, 1980), although such oospore production is generally less regular and less abundant than when the same strain is paired with one of opposite mating type (Savage et al, 1968).

Tsao et al (1980) have reported interesting experiments on formation of oospores in single culture of certain strains of both mating types in *P. parasitica* and *P. palmivora* MF 4. Oospores were produced in aged cultures, 6 mo old or older. Subcultures taken from the parts of the culture containing oospores themselves produced oospores, and this could be repeated for several transfers; however, subcultures from the margin of the colony did not produce oospores. Tsao et al suggested that some substance required for oospore production is produced only slowly but can accumulate to a concentration sufficient to initiate rapid oospore production in the subcultures. Is it possible that we have partial derepression of the synthesis of the postulated steroid hormones controlling antheridium and oogonium development? This and Tsao's interpretation are perhaps unlikely in the light of Ko's (1981) observation that both mating types can be recovered from the area producing oospores; we are looking, rather, at a case of mutation of the mating type genes.

TABLE 7. Possible types of chemically regulated sexuality among members of the genus *Phytophthora*

| Sexuality type | Production of hormones[a] | | Responsiveness to hormones[b] | |
|---|---|---|---|---|
| | $\alpha^1$ | $\alpha^2$ | $\alpha^1$ | $\alpha^2$ |
| Group I: Cross-induction ("heterothallic") | | | | |
| 1 | + | − | − | + |
| 2 | + | − | − | − |
| 3 | − | − | − | + |
| 4 | − | + | + | − |
| 5 | − | + | − | − |
| 6 | − | − | + | − |
| 7 | + | + | − | − |
| 8 | − | − | + | + |
| Group II: Self-induction ("homothallic") | | | | |
| 9 | + | − | + | − |
| 10 | − | + | − | + |
| 11 | + | − | + | + |
| 12 | − | + | + | + |
| 13 | + | + | + | − |
| 14 | + | + | − | + |
| 15 | + | + | + | + |
| Group III: Neuter | | | | |
| 16 | − | − | − | − |

[a]+ = Productive; − = nonproductive.
[b]+ = Responsive; − = nonresponsive.

*Source*: Ko (1980).

## Nature of the $\alpha$ Hormones

What is the chemical nature of the $\alpha 1$ and $\alpha 2$ hormones? Are they themselves steroids? We cannot make any deductions on this matter from Ko's experiments, because V-8 juice was used throughout. One needs to use synthetic medium, with and without sterol.

Pratt and Mitchell (1973) grew strains of *P. capsici* with and without sterol and paired them across a block of plain agar in which oospores developed. Most oospores were formed when both strains were grown on medium containing sterol, slightly fewer were formed when only one strain was on medium containing sterol, and very few were formed when neither strain was supplied with sterol, but it was not determined which strain was the oogonial parent or whether selfing was occurring. They did similar experiments with *Pythium sylvaticum* and found that oospores were formed when the oogonial parent was supplied with sterol, whether or not the antheridial parent had sterol. Furthermore, Child and Haskins (1971) found that in several heterothallic *Pythium* species, adding cholesterol to dilute potato medium greatly stimulated oospore formation in paired cultures of male and female strains but also led to oospore formation in single cultures of the females. These experiments indicate the importance of sterol for female development. Pratt and Mitchell (1973) thought that this could be because of the larger size of the oogonium. I think another possibility is that the sterol supplied to the female parent enhances the production of an antheridiol-like hormone, leading to stronger male development and hence to better oospore development.

In *Pythium,* the strains are simply male and female; mating types as they occur in *Phytophthora* are not found. In Ko's experiments, the point of interest is exchange of material, not between sexes but between mating types, and whether this is steroidal. We, therefore, investigated the transfer of labeled material between strains, one of which was grown with [$^{14}$C]cholesterol (Elliott and Glen, 1982). Cultures of *P. parasitica* strain P991 (mating type A1) and P731 (A2) were grown for 7 days and then paired across a polycarbonate membrane. The cultures were separated after 2 days, and the labeled material in each culture was measured. Most of the label was fixed in the mycelium during the 7 days' growth, but about 1.9% was transferred to the compatible strain growing on unlabeled medium containing sterol; rather more was transferred if the unmarked strain was on sterol-free medium. Analysis of extracts of the two strains indicated that the material transferred was metabolized in the way typical of cholesterol; there was no indication of transfer of anything labeled other than cholesterol or its esters, although our experiments so far would hardly have revealed the transfer of a small quantity of a sterol metabolite.

We have also investigated the development of oospores in *P. parasitica* when the two strains were grown on medium with or without sterol (Elliott and Glen, 1982). Strains P991 and P731 were grown for 7 days and then paired across a polycarbonate membrane. Abundant oospores were formed by P991 when both strains were grown on medium with cholesterol or sitosterol and when P991 was on medium containing sterol and P731 was on medium without sterol. When P731 was on medium containing sterol and P991 was on medium without sterol, P991 formed fewer oogonia, and a smaller proportion of them developed into oospores. When neither strain was on medium containing sterol, only a few oogonia and no oospores were formed. This suggests that production of $\alpha 2$ hormone does not require sterol, but that response to it (or the consequences of response) does. The reproduction of P991 on sterol-free medium when paired with P731 on medium with sterol presumably depends on the sterol transferred from P731. Perhaps, sterol transferred from P991 to P731 could affect production of $\alpha 2$ hormone by the latter strain. In these experiments, the strains were in contact through the membrane for 7 days. We also did an experiment in which 7-day-old cultures were paired for short periods of time; the P731 culture was then removed, and the P991 culture was incubated further, as in the experiments of Yu et al (1981). In our experiment, no oospores were formed by P991 grown on nonsterol medium after 8 hr of contact with P731 grown on medium containing sterol, but P991 on medium with sterol formed abundant oospores after 8 hr of contact with P731 on medium either with or without sterol. Because rather little sterol would have been transferred to P731 in 8 hr, it seems likely that the $\alpha 2$ hormone is not steroidal.

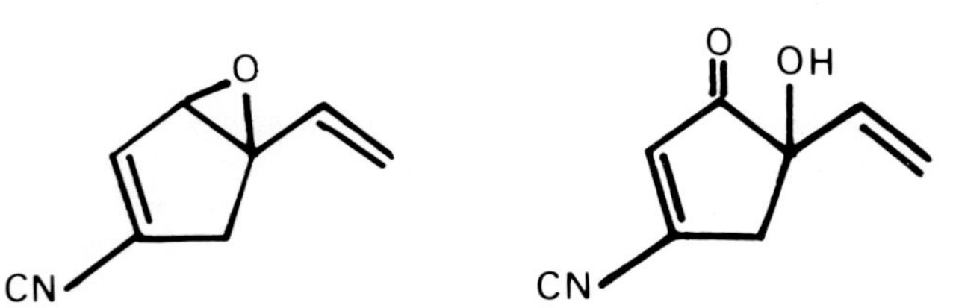


**Fig. 5.** Homothallins I and II, substances produced by *Trichoderma koningii* that induce oospore development in A2 *Phytophthora cinnamomi.* (After Sakata and Rickards, 1980)

### Other Agents Inducing Reproduction

In addition to induction by the opposite mating type, other extrinsic factors have been found that induce oospore development. First, volatile products of *Trichoderma* spp. induce oospore formation in A2 strains (Brasier 1971, 1975; Pratt et al, 1972). Fifty of 62 A2 isolates tested by Brasier (1975), but none of 34 A1 isolates, responded to *T. viride.* All A2 isolates of *P. cinnamomi* responded well; those of *P. cambivora* responded slightly less well. Results were mixed with isolates of *P. arecae, P. infestans, P. nicotianae, P. drechsleri,* and *P. cryptogea,* and several strains of *P. palmivora* gave negative results. Exposure to the *Trichoderma* vapors needed to be for 24 hr or less; sex organs appeared in 2–4 days; and oospores were not obtained in a sterol-free medium (Brasier, 1975). Two novel isocyanides, homothallin I and II, have been identified as the active agent(s) in *T. koningii* vapors, and their synthesis has been reported (Sakata and Rickards, 1980) (Fig. 5).

Zentmyer (1979) showed that extracts of roots of avocado (*Persea americana*) could stimulate oospore production in A2 isolates of *P. cinnamomi, P. capsici,* and *P. drechsleri.* A1 isolates were not stimulated, nor were A2 isolates of *P. cambivora, P. cryptogea, P. infestans, P. meadii, P. nicotianae,* or *P. palmivora.* The active substance was heat stable, dialyzable, and extractable with ether. The fungicide chloroneb was found to induce oospore formation in A2 but not in A1 strains of *P. capsici* (Noon and Hickman, 1974). Just cutting hyphae with a scalpel induced oospore formation in A2 but not in A1 strains of *P. cinnamomi* and *P. cryptogea* (Reeves and Jackson, 1974).

A2 cultures treated with chloroneb produced zoospores of A1 and rarely A1A2 mating types as well as A2 (Ko, 1981); genetic change is evidently induced. Even excluding chloroneb, a remarkable variety of agents apparently can mimic the effects of the $\alpha 1$ hormone. It is not evident what features they have in common with each other and presumably with the $\alpha 1$ hormone that enable them to interact with the $\alpha 1$ receptor in A2 strains. Neither *Trichoderma* vapor nor avocado extract appears to be universally active against A2 isolates. What is striking, however, is that no one has yet found an agent that mimics the effect of the $\alpha 2$ hormone. Yet in Ko's experiments, the A1 strains are generally the more responsive.

## SUMMARY

Temperature, light, oxygen tension, and composition of the medium have all been shown to influence sexual reproduction in *Phytophthora.* Blue light is inhibitory to oospore development, red light stimulatory. The nitrogen source and the carbon-nitrogen ratio are important, and sterols must be added to the medium for reproduction to occur.

The sterol requirement arises because phytophthoras are unable to synthesize sterols. Sterols stimulate vegetative growth and are incorporated into membranes, but sterols that have similar effects on physical properties of model membranes can have different effects on reproduction in

*Phytophthora,* and sterols that have different effects on membranes can have similar effects on reproduction. The effect of sterols on reproduction is, therefore, thought to be the result of their conversion to hormones. In *Achlya,* sterols are converted to steroid hormones that control reproduction. The polar metabolites produced from sterols by *Phytophthora* are thought to include precursors of hormones, and there are quantitative and qualitative differences in polar metabolite production from cholesterol, cholestanol, and sitosterol that may be associated with the different effects of these sterols on reproduction.

In heterothallic species, both crossing and selfing of paired strains of opposite mating type can occur. When strains of opposite mating type are paired but separated by a polycarbonate membrane, one or both strains can reproduce by selfing. Ko postulates that one strain produces a mating type-specific hormone ($\alpha$ hormone) that induces the other to respond by production of sex organs. This suggests that the $\alpha$ hormone derepresses the synthesis of the steroid hormones that control sexual development. Volatile metabolites of *Trichoderma,* an extract from avocado roots, and other agents can induce the reproduction of A2 strains and thus mimic the effect of $\alpha 1$ hormone.

### LITERATURE CITED


Ashby, S. F. 1922. Oospores in cultures of *Phytophthora faberi.* Kew Bull. 1922:257-262.

Ashby, S. F. 1929. Strains and taxonomy of *Phytophthora palmivora* Butler (*P. faberi* Maubl.). Trans. Br. Mycol. Soc. 14:18-38.

Barksdale, A. W. 1960. Inter-thallic sexual reactions in *Achlya,* a genus of the aquatic fungi. Am. J. Bot. 47:14-23.

Barksdale, A. W. 1967. The sexual hormones of the fungus *Achlya.* Ann N.Y. Acad. Sci. 144:313-319.

Barksdale, A. W., and Lasure, L. L. 1973. Induction of gametangial phenotypes in *Achlya.* Bull. Torrey Bot. Club 100:199-202.

Barksdale, A. W., and Lasure, L. L. 1974. Production of hormone B by *Achlya heterosexualis.* Appl. Microbiol. 28:544-546.

Bishop, H. 1940. A study of sexuality in *Sapromyces reinschii.* Mycologia 32:505-529

Boccas, B. R. 1981. Interspecific crosses between closely related heterothallic *Phytophthora* species. Phytopathology 71:60-65.

Brasier, C. M. 1969. The effect of light and temperature on reproduction in vitro in two tropical species of *Phytophthora.* Trans. Br. Mycol. Soc. 52:105-113.

Brasier, C. M. 1971. Induction of sexual reproduction in single $A^2$ isolates of *Phytophthora* species by *Trichoderma viride.* Nature (London) New Biol. 231:283.

Brasier, C. M. 1972. Observations on the sexual mechanism in *Phytophthora palmivora* and related species. Trans. Br. Mycol. Soc. 58:237-251.

Brasier, C. M. 1975. Stimulation of sex organ formation in *Phytophthora* by antagonistic species of *Trichoderma.* I. The effect in vitro. New Phytol. 74:183-194.

Butler, K. W., and Smith, I. C. P. 1978. Sterol ordering effects and permeability regulation in phosphatidylcholine bilayers. A comparison of ESR spin-probe data from oriented multilamellae and dispersions. Can. J. Biochem. 56:117-122.

Butler, K. W., Smith, I. C. P., and Schneider, H. 1970. Sterol structure and ordering effects in spin-labelled phospholipid multibilayer structures. Biochim. Biophys. Acta 219:514-517.

Child, J. J., and Haskins, R. H. 1971. Induction of sexuality in heterothallic *Pythium* spp. by cholesterol. Can. J. Bot. 49:329-332.

Child, J. J., and Haskins, R. H. 1976. Effect of inhibitors of sterol biosynthesis on growth and sexuality of *Pythium* and *Zygorhynchus.* Mycopathologia 59:91-93.

Child, J. J., Défago, G., and Haskins, R. H. 1969. The influence of carbon and nitrogen nutrition on growth and sterol-induced sexuality of *Pythium* sp. PRL 2142. Mycologia 61:1096-1105.

Couch, J. N. 1926. Heterothallism in *Dictyuchus,* a genus of the water moulds. Ann. Bot. 40:849-881.

Demel, R. A., Bruckdorfer, K. R., and Van Deenen, L. L. M. 1972a. Structural requirements of sterols for the interaction with lecithin at the air-water interface. Biochim. Biophys. Acta 255:311-320.

Demel, R. A., Bruckdorfer, K. R., and Van Deenen, L. L. M. 1972b. The effect of sterol structure on the permeability of liposomes to glucose, glycerol and $Rb^+$. Biochim. Biophys. Acta 255:321-330.

Dvornik, D., Kraml, M., and Bagli, J. F. 1966. Agents affecting lipid metabolism. XVIII. A 7-dehydrocholesterol $\Delta^7$-reductase inhibitor (AY-9944) as tool in studies of $\Delta^7$-sterol metabolism. Biochemistry 5:1060-1064.

Elliott, C. G. 1968. Competition and synergism between cholesterol and cholestanol in oospore formation in *Phytophthora cactorum.* J. Gen. Microbiol. 51:137-143.

Elliott, C. G. 1972. Sterols and the production of oospores by *Phytophthora cactorum.* J. Gen. Microbiol. 72:321-327.

Elliott, C. G. 1979. Influence of the structure of the sterol molecule on sterol-induced reproduction in *Phytophthora cactorum.* J. Gen. Microbiol. 115:117-126.

Elliott, C. G., and Glen, B. 1982. Sterol requirements of mating strains of heterothallic phytophthoras. J. Gen. Microbiol. 128:859-863.

Elliott, C. G., and Knights, B. A. 1981a. Uptake and interconversion of cholesterol and cholesteryl esters by *Phytophthora cactorum.* Lipids 16:1-7.

Elliott, C. G., and Knights, B. A. 1981b. Differential metabolism of cholesterol and sitosterol by *Phytophthora cactorum.* Lipids 16:142-145.

Elliott, C. G., and Sansome, E. 1977. The influence of sterols on meiosis in *Phytophthora cactorum.* J. Gen. Microbiol. 98:141-145.

Elliott, C. G., Hendrie, M. R., and Knights, B. A. 1966. The sterol requirement of *Phytophthora cactorum.* J. Gen. Microbiol. 42:425-435.

Galindo A., J., and Gallegly, M. E. 1960. The nature of sexuality in *Phytophthora infestans.* Phytopathology 50:123-128.

Ghosh, D., and Tinoco, J. 1972. Monolayer interactions of individual lecithins with natural sterols. Biochim. Biophys. Acta 266:41-49.

Harnish, W. N. 1965. Effect of light on production of oospores and sporangia in species of *Phytophthora.* Mycologia 57:85-90.

Hendrix, J. W. 1970. Sterols in growth and reproduction of fungi. Annu. Rev. Phytopathol. 8:111-130.

Hendrix, J. W., and Guttman, S.-M. 1968. Annulment of sterol-induced sexual reproduction by estradiol in *Pythium periplocum.* Science 161:1252.

Hendrix, J. W., and Guttman, S.-M. 1969. Resistance to oestradiol in species of *Pythium* and *Phytophthora.* Microbios 2:137-143.

Hendrix, J. W., and Guttman, S.-M. 1972. Steroid structure in relation to induction and inhibition of oospore formation in *Pythium periplocum* and *P. prolatum.* Phytopathology 62:763.

Honour, R. C., and Tsao, P. H. 1974. Production of oospores by *Phytophthora parasitica* in liquid medium. Mycologia 66:1030-1038.

Horlick, L. 1966. Effect of a new inhibitor of cholesterol biosynthesis (AY 9944) on serum and tissue sterols in the rat. J. Lipid Res. 7:116-121.

Hossack, J. A., and Rose, A. H. 1976. Fragility of plasma membranes in *Saccharomyces cerevisiae* enriched with different sterols. J. Bacteriol. 127:67-75.

Huguenin, B. 1973. Application de la technique du marquage fluorescent à la détermination du comportement sexuel du *Phytophthora palmivora.* Cah. ORSTOM Ser. Biol. 20:59-61.

Huguenin, B., and Boccas, B. 1971. Rôle de quelques facteurs dans la formation et la germination des oospores chez le *Phytophthora palmivora* Butl. Ann. Phytopathol. 3:353-371.

Huguenin, B., and Jacques, R. 1973. Établissement du spectre d'action de la lumière sur la sporogenèse et l'oogenèse de souches du *Phytophthora palmivora* Butl. C. R. Hebd. Seances Acad. Sci. Ser. D 276:725-728.

Kleinschmidt, M. G., Chough, K. S., and Mudd, J. B. 1972. Effect of filipin on liposomes prepared with different types of steroids. Plant Physiol. 49:852-856.

Knights, B. A., and Elliott, C. G. 1976. Metabolism of $\Delta^7$- and $\Delta^{5,7}$-sterols by *Phytophthora cactorum.* Biochim. Biophys. Acta 441:341-346.

Ko, W. H. 1978. Heterothallic *Phytophthora:* Evidence for hormonal regulation of sexual reproduction. J. Gen. Microbiol. 107:15-18.

Ko, W. H. 1980. Hormonal regulation of sexual reproduction in

*Phytophthora.* J. Gen. Microbiol. 116:459-463.

Ko, W. H. 1981. Reversible change of mating type in *Phytophthora parasitica.* J. Gen. Microbiol. 125:451-454.

Ko, W. H., and Kunimoto, R. K. 1981. Hormone production and reception among different isolates of *Phytophthora parasitica* and *P. palmivora.* Mycologia 73:440-444.

Langcake, P. 1974. Sterols in potato leaves and their effects on growth and sporulation of *Phytophthora infestans.* Trans. Br. Mycol. Soc. 63:573-586.

Langcake, P. 1975. Uptake of sterols by *Phytophthora infestans,* their intracellular distribution and metabolism. Trans. Br. Mycol. Soc. 64:55-65.

Leal, J. A., and Gomez-Miranda, B. 1967. Effect of amino acids and organic acids on the sexual reproduction of species of *Phytophthora* and *Pythium.* Trans. Br. Mycol. Soc. 50:77-84.

Leal, J. A., Gomez-Miranda, B., and Nicolas, G. 1966. Formation of toxic substances by fungi in the degradation of some amino acids. Can. J. Microbiol. 12:1073-1076.

Leal, J. A., Gallegly, M. E., and Lilly, V. G. 1967. The relation of the carbon-nitrogen ratio in the basal medium to sexual reproduction in species of *Phytophthora.* Mycologia 59:953-964.

Leal, J. A., Lilly, V. G., and Gallegly, M. E. 1970. Some effects of ammonia on species of *Phytophthora.* Mycologia 62:1041-1056.

Lemcke, R. M., and Burrows, M. R. 1980. Sterol requirement for the growth of *Treponema hyodysenteriae.* J. Gen. Microbiol. 116:539-543.

Lilly, V. G. 1966. The effects of sterols and light on the production and germination of *Phytophthora* spores. Pages 259-271 in: The Fungus Spore. M. F. Madelin, ed. Butterworths, London.

McMorris, T. C. 1978. Sex hormones of the aquatic fungus *Achlya.* Lipids 13:716-722.

Mitchell, D. J., and Zentmyer, G. A. 1971. Effects of oxygen and carbon dioxide tensions on sporangium and oospore formation by *Phytophthora* spp. Phytopathology 61:807-811.

Nes, W. D., and Heftmann, E. 1981. A comparison of triterpenoids with steroids as membrane components. J. Nat. Prod. 44:377-400.

Nes, W. D., and Patterson, G. W. 1981. Effects of tetracyclic and pentacyclic triterpenoids on growth of *Phytophthora cactorum.* J. Nat. Prod. 44:215-220.

Nes, W. D., Patterson, G. W., and Bean, G. A. 1979. The effect of steroids and their solubilizing agents on mycelial growth of *Phytophthora cactorum.* Lipids 14:458-462.

Nes, W. D., Patterson, G. W., and Bean, G. A. 1980. Effect of steric and nuclear changes in steroids and triterpenoids on sexual reproduction in *Phytophthora cactorum.* Plant Physiol. 66:1008-1011.

Noon, J. P., and Hickman, C. J. 1974. Oospore production by a single isolate of *Phytophthora capsici* in the presence of chloroneb. Can. J. Bot. 52:1591-1595.

Papa, K. E., Campbell, W. A., and Hendrix, F. F. 1967. Sexuality in *Pythium sylvaticum:* Heterothallism. Mycologia 59:589-595.

Patterson, G. W., Doyle, P. J., Dickson, L. G., and Chan, J. T. 1974. Effects of triparanol and AY-9944 upon sterol biosynthesis in *Chlorella.* Lipids 9:567-574.

Pethybridge, G. H., and Murphy, P. A. 1913. On pure cultures of *Phytophthora infestans* De Bary, and the development of oospores. Sci. Proc. R. Dublin Soc. 13:566-588.

Pratt, B. H., Sedgley, J. H., Heather, W. A., and Shepherd, C. J. 1972. Oospore production in *Phytophthora cinnamomi* in the presence of *Trichoderma koningii.* Aust. J. Biol. Sci. 25:861-863.

Pratt, R. G., and Green, R. J., Jr. 1971. The taxonomy and heterothallism of *Pythium sylvaticum.* Can. J. Bot. 49:273-279.

Pratt, R. G., and Mitchell, J. E. 1973. Differential effects of cholesterol in mating thalli of *Pythium sylvaticum* and *Phytophthora capsici.* Can. J. Bot. 51:595-599.

Raper, J. R. 1939. Sexual hormones in *Achlya.* I. Indicative evidence for a hormonal coordinating mechanism. Am. J. Bot. 26:639-650.

Raper, J. R. 1940. Sexuality in *Achlya ambisexualis.* Mycologia 32:710-727.

Reeves, R. J., and Jackson, R. M. 1974. Stimulation of sexual reproduction in *Phytophthora* by damage. J. Gen. Microbiol. 84:303-310.

Ribeiro, O. K., Zentmyer, G. A., and Erwin, D. C. 1976. The influence of qualitative and quantitative radiation on reproduction and spore germination of four *Phytophthora* species. Mycologia 68:1162-1173.

Rottem, S., Pfendt, E. A., and Hayflick, L. 1971. Sterol requirements of T-strain mycoplasmas. J. Bacteriol. 105:323-330.

Sakata, K., and Rickards, R. W. 1980. Synthesis of homothallin II. Pages 165-172 in: 23rd Symp. Chemistry of Natural Products, Nagoya, Japan.

Sansome, E., Brasier, C. M., and Sansome, F. W. 1979. Further cytological studies on the 'L' and 'S' types of *Phytophthora* from cocoa. Trans. Br. Mycol. Soc. 73:293-302.

Satchuthananthavale, V. 1963. Complementary strains of *Phytophthora palmivora* from Ceylon rubber. Phytopathology 53:729.

Savage, E. J., Clayton, C. W., Hunter, J. H., Brenneman, J. A., Laviola, C., and Gallegly, M. E. 1968. Homothallism, heterothallism, and interspecific hybridization in the genus *Phytophthora.* Phytopathology 58:1004-1021.

Semer, R., and Gelerinter, E. 1979. A spin label study of the effects of sterols on egg lecithin bilayers. Chem. Phys. Lipids 23:201-211.

Shepherd, C. J. 1978. Mating behaviour of Australian isolates of *Phytophthora* species. I. Inter- and intraspecific mating. Aust. J. Bot. 26:123-138.

Sherwood, W. A. 1966. Evidence for a sexual hormone in the water mold *Dictyuchus.* Mycologia 58:215-220.

Smith, P. F. 1964. Relation of sterol structure to utilization in pleuropneumonia-like organisms. J. Lipid Res. 5:121-125.

Smith, P. F., and Lynn, R. J. 1958. Lipid requirements for the growth of pleuropneumonialike organisms. J. Bacteriol. 76:264-269.

Smoot, J. J., Gough, F. J., Lamey, H. A., Eichenmuller, J. J., and Gallegly, M. E. 1958. Production and germination of oospores in *Phytophthora infestans.* Phytopathology 48:165-171.

Stamps, D. J. 1953. Oospore production in paired cultures of *Phytophthora* species. Trans. Br. Mycol. Soc. 36:255-259.

Tsao, P. H., Ugale, R., Hobbs, S., and Farih, A. 1980. Control of homothallic oospore formation in *Phytophthora parasitica* by culture manipulation. Trans. Br. Mycol. Soc. 75:153-156.

Turner, P. D. 1961. Complementary isolates of *Phytophthora palmivora* from cacao and rubber, and their taxonomy. Phytopathology 51:161-164.

Wood, S. G., and Gottlieb, D. 1978a. Evidence from mycelial studies for differences in the sterol biosynthetic pathway of *Rhizoctonia solani* and *Phytophthora cinnamomi.* Biochem. J. 170:343-354.

Wood, S. G., and Gottlieb, D. 1978b. Evidence from cell-free systems for differences in the sterol biosynthetic pathway of *Rhizoctonia solani* and *Phytophthora cinnamomi.* Biochem. J. 170:355-363.

Yu, J.-Y., and Chang, H. S. 1980. Chemical regulation of sexual reproduction in *Phytophthora colocasiae.* Bot. Bull. Acad. Sin. 21:155-158.

Yu, J. Y., Chang, H. S., and Ko, W. H. 1981. Factors affecting the induction of sexual reproduction in *Phytophthora parasitica* by *P. colocasiae.* J. Gen. Microbiol. 123:249-252.

Zentmyer, G. A. 1979. Stimulation of sexual reproduction in the $A^2$ mating type of *Phytophthora cinnamomi* by a substance from avocado roots. Phytopathology 69:1129-1131.

Zentmyer, G. A., Klure, L. J., and Pond, E. C. 1979. The influence of temperature and nutrition on formation of sexual structures by *Phytophthora cinnamomi.* Mycologia 71:55-67.

# 6

# The Cytogenetics and Genetics of *Phytophthora*

D. S. Shaw

CONTENTS



## THE CYTOGENETICS OF *PHYTOPHTHORA*

### Cytogenetics and the Life Cycle

From the earliest times, the interpretation of nuclear figures in the Oomycetes has been surprisingly controversial.

The first drawing of nuclei of *Phytophthora* I can find in the literature is by Anton de Bary (1887). This quite accurate representation of nuclei in sporangia and zoospores of *P. infestans* was published at a time when the significance of mitosis and meiosis was not realized. Interest in nuclear cytology was growing and de Bary predicted that work with fungi would be difficult because they have such small nuclei. However, he found it worth remarking that nuclei in the Saprolegniales were obviously larger than those of other fungi.

By the turn of the century, the nature of divisions of nuclei in gametangia and zygotes was already a well-developed controversy, which continued to dominate cytologic work with these fungi until very recently.

This early work, ably reviewed by Dick and Win-Tin (1973), was done mainly with members of the Saprolegniales but also included *Pythium* and *Albugo*. Murphy (1918) joined in the debate with an excellent description of the nuclei in gametangia and zygotes of *P. erythroseptica*. He identified only one synchronous division in gametangia and concluded that meiosis must be zygotic. In fact, his very large, spaced-out nuclei in oogonia preparing for division are beautiful representations of meiotic prophase. Murphy went so far as to criticize earlier workers, whom he said were obsessed by the idea of reduction division in gametangia. In contrast, Allain (1934) produced evidence of two synchronized divisions of the nuclei in gametangia of *P. cambivora* and sided with those who held that oosporic fungi were haploid in the vegetative state. Textbooks such as Gwynne-Vaughan and Barnes (1927) propagated the interpretation that divisions in gametangia were mitotic and that meiosis must occur in the zygote or during zygote germination. Blackwell (1943) supported this view after her studies with *P. cactorum*.

In the early 1960s, Sansome challenged the accepted dogma with her series of studies on *Pythium, Sclerospora*, and *Phytophthora* (Sansome, 1961, 1963a, 1963b, 1965; Sansome and Harris, 1962). Her interpretations were received with some skepticism by mycologists who had grown up with the doctrine of haploidy. The old controversy was renewed, with proponents on both sides producing new evidence. The great value of this disagreement, as we can see today, was to stimulate not only more cytologic work but also other studies using genetic, biophysical, and biochemical methods designed to settle the argument.

Sansome used the newer squash technique with orcein and carmine stains, and her conclusion that synchronous meiotic divisions occur in gametangia was based on the following observations: 1) large nuclei in oogonia (and antheridia) enter a long prophase; 2) paired chromosomes (bivalents) show chiasmata; 3) multivalents and chromosome rings are seen in some cultures treated with camphor to produce polyploidy; 4) rings and chains (multiple associations) are also seen in natural isolates; 5) second division follows without delay and without increase in size of the nuclei; 6) products of second division are nuclei, often in tetrads, clearly smaller than any vegetative nuclei.

Gametangial meiosis has now been confirmed in a series of studies by other workers with several genera from the Saprolegniales and Peronosporales (Barksdale, 1968; Dick and Win-Tin, 1973; Ellzey, 1974; Flanagan, 1970; Tommerup et al, 1974) and with several *Phytophthora* spp., namely *P. nicotianae* var. *parasitica* (Erselius and Boccas, 1978; Huguenin and Boccas, 1970), *P. drechsleri* (Galindo and Zentmyer, 1967), and *P. capsici* (Maia et al, 1976).

Acceptance of gametangial meiosis, however, came slowly. Much of the chromosomal detail in these small nuclei can barely be resolved by the light microscope. Even the best photographic evidence (e.g., Fig. 1) is always less convincing than direct observation of squashed material, in which optical sections provide clearer details of chromosomal figures. Some of the earliest genetic data for *P. infestans* (Gallegly, 1968) and for *P. capsici* (Timmer et al, 1970) were thought to favor zygotic meiosis and a haploid somatic phase. Cytologic reports on these species followed and continued to argue that gametangial divisions were mitotic whereas zygotic divisions were meiotic (Laviola and Portacci, 1974; Stephenson et al, 1974a, 1974b). Was it still possible that certain species or isolates were haploid and the rest were diploid? Studies of these same species (*P. infestans* by Sansome and Brasier, 1973) and even the same isolates (*P. capsici* by Sansome, 1976) showed that the different conclusions drawn were due to differences in interpretation rather than in the ploidy of the material being examined. This last study by Sansome was of Stephenson's cultures and is an example of a rare cooperative venture that might foster more rapid progress if it were to happen more often.

Sansome and Brasier have continued to provide a wealth of cytologic detail in studies of meiotic chromosomes of various species of *Phytophthora.* A mating of two compatible Mexican isolates of *P. infestans* and a selfing of an A2 isolate by the "*Trichoderma* method" (Brasier, 1971) produced meiotic metaphases in gametangia with five to seven bivalents and an association of six chromosomes (Sansome and Brasier, 1973). The researchers suggested that at least the A2 parent was a structural hybrid for a double interchange involving three pairs of chromosomes and that the high rate of oogonial abortion observed was a natural consequence of nondisjunction from such multiple associations. Failure of meiosis was also suggested as a cause of large restitution nuclei, which were not uncommon. Oogonia with two oospores were also observed.

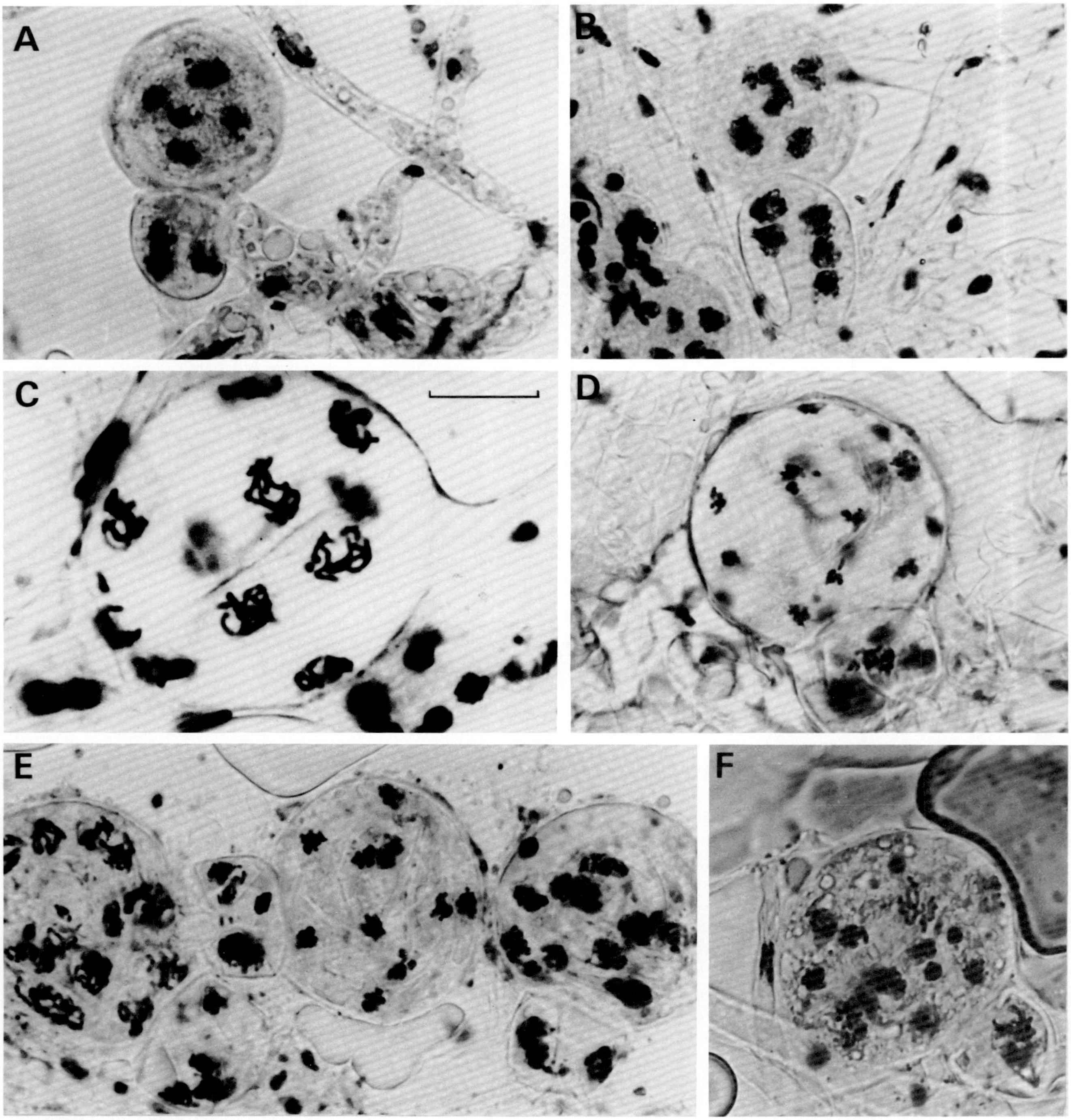


**Fig. 1.** Meioses in gametangia of *Phytophthora nicotianae* var. *parasitica.* Gametangia from matings in pea broth were fixed at intervals and stained with Giemsa solution (Erselius and Boccas, 1978). **A,** Large prophase nuclei in antheridium and in an unusually small oogonium. **B,** Leptotene. **C,** Pachytene. **D,** Metaphase and anaphase. **E** (from left to right), Pachytene, metaphase, leptotene. **F,** Pachytene. Scale bar represents 10 μm, except in C, where it represents 7 μm. (Courtesy L. J. Erselius; parts reprinted, by permission, from Erselius and Boccas, 1978)

By exploiting the observation that A1 isolates of *P. infestans* can be induced to form oogonia when either intraspecific or interspecific pairings are made with suitable isolates of other heterothallic species, Sansome (1977) was able to study the karyotype of Welsh isolates. Oogonial nuclei of *P. infestans* were easily distinguished from the small-chromosome karyotype of *P. palmivora* and from the large-chromosome karyotype of *P. megakarya*, which were used as inducers. She found that several native species of *Phytophthora* could also be used to induce oogonia and antheridia in the cultures of *P. infestans*. Selfed gametangia, if they occurred, must have been rare, as most gametangial combinations that could be identified involving *P. infestans* were hybrid. The surprising result was that meiotic metaphase stages of these Welsh isolates had large numbers of chromosomes—approximately twice the number expected from the study of the Mexican isolates of *P. infestans* (Sansome and Brasier, 1973). Two of the isolates were thought to be tetraploid and a third was thought to carry both diploid and tetraploid nuclei. From the evidence presented, tetraploid nuclei would appear to be functionally diploid; i.e., bivalents rather than tetravalents, or trivalents were formed at meiosis.

Sansome's work indicates that the perfect stage of the single compatibility type A1 of *P. infestans* may occur, although rarely, in the United Kingdom and perhaps in other areas outside Mexico. Recently (D. S. Shaw, *unpublished*), an isolate of *P. drechsleri* of A2 mating type from a blighted potato field in North Wales was found to induce oogonia and oospores in freshly isolated *P. infestans*. Selfing of *P. infestans* induced by this isolate has been confirmed (D. S. Shaw and D. I. Skidmore, *unpublished*) by use of the polycarbonate membrane method (Ko, 1978).

Clearly, the ploidy of populations of *P. infestans* should be determined on a world scale. In a current fluorometric study of nuclei of isolates from the United Kingdom and Italy, preliminary results show that DNA contents of all isolates examined are very similar (C. Evola and D. S. Shaw, *unpublished*).

Studies of karyotype in two taxonomic groups have confirmed divisions less well defined by other criteria. In the first of these, Sansome and Brasier (1974) estimated the chromosome number of *P. megasperma* var. *megasperma* to be in the range $n = 22–27$, whereas those of isolates assigned to *P. megasperma* var. *sojae* were in the range $n = 10–13$. Thus, the former variety seems to be a tetraploid and the latter a diploid. In the second example, isolates formerly assigned to *P. palmivora* were found to fall into two groups based on whether meiotic metaphase had 9–12 small chromosomal configurations (S type) or five to six larger chromosomal configurations (L type) (Sansome et al, 1975). Similar findings were reported by Kaosiri (1978). These karyotypic differences, which were closely correlated with several distinct morphological differences, later became the basis for separating the L type into a new species, *P. megakarya*. The S type was retained within *P. palmivora* (Brasier and Griffin, 1979).

Sansome (1980) has turned her attention to the multiple associations that occur so frequently in meiotic metaphase of heterothallic isolates. A systematic search within 10 heterothallic species revealed that multiple associations occurred in all. Where the parentage of the nuclei undergoing meiosis could be identified (by pairing with a distinctive chromosomal type of another species, by differential orcein staining, or by selfing with the *Trichoderma* method), both A1 and A2 parents showed multiple associations. In contrast, multiple associations were not observed in four homothallic species. However, *P. megasperma* var. *megasperma* proved to be an exception; the rare multiple associations were explained as multivalents—to be expected occasionally in a tetraploid. Sansome's hypothesis is that a reciprocal translocation of segments involving two nonhomologous chromosomes gave rise to a translocation heterozygote. At synapsis, this heterozygote would form a cross-shaped multivalent. The metaphase configuration would depend on the number and positions of chiasmata formed. The most conspicuous figures would be open or closed rings of four chromosomes and chains of four or three chromosomes. Because all heterothallic isolates are translocation heterozygotes and because certain mating associations in *P. drechsleri* are associated with an altered translocation complex, the hypothesis was extended by postulating that the mating type (compatibility type) genes are located on the translocation complex. A corollary of this is that the suppression of mitotic crossing-over expected in segments of the translocation heterozygote would conserve any heterozygosity of their genes. Thus, if one of the mating types were heterozygous (e.g., if A2 were *Aa* and A1 were *aa*) and the gene or genes were located on the complex, then the heterozygous state would be conserved and no somatic segregation of the opposite mating type, *aa*, would be possible. Suppression of mitotic crossing-over depends on imbalance and thus inviability of any homozygotes produced.

Sansome (1980) pointed out that a mechanism must exist to prevent reciprocal translocations becoming homozygous in sexual progeny. The mechanism could involve either selection against structural homozygotes (perhaps by a balanced lethal system) or some form of preferential mating of gametic nuclei that would propagate the structural heterozygote. An attractive hypothesis of the second type, outlined in Figure 2, suggests that nuclei containing chromosomes AB and CD are selected as gametic nuclei in the oogonium, and nuclei having the other combination of chromosomes, BC and DA, are selected to function as antheridial nuclei. The resulting fusion nuclei would be translocation heterozygotes. If we suppose that the A2 mating type is determined by the heterozygous condition of gene *A* located on, say, segment B and the A1

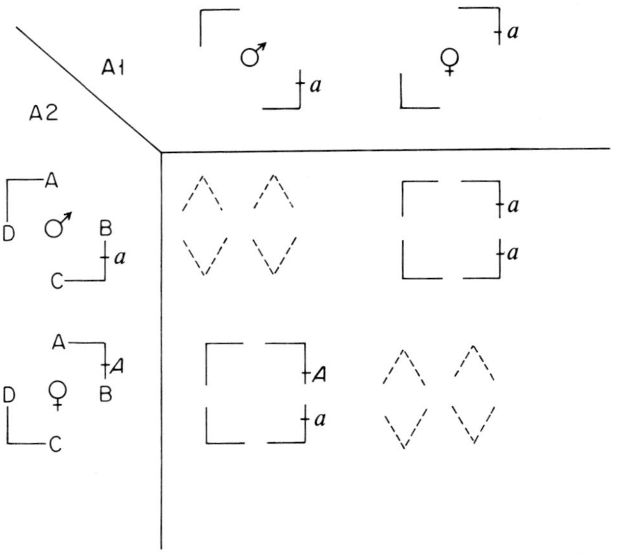


**Fig. 2.** A possible mechanism that retains heterozygosity of the translocation complex. Gametic nuclei are formed by balanced disjunction from a ring of four chromosomes (AB.BC.CD.DA). Of the two possible arrangements, AD.BC and AB.DC, let the former function as a female gamete, the latter as a male. Mating of identical arrangements to produce translocation homozygosity will not occur. The proportion of the two translocation heterozygotes (A1 and A2) will depend on the relative sexual strengths of the parents.

mating type by homozygous *aa*, then the reciprocal crosses will generate different mating types and the segregation of mating type will depend primarily on the frequency of each kind of reciprocal cross. This explanation can account for the range of mating type ratios found in different crosses and even in the same cross made on different occasions.

### Cytology of Nuclei in Hyphae and Asexual Spores

It is hardly surprising that nuclei in asexual stages have been neglected in favor of those in gametangia and oospores or have been mentioned only in passing. Many workers who have studied *Phytophthora* and other oomycetes, among them Bakerspigel (1960), Bosc (1946), Hartog (1895), Sansome (1965), and Stephenson et al (1974a), have commented on the fact that division of nuclei in hyphae is achieved by a gradual pulling apart of the "dumbbell-shaped" predivision nucleus. This kind of division, typified by the absence of well-defined chromosomes on a metaphase plate and by the persistence of the nuclear membrane, has been referred to as a direct, amitotic or promitotic division to distinguish it from the indirect or true mitotic division occurring in higher organisms.

Marks (1965) observed that many nuclei in vegetative structures of *P. infestans* had a prominent Feulgen-positive "cap," which persisted and divided during mitosis. Rarer "granular" nuclei, thought to be entering division, had up to 10 condensed granules of chromatin, which appeared to divide during mitosis. Small, intensely stained, "condensed" nuclei were sometimes observed. These could have been degenerating nuclei. Marks also described some attenuated, almost threadlike hyphal nuclei and proposed that these forms may be typical of nuclei migrating through narrow hyphae. Bakerspigel (1960) observed similar nuclei with a "flame-like appearance" in *Saprolegnia*. It would be interesting to know if these peculiar nuclei, which are particularly common in sporangiophores, are destined for the developing sporangium or if they have another function.

It is now clear that in *P. infestans*, both migration of nuclei (Marks, 1965) and division within the sporangium initial (Graham, 1954) give rise to the multinucleate condition of the sporangium (Laviola, 1975; Sivak, 1973). Similar nuclear migration followed by division within the sporangium initial occurs in *P. capsici* (Maia et al, 1976). A recent, unexpected finding with *P. infestans* is that nuclear degeneration occurs before sporangia become deciduous. Further degeneration coupled with division may occur in detached sporangia before zoospores are cleaved (C. Evola and D. S. Shaw, *unpublished*). In an electron microscope study of detached sporangia of *P. palmivora*, Hemmes and Hohl (1973) described simultaneous division and degeneration of nuclei during dextrin-induced formation of secondary sporangia. The authors proposed that this reorganization was a necessary prelude to the formation of secondary zoosporangia and that the degeneration provided precursors for further division or restored the balance of nucleus and cytoplasm that existed before division. Division coupled with degeneration appears to be part of the normal developmental sequence in the formation of zoosporangia of *Phytophthora*. Hemmes and Hohl's study provides us with some excellent micrographs of dividing nuclei. A small intranuclear spindle of microtubules stretches excentrically across the nucleus between two pairs of centrioles located just outside the nuclear envelope. Chromatin is not obviously condensed, but a persistent nucleolus is clearly visible. As division proceeds, the spindle and nuclear envelope elongate until two lobes are formed. These lobes pinch off into separate daughter nuclei, apparently without any disruption of the nuclear envelope.

### Other Evidence Relating to Ploidy

Detailed electron microscopy of dividing oogonial nuclei has not been conducted in *Phytophthora*, but in *Saprolegnia terrestris*, Howard and Moore (1970) found short sections of paired axial elements, which they interpreted to be precursors of the lateral elements of the synaptonemal complex of nuclei undergoing meiosis. Ellzey and Huizar (1977) reported synaptonemal complexes in antheridial prophase nuclei of *Achlya ambisexualis*. These studies verify the theory that gametangial division is meiotic.

Following the lead of Bryant and Howard (1969) and Howard and Bryant (1971) in comparing amounts of DNA in nuclei of *S. terrestris* and *Apodachlya brachynema* by Feulgen microdensitometry, Mortimer and Shaw (1975) used a microfluorometric method to estimate relative DNA contents of nuclei of *P. drechsleri*. Fluorescence from nuclei stained with the fluorochrome BAO (2,5-bis-3-4[4′-aminophenyl-(1′)]-1,3,4-oxadiazole) was measured with a photomultiplier. Results showed that the basal amount of DNA (1C) occurred only in gametic nuclei; nuclei in hyphae and in young gametangia contained approximately twice or up to four times this amount. Clearly, if gametic nuclei have the 1C amount of DNA and are haploid, then vegetative nuclei with mainly 2C amounts of DNA are diploid, and nuclei in young gametangia are also diploid (2C and replicated to 4C). The divisions in the gametangia resulting in reduction of 4C to 1C must have been meiotic. These results agree closely with those from *Saprolegnia* and *Apodachlya*.

Legrand-Pernot and Pellegrin (1976) used yet another method to define ploidy in nuclei from zoospores of two isolates of *P. palmivora* and one isolate of *P. parasitica*. The viability of zoospores was determined after they had been treated with different doses of $\gamma$-rays. Kill curves for all three isolates were exponential but always had a significant shoulder. Because survival of spores with haploid nuclei might be expected to follow single-hit kinetics, the authors concluded that nuclei in zoospores were diploid. The possibility that the shoulder on the curve was due to nuclei of zoospores being haploid but in the G2 stage of the cell cycle was tested by measuring their DNA contents relative to those of nuclei of germlings and hyphae. The DNA content of nuclei of zoospores and germlings remained constant for 2 hr (the period of irradiation), then increased before the nuclei divided. The basal DNA content of nuclei of zoospores and hyphae was very similar. Zoospore nuclei, therefore, were shown to be at G1 of the cell cycle rather than at G2. The researchers concluded that the shoulder on the kill curve was due to the zoospore nuclei being diploid and not to replication of haploid nuclei.

## GENETIC WORK WITH *PHYTOPHTHORA*

Genetic studies were started in the early 1960s to find out whether patterns of inheritance justified the cytologic claim for a diploid vegetative phase in *Phytophthora* (Table 1). Early optimism was dimmed by data that could be interpreted to indicate either haploidy or diploidy. However, the more recent studies have consistently led to the conclusion that the pattern of inheritance of most traits analyzed is that expected in a diploid fungus. No attempt is made here to discuss the results or the arguments over interpretation that followed. This has been adequately dealt with in several reviews (Caten and Day, 1977; Day, 1974; Dick and Win-Tin, 1973; Erwin et al, 1963; Gallegly, 1970; Webster, 1974). Instead it would be worthwhile to analyze some of the special problems facing the

*Phytophthora* geneticist that have been revealed by these studies.

### Special Problems in Genetic Work with *Phytophthora*

#### Choice of Isolates

Table 1 shows that studies have involved a variety of different species. Frequently, crosses between different isolates have shown variation in vigor of mating, in frequency of gametangial and zygote abortion, in percentage and rate of oospore germination, and in viability of germinated oospores. Interestingly, four of the 10 studies listed were made with homothallic species in which habitual selfing might be expected to have eliminated deleterious alleles and thus to have reduced abortion. Problems related to genome incompatibility will not arise in selfed homothallic isolates but might be expected to contribute to inviability on outcrossing of both homothallic and heterothallic isolates. These problems might be avoided or minimized if closely related isolates from the same field population were used as starting material.

Genetic work with other microorganisms has clearly demonstrated the usefulness of induced or selected mutations as markers that behave in crosses as if they were mutations of single genes. The trend has thus been to use mutagens to induce mutations and to select survivors with a nutrient requirement (Andreoli and Ponchet, 1974; Boccas, 1973; Castro et al, 1971; Elliott and MacIntyre, 1973; Long and Keen, 1977a) or with a large increase in resistance to an inhibitory antibiotic or drug (Dyakov and Vishtynezkaja, 1973; Khaki and Shaw, 1974; Long and Keen, 1977a; Shattock and Shaw, 1975; Shaw and Elliott, 1968; Swiezynski and Jankowska, 1970; Timmer et al, 1970). Because frequencies of recovery of mutants were very low, it is unclear in most cases whether mutations were induced by the mutagen or arose spontaneously.

The dangers of inducing too many nonspecific or semilethal mutations while using a potent mutagen are well known in work with other groups of microorganisms. In *Phytophthora*, Khaki's induced drug-resistant mutants were fertile in crosses to wild-type but in all cases sterile when intercrossed (Khaki, 1974); some mutants had abnormal morphological features, such as the grotesque sporangia of a chloramphenicol mutant shown in Figure 3 (Khaki, 1974; Khaki and Shaw, 1974). Long and Keen's auxotrophs were all self-sterile, and many were deficient in sporulation or pathogenicity (Long and Keen, 1977a). Some of these abnormal phenotypes, particularly those that did not segregate in sexual progeny, might have been the result of the pleiotropic action of a single mutant gene, whereas others were probably due to additional undesirable mutations. Selection of spontaneous mutations or purification by repeated backcrossing or selfing followed by selection would avoid problems caused by unwanted mutations.

The selection of recessive mutants such as auxotrophs presents an obvious problem in a diploid organism, because the mutation will not be expressed until it occurs in the homozygous condition. Elliott and MacIntyre (1973) solved this problem by selfing the survivors after mutagen treatment and selecting auxotrophs from segregating progenies.

In general, attempts should be made to screen the largest possible number of spores for mutant phenotypes so that only the ones showing a clearly defined, stable phenotype need be taken further. Those mutants showing evidence of polygenic inheritance, instabilities, or abnormal phenotypes in the sexual progeny should also be eliminated. Such a rigorous procedure can be conducted if selection is easy (e.g., for drug resistance) but is much more difficult when negative selection (e.g., for auxotrophy) is used and is nearly impossible when wild-types will not produce abundant zoospores. Recovery of induced mutations may be aided in any screening method by ensuring that potential mutants grow enough to express their mutant phenotype before selection is applied. Reconstruction experiments in which a few mutant spores are added to a population of wild-type spores can be useful in perfecting techniques for efficient isolation of mutants (Hopwood, 1970; Shaw, 1965).

#### Selfing in Matings of Heterothallic Species

Because both parents of a mating in a heterothallic species have the potential to form antheridia and oogonia, hybrid oospores may be formed by the combination of an A1 oogonium and an A2 antheridium or by the reciprocal combination, an A2 oogonium and an A1 antheridium. In addition, the presence of a compatible isolate may induce either one or both of the parents to form selfed oospores. The implications of this for the interpretation of genetic data were made clear by Sansome (1970) and Brasier (1972). Obviously,

TABLE 1. Genetic studies of *Phytophthora*

| Date | Authors | Species |
|---|---|---|
| Early work fueling ploidy debate: | | |
| 1967 | Galindo and Zentmyer | *P. drechsleri* |
| 1968 | Satour and Butler | *P. capsici* |
| 1968 | Shaw and Elliott | *P. cactorum* |
| 1968 | Laviola | *P. infestans* |
| 1969 | Romero and Erwin | *P. infestans* |
| 1970 | Timmer et al | *P. capsici* |
| Later work clearly indicating diploidy: | | |
| 1971 | Shaw and Khaki | *P. drechsleri* |
| 1972 | Boccas | *P. syringae* |
| 1973 | Elliott and MacIntyre | *P. cactorum* |
| 1977b | Long and Keen | *P. megasperma* var. *sojae* |

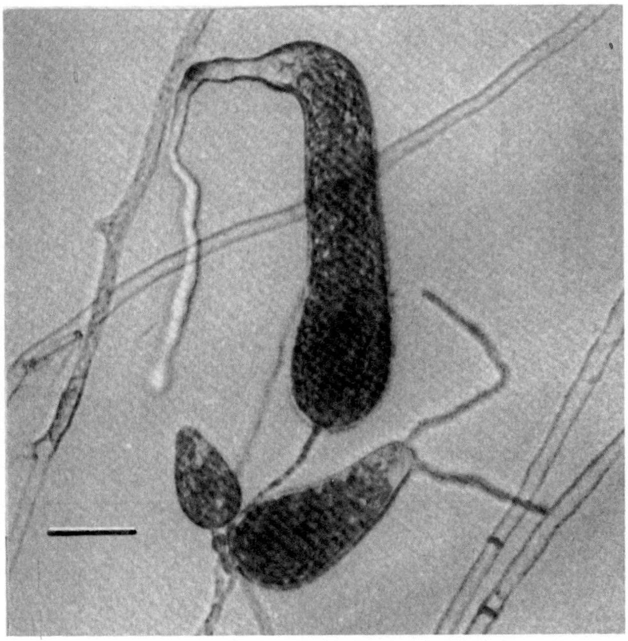

**Fig. 3.** Morphology of sporangia of a resistant mutant of *Phytophthora drechsleri*. Note elongate sporangia with branched and distended papillae. Scale bar represents 30 $\mu$m. (Courtesy I. A. Khaki)

if two kinds of selfed progeny are produced in addition to hybrid individuals, the proportions of the different phenotypes in the progeny of a cross of marked strains will depend on the relative frequencies of the two selfings and the crossing as well as on the ability of the three kinds of zygote to germinate. The frequencies of selfed combinations can be estimated by tracing the hypha subtending each gametangium back to its origin (Stamps, 1953b). Huguenin (1973) claimed that this laborious task may be avoided by labeling the hyphae and gametangia of the one or the other parent with a fluorochrome. Natural differences in the uptake of the stain, aceto-orcein, by gametangia of the two parents of some crosses were used by Brasier and Sansome (1975) to estimate selfing frequency.

These methods, of course, only estimate the proportions of the different possible gametangial unions and are not a measure of selfing in the viable oospore progeny. Perhaps the best way to assess parentage in the sexual progeny would be to make sure that parents were always homozygous for two known markers: e.g., + + *met met* × *arg arg* + +. Hybrid progeny would be prototrophic but heterozygous; selfed progeny would all be auxotrophic, either $Met^-$ or $Arg^-$. A simpler system might use a single semidominant gene conferring drug resistance. The mating of sensitive (*rr*) × resistant (*RR*) would produce hybrids, *Rr*, of intermediate resistance and selfed progeny that were resistant or sensitive.

Although, as noted above, selfed oogonia occur in heterothallic matings within a species, there is no firm evidence that selfed individuals are recovered in progenies of these matings. We do know that single-parent families can be produced by selfing using the "*Trichoderma* method" (Brasier and Sansome, 1975) or by inducing with a compatible mating type of a different species (Boccas, 1981), but we do not know if selfed progeny are recovered in intraspecies crosses when hybrid oospores are viable. Some indirect evidence favoring selfed progeny comes from the fact that certain aberrant ratios found in crosses of *P. drechsleri* could be easily explained on a diploid model if selfing had occurred (Sansome, 1970). However, selfing in several other crosses involving strains derived from the same isolates of *P. drechsleri* would result in a worse fit of the data to the expected than was found (Khaki and Shaw, 1974). Indeed, matings of Khaki's FA4 resistant mutant gave a uniform $F_1$, all of intermediate resistance. Because it is unlikely that these offspring (all A2) were the product of selfing of FA4 (A1), the result indicates that, in this cross at least, all progeny were hybrid. Possibly, in crosses of some isolates, viable progeny are always hybrid whereas in others, progeny are selfed and crossed. Alternatively, tendency to self may vary among different isolates of a species or even within a population, as indicated by recent work with polycarbonate membranes (Ko and Kunimoto, 1981). It would be very useful to know whether the ability of an isolate to respond by selfing in a polycarbonate membrane test is related to its tendency to generate selfed progeny in crosses (Khaki's isolates of *P. drechsleri* have consistently refused to self in such tests). If this is true, then polycarbonate membrane tests might be useful in choosing isolates with little or no selfing tendency. However, some selfing tendency might be regarded as a desirable attribute in a wild-type strain, because rapid inbreeding would then be inducible.

#### Conclusions from the Genetic Evidence

Today we can look back on 15 yr of genetic data with some insight gained from experience. We have to conclude that almost all of the data, particularly those more recently obtained with improved techniques, are consistent with the mass of cytologic evidence and that both confirm that *Phytophthora* spp. have gametangial meiosis and a diploid somatic phase.

## SOMATIC VARIATION IN A DIPLOID FUNGUS

Because *Phytophthora* spp. are asexually reproducing diploid fungi for a large part of their existence, recessive deleterious mutations, which would be eliminated from haploid fungi, must accumulate in heterozygous condition. Muller has pointed out that an asexual diploid organism will become less and less fit as these mutations accumulate (Muller's ratchet; see Maynard Smith, 1978). Occasional inbreeding, as found in homothallic species, would serve to eliminate deleterious recessives that become homozygous (seen as oospore abortion and subviable progeny) and to retain the fitter homozygotes. In practical terms, this means that field isolates may be heterokaryotic due to mutation and also highly heterozygous, and thus every gamete produced by meiosis would be of a different genotype. All sorts of variants would be expected even in selfed progeny. Boccas (1972) shows such a release of variation on selfing of *P. syringae*. Long et al (1975) prudently inbred their field isolates of *P. megasperma* var. *sojae* to reduce the variation in growth rate, colony morphology, and aggressiveness. The inbred derivatives were then used as wild types for mutagenesis and genetic analysis. Their results clearly show how variation in growth rate, colony morphology, and aggressiveness decreased during inbreeding. Other genetic studies using unselected field isolates have shown some evidence of segregation of mutant genes carried in heterozygous parents; e.g., auxotrophic segregants were found in progeny of matings of *P. capsici* (Timmer et al, 1970). Some attempts have been made to inbreed field isolates of the heterothallic species *P. drechsleri* by sib-matings, but problems of loss of vigor were encountered (D. S. Shaw and I. A. Khaki, *unpublished*) that may be analogous to inbreeding depression found in higher organisms. Interestingly, selfing of *P. megasperma* var. *sojae* (a habitual inbreeder) does not decrease vigor (Long et al, 1975).

### Somatic Segregation from a Monokaryon

The tendency for uninucleate, single-spore cultures of *Phytophthora* to produce variants during vegetative growth and asexual reproduction is well known and has been studied in many species. Changes in colony morphology are commonly observed among subcultures of both field isolates and laboratory strains started from single zoospores. The genetic bases of these changes are often discussed, but little evidence is ever available to allow definite conclusions to be reached. Recombination mechanisms involving fusion of nuclei of different genotypes can be excluded as a cause, but mutation and somatic segregation of chromosomal genes or extrachromosomal elements cannot.

#### Mutation

It is difficult to believe that mutations of chromosomal genes in a diploid fungus could result in the high frequencies of segregation so often observed. Experiments using potent mutagens and highly selective methods of mutant recovery have, in almost every case, produced very low frequencies of mutants even in instances where mutations have subsequently been shown to be dominant.

#### Mitotic Crossing-Over

Crossing-over during mitotic prophase in heterozygous diploid nuclei in the diploid yeasts (e.g., *Saccharomyces cerevisiae*) and in transient diploids of the Fungi Imperfecti is well understood. It leads to the formation of diploid segregants if recessive and dominant alleles become homozygous. Distal markers recombine more frequently than proximal ones (Mortimer and Hawthorne, 1969; Roper, 1966). The evidence for such a mechanism in *Phytophthora* is meager and at best circumstantial. Khaki and Shaw (1974) argued that the unexpected occurrence of mutant genes in the homozygous condition must result from somatic crossing-over in heterozygotes during the recovery of mutants, but this is a less convincing explanation of the homozygosity of Long and Keen's auxotrophs. In this latter case, unless the mutant had become homozygous immediately after induction, the heterozygote would have been eliminated. This could be taken as evidence of a spontaneous origin for the mutation in the mycelium, where mitotic crossing-over could have taken place before sporulation. Sansome (1965) suggested that the high frequency of "mutants" induced after ultraviolet (UV) treatment of *P. cactorum* (Buddenhagen, 1958) is a result of a high rate of mitotic crossing-over, which is known to be inducible in other fungi by UV treatment.

Recent evidence shows that the production of type II trisomics by mitotic crossing-over is responsible for the different karyotypes found in self-fertile strains of *P. drechsleri* (Sansome, 1980; E. Sansome, A. M. Mortimer, and D. S. Shaw, *unpublished*).

The theoretical argument for mitotic crossing-over in *Phytophthora* is attractive. Variation provided by mutation would be released from the heterozygous condition and exposed to selection from time to time during extended periods of vegetative growth. Mitotic crossing-over would also allow the predicted accumulation of deleterious genes to be halted by the elimination of these genes in unfit homozygous segregants. This last may be the more important outcome of the mechanism in species thought to be virtually asexual. Mitotic crossing-over may be more frequent in those species to compensate for the scarcity of the sexual stage, which would achieve the same result.

Clearly, studies of mitotic crossing-over are urgently needed. Strains with several linked heterozygous markers are required, but even experiments with a range of single heterozygotes, perhaps using UV or other agents to promote mitotic crossing-over, could provide valuable preliminary data.

### Extrachromosomal Genes

Some characteristics of somatic variation in *Phytophthora*, such as the persistence of somatic segregation in progenies from single zoospores (monokaryons), are similar to those shown by other fungi that are known to be under the control of nonchromosomal genetic elements (Caten, 1971; Caten and Jinks, 1968; Leonian, 1925; Shaw and Elliott, 1968; Stamps, 1953a; Upshall, 1969). The occurrence of several copies of "mutant" and "wild-type" self-replicating extrachromosomal determinants in a heteroplasmon could account for most observed kinds of segregation from monokaryons. MacIntyre and Elliott's detailed selection experiments for growth rate over several sexual and asexual generations indicate polygenic control of the continuous variation observed and cytoplasmic control of the discontinuous variation (similar to that occurring in progenies from single zoospores). In sexual progenies, the discontinuous variation actually increased with inbreeding—a finding inconsistent with chromosomal control (MacIntyre and Elliott 1974).

Although, as discussed above, the high frequency of UV-induced morphological "mutants" in *P. cactorum* (Buddenhagen, 1958) might be explained by mitotic crossing-over, an alternative and equally attractive explanation is that mutations occurred in an extrachromosomal genome that was particularly sensitive to UV. It is well known that mitochondrial mutants, such as "petite" mutants of *S. cerevisiae*, can be induced at high frequency by UV and by other mutagens (Roodyn and Wilkie, 1968).

Mitochondrial DNA is a prominent feature of the zoospore stained with a fluorochrome specific for adenine-thymine base pairs in Figure 4. The mDNA must be a strong candidate for the location of mutations showing nonchromosomal behavior and perhaps particularly for those induced in the zoospore with abundant mitochondria. We might expect to find mitochondrial mutants particularly among a collection of mutants with defects in mitochondrial function. Some mutations conferring resistance to antibacterial antibiotics

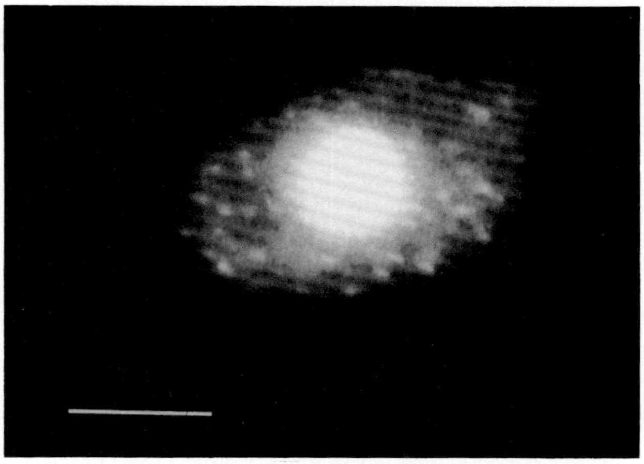

**Fig. 4.** Zoospore of *Phytophthora infestans* stained with fluorochrome DAPI. Epi-illuminated with light of wavelength 360 nm. Note single large nucleus and peripheral DNA within the mitochondria. Scale bar represents 5 $\mu$m.

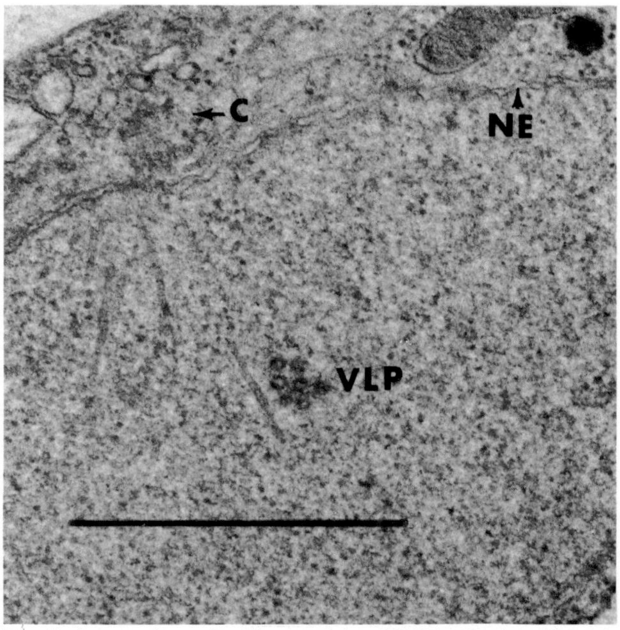


**Fig. 5.** Intranuclear viruslike particles (VLPs). Section through nucleus in 67.5-min germling of *Phytophthora drechsleri* (6503). Note cluster of VLPs, intranuclear microtubules of nucleus entering mitosis, centriole (C), and nuclear envelope (NE). Scale bar represents 1.0 $\mu$m. (Courtesy Urs-Peter Roos)

might well have a location on the mitochondrial genome. Both chromosomal and mitochondrial resistant mutants are well known in *S. cerevisiae* (Roodyn and Wilkie, 1968).

Virus particles may carry nonchromosomal genes in *Phytophthora* spp. Viruslike particles have been found in isolates of *P. infestans* (Corbett and Styer, 1976) and more recently in an isolate of *P. drechsleri* (U.-P. Roos and D. S. Shaw, *unpublished*). In *P. drechsleri*, viruslike particles were found within nuclei of zoospores, cysts, and germlings of a slow-growing variant of strain 6503 (Fig. 5). This strain has shown persistent somatic segregation of colony morphology. High concentrations of nucleic acid, probably double-stranded RNA, may thus be found in some strains. Different rates of virus replication in different nuclei could yield a zoospore population with a range of virus concentration. Conceivably, such zoospore isolates would show differing growth rates related to the frequency of virus particles in the zoospore nucleus.

Critical genetic evidence for extrachromosomal control of some somatic variation can only be obtained by using nuclear markers. Identification of nonchromosomal determinants should be possible by observing the phenotypes of marked segregants derived from a heterokaryon synthesized with a variant suspected of being nonchromosomally determined, namely the "heterokaryon test" of Arlett et al (1962).

## Somatic "Recombination" Between Monokaryons

### Heterokaryosis

Several attempts have been made to identify a mechanism that would result in new phenotypes when two "parental" isolates are mixed during vegetative growth. Only one attempt has been completely successful in identifying the mechanism responsible for this new phenotype—the study by Long and Keen (1977a), who established a prototrophic culture of *P. megasperma* var. *sojae* from a mixture of two auxotrophic mutants. The prototroph was then shown to be a heterokaryon. Heterokaryons were synthesized both within and between inbred lines by growing pairs of auxotrophs on minimal medium. Single hyphal tips from prototrophic growths were established on minimal agar, and zoospore progeny were shown to segregate clearly into the two precise phenotypes of the "parental" monokaryons.

Fortunately, sterility of both monokaryons was complemented in the heterokaryon, allowing sexual progeny to be analyzed. Their results provided further genetic evidence of diploidy (Long and Keen, 1977b). Long and Keen (1977a) remarked on the low frequency of formation of their heterokaryons, even when strong selection pressures were applied. This may mean that heterokaryons are rarely formed in the field and may persist only when selected. Heterokaryons composed of haploid nuclei can provide a considerable variety of different phenotypes from two monokaryons, due to the differing ratios of the component nuclei that are selected in different situations. In diploids, the situation is probably similar but even more open to variation, as allele doses could vary in different heterokaryons composed of the same alleles, even if this nuclear ratio were a constant 1:1. For example, phenotypes of heterokaryon *Aabb* + *aaBb* might be quite different from heterokaryon *AAbb* + *aaBB* if alleles *A* or *B* were not sufficiently dominant to express themselves fully in a 1:3 ratio with *a* or *b*.

### Somatic "Recombination" in *P. infestans*

The absence of a sexual cycle in *P. infestans* over much of its range has stimulated a series of projects designed to detect new phenotypes in the asexual products of a mixture of two isolates differing in race. Wilde (1961), Leach and Rich (1969), and Malcolmson (1970) all used different techniques and in each case found that new races arose that combined the race characteristics of the "parents." Malcolmson, who selected her new race on resistant potato leaves, used complex races in her mixtures; e.g., 3,4,10 + 1,2,3,7. Although possible, it is unlikely that the new race—in this case 1,2,3,4,7,10—arose by mutation or mitotic crossing-over in one of the "parents," because controls never yielded the new phenotype. Leach and Rich (1969) used simpler races, which were mixed in pure culture. "Recombinants," which combined the race characteristics of both parents, arose at high frequency without selection and were identified on leaf tissue. In some combinations, a new virulence characteristic arose that was absent in controls of the parental isolates grown singly; e.g., race 0 + race 1,4 gave rise to race 1,2,4. A similar phenomenon was observed by Denward (1970). The "recombinant" strains were in all cases shown to be monokaryons, but in certain instances a single sporangium was shown to give rise to "recombinant" zoospores as well as to both kinds of parental zoospores.

The mechanism involved in such "recombinations" remains obscure. Mutation or mitotic crossing-over in one parent could have given rise to the new virulence and to the "recombinant," but this possibility seems to be excluded by the controls. In view of the fact that the genetic determination of virulence is unknown, it is not possible to say that heterokaryons were formed, but it does seem likely that hyphal fusions took place between parental races. The possibilities after such an event include cytoplasmic mixing (maybe resulting in a complex nucleocytoplasmic interaction) and true heterokaryon formation followed by nuclear fusion, which may or may not have been followed by chromosome loss and a return to diploidy. Much more could obviously be learned if such experiments could be repeated using chromosomal genes as markers.

A step in this direction has been taken by Dyakov and Kuzovnikova (1974) and Shattock and Shaw (1976), who mixed pairs of isolates of different race on agar medium containing two drugs. Each isolate was labeled by a drug-resistant mutation. Although the groups used different drugs and different techniques, they both found that very vigorous, doubly resistant "recombinants" were selected and that the "recombinant" phenotype was transmitted by all single zoospores. Fluorometric measurements of Shattock and Shaw's material did not reveal differences in DNA content of nuclei between parents and recombinants (A. M. Mortimer, *personal communication*; Shattock and Shaw, 1976).

Dyakov and Kulish (1979) and Kulish and Dyakov (1979) went on to show that one of their new strains was more resistant to UV than its "parents" and that UV induced a segregation of resistance and virulence in zoospore progeny. Because both race characteristics of the "parents" reappeared among the populations of segregants, the "recombinant" is unlikely to have been derived from only one parent by mutation. Both the volume and relative DNA content of zoospores and hyphal nuclei of the "recombinant" and a derived segregant were claimed to be twice those of the "parental" strains.

These results, together with the exceptional vigor of both the Russian and the Welsh "recombinants," argue for the formation of a polyploid (tetraploid if the wild-type is taken as a diploid) by the fusion of a diploid nucleus from each "parental" strain. If all segregants turn out to be tetraploid, then mitotic crossing-over stimulated by UV could account for the observed phenotypes.

What we need now is more data, if possible from

experiments using strains with several known chromosomal markers, and a more extensive study of ploidy levels in wild-type, mutant, and recombinant strains. From the data we have available, we can conclude that parasexual processes possibly occur, but evidence is fragmentary for a cycle similar to the one found in Fungi Imperfecti involving nuclear fusion and stages analogous to Pontecorvo's haploidization (Pontecorvo, 1956). A parasexual mechanism with completely new features may yet be found.

In addition to the evidence for "recombination" in *P. infestans* referred to above, a study of the frequencies of the different races occurring in populations of *P. infestans* from potato fields in North Wales shows that the frequencies observed are almost exactly those expected if free recombination occurs among all single virulence factors within the area (Shattock et al, 1977). Thus, an unrestricted recombination among races is predicted to take place during vegetative growth.

A possible alternative mechanism of recombination in the field might involve the induced selfing of *P. infestans* by A2 mating types of other species, a phenomenon mentioned earlier. Anastomoses of hyphae of different physiologic races could allow heterokaryon formation; rare selfing, meiosis, and karyogamy would then result in outbreeding comparable to that between A1 and A2 mating types. One may ask how populations of the fungus outside Mexico have retained sexuality and normal meiosis even though these processes were made redundant by mating type separation centuries ago. The answer may be that sexuality is not redundant but that survival depends on occasional selfing induced by chance meetings with A2 isolates of other species.

Any somatic "recombination" between strains presumably requires some form of fusion between somatic structures from each strain. Several of those workers who produced race recombinants in *P. infestans* observed fusions between hyphae in mixed cultures. Leach and Rich (1969) claim to have seen anastomosis only in mixtures yielding new races. Wilde (1961) provides a drawing of anastomosis, whereas Dyakov and Kuzovnikova-Vishtynetskaya (1973) produce several photographs of bridges between trunk hyphae. The latter also claim to have seen hyphal contents pass through bridges. In their search for hyphal fusions in mixed A1 and A2 cultures of *P. capsici, P. drechsleri*, and *P. infestans*, Stephenson et al (1974c) produced evidence for bridges between hyphae of different parentage and also between hyphae of two different A1 mating types of *P. capsici.*

It is fairly clear that anastomosis does take place, but no estimates have been provided of the frequency of its occurrence. Perhaps in mixture combinations where "recombinants" are common even without selection, hyphae of "parental" isolates anastomose freely, whereas in combinations in which strong selection pressure yields only rare recombinants, anastomosis is rare or the unions produced are incompatible.

Fusion of zoospores has been reported in the laboratory for *P. infestans* (Hiddema and Kole, 1954) and therefore is a possible means for heterokaryon synthesis in the field, where zoospores of different genotype find themselves in the same drop of water. It will be interesting to find out whether zoospores or their germ tubes fuse when dense aggregates form spontaneously or in response to attractants from the host (Hickman and Ho, 1966; Porter and Shaw, 1978).

From an examination of the structure of hyphae and nuclei, one might predict that heterokaryons, if formed by fusion, mutation, or mitotic crossing-over, would tend to be unstable and transient. Mixing of the large nuclei would be difficult or impossible except in the largest trunk hyphae, whose nuclei can be seen "two abreast." In smaller hyphae, the nucleus occupies most of the lumen of the hypha, with the result that side branches would almost certainly become monokaryotic unless mass flow of cytoplasm and nuclei occurred. We still know little or nothing about the behavior of nuclei in a mycelium; this is an area in which careful study would be rewarding.

## HYBRIDIZATION BETWEEN SPECIES

Most self-sterile isolates of heterothallic species, as well as forming oospores when mated with isolates of the opposite mating type belonging to the same species, also form oospores when paired with compatible isolates of other species (Apple, 1959; Barrett, 1948; Brasier, 1972; Galindo and Gallegly, 1960; Haasis and Nelson, 1963; Huguenin, 1973; Ribeiro et al, 1975; Savage et al, 1968; Shepherd, 1978). This observation has allowed mating types to be standardized throughout the genus (Savage et al, 1968). Clinton (1911) noted oospores in pairings of *P. infestans* and *P. phaseoli* and claimed that they were the result of hybridization. Even as early as 1918, Murphy suggested an alternative explanation, that the normally self-sterile isolates were encouraged to self in the presence of each other. More recent work (Boccas, 1981; Boccas and Zentmyer, 1976; Erselius, 1980; Kaosiri and Zentmyer, 1980; Kaosiri et al, 1980) shows that the viability of oospores produced in interspecies pairings is low but that reasonable numbers of oospores can be germinated. Protein electrophoresis enables accurate identification of the parentage of the single-oospore isolates. In most progenies, segregation was found for growth rate, temperature sensitivity, pathogenicity, and occasionally for mating type. Protein patterns were always identical in most respects to one or the other parent. The interpretation offered is that although hybrid gametangial combinations may occur frequently (Sansome et al, 1979; Savage et al, 1968), abortion of hybrid oospores occurs before or during germination; the viable oospores are the result of selfing of one or both parents. Segregants are most probably due to selfing of parents that are heterozygous for genes influencing those characters that segregate.

A rare exception to selfing was detected in only one oospore culture from a *capsici* × *palmivora* mating (Boccas, 1981). From its intermediate protein pattern, this isolate appears to have been a true hybrid. In general, then, interspecific hybridization, even between species thought to be rather closely related (e.g., *P. palmivora* and *P. nicotianae* var. *parasitica*), appears to be unlikely to contribute to variation seen in the field. However, rare hybrids might possibly survive and mate further with parental types, causing introgression, or new allopolyploid species might arise, initially by hybridization of distinct genomes, and then by chromosome doubling.

## INHERITANCE AND DETERMINATION OF MATING TYPE

One of the most readily available characters for study of inheritance pattern is mating type (compatibility type). In all the genetic work done with heterothallic isolates, mating type data have been collected and an interpretation attempted. The following features of mating type inheritance emerge:

1) In some studies, the ratios of A1 and A2 types in individual progenies rarely deviate significantly from 1:1, but a consistent excess of one mating type may be found in most of the crosses analyzed (e.g., Timmer et al, 1970).

2) In other studies, a large excess of one mating type is usual; that is, deviation from 1:1 is significant (e.g., Khaki and Shaw, 1974).
3) With some exceptions, the mating type of an isolate is stable indefinitely during vegetative growth and asexual reproduction. The A2 type of *P. infestans* is unknown outside Mexico.
4) In interspecific pairings where selfing is induced, the A2 parent, but not the A1 parent, can give rise to progeny of both mating types (Boccas, 1981; Boccas and Zentmyer, 1976).
5) In matings within many heterothallic species, a proportion of single-oospore colonies may be self-fertile (homothallic) (e.g., Erselius, 1980; Khaki and Shaw, 1974). Self-fertility may be stable and be transmitted by the single zoospore (Laviola, 1968; Niederhauser, 1961; Webster, 1974). In other cases, A1 and A2 may segregate at zoosporogenesis in the germ sporangium (Laviola, 1968) or later during growth and zoosporogenesis when mating type mosaics are produced (Mortimer et al, 1977).
6) Single-zoospore progeny from mating type mosaics of *P. drechsleri* may be A1 or A2 or self-fertile (A1A2). Further zoospore analysis of these progeny shows that A1A2 colonies may be stable or may segregate sterile A1 and A2 colonies. Some A2 colonies are also unstable and segregate A1A2 and A1 types (Mortimer et al, 1977).
7) Analyses of meiotic metaphases of self-fertile isolates derived from single oospores of *P. drechsleri* show that the nuclei are trisomic. The extra chromosome is associated with the four chromosomes showing translocation heterozygosity that have been found in all heterothallic species examined (E. Sansome, A. M. Mortimer, and D. S. Shaw, *unpublished*).

These data are most easily interpreted as follows. The trisomy in fertile and in sterile subcultures recently derived from self-fertile single-oospore cultures accounts for the observation that both the A1 and A2 characters may reside within a uninucleate zoospore. (A heterokaryon of A1 and A2 nuclei can be excluded as an explanation here.) The fact that trisomics are tertiary, i.e., involve duplication of a chromosome of the translocation complex, firmly establishes that mating type determinants are associated with segments of this complex. Because A1 and A2 segregants are still trisomic and may (at least in the case of A2 segregants) give rise to the opposite mating type or the self-fertile A1A2 type during further vegetative growth, whole chromosome loss can be ruled out as being responsible for the instability. Instead, there is evidence of chromosomal rearrangements in the segregants, which could be easily generated by mitotic crossing-over of segments carrying mating type genes.

This hypothesis, outlined by Sansome (1980), suggests that the A2 mating type is heterozygous for a gene (or several tightly linked genes) located on a segment of the translocation heterozygote and that the A1 mating type is homozygous at this locus. Thus, selfed progeny of an A2 parent include A1 mating types, but all the progeny of selfed A1 mating types are A1. The stability of the homozygous A1 mating type is assured, but if the A2 mating type is heterozygous, then mitotic crossing-over yields the A1 homozygote unless the gene(s) is located by the centromere or, as seems to be the case in *Phytophthora*, on a segment involved in reciprocal translocation. Mitotic crossing-over within a translocation complex would be suppressed because products would be imbalanced and thus nonviable. In contrast, the presence of extra segments in a tertiary trisomic results in the survival of some products of mitotic crossing-over and thus the instability of such cultures. The varying ratios of A1 and A2 mating types so often observed in crosses of wild-type isolates of many species could result from a mechanism that retains heterozygosity of the translocation, such as the one outlined in Figure 2.

One might expect that when the four chromosomes of the translocation heterozygote associate at meiosis, some disjunctions would result in unbalanced disomic meiotic products; others would result in normal balanced products. Fusion of a disomic gamete with a normal gamete would then produce a trisomic zygote, and thus trisomy in the progeny might be a natural consequence of translocation heterozygosity.

Although wild isolates of a single mating type are usually stable during vegetative growth and sporulation, instabilities have been noted (Table 2). These investigations suggest that aged inocula of some A1 isolates and some A2 isolates may give rise to colonies bearing abundant oospores. Tsao et al (1980) found that their self-fertile colonies of *P. parasitica* were unstable and reverted to the original mating type, A1 or A2, during vegetative growth. In *P. capsici* isolated from pumpkin by W. Kamjaipai (*personal communication*), self-fertile colonies grown from aged inocula were derived from most A2 isolates but not from A1 isolates. Single-zoospore propagations from these self-fertile colonies were mostly A2 but included a proportion of A1 types. Ko (1981) analyzed zoospores of his self-fertile colonies from aged inocula of both A1 and A2 isolates of *P. parasitica*. Here again, most single-zoospore colonies were of the mating type of the original isolate before aging, but some were of the opposite mating type. Very rare self-fertile colonies (A1A2) were recovered in one case.

The change to self-fertility may also be induced by the fungicide chloroneb in *P. capsici*, apparently only in the A2 mating type (Noon and Hickman, 1974). Treated cultures continued to produce oospores even in the absence of chloroneb. Ko (1981) induced similar self-fertility in A2 isolates of *P. parasitica* grown on chloroneb and showed that the fertility was associated with the somatic segregation of the A1 mating type; most zoospore colonies propagated from induced fertile sectors were A2, but A1 and very rare self-fertile (A1A2) types were found.

The above data are fully consistent with the hypothesis of a mating type complex derived from the cytologic studies of heterothallic species and the cytologic and genetic studies of mating type instability in *P. drechsleri*. The appearance or induction of A1 segregants in some A2 cultures can be accounted for by a more distal location of the chromosomal segment bearing mating type gene(s) in these isolates, allowing mitotic crossing-over in *Aa* genotypes to produce *aa* (A1) genotypes. Alternatively, cultures giving rise to the opposite mating type may be trisomic, *Aaa* (A1) and *AAa* (A2). Rare crossing-over favored by aging or chloroneb treatment would result in the segregation of the opposite mating type, giving a mating type heterokaryon or a local physical mixture of compatible hyphae, either of which would produce a fertile sector. If the two genotypes became physically separated, vigorous vegetative growth would continue, resulting in overgrowth of the fertile segment by sterile hyphae of a single mating type, as observed by Tsao et al (1980). Clearly, more cytologic and genetic work is required before the details of the hypothesis can be developed further.

Further recent support for the existence of a "mating type complex" of chromosomes has come from the finding of reciprocal translocation heterozygosity in *Bremia lactucae* and its association with trisomy in unstable self-fertile isolates; these naturally occurring self-fertile isolates behave almost identically to those of *P. drechsleri* isolated as single oospores

TABLE 2. Occurrence of isolates bearing abundant oospores derived from single mating types of heterothallic species of *Phytophthora*

| Species | Mating type of parent isolate | Behavior | Mating types of zoospore progeny | Treatment | Study |
|---|---|---|---|---|---|
| *P. infestans* | Presumably A1 | One isolate highly fertile | ...[a] | None | Tucker (1931) |
| *P. parasitica* | ... | 35 of 57 strains self-fertile | ... | None | Tucker (1931) |
| *P. parasitica* (and others) | ... | Patches of oospores in culture; subcultures "homothallic" but have strong A2 reaction | ... | None, but oospores most abundant on dilute or thin agar medium | Savage et al (1968) |
| "*Phytophthora* MF4" (also *P. palmivora*) | Some A1 and some A2 isolates became fertile | New cultures from old stock were most fertile; subcultures from these became self-fertile | ... | Aging | Brasier (1972) |
| *P. capsici* | A2 not A1 | Treated colonies showed fertile sectors | ... | Growth on medium with chloroneb | Noon and Hickman (1974) |
| *P. parasitica* | A1<br>A2 | Colonies showed fertile sectors | Self-fertile and A1<br>Self-fertile and A2 | Aging? | Shepherd (1978) |
| *P. capsici* | A2 not A1 | Colonies showed fertile sectors | Progeny from fertile sectors A2 but also A1 | Aging | W. Kamjaipai (*personal communication*) |
| *P. parasitica* (also "*Phytophthora* MF4") | A1 and A2 | Young subcultures from old stock fertile; subcultures from periphery sterile and same mating type as parental isolate | ... | Aging more than 6 mo | Tsao et al (1980) |
| *P. parasitica* | A1 and A2 | Colonies showed fertile sectors | A1, A2, and rarely, self-fertile | Aging | Ko (1981) |
| | A2 | | A1, A2, and rarely, self-fertile | Growth on medium with chloroneb | Ko (1981) |

[a]... = Not tested.

(Michelmore and Ingram, 1982; Michelmore and Sansome, 1982).

## SUMMARY

After nearly a century of study and the accumulation of a considerable amount of data, general agreement has been reached that a classical meiosis takes place in the antheridium and the oogonium and that *Phytophthora* is a diploid fungus. Although recent work has concentrated on detailed comparison of some karyotypes within the genus, many more comparisons both within and among present taxonomic divisions remain to be made. Chromosome size and number are now accepted as valuable criteria in delimiting taxa. The significance of chromosomal rearrangements identified at meiotic metaphase in heterothallic species is now beginning to be realized. A complex of chromosomes, "the mating type complex," is thought to be the carrier of mating type determinants.

The activities and fates of nuclei in the coenocytic mycelium and in the development of sporangium and zoospore are still mysteries. Cytologic studies would complement those made using genetic techniques to learn more about the synthesis, maintenance, and breakdown of heterokaryons and the compatibility of somatic unions between lines with genetically distinct nuclei.

Genetic investigations of *Phytophthora* have not been easily accomplished but have provided evidence consistent with gametangial meiosis. Much has now been learned about the pitfalls likely to be encountered on the pathway to reliable genetic data. Early evidence, interpreted at the time to support a haploidy hypothesis, can be reexamined in the light of experience. Expectations in sexual progenies may be modified not only in crosses showing low oospore viability but also by the occurrence of selfing, by the presence of natural and induced recessive mutations, and by the polygenic determination of natural markers.

Rapid progress still depends on much more spadework in refining and controlling chosen model genetic systems and in the isolation and careful preservation of a range of the best possible mutants for use in that system. Then we should be able to do the critical experiments needed to unequivocally identify mechanisms such as those underlying somatic variation. Perhaps the occurrence of mitotic crossing-over and the segregation of extrachromosomal genes proposed by so many will soon be verified.

Hopefully, improved systems will also allow a reexamination of the determination of physiologic race in *P. infestans* and subsequently lead to an understanding of the mechanisms resulting in somatic "recombination" and a fuller knowledge of the determination of mating type and sexuality.

*Phytophthora* genetics is passing from its infancy to an awkward adolescence. No one pretends that work in this field is not slow and frustrating, but findings over the last few years have shown how more rapid progress can be made in developing substantial genetic analyses that will eventually allow the solution of many problems of a more applied nature.

## ACKNOWLEDGMENTS

I wish to thank E. R. Sansome for her stimulating discussions and her critical reading of the manuscript. Thanks are also due to colleagues at the School of Plant Biology, who helped in so many ways.

## LITERATURE CITED


Allain, A. 1934. Cytologie de l'oeuf du *Phytophthora cambivora*. C. R. Seances Soc. Biol. 117:778-780.

Andreoli, C., and Ponchet, J. 1974. Recherche de mutants auxotrophes de *Phytophthora nicotianae* F. *parasitica* (Dastur) Waterhouse. Ann. Phytopathol. 6:255-264.

Apple, J. L. 1959. Sexuality of *Phytophthora parasitica* var. *nicotianae*. Phytopathology 49:37-43.

Arlett, C. F., Grindle, M., and Jinks, J. L. 1962. The "red" cytoplasmic variant of *Aspergillus nidulans*. Heredity 17:197-209.

Bakerspigel, A. 1960. Nuclear structure and division in the vegetative mycelium of the Saprolegniaceae. Am. J. Bot. 47:94-100.

Barksdale, A. W. 1968. Meiosis in the antheridium of *Achlya ambisexualis* E87. J. Elisha Mitchell Sci. Soc. 84:187-194.

Barrett, J. T. 1948. Induced oospore production in the genus *Phytophthora*. (Abstr.) Phytopathology 38:2.

Blackwell, E. 1943. The life history of *Phytophthora cactorum* (Leb. and Cohn.) Schroet. Trans. Br. Mycol. Soc. 26:71-89.

Boccas, B. 1972. Contribution à l'étude du cycle chez les *Phytophthora*. Analyse du mode de transmission d'un caractère génétique quantitatif chez une espèce homothallique, le *Phytophthora syringae* Kleb. C. R. Hebd. Seances Acad. Sci. Ser. D 275:663-666.

Boccas, B. 1973. Observations preliminaires sur l'hérédité du pouvoir pathogène chez le *P. palmivora* (Butl.) Butl. Cah. ORSTOM Ser. Biol. 20:51-56.

Boccas, B. R. 1981. Interspecific crosses between closely related heterothallic *Phytophthora* species. Phytopathology 71:60-65.

Boccas, B., and Zentmyer, G. A. 1976. Genetical studies with interspecific crosses between *Phytophthora cinnamomi* and *Phytophthora parasitica*. Phytopathology 66:477-484.

Bosc, M. 1946. Sur la structure des noyaux et la méiose de *Plasmopara viticola* (Berk. et Curt.) Berl. et de Toni. C. R. Hebd. Seances Acad. Sci. Ser. D 223:584-586.

Brasier, C. M. 1971. Induction of sexual reproduction in single A2 isolates of *Phytophthora* species by *Trichoderma viride*. Nature (London) New Biol. 231:283.

Brasier, C. M. 1972. Observations on the sexual mechanism in *Phytophthora palmivora* and related species. Trans. Br. Mycol. Soc. 58:237-251.

Brasier, C. M., and Griffin, M. J. 1979. Taxonomy of *Phytophthora palmivora* on cocoa. Trans. Br. Mycol. Soc. 72:111-143.

Brasier, C. M., and Sansome, E. 1975. Diploidy and gametangial meiosis in *Phytophthora cinnamomi*, *P. infestans* and *P. drechsleri*. Trans. Br. Mycol. Soc. 65:49-65.

Bryant, T. R., and Howard, K. L. 1969. Meiosis in the Oömycetes: I. A microspectro-photometric analysis of nuclear deoxyribonucleic acid in *Saprolegnia terrestris*. Am. J. Bot. 56:1075-1083.

Buddenhagen, I. W. 1958. Induced mutations and variability in *Phytophthora cactorum*. Am. J. Bot. 45:355-365.

Castro, J., Zentmyer, G. A., and Belser, W. L., Jr. 1971. Induction of auxotrophic mutants in *Phytophthora* by ultraviolet light. Phytopathology 61:283-289.

Caten, C. E. 1971. Single zoospore variation in *Phytophthora infestans* and attenuation of strains in culture. Trans. Br. Mycol. Soc. 56:1-7.

Caten, C. E., and Day, A. W. 1977. Diploidy in plant pathogenic fungi. Annu. Rev. Phytopathol. 15:295-318.

Caten, C. E., and Jinks, J. L. 1968. Spontaneous variability of single isolates of *Phytophthora infestans*. I. Cultural variation. Can. J. Bot. 46:329-348.

Clinton, G. P. 1911. Oospores of potato blight, *Phytophthora infestans*. Pages 753-774 in: Conn. Agric. Exp. Stn. Bienn. Rep. 1909–10 Part 10.

Corbett, M. K., and Styer, E. L. 1976. Intranuclear viruslike particles in *Phytophthora infestans*. Proc. Am. Phytopathol. Soc. 3:332.

Day, P. R. 1974. Genetics of Host-Parasite Interaction. W. H. Freeman and Co., San Francisco. 238 pp.

de Bary, A. 1887. Comparative Morphology and Biology of the Fungi, Mycetozoa and Bacteria. Clarendon Press, Oxford. 525 pp.

Denward, T. 1970. Differentiation in *Phytophthora infestans*. II. Somatic recombination in vegetative mycelium. Hereditas 66:35-48.

Dick, M. W., and Win-Tin. 1973. The development of cytological theory in the Oomycetes. Biol. Rev. 48:133-158.

Dyakov, Y. T., and Kulish, V. B. 1979. Genetic investigation of strains of *Phytophthora infestans* (Mont.) de Bary obtained as a result of paired cultivations of mutant cultures. Genetika 15:49-56.

Dyakov, Y. T., and Kuzovnikova, T. A. 1974. Heterokaryosis in *Phytophthora infestans* (Mont.) de Bary. II. Genetic investigations. Mikol. Fitopatol. 8:81-89. (In Russian)

Dyakov, Y. T., and Kuzovnikova-Vishtynetskaya, T. A. 1973. Heterocaryosis of *Phytophthora infestans* (Mont.) de Bary. I. Cytological investigations. Mikol. Fitopatol. 7:479-485. (In Russian)

Dyakov, Y. T., and Vishtynezkaja, T. A. 1973. Mutants of *Phytophthora infestans* (Mont.) de Bary resistant to medicinal substances. Vestn. Mosk. Univ. Ser. VI. 28:57-60. (In Russian, English summary)

Elliott, C. G., and MacIntyre, D. 1973. Genetical evidence on the life-history of *Phytophthora*. Trans. Br. Mycol. Soc. 60:311-316.

Ellzey, J. T. 1974. Ultrastructural observations of meiosis within antheridia of *Achlya ambisexualis*. Mycologia 66:32-47.

Ellzey, J. T., and Huizar, E. H. 1977. Synaptonemal complexes in antheridia of *Achlya ambisexualis* E87. Arch. Microbiol. 112:311-313.

Erselius, L. J. 1980. A study of mating in *Phytophthora* species. Ph.D. thesis, University of Wales, Bangor. 82 pp.

Erselius, L. J., and Boccas, B. 1978. Gametangial meiosis in cocoa black pod *Phytophthora* isolates. Ann. Phytopathol. 10:287-294.

Erwin, D. C., Zentmyer, G. A., Galindo, J., and Niederhauser, J. S. 1963. Variation in the genus *Phytophthora*. Annu. Rev. Phytopathol. 1:375-396.

Flanagan, P. W. 1970. Meiosis and mitosis in Saprolegniaceae. Can. J. Bot. 48:2069-2076.

Galindo A., J., and Gallegly, M. E. 1960. The nature of sexuality in *Phytophthora infestans*. Phytopathology 50:123-128.

Galindo A., J., and Zentmyer, G. A. 1967. Genetical and cytological studies of *Phytophthora* strains pathogenic to pepper plants. Phytopathology 57:1300-1304.

Gallegly, M. E. 1968. Genetics of pathogenicity of *Phytophthora infestans*. Annu. Rev. Phytopathol. 6:375-396.

Gallegly, M. E. 1970. Genetics of *Phytophthora*. Phytopathology 60:1135-1141.

Graham, K. M. 1954. Nuclear behaviour in *Phytophthora infestans* (Mont.) de Bary. (Abstr.) Phytopathology 44:490.

Gwynne-Vaughan, H. C. I., and Barnes, B. 1927. The Structure and Development of the Fungi. Cambridge University Press, London. 384 pp.

Haasis, F. A., and Nelson, R. R. 1963. Studies on the biological relationship of species of *Phytophthora* as measured by oospore formation in intra- and interspecific crosses. Plant Dis. Rep. 47:705-709.

Hartog, M. 1895. On the cytology of the vegetative and reproductive organs of the Saprolegnieae. Trans. R. Ir. Acad. 30:649-708.

Hemmes, D. E., and Hohl, H. R. 1973. Mitosis and nuclear degeneration: Simultaneous events during secondary sporangia formation in *Phytophthora palmivora*. Can. J. Bot. 51:1673-1675.

Hickman, C. J., and Ho, H. H. 1966. Behaviour of zoospores in plant-pathogenic Phycomycetes. Annu. Rev. Phytopathol. 4:195-220.

Hiddema, J., and Kole, A. P. 1954. Enkele waarnemingen over versmelten van zoösporen bij *Phytophthora infestans* (Mont.) de Bary. Tijdschr. Plantenziekten 60:138-139.

Hopwood, D. A. 1970. The isolation of mutants. Pages 363–433 in: Methods in Microbiology, Vol. 3A. J. R. Norris and D. W. Ribbons, eds. Academic Press, London and New York. 505 pp.

Howard, K. L., and Bryant, T. R. 1971. Meiosis in the Oömycetes: II. A microspectrophotometric analysis of DNA in *Apodachlya brachynema*. Mycologia 63:58-68.

Howard, K. L., and Moore, R. T. 1970. Ultrastructure of oogenesis in *Saprolegnia terrestris*. Bot. Gaz. 131:311-336.

Huguenin, B. 1973. Application de la technique du marquage fluorescent à la détermination du comportement sexuel du *Phytophthora palmivora*. Cah. ORSTOM Ser. Biol. 20:59-61.

Huguenin, B., and Boccas, B. 1970. Etude de la caryocinèse chez le *Phytophthora parasitica* Dastur. C. R. Hebd. Seances Acad. Sci. Ser. D 271:660-663.

Kaosiri, T. 1978. Morphological, taxonomic and cytological studies of *Phytophthora palmivora*. Ph.D. dissertation, University of

California, Riverside. 148 pp.

Kaosiri, T., and Zentmyer, G. A. 1980. Protein, esterase, and peroxidase patterns in the *Phytophthora palmivora* complex from cacao. Mycologia 72:988-1000.

Kaosiri, T., Zentmyer, G. A., and Erwin, D. C. 1980. Oospore morphology and germination in the *Phytophthora palmivora* complex from cacao. Mycologia 72:888-907.

Khaki, I. A. 1974. Genetical studies of a species of *Phytophthora* pathogenic to pepper. Ph.D. thesis. University of Wales, Bangor. 92 pp.

Khaki, I. A., and Shaw, D. S. 1974. The inheritance of drug resistance and compatibility type in *Phytophthora drechsleri*. Genet. Res. 23:75-86.

Ko, W. H. 1978. Heterothallic *Phytophthora*: Evidence for hormonal regulation of sexual reproduction. J. Gen. Microbiol. 107:15-18.

Ko, W. H. 1981. Reversible changes of mating type in *Phytophthora parasitica*. J. Gen. Microbiol. 125:451-454.

Ko, W. H., and Kunimoto, R. K. 1981. Hormone production and reception among different isolates of *Phytophthora parasitica* and *P. palmivora*. Mycologia 73:440-444.

Kulish, V. B., and Dyakov, Y. T. 1979. Isolation of diploid strains from mixed cultures of two physiological races of *Phytophthora infestans* (Mont.) de Bary. Dokl. Akad. Nauk SSSR Ser. Biol. 244:735-738.

Laviola, C. 1968. Studies on the genetics of *Phytophthora infestans*. Ph.D. dissertation, W. Va. Univ., Morgantown. 72 pp.

Laviola, C. 1975. On the origin of multinuclearity of sporangia in *Phytophthora infestans* (Mont.) de Bary. Phytopathol. Mediterr. 14:30-31.

Laviola, C., and Portacci, M. 1974. Osservazioni citologiche sulle strutture sessuali di *Phytophthora infestans* (Mont.) de Bary. Phytopathol. Mediterr 13:87-92.

Leach, S. S., and Rich, A. E. 1969. The possible role of parasexuality and cytoplasmic variation in race differentiation in *Phytophthora infestans*. Phytopathology 59:1360-1365.

Legrand-Pernot, F., and Pellegrin, F. 1976. Nature du cycle végétatif des *Phytophthora parasitica* Dastur et *palmivora* (Butler) Butler. Ann. Phytopathol. 8:379-388.

Leonian, L. H. 1925. Physiological studies on the genus *Phytophthora*. Am. J. Bot. 12:444-498.

Long, M., and Keen, N. T. 1977a. Evidence for heterokaryosis in *Phytophthora megasperma* var. *sojae*. Phytopathology 67:670-674.

Long, M., and Keen, N. T. 1977b. Genetic evidence of diploidy in *Phytophthora megasperma* var. *sojae*. Phytopathology 67:675-677.

Long, M., Keen, N. T., Ribeiro, O. K., Leary, J. V., Erwin, D. C., and Zentmyer, G. A. 1975. *Phytophthora megasperma* var. *sojae*: Development of wild-type strains for genetic research. Phytopathology 65:592-597.

MacIntyre, D., and Elliott, C. G. 1974. Selection for growth rate during asexual and sexual propagation in *Phytophthora cactorum*. Genet. Res. 24:295-309.

Maia, N., Venard, P., and Lavrut, F. 1976. Etude des divisions mitotiques et méiotiques du cycle de *P. capsici*. Ann. Phytopathol. 8:141-146.

Malcolmson, J. F. 1970. Vegetative hybridity in *Phytophthora infestans*. Nature (London) 225:971-972.

Marks, G. E. 1965. The cytology of *Phytophthora infestans*. Chromosoma 16:681-692.

Maynard Smith, J. 1978. The Evolution of Sex. Cambridge University Press, Cambridge. 222 pp.

Michelmore, R. W., and Ingram, D. S. 1982. Secondary homothallism and heterokaryosis in *Bremia lactucae* Regel (Peronosporaceae). Trans. Br. Mycol. Soc. 78:1-9.

Michelmore, R. W., and Sansome, E. R. 1982. Cytological studies of heterothallism and secondary homothallism in *Bremia lactucae*. Trans. Br. Mycol. Soc. (In press)

Mortimer, A. M., and Shaw, D. S. 1975. Cytofluorimetric evidence for meiosis in gametangial nuclei of *Phytophthora drechsleri*. Genet. Res. 25:201-205.

Mortimer, A. M., Shaw, D. S., and Sansome, E. R. 1977. Genetical studies of secondary homothallism in *Phytophthora drechsleri*. Arch. Microbiol. 111:255-259.

Mortimer, R. K., and Hawthorne, D. C. 1969. Yeast genetics. Pages 386-460 in: The Yeasts. Vol. 1. A. H. Rose and J. S. Harrison, eds. Academic Press, London. 508 pp.

Murphy, P. A. 1918. The morphology and cytology of the sexual organs of *Phytophthora erythroseptica* Pethyb. Ann. Bot. (London) 32:115-153.

Niederhauser, J. S. 1961. Genetic studies of *Phytophthora infestans* and *Solanum* species in relation to late-blight resistance in potato. Pages 491-497 in: Recent Advances in Botany. Proc. Int. Bot. Congr. 9th. Univ. Toronto Press. 1,766 pp.

Noon, J. P., and Hickman, C. J. 1974. Oospore production by a single isolate of *Phytophthora capsici* in the presence of chloroneb. Can. J. Bot. 52:1591-1595.

Pontecorvo, G. 1956. The parasexual cycle in fungi. Annu. Rev. Microbiol. 10:393-400.

Porter, J. R., and Shaw, D. S. 1978. Aggregation of *Phytophthora drechsleri* zoospores; pattern analysis suggests a taxis. Trans. Br. Mycol. Soc. 71:515-518.

Ribeiro, O. K., Erwin, D. C., and Zentmyer, G. A. 1975. An improved synthetic medium for oospore production and germination of several *Phytophthora* species. Mycologia 67:1012-1019.

Romero, S., and Erwin, D. C. 1969. Variation in pathogenicity among single-oospore cultures of *Phytophthora infestans*. Phytopathology 59:1310-1317.

Roodyn, D. B., and Wilkie, D. 1968. The Biogenesis of Mitochondria. Methuen and Co. Ltd., London. 123 pp.

Roper, J. A. 1966. Mechanisms of inheritance. 3. The parasexual cycle. Pages 589-617 in: The Fungi. Vol. 2. G. C. Ainsworth and A. S. Sussman, eds. Academic Press, New York. 805 pp.

Sansome, E. 1961. Meiosis in the oogonium and antheridium of *Pythium debaryanum* Hesse. Nature (London) 191:827-828.

Sansome, E. 1963a. Meiosis in *Pythium debaryanum* Hesse and its significance in the life history of the Biflagellatae. Trans. Br. Mycol. Soc. 46:63-72.

Sansome, E. 1963b. Cytological observations on *Sclerospora sorghi* in relation to the life history of the Oomycetes. Proc. Sci. Assoc. Nigeria 6:48-52.

Sansome, E. 1965. Meiosis in diploid and polyploid sex organs of *Phytophthora* and *Achlya*. Cytologia 30:103-117.

Sansome, E. 1970. Selfing as a possible cause of disturbed ratios in *Phytophthora* crosses. Trans. Br. Mycol. Soc. 54:101-107.

Sansome, E. 1976. Gametangial meiosis in *Phytophthora capsici*. Can. J. Bot. 54:1535-1545.

Sansome, E. 1977. Polyploidy and induced gametangial formation in British isolates of *Phytophthora infestans*. J. Gen. Microbiol. 99:311-316.

Sansome, E. 1980. Reciprocal translocation heterozygosity in heterothallic species of *Phytophthora* and its significance. Trans. Br. Mycol. Soc. 74:175-185.

Sansome, E., and Brasier, C. M. 1973. Diploidy and chromosomal structural hybridity in *Phytophthora infestans*. Nature (London) 241:344-345.

Sansome, E., and Brasier, C. M. 1974. Polyploidy associated with varietal differentiation in the *megasperma* complex of *Phytophthora*. Trans. Br. Mycol. Soc. 63:461-467.

Sansome, E., and Harris, B. J. 1962. Use of camphor-induced polyploidy to determine the place of meiosis in fungi. Nature (London) 196:291-292.

Sansome, E., Brasier, C. M., and Griffin, M. J. 1975. Chromosome size differences in *Phytophthora palmivora*, a pathogen of cocoa. Nature (London) 255:704-705.

Sansome, E., Brasier, C. M., and Sansome, F. W. 1979. Further cytological studies on the 'L' and 'S' types of *Phytophthora* from cocoa. Trans. Br. Mycol. Soc. 73:293-302.

Satour, M. M., and Butler, E. E. 1968. Comparative morphological and physiological studies of the progenies from intraspecific matings of *Phytophthora capsici*. Phytopathology 58:183-192.

Savage, E. J., Clayton, C. W., Hunter, J. H., Brenneman, J. A., Laviola, C., and Gallegly, M. E. 1968. Homothallism, heterothallism and interspecific hybridization in the genus *Phytophthora*. Phytopathology 58:1004-1021.

Shattock, R. C., and Shaw, D. S. 1975. Mutants of *Phytophthora infestans* resistant to and dependent upon antibiotics. Trans. Br. Mycol. Soc. 64:29-41.

Shattock, R. C., and Shaw, D. S. 1976. Novel phenotypes of *Phytophthora infestans* from mixed cultures of antibiotic resistant mutants. Trans. Br. Mycol. Soc. 67:201-206.

Shattock, R. C., Janssen, B. D., Whitbread, R., and Shaw, D. S. 1977.

An interpretation of the frequencies of host specific phenotypes of *Phytophthora infestans* in North Wales. Ann. Appl. Biol. 86:249-260.

Shaw, D. S. 1965. Drug resistance and morphological variation in *Phytophthora cactorum*. Ph.D. thesis, University of Glasgow, Scotland. 146 pp.

Shaw, D. S., and Elliott, C. G. 1968. Streptomycin resistance and morphological variation in *Phytophthora cactorum*. J. Gen. Microbiol. 51:75-84.

Shaw, D. S., and Khaki, I. A. 1971. Genetical evidence for diploidy in *Phytophthora*. Genet. Res. 17:165-167.

Shepherd, C. J. 1978. Mating behaviour of Australian isolates of *Phytophthora* species. I. Inter- and intra-specific mating. Aust. J. Bot. 26:123-138.

Sivak, B. 1973. The nature of the "nuclear cap" and nuclear behaviour in *Phytophthora infestans*. Can. J. Bot. 51:1983-1984.

Stamps, D. J. 1953a. Variation in a strain of *Phytophthora cactorum*. Trans. Br. Mycol. Soc. 36:248-254.

Stamps, D. J. 1953b. Oospore production in paired cultures of *Phytophthora* species. Trans. Br. Mycol. Soc. 36:255-259.

Stephenson, L. W., Erwin, D. C., and Leary, J. V. 1974a. Cytology of somatic and gametangial nuclei in *Phytophthora capsici* and *Phytophthora megasperma* var. *sojae*. Can. J. Bot. 52:2055-2060.

Stephenson, L. W., Erwin, D. C., and Leary, J. V. 1974b. Meiotic configurations in the oospore of *Phytophthora capsici*. Can. J. Bot. 52:2141-2143.

Stephenson, L. W., Erwin, D. C., and Leary, J. V. 1974c. Hyphal anastomosis in *Phytophthora capsici*. Phytopathology 64:149-150.

Swiezynski, K. M., and Jankowska, H. 1970. Drug-resistant isolates of *Phytophthora infestans*. Genet. Pol. 11:27-36.

Timmer, L. W., Castro, J., Erwin, D. C., Belser, W. L., and Zentmyer, G. A. 1970. Genetic evidence for zygotic meiosis in *Phytophthora capsici*. Am. J. Bot. 57:1211-1218.

Tommerup, I. C., Ingram, D. S., and Sargent, J. A. 1974. Oospores of *Bremia lactucae*. Trans. Br. Mycol. Soc. 62:145-150.

Tsao, P. H., Ugale, R., Hobbs, S., and Farih, A. 1980. Control of homothallic oospore formation in *Phytophthora parasitica* by culture manipulations. Trans. Br. Mycol. Soc. 75:153-156.

Tucker, C. M. 1931. Taxonomy of the genus *Phytophthora* de Bary. Mo. Agric. Exp. Stn. Res. Bull. 153. 208 pp.

Upshall, A. 1969. Variation among monosporous progeny of *Phytophthora infestans*. Can. J. Bot. 47:1171-1174.

Webster, R. K. 1974. Recent advances in the genetics of plant pathogenic fungi. Annu. Rev. Phytopathol. 12:331-353.

Wilde, P. 1961. Ein Beitrag zur Kenntnis der Variabilität von *Phytophthora infestans* (Mont.) de Bary. Arch. Microbiol. 40:163-195.

# 7

# Motility, Taxis, and Tropism in *Phytophthora*

Michael J. Carlile

CONTENTS



## INTRODUCTION

The zoospores of *Phytophthora* can be moved passively by water currents but are also motile, swimming by means of flagella. If motile cells possess sensory systems, they are able to respond to environmental factors so as to accumulate at some locations and avoid others. A variety of such tactic responses or taxes (singular taxis) are known in *Phytophthora.*

Fungal hyphae show protoplasmic streaming and polarized growth. Such polarized growth, if accompanied by biased protoplasmic streaming, has some attributes of motility, because a net translocation of living matter occurs. Moreover, fungal hyphae sometimes possess sensory systems that permit growth to be precisely oriented with respect to environmental stimuli. Such tropic responses (tropisms) have so far received little attention in *Phytophthora* but may be important.

I have recently discussed the mechanisms of sensory transduction (Carlile, 1980a) and the ecological significance of motility, taxis, and tropism in microorganisms (Carlile, 1980b). In this chapter, the motility, taxis, and tropism of *Phytophthora* are reviewed in more detail, and reference is made to work on other microorganisms, especially other oomycetes, where information on *Phytophthora* is sparse or comparison helpful. Earlier reviews covering a similar area are those by Gooday (1975), Hickman (1970), Hickman and Ho (1966), and Zentmyer (1970).

## ZOOSPORES AND THEIR MOTILITY

### Preparation of Zoospore Suspensions

Zoospore suspensions are commonly prepared by allowing sporangia to discharge into distilled water. However, pure water is not a very satisfactory suspending medium for protoplasm unprotected by cell walls. Gooding and Lucas (1959) found that zoospore suspensions of *P. parasitica* var. *nicotianae* in distilled water remained motile for 4 hr at cell densities of $10^6$/ml but for only 5 min at $10^3$ cells per milliliter. The protective effect of high cell densities could be due to the "sharing" of the toxic effects of traces of heavy metals or to the provision of essential ions, such as $Mg^{2+}$ and $Ca^{2+}$, in adequate amounts from the high densities of sporangia.

Cameron (1979) found that dilute suspensions of zoospores of *P. palmivora* swam well for 10 hr in the dilute salts solution that Machlis (1958) devised for *Allomyces* gametes but mostly lysed within 30 min in distilled water. By controlling the levels of free $Mg^{2+}$ and $Ca^{2+}$ with chelating agents, Cameron (1979) found that both these ions were needed at about 1 $\mu M$ for normal and prolonged motility. At a free $Ca^{2+}$ level of 0.01 $\mu M$, motility ceased in a few minutes, and at 0.1 $\mu M$, zoospores swam in circles with a radius of about 100 $\mu$m. A satisfactory suspending medium based on that of Machlis (1958) contained potassium phosphate buffer (pH 5.8, 0.5 m$M$) and 0.05 m$M$ $MgCl_2$ and $CaCl_2$ (Cameron and Carlile, 1978, 1980). The pH range over which zoospores show good motility is wide—pH 5–9.3 for *P. megasperma* var. *sojae* (Ho and Hickman, 1967a)—but the use of pH buffers and added $Mg^{2+}$ and $Ca^{2+}$ seems prudent if prolonged motility by low cell densities is required.

### Zoospore and Flagellum Form

The form and dimensions of the zoospores and flagella of *Phytophthora* have been studied by several groups, as indicated in the review by Bartnicki-Garcia and Hemmes (1976). The different groups, however, were often concerned with different features, used different methods, and studied different species. Information on zoospore form has been obtained by light microscopy (Ho and Hickman, 1967a; Ho et al, 1968b), electron microscopy of thin sections (Bimpong and Hickman, 1975; Ho et al, 1968a; Reichle, 1969), and electron microscopy of whole zoospores (Desjardins et al, 1973a). Studies of motile zoospores with dark-field microscopy (Ho and Hickman, 1967a) and cinefilming with phase contrast microscopy (Ho et al, 1968b) have helped to relate form to direction of movement. Details of flagellum form have been studied with electron microscopy by Vujičić et al (1968) and Desjardins et al (1969, 1973b). These and other studies show substantial agreement, but no single study provides a full account of the form and dimensions, including the variation in these, of the zoospores and flagella of a single species. The description provided below is hence a composite account, based on work on a range of species.

Zoospores are biflagellate. The point of insertion of the two flagella lies in a deep groove (Fig. 1) in what, for convenience, may be termed the ventral surface. The anterior flagellum is tinsel and the posterior flagellum is nearly smooth (Fig. 2). Zoospores are longer than they are wide, are narrower at the anterior than at the posterior end, and are flattened dorsoventrally. The anterior flagellum bears a row of mastigonemes on each side (Fig. 3). These mastigonemes are 2 μm long and 20 nm wide except for the terminal 0.5 μm, which is 13 nm wide (Desjardins et al, 1969). The posterior flagellum carries fine hairs 0.5 μm long and 15 nm wide. The zoospore of *P. palmivora* is 12.9 ± 0.8 μm long and 8.9 ± 0.6 μm wide (standard deviations indicated) (M. Sanchez-Alvarez and M. E. J. Holwill, *personal communication*). The lengths of the flagella, measured from the electron micrographs of Desjardins et al (1973a), are 15.6 μm (anterior) and 23.9 μm (posterior). Sanchez-Alvarez and Holwill found that the anterior flagella extended 14.2 ± 1.9 μm and the posterior flagella extended 16.9 ± 2.3 μm beyond the soma. The combined data of the two groups suggest that the flagella are inserted toward the anterior end of the soma and 4.5 μm apart. The measurements of Sanchez-Alvarez and Holwill give an overall length of 44 μm for the zoospore with flagella extended; Allen and Newhook (1973) gave a minimum estimate of 30 μm for *P. cinnamomi.*

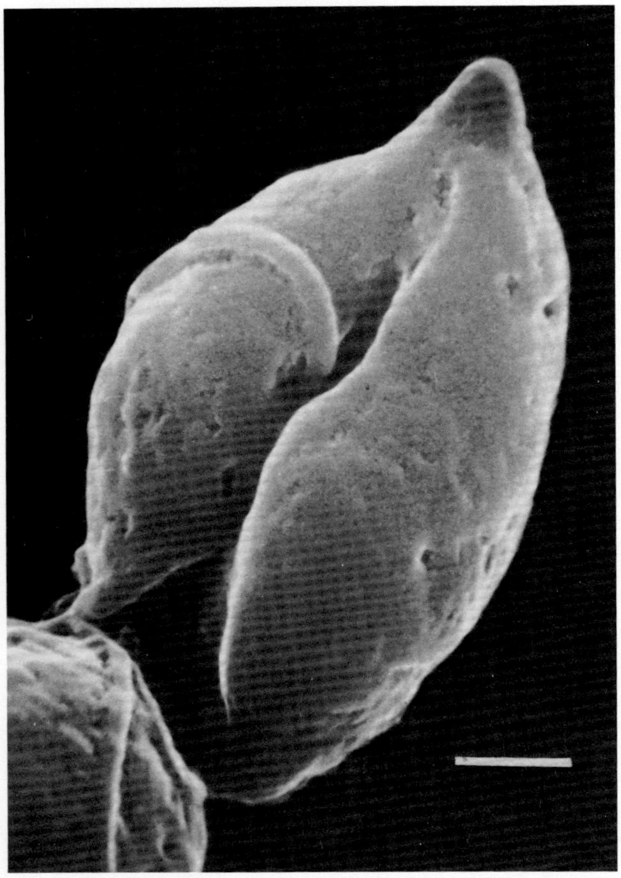

**Fig. 1.** Scanning electron micrograph of a glutaraldehyde-fixed zoospore of *Phytophthora palmivora,* showing the deep central groove from which the flagella emerge. The anterior tip of the zoospore is at the top of the figure. The scale bar represents 1 μm. (Reprinted, by permission, from Desjardins et al, 1973a)

### Flagellum Action and Zoospore Movement

Flagellum action in *P. palmivora* has recently been the subject of cinefilming and hydrodynamic analysis by M. Sanchez-Alvarez and M. E. J. Holwill (*unpublished*); some of the results are discussed in a review of the dynamics of eukaryotic flagellum action by Holwill (1982). They found that both flagella undulate in a plane, with waves moving from base to apex. This confirms the conclusions of Hickman (1970). It is intuitively clear how waves moving from base to tip of a posterior flagellum can produce a thrust moving the zoospore forward, but not how forward-moving waves on the anterior flagellum can contribute to this thrust. The explanation lies in the way the mastigonemes, the stiff lateral hairs on the anterior flagellum, move in response to the flagellum undulations (Holwill, 1982). The thrust generated by each flagellum can be calculated from the amplitude, wavelength, and frequency of the waves passing along the flagellum; the anterior flagellum contributes about 10 times the thrust of the posterior flagellum. The action of the posterior flagellum is, however, responsible for producing changes in the direction of swimming. The posterior flagellum ceases to beat for a short time (50–100 msec), during which it bends almost perpendicular to the direction of movement. The anterior flagellum continues to beat, so that the rear flagellum acts as a rudder, causing a swift but smooth change in the direction of zoospore movement. The calculated thrust from the flagellum agrees well with the speed at which the zoospore moves.

*Phytophthora* zoospores often follow a helical path, rotating about their long axis as they swim. The helical path appears as a wavy line in dark-field photomicrographs (Fig. 4) taken with exposures of a few seconds (Allen and Newhook, 1973; Ho and Hickman, 1967a). Data on the dimensions of the helix for *P. cinnamomi* are provided by Allen and Newhook (1973, 1974). Information is lacking as to whether zoospores maintain a constant configuration with respect to the axis of the zoospore, such as the ventral surface facing the axis, but intuitively this seems probable. Couch (1941), studying *Saprolegnia ferax* zoospores, noted that at any moment, obvious undulations could be observed on only one of the two flagella. We (J. N. Cameron and M. J. Carlile, *unpublished*) find with *P. palmivora* that waves are commonly observed on the two flagella alternately and not simultaneously. Such observations are consistent with the two flagella undulating in planes perpendicular to each other and thus being observable alternately as the zoospore rotates. A microorganism that does not rotate may yaw and gradually change direction. Rotation converts a tendency to yaw into a helical path with a persisting direction and is hence advantageous in enabling zoospores to maintain a constant direction in unobstructed surroundings.

However, zoospores sometimes swim smoothly instead of rotating and following a helical path (Ho and Hickman, 1967a). This seems to apply in the study of M. Sanchez-Alvarez and M. E. J. Holwill (*unpublished*), who do not remark upon the alternation of waves on anterior and posterior flagella or rotation. The shallow depth of suspending medium required for cinefilming might well constrain the zoospores to swim smoothly. Such constraints often apply in nature, where thin films of liquid occur and numerous obstacles may prevent a persisting helical path. Hence, both smooth and helical swimming are probably natural and appropriate to different situations.

## Zoospore Performance

The distance a zoospore can travel depends on the duration of swimming, the speed, and the path taken.

**Duration of Swimming**

Most published work on how long zoospores swim is concerned with the maximum rather than the average duration of swimming, which is likely to give an exaggerated impression of the average duration of zoospore motility. Bimpong and Clerk (1970), for example, claimed 84 hr for the maximum duration of motility of zoospores of *P. palmivora* in distilled water at 17° C but noted that 50% had encysted in 24 hr. The lack of data on percentage of zoospores remaining motile after successive time intervals is unfortunate. Studies on the way different factors influence the duration of zoospore motility, such as those of Bimpong and Clerk (1970) on *P. palmivora* and Ho and Hickman (1967a) on *P. megasperma* var. *sojae,* nevertheless yield reasonably consistent results.

Zoospores typically swim for many hours. The duration of swimming is greatest in distilled water or perhaps, as suggested earlier, in very dilute solutions of suitable ions. The addition of nutrients with high concentrations of ions to the suspending medium generally shortens the period of motility through encystment or sometimes by lysis. The optimal temperature for prolonged motility is relatively low—Bimpong and Clerk (1970), for example, found that zoospores of *P. palmivora* swam longest at 17° C, although motility occurred up to 33° C. An abundance of obstacles, whether brought about by a very dense zoospore suspension, added microbeads, or shallow suspending medium, can hasten encystment (Ho and Hickman, 1967a). The effectiveness of agitation (Bartnicki-Garcia, 1973) in causing encystment suggests that these factors may act by increasing the frequency of collision between zoospores or between zoospores and other objects. However, Cameron and Carlile (1980) noted that zoospores have some ability to avoid collisions with other zoospores, and presumably with other similarly charged particles, and Young et al (1979) reported that glass microbeads that had been appropriately treated had a greatly reduced tendency to cause encystment. Earlier studies may have exaggerated the extent to which obstacles cause encystment.

The maximum duration of zoospore swimming may be determined by the availability of energy (for motility, osmoregulation, maintenance of ionic balance) from food reserves within the zoospore. Bimpong (1975) demonstrated substantial lipid reserves within the zoospores of *P. palmivora,* and Wang and Bartnicki-Garcia (1980) showed the presence of $(1 \rightarrow 3)$-$\beta$-D-glucan (mycolaminaran). Bimpong (1975) showed that after zoospores had been swimming for 6 hr, their lipid content had fallen to about 46% and suggested that endogenous lipids were the main source of energy for zoospore motility. The metabolism of $^{14}$C-labeled glucose, acetate, serine, glutamate, and aspartate reported by Barash et al

**Figs. 2 and 3. 2,** Transmission electron micrograph, glutaraldehyde-fixed and rotary shadow cast, of a zoospore of *Phytophthora palmivora.* The anterior tinsel flagellum is on the right, and the nearly smooth posterior flagellum is on the left. The scale bar represents 1 $\mu$m. **3,** Transmission electron micrograph, glutaraldehyde-fixed and unidirectionally shadow cast with platinum-palladium alloy, of the flagella of a zoospore of *Phytophthora palmivora.* The anterior flagellum with prominent mastigonemes is at the top; the posterior flagellum with fine hairs is below. The scale bar represents 1 $\mu$m. (Reprinted, by permission, from Desjardins et al, 1969)

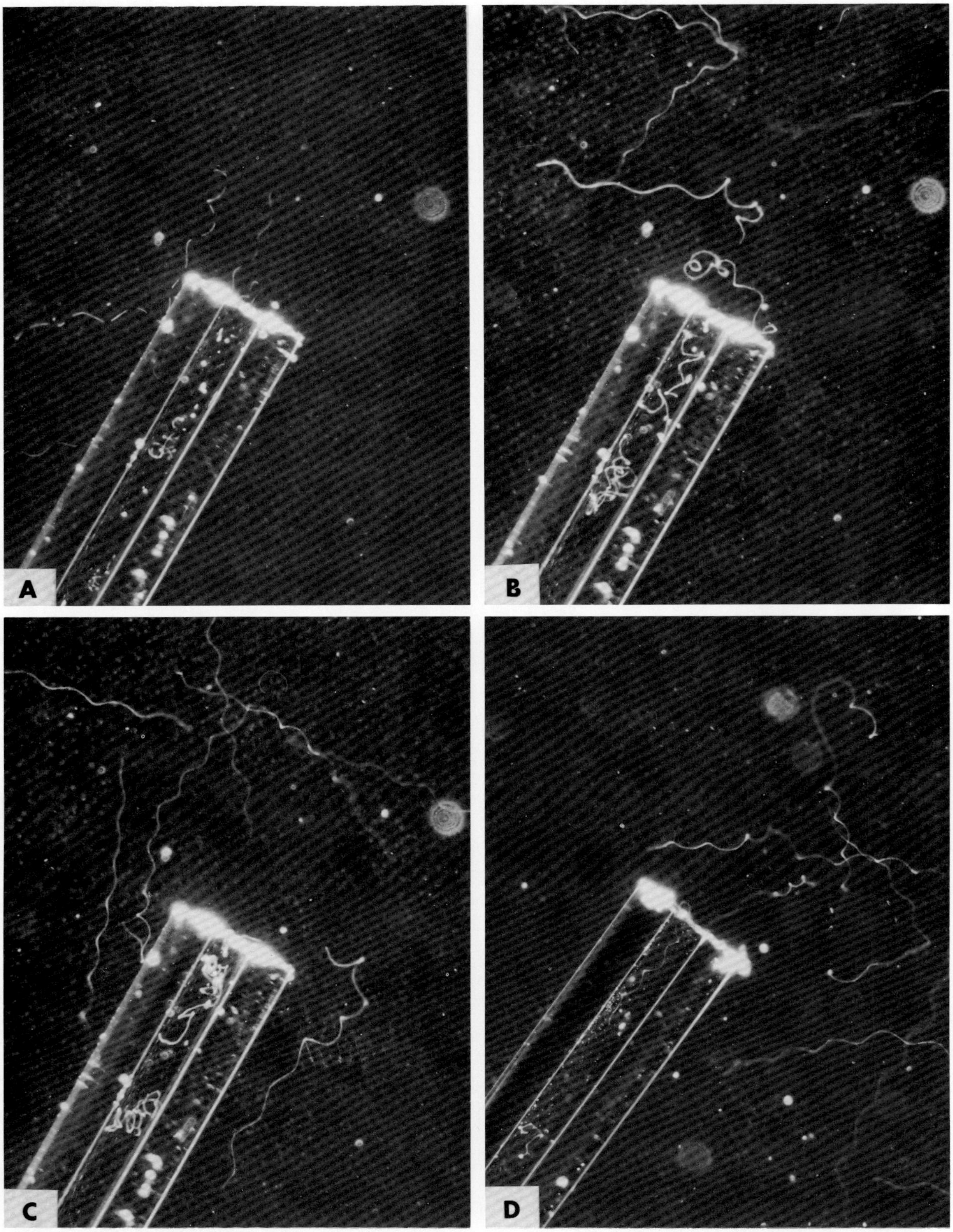


**Fig. 4.** Dark-field photomicrographs showing tracks of zoospores of *Phytophthora cinnamomi* in the presence of a capillary tube (190 $\mu$m internal diameter) containing 5 m*M* ethanol. **A,** Illumination by stroboscopic cycles of 3 sec with successive flashes of 1, 0.67, and 0.33 sec. This enables the speed and direction of swimming to be deduced. Part of the track is, however, not in the plane of focus. **B,** Illumination of 15 sec without strobe. One zoospore swimming in a path that would have taken it past the mouth of the tube has undergone turning reactions that have reoriented the path toward the tube. Another zoospore exhibits continuous turning near the mouth of the tube, and others do so within the tube. **C,** A further 15-sec exposure starting 2 sec after that shown in B. The zoospore that had reoriented itself toward the tube has now entered, and the one that had been continuously turning near the mouth of the tube has now encysted. A further zoospore that had been swimming parallel to the tube has turned toward its mouth. **D,** A track can be followed toward and into a capillary tube. (Reprinted, by permission, from Allen and Newhook, 1973)

(1965) appears to be related mainly to cyst germination rather than to zoospore motility. The good performance of zoospores in the absence of added nutrients, the low levels of nutrients in soil water generally, and the high levels of sugars and amino acids in the rhizosphere of hosts suggest that zoospore swimming is based on the metabolism of endogenous reserves and that nutrients may serve as signals for encystment rather than as energy sources for the maintenance of the zoospore phase. Cameron (1979) estimated, from measurements of oxygen utilization by zoospores, that the lipid reserves demonstrated by Bimpong and Hickman (1975) would support zoospore activity for at least 10 hr.

#### Swimming Speeds

The rate of movement of zoospores can be determined with respect to either the helical path followed or the axis of the helix. The former values give the actual swimming speed of the zoospores; the latter give the speed with which they travel from place to place. Ho and Hickman (1967a) found that zoospores of *P. megasperma* var. *sojae* traveled at 88 ± 17 μm/sec at 15°C and 157 ± 23 μm/sec at 26°C; the corresponding speeds along the helical path were 122 ± 32 μm/sec and 232 ± 43 μm/sec. Allen and Newhook (1973), studying *P. cinnamomi* at intermediate temperatures (20–24°C), found intermediate values; they measured 121 ± 12 μm/sec for the rate of travel and calculated 182 ± 14 μm/sec for the speed along the helical path. Higher metabolic rates presumably account for faster swimming at high temperatures; a consequently more rapid depletion of reserves may account for the shorter duration of swimming mentioned earlier.

#### Path Taken and Distance Traveled

The above rates of travel cannot usefully be converted into millimeters per minute or centimeters per hour because zoospores change direction frequently. Allen and Newhook (1973) found that the rate of change of direction of the axis of the helical path for zoospores of *P. cinnamomi* was 0.14 ± 0.05 rad/sec (8 ± 3°/sec). Information on the mean frequency and mean angle of directional change would also be useful for comparison with the extensive studies on the random walk of bacterial cells (Berg, 1975; Koshland, 1979, 1980; Macnab, 1978, 1979), but the published results clearly show that swimming spreads zoospores much less rapidly than their speed in micrometers per second would suggest.

Duniway (1976) found that zoospores of *P. cryptogea* swam 2.5–3.5 cm in surface water over flooded soils or through coarse-textured soils. Newhook et al (1981) studied the swimming of zoospores of *P. cinnamomi* through coarse wet sand from a reservoir. The numbers of zoospores trapped by inserting capillary tubes containing an attractant at 3, 6, and 10 hr at various distances from the reservoir were compared with the numbers trapped at the same time in the reservoir. Zoospores had reached 4 cm in 3 hr, 5 cm in 6 hr, and 6 cm in 10 hr, although the numbers trapped were only 0.1–0.2% of those trapped in the reservoir. Numbers were correspondingly higher near the reservoir and diminished with distance from the reservoir in a manner reminiscent of diffusion. A diffusion coefficient (perhaps better termed a motility coefficient) of about 0.004 $mm^2$/sec was calculated; this is a little higher than the diffusion coefficient for HCl. Hence zoospores, through their motility, may reach 3–6 cm from their source, dispersing a little faster than would small molecules by diffusion but much faster than would large molecules or inanimate objects the size of zoospores. Whether temperature influences the distance that can be reached is not clear—although zoospores swim more slowly at low temperatures, their period of motility is longer.

### Motility in Nature

As indicated earlier, abundant obstacles may lead to more rapid encystment of zoospores, presumably through collision with these obstacles. Zoospores swimming in soil water are likely to encounter numerous obstructions, and the possibility of collision will be enhanced if zoospores are carried through soil by rapidly flowing water. The question therefore arises as to whether zoospores are able to remain motile in soil water long enough for swimming to play a useful role.

Young et al (1979) studied the effects of "ideal soils" composed of glass microbeads of various sizes on motility of zoospores of *P. cinnamomi.* Different sized microbeads lead to different sized passages ("pore necks") connecting the voids between the beads. Zoospore suspensions were passed through columns of beads of various pore neck sizes at different speeds. With a pore neck size of 150–250 μm, a zoospore suspension could be passed at 50 cm/min without affecting subsequent motility. Even with pore necks of 25–35 μm, flow rates of 2.5 cm/min had no adverse effect. Zoospores failed to pass through columns with pore necks of 5–10 μm, dimensions similar to those of the zoospores. Static confinement of zoospores in columns of glass beads (pore neck sizes 25–35 or 150–200 μm) had almost no effect on the proportion of zoospores encysted at 1 hr (25–29%) and only a moderate effect on the numbers encysted at 6 hr (about 80% compared with 70%). Some other *Phytophthora* spp., however, have zoospores more susceptible to the induction of encystment by glass microbeads, although not to such an extent as to suggest that motility is of no value in the soil (Benjamin and Newhook, 1982).

Newhook et al (1981) extended observations from "ideal" soils of glass microbeads to coarse sand, sand-loam mixtures, and cores of undisturbed soil. Their results on the movement of zoospores through sand were discussed earlier, and experiments with sand-loam mixtures showed that chemotactic trapping, and hence presumably zoospore motility, was more effective where the proportion of sand was high and hence many large pore necks occurred. Zoospore suspensions were allowed to pass by gravity into saturated cores of natural soil. Drainage of the soil recovered about a quarter of the zoospores initially added, and further flushings brought the total of zoospores retrieved, still motile, to about three-quarters of the original number. The unrecovered quarter may have entered water-filled voids that were not part of the free draining system.

The conclusion seems probable, therefore, that zoospores can swim effectively in many natural soils and can be transported rapidly by water movements in soils but still retain their motility.

## TAXIS OF ZOOSPORES

Microorganisms, including zoospores, may accumulate at (positive taxis) or avoid (negative taxis) a site through modification of their motility in three basic ways. First, they may become oriented so as to swim toward or away from the site; this is referred to as topotaxis, and many workers, especially photobiologists (Diehn et al, 1977), wish to restrict use of the term taxis to this phenomenon. Second, the organisms may perform a biased random walk, delaying a change in direction when they happen to be moving in the "right" direction or hastening it if they are going the "wrong" way. This is known as klinokinesis; the "chemotaxis" now intensively studied (Adler, 1975; Berg, 1975; Koshland, 1979, 1980; Macnab, 1978, 1979) in *Escherichia coli* and *Salmonella typhimurium* is in fact chemoklinokinesis. Finally, the organisms may show orthokinesis; that is, change their speed as they approach the site. If they slow down, they will spend more time near the site than away from it, and if they speed up, the reverse will apply.

The various forms of taxis (using the term in the broad sense) can all contribute to the arrival of microorganisms at a site or their dispersal from it. An extreme form of orthokinesis, immobilization, may bring a positive taxis to an appropriate conclusion. I have discussed terminological and semantic problems (Carlile, 1980a) and the interaction of orthokinesis, klinokinesis, and topotaxis (Carlile, 1980a, 1980b) in more detail elsewhere. The relative contribution of these processes in the positive and negative taxis of *Phytophthora* zoospores is unclear, but much information, reviewed below, is available on the factors that bring about accumulation or dispersal.

## Pattern Swimming

Cameron (1979) observed that petri dishes containing very dense suspensions (e.g., $10^6$/ml) of zoospores of *P. palmivora* developed a tessellated appearance, in which areas densely populated with cells were separated from each other by narrow zones in which few zoospores occurred. Microscopic examination showed that zoospores were swimming upward at the center of each area; then they moved outward, descended, and moved to the center once more. This phenomenon of pattern swimming is well known in algae and protozoa. The most likely explanation (Plesset et al, 1975) is that cells, slightly denser than water, swim upward as a result of a taxis. The upper layer of water plus cells then becomes denser than the lower layer, and Rayleigh-Taylor instability leads to downward movement. We (J. N. Cameron and M. J. Carlile, *unpublished*) have also observed pattern swimming in *P. infestans.* Ko and Chase (1973) and Ko and Chan (1974) described and illustrated pattern swimming, which they termed "aggregation," in *P. capsici* and *P. palmivora.*

The extremely high cell densities needed to produce pattern swimming suggest that it is a laboratory artifact. Its presence, however, alerts the investigator to the occurrence of a taxis-producing upward movement. Such responses could be to light (positive phototaxis), air (positive aerotaxis), or gravity (negative geotaxis). The only claim for phototaxis in *Phytophthora* (Petri, 1925) has never been confirmed, and no evidence exists for aerotaxis in *Phytophthora* (Ho and Hickman, 1967b). Negative geotaxis, however, occurs (Cameron and Carlile, 1977) and would suffice to produce the upward movement needed to initiate pattern swimming.

## Geotaxis

Cameron and Carlile (1977) found that zoospores of *P. cactorum, P. palmivora,* and *P. nicotianae* tend to swim upward. Such negative geotaxis is widespread in eukaryotic microorganisms (Carlile, 1980a). The precise mechanism of the response is controversial, but it seems clear that the orientation that produces upward swimming is a consequence of cell shape, location of the center of gravity, and propulsive effort in swimming microorganisms rather than the result of any involvement of a sensory system. The significance of negative geotaxis for *Phytophthora* may be that it helps to keep zoospores near the surface of soils, where host rootlets are more abundant than at greater depths.

## Attraction to Roots

Hickman (1970) lists 12 species of *Phytophthora* for which accumulation of zoospores behind the root tips of plants has been demonstrated. The method usually employed is to immerse severed roots, or sometimes entire root systems, in zoospore suspensions. Massive accumulation commonly occurs fast enough to indicate that attraction is involved and not merely the immobilization of zoospores arriving at the root surface by random movement. Accumulation is commonly in the zone of elongation just behind the root tip, but it can be more widespread or can be limited to wounds. Different accumulation patterns have been reported for the same parasite and host species by different workers, perhaps because of differences in the nutritional status or age of the plants. Attraction is generally reported to be of low specificity, in that zoospores of a *Phytophthora* species are usually attracted to a wide variety of nonhost as well as host plants, although *P. cinnamomi* seems to respond more effectively to hosts than to nonhosts (Zentmyer, 1961, 1966, 1970). Zoospores are often attracted not only to living roots but also to killed roots—again *P. cinnamomi* is an exception, failing to respond to roots of avocado killed by boiling or with propylene oxide (Zentmyer, 1961). Root exudates contain a wide range of low-molecular-weight compounds, so it is likely that attraction to roots is due to positive chemotaxis to substances released from the root and diffusing through the aqueous phase in the soil. Positive chemotaxis to root exudates and to substances known to be or likely to be present in root exudates is reviewed below.

The studies above were carried out with root systems or severed roots placed in zoospore suspensions. Mehrotra (1970), however, was able to study attraction of zoospores to undisturbed roots growing in soil. He grew his plants in sterile soil in polypropylene tubes with cellulose acetate "windows" (millipore membranes with 1-$\mu$m pores) so that the roots came into contact with the window. Outside the tube was nonsterile soil. Either the tubes were immersed in suspensions of zoospores of *P. megasperma* var. *sojae* or *P. drechsleri,* or the zoospore suspensions were added to the outer soil. After a suitable interval, the acetate membrane was examined for encysted zoospores, which were found to have been attracted in large numbers to points on the outer side of the membrane corresponding to the position of root tips on the inner side. In other experiments, zoospore suspensions were added to soil containing pea seedlings. Encysted zoospores could later be detected on roots by staining the roots with a fluorescent dye and examining them with fluorescence microscopy.

## Positive Chemotaxis

### Methodology

Most quantitative studies of positive chemotaxis of *Phytophthora* zoospores have been conducted with an attractant diffusing from a capillary tube into a "pond" containing zoospores (Fig. 4). Equations describing the concentration gradient produced by such diffusion have been provided by Brokaw (1958) and Futrelle and Berg (1972). Tests with HCl diffusing into indicator solution (Cameron, 1979) showed that for the first few minutes Brokaw's equation, but later that of Futrelle and Berg, most accurately described diffusion from a capillary tube. Experiments with dyes established that diffusion from large-bore capillary tubes is too rapid to maintain a satisfactory gradient; small tubes (e.g., 190 $\mu$m internal diameter) are preferable.

Most workers fill the capillary tubes by capillarity with test solution mixed with molten agar. Under these circumstances, chemotaxis is assessed by photographing the area next to the mouth of the tube after a suitable time interval and comparing zoospore numbers with those in a photomicrograph of a control area. Khew and Zentmyer (1973) used as their control an area well away from the capillary tube. Halsall (1976) compared the experimental counts with those in a separate control chamber. The respective merits of the two approaches depend on the design of the experiment and the information sought.

Allen and Newhook (1973) omitted agar from their capillary tubes. Such tubes can conveniently be filled with test solutions by heat-sealing one end of the tube and allowing the test solution to be drawn in as the tube cools (Cameron and Carlile, 1978). Experimental and control tubes are placed in zoospore suspensions, and after an appropriate interval, the tubes are removed from the test chamber and the numbers of zoospores that have entered each are determined. The zoospores are immobilized by exposure to formalin vapor before the tubes are removed (Allen and Newhook, 1973) or by passing the tubes swiftly through a flame on removal (Cameron and Carlile, 1978).

The swim-in test is probably a valid measure of *attraction,* in that the numbers that have swum into a narrow aperture under the influence of attractant are compared with the numbers that have entered a control tube by random swimming. The number of zoospores next to an agar-filled capillary, on the other hand, will be determined not only by attraction but also by any orthokinetic trapping and immobilization brought about by the substances diffusing from the tube. This may enhance the performance of weak attractants and give apparent attraction by substances that lack any topotactic or klinokinetic effect. The swim-in test is probably the more useful for detailed studies on known attractants, whereas the test with agar-filled capillaries may well give fuller information about substances that are likely to cause accumulation in nature. In a variant of the swim-in test, the "rhizosphere model" of Allen and Newhook (1973), capillary tubes connect a large attractant source with a "pond" of zoospores. Under these circumstances, a steady state can be achieved and a uniform gradient set up and maintained.

Whichever assay method is preferred, statistical comparison of experimental and control zoospore numbers is desirable. With some attractants, pH adjustment or buffering may be needed. Here it should be borne in mind that $H^+$ and to a lesser extent other cations are repellents (Allen and Newhook, 1973; Cameron and Carlile, 1980), which, if present at too high a concentration, may interfere with positive chemotaxis.

**Attraction by Root Exudates**

Because zoospores are attracted to roots and roots are known to produce exudates containing a wide variety of substances, it is reasonable to infer that attraction to roots is a consequence of positive chemotaxis to root exudates. This inference has led to extensive tests of the chemotactic effects of compounds known to occur, or likely to occur, in root exudates. However, surprisingly little work has been performed on chemotaxis to actual root exudates.

Ho and Hickman (1967a) described a method for obtaining root exudates from pea plants grown in sterile sand. They found that the period of motility for zoospores of *P. megasperma* var. *sojae* was greatly shortened by the addition of root *exudates* but was prolonged by the addition of *extracts* from the same roots. This finding indicates that root extracts are not an adequate substitute for root exudates in the study of zoospore motility and taxis. Ho and Hickman (1967b) demonstrated chemotaxis of zoospores of *P. megasperma* var. *sojae* to a filter paper disk impregnated with root exudate.

The most detailed studies on the chemotaxis of zoospores to root exudates are those of Chang-Ho and Hickman (1970), extending the work of Royle and Hickman (1964a, 1964b) on the attraction of *Pythium aphanidermatum* zoospores to pea roots and pea root exudates. Chang-Ho and Hickman (1970) used the capillary tube method to investigate the attraction of root exudates and of cationic, anionic, and neutral fractions from the exudate. The cationic fraction, shown to contain a wide range of amino acids, was the most attractive, followed by the neutral fraction, which contained sugars. Weak attraction was observed to the anionic fraction, which contained low levels of carboxylic acids. Gaworzewska and Carlile (1982), studying attraction of the symbiotic bacterium *Rhizobium leguminosarum* to pea root exudate, found the same order of attractiveness with respect to cationic, neutral, and anionic fractions. Therefore, a bacterium and a fungus with the same host appear to have evolved a similar pattern of chemotactic sensitivity.

The discovery (discussed below) that ethanol and other volatile compounds can be attractants will need to be considered in future studies on root exudates, so that such substances are not lost during the collection or concentration of exudates.

**Attraction by Pure Substances**

A variety of amino acids and sugars attract *Phytophthora* zoospores (Hickman, 1970), as would be expected from studies on root exudates. Khew and Zentmyer (1973) studied the attractiveness of 12 amino acids for zoospores of five *Phytophthora* species. Most of the amino acids had some attraction for all five species, but aspartic acid, glutamic acid, arginine, and methionine were the most effective. The threshold for attraction by these amino acids was about 10–40 $\mu M$ for *P. cactorum, P. capsici,* and *P. palmivora* and about 100–300 $\mu M$ for *P. citrophthora* and *P. cinnamomi.* Halsall (1976) also studied the chemotactic response of *P. cinnamomi* and four species not studied by Khew and Zentmyer. She found strong chemotactic responses to a number of amino acids by *P. cryptogea, P. drechsleri,* and *P. citricola* and weaker responses by *P. nicotianae* var. *parasitica* and *P. nicotianae.* Responses to sugars were also shown.

An important development was the demonstration by Allen and Newhook (1973) that ethanol, which is produced by roots in waterlogged soils, is an attractant for *P. cinnamomi* zoospores in a range of about 0.2–20 m$M$. Halsall (1976) found that ethanol was a better attractant for *P. cinnamomi* than any other substance she tested but was not the most effective attractant for her other species. Cameron and Carlile (1978) found that ethanol at 5 m$M$ attracted *P. palmivora.* M. Benjamin and F. J. Newhook (*personal communication*) found thresholds of 50 $\mu M$ for ethanol attraction of *P. megasperma* and 200 $\mu M$ for *P. citrophthora* and *P. cinnamomi. P. palmivora* was less sensitive, and *P. megasperma* var. *sojae* and *P. cactorum* did not respond to ethanol.

Cameron and Carlile (1978, 1981) found that zoospores of *P. palmivora* are attracted by a wide range of fatty acids, aldehydes, and alcohols; isovaleraldehyde had a chemotactic threshold at 1 $\mu M$. Some of these attractants, including isovaleraldehyde, have now been shown to attract zoospores of *P. cactorum* (C. A. Duggan and M. J. Carlile, *unpublished*).

**The Mechanism of Positive Chemotaxis**

The first step in a chemotactic response is the interaction of the attractant or repellent with the cell surface. Studies on the positive chemotaxis of bacteria have led to the recognition of specific receptors for sugars and amino acids, which are located in the plasma membrane or the adjacent periplasmic space between wall and membrane (Adler, 1975; Koshland, 1979, 1980; Macnab, 1978, 1979). These receptors are proteins for which other functions—usually transport of the sugar or amino acid into the cell—are known. So far, little comparable work has been done for eukaryote chemotaxis, although

Cameron and Carlile (1981) have shown that zoospores of *P. palmivora* specifically bind the attractant isovaleraldehyde, which can be displaced by chemically related attractants such as anteisovaleraldehyde or valeraldehyde but not by distantly related attractants such as isoamyl alcohol or unrelated ones such as asparagine. This indicates a specific receptor for isovaleraldehyde that also binds some closely related compounds. The dissociation constants for these compounds, calculated from their ability to displace isovaleraldehyde, parallel their chemotactic effectiveness. The fact that a number of compounds can interact with a single receptor means that not all the compounds shown to be chemotactic in the laboratory may be significant in nature.

Studies on bacteria have shown that although binding to a specific receptor is limited to closely related compounds, with respect to chemotaxis, competition may occur between distantly related compounds. At the molecular level, this results from competition between different ligand-receptor complexes interacting with a common secondary receptor. Several such groups of competing compounds have been demonstrated in bacterial chemotaxis, and a similar effect has been found in the chemotaxis of a eukaryote, the myxomycete *Physarum polycephalum* (Knowles and Carlile, 1978). The ability of a background of one amino acid to inhibit chemotaxis to a second amino acid (Khew and Zentmyer, 1973) indicates that here too, competition between attractants with different receptors may occur.

Adaptation to an attractant is an essential part of the klinokinetic mechanism of attraction and also permits sensitivity over a wide range of concentrations (Carlile, 1980a). The decreasing ability of zoospores to bind isovaleraldehyde following exposure to this ligand (Cameron and Carlile, 1981) may constitute an adaptation mechanism. Sensory adaptation in bacteria involves protein methylation, and mutants auxotrophic for methionine may fail to show chemotaxis. If a similar mechanism occurs in *Phytophthora*, it may account for the poor chemotactic performance of a methionine auxotroph (Khew and Zentmyer, 1973), although one would still need to account for the poor performance of their arginine auxotroph.

Allen and Newhook (1973) followed the path of zoospores of *P. cinnamomi* attracted to ethanol in capillary tubes by means of long-exposure (15-sec) dark-field photomicrography (Fig. 4). Their photographs suggest a form of klinokinesis, with zoospores continuing in their helical path if they are experiencing increasing ethanol concentrations, but turning sharply if they are going in the "wrong" direction. Allen and Newhook (1974), reanalyzing their earlier data with a more sophisticated statistical approach, found that the rate of change of direction shown by zoospores is less in isotropic solutions of ethanol than in distilled water. In studies of bacterial klinokinesis in which stopped flow devices were used to rapidly mix test solutions and bacterial suspensions, turning due to attractants was markedly suppressed before adaptation occurred and normal movement resumed. More gentle mixing procedures are necessary to maintain zoospore motility, and adaptation will probably diminish the effect of ethanol considerably before observations can be made. Not surprisingly, therefore, the effects detected by Allen and Newhook were small. The evidence now available does not make clear whether topotaxis and orthokinesis contribute to the attraction of zoospores, nor do we have any information on the sensory transduction path linking chemoreception with control of the flagellum movement that brings about directional changes.

**Significance of Positive Chemotaxis in Nature**

As indicated earlier, Young et al (1979) and Newhook et al (1981) found that in soils with adequate pore size, zoospores can remain motile and even be transported passively by water currents and retain their motility. They also found that chemotaxis could occur in "ideal" soils of glass microbeads, even if zoospores were being passively transported by water flow at speeds similar to their swimming speed (Young et al, 1979). Chemotaxis in coarse sand and sand-soil mixtures was also demonstrated (Newhook et al, 1981). Therefore, zoospores that have been passively transported over long distances can probably show active motility and, if close to a host, reach it through positive chemotaxis.

Positive chemotaxis, at least for isovaleraldehyde, is specific in the sense that it is brought about through the interaction of a ligand with a receptor that can interact only with it and a few other molecular species with similar configurations. In the ecological sense, however, positive chemotaxis is of low specificity, in that it can be brought about by a wide variety of compounds common in the root exudates of a wide range of plants. This is consistent with the wide host range of many *Phytophthora* spp. Different species do, however, show some differences in the effectiveness of their response to different groups of attractants; *P. cinnamomi,* for example, is most sensitive to ethanol and *P. drechsleri* to amino acids (Halsall, 1976). Such differences may confer a limited specificity and lead to species differences in the circumstances under which infection is most likely.

## Negative Chemotaxis

Allen and Harvey (1974) showed that zoospores of *P. cinnamomi* were repelled by HCl, KCl, $NH_4Cl$, and NaCl. They filled capillary tubes with zoospore suspensions and placed the tubes in test solutions. The zoospores were observed to retreat down the capillary tubes as the ions being tested diffused into the tubes. Negative chemotaxis appeared to be caused by abrupt turning when zoospores encountered a critical concentration of repellent. The most effective cation was $H^+$, which repelled at 37 $\mu M$ (i.e., pH 4.4).

Cameron and Carlile (1980) extended these results with zoospores of *P. palmivora,* demonstrating repulsion by other monovalent cations and by $Mg^{2+}$, $Ca^{2+}$, and $La^{3+}$. The most effective monovalent cation was again $H^+$, which repelled at 150 $\mu M$ (i.e., pH 3.8). The critical concentrations for monovalent cations approximately parallel their ionic conductivities, which indicates that an explanation should be sought not in specific receptors but in interaction of cations with the negative charge (Khew and Zentmyer, 1974) at the zoospore surface. Cameron and Carlile (1980) suggested that the approach of cations to the cell surface reduced the surface potential and hence changed the transmembrane potential, resulting in changed flagellum activity and turning.

Negative chemotaxis may have evolved as a way of avoiding excessively acidic conditions, and the response to other cations may be a fortuitous consequence of the mechanism of negative chemotaxis. On the other hand, some cations can occur at high concentrations in soils and are of metabolic importance, so here too, repulsion may have ecological significance. Experiments with anions other than chloride as the coionic species would be of interest.

## Electrotaxis

The effect of electric currents on zoospore motility was studied by Troutman and Wills (1964), Ho and Hickman (1967b), Katsura and Miyata (1971), and Katsura et al (1966). Discrepancies among the results obtained by these different authors led Khew and Zentmyer (1974) to study the response of zoospores of seven *Phytophthora* species to a range of current intensities. Differences among species were slight. At low current intensities, usually those below 0.5 μA, zoospores swam toward the anode and ultimately encysted around the electrode. At higher intensities, usually those above 0.5 μA, zoospores swam toward the anode but were repulsed near the anode. Hence a zoospore-free zone, its margin corresponding to an equipotential line and its width depending on the current intensity, surrounded the anode. Zoospores ultimately encysted just outside the zone of repulsion. At these high current intensities, zoospores swimming near the cathode were immobilized and thus accumulated. These trapped zoospores often burst and rarely germinated.

Khew and Zentmyer (1974) published dark-field photomicrographs of the tracks of zoospores showing electrotaxis. Precise orientation of zoospores does not seem to occur, but attraction seems to be due to the suppression of turning in zoospores swimming toward the anode. At high intensities, sharp turning occurred at the boundary of the zone of repulsion. Electric currents probably suppress or cause turning by altering transmembrane potential. The role of electrotaxis in nature is unclear, but weak electric currents do occur in the vicinity of roots, so electrotaxis may perhaps contribute to zoospore accumulation at the root surface.

### Rheotaxis

Katsura and Miyata (1971) found that zoospores of *P. capsici* showed positive rheotaxis, swimming upstream in a water current. Zoospores would swim toward the mouth of a capillary tube from which a water current was emerging, but accumulated at a distance from the tube corresponding to a current velocity of 130–140 μm/sec. Because zoospores swim at about this speed, a fairly precise upstream orientation of zoospores during positive rheotaxis is implied.

In natural soils, water will often gradually percolate downward. Hence, positive rheotaxis against gentle water currents may tend (as perhaps negative geotaxis does) to keep zoospores near the surface of the soil where roots occur.

### Zoospore Aggregation

The production of zoospore clumps has been observed when dense concentrations are spread on an agar surface (Porter and Shaw, 1978). Because only a film of liquid is present, pattern swimming seems to be excluded as the cause of the aggregation. A possible explanation is that small clumps are formed as a result of random collisions of zoospores and that these small clumps then become the source of a diffusible factor that attracts other zoospores and hence prevents the formation of further clumps in the vicinity.

## ENCYSTMENT AND CYST GERMINATION

### Encystment

Provided that highly unfavorable conditions do not cause lysis, zoospores ultimately encyst. They become sluggish, travel shorter distances between turns, show irregular jerky movements, or travel in small circles. Motility ceases, and the flagella are withdrawn or shed, both processes having been observed in the same population (Reichle, 1969). Rounding off occurs and the cyst wall is synthesized, which takes about 10 min. The encystment process has been reviewed by Bartnicki-Garcia (1973) and Bartnicki-Garcia and Hemmes (1976).

Ho and Hickman (1967a, 1967b) studied the circumstances that bring about the encystment of zoospores of *P. megasperma* var. *sojae*. The longer the zoospores had been swimming, the more sensitive they were to factors that induce encystment. These factors include extremes of temperature or pH, root exudates, a range of nutrients, and circumstances (such as proximity of surfaces or agitation) likely to cause collisions. Similar results have been obtained in less detailed studies by other researchers. In general, encystment appears to be caused either by unfavorable conditions or by proximity to the root surface.

Studies on the induction of encystment by potentially harmful agents do not, however, always discriminate between production of viable and nonviable cysts by testing for subsequent germination; results are often based only on the observation of immobilization or rounding up of zoospores. Mildly adverse conditions probably induce the production of viable cysts, whereas sudden exposure to very unfavorable conditions causes death before encystment is complete. The response to unfavorable conditions (extremes of pH and temperature, high concentrations of ions) perhaps permits escape from harmful agents and survival until conditions are more favorable. Zoospores that approach a root through chemotaxis to root exudates will experience high concentrations of root exudates and may collide with the root surface or each other. It is appropriate that such factors should induce encystment because the zoospore's role is completed when the root surface is reached. Many published photomicrographs show zoospore accumulation at root surfaces. Many of these were obtained with cut roots in zoospore suspensions, but some are from experiments in which zoospores were added to soils containing growing plants (e.g., Hickman, 1970).

Zoospores become adhesive early in encystment (Sing and Bartnicki-Garcia, 1975), and if they come into contact with a solid surface at this time, they become firmly attached. In nature, this probably facilitates invasion of plant roots. Adhesiveness is soon lost, so zoospores that fail to contact a solid surface during the adhesive phase produce cysts that remain unattached. In nature such cysts, perhaps formed far from roots as a result of unfavorable conditions, are free to be moved to other sites by water currents or other agents.

### Cyst Germination

Information on the factors that promote or suppress cyst germination is sparse, and published work does not always discriminate between factors that induce cyst formation and those that suppress subsequent germination of cysts. Factors that signal favorable conditions (e.g., root exudates) seem likely to stimulate both encystment and subsequent germination, but unfavorable factors that stimulate encystment may inhibit subsequent germination. Properly designed experiments to test these possibilities would involve transfer of recently encysted zoospores (perhaps attached by their adhesive properties to an artificial surface) from the medium used for encystment to the one for studying germination. Moreover, to study the suppression of germination, researchers would need to further transfer the zoospores from the agent suppressing germination to a medium suitable for germination to ascertain whether the suppression is temporary or whether irrevocable damage has occurred.

Cysts may germinate either by producing germ tubes or by releasing a secondary zoospore. Ho and Hickman (1967a, 1967b) found that germination of cysts that had formed on plant roots began within 30 min and that all cysts had produced germ tubes within 3 hr. They showed that root exudates and nutrients stimulated germination and favored production of germ tubes rather than secondary zoospores. In distilled water, 57% of their cysts produced germ tubes and

15% produced zoospores; on a rich liquid medium, 93% produced germ tubes and less than 1% produced zoospores. Stimulation of germ tube emergence and suppression of zoospore formation are appropriate responses to the proximity of a root, whereas secondary zoospores may be useful if a germinating cyst is located at a distance from roots.

## TROPISM OF HYPHAE

The tropisms of the sporangiophore of *Phycomyces* have been the subject of intensive study (Cerdá-Olmedo, 1977; Foster, 1977; Russo, 1980). Zygomycete sporangiophores are specialized hyphae, which change direction through bending at some distance below the hyphal apex. This mode of directional change is also feasible for other aerial hyphae but not for hyphae in contact with the substratum. Here, directional change must occur by a redistribution of growth at the apex; the apical growing region shifts to one side of the apex, so that with continued extension, the hypha curves until it ultimately points toward an attractant or away from a repellent. Hence precise orientation, which is unlikely with most forms of taxis in the smaller microorganisms, occurs with tropism.

Whereas the *Phycomyces* sporangiophore is an excellent experimental object, studies on the tropism of other types of hyphae have not proved easy, and the literature is relatively sparse. Chemotropism of fungi has been reviewed by Gooday (1975) and phototropism by Carlile (1970). Prospects for future studies on fungal chemotropism have been improved by the development of a method for determining threshold concentrations of attractants (Musgrave et al, 1977).

Germ tube tropism in *Phytophthora* has received some study; tropism may also occur in other phases in the life cycle.

### Orientation of Germ Tubes Toward Roots

When cysts located near roots germinate, the germ tubes emerge from the side nearest the root and grow toward it, as observed for *P. cinnamomi* with avocado roots by Zentmyer (1966, 1970) and for *P. megasperma* var. *sojae* with soybean roots (Ho and Hickman, 1967b). In addition to these in vitro experiments, Mehrotra (1970) observed similar orientation with germ tubes from *P. drechsleri* cysts toward safflower roots growing in soil. It is clearly important that germ tubes should grow toward rather than away from roots. Khew and Zentmyer (1974) found no evidence of electrotropic orientation of germ tubes of *Phytophthora*, so positive chemotropism toward root exudates, which has been demonstrated, is probably responsible for the orientation of *Phytophthora* germ tubes toward roots.

### Positive Chemotropism of Germ Tubes

Positive chemotropism of germ tubes of *P. palmivora* to cocoa pod extract (Bimpong and Clerk, 1970) and of *P. cinnamomi* to avocado root exudate (Zentmyer, 1970) has been observed. In addition, Bimpong and Clerk observed tropism to aspartic acid and Khew and Zentmyer (1973) to a number of amino acids. Cameron (1979) tested singly several potent chemotactic agents on the germ tubes of *P. palmivora* but failed to observe chemotropism. Ho and Hickman (1967b) found that germ tubes of *P. megasperma* var. *sojae* grew toward $H^+$ resin particles; they confirmed with indicators that this was growth up a hydrogen ion gradient. Hickman (1970) speculated that this response might be important in infection, which could be correct if the threshold for the response is lower than that for the negative chemotactic response to hydrogen ions discussed earlier. Porter and Shaw (1978) observed orientation of germ tubes toward clumps of encysted zoospores. C. A. Duggan and M. J. Carlile (*unpublished*) found that cysts of *P. cactorum* germinating close to a cover slip placed on an agar surface produced germ tubes that were oriented toward and grew under the cover slip, suggesting negative aerotropism; such a response might facilitate penetration of nonphotosynthetic plant tissues or even of photosynthetic tissues during darkness.

The most detailed study of germ tube chemotropism in the Oomycetes is that by Musgrave et al (1977) with *Achlya bisexualis*. They observed chemotropism to various amino acid mixtures; one mixture had a threshold at 4 μg/ml and became ineffective at 125 μg/ml. As the number of amino acids in the mixture was reduced, the threshold concentration increased until it approached the value at which chemotropic indifference occurred. Chemotropism was detected with two but not with one amino acid alone. A gradient of a single amino acid could, however, produce chemotropism in the presence of a uniform background of 200 μg of casein hydrolysate per milliliter. Under these circumstances, chemotropism was demonstrated with 15 of the 21 amino acids tested, the lowest threshold being observed with methionine at 8 μ*M* (about 1 μg/ml). The greatest range of concentrations at which chemotropism occurred was obtained with a casein hydrolysate, in which the minimum concentration was 20 μg/ml and the maximum was 10,000 μg/ml. Musgrave et al (1977) also found indications of negative autotropism, with germ tubes growing away from areas occupied by other germ tubes.

Further studies on germ tube chemotropism are needed. The results of Musgrave et al (1977) suggest that chemotropism should be easy to detect with mixtures such as casein hydrolysate but difficult with single amino acids. The growth toward $H^+$ resin particles observed by Ho and Hickman (1967b) indicates that pH gradients should be avoided if other potentially chemotropic agents are being studied. The nature of the chemotropic factors for *Phytophthora* germ tubes remains largely unknown.

### Other Tropisms

Fungal hyphae in agar culture commonly produce a monopodial branching pattern that is optimal for draining an area of nutrients (Leopold, 1971). Such patterns and spacing between hyphae suggest that vegetative growth of fungi involves negative autotropism; i.e., hyphae growing so as to avoid each other (Gooday, 1975), but no repellent responsible for such negative autotropism has yet been isolated. Robinson (1973a, 1973b) suggested that the apparent negative autotropism may be positive chemotropism to oxygen, with hyphae avoiding areas in the immediate vicinity of other hyphae whose metabolic activity will have depleted the area's oxygen. Avoidance reactions by hyphae of *Mucor hiemalis* have been demonstrated by Hutchinson et al (1980), who moved hyphae close to other hyphae by micromanipulation. No studies of hyphal spacing with *Phytophthora* cultures have been reported, but some form of negative autotropism or positive chemotropism presumably plays a role in the efficient invasion and exploitation of plant tissues.

In some *Phytophthora* spp., light promotes sporulation (see Ribeiro, Chapter 4). In view of this photosensitivity, the possibility that phototropic responses occur in *Phytophthora* should be considered. Positive phototropism guides the sporangiophores of many fungi (Carlile, 1970), but so far no reports have been made of phototropic effects on sporangiophore growth in *Phytophthora*. Negative phototropism influences the germination of the spores of some plant pathogens and presumably guides the germ tube into the substratum (Carlile, 1970). Such negative phototropism should be sought in the direct germination (germ tube production) of the sporangia of photosensitive *Phytophthora* spp.

Finally, some form of chemotropism seems to be involved in the growth of the antheridium toward the oogonium in sexual interaction. In *Achlya,* the hormone antheridiol, which is released by a female strain and induces sexual differentiation in the male, is also responsible for the oriented growth of the antheridium toward the oogonium (Gooday, 1975, 1978; Kochert, 1978). Some similar reactions in both heterothallic and homothallic species presumably occur in *Phytophthora.*

## DISCUSSION

There can be little doubt that zoospore motility is important in the survival of *Phytophthora.* Eukaryotic flagella are structures of considerable complexity (Cavalier-Smith, 1978); those of *Chlamydomonas,* for example, contain at least 130 polypeptides. This means that there are numerous genetic loci at which a mutation could disrupt flagellum function. Were flagella unimportant to *Phytophthora,* such mutants could survive, and natural selection would favor further mutations leading to the elimination of an unnecessary organ. The existence of a functioning structure of the complexity of the *Phytophthora* flagella is in itself an argument for an important role. Similar reasoning can be applied to the occurrence of chemotaxis. The sensory systems that influence flagellum action in bacteria require the correct functioning of many genes (Koshland, 1980) and eukaryotic sensory systems are unlikely to be any less complex.

Possibly, the value of motility and chemotaxis to *Phytophthora* could be tested more directly. Zoospores are more dense than water yet accumulate at the top of a column of water as a result of negative geotaxis (Cameron and Carlile, 1977). Hence mutants with diminished motility might be obtained by selecting those that sediment. Negative geotaxis might also be used to obtain chemotaxis mutants, because mutants that remain motile but are not chemotactic should swim upward, ignoring a chemical attractant. Such a procedure has been used to obtain sensory response mutants in *Paramecium* (Kung et al, 1975). Motility and chemotaxis mutants, if successfully produced, could then be compared with the wild-type in their ability to bring about infection.

Although motility and chemotaxis are doubtlessly important to *Phytophthora,* a critical approach to the role and significance of individual responses remains justified. Demonstrating a chemotactic response to a particular chemical in the laboratory, for example, does not establish that a response to this chemical is important in nature; the chemical may not occur in nature but may sufficiently resemble a natural attractant to produce a response. In general, however, one can reasonably conclude that tactic responses persist because they are of importance and that their nature may tell us something about the ecology of the organism. Thus, the threshold values for negative chemotaxis for zoospores of a *Phytophthora* sp. may indicate the levels of different ions that are harmful to that species, and this knowledge may perhaps be applied to control infection. A more remote possibility would be to prevent zoospores from reaching the roots of plants of high value by misleading them with a background of an attractant or terminating their motility prematurely with an inducer of encystment.

## SUMMARY

1) *Phytophthora* zoospores can swim at speeds of about 100 $\mu$m/sec. Most of the thrust is provided by the anterior flagellum, but directional changes are brought about by the posterior flagellum.
2) Zoospores can be carried through soils by water currents and remain motile.
3) Zoospores, as a consequence of frequent directional changes, travel much shorter distances than their swimming speeds would suggest—at the most, a few centimeters in many hours.
4) Zoospores are attracted by plant roots and encyst at the root surface.
5) Root exudates attract zoospores and also bring about encystment.
6) Positive chemotaxis of zoospores to a range of amino acids, ethanol, and a variety of other compounds known or likely to be produced by plants has been demonstrated.
7) Zoospores are repelled by high concentrations of hydrogen ions and other cations.
8) Zoospores show electrotaxis, positive rheotaxis, and negative geotaxis.
9) Germ tubes emerging from encysted zoospores are attracted by plant roots.
10) The dispersal of zoospores is probably due largely to passive transport by water currents. Swimming and taxis are likely to be important close to the host and germ tube tropism important in the penetration of the host.

## ACKNOWLEDGMENTS

I wish to thank Robert Allen, Neil Cameron, Michael Holwill, Frank Newhook, and Marcella Sanchez-Alvarez for access to unpublished work and for helpful discussion, and Robert Allen and Paul Desjardins for providing figures.

## LITERATURE CITED

Adler, J. 1975. Chemotaxis in bacteria. Annu. Rev. Biochem. 44:341-356.

Allen, R. N., and Harvey, J. D. 1974. Negative chemotaxis of zoospores of *Phytophthora cinnamomi.* J. Gen. Microbiol. 84:28-38.

Allen, R. N., and Newhook, F. J. 1973. Chemotaxis of zoospores of *Phytophthora cinnamomi* to ethanol in capillaries of soil pore dimensions. Trans. Br. Mycol. Soc. 61:287-302.

Allen, R. N., and Newhook, F. J. 1974. Suppression by ethanol of spontaneous turning activity in zoospores of *Phytophthora cinnamomi.* Trans. Br. Mycol. Soc. 63:383-385.

Barash, I., Klisiewicz, J. M., and Kosuge, T. 1965. Utilization of carbon compounds by zoospores of *Phytophthora drechsleri* and their effect on motility and germination. Phytopathology 55:1257-1261.

Bartnicki-Garcia, S. 1973. Cell wall genesis in a natural protoplast: The zoospore of *Phytophthora palmivora.* Pages 77-91 in: Yeast, Mould and Plant Protoplasts. J. R. Villaneuva, I. Garcia-Acha, S. Gascon, and F. Uruburu, eds. Academic Press, London.

Bartnicki-Garcia, S., and Hemmes, D. E. 1976. Some aspects of the form and function of oomycete spores. Pages 593-641 in: The Fungal Spore: Form and Function. D. J. Weber and W. M. Hess, eds. John Wiley & Sons, New York.

Benjamin, M., and Newhook, F. J. 1982. Effects of glass microbeads on *Phytophthora* zoospore motility. Trans. Br. Mycol. Soc. 78:43-46.

Berg, H. C. 1975. Chemotaxis in bacteria. Annu. Rev. Biophys. Bioeng. 4:119-136.

Bimpong, C. E. 1975. Changes in metabolic reserves and enzyme activities during zoospore motility and cyst germination in *Phytophthora palmivora.* Can. J. Bot. 53:1411-1416.

Bimpong, C. E., and Clerk, G. C. 1970. Motility and chemotaxis in zoospores of *Phytophthora palmivora* (Butl.) Butl. Ann. Bot. 34:617-624.

Bimpong, C. E., and Hickman, C. J. 1975. Ultrastructural and cytochemical studies of zoospores, cysts, and germinating cysts of *Phytophthora palmivora*. Can. J. Bot. 53:1310-1327.

Brokaw, C. J. 1958. Chemotaxis of bracken spermatozoids: Implications of electrochemical orientation. J. Exp. Biol. 35:197-212.

Cameron, J. N. 1979. Taxes of zoospores of the fungus *Phytophthora*. Ph.D. thesis, University of London. 169 pp.

Cameron, J. N., and Carlile, M. J. 1977. Negative geotaxis of zoospores of the fungus *Phytophthora*. J. Gen. Microbiol. 98:599-602.

Cameron, J. N., and Carlile, M. J. 1978. Fatty acids, aldehydes and alcohols as attractants for zoospores of *Phytophthora palmivora*. Nature (London) 271:448-449.

Cameron, J. N., and Carlile, M. J. 1980. Negative chemotaxis of zoospores of the fungus *Phytophthora palmivora*. J. Gen. Microbiol. 120:347-353.

Cameron, J. N., and Carlile, M. J. 1981. Binding of isovaleraldehyde, an attractant, to zoospores of the fungus *Phytophthora palmivora* in relation to zoospore chemotaxis. J. Cell Sci. 49:273-281.

Carlile, M. J. 1970. The photoresponses of fungi. Pages 309-344 in: Photobiology of Microorganisms. P. Halldal, ed. John Wiley & Sons, London.

Carlile, M. J. 1980a. Positioning mechanisms—The role of motility, taxis and tropism in the life of microorganisms. Pages 55-74 in: Contemporary Microbial Ecology. D. C. Ellwood, J. N. Hedger, M. J. Latham, J. M. Lynch, and J. H. Slater, eds. Academic Press, London.

Carlile, M. J. 1980b. Sensory transduction in aneural organisms. Pages 1-22 in: Photoreception and Sensory Transduction in Aneural Organisms. F. Lenci and G. Colombetti, eds. Plenum Press, New York.

Cavalier-Smith, T. 1978. The evolutionary origin and phylogeny of microtubules, mitotic spindles and eukaryote flagella. BioSystems 10:93-114.

Cerdá-Olmedo, E. 1977. Behavioral genetics of *Phycomyces*. Annu. Rev. Microbiol. 31:535-547.

Chang-Ho, Y., and Hickman, C. J. 1970. Some factors involved in the accumulation of phycomycete zoospores on plant roots. Pages 103-108 in: Root Diseases and Soil-Borne Pathogens. T. A. Toussoun, R. V. Bega, and P. E. Nelson, eds. Univ. Calif. Press, Berkeley.

Couch, J. N. 1941. The structure and action of the cilia in some aquatic phycomycetes. Am. J. Bot. 28:704-713.

Desjardins, P. R., Zentmyer, G. A., and Reynolds, D. A. 1969. Electron microscopic observations of the flagellar hairs of *Phytophthora palmivora* zoospores. Can. J. Bot. 47:1077-1079.

Desjardins, P. R., Wang, M. C., and Bartnicki-Garcia, S. 1973a. Electron microscopy of zoospores and cysts of *Phytophthora palmivora:* Morphology and surface texture. Arch. Mikrobiol. 88:61-70.

Desjardins, P. R., Zentmyer, G. A., Chen, D.-W., DeWolfe, T. A., Klotz, L. J., and Reynolds, D. A. 1973b. Flagellar hairs on zoospores of *Phytophthora* species: Tip hairs on the whiplash flagellum. Experientia 29:240-241.

Diehn, B., Feinleib, M., Haupt, W., Hildebrand, E., Lenci, F., and Nultsch, W. 1977. Terminology of behavioral responses of motile microorganisms. Photochem. Photobiol. 26:559-560.

Duniway, J. M. 1976. Movement of zoospores of *Phytophthora cryptogea* in soils of various textures and matric potentials. Phytopathology 66:877-882.

Foster, K. W. 1977. Phototropism of coprophilous zygomycetes. Annu. Rev. Biophys. Bioeng. 6:419-443.

Futrelle, R. P., and Berg, H. C. 1972. Specification of gradients used for studies of chemotaxis. Nature (London) 239:517-518.

Gaworzewska, E. T., and Carlile, M. J. 1982. Positive chemotaxis of *Rhizobium leguminosarum* and other bacteria towards root exudates from legumes and other plants. J. Gen. Microbiol. 128:1179-1188.

Gooday, G. W. 1975. Chemotaxis and chemotropism in fungi and algae. Pages 155-204 in: Primitive Sensory and Communication Systems: The Taxes and Tropisms of Micro-Organisms and Cells. M. J. Carlile, ed. Academic Press, London.

Gooday, G. W. 1978. Microbial hormones. Pages 207-219 in: Companion to Microbiology. A. T. Bull and P. M. Meadow, eds. Longman, London.

Gooding, G. V., and Lucas, G. B. 1959. Factors influencing sporangial formation and zoospore activity in *Phytophthora parasitica* var. *nicotianae*. Phytopathology 49:277-281.

Halsall, D. M. 1976. Zoospore chemotaxis in Australian isolates of *Phytophthora* species. Can. J. Microbiol. 22:409-422.

Hickman, C. J. 1970. Biology of *Phytophthora* zoospores. Phytopathology 60:1128-1135.

Hickman, C. J., and Ho, H. H. 1966. Behaviour of zoospores in plant-pathogenic phycomycetes. Annu. Rev. Phytopathol. 4:195-220.

Ho, H. H., and Hickman, C. J. 1967a. Asexual reproduction and behavior of zoospores of *Phytophthora megasperma* var. *sojae*. Can. J. Bot. 45:1963-1981.

Ho, H. H., and Hickman, C. J. 1967b. Factors governing zoospore responses of *Phytophthora megasperma* var. *sojae* to plant roots. Can. J. Bot. 45:1983-1994.

Ho, H. H., Zachariah, K., and Hickman, C. J. 1968a. The ultrastructure of zoospores of *Phytophthora megasperma* var. *sojae*. Can. J. Bot. 46:37-41.

Ho, H. H., Hickman, C. J., and Telford, R. W. 1968b. The morphology of zoospores of *Phytophthora megasperma* var. *sojae*. Can. J. Bot. 46:88-89.

Holwill, M. E. J. 1982. Dynamics of eukaryotic flagellar movement. Pages 289-312 in: Eukaryotic and Prokaryotic Flagella. E. B. Amos and J. G. Duckett, eds. Symp. Soc. Exp. Biol. 35. Cambridge Univ. Press, Cambridge.

Hutchinson, S. A., Sharma, P., Clarke, K. R., and MacDonald, I. 1980. Control of hyphal orientation in colonies of *Mucor hiemalis*. Trans. Br. Mycol. Soc. 75:177-191.

Katsura, K., and Miyata, Y. 1971. Swimming behavior of *Phytophthora capsici* zoospores. Pages 107-128 in: Morphological and Biochemical Events in Plant-Parasite Interaction. S. Akai and S. Ouchi, eds. Phytopathol. Soc. Jpn., Tokyo.

Katsura, K., Masago, H., and Miyata, Y. 1966. Movements of zoospores of *Phytophthora capsici*. I. Electrotaxis in some organic solutions. Ann. Phytopathol. Soc. Jpn. 32:215-220.

Khew, K. L., and Zentmyer, G. A. 1973. Chemotactic response of zoospores of five species of *Phytophthora*. Phytopathology 63:1511-1517.

Khew, K. L., and Zentmyer, G. A. 1974. Electrotactic response of zoospores of seven species of *Phytophthora*. Phytopathology 64:500-507.

Knowles, D. J. C., and Carlile, M. J. 1978. The chemotactic response of plasmodia of the myxomycete *Physarum polycephalum* to sugars and related compounds. J. Gen. Microbiol. 108:17-25.

Ko, W. H., and Chan, M. J. 1974. Aggregation of *Phytophthora capsici* zoospores and their interaction with zoospores of *P. palmivora*. J. Gen. Microbiol. 80:549-551.

Ko, W. H., and Chase, L. L. 1973. Aggregation of zoospores of *Phytophthora palmivora*. J. Gen. Microbiol. 78:79-82.

Kochert, G. 1978. Sexual pheromones in algae and fungi. Annu. Rev. Plant Physiol. 29:461-468.

Koshland, D. E. 1979. Bacterial chemotaxis. Pages 111-166 in: The Bacteria: A Treatise on Structure and Function, Vol. 7. J. R. Sokatch and L. N. Ornston, eds. Academic Press, New York.

Koshland, D. E. 1980. Bacterial Chemotaxis as a Model Behavioral System. Raven Press, New York. 193 pp.

Kung, C , Chang, S.-Y., Satow, Y., van Houten, J., and Hansma, H. 1975. Genetic dissection of behavior in *Paramecium*. Science 188:898-904.

Leopold, L. B. 1971. Trees and streams: The efficiency of branching patterns. J. Theor. Biol. 31:339-354.

Machlis, L. 1958. Evidence for a sexual hormone in *Allomyces*. Physiol. Plant. 11:181-192.

Macnab, R. M. 1978. Bacterial motility and chemotaxis: The molecular biology of a behavioral system. CRC Crit. Rev. Biochem. 5:291-341.

Macnab, R. M. 1979. Chemotaxis in bacteria. Pages 310-334 in: Physiology of Movements. W. Haupt and M. E. Feinleib, eds. Springer-Verlag, Heidelberg.

Mehrotra, R. S. 1970. Techniques for demonstrating accumulation of zoospores of *Phytophthora* species on roots in soil. Can. J. Bot. 48:879-882.

Musgrave, A., Ero, L., Scheffer, R., and Oehlers, E. 1977.

Chemotropism of *Achlya bisexualis* germ hyphae to casein hydrolysate and amino acids. J. Gen. Microbiol. 101:65-70.

Newhook, F. J., Young, B. R., Allen, S. D., and Allen, R. N. 1981. Zoospore motility of *Phytophthora cinnamomi* in particulate substrates. Phytopathol. Z. 101:202-209.

Petri, L. 1925. Osservazioni biologiche sulla '*Blepharospora cambivora*.' Reprinted from Ann. R. Ist. Sup. Agrario e Forestale, Ser. 2a, in Rev. Appl. Mycol. 10:122-123 (1931).

Plesset, M. S., Whipple, C. G., and Winet, H. 1975. Analysis of the steady state of bioconvection in swarms of swimming micro-organisms. Pages 339-360 in: Swimming and Flying in Nature, Vol. 1. T. Y.-T. Wu, C. J. Brokaw, and C. Brennan, eds. Plenum Press, New York.

Porter, J. R., and Shaw, D. S. 1978. Aggregation of *Phytophthora drechsleri* zoospores; pattern analysis suggests a taxis. Trans. Br. Mycol. Soc. 71:515-518.

Reichle, R. E. 1969. Fine structure of *Phytophthora parasitica* zoospores. Mycologia 61:30-51.

Robinson, P. M. 1973a. Autotropism in fungal spores and hyphae. Bot. Rev. 39:367-384.

Robinson, P. M. 1973b. Chemotropism in fungi. Trans. Br. Mycol. Soc. 61:303-313.

Royle, D. J., and Hickman, C. J. 1964a. Analysis of factors governing in vitro accumulation of zoospores of *Pythium aphanidermatum* on roots. I. Behaviour of zoospores. Can. J. Microbiol. 10:151-162.

Royle, D. J., and Hickman, C. J. 1964b. Analysis of factors governing in vitro accumulation of zoospores of *Pythium aphanidermatum* on roots. II. Substances causing response. Can. J. Microbiol. 10:201-219.

Russo, V. A. 1980. Sensory transduction in phototropism: Genetic and physiological analysis in *Phycomyces*. Pages 373-395 in: Photoreception and Sensory Transduction in aneural organisms. F. Lenci and G. Colombetti, eds. Plenum Press, New York.

Sing, V. O., and Bartnicki-Garcia, S. 1975. Adhesion of *Phytophthora palmivora* zoospores: Electron microscopy of cell attachment and cyst wall fibril formation. J. Cell Sci. 18:123-132.

Troutman, J. L., and Wills, W. H. 1964. Electrotaxis of *Phytophthora parasitica* zoospores and its possible role in infection of tobacco by the fungus. Phytopathology 54:225-228.

Vujičić, R., Colhoun, J., and Chapman, J. A. 1968. Some observations on the zoospores of *Phytophthora erythroseptica*. Trans. Br. Mycol. Soc. 51:125-127.

Wang, M. C., and Bartnicki-Garcia, S. 1980. Distribution of mycolaminarans and cell wall $\beta$-glucans in the life cycle of *Phytophthora*. Exp. Mycol. 4:269-280.

Young, B. R., Newhook, F. J., and Allen, R. N. 1979. Motility and chemotactic response of *Phytophthora cinnamomi* zoospores in 'ideal soils.' Trans. Br. Mycol. Soc. 72:395-401.

Zentmyer, G. A. 1961. Chemotaxis of zoospores for root exudates. Science 133:1595-1596.

Zentmyer, G. A. 1966. Role of amino acids in chemotaxis of zoospores of three species of *Phytophthora*. (Abstr.) Phytopathology 56:907.

Zentmyer, G. A. 1970. Tactic responses of zoospores of *Phytophthora*. Pages 109-111 in: Root Diseases and Soil-Borne Pathogens. T. A. Toussoun, R. V. Bega, and P. E. Nelson, eds. Univ. Calif. Press, Berkeley.

# 8

# Biophysical Aspects of the Development of *Phytophthora*

U. Gisi

CONTENTS



## INTRODUCTION

The development of fungi is governed by different kinds of biological, physical, and chemical mechanisms. It is not easy to separate these three mechanisms clearly from one another. Many ecological problems in soil can be solved by the study of antagonistic or synergistic interactions: two examples in *Phytophthora* research are the stimulation of sporangium production of *P. cinnamomi* by certain soil bacteria (Zentmyer, 1965) and the possibility of biological control of *Phytophthora* diseases (Shea and Broadbent, Chapter 26). Looking more closely, the reason for the stimulation of sporangium production of *P. cinnamomi* in soil seems to be the production of exudates by certain soil bacteria, which is a chemical mechanism (Ayers and Zentmyer, 1971), while sporangium production per se is governed by soil matric potential, which is a physical mechanism (Gisi et al, 1980).

In this chapter, some of the most important physical factors governing fungal development (i.e., biophysical aspects) are described; however, it must be kept in mind that many chemical factors such as nutrition, pH, redox potential, and enzyme activity are of equal importance and interact with physical factors.

### Temperature and Light

As described for many fungi, temperature is an important parameter for mycelial growth, sporulation, and spore germination in *Phytophthora*; these aspects are discussed in more detail by Ribeiro (Chapter 4). Perhaps less is known about the correlation between temperature and zoospore motility; Ho and Hickman (1967a) observed velocities of zoospore movement in *P. megasperma* var. *sojae* of 87 and 156 $\mu$m/sec at 18 and 26° C, respectively. The mechanisms of such correlations are not clear; probably the viscosity of the fluid in which they swim and the production of energy needed for swimming are involved.

The influence of light on different phases in the life cycle of *Phytophthora* has been described by several authors. Among the most interesting events are the stimulation of sporangium production on young hyphae by continuous light and the stimulation of oospore production in darkness (Englander and Roth, 1980; Gisi, 1975; Zentmyer and Ribeiro, 1977). The influence of light on the saprophytic phase of *Phytophthora* is probably less important, because for many species the production of sporangia and oospores in soil can take place in the dark.

### Electricity, Viscosity, and Mechanical Forces

Every biological surface, especially membranes, is charged electrically. In mycology, there is little information on the effect of electric charges on transport toward or through surfaces. Slayman and Slayman (1962) described the migration of vesicles toward the hyphal tip along an electrochemical gradient in *Neurospora*. Ho and Hickman (1967b), investigating the behavior of zoospores of *P. megasperma* var. *sojae* in an electric field, observed a rapid trapping of the zoospores around the negative pole. Similar results were found by Singh (1975) on the response of *Phytophthora* zoospores to ionically charged resins. A surface charge of about −25 mV has been demonstrated for certain plant protoplasts (Nagata and Melchers, 1978); the success of fusion experiments with protoplasts depends on this charge. Because zoospores of *Phytophthora* are free of a cell wall, surface charges of the membrane may be involved during aggregation as well as cleavage of zoospores. The autoaggregation of zoospores in dense suspensions (Ko and Chase, 1973; Porter and Shaw, 1978) could be explained by electrical effects. Electrical forces may participate during cleavage of zoospores and prevent them from fusing; electrical repulsion might even help explain zoospore release from the sporangium. In fact, investigations of zoospore release from sporangia of *Saprolegnia monoica* support such an assumption (Borkowski, 1968).

Movement of zoospores in a liquid is governed by the liquid's viscosity; the higher the viscosity, the lower the speed. Tactic responses and zoospore release are affected by the viscosity of the medium. Two different factors are probably involved: more energy is consumed by zoospores swimming in viscous media, and water transport into sporangia or zoospores is restricted.

The influence of other physical factors on the development and behavior of *Phytophthora* spp. has been observed, but it is difficult to explain the mechanisms. Mechanical forces such as the streaming of fluids (probably the streaming resistance) are known to induce positive rheotaxis of zoospores (Katsura and Miyata, 1971). Contact stimulus, in culture or in soil, was reported to induce early encystment of zoospores (Bimpong and Clerk, 1970; Ho and Hickman, 1967a; MacDonald and Duniway, 1978c); in "ideal soils," this phenomenon seems to be less important (Young et al, 1979). Gravity influences zoospore motility; as reported by Cameron and Carlile (1977), zoospores exhibit negative geotaxis (see Carlile, Chapter 7).

### Aeration and Moisture

Little work has been done on the effect of oxygen, carbon dioxide, and other gases on the development of *Phytophthora* spp. except that of Mitchell and Zentmyer (1971a, 1971b), perhaps because in nature, the influence of aeration is not readily separated from that of moisture. The amount and form of water in the environment are expressed best by its availability. In thermodynamic terms, this is called water activity. For physiologic purposes, water activity may be replaced by the related term water potential. As a simplification, water potential indicates the energy needed to remove a certain quantity of water from a given system. Thus, by definition, chemically pure water under standard conditions has a water potential of 0 bar (1 bar = 1,022 cm water column = 0.987 atm = 97.7 kPa).

The total water potential $\Psi$ of a biological system has various components—a matric term $\psi_m$ due to water associated with the matrix (imbibition by solids, colloids, wall); an osmotic term $\psi_\pi$ due to particles in solution (also called solute potential $\psi_s$); and a pressure term $\psi_p$ (usually termed turgor pressure). The first two terms are negative, and the third is positive. The relative humidity of the vapor phase in equilibrium with a body of water is also an indirect measure of the total water potential (Griffin, 1977). As examples of magnitudes of water potential, soils of medium water content (field capacity) have matric potentials of about −0.3 bar; a physiologic NaCl solution (0.85% w/v) has an osmotic potential of about −6.8 bars; the turgor pressure of a plant or fungal cell normally ranges from 5 to 10 bars (for comparison, automobile tires are usually inflated to a pressure of about 2 bars); and a value of 95% relative humidity corresponds to a water potential of −71 bars (Griffin, 1977).

Why is water potential of such immense importance for *Phytophthora*? One answer is that *Phytophthora* is positioned between water and land life and as a consequence is extremely sensitive to all components of water potential in habitats such as free water (osmotic water component) as well as to soil (matric water component) and to air (relative humidity).

In the following sections, the physical importance of water in the development of the nonparasitic phases of *Phytophthora* is reviewed. Duniway (Chapter 14) describes in more detail the effect of soil physical factors on the development of diseases caused by *Phytophthora*.

## GROWTH AND WATER RELATIONS

Growth on agar or liquid media is sensitive to osmotic water potential as well as to nutrient concentration. Thus, it is difficult to distinguish between the effects of those two factors on mycelial growth. Sommers et al (1970) demonstrated that the lower limit for mycelial growth of *P. cinnamomi*, *P. megasperma*, and *P. parasitica* on different agar media was about −30 to −50 bars. On the other hand, Sterne et al (1976) found that in liquid culture, growth of *P. cinnamomi* was limited at osmotic water potentials between −15 and −25 bars, depending on the osmoticum used; similarly, Gisi et al (1977) determined the lower limit of growth for *P. cactorum* to be at about −35 bars.

The apparent discrepancy between the results obtained on agar vs liquid cultures may be explained by considering the following points. First, species differ slightly in their sensitivity to water potential, analogous to the situation in sporangium formation (Gisi et al, 1980). Second, an agar culture differs from a liquid one in having a matric component in addition to an osmotic component of water potential (Wiebe, 1966). Third, in liquid culture, growth is largely submerged and adapts continuously to the osmotic conditions of the liquid. A higher solute transport into a liquid culture mycelium (Duniway, 1979) results in a higher sensitivity to water potential or to specific solute effects.

Solute effects must be considered when comparing the influence of osmotic vs matric potential on mycelial growth of *Phytophthora*. Low matric potential values are reported to be more limiting to growth than are equivalent low values of osmotic potential (Adebayo and Harris, 1971; Duniway, 1979). Except for *P. cinnamomi*, most *Phytophthora* species show a very low rate of growth in nonsterile soils (Gisi, 1975; Tsao, 1969), and even in sterile soils amended with glucose, the maximum growth rate of *P. cactorum* is lower by a factor of about 100 compared with *P. cinnamomi* (Table 1). Nutrient conditions of the inoculum and chemical factors of the test soils probably obscure the growth response to matric potential. The lower limit of growth in sterile soils amended with glucose is at matric potentials of about −40 bars for *P. cinnamomi* (Malajczuk and Theodorou, 1979) and −5 bars for *P. cactorum* (Gisi, 1975). Optimum growth conditions are at −11 bars and 0 bars for *P. cinnamomi* and *P. cactorum*, respectively (Table 1). Growth responses of *P. cinnamomi*, in nutrient-rich soils, to matric potential adjusted by adding increasing amounts of KCl (Malajczuk and Theodorou, 1979) seem to be similar to those in agar cultures adjusted osmotically (Sommers et al, 1970); however, with *P. cactorum*, there is a clear discrepancy between the influence of matric and osmotic potential. Optimum growth of *P. cactorum* in sterile soils at matric potentials around 0 bars (saturated soil, Table 1) seems to be typical for species with a low growth rate. Kuan and Erwin (1982) found that mycelial fragments of *P. megasperma* f. sp. *medicaginis* developed best at matric potentials higher than −0.01 bar. Thus, among various *Phytophthora* species there are different levels of sensitivity to matric potential.

TABLE 1. Growth of *Phytophthora cactorum* and *P. cinnamomi* in sterile soils at different water potentials

| | Growth rate of hyphae (μm/day) | | |
|---|---|---|---|
| | *P. cactorum* in soil[a] | | *P. cinnamomi* in soil[b] |
| Water potential (bars) | Without glucose | +1.5% Glucose | + About 0.5% glucose |
| Flooded | 0 | 75 | ... |
| Saturated (= 0 bars) | 2 | 88 | ... |
| −0.05 | 23 | 12 | ... |
| −0.90 | ...[c] | ... | 2,570 |
| −1.8 | ... | ... | 7,140 |
| −5 | 12 | 0 | ... |
| −10.8 | ... | ... | 7,430 |
| −20.8 | ... | ... | 6,715 |
| −30.3 | ... | ... | 3,430 |
| −38 | ... | ... | 145 |
| −49 | ... | ... | 0 |

[a]Water potential adjusted matrically. Data calculated from Gisi (1975).

[b]Water potential adjusted by adding increasing amounts of KCl. Data calculated from Malajczuk and Theodorou (1979).

[c]No observation.

## SPORULATION AND WATER RELATIONS

### Zoosporangia

In his excellent review on the "water relations of water molds," Duniway (1979) stated that "water status is one of the most determining factors in the formation and behavior of sporangia and zoospores." Formation of sporangia in soils at different matric potentials has been described for a variety of *Phytophthora* species (Gisi et al, 1980), but little information is available on the effect of osmotic potential on sporangium formation. When washed mycelium is used as inoculum, production of sporangia in soil ceases at matric potentials between −2 and −5 bars, the exact values depending on the inoculation method as well as on the species (Fig. 1A–E). By using infected plant material as inoculum, the lower limit of sporangium formation is at about −0.2 bar, whereas optimum sporulation occurs near saturated soil conditions (Kuan and Erwin, 1982; Pfender et al, 1977; Sugar, 1977). Because infected plant material and washed mycelium differ in their nutrient status, water potential, and microbial degradability, it is not surprising that different intensities of sporangium production were found by Kuan and Erwin (1982) with these two types of inoculum from the same isolate of *P. megasperma* (Fig. 1C and F).

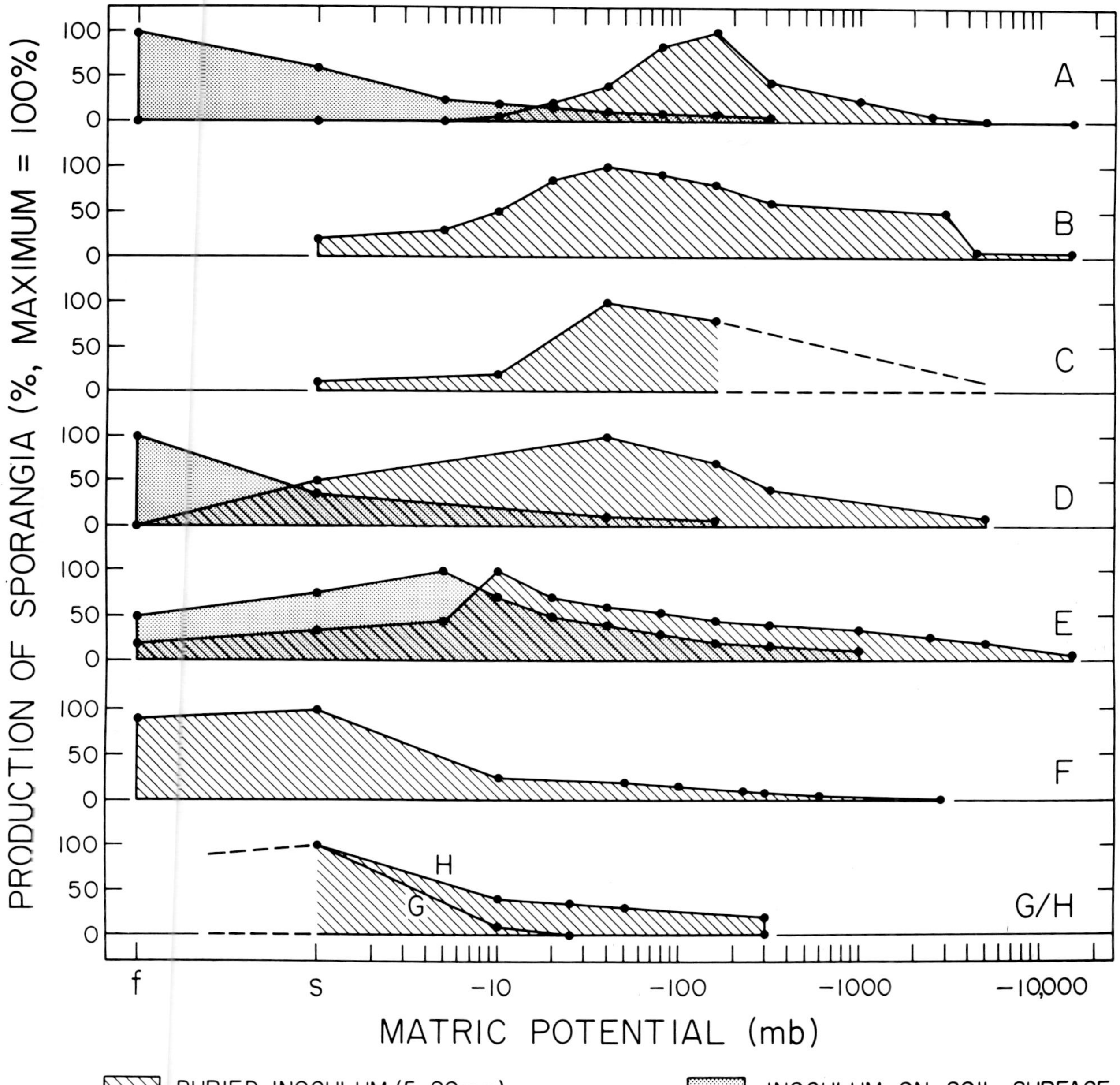


**Fig. 1.** Effect of soil matric potential on production of sporangia by *Phytophthora* species, expressed as a percentage of the maximum number (= 100%) observed after 2–4 days (f = flooded soil, s = saturated soil). **A,** *P. cinnamomi* (Gisi et al, 1980). **B,** *P. cryptogea* (Duniway, 1975a, 1975b). **C,** *P. megasperma* (Kuan and Erwin, 1982; MacDonald, 1977). **D,** *P. cactorum* (Gisi, 1975 [surface]; Meyer and Schönbeck, 1975 and Sneh and McIntosh, 1974 [buried]). **E,** *P. palmivora* (Gisi et al, 1980). **F,** *P. megasperma* (Kuan and Erwin, 1982; Pfender et al, 1977). **G** and **H,** *P. cambivora* (G) and *P. drechsleri* (H) (Sugar, 1977). (Data from Gisi et al, 1980)

Different intensities of sporangium production are observed at very high[1] matric potentials when using different inoculation techniques. Buried mycelium is affected adversely by poor aeration, resulting in a distinct decline in sporangium numbers at matric potentials higher than −0.01 bar (Fig. 1). When mycelium is exposed on the soil surface (optimum conditions for aeration) or when infected plant material is used as inoculum, sporangium formation increases markedly at matric potentials higher than −0.01 bar, with highest numbers under saturated or flooded[2] soil conditions (Fig. 1). This behavior indicates that other factors besides matric potential, such as interaction with soil microorganisms and their exudates, are also important for sporangium production (Ayers and Zentmyer, 1971), especially in situations where solute diffusion through soil pores is high (see below). Water potential may also affect *Phytophthora* indirectly by changing the competitive behavior of some important soil microorganisms.

### Chlamydospores and Oospores

Very little is known about the influence of water potential on the formation of chlamydospores and oospores. These structures function as survival propagules, and their formation appears to be a response to an unfavorable environment. Chlamydospore production by *P. cinnamomi* occurred at matric potentials between −0.5 and −20 bars (Malajczuk and Theodorou, 1979; Reeves, 1975; Weste and Vithanage, 1979). Oospore production on mycelium of *P. cactorum* occurred at matric potentials lower than about −1 bar[1] (Meyer and Schönbeck, 1975; Sneh and McIntosh, 1974). Thus, chlamydospore and oospore formation is adapted to water conditions considerably drier than those that permit sporangium production (Fig. 1).

### Survival of Spores

Survival of fungal material in soil depends on two main conditions: microbial activities (competition, antagonism, parasitism) and the physiologic ability of the propagule to remain viable for a certain time in adverse physicochemical situations. Since both conditions are linked to water relations, spore survival can be correlated indirectly with different soil water potentials. It has consistently been reported that no spore types (sporangia, zoospores, chlamydospores, oospores) of different *Phytophthora* species survive more than a few days at matric potentials higher than about −0.1 bar (Gisi, 1975; McIntosh, 1972; Meyer and Schönbeck, 1975; Sneh and McIntosh, 1974). In some cases, breakdown of *Phytophthora* propagules by soil bacteria (Broadbent and Baker, 1974; Malajczuk et al, 1977) or by the parasitic activity of other soil fungi (Sneh et al, 1977) has been demonstrated in relatively wet soils. However, *Phytophthora* spores generally tolerate matric potentials as low as −5 to −30 bars for 4–8 mo without much loss of viability, thus suggesting that at these water potentials, microbial degradation is low. An interesting short-term experiment by MacDonald and Duniway (1978b) showed that only 50% of the rewetted sporangia of *P. cryptogea* and *P. megasperma* lost their viability after being dried to matric potentials as low as −45 bars for 2 days—a surprising physiologic tolerance to low water potential.

## SPORE GERMINATION AND WATER RELATIONS

### Zoosporangia and Zoospore Cysts

There is no doubt that water conditions and temperature are the main factors that govern the germination of zoosporangia (Duniway, 1979). Chemically pure water induces zoospore release[3], but at lower water potentials, germ tube formation replaces zoospore release. The precise value of water potential that allows zoospore release from sporangia is difficult to determine, but it is definitely higher than about −3.5 bars for osmotic potential (Gisi et al, 1979a) and higher than about −0.05 bar for matric potential (Duniway, 1976; Pfender et al, 1977). Since zoospore release is not strictly a germination process but a type of spore liberation, this topic is discussed further in a later section.

Germination of sporangia of *Phytophthora* can be divided into two steps: 1) induction and formation of the germ tube and 2) growth of the germ tube with or without formation of a new sporangium. It seems that growth of germ tubes is best in water-saturated soil (Gisi, 1975) or at values higher than −0.3 bar (Sneh and McIntosh, 1974). This behavior is typical for species with a low growth rate and reflects a situation similar to that described for mycelial growth of *P. cactorum* in soil. Formation of germ tubes terminated by a new sporangium was observed at much drier conditions (Meyer and Schönbeck, 1975).

In encysted zoospores, germ tubes may be produced at very high water potentials, but growth beyond about 6 μm is possible only when a carbon source like sucrose is added at concentrations of 5% (Hemmes and Hohl, 1971). At this concentration, the osmotic potential is about −3.5 bars and it prevents zoospore cysts from germinating by formation of secondary zoospores. As mentioned earlier, −3.5 bars is also the level at which zoospore release from sporangia is prevented. Highest numbers of germinated zoospores of *P. cactorum* were found under saturated soil conditions (Meyer and Schönbeck, 1975).

### Chlamydospores and Oospores

Infection of avocado roots in soils infested with chlamydospores of *P. cinnamomi* is limited at a matric potential of −0.25 bar; the highest infection rates occurred at 0 bars (Sterne et al, 1977). This observation is an indication of the water potential range that permits germination and growth of chlamydospore germ tubes. Kuan and Erwin (1982) also reported that the highest incidence of root rot of alfalfa caused by germinating oospores of *P. megasperma* f. sp. *medicaginis* occurred at 0 bars with a proportional reduction in root rot to nearly zero as matric potential values were reduced to −0.1 bar. Both reports support the hypothesis that at matric potentials higher than about −0.25 bar, more nutrients become available, favoring the germination of spores and the growth of germ tubes. Further support comes from two additional findings. In saturated soil, a much higher nutrient leakage from alfalfa roots predisposed the roots to infection (Kuan and Erwin, 1980). A reduction of soil matric potential decreased hydraulic

[1]Since matric and osmotic water potentials are negative values, the highest values are those closest to 0 bars; e.g., higher than −0.01 bar means between −0.01 bar and 0 bars; a value lower than −1 bar is a value of −2 bars. At saturated and flooded soil conditions, the matric potential is 0 bars; the osmotic potential is lower than 0 bars, depending on the quantity of water and the quantity of solutes.

[2]At saturated conditions, all soil pores are filled with water; at flooded conditions, there is an excess of water.

[3]Zoospore release from sporangia is often called "indirect germination" to distinguish it from germ tube formation, termed "direct germination."

conductivity, K, and thus restricted diffusion of solutes through soil pores (Marshall and Holmes, 1979).

## CYTOPLASMIC DISCHARGE (ZOOSPORE RELEASE) AND WATER RELATIONS

### Preconditions

Zoospore release can be subdivided into two independent steps: 1) zoosporogenesis or cleavage of cytoplasm and 2) zoospore release or cytoplasmic discharge. The cleavage of cytoplasm by fusion of cleavage vesicles has been clearly demonstrated on an ultrastructural level (Hohl and Hamamoto, 1967; see Hemmes, Chapter 2), but there is no information on the physiologic reasons why and how these vesicles migrate in the cytoplasm and what triggers the sporangium to form zoospores. Sporangia of *Phytophthora* have a strong tendency to release zoospores, but this trend is reversed when conditions favor germ tube formation (Hemmes and Hohl, 1972). Zoospore differentiation is not a prerequisite for cytoplasmic discharge, although this is the most frequent case (Gisi et al, 1979a). Transfer of sporangia to pure water initiates zoospore release after a short time; this demonstrates that very few conditions are needed for triggering this process. The most important preconditions for zoospore release are a turgor pressure in the sporangium high enough to initiate the process and an osmotic gradient between sporangium and external medium high enough to allow a significant water uptake into the sporangium. Biochemical experiments are needed to elucidate this first step of zoospore release. The second step, cytoplasmic discharge or the liberation of zoospores, will be described in detail with special reference to water relations and the motive forces for cytoplasmic transport.

### Significance of the Discharge Vesicle

When sporangia are transferred from the sporulation medium to water or to hypotonic solutions, the osmotic gradient between a sporangium and the external medium increases. As a result water is taken up by the sporangium. Commonly used sporulation media such as bean meal broth, diluted salt solutions, and soil extracts have osmotic potentials of about −0.1 to −1.0 bar, whereas sporangium cytoplasm has values of about −6 bars (Gisi and Zentmyer, 1980). Since sporangium walls are not very distensible, water uptake does not increase the volume of the sporangium; instead the turgor pressure increases. On the other hand, the papillar material located between the cytoplasm and the primary (outer) sporangial wall is highly distensible (Gisi et al, 1979a) and is probably a gel-like substance (MacDonald and Duniway, 1978a). The turgor pressure increases to the point where the limit of elasticity of the primary wall is exceeded. This is a plausible explanation for the observation that zoospore release is initiated either by ejection of a substantial part of the primary wall covering the papilla or by a rupture and curling back of the wall at the edge of the exit pore. The first possibility is frequently observed in *P. cactorum* (Figs. 2–4); the second one has been described by Gisi et al (1979a) for *P. palmivora* (Fig. 5).

After the rupture of the primary wall, the papillar material underneath (= plug, = vesiculogen; Gisi et al, 1979a) is forced through the exit pore. As this plastic material distends, it creates a large spherical "vesicle." The vesicle "wall" is plastic and consists of thin fibrillar elements of undetermined lengths. In my opinion, the inflation of papillar material of a sporangium to form a discharge vesicle equal in volume to the sporangium itself can be explained only by the action of substantial internal pressure. Within certain limits, the diameter as well as the surface area of the discharge vesicle increases linearly with time; the velocity of enlargement of the surface of the discharge vesicle in *P. cactorum* is about 200 $\mu m^2/sec$ (Gisi et al, 1979a). Furthermore, the rate of volume increase of the vesicle per unit time is nearly constant, with values of about 80 $\mu m^3/sec^2$. We calculated (Gisi et al, 1979b) that the rate of volume increase (dV/dt) of the vesicle was proportional to the decrease in hydrostatic pressure (dp) of the sporangium and that it decreased hyperbolically with time (t):

$$\frac{dV}{dt} = -K_1 \cdot dp = -K_2 \cdot dt^{-1}. \qquad (1)$$

Thus, the enlargement of the discharge vesicle corresponds to the change in hydrostatic pressure in the sporangium. Since up to 70% of the total number of zoospores per sporangium are released into the growing discharge vesicle before the vesicle bursts, a good part of zoospore release is also governed by changes in hydrostatic pressure. The velocity of zoospore release showed a similar correlation to the change in hydrostatic pressure as the enlargement of the discharge vesicle did (see below). The establishment of a distinct discharge vesicle during zoospore release could be observed in our experiments with many *Phytophthora* species, including *P. cactorum, P. capsici, P. cinnamomi, P. citricola, P. cryptogea, P. infestans, P. megasperma,* and *P. palmivora.* This suggests that a pressure-driven zoospore release is valid for the entire genus. We can even postulate that all genera of fungi in which a discharge vesicle occurs could have this mechanism for cytoplasmic discharge; this would include *Pythium* as well as many members of the Chytridiales and Hyphochytridiales.

Further observations support a pressure-driven mechanism of cytoplasmic discharge. Normally, velocities of vesicle enlargement and increase in diameter of the expelled zoospore mass are about equal. The vesicle wall can be seen only by careful observation. When cytoplasmic discharge of *P. palmivora* sporangia was studied using $CaCl_2$ or polyethylene glycol (PEG) at osmotic potentials of about −3 bars, instead of double-distilled water, the process slowed down to a level where only the enlargement of the discharge vesicle occurred (Gisi et al, 1979a). The maximum diameter of the vesicle was nearly reached by the time the first two zoospores were released (Figs. 6–10). Probably, the turgor pressure or at least

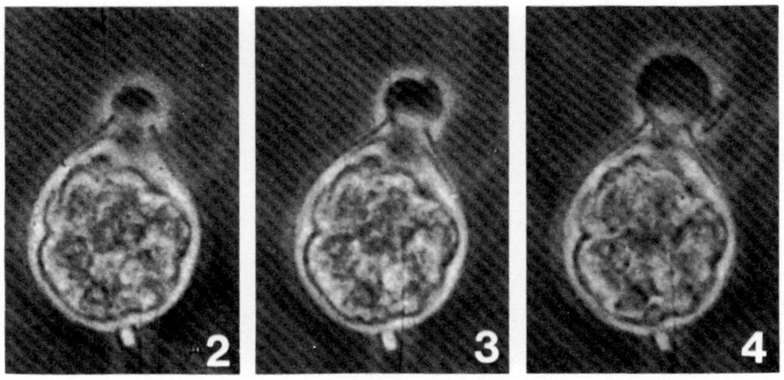


**Figs. 2–4.** Sequence of opening of *Phytophthora cactorum* sporangium and ejection of a cover from the apex of the sporangium before enlargement of the discharge vesicle (×550). (Reprinted, by permission, from Gisi et al, 1979b)

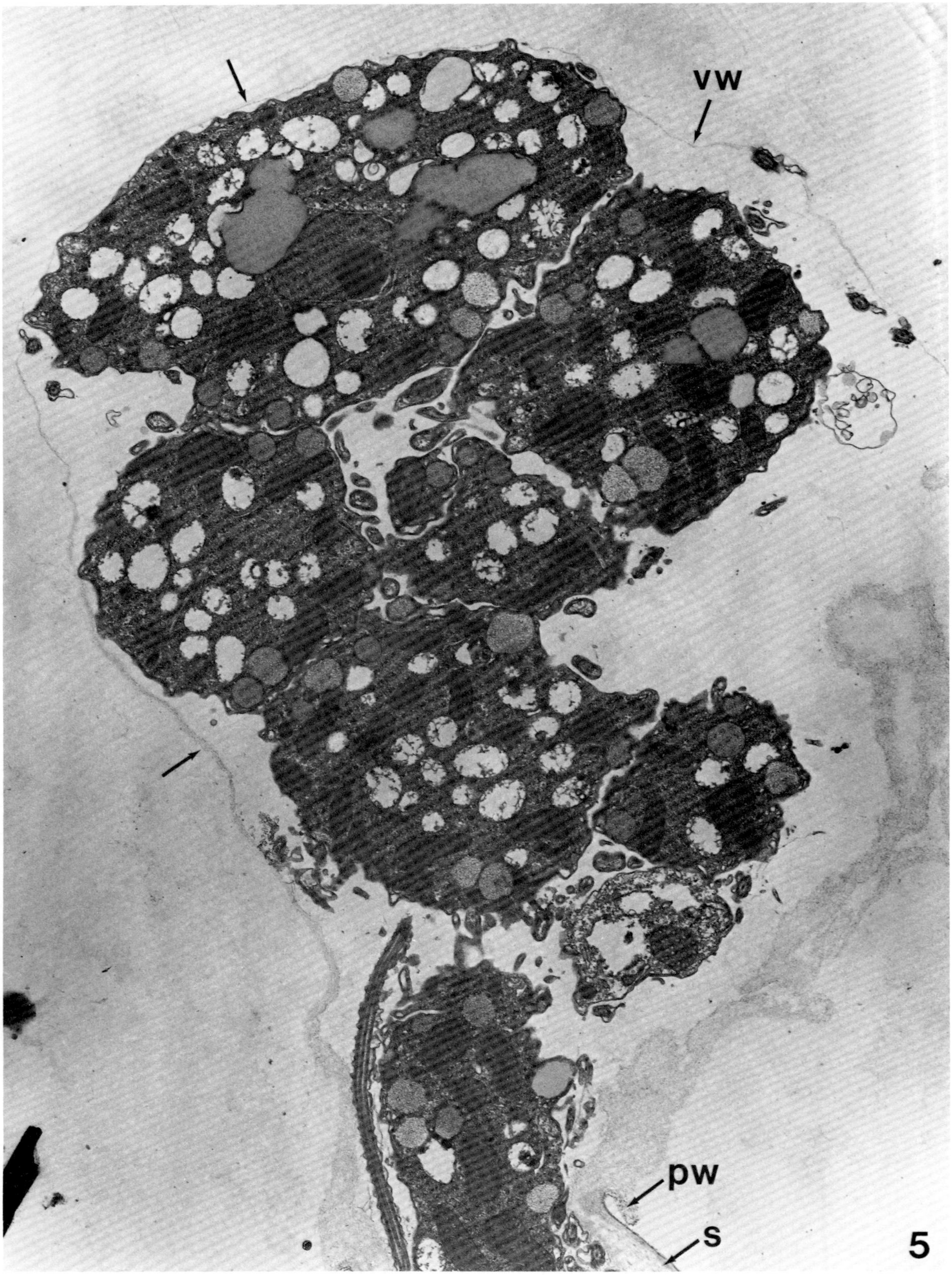


**Fig. 5.** Median longitudinal section through the discharge vesicle of *Phytophthora palmivora* sporangium (×7,500). Note the continuous vesicle wall (vw, arrows) diminishing in thickness from the bottom to the top, the material of the primary wall (pw) curled back from the apex of the sporangium and forming a collar around the exit pore, and the sporangium (s). (Reprinted, by permission, from Gisi et al, 1979a).

the gradient of the osmotic water potential between sporangium and outside was not sufficient for, or was equilibrated prior to, zoospore release, since release did not continue further.

With regard to the energy relations of cytoplasmic discharge, at least three factors can retard zoospore release. First, since a sporangium and its full-size discharge vesicle are one "chamber," about half of the initial turgor pressure is lost because of the doubling of the initial sporangial volume (Boyle-Mariotte's law). Second, since the exit pores of most sporangia are smaller in diameter than the zoospores (5–8 $\mu$m vs about 10 $\mu$m), except in *P. cinnamomi*, zoospores must be squeezed through the exit pore and would thus consume much of the initial pressure. Third, the papillar material is forced through the exit pore and is continuously deformed and inflated into a vesicle with a wall about 0.1 $\mu$m thick or thinner. Probably most of the initial pressure is consumed in vesicle formation. If one considers the discharge vesicle as a "walled cell," changes in turgor pressure may be related to changes in volume by the elastic (and plastic?) properties of the wall, described by the volumetric elastic modulus $\epsilon$ as follows (Zimmerman, 1978):

$$dp = \epsilon \frac{dV}{V} . \qquad (2)$$

For high polymers, the modulus is known to be directly proportional to the acting pressure (Hall, 1968). A change of pressure on the plastic vesicle wall would change the modulus; at a critical increase in volume, the wall will rupture (Gisi et al, 1979a). These three factors indicate an extremely high resistance to cytoplasmic discharge. I would expect that in those aquatic fungi where sporangium discharge occurs without a discharge vesicle, the velocity of zoospore release would be much higher.

Two further observations lend support to the hypothesis of a pressure-driven enlargement of the discharge vesicle. First, from base to top, the vesicle wall diminishes gradually in thickness (Fig. 5), indicating that the enlargement velocities of the vesicle are unequal on its two hemispheres, the main flow contribution being along the major axis of the sporangium. After a critical increase in vesicle volume, the wall in the distal half of the vesicle becomes so thin that it cannot withstand a further increase in volume. Second, the sharpness of the convergence angle (i.e., the decreasing diameter of the sporangial apex) greatly influences the behavior of the flow coming out from the exit pore. In *P. cinnamomi* and *P. cactorum*, the angles are about 40° and 10°, respectively (Gisi et al, 1979a); the vesicle of *P. cinnamomi* seems to be more frail (Blackwell, 1949) than that of *P. cactorum*. The reason for this may be that increasing convergence angles result in more unfavorable hydrodynamic properties during discharge, namely higher discharge velocities and turbulence (Gisi et al, 1979a).

### Velocity of Cytoplasmic Discharge

To evaluate a pressure-driven mechanism for zoospore release, the velocity with which the zoospores pass through the exit pore was measured. There are physical equations describing the correlation between pressure gradient and velocity of a fluid flow based on Bernoulli's energy balance and Hagen-Poiseuille's law of streaming fluids. The pressure gradient dp/dl (change in pressure per unit length) is a function of viscosity, $\eta$, velocity, v, and radius of the exit pore, r (Gisi and Zentmyer, 1980):

$$-\frac{dp}{dl} = 8\eta \frac{v}{r^2} \qquad (3)$$

The passage time of one zoospore through the exit pore increases linearly with increasing numbers of released zoospores (Table 2). Since the number of zoospores in a sporangium is an expression of decreasing volume, or pressure during discharge, there is an inverse relationship between decreasing pressure and the passage time of zoospores, a situation that is described by Equation 3 (v is proportional to 1/t). For three *Phytophthora* species (Table 2), the velocities of zoospore release are lowest in *P. cactorum*, medium in *P. palmivora*, and highest in *P. cinnamomi*. This behavior can be explained using again Equation 3, in which the velocity is proportionally related to the square of the radius. In fact, *P. cinnamomi* sporangia have the largest exit pores compared to *P. palmivora* and *P. cactorum* (Table 3).

Zoospores are deformed considerably during their passage through the exit pore, since their diameters are larger than

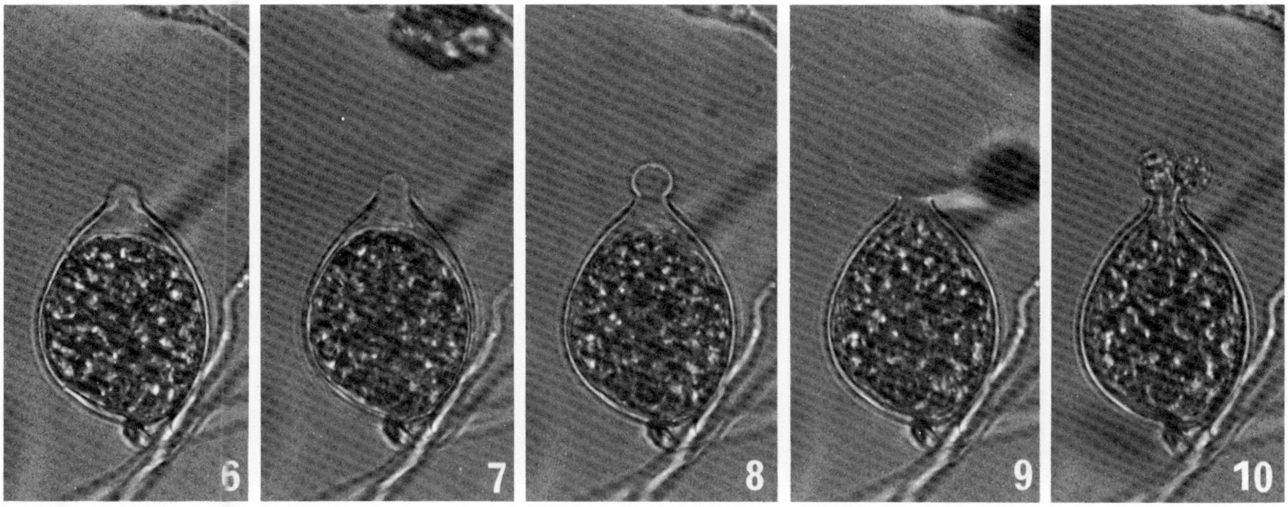


**Figs. 6–10.** Sequence showing the enlargement of the discharge vesicle before zoospore release from sporangia of *Phytophthora palmivora* exposed to polyethylene glycol (×530). (Reprinted, by permission, from Gisi et al, 1979a).

those of the exit pores; even the zoospores of *P. cinnamomi* are deformed, since commonly two zoospores squeeze through the pore at the same time. Considering the deformation of the zoospores and using the measured values of the mean passage time per zoospore, we may calculate the maximum flow velocity, v, for a zoospore (Table 3). The values for *P. cactorum* are lower than those for the other two species. Flow velocity values of about 10–30 μm/sec are typical for zoospore release in *Phytophthora*; they are about 10 times lower than the speeds of freely swimming zoospores: 160 μm/sec (Ho and Hickman, 1967a), 120 μm/sec (Allen and Newhook, 1974), and 140 μm/sec (Gisi, 1977). On the other hand, they are about 10 times higher than reported velocities of amoeboid movement (e.g., Holloway and Heath, 1977).

An additional condition influencing the velocity of zoospore release is the initial pressure immediately before the opening of the sporangium. This pressure can be determined by measuring the isotonic point of the sporangium, i.e. the concentration of the surrounding osmoticum at which 50% of the sporangia are plasmolyzed (Gisi and Zentmyer, 1980). Sporangia that are ready to release zoospores show turgor pressures between 4.8 and 7 bars (Table 3), values that are in the neighborhood of known turgor pressures of higher plant cells. The values of *P. cactorum* are significantly lower than those of *P. palmivora* and the highest values are reached in *P. cinnamomi* sporangia. With this information, we calculated the pressure gradient, dp/dl, of cytoplasmic discharge according to Equation 3. The values for *P. cactorum* and *P.* *palmivora* (Table 3) are about 10 times higher than, and those of *P. cinnamomi* about equal to, those for phloem transport in sieve tubes of herbaceous plants (Milburn, 1975). This indicates a higher resistance to cytoplasmic flow in zoospore release than in phloem transport. According to studies on blood flow (Fung, 1981), there is an additional pressure drop (resistance) caused by the presence of red blood cells in the suspending fluid (plasm). The resistance is proportional to the number of cells per unit length. The high resistance to cytoplasmic flow in *Phytophthora* is probably due to a similar effect, since zoospores are transported within this fluid flow. Also, the inflation of the discharge vesicle contributes much to this resistance.

TABLE 2. Passage time of zoospores released from sporangia of *Phytophthora cactorum, P. palmivora*, and *P. cinnamomi* exposed to double-distilled water at 25–30° C[a]

| Zoospore release[b] (%) | Passage time (sec) | | |
|---|---|---|---|
| | *P. cactorum* | *P. palmivora* | *P. cinnamomi* |
| 20 | 0.79 | 0.67 | 0.35 |
| 40 | 1.12 | 0.79 | 0.43 |
| 60 | 1.44 | 0.91 | 0.51 |
| 80 | 1.76 | 1.03 | 0.59 |
| 100 | 2.08 | 1.15 | 0.67 |

[a] Sporangia were produced in a chloride salt solution (CSS) within 24–48 hr. The composition of CSS was 0.01 *M* $CaCl_2$, 0.005 *M* KCl, 0.004 *M* $MgSO_4$, and 1 ml of chelated iron solution per liter of CSS (Gisi et al, 1980). Values have been recalculated from Gisi and Zentmyer (1980).

[b] Percentage of initial number of zoospores in sporangia.

TABLE 3. Characteristics of zoospore release from sporangia of *Phytophthora cactorum, P. palmivora*, and *P. cinnamomi* exposed to double-distilled water at 25–30° C[a]

| Characteristic | Unit | *P. cactorum* | *P. palmivora* | *P. cinnamomi* |
|---|---|---|---|---|
| Mean diameter of exit pores | μm | 4.68 | 5.37 | 15.90 |
| Mean passage time of a zoospore | sec | 1.38 | 0.86 | 0.47 |
| Maximum velocity of a zoospore | μm/sec | 12.8 | 19.7 | 29.2 |
| Initial turgor pressure | bars | 4.8 | 5.9 | 7.0 |
| Pressure gradient (−dp/dl) | mb/μm | 0.19 | 0.21 | 0.04 |

[a] Sporangia were produced in a chloride salt solution within 24–48 hr (see Table 2).

*Source*: Gisi and Zentmyer (1980).

## Influence of Osmotics and Ions

By transferring sporangia from sporulation media to hypertonic solutions, it is possible to retard or even stop the enlargement of the discharge vesicle and subsequent zoospore release. PEG prevents zoospore release from *P. cactorum, P. palmivora,* and *P. cinnamomi* sporangia almost completely at osmotic water potentials of −2.4, −3.1, and −3.6 bars, respectively (Gisi and Zentmyer, 1980); similar results were obtained with $CaCl_2$. The exact value of complete inhibition of zoospore release depends on the species as well as on the physiologic conditions of the sporangia, e.g. method of sporangium production, age, and prior exposure to solutes. With PEG, the lower limit of zoospore release is at about −0.2 bar for *P. cactorum* (Gisi et al, 1977) and −1.2 bars for *P. megasperma* (MacDonald and Duniway, 1978a). With 10 m*M* NaCl, the frequency of zoospore release from sporangia of *P. cactorum* was found to be 100, 60, and 10% for sporangia 7, 13, and 17 days old, respectively (Gisi et al, 1977).

In a variety of salt and sugar solutions, zoospore release is influenced not only by osmotic water potential but also by specific ions. Table 4 lists the osmotic water potentials at which zoospore release from sporangia of *P. cactorum* is inhibited 50% (GR 50). Alkali and earth-alkali chlorides as well as two sulfates, sucrose, and PEG were tested. All compounds except $BeCl_2$, $MgCl_2$, and $BaCl_2$ show GR 50 values that are significantly lower than that for PEG. PEG values are used as reference as this substance probably produces a minimum of side effects since it has no charge, it is not metabolizable, and it is not readily taken up. Since we are comparing mainly chlorides with one another, the specific ion effect is due to the cation. $BeCl_2$ is very acidic and therefore unsuitable for studying zoospore release. The behavior of $Ba^{2+}$ is roughly similar to that of PEG (Gisi et al, 1977), whereas $Mg^{2+}$ clearly inhibits zoospore release (Table 4). The sequence of GR 50 values from $Li^+$ (lowest) to $Cs^+$ (highest) parallels the crystal radii (lyotropic series) of elements. This behavior strongly points to a differential solute uptake into the sporangium before zoospore release. Ions with small radii are obviously taken up more intensively and faster than those with large radii. This greater solute uptake results in a further increase in turgor pressure; zoospores are then released at a lower level of osmotic potential (at higher osmolarities of the medium). The differential influence of specific ions on zoospore release is due to a different reflection coefficient δ during uptake, according to the well-known equation

$$J = L(d\psi_p - \delta d\psi_\pi), \qquad (4)$$

where J is the volume flow (numerically equal to the flow velocity), L is the hydraulic conductivity, $d\psi_p$ the change in turgor pressure, and $d\psi_\pi$ the change in osmotic potential. A similar sequence of differential solute uptake before zoospore release from *P. megasperma* sporangia was found by MacDonald and Duniway (1978a), but the values are somewhat lower than those of Gisi et al (1977).

The hypothesis of a differential solute uptake before zoospore release is supported by the analogous behavior of *P. cactorum* sporangia when exposed to the same salts but at much higher concentrations than those causing zoospore release (Table 4). The osmotic water potential at which 50% of the sporangia are plasmolyzed (PR 50, isotonic point) is very low for LiCl and $MgCl_2$ and successively increases: LiCl << KCl ≈ RbCl ≈ CsCl; and $MgCl_2 < CaCl_2 < SrCl_2 < BaCl_2$; the highest values are reached with PEG (about −11 bars). The results of zoospore release, as well as those of plasmolysis of sporangia in different salts, suggest that the dynamics of ion uptake are related to the crystal and not to the hydrated radius of the cations (Table 4), but this has to be confirmed with additional experiments.

Another observation supports the occurrence of differential ion uptake into the sporangium. In 1*M* salt solutions, plasmolysis of all sporangia occurs after a short time (about 20 min) but, with increasing exposure time, the proportion of plasmolyzed sporangia decreased continuously to about 15% (LiCl) and 60% (NaCl) after 7 hr (Gisi et al, 1977). This observed deplasmolysis and the different behavior in LiCl vs NaCl point to differential ion uptake according to the crystal radii of the cations. The sequence of solute uptake in sporangia is $Li^+ >> Na^+ > K^+ > Ca^{2+} >$ PEG $>> Mg^{2+}$. The same sequence obtains for equimolar and equiosmolar solutions, and it does not correspond to that frequently found in higher plant cells: $K^+ > Na^+ > Li^+ > Ca^{2+} > Mg^{2+}$. The same sequence of cation uptake by sporangia was found for chloride or sulfate salts (Table 4). Besides cation radius, there are differences in uptake between single and double charged ions; this can be seen best by comparing $Na^+$ and $Ca^{2+}$: the crystal radii are about equal ($9.7$ and $9.9 \times 10^{-2}$ nm, respectively), but the values of "GR 50" and "PR 50" differ clearly; zoospore release (GR 50) occurred at −0.55 bar (NaCl) vs −0.17 bar ($CaCl_2$) and plasmolysis (PR 50) at −44.0 bars (NaCl) vs −24.5 bars ($CaCl_2$).

It is not known at which stage $Mg^{2+}$ inhibits zoospore release during cleavage or during discharge of the zoospores. The same uncertainty holds for $NH_3$ and $HNO_2$, which were recently described by Tsao and Oster (1981) as toxic at very low concentrations. $Mg^{2+}$ is a cofactor in numerous biochemical reactions and plays a role in the integrity of cellular organelles; a concentration greater than 5 mmol might be toxic for important processes. After cleavage of the cytoplasm, the semipermeable membrane (plasmalemma) of the sporangium becomes part of the plasmalemma of the peripheral zoospores; as a consequence, inside and outside concentrations will equilibrate, and turgor pressure begins to decrease. This is a plausible explanation of why cleaved sporangia release zoospores only during a critical exposure time and then lose their ability to expel zoospores.

TABLE 4. Effect of osmotic water potential on zoospore release from sporangia (GR 50) and on plasmolysis of sporangia (PR 50)[a]

| | Osmotic potential[b] (−bars) | | Cation radius ($nm \times 10^{-2}$) | |
|---|---|---|---|---|
| Osmotics | GR 50 | PR 50 | Crystal[c] | Hydrated[d] |
| LiCl | 1.29 | 73.8 | 6.8 | 38 |
| NaCl | 0.55/0.28[e] | 44.0 | 9.7 | 31 |
| KCl | 0.40/0.20[e] | 34.0 | 13.3 | 28 |
| RbCl | 0.31 | 33.0 | 14.7 | 25 |
| CsCl | 0.26 | 37.0 | 16.7 | ...[g] |
| $BeCl_2$ | 0[f] | ... | 3.5 | ... |
| $MgCl_2$ | 0.03 | 26.8 | 6.6 | 46 |
| $CaCl_2$ | 0.17 | 24.5 | 9.9 | 44 |
| $SrCl_2$ | 0.11 | 22.4 | 11.2 | 43 |
| $BaCl_2$ | 0.07 | 19.5 | 13.4 | ... |
| $Na_2SO_4$ | 0.57/0.40[e] | ... | ... | ... |
| $K_2SO_4$ | 0.25/0.17[e] | ... | ... | ... |
| Sucrose | 0.24 | ... | ... | 53 |
| PEG | 0.10 | 10.9 | ... | ... |

[a] Sporangia of *Phytophthora cactorum* were produced in bean meal decoction within 10–14 days. Zoospore release or plasmolysis was observed after a 2-hr exposure of the sporangia to the osmoticum.

[b] Values were obtained by measuring the real osmolarity (mosm) cryoscopically and converting to values of osmotic potential (41 mosm = −1 bar; at 20° C) (data recalculated from Gisi et al, 1977). GR 50 = 50% inhibition of zoospore release; PR 50 = 50% of sporangia are plasmolyzed.

[c] From CRC Handbook of Chemistry and Physics, 62nd ed., 1981–1982. R. C. Weast and M. J. Astle, eds. CRC Press, Boca Raton, FL.

[d] From McFarlane and Berry (1974).

[e] Osmotic water potential calculated from cation concentration only.

[f] At concentrations greater than 0.5 m*M*, $BeCl_2$ was toxic because of a pH value of 3.5; all other compounds could be adjusted to a pH of 7.0, at which all experiments were done.

[g] No observation.

## Significance of Water Potential

The data suggest a mechanism of zoospore release from sporangia governed by a pressure-driven fluid flow. The main observations supporting this hypothesis are 1) the manner of the opening of the sporangium and of the inflation of the discharge vesicle, 2) the increase in passage time during zoospore release, 3) the possibility of osmotically retarding and stopping the enlargement of the discharge vesicle as well as the zoospore release, 4) the differential influence of alkali and earth-alkali ions on zoospore release and on plasmolysis/deplasmolysis of sporangia, and 5) the measurements of initial turgor pressure and of important hydrodynamic characteristics of cytoplasmic discharge, such as the diameter of the exit pore, the convergence angle of the sporangial apex, and the pressure gradient. According to Equation 4, the driving forces for this flow are the differences in hydrostatic and osmotic water potential.

In soil, we should consider the additional action of matric water potential. As described for mycelial growth, zoospore release seems to be much more sensitive to matric than to osmotic potentials. Several authors (e.g., Duniway, 1976; MacDonald and Duniway, 1978a; Pfender et al, 1977) found the lower limit for zoospore release in *P. cryptogea* and *P. megasperma* to be about −0.05 bar. The different sensitivity of zoospore release to the two components of water potential agrees with the observation that different osmotics can change the values of "GR 50" over an almost 10-fold range, e.g., LiCl vs PEG (Table 4). The response to osmotic potential is mostly connected with a differential solute uptake into the sporangium, allowing zoospore release at lower osmotic potentials than the limitations of matric potential would suggest.

The high sensitivity of zoospore release to matric and to osmotic water potential indicates the need for freely available water in the environment for this process. There is a need for a continuous influx of water into the sporangium not only before but also during zoospore release. This osmotically governed and pressure-driven fluid flow is responsible for the transport of the zoospores out of the sporangium. Only the last few zoospores may swim out actively using their flagella. Under special conditions, especially in near-isotonic solutions, the cleavage of zoospores may be incomplete and the cytoplasm discharges as a whole mass, proving that cleavage is not a precondition and that the action of flagella is not necessary for zoospore release.

Chilling the sporangia has been described as a precondition for zoospore release, but I have never had difficulty in obtaining 100% zoospore release, in double-distilled water, from sporangia of various *Phytophthora* species, including *P. cinnamomi* and *P. infestans*, without chilling. Interestingly, there was a critical time after transferring sporangia to water before zoospore release started: 12 min for *P. cactorum*, 16 min for *P. palmivora*, 55 min for *P. cinnamomi* and *P. cryptogea*, and about 80 min for *P. infestans*. Temperature has a distinct influence on zoospore release. For *P. infestans*, the optimal temperature is about 12–16° C (Bohnen, 1963; U. Gisi, *unpublished*), and it is about 18–28° C for most other species. The advantage of chilling sporangia is that a more synchronous zoospore release can be obtained. A possible explanation for this phenomenon is derived from Equation 3, where flow velocity and viscosity are inversely correlated. Since the viscosity is dependent on temperature, cooling results in an increase and rewarming in a decrease in viscosity; the latter causes an influx of water into the sporangium. The rapid increase in turgor pressure bursts the sporangial apex, and zoospores are released after a short time (about 15 min).

Future experiments on the influence of temperature on the velocity of zoospore release, exact measurements of the diameters of zoospores during and after zoospore release, the action of certain inhibitors, direct measurement of turgor pressure of mature sporangia vs that of free zoospores, and studies on solute uptake will test the validity of the hypothesis of a pressure-driven zoospore release. A comparison with other fungi that release zoospores with and without the formation of a discharge vesicle may indicate whether this type of cytoplasmic discharge is governed by the same basic mechanism. Preliminary support comes from our observations on *Pythium* (Gisi and Zentmyer, 1980) and investigations on *Aphanomyces* by Hoch and Mitchell (1973).

## SUMMARY

The influence of physical factors, especially water potential, on the development of *Phytophthora* is described. Different stages in the life cycle of *Phytophthora* show different sensitivities to water potential. The lower limits of matric potential for mycelial growth are at about −30 bars, for production of sporangia at about −5 bars, and for zoospore release at about −3.5 to −0.05 bar. The water relations of sporangia are discussed in detail in order to elucidate the motive forces for zoospore release. After a critical time following the transfer of sporangia from the sporulation medium to hypotonic solutions, zoospore release is initiated. The outer sporangial wall covering the apex of the sporangium is ruptured and either curled back or ejected; the papillar material is then forced through the exit pore and inflated to form a discharge vesicle with a continuous and plastic wall. Full-size discharge vesicles contain up to 70% of the total number of zoospores and reach a volume about equal to that of the sporangium before rupture. The passage time of zoospores released from sporangia increases linearly with increasing number of released zoospores; the mean passage time ranges from 0.5 to 1.4 sec and the mean flow velocity from 10 to 30 μm/sec, depending on the diameter of the exit pore of the sporangium and on the initial turgor pressure. The opening of the sporangium, the enlargement of the discharge vesicle, and the subsequent zoospore release can be retarded or stopped osmotically by transferring sporangia into solutions with osmotic water potentials of −2.4 to −3.6 bars. Zoospore release as well as plasmolysis of sporangia can be manipulated by placing sporangia in salt and sugar solutions; both processes are decreased by alkali and earth-alkali ions in proportion to the crystal radii of the cations. This suggests a differential solute uptake into the sporangium. All of the described characteristics of zoospore release suggest a pressure-driven and osmotically governed cytoplasmic discharge that can be explained physically by the dynamics of fluid flow.

## LITERATURE CITED

Adebayo, A. A., and Harris, R. F. 1971. Fungal growth responses to osmotic as compared to matric potential. Soil Sci. Soc. Am. Proc. 35:465-469.

Allen, R. N., and Newhook, F. J. 1974. Suppression by ethanol of spontaneous turning activity in zoospores of *Phytophthora cinnamomi*. Trans. Br. Mycol. Soc. 63:383-385.

Ayers, W. A., and Zentmyer, G. A. 1971. Effect of soil solution and two soil pseudomonads on sporangium production by *Phytophthora cinnamomi*. Phytopathology 61:1188-1193.

Bimpong, C. E., and Clerk, G. C. 1970. Motility and chemotaxis in zoospores of *Phytophthora palmivora*. Ann. Bot. 34:617-624.

Blackwell, E. 1949. Terminology in *Phytophthora*. Mycol. Pap. 30. Commonw. Mycol. Inst., Kew, Surrey, England. 24 pp.

Bohnen, K. 1963. Ein Verfahren zur Anzucht in vitro von *Phytophthora infestans* und die Verwendung des Pilzes im Sporenkeimtest zur Mittelprüfung. Phytopathol. Z. 46:199-206.

Borkowski, M. 1968. Zur Entleerungsweise der Zoosporangien bei *Saprolegnia monoica* (Pringsheim, DeBary). Z. Allg. Mikrobiol. 8:339-354. (In German)

Broadbent, P., and Baker, K. F. 1974. Association of bacteria with sporangial formation and breakdown of sporangia in *Phytophthora* species. Aust. J. Agric. Res. 25:139-145.

Cameron, J. N., and Carlile, M. J. 1977. Negative geotaxis of zoospores of the fungus *Phytophthora*. J. Gen. Microbiol. 98:599-602.

Duniway, J. M. 1975a. Formation of sporangia by *Phytophthora drechsleri* in soil at high matric potentials. Can. J. Bot. 53:1270-1275.

Duniway, J. M. 1975b. Limiting influence of low water potential on the formation of sporangia by *Phytophthora drechsleri* in soil. Phytopathology 65:1089-1093.

Duniway, J. M. 1976. Movement of zoospores of *Phytophthora cryptogea* in soils at various textures and matric potentials. Phytopathology 66:877-882.

Duniway, J. M. 1979. Water relations of water molds. Annu. Rev. Phytopathol. 17:431-460.

Englander, L., and Roth, L. F. 1980. Interaction of light and sterol on sporangium and chlamydospore production by *Phytophthora lateralis*. Phytopathology 70:650-654.

Fung, Y. C. 1981. Biomechanics; Mechanical Properties of Living Tissues. Springer, New York. 433 pp.

Gisi, U. 1975. Investigations on the soil phase of *Phytophthora cactorum* (Leb. et Cohn) Schroet. with fluorescent optical direct observation. Z. Pflanzenkr. Pflanzenschutz 82:355-377. (In German)

Gisi, U. 1977. Indirect germination of sporangia of *Phytophthora cactorum* (Oomycetes). (Motion picture) Schweizerische Gemeinschaft für den Hochschul- und Forschungsfilm, Basel, Switzerland.

Gisi, U., and Zentmyer, G. A. 1980. Mechanism of zoospore release in *Phytophthora* and *Pythium*. Exp. Mycol. 4:362-377.

Gisi, U., Oertli, J. J., and Schwinn, F. J. 1977. Water and salt relations of sporangia of *Phytophthora cactorum* in vitro. Phytopathol. Z. 89:261-284. (In German)

Gisi, U., Hemmes, D. E., and Zentmyer, G. A. 1979a. Origin and significance of the discharge vesicle in *Phytophthora*. Exp. Mycol. 3:321-339.

Gisi, U., Schwinn, F. J., and Oertli, J. J. 1979b. Dynamics of indirect germination in *Phytophthora cactorum* sporangia. Trans. Br. Mycol. Soc. 72:437-446.

Gisi, U., Zentmyer, G. A., and Klure, L. J. 1980. Production of sporangia by *Phytophthora cinnamomi* and *P. palmivora* in soils at different matric potentials. Phytopathology 70:301-306.

Griffin, D. M. 1977. Water potential and wood-decay fungi. Annu. Rev. Phytopathol. 15:319-329.

Hall, I. H. 1968. Deformation of Solids. Thomas Nelson and Sons, London.

Hemmes, D. E., and Hohl, H. R. 1971. Ultrastructural aspects of encystation and cyst-germination in *Phytophthora parasitica*. J. Cell Sci. 9:175-191.

Hemmes, D. E., and Hohl, H. R. 1972. Flagellum degeneration in the fungus *Phytophthora palmivora* (formerly *Phytophthora parasitica*). J. Gen. Microbiol. 73:345-351.

Ho, H. H., and Hickman, C. J. 1967a. Asexual reproduction and behavior of zoospores of *Phytophthora megasperma* var. *sojae*. Can. J. Bot. 45:1963-1981.

Ho, H. H., and Hickman, C. J. 1967b. Factors governing zoospore responses of *Phytophthora megasperma* var. *sojae* to plant roots. Can. J. Bot. 45:1983-1994.

Hoch, H. C., and Mitchell, J. E. 1973. The effects of osmotic water potentials on *Aphanomyces euteiches* during zoosporogenesis. Can. J. Bot. 51:413-420.

Hohl, H. R., and Hamamoto, S. T. 1967. Ultrastructural changes during zoospore formation in *Phytophthora parasitica*. Am. J. Bot. 54:1131-1139.

Holloway, S. A., and Heath, I. B. 1977. Morphogenesis and the role of microtubules in synchronous populations of *Saprolegnia* zoospores. Exp. Mycol. :9-29.

Katsura, K., and Miyata, Y. 1971. Swimming behavior of *Phytophthora capsici* zoospores. Pages 107-128 in: Morphological and Biochemical Events in Plant-Parasite Interaction. S. Akai and S. Ouchi, eds. Phytopathol. Soc. Jpn., Tokyo. 415 pp.

Ko, W. H., and Chase, L. L. 1973. Aggregation of zoospores of *Phytophthora palmivora*. J. Gen. Microbiol. 78:79-82.

Kuan, T.-L., and Erwin, D. C. 1980. Predisposition effect of water-saturation of soil on Phytophthora root rot of alfalfa. Phytopathology 70:981-986.

Kuan, T.-L., and Erwin, D. C. 1982. Effect of soil matric potential on Phytophthora root rot of alfalfa. Phytopathology 72:543-548.

MacDonald, J. D. 1977. Influence of soil water on the sporangia and zoospores of three root-infecting species of *Phytophthora*. Ph.D. thesis, Univ. Calif., Davis. 109 pp.

MacDonald, J. D., and Duniway, J. M. 1978a. Influence of the matric and osmotic components of water potential on zoospore discharge in *Phytophthora*. Phytopathology 68:751-757.

MacDonald, J. D., and Duniway, J. M. 1978b. Temperature and water stress effects on sporangium viability and zoospore discharge in *Phytophthora cryptogea* and *P. megasperma*. Phytopathology 68:1449-1455.

MacDonald, J. D., and Duniway, J. M. 1978c. Influence of soil texture and temperature on the motility of *Phytophthora cryptogea* and *P. megasperma* zoospores. Phytopathology 68:1627-1630.

Malajczuk, N., and Theodorou, C. 1979. Influence of water potential on growth and cultural characteristics of *Phytophthora cinnamomi*. Trans. Br. Mycol. Soc. 72:15-18.

Malajczuk, N., Nesbitt, H. J., and Glenn, A. R. 1977. A light and electron microscope study of the interaction of soil bacteria with *Phytophthora cinnamomi* Rands. Can. J. Microbiol. 23:1518-1525.

Marshall, T. J., and Holmes, J. W. 1979. Soil Physics. Cambridge Univ. Press, Cambridge. 345 pp.

McFarlane, J. C., and Berry, W. L. 1974. Cation penetration through isolated leaf cuticles. Plant Physiol. 53:723-727.

McIntosh, D. L. 1972. Effects of soil water suction, soil temperature, carbon and nitrogen amendments, and host rootlets on survival in soil of zoospores of *Phytophthora cactorum*. Can. J. Bot. 50:269-272.

Meyer, D., and Schönbeck, F. 1975. Investigations on the development of *Phytophthora cactorum* (Leb. et Cohn) Schroet. in soil. Z. Pflanzenkr. Pflanzenschutz 82:337-354. (In German)

Milburn, J. A. 1975. Pressure flow. Pages 328-353 in: Transport in Plants. I. Phloem Transport. M. H. Zimmerman and J. A. Milburn, eds. Encyclopedia of Plant Physiology, New Series, Vol. 1. Springer-Verlag, Berlin.

Mitchell, D. J., and Zentmyer, G. A. 1971a. Effects of oxygen and carbon dioxide tensions on growth of several species of *Phytophthora*. Phytopathology 61:787-791.

Mitchell, D. J., and Zentmyer, G. A. 1971b. Effects of oxygen and carbon dioxide tensions on sporangium and oospore formation by *Phytophthora* spp. Phytopathology 61:807-812.

Nagata, T., and Melchers, G. 1978. Surface charge of protoplasts and their significance in cell-cell interaction. Planta 142:235-238.

Pfender, W. F., Hine, R. B., and Stanghellini, M. E. 1977. Production of sporangia and release of zoospores by *Phytophthora megasperma* in soil. Phytopathology 67:657-663.

Porter, J. R., and Shaw, D. S. 1978. Aggregation of *Phytophthora drechsleri* zoospores; pattern analysis suggests a taxis. Trans. Br. Mycol. Soc. 71:515-518.

Reeves, R. J. 1975. Behavior of *Phytophthora cinnamomi* Rands in different soils and water regimes. Soil Biol. Biochem. 7:19-24.

Singh, N. 1975. Responses of fungal zoospores to ionic forms of resins. Trans. Br. Mycol. Soc. 64:521-524.

Slayman, C. L., and Slayman, C. W. 1962. Measurement of membrane potentials in *Neurospora*. Science 136:876-877.

Sneh, B., and McIntosh, D. L. 1974. Studies on the behavior and survival of *Phytophthora cactorum* in soil. Can. J. Bot. 52:795-802.

Sneh, B., Humble, S. J., and Lockwood, J. L. 1977. Parasitism of oospores of *Phytophthora megasperma* var. *sojae, P. cactorum, Pythium* sp., and *Aphanomyces euteiches* in soil by oomycetes, chytridiomycetes, hyphomycetes, actinomycetes, and bacteria. Phytopathology 67:622-628.

Sommers, L. E., Harris, R. F., Dalton, F. N., and Gardner, W. R. 1970. Water potential relations of three root-infecting *Phytophthora* species. Phytopathology 60:932-934.

Sterne, R. E., Zentmyer, G. A., and Bingham, F. T. 1976. The effect of osmotic potential and specific ions on growth of *Phytophthora cinnamomi*. Phytopathology 66:1398-1402.

Sterne, R. E., Zentmyer, G. A., and Kaufmann, M. R. 1977. The influence of matric potential, soil texture, and soil amendment on root disease caused by *Phytophthora cinnamomi*. Phytopathology 67:1495-1500.

Sugar, D. 1977. The development of sporangia of *Phytophthora cambivora, P. megasperma*, and *P. drechsleri*, and severity of root and crown rot in *Prunus mahaleb* as influenced by soil matric potential. M.S. thesis, Univ. Calif., Davis. 56 pp.

Tsao, P. H. 1969. Studies on the saprophytic behavior of *Phytophthora parasitica* in soil. Proc. 1st Int. Citrus Symp. 3:1221-1230.

Tsao, P. H., and Oster, J. J. 1981. Relation of ammonia and nitrous acid to suppression of *Phytophthora* in soils amended with nitrogenous organic substances. Phytopathology 71:53-59.

Weste, G., and Vithanage, K. 1979. Survival of chlamydospores of *Phytophthora cinnamomi* in several non-sterile, host-free forest soils and gravels at different soil water potentials. Aust. J. Bot. 27:1-9.

Wiebe, H. H. 1966. Matric potential of several plant tissues and biocolloids. Plant Physiol. 41:1439-1442.

Young, B. R., Newhook, F. J., and Allen, R. N. 1979. Motility and chemotactic responses of *Phytophthora cinnamomi* zoospores in "ideal soils." Trans. Br. Mycol. Soc. 72:395-401.

Zentmyer, G. A. 1965. Bacterial stimulation of sporangium production in *Phytophthora cinnamomi*. Science 150:1178-1179.

Zentmyer, G. A., and Ribeiro, O. K. 1977. The effect of visible and near-visible radiation on sporangium production by *Phytophthora cinnamomi*. Phytopathology 67:91-95.

Zimmerman, U. 1978. Physics of turgor- and osmoregulation. Annu. Rev. Plant Physiol. 29:121-148.

# 9

# Biochemical Aspects of Morphogenesis in *Phytophthora*

S. Bartnicki-Garcia and M. C. Wang

CONTENTS



## INTRODUCTION

Compared to the wealth of information available on some fungi, our knowledge of the biochemistry of the genus *Phytophthora* is very limited—a surprising void given the agricultural importance of these fungi and their facility of cultivation. Besides these practical considerations, *Phytophthora* has the attractiveness of being an oomycete—one of a group of fungi whose biochemical uniqueness separates them quite distinctly from the rest of the fungi.

This chapter is a summary of those biochemical aspects of the genus *Phytophthora* that appear to be immediately significant in morphogenesis. Considerable attention will be paid to cell walls. As discussed elsewhere (Bartnicki-Garcia, 1968, 1969, 1973a, 1973b), cell shape of fungi is mainly dictated by the shape of the wall. Consequently, the study of morphogenesis can be approached experimentally as a question of cell wall differentiation.

Except for the brief zoospore stage, all stages in the life cycle of *Phytophthora* (cysts, mycelia, sporangia, chlamydospores, gametangia, oospores) have well-defined cell walls. Since the cell walls of *Phytophthora* are composed mainly of β-glucans, elucidation of the structure and metabolism of these polymers is a major preoccupation in the quest for understanding the biochemistry of development of these fungi.

## CELL WALLS

### Role

The cell wall has two major roles: protection and morphogenesis. In addition to allowing a fungus to safely generate high metabolite concentrations in its cytoplasm to sustain high growth rates, the wall can be molded into the diversity of shapes that characterize the vegetative and reproductive structures found in the life cycle of a fungus. This morphological plasticity is a key part of an adaptive capacity with which a fungus can successfully meet a wide assortment of ecological challenges.

The importance of the wall may be demonstrated in a rather simple but drastic manner by removing it selectively. This can be accomplished readily with mycelium of *Phytophthora* spp. by digesting the walls with mixtures of β-glucanases in an isotonic or hypertonic medium (Bartnicki-Garcia and Lippman, 1966). The resulting spherical protoplasts lack the characteristic tubular shape of the hyphae from which they emerged (Fig. 1). The protoplasts would readily lyse if placed back in the same medium where the mycelium was grown or in other hypotonic media. Even in a hypertonic nutrient medium, protoplasts are unable to survive for long unless they regenerate a cell wall.

In the zoospore stage, the fungus trades the advantages of a wall for the freedom of self-propulsion, a trade that in aquatic habitats must confer upon the organism a distinct advantage in dispersal. The wall-less stage is, however, ephemeral. A zoospore cannot grow or multiply; the continued existence of the organism requires a prompt return to a walled existence, namely encystment. All subsequent development requires walled cells.

### Chemical Composition

#### General

The cell walls of *Phytophthora* are made principally of β-linked glucose polymers. These β-glucans comprise about 80–90% of the wall dry weight. The walls also contain small amounts of protein, lipid, and other polysaccharides, in which mannose and glucosamine appear to be main sugar components. A trace of galactose has also been found in some wall hydrolyzates (Lippman et al, 1974; Novaes-Ledieu et al, 1967), but whether or not this sugar is a bona fide wall component is open to question. Table 1 shows the overall cell wall composition of mycelium of several species and of other life-cycle stages of selected species. The data on mycelial walls

TABLE 1. Chemical composition of the cell walls of *Phytophthora* species[a]

| | Mycelium | | | | | Sporangium | Cyst | Oogonium-oospore |
|---|---|---|---|---|---|---|---|---|
| Component | *P. cinnamomi*[b] | *P. parasitica*[b] | *P. palmivora*[c] | *P. heveae*[d] | *P. megasperma* var. *sojae*[e] | *P. palmivora*[c] | *P. palmivora*[c] | *P. megasperma* var. *sojae*[f] |
| Hexose | | | 90.1 | | | 93.5 | 92.9 | |
| Glucose | 88.0 | 86.0 | | 90.0 | ~86.0 | | | 75.9 |
| Mannose | 0.6 | 0.7 | | $<$1.0 | ~3.0 | | | 1.9 |
| Glucosamine | 0.3 | 0.3 | 0.6 | 2.3 | $<$1.0 | 0.2 | 0.1 | 0.4 |
| Protein | 3.5 | 4.2 | 10.5 | 5.6 | ~9.0 | 1.7 | 2.4 | 11.4 |
| Lipid | | | | | | | | |
| Free | 0.3 | 0.2 | ND | 1.2 | ND | ND | ND | 5.7 |
| Bound | 2.1 | 0.9 | ND | 1.3 | ND | ND | ND | 5.6 |
| Phosphorus | 0.13 | ND | ND | 0.26 | ND | 0.25 | 0.1 | 0.55 |

[a] Composition expressed in percentage of wall dry weight. Sugar values correspond to anhydrous form. ND = Not determined.
[b] Bartnicki-Garcia (1966).
[c] Tokunaga and Bartnicki-Garcia (1971b).
[d] Novaes-Ledieu et al (1967).
[e] Ayers et al (1976b).
[f] Lippman et al (1974).

indicate that wall composition is fairly constant among the different species examined.[1] Two β-glucans are the major wall components throughout the life cycle, although their relative proportions vary significantly from stage to stage (Table 2).

### Cell Wall β-Glucans

Enzymatic and physical studies (Bartnicki-Garcia, 1966; Bartnicki-Garcia and Lippman, 1967; Novaes-Ledieu and Jiménez-Martínez, 1969) have shown that the glucans in the cell walls of *Phytophthora* are β-linked. Cell wall β-glucans are highly insoluble polymers. They resist extraction by ordinary mineral acids or alkalis at 20–30° C. By means of cuprammonium hydroxide (Schweizer reagent—the classical cellulose solvent), two different types of glucans can be distinguished in the walls of *Phytophthora*: a cuprammonium-soluble glucan or cellulose and an insoluble, noncellulosic glucan. The noncellulosic glucan is much more abundant, nearly two to tenfold more than the cellulosic glucan (Table 2). The cellulosic fraction contains the typical β-1,4-linked chains of cellulose, but this cellulose is poorly crystalline and does not give any sharp reflections upon X-ray diffraction (Bartnicki-Garcia, 1966).

The structure of the major noncellulosic glucan from hyphal walls of *P. cinnamomi* was elucidated, in part, by enzymatic degradation (Bartnicki-Garcia and Lippman, 1967) and by methylation analysis (Zevenhuizen and Bartnicki-Garcia, 1969). The products of digestion and methylation provided the evidence (Table 3) to conclude that the noncellulosic glucan is a highly branched β-1,3-glucan with β-1,6 links at the branching residues. There are about four chain residues per branching point. An idealized, purely speculative model of the glucan is shown in Figure 2. In addition, about 10% of the glucose units in the noncellulosic glucan were found to be joined by 1,4-linkages. The relationship of these units to the rest of the molecule was not fully elucidated. Conversely, methylation analysis of the cellulosic glucan from cyst or sporangial walls of *P. palmivora* showed that about 6–8% of

[1] The only exception to a wall composition consisting principally of β-glucans is in a report by Miyazaki et al (1974). These authors claimed that the major component of hyphal walls of *P. infestans* was an alkali-soluble α-1,3-glucan. There is no evidence in the literature for such a polymer being in the cell walls of *Phytophthora* spp. or related genera. Our own investigations showed that the hyphal walls of *P. infestans* contain mainly alkali-*insoluble* glucan, as do the hyphal walls of other species of *Phytophthora* (Wang and Bartnicki-Garcia, 1980). Clearly, the unusual presence of α-1,3-glucan requires confirmation. The possibility must be ruled out that the α-1,3-glucan originated from a contaminating higher fungus whose walls do contain this type of polysaccharide.

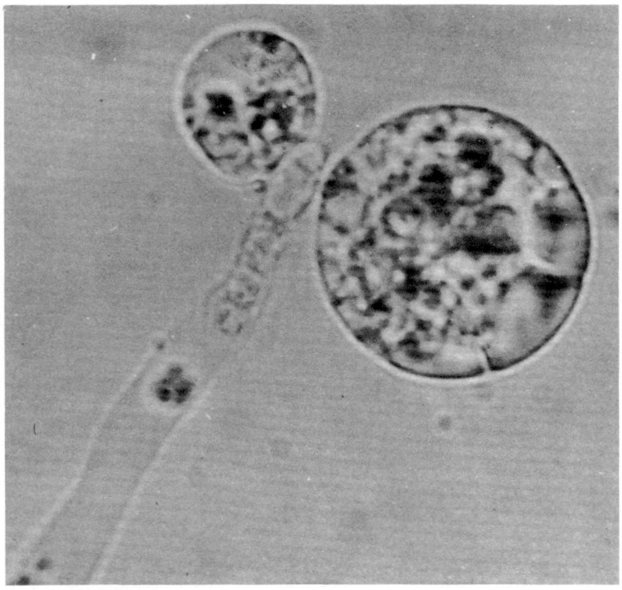

**Fig. 1.** Release of two protoplasts from a hyphal tip of *Phytophthora parasitica* after digestion with crude glucanases from *Streptomyces* sp. in 0.6 *M* mannitol (×2,000). (Reprinted, by permission, from Bartnicki-Garcia and Lippman, 1966).

TABLE 2. β-Glucan content of the cell walls of *Phytophthora* species[a]

| β-Glucan | Mycelium of *P. cinnamomi*[b] | Sporangium of *P. palmivora*[c] | Cyst of *P. palmivora*[c] | Oospore-oogonium of *P. megasperma* var. *sojae*[d] |
|---|---|---|---|---|
| Noncellulosic | 66 | 54 | 48 | 68 |
| Cellulosic | 11 | 27 | 30 | 7 |

[a] Cellulosic and noncellulosic glucans were separated by solubility in cuprammonium reagent and determined colorimetrically with anthrone reagent. Values are percentage of wall dry weight.
[b] Bartnicki-Garcia (1966).
[c] S. Bartnicki-Garcia and M. C. Wang (*unpublished*).
[d] Lippman et al (1974).

TABLE 3. Structural analyses of hyphal wall β-glucans of *Phytophthora cinnamomi* by enzymatic digestion and methylation

| | | Methylation fragments | | | |
|---|---|---|---|---|---|
| Fraction | Digestion product | -Methyl glucoside | Abundance | Structural feature | References |
| Whole hyphal wall | Laminaribiose<br>Laminaritriose<br>Gentiobiose<br>Cellobiose | | | β-1,3 link<br>β-1,3 link<br>β-1,6 link<br>β-1,4 link | Bartnicki-Garcia and Lippman (1967) |
| Hyphal wall, noncellulosic fraction | Glucose<br>Laminaribiose<br>Gentiobiose<br>Cellobiose | | | β-1,3 link<br>β-1,3 link<br>β-1,6 link<br>β-1,4 link | S. Bartnicki-Garcia and M. C. Wang (*unpublished*)[a] |
| Hyphal wall, noncellulosic fraction | | 2,3,4,6-tetra-*O*-<br>2,4,6-tri-*O*-<br>2,3,6-tri-*O*-<br>2,4-di-*O*- | 15–20%<br>45–50%<br>10–13%<br>12–20% | Nonreducing termini<br>C-3 linked residues<br>C-4 linked[b] residues<br>C-3,C-6 branches | Zevenhuizen and Bartnicki-Garcia (1969) |

[a] Digested with an exo-β-1,3-glucanase from *Sporotrichum dimorphosporum* QM 806.
[b] Determined as glucitol acetate (M. C. Wang and S. Bartnicki-Garcia, *unpublished*).

the glucose units are not 1,4- but 1,3-linked (S. Bartnicki-Garcia and M. C. Wang, *unpublished*). Methylation analysis of hyphal walls of *P. megasperma* var. *sojae* (Ayers et al, 1976a) revealed that the structural features of the wall glucans were similar to those found in *P. cinnamomi* (Zevenhuizen and Bartnicki-Garcia, 1969).

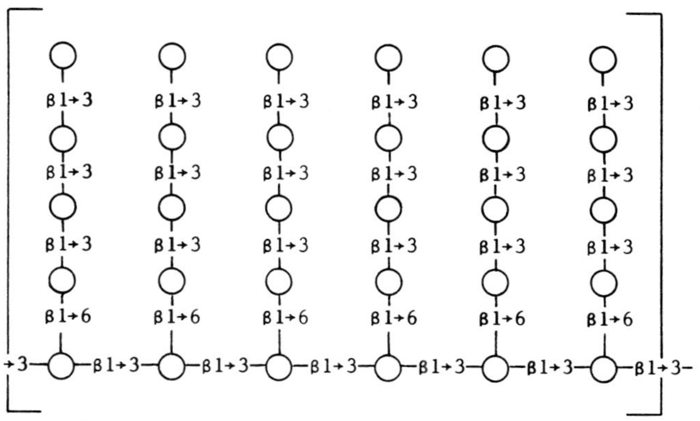

**Fig. 2.** Hypothetical structure of the noncellulosic β-1,3-β-1,6-glucan of *Phytophthora* cell walls. This is only one of the many possible structures based on the methylation data of Zevenhuizen and Bartnicki-Garcia (1969).

### Mannose-Containing Polysaccharides

Most of the mannose in the mycelial walls of *Phytophthora* spp. is probably part of surface glycoproteins that can be extracted by hot water, alkali, or digestion with pronase (Ayers et al, 1976b; Keen and Legrand, 1980). These mannoproteins contain much of the glucosamine found in the wall (see next section) and may be part of more complex glucomannoproteins.

The mannoproteins are important in various respects. They may be part of the cementing material that rigidifies the β-glucan fabric. They are probably the adhesive substance with which zoospores attach themselves to solid surfaces (see below). In addition, the carbohydrate moieties of these glycoproteins seem to function as elicitors of phytoalexin production and determinants of race specificity in host-pathogen interactions (see Keen and Yoshikawa, Chapter 21). Mannose and glucose, together with a small amount of galactose, are also conspicuous components of the invertases isolated from *P. megasperma* var. *sojae* by West et al (1980). Although the localization of these enzymes in the wall has not been proven, the data suggest that they are bound to a cell structure (wall?) by disulfide bridges.

### Glucosamine-Containing Polysaccharides

Hydrolyzates from purified cell walls of *Phytophthora* spp. contain small amounts of glucosamine (usually less than 1%) (Ayers et al, 1976b; Bartnicki-Garcia, 1966; Keen and Legrand, 1980; Lippman et al, 1974; Novaes-Ledieu et al, 1967; Tokunaga and Bartnicki-Garcia, 1971b). Most of this glucosamine is probably part of cell wall glycoproteins; at best, only a minute trace could be chitin. Thus, Novaes-Ledieu and Jiménez-Martínez (1969) found that glucosamine was extracted from the walls of *P. heveae* (and *Pythium butleri*) by hot 1 *M* HCl. This acid-soluble fraction contained most of the glucan and protein of the wall. Ayers et al (1976b) found that nearly half of the total glucosamine present in the mycelial walls of *P. megasperma* var. *sojae* was extractable with hot water. The resulting fraction, a polysaccharide rich in glucose, mannose, and glucosamine, was studied for its phytoalexin-eliciting properties. These glucosamine-containing glycoproteins are similar, if not identical, to the cell surface glycoproteins analyzed by Keen and Legrand (1980). The observation that tunicamycin, at low concentration (1 μg/ml), inhibits mycelial growth of *P. megasperma* var. *sojae* is an indirect piece of evidence for the existence of glycoproteins whose carbohydrate moiety is linked to protein via GlcNAc residues (tunicamycin acts by blocking such a linking reaction [Waechter and Lennarz, 1976]).

The amount of chitin in the wall, if any, is probably too small to have any structural significance. The presence of traces of chitin in the walls of *Phytophthora* should not be construed as validation of the old claim that the walls of these fungi are chitinous (Thomas, 1943).

### Proteins

The protein content of the walls of *Phytophthora* (Table 1) is in the range found for other fungi. The proteins of the mycelial wall of *P. cinnamomi, P. parasitica* (Bartnicki-Garcia, 1966), and *P. heveae* (Novaes-Ledieu et al, 1967) and the oospore/oogonium walls of *P. megasperma* var. *sojae* (Lippman et al, 1974) contain hydroxyproline. This amino acid is a distinctive component of oomycetous cell walls (Lin et al, 1976; Novaes-Ledieu et al, 1967; Pao and Aronson, 1970).

**Fig. 3.** Detail of the microfibrillar fabric of a mechanically isolated cyst wall of *Phytophthora palmivora* (×60,000).

### Wall Architecture

Electron microscopic examination of isolated cell walls of *Phytophthora* reveals the presence of two different structural elements: long microfibrils and nonfibrillar or amorphous material. The architectural design of the walls of *Phytophthora* is therefore similar to that of other fungal walls, namely, a microfibrillar meshwork embedded in a cementing matrix (for reviews on fungal walls, see Aronson [1965], Bartnicki-Garcia [1968], and Wessels and Sietsma [1981]).

In shadow cast preparations, microfibril diameter was estimated to be about 10–15 nm (Hunsley and Burnett, 1968; Tokunaga and Bartnicki-Garcia, 1971b) or 15–25 nm (Hegnauer and Hohl, 1973). Microfibril length is very difficult to estimate because the microfibrils are tightly interwoven (Fig. 3). In favorable specimens, e.g. cyst walls of *P. palmivora*, uninterrupted lengths of several micrometers can be discerned (Fig. 4).

The composition of microfibrils has not been fully elucidated. The microfibrils of *Phytophthora* are generally considered to be cellulosic and, indeed, they disappear from the mycelial walls of *P. parasitica* upon digestion with cellulase (following a previous digestion with $\beta$-1,3-glucanase) (Hunsley and Burnett, 1970). Likewise, extraction of cellulose from the sporangial walls of *P. palmivora*, by treatment with

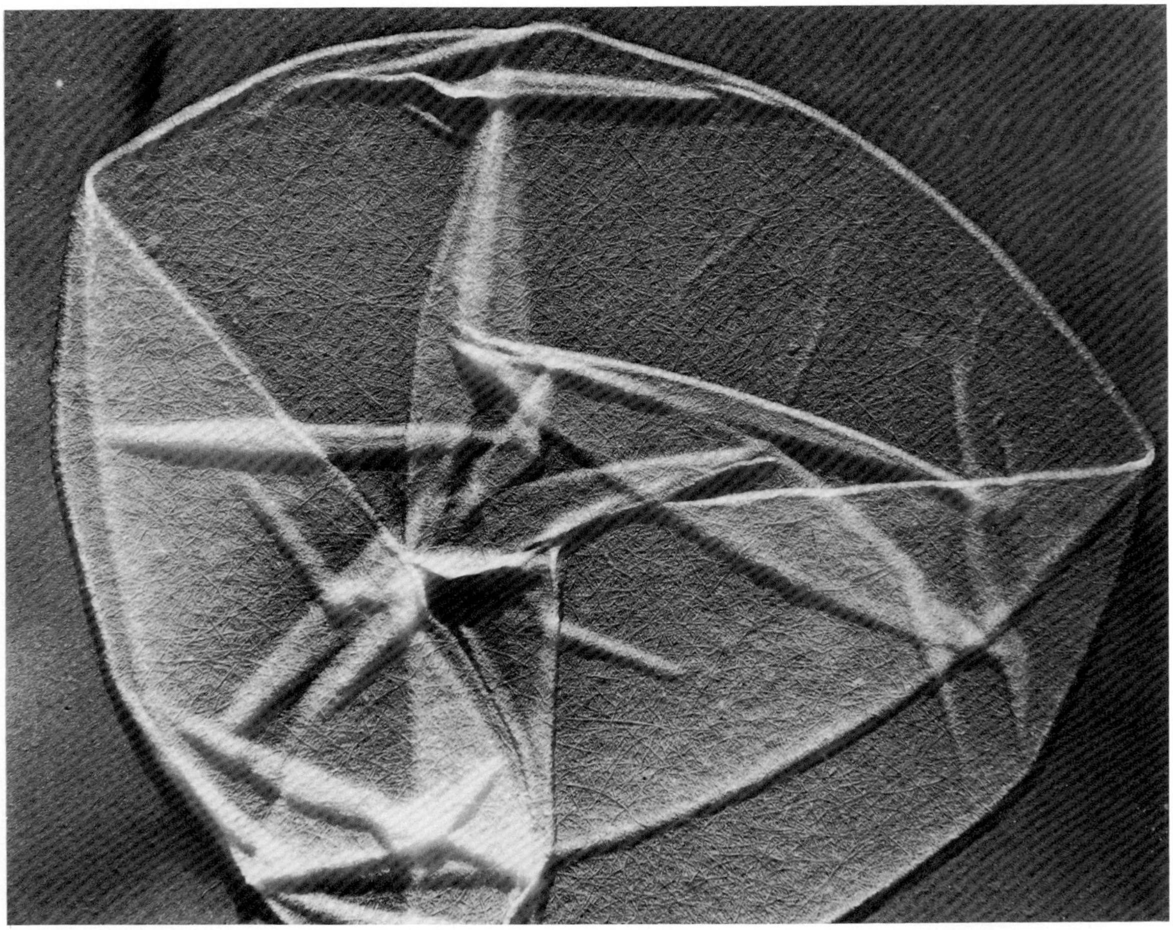

**Fig. 4.** Cyst wall of *Phytophthora palmivora*. The wall was prepared by extracting cysts with hot ethanolic-KOH (×18,000). (Reprinted, by permission, from Tokunaga and Bartnicki-Garcia, 1971b).

Schweizer reagent, removed the microfibrillar component (Hegnauer and Hohl, 1978). However, we have reason to suspect that the microfibrils are made of both cellulosic and noncellulosic glucan. Thus, purified cyst walls of *P. palmivora* exhibit only microfibrils without any ostensible amorphous material (Fig. 4), yet most of the glucan in these walls is noncellulosic (Tokunaga and Bartnicki-Garcia, 1971b).

Two types of amorphous components can be detected in cell walls of *Phytophthora* and distinguished by their solubility: 1) loosely bound amorphous material that can be readily removed by sonication or by washings with hot water or alkali; for instance, the amorphous "fluffy coat" secreted at the time of encystment (Desjardins et al, 1973; Sing and Bartnicki-Garcia, 1975a, 1975b); and 2) firmly bound amorphous components such as the noncellulosic glucan material on the outer surface of hyphal walls, which resists extraction with hot alkali (Tokunaga and Bartnicki-Garcia, 1971b).

## Structure and Morphogenesis

During the life cycle of *Phytophthora*, there are major changes in cell wall structure. Thickness, texture, and layering change substantially, yet the basic chemical composition of the wall is maintained, i.e. insoluble $\beta$-glucans are the main ingredients of these walls, with noncellulosic glucan predominating over the cellulosic one. The life cycle encompasses a wide range of cell wall differentiation—from cells where there is no wall (zoospores) to cells with complex multilayered walls (oospores), where the wall mass comprises nearly half the total weight of the cell.

Although zoospores of *P. palmivora* have no cell wall, their plasma membrane is not a simple bilayer but displays on the external surface a "fuzzy coat" (glycocalyx?), about 20–25 nm thick, of unknown composition. This coat can be best seen in freeze-fracture views (Fig. 5) (Nogueira et al, 1977). Its role is not known, but it may serve to strengthen the plasma membrane of this wall-less cell.

A cell wall is produced de novo during encystment. This cyst wall is a very thin (50–100 nm) fabric made almost entirely of microfibrils. The fabric is covered with patches of amorphous material (Desjardins et al, 1973; Hegnauer and Hohl, 1973; Tokunaga and Bartnicki-Garcia, 1971b). This amorphous material, also called the "fluffy" coat (Hegnauer and Hohl, 1973), is readily lost during the wall isolation procedures that employ either physical disruption (sonication) or chemical extraction (hot ethanolic alkali) (Tokunaga and Bartnicki-Garcia 1971b). The isolated cyst wall exhibits no amorphous component; both the inner and outer surfaces have a sharp microfibrillar texture. The microfibrils are tightly knit and randomly oriented (Fig. 4).

The germ tube that emerges from the cyst has a more elaborate wall structure with two distinct textures (Fig. 6). On the inside is a microfibrillar texture that is essentially a direct extension of the cyst wall. On the outside is a layer of insoluble, tightly bound, amorphous glucan that covers the microfibrils completely. This layer can be selectively removed by digestion with an exo-$\beta$-1,3-glucanase to reveal the underlying fibrillar skeleton (compare Fig. 6A and B with 6C and D). The layering is not sharply defined. The two components are intermixed. There is probably a gradient of wall components, with microfibrils predominating on the inside surface and nonfibrillar glucan on the outer surface. In cross sections of germinating cysts of *P. palmivora*, Hegnauer and Hohl (1973) observed a distinct line of demarcation between cyst and germ tube walls. According to them, the germ tube wall is a separate entity built by apposition under the cyst wall. However, the images shown by Tokunaga and Bartnicki-Garcia (1971b) (see Fig. 6C) indicate clearly an uninterrupted microfibrillar fabric from cyst to germ tube. Hence, we must conclude that during germination, the newly forming microfibrils of the germ tube are interwoven with the old microfibrils of the cyst wall to produce a continuous seamless fabric. The demarcation seen between two walls is probably an indication of the abrupt change in texture and composition between the two walls, which are, nevertheless, continuous in their main skeletal component.

The dual texture of the germ tube wall is clearly evident from base to apex (Fig. 6). It is also present in mycelial hyphae (Fig. 7) and seems to be a general feature of hyphal walls of fungi (Hunsley and Burnett, 1970). The differences in wall organization between cysts and germ tubes may be a direct result of their morphogenesis. Whereas the cyst wall is "morphogenetically passive" (Bartnicki-Garcia, 1973b), i.e. it is made on an already existing cell, the germ tube wall is "morphogenetically active," i.e. it is produced as the germ tube grows. As this wall grows, it dictates the cylindric shape of the expanding cell. The dual wall texture on hyphal tubes may be required for extension growth and may be related to the need for controlled plasticity of the apical zone where expansion takes place (Bartnicki-Garcia, 1973a). If cysts of *P. palmivora* are germinated in the presence of exo-$\beta$-1,3-glucanase, the outer wall layer is eroded and the germ tubes have distorted, swollen shapes and unusual apical branching (Meyer et al, 1976). This observation may be adduced as evidence for the importance of the outer wall layer in hyphal wall morphogenesis.

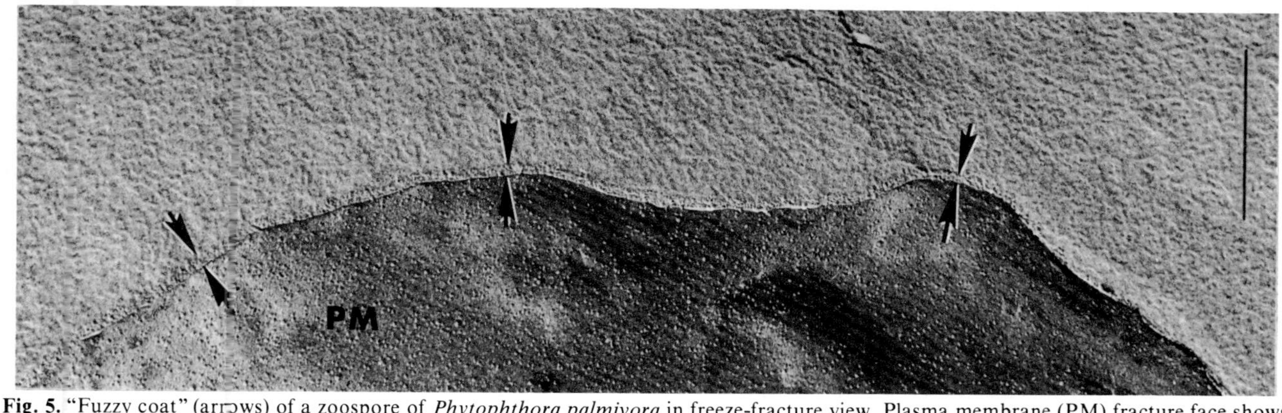


**Fig. 5.** "Fuzzy coat" (arrows) of a zoospore of *Phytophthora palmivora* in freeze-fracture view. Plasma membrane (PM) fracture face shows typical granules. Bar represents 500 nm. (Reprinted, by permission, from Nogueira et al, 1977)

these vesicles are responsible for the synthesis of the insoluble, amorphous glucan component of hyphal walls.

**Chlamydospores**

The asexual resting spore, or chlamydospore, resembles the sporangium in that it stores a significant amount of mycolaminaran (Fig. 13) but with little or no mycolaminaran phosphate. The chemical composition of chlamydospore walls has not been described in the literature; our own unpublished data on *P. parasitica* indicate that chlamydospore walls are made principally of β-glucans. These glucans comprise 24% of the dry weight of the chlamydospore.

**Oospores**

Sexual sporulation depends largely on the mycolaminaran accumulated during mycelial growth. This conclusion stems from the observation that developing gametangia, particularly the expanding oogonium, are filled by a massive influx of mycelial cytoplasm. The mycolaminaran thus mobilized is probably all utilized for the complex sequence of sexual differentiation; the cytoplasm of mature oospores contains only a trace of neutral mycolaminaran and no mycolaminaran phosphate (Wang and Bartnicki-Garcia, 1980). Presumably, the glucans of the thick oospore/oogonium wall complex, which amounts to 35% of the cell dry weight (Fig. 13) (Lippman et al, 1974), are largely or exclusively derived from mycelial mycolaminaran.

The very small amount of mycolaminaran in mature oospores is surprising considering that oospores are capable of germinating in the absence of nutrients to give rise to a mycelium and/or a zoosporangium—processes that require extensive utilization of endogenous reserves. The dormant oospore does store a large amount of glucose to be used later in germination. This glucose is not stored in the cytoplasm in the form of soluble mycolaminarans but in the oospore wall as insoluble β-glucans. This conclusion is based on the observation (Blackwell, 1943; Zentmyer and Erwin, 1970) that the thick inner oospore wall of *Phytophthora* is markedly eroded during germination (see Hemmes, Chapter 2, Fig. 40). For some unknown reason, perhaps insolubility and attendant resistance to hydrolysis, dormant oospores rely on an insoluble glucan as a more suitable form of glucose for long-term storage.

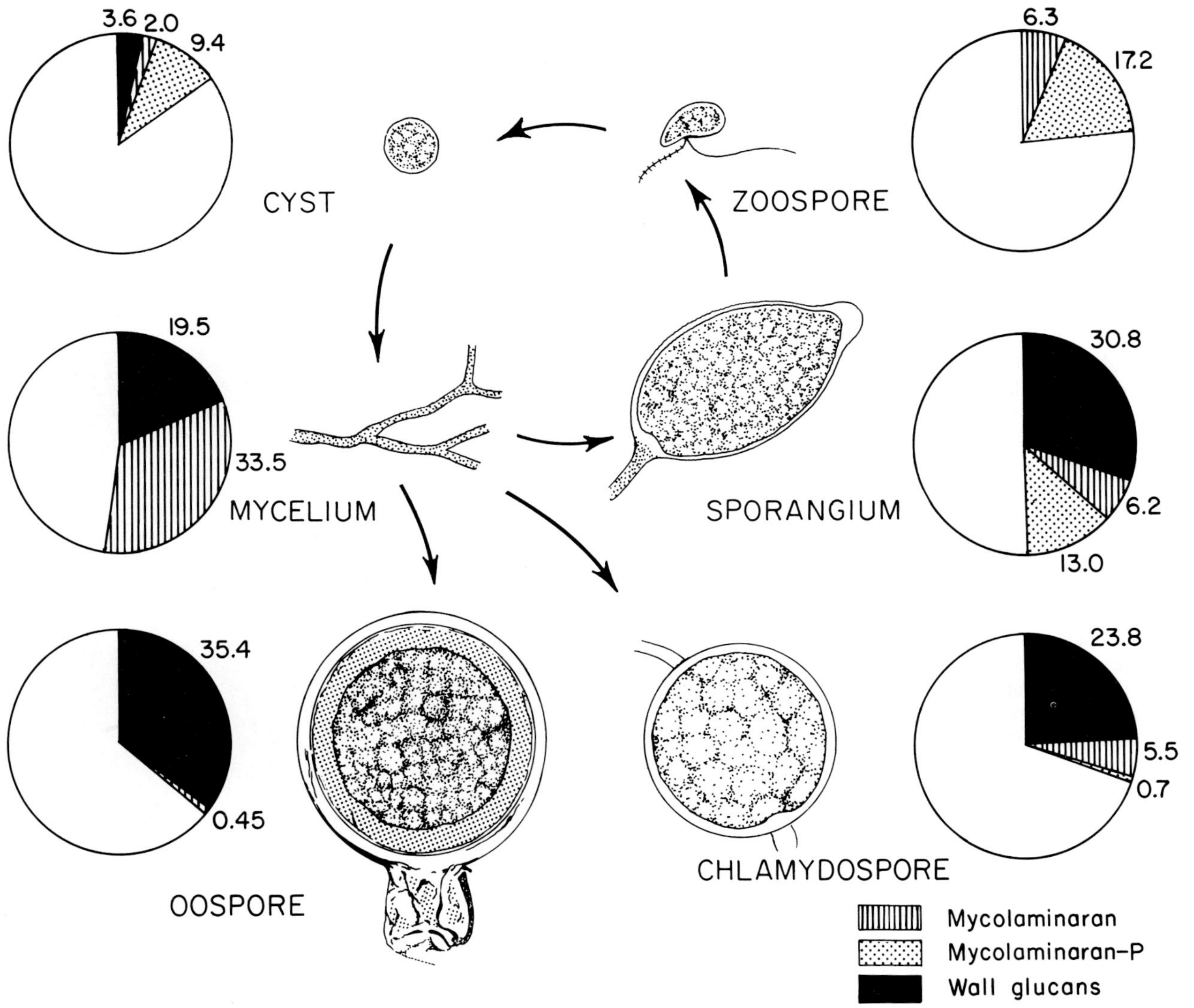


**Fig. 13.** Flux of cell wall and cytoplasmic β-glucans during the life cycle of *Phytophthora*. Circles represent cell dry weight. Numbers are the actual percentage values. The scheme was assembled from data for zoospores, cysts, mycelia, and sporangia of *P. palmivora*; chlamydospores of *P. parasitica*; and oospores of *P. megasperma* var. *sojae*. (Redrawn from data in Wang and Bartnicki-Garcia, 1980).

## COMPARATIVE BIOCHEMISTRY

Information on wall composition and storage polysaccharides of other genera of Oomycetes is rather limited. Nevertheless, the available evidence indicates that the β-glucan scheme described for the life cycle of *Phytophthora* may be representative of Oomycetes in general. The two β-glucans (cellulosic and noncellulosic β-1,3-β-1,6-glucan) characteristic of the walls of *Phytophthora* spp. also occur in the walls of all genuine members of the Oomycetes examined. They include representatives of the Peronosporales, Saprolegniales, and Leptomitales (Aronson and Lin, 1978; Aronson et al, 1967; Bertke and Aronson, 1980; Novaes-Ledieu et al, 1967; Pao and Aronson, 1970; Sietsma et al, 1969). Glycogen, the typical storage polysaccharide of fungi (Blumenthal, 1976), has not been found in Oomycetes. Instead, soluble glucans similar to the mycolaminarans of *Phytophthora* (Peronosporales) have been described for two other orders of Oomycetes, Saprolegniales (Faro, 1972a, 1972b) and Leptomitales (Coulter and Aronson, 1977), and appear to play the same role. Thus, Faro (1972a) found that the soluble glucan from *Achlya* was extensively utilized in the formation of sex organs.

In addition to this unique reliance on β-glucans as major cytoplasmic and wall polysaccharides, oomycetous fungi have various other biochemical idiosyncrasies, notably making lysine by the diaminopimelic acid pathway and not by the aminoadipic acid route typical of most fungi. These and other biochemical peculiarities (see review by Bartnicki-Garcia [1970]) support the belief that Oomycetes (and Hyphochytridiomycetes) are a unique group of fungi that evolved independently of the major trunk of fungal evolution (Chytridiomycetes, Zygomycetes, Ascomycetes, and Basidiomycetes).

## SUMMARY

Our limited knowledge of the biochemistry of morphogenesis of *Phytophthora* is discussed. The main emphasis of this chapter is on the structure and biosynthesis of the cell walls; except for zoospores, the shape of the fungus resides in the shape of its wall.

About 80–90% of the wall is made of β-glucans. Two different β-glucans are present. The most abundant is a noncellulosic β-1,3-β-1,6-glucan. A poorly crystalline cellulose is the minor glucan component. These two β-glucans are the main components of the wall in all species and life cycle stages examined. Other sugars present in small amounts are mannose and glucosamine; these form part of complex glycoproteins.

The wall composition of *Phytophthora* spp. is typical of an oomycete (β-1,3-β-1,6-glucan + cellulose) and includes hydroxyproline, an unusual component of fungal wall proteins.

The architecture and texture of the wall change drastically during the life cycle, from a thin, microfibrillar envelope in the cyst wall, to the double texture (internally microfibrillar, externally granular) of the hyphal walls, to the thicker and more elaborate cell walls of sporangia, chlamydospores, and oospores.

Alkali-insoluble β-1,3-β-1,6-glucans similar to those present in the wall can be synthesized in vitro from UDP-Glc by particulate preparations. The organellar nature of these particles has not been elucidated. No evidence for cellulose (β-1,4-glucan) biosynthesis in vitro has been gathered; the reason is still elusive.

β-Glucans are also the main reserve polysaccharides. Unlike the highly insoluble glucans in the wall, these cytoplasmic glucans, termed mycolaminarans, are water-soluble. They consist of a β-1,3-linked glucan chain with one or two branches at C-6. Mycolaminarans accumulate during vegetative (mycelium) development and are consumed in later stages. During differentiation, i.e., asexual sporulation, there is extensive formation of mycolaminaran phosphate.

Sporulation is usually triggered by starvation of the vegetative mycelium. Thus, both asexual and sexual development rely on endogenous nutrient reserves. Mycolaminarans probably play a key role as a C and energy source for these reproductive processes. In the oospore, however, little if any mycolaminaran is stored. The main carbohydrate reserve for long-term storage seems to be the insoluble β-glucans deposited as a conspicuous, thick, inner wall layer of the oospore. During oospore germination, this wall layer is largely consumed.

*Phytophthora* spp. are a good example of the importance of polysaccharide biochemistry in the life cycle of fungi. A single group of polysaccharides, the β-glucans, plays a preponderant role in the metabolic activities of oomycetous fungi. The importance of β-glucans is both quantitative and qualitative. β-Glucans constitute a major part of the cell mass. They are the principal wall constituents and reserve material of these fungi. In some stages of the life cycle, β-glucans comprise one half of the total cell dry weight. β-Glucans are the central ingredient for cell wall construction, i.e. cell morphogenesis. Given the abundance of β-glucan in *Phytophthora* cells, one must conclude that these organisms spend a major portion of their metabolic activities making and breaking β-glucosidic linkages.

## ACKNOWLEDGMENTS

Our own experimental work described herein was supported in part by research grants (GB-32165, BMS-72-01862, and OIP75-08378) from the National Science Foundation.

## LITERATURE CITED

Aragaki, M. 1966. Deposition and function of callose in the Peronosporales. Phytopathology 56:145.

Aronson, J. M. 1965. The cell wall. Pages 49-76 in: The Fungi. Vol. 1. G. C. Ainsworth and A. S. Sussman, eds. Academic Press, New York.

Aronson, J. M., and Lin, C. C. 1978. Hyphal wall chemistry of *Leptomitus lacteus*. Mycologia 70:363-369.

Aronson, J. M., Cooper, B. A., and Fuller, M. S. 1967. Glucans of oomycete cell walls. Science 155:332-335.

Ayers, A. R., Ebel, J., Finelli, F., Berger, N., and Albersheim, P. 1976a. Host-pathogen interactions. IX. Quantitative assays of elicitor activity and characterization of the elicitor present in the extracellular medium of cultures of *Phytophthora megasperma* var. *sojae*. Plant Physiol. 57:751-759.

Ayers, A. R., Ebel, J., Valent, B., and Albersheim, P. 1976b. Host-pathogen interactions. X. The fractionation and biological activity of an elicitor isolated from the mycelial walls of *Phytophthora megasperma* var. *sojae*. Plant Physiol. 57:760-765.

Barash, I., Klisiewicz, J. M., and Kosuge, T. 1965. Utilization of carbon compounds by zoospores of *Phytophthora drechsleri* and their effect on motility and germination. Phytopathology 55:1257-1261.

Bartnicki-Garcia, S. 1966. Chemistry of hyphal walls of *Phytophthora*. J. Gen. Microbiol. 42:57-69.

Bartnicki-Garcia, S. 1968. Cell wall chemistry, morphogenesis, and taxonomy of fungi. Annu. Rev. Microbiol. 22:87-108.

Bartnicki-Garcia, S. 1969. Cell wall differentiation in the Phycomycetes. Phytopathology 59:1065-1071.

Bartnicki-Garcia, S. 1970. Cell wall composition and other biochemical markers in fungal phylogeny. Pages 81-102 in: Phytochemical Phylogeny. J. B. Harborne, ed. Academic Press, London.

Bartnicki-Garcia, S. 1973a. Fundamental aspects of hyphal morphogenesis. Pages 245-267 in: Microbial Differentiation. Vol. 23. Symp. Soc. Gen. Microbiol. J. M. Ashworth and J. E. Smith, eds. Cambridge University Press, Cambridge.

Bartnicki-Garcia, S. 1973b. Cell wall genesis in a natural protoplast: The zoospore of *Phytophthora palmivora*. Pages 77-91 in: Yeast, Mould and Plant Protoplasts. J. R. Villanueva, I. García-Acha, S. Gascón, and F. Uruburu, eds. Academic Press, London.

Bartnicki-Garcia, S. 1981. Cell wall construction during spore germination in Phycomycetes. Pages 533-556 in: The Fungal Spore: Morphogenetic Controls. G. Turian and H. R. Hohl, eds. Academic Press, London.

Bartnicki-Garcia, S., and Hemmes, D. E. 1976. Some aspects of the form and function of oomycete spores. Pages 593-641 in: The Fungal Spore: Form and Function. D. J. Weber and W. M. Hess, eds. John Wiley & Sons, New York.

Bartnicki-Garcia, S., and Lippman, E. 1966. Liberation of protoplasts from the mycelium of *Phytophthora*. J. Gen. Microbiol. 42:411-416.

Bartnicki-Garcia, S., and Lippman, E. 1967. Enzymic digestion and glucan structure of hyphal walls of *Phytophthora cinnamomi*. Biochim. Biophys. Acta 136:533-543.

Bartnicki-Garcia, S., and Lippman, E. 1969. Fungal morphogenesis: Cell wall construction in *Mucor rouxii*. Science 165:302-304.

Bartnicki-Garcia, S., Bracker, C. E., Reyes, E., and Ruiz-Herrera, J. 1978. Isolation of chitosomes from taxonomically diverse fungi and synthesis of chitin microfibrils in vitro. Exp. Mycol. 2:173-192.

Bertke, C. C., and Aronson, J. M. 1980. Hyphal wall composition in *Apodachlyella completa*. Curr. Microbiol. 4:235-238.

Bimpong, C. E. 1975. Changes in metabolic reserves and enzyme activities during zoospore motility and cyst germination in *Phytophthora palmivora*. Can. J. Bot. 53:1411-1416.

Bimpong, C. E., and Clerk, G. C. 1970. Motility and chemotaxis in zoospores of *Phytophthora palmivora* (Butl.) Butl. Ann. Bot. 34:617-624.

Blackwell, E. 1943. On germinating the oospores of *Phytophthora cactorum*. Trans. Br. Mycol. Soc. 26:93-103.

Blumenthal, H. J. 1976. Reserve carbohydrates in fungi. Pages 292-307 in: The Filamentous Fungi. Vol. 2. J. E. Smith and D. R. Berry, eds. Edward Arnold, London.

Bokhary, H. A., and Cooke, R. C. 1974. Translocation of [$^{14}$C] glucose by *Phytophthora cactorum*. Trans. Br. Mycol. Soc. 63:535-540.

Bracker, C. E., Ruiz-Herrera, J., and Bartnicki-Garcia, S. 1976. Structure and transformation of chitin synthetase particles (chitosomes) during microfibril synthesis in vitro. Proc. Natl. Acad. Sci. USA 73:4570-4574.

Cabib, E., and Farkas, V. 1971. The control of morphogenesis: An enzymatic mechanism for the initiation of septum formation in yeast. Proc. Natl. Acad. Sci. USA 68:2052-2056.

Chapman, J. A., and Vujičić, R. 1965. The fine structure of sporangia of *Phytophthora erythroseptica* Pethyb. J. Gen. Microbiol. 41:275-282.

Clark, M. C., and Melanson, D. L. 1978. Purine metabolism and differential inhibition of spore germination in *Phytophthora infestans*. Can. J. Microbiol. 24:1032-1038.

Coulter, D. B., and Aronson, J. M. 1977. Glycogen and other soluble glucans from Chytridiomycete and Oomycete species. Arch. Microbiol. 115:317-322.

Desjardins, P. R., Wang, M. C., and Bartnicki-Garcia, S. 1973. Electron microscopy of zoospores and cysts of *Phytophthora palmivora*: Morphology and surface texture. Arch. Mikrobiol. 88:61-70.

Doke, N., Garas, N. A., and Kuć, J. 1979. Partial characterization and aspects of the mode of action of a hypersensitivity-inhibitory factor (HIF) isolated from *Phytophthora infestans*. Physiol. Plant Pathol. 15:127-140.

Ehrlich, M. A., and Ehrlich, H. G. 1966. Ultrastructure of the hyphae and haustoria of *Phytophthora infestans* and hyphae of *P. parasitica*. Can. J. Bot. 44:1495-1504.

Faro, S. 1972a. The role of a cytoplasmic glucan during morphogenesis of sex organs in *Achlya*. Am. J. Bot. 59:919-923.

Faro, S. 1972b. A soluble beta-1,3-glucan found in selected genera of Oomycetes. J. Gen. Microbiol. 72:393-394.

Gooday, G. W. 1971. An autoradiographic study of hyphal growth of some fungi. J. Gen. Microbiol. 67:125-133.

Gooday, G. W., and Hunsley, D. 1971. Cellulose wall ingrowths in *Phytophthora parasitica*. Trans. Br. Mycol. Soc. 57:178-179.

Grove, S. N., and Bracker, C. E. 1978. Protoplasmic changes during zoospore encystment and cyst germination in *Pythium aphanidermatum*. Exp. Mycol. 2:51-98.

Hegnauer, H., and Hohl, H. R. 1973. A structural comparison of cyst and germ tube wall in *Phytophthora palmivora*. Protoplasma 77:151-163.

Hegnauer, H., and Hohl, H. R. 1978. Cell wall architecture of sporangia, chlamydospores, oogonia and oospores in *Phytophthora*. Exp. Mycol. 2:216-233.

Hemmes, D. E., and Bartnicki-Garcia, S. 1975. Electron microscopy of gametangial interaction and oospore development in *Phytophthora capsici*. Arch. Microbiol. 103:91-112.

Hemmes, D. E., and Hohl, H. R. 1971. Ultrastructural aspects of encystation and cyst germination in *Phytophthora parasitica*. J. Cell Sci. 9:175-191.

Hepler, P. K., and Palevitz, B. A. 1974. Microtubules and microfilaments. Annu. Rev. Plant Physiol. 25:309-362.

Hickman, C. J. 1970. Biology of *Phytophthora* zoospores. Phytopathology 60:1128-1135.

Hinch, J. M., and Clarke, A. E. 1980. Adhesion of fungal zoospores to root surfaces is mediated by carbohydrate determinants of the root slime. Physiol. Plant Pathol. 16:303-307.

Ho, H. H. 1979. Electron microscope studies of oogonial wall and elemental composition of oogonia in *Phytophthora*. Mycopathologia 68:17-21.

Ho, H. H., and Hickman, C. J. 1967. Asexual reproduction and behavior of zoospores of *Phytophthora megasperma* var. *sojae*. Can. J. Bot. 45:1963-1981.

Ho, H. H., Zachariah, K., and Hickman, C. J. 1968. The ultrastructure of zoospores of *Phytophthora megasperma* var. *sojae*. Can. J. Bot. 46:37-41.

Hohl, H. R., and Hamamoto, S. T. 1967. Ultrastructural changes during zoospore formation in *Phytophthora parasitica*. Am. J. Bot. 54:1131-1139.

Holten, V. Z., and Bartnicki-Garcia, S. 1972. Intracellular $\beta$-glucanase activity of *Phytophthora palmivora*. Biochim. Biophys. Acta 276:221-227.

Honour, R. C. 1973. Lysis of *Phytophthora parasitica* mycelium and oospores in soil. Ph.D. thesis. Univ. Calif., Riverside. 204 pp.

Hunsley, D. 1973. Apical wall structure in hyphae of *Phytophthora parasitica*. New Phytol. 72:985-990.

Hunsley, D., and Burnett, J. H. 1968. Dimension of microfibrillar elements in fungal walls. Nature (London) 218:462-463.

Hunsley, D., and Burnett, J. H. 1970. The ultrastructural architecture of the walls of some hyphal fungi. J. Gen. Microbiol. 62:203-218.

Keen, N. T., and Legrand, M. 1980. Surface glycoproteins: Evidence that they may function as the race specific phytoalexin elicitors of *Phytophthora megasperma* f. sp. *glycinea*. Physiol. Plant Pathol. 17:175-192.

Leonian, L. H. 1934. Identification of *Phytophthora* species. W. Va. Agric. Exp. Stn. Bull. 262. 36 pp.

Lin, C. C., Sicher, R. C., Jr., and Aronson, J. M. 1976. Hyphal wall chemistry in *Apodachlya*. Arch. Microbiol. 108:85-91.

Lippman, E., Erwin, D. C., and Bartnicki-Garcia, S. 1974. Isolation and chemical composition of oospore-oogonium walls of *Phytophthora megasperma* var. *sojae*. J. Gen. Microbiol. 80:131-141.

Menyonga, J. M., and Tsao, P. H. 1966. Production of zoospore suspensions of *Phytophthora parasitica*. Phytopathology 56:359-360.

Meyer, R., Parish, R. W., and Hohl, H. R. 1976. Hyphal tip growth in *Phytophthora*. Gradient distribution and ultrahistochemistry of enzymes. Arch. Microbiol. 110:215-224.

Miyazaki, T., Yamada, M., and Ohno, T. 1974. Isolation and structure of $\alpha$-1,3-linked glucan from the hyphal wall of *Phytophthora infestans*. Chem. Pharm. Bull. 22:1666-1669.

Nogueira, M. L. B., Pinto da Silva, P., and Bartnicki-Garcia, S. 1977. Freeze-fracture and freeze-etching study of encystment of *Phytophthora palmivora* zoospores. J. Gen. Microbiol. 102:149-155.

Novaes-Ledieu, M., and Jiménez-Martínez, A. 1969. The structure of cell walls of Phycomycetes. J. Gen. Microbiol. 54:407-415.

Novaes-Ledieu, M., Jiménez-Martínez, A., and Villanueva, J. R. 1967. Chemical composition of hyphal wall of Phycomycetes. J. Gen. Microbiol. 47:237-245.

Pao, V. M., and Aronson, J. M. 1970. Cell wall structure of *Sapromyces elongatus*. Mycologia 62:531-541.

Powell, M. J., and Bracker, C. E. 1977. Isolation of zoospore organelles from *Phytophthora palmivora*. Page 533 in: Proc. 2nd Int. Mycol. Congr., Tampa, FL.

Reichle, R. E. 1969a. Retraction of flagella by *Phytophthora parasitica* var. *nicotiana* zoospores. Arch. Mikrobiol. 66:340-347.

Reichle, R. E. 1969b. Fine structure of *Phytophthora parasitica* zoospores. Mycologia 61:30-51.

Ruiz-Herrera, J., and Bartnicki-Garcia, S. 1974. Synthesis of cell wall microfibrils in vitro by a "soluble" chitin synthetase from *Mucor rouxii*. Science 185:357-359.

Schoulties, C. L., Baker, K. F., and Sabersky-Lehmann, C. 1980. Factors influencing zoospore production by *Phytophthora cinnamomi* in axenic culture. Can. J. Bot. 58:2117-2122.

Sietsma, J. H., Eveleigh, D. E., and Haskins, R. H. 1969. Cell wall composition and protoplast formation of some oomycete species. Biochim. Biophys. Acta 84:306-317.

Sing, V. O. 1974. Cytological and biochemical basis of cellular adhesion in zoospores of *Phytophthora palmivora*. Ph.D. thesis. Univ. Calif., Riverside. 114 pp.

Sing, V. O., and Bartnicki-Garcia, S. 1972. Adhesion of zoospores of *Phytophthora palmivora* to solid surfaces. Phytopathology 62:790.

Sing, V. O., and Bartnicki-Garcia, S. 1975a. Adhesion of *Phytophthora palmivora* zoospores: Electron microscopy of cell attachment and cyst wall fibril formation. J. Cell Sci. 18:123-132.

Sing, V. O., and Bartnicki-Garcia, S. 1975b. Adhesion of *Phytophthora palmivora* zoospores: Detection and ultrastructural visualization of concanavalin A-receptor sites appearing during encystment. J. Cell Sci. 19:11-20.

Thomas, R. C. 1943. Composition of fungus hyphae. IV. *Phytophthora*. Ohio J. Sci. 43:135-138.

Tokunaga, J., and Bartnicki-Garcia, S. 1971a. Cyst wall formation and endogenous carbohydrate utilization during synchronous encystment of *Phytophthora palmivora* zoospores. Arch. Mikrobiol. 79:283-292.

Tokunaga, J., and Bartnicki-Garcia, S. 1971b. Structure and differentiation of the cell wall of *Phytophthora palmivora*: Cysts, hyphae and sporangia. Arch. Mikrobiol. 79:293-310.

Vujičić, R., Chapman, J. A., and Colhoun, J. 1965. Ultra-structural studies of sporangia and zoospores of *Phytophthora erythroseptica*. Trans. Br. Mycol. Soc. 48:153-154.

Waechter, C. J., and Lennarz, W. J. 1976. The role of polyprenol-linked sugars in glycoprotein synthesis. Annu. Rev. Biochem. 45:95-112.

Wang, M. C. 1968. Biosynthesis of cell-wall polysaccharide of *Phytophthora cinnamomi*. Ph.D. thesis. Univ. Calif., Riverside. 131 pp.

Wang, M. C., and Bartnicki-Garcia, S. 1966. Biosynthesis of $\beta$-1,3- and $\beta$-1,6-linked glucan by *Phytophthora cinnamomi* hyphal walls. Biochem. Biophys. Res. Commun. 24:832-837.

Wang, M. C., and Bartnicki-Garcia, S. 1973. Novel phosphoglucans from the cytoplasm of *Phytophthora palmivora* and their selective occurrence in certain life cycle stages. J. Biol. Chem. 248:4112-4118.

Wang, M. C., and Bartnicki-Garcia, S. 1974. Mycolaminarans: Storage (1 → 3) -$\beta$-D-glucans from the cytoplasm of the fungus *Phytophthora palmivora*. Carbohydr. Res. 37:331-338.

Wang, M. C., and Bartnicki-Garcia, S. 1980. Distribution of mycolaminarans and cell wall $\beta$-glucans in the life cycle of *Phytophthora*. Exp. Mycol. 4:269-280.

Wang, M. C., and Bartnicki-Garcia, S. 1982. Synthesis of non-cellulosic cell wall $\beta$-glucan by cell free extracts from zoospores and cysts of *Phytophthora palmivora*. Exp. Mycol. 6:125-135.

Waterhouse, G. M. 1962. The zoospore. Trans. Br. Mycol. Soc. 45:1-20.

Wessels, J. G. H., and Sietsma, J. H. 1981. Fungal cell walls: A survey. Pages 352-394 in: Encyclopedia of Plant Physiology, New Series. Vol. 13. B. Plant Carbohydrates II. W. Tanner and F. A. Loewus, eds. Springer-Verlag, Berlin.

West, C., Wade, M., McMillan, C., and Albersheim, P. 1980. Purification and properties of invertases extractable from *Phytophthora megasperma* var. *sojae* mycelia. Arch. Biochem. Biophys. 201:25-35.

Wood, F. A., Singh, R. P., and Hodgson, W. A. 1971. Characterization of a virus-inhibiting polysaccharide from *Phytophthora infestans*. Phytopathology 61:1006-1009.

Yamamoto, M., and Tanino, J. 1961. Physiological studies on the formation and germination of sporangia of *Phytophthora infestans* (Mont.) de Bary. Forsch. Geb. Pflanzenkr. (Shokubutsu Byogai Kenkyu) 7:7-22.

Yoshikawa, M., and Masago, H. 1977. Effect of cyclic AMP on catabolite-repressed zoosporangial formation in *Phytophthora capsici*. Can. J. Bot. 55:840-843.

Yoshikawa, M., Tanaka, H., Miyata, Y., and Masago, H. 1976. Evaluation by specific inhibitors of metabolic dependence of asexual differentiation processes in *Phytophthora capsici*. Bull. Kyoto Prefect. Univ. 280:57-62.

Zentmyer, G. A. 1970. Tactic responses of zoospores of *Phytophthora*. Pages 109-111 in: Root Diseases and Soil-Borne Pathogens. T. A. Toussoun, R. V. Bega, and P. E. Nelson, eds. Univ. Calif. Press, Berkeley.

Zentmyer, G. A., and Erwin, D. C. 1970. Development and reproduction of *Phytophthora*. Phytopathology 60:1120-1127.

Zevenhuizen, L. P. T. M., and Bartnicki-Garcia, S. 1969. Chemical structure of the insoluble hyphal wall glucan of *Phytophthora cinnamomi*. Biochemistry 8:1496-1502.

Zevenhuizen, L. P. T. M., and Bartnicki-Garcia, S. 1970. Structure and role of a soluble cytoplasmic glucan from *Phytophthora cinnamomi*. J. Gen. Microbiol. 61:183-188.

# 10

# Present Criteria for Classification of *Phytophthora*

Grace M. Waterhouse, F. J. Newhook, and D. Jean Stamps

CONTENTS



## HISTORICAL INTRODUCTION

The features that de Bary (1887) singled out as characterizing *Phytophthora infestans* were the branching of the sporangiophore (at first monopodial, secondary branches sympodial), the shedding of the sporangia, the formation of zoospores in the sporangium, and the germination of the sporangium by zoospores or by a tube. At that time, the sex organs were not known.

During the 41 yr following the establishment of the genus, 12 more species were described under the generic name, but there were no established criteria. This led Rosenbaum (1917) to embark on a comparative study of 11 of these species with a view to constructing a key to the species. He clearly understood the problems, as in his opening paragraph he wrote,

> There was a great need of a comparative study under the same conditions of all the described species of the genus. Hitherto, the various species had never been gathered and grown by any one investigator; nor had comparisons been made when grown under similar conditions; nor had any general survey of the genus been made since it was established. . . . The cytology of the genus has likewise been almost entirely neglected.

Rosenbaum was able to obtain all the described species except *P. thalictri* and *P. colocasiae*. He did not consider *P. citrophthora*, then in the genus *Pythiacystis*, or *P. cambivora*, then in the genus *Blepharospora*. He had only one isolate of each species except *P. cactorum*, for which he had two.

The main criteria Rosenbaum selected for the identification and separation of the species and for his key were: 1) antheridia paragynous, amphigynous, or unknown (if no sex organs), papilla prominent or not ("good"); 2) sporangial length-breadth ratio (l/b) and the mean or mode; 3) chlamydospore presence or absence; 4) chlamydospore size ("good"); 5) oospore size (he did not measure oogonia); and 6) oospore abundance (for two species only). Because Rosenbaum had only papillate species, he did not use the nature of the papilla as a diagnostic character.

Rosenbaum concluded that cultures varied with the medium used and temperature; the proportionate value attached to each character in the separation and relationship of species must vary. He considered growth on a given medium to be of some value. On account of the variability in the size of the spore forms, a more ready comparison could be made by the use of biometric constants that further confirmed the identity and relationship of the species. This, then, was really the start of the establishment of criteria for identification of species.

By the time the next monograph (Tucker, 1931) appeared, 13 species (as now recognized) had been added as well as several that were later placed in synonymy. Tucker obtained 150 isolates, including many of *P. palmivora, P. parasitica*, and *P. cactorum*, but only one or a few of the remainder of the species he observed. He used in his key 11 criteria that he considered to be most valuable for differentiation of species: 1) ability to grow on certain media; 2) type of antheridium; 3) presence (abundance) or absence of sex organs; 4) cardinal temperatures, especially optimum and maximum; 5) host (or host family) that was parasitized; 6) nature of the papilla; 7) presence or absence of chlamydospores; 8) abundance of sporangia and chlamydospores; 9) pathogenicity determined by inoculation; 10) oospore size; and 11) hyphal swelling (for *P. cinnamomi*).

In Tucker's discussion of individual species, eight additional criteria that he had investigated were mentioned as useful for certain species. He reduced a number of species and varieties to synonymy with earlier species, and all of these reductions (except four) are accepted today; also the species he confirmed are, on the whole, well established still. He did not have cultures or specimens of *P. heveae* or *P. cyperi-rotundati*.

About this time, Leonian (1925) was studying the physiology of various isolates, the effect of host on sporangial form (Leonian, 1927), and the value of size of sporangia obtained under standard conditions (Leonian and Geer, 1929). He examined 101 isolates of 22 species and produced a key (Leonian, 1934), using the same criteria as Tucker (1931) except for pathogenicity after inoculation, but adding pedicel, growth and spore production in distilled water following growth in pea broth, and differential growth in media containing malachite green at different concentrations. He found that the ability to produce sporangia, oogonia, and/or chlamydospores was a valuable criterion. Although he keyed out 22 species and admitted that they could be identified by the key, he concluded that there were not more than three "good" species, namely *P. infestans, P. cactorum*, and *P. palmivora*.

Tucker's monograph (1931) remained the standard work for more than 30 yr. During this time, occasional studies appeared on *Phytophthora* species present in individual countries, either with a key (Frezzi, 1950) or without (Schwinn, 1959; Wager, 1940). Frezzi amassed many isolates from Argentina but studied only 57 in detail, finally identifying 12 species. He concluded with a synoptic key but did not specify which characters he thought were most important; he combined those of both Tucker (1931) and Leonian (1934).

Schwinn (1959) used 37 isolates (listed under 25 names) from Germany, one isolate of each except *P. cactorum* and *P. parasitica* (for which he had eight isolates each), and reduced these names to 18 "good" species. Though he did not produce a key, he examined 14 criteria in comparative studies. Those that gave some distinction between species were hyphal thickness, hyphal swellings (*P. cinnamomi* and *P. syringae*), cultural characters, the antheridium, abundance of both sex organs and sporangia on solid media, oospore and chlamydospore size (for occasional species), l/b ratio, growth in relation to temperature, and the papilla, which, with the pedicel, was thought to be of only secondary importance.

By 1960, the number of specific epithets attached to *Phytophthora* had reached over 60, and opinions varied as to how many "good" species there were. It was felt that the time had come to attempt to produce a "world" key. The new key (Waterhouse, 1963) was based on studies of cultures received at the Commonwealth Mycological Institute over a period of 17 yr and on slides left by S. F. Ashby. Type or authentic cultures were obtained where available. As a result, 43 species and subspecies taxa were recognized, and new descriptions were given where the original ones were deemed inadequate.

To assist in the identification of species, the genus was divided into six groups on the basis of the following criteria: 1) sporangial apex (extent of apical thickening and width of the exit pore); 2) abundance of sporangia on solid media; 3) deciduous or nondeciduous nature of sporangia; 4) internal proliferation of the sporangiophore; 5) nature of the antheridium; and 6) the presence (abundance) or absence of oospores on the host and in culture. Other characters featured in the key for certain species were: 7) sporangial dimensions, particularly maximum length; 8) l/b ratio; 9) sporangiophores; 10) hyphal swellings; 11) presence or absence of chlamydospores; 12) oogonial size; 13) ornamentaion of the oogonial wall; 14) host specificity; and 15) cardinal temperatures (optimum and maximum). This monograph removed a number of superfluous species that had been described since Tucker's monograph but revived four that he had placed in synonymy (*P. meadii, P. arecae, P. citricola*, and *P. hibernalis*).

In a sense, this key was too simple; it brought to light, but did not resolve, the complexities of some so-called species (namely, the "*palmivora* group" and *nicotianae-parasitica*) and also the very closely related nonpapillate species. In another sense, it was too difficult—*Phytophthora* has always been recognized as a notoriously difficult genus—especially for those who wished only to identify an isolate and were not concerned with taxonomic problems and/or minute details.

These considerations, plus the fact that six new species had been described and that the "*palmivora* group" was being resolved into distinguishable taxa, prompted a new type of key (Newhook et al, 1978). This new key was designed to aid both plant pathologists and others who wanted to know as nearly as possible the identity of a particular isolate and taxonomists who were endeavoring to achieve a better classification. It took the form of a tabular presentation of most of the traditional characters used in identifications and descriptions. It was not possible to include details of cultural characters or of temperature relationships, although this information is pointed out in some instances where such characters are diagnostically important. It was hoped that the tables would assist rapid and accurate identification, highlight the gaps in our knowledge (the gaps in the tables to be filled), and lead to better and much more comprehensive descriptions of future new species. Species were again arranged in the six groups of Waterhouse (1963), and at the time of publication of the 1978 key, included the species shown in Table 1.

## PRESENT CRITERIA

In this section, we will comment on the relative usefulness of the characters used in the tabular key (Newhook et al, 1978). In particular, we will consider their stability in relation to classification, identification, and description in view of the variability of *Phytophthora* species. The range of variability within a single species must be taken into account, and overlapping between species of ranges of individual characters must be accepted. Not all characters listed for a species need be present in a single isolate. This last feature, perhaps, has the most influence on our assessment of relative stability. For these purposes, characters are divided into those good for main groups; those good at species level; those useful for some species only; those important for one or two species; supporting characters; and characters needing further examination.

### Good for Main Groups

The characters that Waterhouse (1963) used to segregate the species into six main groups (I-VI), which were retained by Newhook et al (1978) in the tabular key, are: 1) apical thickening of sporangia and width of the exit pore; 2) caducity of sporangia and pedicel length; and 3) antheridia amphigynous, paragynous, or both.

#### Apical Thickening of Sporangia and Width of the Exit Pore

The structure of the sporangium apex is a marked feature of *Phytophthora* species. There is always apical thickening, which may be hemispheric, shallow, or inconspicuous and which is always covered by a thin continuation of the outer sporangial wall, and there is always a clear area between the base of the thickening and the surface of the protoplasm.

The events that occur from beginning to end of emission of zoospores were clearly established by Gisi et al (1979). Zoospores are normally differentiated from the protoplasm before emission. The thin, ephemeral covering of the apex rapidly dissolves or breaks. This is caused by the rapid swelling of the substance of the apical thickening into an evanescent vesicle into which the zoospores pass.

This apical structure and sequence of events for zoospore emission are outstanding features that help to distinguish *Phytophthora* from other closely related genera. We would exclude *P. vesicula* (Anastasiou and Churchland, 1969) and the later described marine species (Fell and Master, 1975) from the genus on these grounds alone. Although the sequence of events in zoospore emission can be speeded or slowed, the process is fundamentally the same for each species. The feature that is not subject to variation within species is the width of the exit pore and depth of apical thickening. Species with hemispheric or shallow thickening have a narrow exit pore (5–7 $\mu$m), and those with inconspicuous thickening have a broader pore (12 $\mu$m). This seems to be an unvarying feature good for an initial grouping of species.

**Caducity of Sporangia and Pedicel Length**

In the type species *P. infestans* and some others, most or many sporangia are shed at maturity. This condition is termed caducity. In the noncaducous species, which include all nonpapillate species (Waterhouse, 1963), shedding does not occur in any isolate, i.e., there is no provision in the structure and physiology of the fungus for this. This noncaducity or persistence is invariable for those species. In species where there is some shedding, the percentage of sporangia liberated appears to be variable (Al-Hedaithy and Tsao, 1979), and there may be some noncaducous isolates of species that are characteristically caducous (Zentmyer et al, 1977). We need to know more about caducity (e.g., structure of the abscission point and effect of environmental conditions on caducity), but even so this character is a good guide where it occurs. Erwin (Chapter 11) reviews the paper on caducity by Al-Hedaithy and Tsao (1979) and reproduces their figures.

The sporangium is shed with a length of hypha between the breaking-off point and the sporangium base, which is called the pedicel. The pedicel is of variable length in different species, because the length depends to a certain extent on how much the hypha grows before the apex starts to swell to form the sporangium. Caducous species fall into three pedicel-length categories (Al-Hedaithy and Tsao, 1979; Zentmyer et al, 1977): 1) short (<5 $\mu$m), wide, and occluded, e.g. *P. palmivora* morphological form (MF) 1, *P. infestans,* and *P. cactorum*; 2) intermediate (5–20 $\mu$m), thin, and with protoplasmic contents, e g. *P. botryosa, P. colocasiae,* and *P. megakarya*; and 3) long (>20 $\mu$m), thin, and with protoplasmic contents, e g. *P. hibernalis, P. palmivora* MF 4, and *P. capsici.* These categories seem to be fairly well substantiated, allowing for variability within them.

**Antheridia Amphigynous, Paragynous, or Both**

Groups II, IV, and VI are characterized by amphigynous antheridia. In group II, all the antheridia are amphigynous, with apparently no records of paragyny; in groups IV and VI, the only species exhibiting occasional paragyny are *P. hibernalis* and *P. richardiae.* Therefore, one can regard amphigyny as a feature that varies little. On the other hand, in species with mostly paragynous antheridia, there is usually a small proportion (about 1%) of amphigynous antheridia. Thus, paragyny is less consistent but not unstable enough to affect the reliability of this criterion for those species with predominantly paragynous antheridia. In the remaining species, which have large proportions of both types of antheridia, the numbers of each are variable for unknown reasons; however, this category has only about five species, and these are easily delimited from the others by other characters.

## Good at the Species Level

**Cardinal Temperatures**

Since Tucker (1931), all keys have used growth at minimum, optimum, and maximum temperatures as a useful criterion, although there is usually a range of variability of ±1–2°C among isolates of one species. All but seven of the 57 isolates of *P. nicotianae* var. *parasitica* studied by Tucker (1931) grew well at 35°C, and these seven grew more than 0.5 cm in 96 hr. High maximum temperature is also useful for *P. drechsleri, P. capsici,* and *P. melonis.* A low maximum temperature distinguishes *P. syringae, P. hibernalis, P. infestans, P. lateralis, P. ilicis,* and *P. primulae.* Occasionally, an isolate conforms to a particular specific description except for growth at a temperature higher than the accepted maximum. Such an isolate might be designated a high-temperature form of the species.

Optimum range is more variable but falls within certain limits, and one can delineate species with a high or a low range (Table 2). Minimum temperatures have not been worked out in great detail, nor has the distinction been made between no growth and no life. The lethal effect of temperature on growth

TABLE 1. Grouping of *Phytophthora* species (Newhook et al, 1978)

| Group | Species |
|---|---|
| I | *cactorum* (Lebert & Cohn) Schröter |
| | *iranica* Ershad |
| II | *palmivora* (Butler) Butler |
| | MF 1 |
| | MF 2 |
| | MF 3 (*P. megakarya* Brasier & Griffin) |
| | MF 4 |
| | *arecae* (Coleman) Pethybr. |
| | *boehmeriae* Saw. |
| | *botryosa* Chee |
| | *heveae* Thompson |
| | *meadii* McRae |
| | *nicotianae* B. de Haan |
| | var. *nicotianae* |
| | var. *parasitica* (Dastur) Waterh. |
| | *capsici* Leonian |
| | *citrophthora* (Sm. & Sm.) Leonian |
| | *mexicana* Hotson & Hartge |
| | *castaneae* Katsura & Uchida (now *katsurae*) |
| III | *citricola* Saw. |
| | *syringae* (Kleb.) Kleb. |
| | *porri* Foister |
| | *primulae* Tomlinson |
| | *cyperi* (Ideta) Ito |
| | *cyperi-bulbosi* Seeth. & Ramakr. |
| | *inflata* Caroselli & Tucker |
| | *lepironiae* Saw. |
| | *macrospora* (Sacc.) Ito & Tanaka |
| | *vesicula* Anastasiou & Churchland |
| IV | *colocasiae* Racib. |
| | *hibernalis* Carne |
| | *ilicis* Buddenhagen |
| | *infestans* (Mont.) de Bary |
| | *phaseoli* Thaxter |
| | *melonis* Katsura |
| V | *fragariae* Hickman |
| | *megasperma* Drechsler |
| | var. *megasperma* |
| | var. *sojae* Hildebrand |
| | *quininea* Crandall |
| | *verrucosa* Alcock & Foister |
| VI | *cambivora* (Petri) Buisman |
| | *cinnamomi* Rands |
| | *cryptogea* Pethybr. & Laff. |
| | *drechsleri* Tucker |
| | var. *drechsleri* |
| | var. *cajani* Pal, Grewal & Sarbhoy |
| | *erythroseptica* Pethybr. |
| | var. *erythroseptica* |
| | var. *pisi* Bywater & Hickman |
| | *richardiae* Buisman |
| | *vignae* Purss |
| | *lateralis* Tucker & Milbrath |
| | *gonapodyides* (Petersen) Buisman |
| | *japonica* Waterh. |

of mycelium should be differentiated from the effect on germination of spores, since these could be more or less sensitive.

The production of organs in culture in relation to temperature has proved useful for separation of some species (Brasier, 1969). In *P. lateralis*, the optimum temperature for chlamydospore production is 24–25°C, which is also approximately the maximum for linear growth and dry weight increase. However, the optimum for sporangial production is 14–16°C (Englander and Roth, 1980).

### Growth Rates

Although growth rates vary both among isolates of one species and in different conditions, in five species—*P. syringae, P. porri, P. primulae, P. infestans,* and *P. phaseoli*—very slow growth is a useful indicator.

### Chromosomes

The chromosomes in most species so far studied are small and difficult to count. About 9 to 10 or 12 chromosomes are present in many species, but *P. megasperma* var. *megasperma* has twice the usual number, whereas *P. megasperma* var. *sojae* has 10–15 (Sansome and Brasier, 1974); British tetraploid isolates of *P. infestans* have double the number present in Mexican diploid isolates (Brasier and Sansome, 1975). Table 3 summarizes the present state of knowledge of chromosome numbers in *Phytophthora*. The form is sometimes diagnostic; for example, *P. megakarya* has larger and fewer chromosomes (5–6) than most species, and chromosomes of *P. meadii* are much smaller than those of *P. palmivora*. Configurations at diakinesis caused by associations of chromosomes are sometimes characteristic of species (Sansome, 1980).

### Abundance of Oogonia in Agar Media

This character was used by Rosenbaum (1917), Tucker (1931), Leonian (1934), Frezzi (1950), and Waterhouse (1963) for a number of species, e.g. *P. cactorum*. It appears to be consistent enough for most isolates for establishment of the

TABLE 2. Grouping of *Phytophthora* species according to optimum temperatures for growth

| Optimum temperatures (°C)[a] | | | | | |
|---|---|---|---|---|---|
| High | | Intermediate | | Low | |
| Species | Temperature | Species | Temperature | Species | Temperature |
| *parasitica* | 30–32 | *oryzae* | 26–28 | *fragariae* | 20–22 |
| *melonis* | 28–32 | *citricola* | 25–28 | *phaseoli* | 15–20 |
| *palmivora* MF 1 | 27–32 | *cinnamomi* | 24–28 | *primulae* | 15–20 |
| *capsici* | 26–32 | *citrophthora* | 24–28 | *syringae* | 20 |
| *drechsleri* | 28–31 | *macrospora* | 28 | *hibernalis* | 20 |
| *vignae* | 28–30 | *erythroseptica* | 27.5 | *infestans* | 20 |
| *palmivora* MF 4 | 27–30 | *mexicana* | 27.5 | *lateralis* | ≈20 |
| *colocasiae* | 27–30 | *cactorum* | 20–28 | *ilicis* | 20 |
| *arecae* | 27–30 | *megasperma* | 15–27.5 | | |
| *meadii* | 25–30 | *botryosa* | 26 | | |
| *nicotianae* | 25–30 | *boehmeriae* | 25 | | |
| *inflata* | 25–30 | *heveae* | 25 | | |
| | | *quininea* | 25 | | |
| | | *porri* | 25 | | |
| | | *cryptogea* | 22–25 | | |
| | | *megasperma* var. *sojae* | 25 | | |
| | | *cambivora* | 22–24 | | |
| | | *erythroseptica* var. *pisi* | 24 | | |

[a]Data from Waterhouse (1963) except for var. *sojae*, which is from Ho (1971), and *lateralis*, which is from Englander and Roth (1980).

TABLE 3. Chromosome complements of *Phytophthora* species, showing multiple associations in heterothallic species

| | | Chromosome complements | | |
|---|---|---|---|---|
| Species | Ploidy | *n* | 2*n* (pairs + associations) | Sexuality |
| *megakarya* | diploid | 5 | 3 + 4 | heterothallic |
| *capsici* | diploid | 8–10 | 7–8 + 4 | heterothallic |
| *cambivora* | diploid | 8–10 | 7–9 + 4 | heterothallic |
| *cinnamomi* | diploid | 9±1 | 7–8 + 4 | heterothallic |
| *palmivora* MF 1 | diploid | 9–12 | 7–10 + 4 | heterothallic |
| *palmivora* MF 4 | diploid | 9–12 | 7–10 + 4 | heterothallic |
| *drechsleri* | diploid | 9–12 | 7–10 + 4 | heterothallic |
| *cryptogea* | diploid | 9 | 7 + 4 | heterothallic |
| *nicotianae* var. *parasitica* | diploid or trisomic | 7–9 | 6–8 + 4 or 5 | homothallic and heterothallic isolates |
| *infestans* (Mexican isolates) | diploid | 9–10 | 6–7 + 6 | heterothallic |
| *infestans* (British isolates) | tetraploid | 18–20 | multiples present | heterothallic (A1 only) |
| *meadii* | tetraploid | 16–18 | multiples present | homothallic (secondary) |
| | diploid | 8–10 | | heterothallic |
| *megasperma* var. *megasperma* | tetraploid | 22–27 (–30) | no multiples present | homothallic |
| *megasperma* var. *sojae* | diploid | 10–15 | no multiples present | homothallic |

*Source:* Sansome (1980), with additions by E. Sansome.

main groups, provided the right type of agar medium is used. However, some rare isolates do not always conform, and the ability to produce oogonia may be lost after prolonged culture.

## Useful for Some Species Only

### Sporangial Size

Tucker (1931) concluded that "dimensions of sporangia considered independently . . . cannot be accorded much importance taxonomically." There are problems in standardizing methods for producing sporangia, because the best conditions for one species may be unsuitable for another. Unless measurements are made during the first flush of growth, the presence of many atypically small sporangia may distort the size range for an isolate. Many species share similar or overlapping ranges. However, it is still possible to separate one group with small sporangia (rarely more than 45 μm long), which includes *P. cactorum, P. iranica, P. botryosa*, and *P. katsurae* (new name for *P. castaneae*) from another group with large sporangia (often more than 75 μm long), which includes *P. palmivora* MF 1 and MF 4, *P. capsici, P. citrophthora, P. mexicana, P. porri, P. primulae, P. cyperi, P. fragariae, P. cambivora, P. cinnamomi*, and *P. japonica*.

### Sporangial Length-Breadth Ratio

The relative narrowness or broadness of sporangia depends to some extent on external conditions and substrate. Brasier and Griffin (1979) found a higher l/b ratio for *P. palmivora* MF 1 sporangia harvested from cacao pods than from agar plates. Light or darkness, substrate (host, nature of agar medium), age of culture, and presence of free water may all have an effect. Species with a distinctively large l/b ratio include *P. palmivora* MF 4, *P. mexicana, P. cyperi*, and *P. hibernalis*. Of 18 species in which broad sporangia (l/b ratio $< 1.6$) are common, 12 have some sporangia that are spherical or nearly so, and all except one of these (*P. drechsleri* f. sp. *cajani*) are papillate or semipapillate.

### Oogonial Size

The range of oogonial sizes appears fairly consistent among isolates of a species. Restriction to small sizes is especially important for classification of *P. botryosa* and *P. heveae* and for separation of *P. nicotianae* var. *parasitica* from *P. nicotianae* var. *nicotianae*. On the other hand, large oogonia are conspicuous for *P. quininea, P. cambivora*, and *P. cinnamomi*, and the size range extends to above 50 μm in *P. megasperma* var. *megasperma* but never in *P. megasperma* var. *sojae*.

### Oogonial Wall

We do not yet know enough about properties of the oogonial wall for the genus as a whole, but there are indications of differences in staining reaction of some species (Table 4). The young wall stains darkly with aceto-orcein in *P. megakarya* and *P. palmivora* MF 4 but not in *P. palmivora* MF 1 (Brasier and Medeiros, 1978). A wall thicker than usual has been noted for *P. megakarya, P. palmivora* MF 4, *P. capsici, P. lepironiae, P. verrucosa*, and *P. erythroseptica* var. *pisi*. It is essential to examine staining reactions with young walls, because thickening and maturation begin as soon as oogonia have reached full size.

### Hyphal Swellings in Water

Rounded or angular swellings, in chains or clusters, which vary in frequency among isolates, may indicate any one of a number of species. Alternatively, the absence of such structures rather than their presence characterizes other species. In group III (Newhook et al, 1978), such swellings are associated with *P. syringae, P. porri*, and *P. primulae* but not with *P. citricola*.

### Chlamydospores

Chlamydospores are defined as being completely separated from the mycelium by a basal septum with an inner thickening of the wall. Since in some species chlamydospores are formed only by certain isolates, the significance of chlamydospores for classification in the genus as a whole is restricted. However, for separating some individual species, such as *P. palmivora* MF 1 from *P. megakarya* or *P. meadii*, their presence or absence, respectively, is a useful indicator.

### Pathogenicity

Tucker (1931) used 18 different hosts, including fruits, tubers, seedlings, and stems. For the few host-restricted *Phytophthora* spp. such as *P. fragariae, P. infestans*, and *P. ilicis*, a record of the source host is obviously still an aid to classification. Recently, too, further use has been made of specific pathogenicity for classifying isolates as formae speciales within species, such as *P. drechsleri* f. sp. *cajani* on pigeon pea (Kannaiyan et al, 1980), *P. megasperma* f. sp. *medicaginis* on alfalfa, and *P. megasperma* f. sp. *glycinea* on soybean (Kuan and Erwin, 1980).

### Culture Pattern on Agar Media

Colony morphology is determined by growth and branching pattern of the hyphae; however, we know little about how hyphal branching is regulated. Ho (1978) suggested that the branching characteristics at the periphery of a colony might be useful in taxa delineation. Culture pattern depends to a certain extent on the media used; examples of media that give distinctive patterns are potato-dextrose agar, cornmeal agar, and carrot agar. The main types include: 1) radiate (uniformly radiate hyphae), e.g. *P. citrophthora*; 2) chrysanthemum (narrow petaloid sectors), e.g. *P. citricola*; 3) rose or camellia (broad, rounded petaloid sectors), e.g. *P. cinnamomi* and *P. syringae*; 4) uniform (without any obvious pattern), e.g. *P. nicotianae, P. capsici*, and *P. infestans*; and 5) stellate (may be intermediate between radiate and chrysanthemum), e.g. *P. palmivora* MF 1.

## Intermediate Characters

Some characters are noticeable for certain species but have particular importance in only a few.

*Sporangial base tapered.* Although a variable proportion of sporangia with tapered bases may be seen in 15 species or

TABLE 4. Depth of staining reaction of walls of young oogonia of *Phytophthora* species to aceto-orcein and aceto-carmine

| Species | Stain | Staining reaction |
|---|---|---|
| *drechsleri* | aceto-orcein | A1 pale, A2 deep |
| *cinnamomi* | aceto-orcein | A1 pale, A2 deep (also antheridium and cytoplasm) |
| *capsici* | aceto-orcein | deep |
| *megakarya* | aceto-orcein | deep |
| *palmivora* MF 1 | aceto-orcein | pale |
| *cryptogea* | aceto-orcein | deep |
| *palmivora* MF 4 | aceto-orcein (especially) and aceto-carmine | deep (also cytoplasm) |
| *infestans* | aceto-orcein | deep |

*Source:* Brasier and Sansome (1975).

varieties, the regularity of occurrence of this feature is outstanding for *P. hibernalis* and *P. palmivora* MF 4, which also have the broadest dimension closer to the apex than to the basal septum.

*More than one apex.* Although at least 12 species or varieties have been recorded as producing some sporangia with more than one apex, we consider that the character has particular importance only for *P. citrophthora* and *P. citricola.* The sporangia of these species may be deeply cleft compared with those of *P. nicotianae* var. *nicotianae*, where there may be two or more papillae but where the shape of the sporangium is otherwise not changed. It may or may not be significant that the character has not been observed in association with nonpapillate sporangia.

*Oospores markedly aplerotic.* The relationship of the size of the oospore to the size of the oogonium has not been used in *Phytophthora* as it has in *Pythium* (Middleton, 1943). Whereas in most species the oospore nearly fills the oogonium (plerotic), 15 species or varieties are listed (Newhook et al, 1978) in which a variable proportion of oospores can be considered aplerotic. Among these, the space between oogonial and oospore walls is particularly large in many sex organs of *P. heveae, P. porri,* and *P. vignae.*

*Oogonia with tapered base.* While we consider that 10 species or varieties have sufficient oogonia with tapered bases to warrant their being so listed (Newhook et al, 1978), the character is especially notable in degree and frequency in *P. heveae* and *P. megakarya*, where this character contributes markedly to ease of identification. It occurs only in species with amphigynous antheridia. It is important to avoid confusing tapered oogonial bases, where the antheridium is largely below the tapered region, with tapered oogonial stalks, where the tapering occurs within the amphigynous antheridium.

### Important for One or Two Species

*Sporangial shape often distorted.* Ignoring slight asymmetry, some degree of distortion occurs in 14 species in the key (Newhook et al, 1978). In particular, in *P. citrophthora, P. citricola,* and *P. primulae*, the character is sufficiently common to give it special diagnostic value.

*Sporangial collapse after dehiscence.* This character has been noted for *P. cinnamomi* and *P. drechsleri* and is useful without being invariably present. Rands' original description for *P. cinnamomi* includes sporangial collapse, but the character seems to be encountered in only a few isolates.

*Sporangiophore compound. P. infestans* and *P. phaseoli* have this very distinctive feature. The fact that compound sporangiophore production is displayed more characteristically on the host than in culture is evidence for the need to study *Phytophthora* spp. under natural conditions and to investigate the effects of cultural conditions on morphology.

*Oogonial wall ornamented.* This character is particularly important for *P. cambivora* and *P. katsurae*, which produce hollow bullate protuberances. Solid wartlike protrusions are found in *P. verrucosa.* In all cases, the ornamentation is confined to the oogonial wall, although an optical illusion may suggest erroneously that the oospore wall is also involved. A single oogonium may have from one to many protuberances of varying sizes.

*Oogonial stalk eccentric.* This character occurs in some amphigynous antheridia and is a feature that contributes to the maintenance of *P. nicotianae* var. *nicotianae* as a separate taxon. Without careful observation, the condition could pass for a paragynous antheridium in close contact with an oogonial stalk.

*Chlamydospores lateral, sessile.* Although all chlamydospores in a slide preparation of *P. lateralis* will not be lateral, it is such a prominent and unique feature that its presence makes positive identification easy.

*Hyphal ends coiling.* Although coiling of hyphae is not uncommon, perhaps especially on solid media low in nutrients and when bacterial contaminants are present, it is so well developed in *P. porri* that it constitutes a worthwhile diagnostic character.

*Hyphal swellings on agar.* Coralloid hyphae on agar, especially noticeable at the margin of the colony, are diagnostic for *P. cinnamomi.* These scarcely involve an increase in hyphal diameter and might be described simply as extremely irregular hyphal walls. Some cultures have large, irregular hyphal swellings like those found in some isolates of *P. cambivora,* but the character "hyphal swellings large, spherical, botryose" is unique to *P. cinnamomi.* Such spherical swellings are distinguishable from chlamydospores in that they lack a basal septum and do not have a continuous inner-wall thickening.

*Hyphal swellings spherical, with radiating hyphae.* On agar or in water, these swellings are characteristic of *P. nicotianae* var. *nicotianae* and are shown in van Breda de Haan's original drawings. They reach a diameter of 40 $\mu$m, equivalent to that of the chlamydospores, which they resemble except for the presence of as many as four or five radiating hyphae that arise directly from the nonthickened wall. The appearance of these hyphae does not suggest direct germination. They may terminate in similar spherical swellings or, perhaps, chlamydospores. Analogous radial projections, as outgrowths of the wall, occur as a unique character on sporangia of many isolates of *P. nicotianae* var. *nicotianae.*

*Mycelium tough.* Although identifications should be made on the basis of characters less subjective than this, the extraordinary toughness toward a needle or scalpel blade of the aerial mycelium of *P. cinnamomi* growing on rich media is notable. Lack of toughness distinguishes *P. cambivora* from *P. cinnamomi*, which not infrequently occurs with it.

*Sporangia in clumps.* This character is unique to *P. botryosa.* Chee (1969) recorded clumps of sporangia in culture, visible to the naked eye, with up to 970 sporangia per square millimeter of agar surface. Clumping is due to the large number of sporangia that are produced on a sporangiophore; the branching habit is a simple monochasial sympodium that is usually close.

*Oogonia clustered.* The abundant oogonia formed in culture by *P. heveae* often occur in close groups, with several oogonia apparently radiating from a single point of origin.

*Absence of antheridia.* A new species recently described, *P. insolita* Ann & Ko (1980), produced oogonia and oospores without antheridia. It is the only known species to do so.

### Supporting Characters

Supporting characters cannot by themselves be used as primary diagnostic features, but they help to confirm the identification of certain species.

*Hyphal diameter.* Although the range of hyphal diameter (of mature hyphae) is not large, there are specific differences in the fineness or coarseness. For example, *P. cinnamomi, P. cambivora,* and *P. capsici* have comparatively wide hyphae (mostly 7 or 8 $\mu$m), whereas in others, such as *P. drechsleri* and *P. palmivora* MF 1, hyphal width rarely exceeds 5 $\mu$m.

*Sporangiophores.* Apart from the very characteristic erect, compound sympodia of *P. infestans* and *P. phaseoli*, the close sympodia of species such as *P. cactorum, P. palmivora* MF 1, and *P. syringae* are noticeable features contributing to our idea of these species. The loose, irregularly branched sporangiophores common in the nonpapillate species are rather characterless and usually very narrow (2–4 μm wide). Another distinguishing feature sometimes noted is the presence of swellings at the points of branching (*P. citrophthora, P. citricola, P. colocasiae, P. infestans, P. meadii, P. phaseoli*) or along the sporangiophore itself (*P. nicotianae* var. *nicotianae, P. richardiae*). Form depends very much on conditions; the sporangial branching of *P. infestans* is much more characteristic on the host than in culture. The perfection of the sympodia is disturbed by free moisture, in which sporangiophores become more elongated.

*Abundance of sporangia on solid media.* Species with nonpapillate sporangia usually do not form them on solid media. In contrast, most of the species that form papillate sporangia can produce them abundantly on a wide variety of common agar media (e.g., *P. palmivora, P. cactorum,* and *P. boehmeriae*). However, individual isolates can vary in their potential for sporangial production.

*Antheridial size.* The antheridia of many species are similar in size and shape, but some species have characteristically small (*P. heveae, P. syringae*) or large (*P. fragariae, P. cambivora, P. richardiae*) antheridia.

*Sporangial apex flattening.* When sporangia of some nonpapillate species (*P. cryptogea, P. drechsleri* var. *drechsleri, P. erythroseptica, P. lateralis,* all in group VI) (Newhook et al, 1978) are mounted in lactophenol blue, the apex flattens. Why this happens in some species but not in others is not known.

*Oospore wall thick relative to diameter.* The form of the oospore is very consistent throughout the genus, allowing for some variation in size dictated by the size of the oogonium, slight differences in color (sometimes induced by the medium), and some differences in wall thickness that are not necessarily related to oospore size. Some species may have a small spore with a relatively thick wall and others a larger spore with a relatively thin wall; for example, *P. cryptogea* with oospores 25 μm in diameter has a wall 3.5 μm thick, whereas *P. cinnamomi* with oospores about 38 μm in diameter has a wall only 2–3 μm thick.

### Characters Needing Further Examination

The characters included in this section have been described in some species but have not always been indicated as criteria; some have received little consideration. They should always be included in descriptions when present.

*Basal plug of the sporangium.* At the base of the sporangium, on the cross wall separating it from the sporangiophore, there is often a plug bulging into the sporangium. It is possible that specific differences might be identified here.

*Sporangia intercalary.* Intercalary sporangia appear infrequently in cultures of about 10 species, all in the papillate groups. Very little is known about their occurrence or stability as a diagnostic character.

*Central vacuole in sporangium before zoospore emergence.* This character has been reported in a few species (*P. cryptogea* and *P. erythroseptica*), but again the constancy of its occurrence is uncertain.

*Zoospore secondary emergence.* Ho and Zentmyer (1977) found this to be a feature of *P. cinnamomi.* Drechsler (1930) made much of it, particularly for nonpapillate species. Occurrence or nonoccurrence may be due to conditions when germination is about to take place.

*Two-celled elongated antheridia.* Elongation is a feature of some species (e.g. *P. cambivora* or *P. cinnamomi*), but not all the elongated antheridia in any one isolate are two-celled (Ho and Zentmyer, 1977). More investigation seems warranted to determine how specific this character is.

*Subterminal antheridia.* Not uncommonly, the antheridial hypha swells to form the antheridium a short distance from the apex, leaving a length of hypha projecting from the antheridium, as for example in *P. richardiae.*

*Residual material in antheridium.* After fertilization, the antheridium is usually empty, but in *P. boehmeriae* and *P. primulae*, residual globules are left behind. They are such a noticeable and constant feature that the character is worth recording.

*Antheridia monoclinous or diclinous.* Observation of these conditions has not been recorded as much as it has in *Pythium* (Middleton, 1943). Probably for amphigynous species, dicliny is the rule. In paragynous species both occur, and dicliny in *P. citricola* was used as one of the features separating this species from *P. cactorum* (Waterhouse, 1957). Further observation of this character is needed.

*Delayed nuclear fusion in the oospores.* Normally, fusion occurs soon after the fusion nuclei meet. However, delayed fusion has recently been pointed out as a feature of *P. capsici* (Brasier and Brasier, 1978).

*Oospore germination.* A different type of germination distinguishes *P. palmivora* MF 4 from MF 1. Kaosiri et al (1980) observed that crosses of A1 and A2 isolates of *P. palmivora* MF 1 germinated by a single germ tube on which nondeciduous sporangia formed. MF 4 A1 and A2 crosses produced oospores that germinated by as many as 14 germ tubes, each terminated by a deciduous sporangium with a long stalk.

*Oogonial wall structure.* There are preliminary reports of differences among species in regard to wall ultrastructure. Ho (1979), in scanning electron micrograph studies, recorded smooth oogonial wall surfaces for *P. palmivora* MF 4, finely granular surfaces for *P. megasperma* var. *megasperma* and var. *sojae*, and coarsely granular surfaces for *P. nicotianae* var. *parasitica*. Further scanning electron microscope and transmission electron microscope studies may well visualize other differences that may prove useful in diagnosis.

*Chlamydospores in soil.* Some species, such as *P. drechsleri*, that do not normally produce chlamydospores in culture did so in soil (Cother and Griffin, 1973). Chlamydospores smaller than usual were formed in soil by other species.

*Chlamydospore wall thickness.* It might be worthwhile checking if chlamydospore wall thickness is more reliable (less variable) as a specific difference than size.

*Amount of aerial mycelium.* This has been used as one of the characters to separate pairs of closely related taxa (*P. erythroseptica* from its var. *pisi* and *P. porri* from *P. syringae*).

*Color.* Sex organs, particularly oogonia, of *Phytophthora* have a tendency to become brown, but this is sometimes due to the medium. The feature should be taken as a criterion only if it occurs on all media (Shepherd, 1976).

*Growth in malachite green.* Ability to grow in media incorporating malachite green was a character used by Leonian (1934), Urquijo-Landaluze (1947) (who was thus able to distinguish between *P. cambivora* and *P. cinnamomi*), Frezzi (1950), and Satour and Butler (1968). Despite records of some specific differences, the approach does not seem to be used frequently.

*The* Trichoderma *effect.* Brasier (1971) showed that volatile metabolites of *Trichoderma* sp. stimulated development of sex organs in unpaired A2 cultures of some heterothallic species. Brasier and Griffin (1979) pointed out that *P. palmivora* MF 4 responded, but *P. megakarya* and *P. palmivora* MF 1 did not.

## CONCLUSIONS

Present criteria would include all characters that are being used to differentiate taxa in the genus. We have been restricted to the classic criteria; some more modern aspects are dealt with by Gallegly (Chapter 12) and Brasier (Chapter 27). Our survey has attempted to define and evaluate these criteria. They have been grouped according to their usefulness in the classification, identification, and description of species.

A distinction should be made between classification and identification. For a better classification, all the characters available are needed, and it is expected that new characters identified in the fields of physiology, cytology, histology, and electron microscopy will be added. If a key for identifying isolates is to be practical, it must include clearly defined features that can be demonstrated and recognized easily. The relevance of these features to natural systematic relationships is not of prime importance. Conversely, some more basic criteria such as chromosome complements and complex biochemical differences, presumably of greater biological significance, may be more difficult to determine and thus less useful for the practical purposes of identification. We hope that their main usefulness might be to strengthen specific relationships based on morphological grounds.

The known criteria serve different purposes and have different degrees of usefulness in separating species; each should be applied according to its merits. Those that are fundamental to classification are not necessarily most accessible to the observer. Some appear relatively trivial and are not always seen, yet they add weight to the evidence for placement of an isolate or a species. Certain characters are consistently associated in some section of the genus. For the genus as a whole, there are large gaps in information on some of the basic aspects, and much more research is needed to determine linkages between these and more easily observable morphological criteria.

Despite recommendations that go back at least as far as Rosenbaum (1917), standardization of media and growing conditions remains an objective that is far from realized. We accept that there may be ultimate limits to achieving uniformity because of the differing requirements of the various species.

Our main conclusions are as follows:

1) Morphological characters will probably remain the chief criteria for interspecific distinctions.
2) Simple or sophisticated chemical or physical tests may come to have an increasing, but perhaps mainly supportive, role in identification.
3) The extent of the range for a single criterion is important and is probably more important than the mean. Interpretation of measurements should take heed of variability among as well as within isolates of a species.
4) The taxonomy of the genus is still beset with difficulties that stem from original descriptions based on a single type isolate that, by chance, might be far from representative of the species as a whole.
5) Further efforts should be directed toward standardizing conditions under which observations are made.
6) Further research is needed on methods for inducing production of sex organs in single cultures of the so-called "heterothallic" species and on any possible taxonomic significance of mating reactions.
7) As concluded by Rosenbaum more than 60 yr ago, an aggregate of as many characters as possible should be considered if individual species are to be delimited satisfactorily. We still feel there is merit in our suggestion (Newhook et al, 1978) that, ideally, species should be redescribed from material grown under standard growing conditions, with establishment of new types where originals are not available. Only then will many of the difficulties inherent in the present situation be resolved.

## LITERATURE CITED


Al-Hedaithy, S. S. A., and Tsao, P. H. 1979. Sporangium pedicel length in *Phytophthora* species and the consideration of its uniformity in determining sporangium caducity. Trans. Br. Mycol. Soc. 72:1-13.

Anastasiou, C. J., and Churchland, L. M. 1969. Fungi on decaying leaves in marine habitats. Can. J. Bot. 47:251-257.

Ann, P. J., and Ko, W. H. 1980. *Phytophthora insolita*, a new species from Taiwan. Mycologia 72:1180-1185.

Brasier, C. M. 1969. The effect of light and temperature on reproduction in vitro of two tropical species of *Phytophthora*. Trans. Br. Mycol. Soc. 52:105-113.

Brasier, C. M. 1971. Induction of sexual reproduction in single $A_2$ isolates of *Phytophthora* species by *Trichoderma viride*. Nature (London) New Biol. 231:283.

Brasier, C. M., and Brasier, C. 1978. Gametic and zygotic nuclei in oospores of *Phytophthora*. Trans. Br. Mycol. Soc. 70:297-302.

Brasier, C. M., and Griffin, M. J. 1979. Taxonomy of '*Phytophthora palmivora*' on cocoa. Trans. Br. Mycol. Soc. 72:111-143.

Brasier, C. M., and Medeiros, A. G. 1978. Karyotype of *Phytophthora palmivora* morphological form 4. Trans. Br. Mycol. Soc. 70:295-297.

Brasier, C. M., and Sansome, E. 1975. Diploidy and gametangial meiosis in *Phytophthora cinnamomi, P. infestans* and *P. drechsleri*. Trans. Br. Mycol. Soc. 65:49-65.

Chee, K. H. 1969. Variability of *Phytophthora* species from *Hevea brasiliensis*. Trans. Br. Mycol. Soc. 52:425-436.

Cother, E. J., and Griffin, D. M. 1973. Formation of chlamydospores by *Phytophthora drechsleri*. Trans. Br. Mycol. Soc. 61:379-402.

de Bary, A. 1887. Comparative Morphology and Biology of the Fungi, Mycetozoa and Bacteria. Oxford, Clarendon Press. 525 pp.

Drechsler, C. 1930. Repetitional diplanetism in the genus *Phytophthora*. J. Agric. Res. 40:557-563.

Englander, L., and Roth, L. F. 1980. Interaction of light and sterol on sporangium and chlamydospore production by *Phytophthora lateralis*. Phytopathology 70:650-654.

Fell, J. W., and Master, I. M. 1975. Phycomycetes (*Phytophthora* spp. nov. and *Pythium* sp. nov.) associated with degrading mangrove (*Rhizophora mangle*) leaves. Can. J. Bot. 53:2908-2922.

Frezzi, M. J. 1950. Las especies de '*Phytophthora*' en la Argentina. Rev. Invest. Agric. 4:47-133.

Gisi, U., Hemmes, D. E., and Zentmyer, G. A. 1979. Origin and significance of the discharge vesicle in *Phytophthora*. Exp. Mycol. 3:321-339.

Ho, H. H. 1971. A study on the growth and sporulation of *Phytophthora megasperma* var. *sojae*. Sydowia 24(1970):51-58.

Ho, H. H. 1978. Hyphal branching systems in *Phytophthora* and other phycomycetes. Mycopathologia 64:83-86.

Ho, H. H. 1979. Electron microscope studies of oogonial wall and elemental composition of oogonia in *Phytophthora*. Mycopathologia 68:17-21.

Ho, H. H., and Zentmyer, G. A. 1977. Morphology of *Phytophthora cinnamomi*. Mycologia 69:701-713.

Kannaiyan, J., Ribeiro, O. K., Erwin, D. C., and Nene, Y. L. 1980. Phytophthora blight of pigeon pea in India. Mycologia 72:169-181.

Kaosiri, T., Zentmyer, G. A., and Erwin, D. C. 1980. Oospore morphology and germination in the *Phytophthora palmivora* complex from cacao. Mycologia 72:888-907.

Kuan, T.-L., and Erwin, D. C. 1980. Formae speciales differentiation of *Phytophthora megasperma* isolates from soybean and alfalfa. Phytopathology 70:333-338.

Leonian, L. H. 1925. Physiological studies on the genus *Phytophthora*. Am. J. Bot. 12:444-495.

Leonian, L. H. 1927. The effect of different hosts upon the sporangia of some phytophthoras. Phytopathology 17:483-490.

Leonian, L. H. 1934. Identification of *Phytophthora* species. W.Va.

Agric. Exp. Stn. Bull. 262. 36 pp.

Leonian, L. H., and Geer, H. L. 1929. Comparative value of the size of *Phytophthora* sporangia obtained under standard conditions. J. Agric. Res. 39:293-311.

Middleton, J. T. 1943. The taxonomy, host range and geographical distribution of the genus *Pythium*. Mem. Torrey Bot. Club 20:1-171.

Newhook, F. J., Waterhouse, G. M., and Stamps, D. J. 1978. Tabular key to the species of *Phytophthora* de Bary. Mycol. Pap. 143. Commonw. Mycol. Inst., Kew, Surrey, England. 20 pp.

Rosenbaum, J. 1917. Studies of the genus *Phytophthora*. J. Agric. Res. 8:233-276.

Sansome, E. 1980. Reciprocal translocation heterozygosity in heterothallic species of *Phytophthora* and its significance. Trans. Br. Mycol. Soc. 74:175-185.

Sansome, E., and Brasier, C. M. 1974. Polyploidy associated with varietal differentiation in the *megasperma* complex of *Phytophthora*. Trans. Br. Mycol. Soc. 63:461-467.

Satour, M. M., and Butler, E. E. 1968. Comparative morphological and physiological studies of the progenies from intraspecifc matings of *Phytophthora capsici*. Phytopathology 58:183-192.

Schwinn, F. J. 1959. Untersuchungen zur Systematik der Gattung *Phytophthora* de Bary. Arch. Mikrobiol. 33:223-252.

Shepherd, C. J. 1976. Pigment production from tyrosine by Australian isolates of *Phytophthora* species. Aust. J. Bot. 24:607-617.

Tucker, C. M. 1931. Taxonomy of the genus *Phytophthora* de Bary. Mo. Agric. Exp. Stn. Res. Bull. 153. 208 pp.

Urquijo-Landaluze, P. 1947. Revisíon taxonómica de los hongos productores de la enfermedad del Castaño llamada de la 'tinta'. Bol. Patol. Veg. Entomol. Agric. 16:253-270.

Wager, V. A. 1940. Descriptions of South African Pythiaceae, with records of their occurrence. Bothalia 4:3-35.

Waterhouse, G. M. 1957. *Phytophthora citricola* Sawada (syn. *P. cactorum* var. *applanata* Chester). Trans. Br. Mycol. Soc. 40:349-357.

Waterhouse, G. M. 1963. Key to the species of *Phytophthora* de Bary. Mycol. Pap. 92. Commonw. Mycol. Inst., Kew, Surrey, England. 22 pp.

Zentmyer, G. A., Kaosiri, T., and Idosu, G. 1977. Taxonomic variants in the *Phytophthora palmivora* complex. Trans. Br. Mycol. Soc. 69:329-332.

# 11

# Variability Within and Among Species of *Phytophthora*

D. C. Erwin

CONTENTS



## INTRODUCTION

Description of variability and definition of the possible limits of variability of *Phytophthora* will be the concerns of this chapter. Variability in morphology, cultural and physiologic characteristics, pathogenicity, and resistance to fungicides will be described. Much of the information will be related to the discussion of control by resistant cultivars by Umaerus et al (Chapter 24), the current approach to epidemiology by MacKenzie et al (Chapter 23), and the genetic mechanisms of variability discussed by Shaw (Chapter 6).

The most notable advance in recent years has been the establishment of the concept that the Oomycetes, of which *Phytophthora* is a member, are diploid in their vegetative state (Sansome, 1965, 1966) (see Shaw, Chapter 6). Because almost all other fungi are haploid, concepts of variability and how it occurs must be rethought, especially in relation to the development of new races and the response of *Phytophthora* to the new narrow-spectrum fungicides. *Phytophthora* species must be heterozygous for many factors.

The literature on variation of *Phytophthora* species has increased so rapidly since a previous review (Erwin et al, 1963) that many important subtopics and papers have had to be omitted. Some other excellent sources of information on variation include articles by Buddenhagen (1958), Cameron (1966), Caten and Day (1977), Chee (1969), Denward (1967, 1970), Gallegly (1968), Toxopeus (1956), and Webster (1974).

## TYPES OF VARIABILITY

### Morphological and Cultural Variability

Identification of species within the fungi has been based more on morphology than on any other criterion. Chesters (1968) stated, "That these criteria should be morphological for the vast majority of fungi cannot be disputed, but it must never be forgotten that the morphology of an individual is the ultimate expression of its growth processes, the final display of all its complex relationships with its normal habitat." Many species can be readily recognized and identified. However, there are many overlapping characteristics that cause problems in identification. There is an inherent danger in accepting the concept that all isolates must conform to the exact morphological description given for the so-called "type species." Leonian (1934) questioned the wisdom of this restriction: "The accident of a first discovery and description determines the typical strain, while a subsequent discovery and observation of another strain, which may sharply differ from the original, determines the atypical strain. Thus, nothing more tangible than a mere state of mind is brought forth." Leonian also stated, "There is no such thing as typical or atypical but merely variability of living things" (Leonian, 1934). Morphological variability of sporangial sizes from one single-zoospore isolate of the pathogen *P. parasitica* var. *rhei* is shown in Figure 1.

In many species, such as *P. cinnamomi*, certain isolates appear to remain stable in growth rate, morphology, and pathogenicity; however, when numerous isolates from many regions of the world are observed, there are many variants

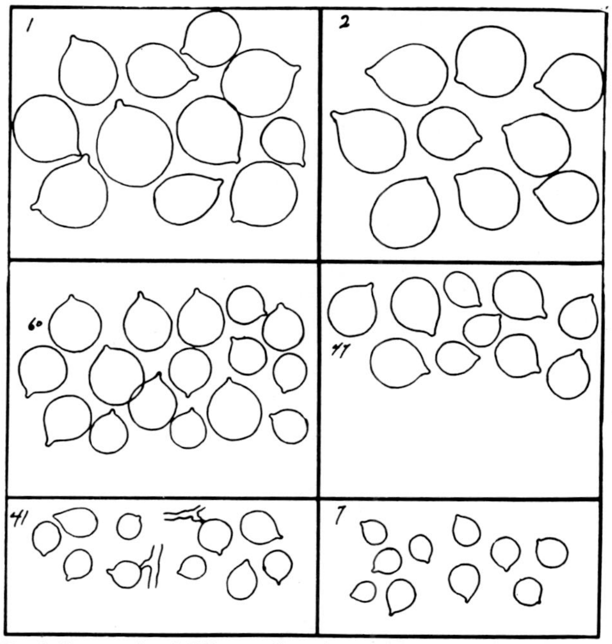


**Fig. 1.** Sporangia of different dissociants (numbers indicate isolates) of *Phytophthora parasitica* var. *rhei*. Note the extreme variation in size. (Reprinted, by permission, from Leonian, 1934)

from the original type species described by Rands (1922). Research in both the northern and southern hemispheres indicates considerable variability in response to temperature (Shepherd and Pratt, 1974; Shepherd et al, 1974; Zentmyer et al, 1976). Zentmyer et al (1976) showed that colony diameters of 187 isolates from different regions of the world varied from 47.5 to 82.5 mm (4 days), with a mean of 67.5 mm (Fig. 2). The rate of growth of 16 A1 and 16 A2 isolates also varied with the type of medium employed (Fig. 3). Cardinal temperatures for growth of *P. cinnamomi* varied from 5 to 16° C (minimum), 20 to 32° C (optimum), and 30 to 36° C (maximum). These ranges were considerably wider than those in Waterhouse's key (1963), which were 5° C (minimum), 24–28° C (optimum), and 32–34° C (maximum). This is but one example of the need to reevaluate and perhaps redescribe this species based on evidence from a large number of isolates from many regions of the world. Zentmyer's (1980) monograph on *P. cinnamomi* summarizes much of this information.

Shepherd and Pratt (1974) studied 361 isolates of *P. cinnamomi* from Australia and noted considerable variation in growth rate on cornmeal agar of single-zoospore progeny (considered to be almost 100% uninucleate) from parent cultures. Cultures from single sporangia and from hyphal tips (multinucleate) also varied in growth rate from that of the parent but less markedly. Shepherd and Pratt concluded that the parental cultures were most likely heterokaryotic, as were the hyphal tip and sporangial cultures, and that the single-zoospore isolates were homokaryotic, since there was little variation in second- and third-generation single-zoospore cultures (Fig. 4).

Uniform standards are necessary for correct interpretation of morphological differences. As an example, the term caducity describes the capability of sporangia to become readily detached from the mycelium. Undoubtedly, those *Phytophthora* species with caducous sporangia have a selective advantage in aerial or water-splash dissemination. Waterhouse (1974b) reviewed the status of caducity in the genus, explaining that in some species, sporangia are shed in dry air and in water (only *P. infestans* and *P. phaseoli*) and in others only in water (*P. palmivora* and 17 other species). Kaosiri et al (1978) recently reported on stalk length as a stable taxonomic criterion for *P. palmivora* isolates from cacao. Al-Hedaithy and Tsao (1979a, 1979b) showed the importance of using a uniform technique to determine caducity. They (1979b) reported that the age of the culture greatly affected the degree of caducity, because sporangia separate from the mycelium only when mature. Although the percentage of caducous sporangia was low in some species and often varied on different media, the length of the pedicel was always remarkably uniform. Some sporangia of *P. cinnamomi,* which produces noncaducous sporangia, often become detached accidentally, but the length of the stalk of these sporangia was always variable. Caducous species could be divided into those with short pedicels (<5 μm), e.g. *P. cactorum, P. infestans,* and *P. palmivora;* those with pedicels of intermediate length

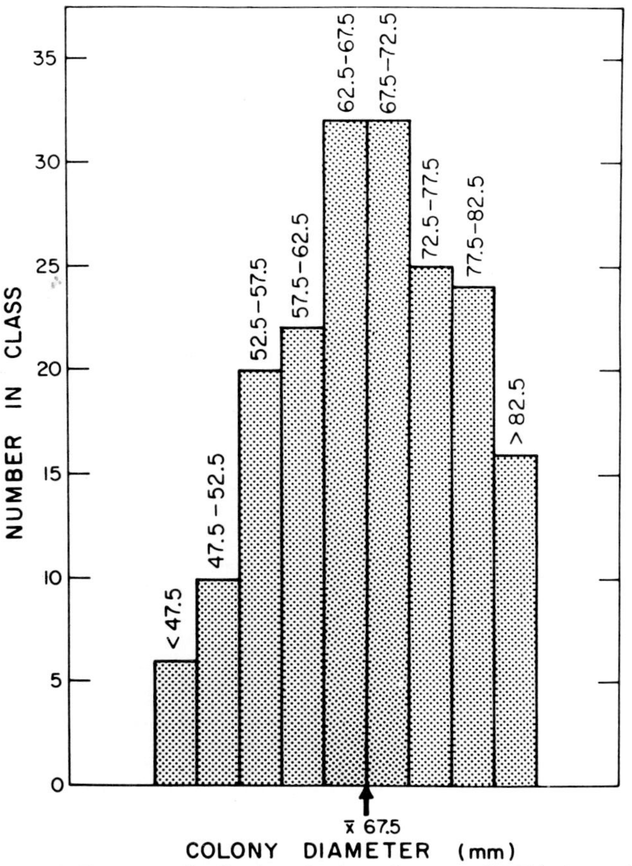


**Fig. 2.** Frequency distribution of colony diameters of 187 isolates of *Phytophthora cinnamomi* grown on potato-dextrose agar for 4 days at 25° C. (Reprinted, by permission, from Zentmyer et al, 1976)

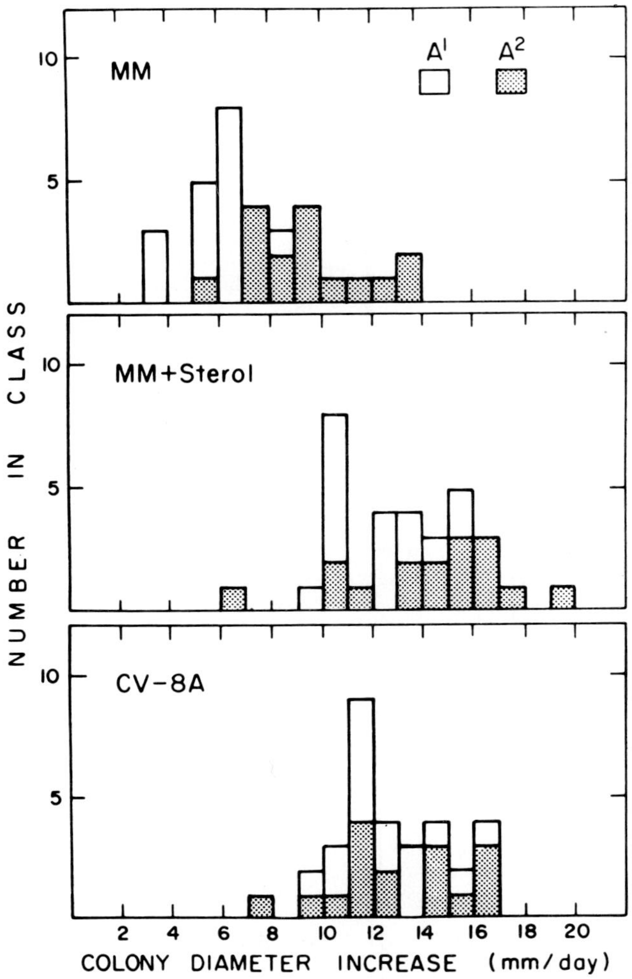


**Fig. 3.** Frequency distributions of growth rates of 16 A1 and 16 A2 isolates of *Phytophthora cinnamomi* grown on minimal medium (MM), MM + β-sitosterol, and clear V-8 agar (CV-8A) at 25° C. (Reprinted, by permission, from Zentmyer et al, 1976)

(5–20 μm), e.g., *P. botryosa, P. colocasiae,* and *P. megakarya*; and those with long pedicels (>20 μm), e.g., *P. hibernalis* and *P. palmivora* "MF 4" (Fig. 5). To obtain dependable data, uniform shaking treatment and uniform media were necessary to draw the proper conclusions. Al-Hedaithy and Tsao (1979a, 1979b) concluded that pedicel length was a highly stable diagnostic character, but that an isolate should be considered caducous only if a certain percentage of sporangia were detached at maturity and only if the pedicel lengths were uniform. Although the percentage of caducous sporangia varied with age, temperature, and medium, the length of the pedicel was apparently independent of these environmental factors.

### Physiologic Variability

Temperature and nutrition are extremely important factors that affect size of various structures and mycelial growth. Zentmyer et al (1979) showed that oospores from pairing an A1 and an A2 isolate of *P. cinnamomi* from avocado and from the A2 isolate incubated in avocado root extract formed at temperatures between 12 and 30° C, with optimal production at 15–2_° C for the isolate from the paired cultures and at 18–24° C for the A2 homothallic isolate produced in root extract. Although temperature did not markedly affect size of oogonia or oospores, nutrients provided by different media did. The oospores from root extract and those from the cross on a synthetic medium (Ribeiro et al, 1975) were consistently larger than those from the cross on carrot agar, cornmeal agar, clear V-8 juice agar, or clear V-8 juice broth. This might have been due to the formation of more oospores on the natural media than on the synthetic medium. Thus, size of organs for

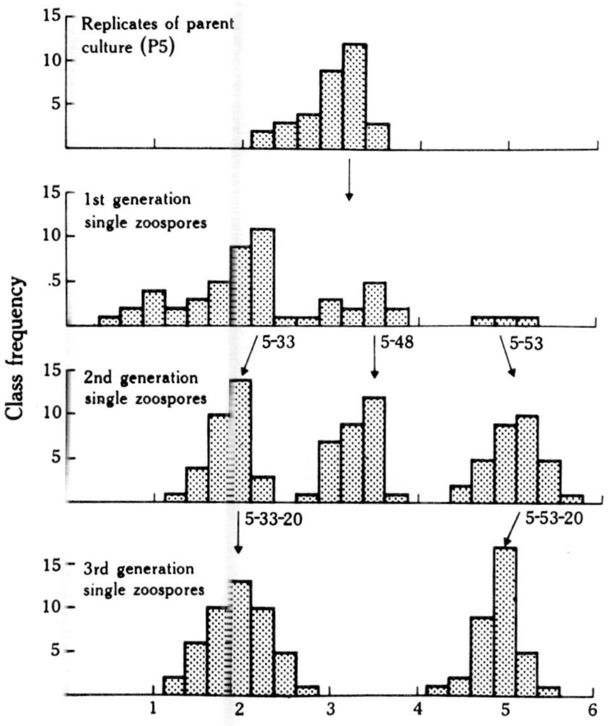


**Fig. 4.** Growth rate distributions among successive generations of single-zoospore progenies of *Phytophthora cinnamomi* field isolate P5. (Numbers below certain bars indicate subisolates from P5 that were single-zoospored and retested.) (Reprinted, by permission, from Shepherd and Pratt, 1974)

taxonomic purposes must be viewed with caution. Also, Zentmyer et al (1979) noted that the size of sex organs in the Waterhouse key (1963) was based on pairing *P. cinnamomi* with *P. cryptogea* on cornmeal and oatmeal agars. Cornmeal agar supported production of the lowest number of gametangia and the smallest percentage of normal oospores. A point raised in the discussion is apt: "For taxonomic purposes, media should be used which support normal maturation and germination of oospores and facilitate observations. . . . A single set of conditions cannot suffice for all species; the conditions used in each case must be described in detail if the characteristics of sex organs are used in species identification" (Zentmyer et al, 1979). Shepherd and Pratt (1973) reported that an ecotype of *P. drechsleri* from northern Australia had an upper temperature limit of 36–37.5° C, whereas the limit for the southern ecotype was 33–36° C.

Use of gel electrophoresis (Boccas, 1981; Boccas and Zentmyer, 1976) and serology (Halsall, 1976) to differentiate protein patterns of species and isolates of *Phytophthora* will be cited later in this chapter. A more refined method, isoelectric focusing, is able to differentiate biotypes within species (L. Erselius, *personal communication*). These tools are of great value for the consideration of relationships between similar species. Gallegly (Chapter 12) discusses these approaches more fully.

Pigment formation in media has significance in taxonomy of some groups of actinomycetes but appears to be too variable to be used for identification of *Phytophthora* spp. Pigment production was first used to differentiate variants within progeny of crosses of A1 and A2 isolates of *P. drechsleri* (Galindo and Zentmyer, 1964). In a synthetic medium containing tyrosine, Timmer et al (1970b) reported variation in production of a melanin pigment among several single-oospore cultures of *P. capsici.* A cross of two nonpigment ($Br^-$) producers, J-47 × P-505-6, resulted in some single-oospore offspring with $Br^{3+}$ (maximum brown pigment). $Br^-$ single-oospore offspring were recovered from a cross of K-10 × K-19, both of which were $Br^{2+}$ producers. Pigment was produced by some species but not by others (Timmer et al, 1970b). Shepherd (1976) found, as Timmer et al (1970b) had, that *P. cinnamomi* produced little or no pigment on a medium containing tyrosine, but that pigment production of A1 compatibility types of *P. parasitica* was double that of the A2 isolates. Shepherd concluded that pigment production might be a poor taxonomic criterion but might be useful as a primary screening technique.

### Pathogenic Variability

Variation in pathogenicity among isolates within a species has long been recognized. Loss of virulence with continued culturing is not an uncommon phenomenon (Caten, 1971; Erwin, 1966; Jeffrey et al, 1962). Recognition of this type of variation is of paramount importance to plant pathologists and to plant breeders. Either the type of pathogenic race or the degree of aggressiveness may vary. Aggressiveness is a term applied to the pathogen to account for different degrees of ability of the pathogen to parasitize the host. It has a quantitative connotation and is considered to be the counterpart of horizontal resistance in the host (Robinson, 1969).

Vanderplank (1963) coined the term "vertical pathogenicity," which is based on a bar graph that showed that certain populations of *P. infestans* were pathogenic to one or more potato host cultivars but not to cultivars with "vertical resistance" usually inherited as single R-genes. Robinson (1969) called these "vertical pathodemes" and the population of the pathogen "vertical pathotypes." However, the term "race," which Robinson would avoid, is firmly entrenched in the literature as physiologic race or pathogenic race, meaning a population that has differential pathogenicity to different host cultivars. Vanderplank's (1963) term "horizontal resistance" (also referred to by different authors as general resistance, multigenic resistance, polygenic resistance, and field resistance) depicts a cultivar that does not show a differential cultivar × race interaction. Horizontal resistance is considered to be effective against all races. The importance of horizontal resistance is that it reduces the rate of disease progress and is related to various components of the pathogenic process (MacKenzie et al, Chapter 23; Umaerus et al, Chapter 24).

Epidemiologically, vertical resistance implies only limited infection of the host and little or no reproduction of the pathogen. Vertical resistance acts on the epidemic by restricting the production of the initial inoculum (genetic sanitation; MacKenzie et al, Chapter 23), whereas horizontal resistance reduces the rate of disease by slowing the increase in the reproduction of the pathogen and the size of lesions. The net effect of horizontal resistance is a reduction of the rate of disease increase (MacKenzie, 1980) (see also MacKenzie et al, Chapter 23).

At least six ways by which variation occurs outside the sexual stage have been postulated: 1) random mutations followed by selection (Gallegly and Eichenmuller, 1959; Graham and Hodgson, 1965); 2) parasexuality (Leach and Rich, 1969; Stephenson et al, 1974; Wilde, 1961); 3) somatic recombination (Denward, 1967, 1970; Malcolmson, 1970); 4) cytoplasmic factors (Caten, 1971; Caten and Jinks, 1968); 5) heterokaryosis (Graham, 1954; Long and Keen, 1977a; Schultz, 1962); and 6) physiologic adaptation or "adaptive parasitism" (Graham and Hodgson, 1965; Mills, 1940; Mills and Peterson, 1952). It does not seem possible with the evidence at hand to categorically state which of these mechanisms is most important.

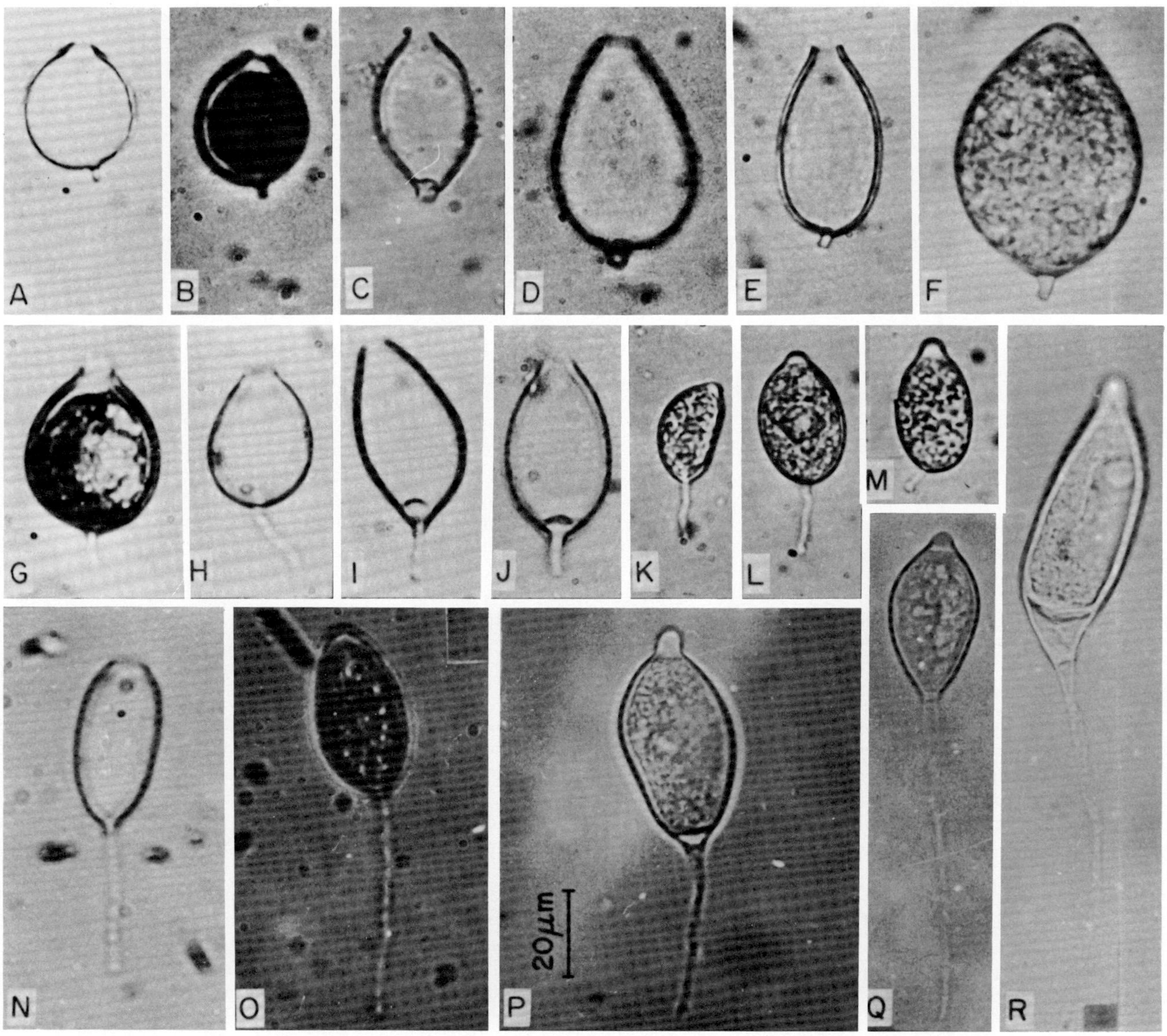


**Fig. 5.** Caducous sporangia of *Phytophthora cactorum* (A and B); *P. infestans* (C); *P. palmivora* sensu stricto, or morphological form (MF) 1 (D–F); "*P. palmivora*" MF 3 from cocoa (G and H); *P. colocasiae* (I and J); *P. botryosa* (K–M); *P. hibernalis* (N and O); and "*P. palmivora*" MF 4 from black pepper (P–R). The pedicel lengths were considered to be short (<5 μm) (A–F); intermediate (5–20 μm) (G–M); or long (>20 μm) (N–R). All figures share the same scale bar, which represents 20 μm. (Reprinted, by permission, from Al-Hedaithy and Tsao, 1979b).

***P. infestans***

Pathogenicity of *P. infestans* was reviewed by Gallegly (1968) and is discussed by Umaerus et al (Chapter 24). Races of *P. infestans* were first noted by Schick (1932) some 7 yr after the introduction of resistant hybrid potato cultivars from the cross of the wild potato, *Solanum demissum,* with *S. tuberosum.* Later, new races also developed in England, Scotland, and the United States. R-genes eventually became of little agricultural use because of the advent of new races.

In 1953, four races of *P. infestans* using potato differentials with genes R1, R2, R3, and R4 were differentiated from the wild potato, *S. demissum,* and designated races *1, 2, 3,* and *4* (Black et al, 1953). Within some isolates of *P. infestans,* vertical pathogenicity to cultivars with as many as four R-genes for resistance was found, and these were labeled as the complex race *1,2,3,4* (Black et al, 1953). Later, Malcolmson and Black (1966) reported three new major genes for hypersensitivity from *S. demissum,* designated R7, R8, and R9. By inoculating potato differential cultivars, Malcolmson and Black (1966) identified "nine new races, each capable of overcoming the hypersensitivity induced by at least one of the new genes." All originated from race *1,2,3,4.*

Malcolmson (1970) posed the question of how new and often complex races originated from the functionally asexual *P. infestans* (in the United Kingdom). Besides mutation, another system regulating variability was postulated. She showed "that mixtures of races with different specificities, say *a* and *b,* grown together on a susceptible host yield new races pathogenic to $R_aR_b$ hosts which are themselves immune to the original *a* and *b* races." She found that by mixing the complex race *1,2,3,7* with race *4,* the complex race *1,2,3,4,7* was obtained, and that by mixing the complex races *3,4,10* and *1,2,3,7,* the complex race *1,2,3,4,7,10* was obtained. She concluded, "It is therefore established that mixtures of races grown together on a susceptible host combine genetically to produce new races that have the sum of the parental specificities. The mechanism is obscure but, because the zoospores of *P. infestans* are uninucleate, nuclear fusion is implied." She termed this phenomenon "vegetative hybridity."

These and other examples became typical of the pathological variability within *P. infestans* and were once considered to be solely the result of selection pressure on the field population caused by the introduction of potato cultivars with single R-gene vertical resistance. The widespread prevalence of race *4* even in the absence of an R4 genotype has been puzzling. Gallegly and Eichenmuller (1959) showed that race *4* could arise from every race from which they made single-zoospore or single-sporangium isolations. They postulated that the occurrence of the race *4* character was due to mutation. Gallegly (1968) cited other examples of this phenomenon.

Malcolmson (1969) discounted the contention that new races arose only in response to the selective effect of host genotypes containing R-genes for vertical resistance:

> This suggested influence of breeders' material is not borne out by the present studies, for, of the 70 races recorded, 16 occurred among clones used for breeding purposes but only seven were confined to such material. Moreover, race *4* has long been regarded as the common race even though none of the popular commercial varieties carries the gene $R_4$; further, it is noteworthy that races complementing the newer R genes ($R_5$, $R_6$, $R_7$, $R_8$, $R_9$, $R_{10}$, and $R_{11}$) are widely distributed (Tables 1 and 2; Malcolmson & Black, 1966) though these genes are not present in common commercial varieties.
>
> That the presence of R genes is unnecessary for the increase in complex races is clearly illustrated by the races obtained from Andigena Group clones at Rosewarne. Only extremely complex races (indeed, the most complex races isolated) were obtained from these clones, which have no R genes (Simmonds & Malcolmson, 1967) and are grown in isolation from clones with R genes, yet from neighbouring material of diploid *S. tuberosum* the comparatively simple race *4,7,10,11* was isolated.

Shattock et al (1977) agreed with Malcolmson (1969). They pointed out that "the increase in the frequency of the specific virulence characters for *1, 2,* and *3* (races) in populations of *P. infestans* appears to be the result of host selection following the introduction of cultivars with R genes like Pentland Dell R1R2R3 and Maris Peer R1R2." However, they noted that R-genes 5–11 have never been incorporated into commercial potato cultivars and that the frequent isolation of races with virulence for cultivars with these genes cannot be explained by host selection:

> The determinants of specific virulence whilst still unidentified behave as genes which recombine giving all combinations possible and with minor exceptions, recombinants are not at selective disadvantage. The high frequencies of widely virulent races revealed by these two surveys [theirs and Malcolmson, 1969] therefore do not support the theory of selection against races with unnecessary specific virulence as proposed by Vanderplank (1968).

Another statement by Shattock et al (1977) presents a useful hypothesis:

> If specific virulence is determined by nuclear characters, then presumably nuclear fusion occurs between the vegetative diploid nuclei after heterokaryon formation, allowing reassortment of race determinants and producing the wide variety of different host-specific phenotypes evident in the population of *P. infestans.*

Although races have been described within only a few species, differences in degree of horizontal pathogenicity, which include aggressiveness, rate of sporulation, and rates of growth, are expressed within *P. infestans* and in all species of *Phytophthora.* Isolates of *P. infestans* may also increase in horizontal pathogenicity (aggressiveness) by continuous passage through potato tuber tissue (Caten, 1971; Leach and Rich, 1969). The mechanism of these changes is not known with certainty, but several explanations, including cytoplasmic inheritance, have been postulated (Caten, 1970, 1971, 1974).

Jeffrey et al (1962) isolated strains of *P. infestans* of the same pathogenic race and demonstrated differences between the rate of growth on and depth of penetration of potato tubers. Speed of sporulation also differed. It is interesting that these differences were modified or lost after prolonged maintenance on laboratory media. Jeffrey et al favored the explanation that this change was based on selection of nuclear differences.

These papers reveal the apparent genetic diversity in the genome of *P. infestans* and, if we may extrapolate, probably of all other species as well. Caten (1971) suggested that these increases in aggressiveness may be explained by changes in selection pressures acting on the variable cytoplasmic system of the fungus. Caten (1974) reported an increase in growth rate of isolates of *P. infestans* by passage through potato cultivars with horizontal resistance, thus indicating that even horizontal resistance could be susceptible to adaptation by *P. infestans* isolates. Although this danger is potentially important, horizontal resistance has held up much better agriculturally than vertical resistance (see Umaerus et al, Chapter 24). According to Vanderplank (1971), the field evidence supports the supposition that horizontal resistance is stable.

Malcolmson (1976) devised a scoring system for horizontal resistance (she uses the term "field resistance") and found a reasonable degree of correlation between results in greenhouse experiments and those in the field. Studies of this type are important, because the main problem with manipulating horizontal resistance in a breeding system is the time-consuming evaluation technology. Vanderplank (1963) pointed out that horizontal resistance had been neglected because vertical resistance could be so much more easily manipulated in a breeding program.

The net result of the recognition of such a wide range of new races in the field has been the de-emphasis by potato breeders of the use of R-genes for vertical resistance. Both breeders and epidemiologists are now pursuing the use of horizontal (general) resistance to a greater degree than before. For more detail, see Umaerus et al (Chapter 24) and MacKenzie et al (Chapter 23).

### *P. megasperma* f. sp. *glycinea*

Physiologic races are common in *P. megasperma* f. sp. *glycinea,* the pathogen that causes root rot of soybean, which was described as *P. megasperma* var. *sojae* by Hildebrand (1959) in Ontario. The name change to *P. megasperma* f. sp. *glycinea* was proposed by Kuan and Erwin (1980) because the forma specialis more correctly designated its host-specific pathogenicity to soybean. The first report of the disease by Suhovecky and Schmitthenner (1955) showed that the cultivars Blackhawk and Monroe were resistant. Kaufmann and Gerdemann (1958) described the disease, named the causal agent *P. sojae,* and described the stem puncture inoculation technique used to delineate resistant cultivars. Hildebrand (1959) not only noted different levels of virulence among isolates but also showed that the variety Mukden was resistant to all isolates. In 1965, Morgan and Hartwig (1965) differentiated race 1 from race 2 by using the soybean cultivar D60-9674. Since then, a new race has been identified in the field each time a new vertical gene for resistance has been introduced in soybean. There were nine vertical pathotypes in 1977 (Laviolette and Athow, 1977), 16 in 1980, and 20 in 1982 (Keeling, 1980, 1982).

Athow et al (1980) reported that the cultivar PI 86050 possesses a gene, $Rps_{1c}$, that conveys resistance to races 1, 2, 3, 6, 7, 8, and 9. Their group at Purdue University has produced a series of definitive papers on physiologic races of *P. megasperma* f. sp. *glycinea* (Athow et al, 1974, 1979, 1980; Laviolette and Athow, 1977, 1981; Laviolette et al, 1979; Mueller et al, 1978; Wilcox, 1976) that are too detailed to cover here. Lockwood and Cohen (1978) developed a flooded soil assay using unwounded seedlings of differential soybean cultivars to type races. They found that a natural soil assay was quicker than and gave results similar to isolation and subsequent inoculation of plants with zoospores.

Inoculum concentration can be an important variable in typing races. Eye et al (1978) reported that inoculation of soybean seedlings with 500 zoospores per seedling did not induce disease, inoculation with $10^4$ zoospores induced disease on susceptible but not on resistant plants, and inoculation with $10^5$ zoospores induced disease on some resistant as well as susceptible plants.

Hilty and Schmitthenner (1962) noted that single-zoospore isolates ranged in pathogenicity from nonpathogenic to virulent. They found one single-spore isolate that caused disease on the soybean cultivar that was resistant to the parental culture. There was no relationship between cultural characteristics and pathogenicity.

Because *P. megasperma* f. sp. *glycinea* is soilborne, the rise of new races with the introduction of resistant cultivars seems to be too rapid to be accounted for by chance mutation, dispersal, and subsequent natural reselection by the resistant soybean cultivar. The work of Hobe (1981) (courtesy A. F. Schmitthenner) suggests the explanation that a wide range of races may exist in the natural wild-type population that may be expressed when a resistant soybean cultivar is introduced. Isolations of single-zoospore isolates from soil, baited with leaf disks of a susceptible cultivar, yielded a wide range of races in soybean fields. Of the 58 single-zoospore isolates obtained from soil and plant cultures, 59% were a different race than the original mass culture. Twelve new races were isolated, and 10 additional races were identified when two more differential cultivars of soybean, Kingwa and PI 68708, were inoculated. The population of races from soil was more diverse than that from diseased plants.

Agriculturally, vertical resistance is still a viable control method for Phytophthora root and stem rot of soybean. This may seem, in light of the experience previously described for late blight of potato, to be an anomaly. However, new races of *P. infestans* might be more limiting than new races of *P. megasperma* f. sp. *glycinea* because the latter is a soilborne pathogen and would not be likely to spread as rapidly. As in *P. infestans,* it seems highly likely that genetic factors for vertical pathogenicity could reside in the wild-type population but not be expressed until the host genotype is established as a dominant cultivar.

### Other Narrow-Host-Range *Phytophthora* spp.

Other *Phytophthora*-host systems in which vertical pathotypes occur include *P. parasitica* var. *nicotianae* sensu Tucker on tobacco (Apple, 1957; Hendrix and Flowers, 1967; Maronek and Hendrix, 1978; McIntyre and Taylor, 1978), *P. drechsleri* f. sp. *cajani* on pigeon pea (Kannaiyan et al, 1980), *P. vignae* on cowpea (Partridge and Keen, 1976; Purss, 1972), and *P. fragariae* on strawberry (Converse, 1967; Hickman, 1962; Maas, 1972, 1973; Montgomerie and Kennedy, 1975; Pepin and Daubeny, 1964). Polach and Webster (1972) defined 14 pathogenic strains of *P. capsici* based on pathogenicity to different hosts. In this research, based on 391 single-oospore progenies from crosses of A1 and A2 types, recombination between pathogenicity and mating type occurred. Pathogenicity to pepper was controlled by at least two genes and to other hosts by a "single gene or gene system." This work also confirmed that pathogenicity to pepper was an inadequate taxonomic criterion (Tucker, 1931), as previously observed by Satour and Butler (1968).

### *P. cinnamomi*

*P. cinnamomi* is distributed worldwide and has a wide host range (Zentmyer, 1980). Because *P. cinnamomi* has such a wide host range and is the cause of economically destructive epidemics, such as avocado root rot in the United States, South Africa, Australia, and other parts of the subtropics and the jarrah "dieback" in Western Australia (probably the most devastating of all *Phytophthora* epidemics except late blight of potato) (see Shea and Broadbent, Chapter 26, and Weste, Chapter 18), much research has been done on this species.

There is no evidence for physiologic races of *P. cinnamomi*, but several reports indicate varying degrees of virulence of isolates on different hosts (Crandall et al, 1945; Manning and Crossan, 1966; Torgeson, 1954). As a striking example of isolates different in pathogenicity, an A1 isolate from *Camellia japonica* caused a severe reaction on both avocado and camellia, whereas an A2 isolate from avocado was highly pathogenic to avocado but not to camellia. In nature, all isolates from avocado, with one exception, have been A2 (Zentmyer and Guillemet, 1981). This indicates that the avocado root disease, which markedly affects production, could be introduced as A1 types on diseased ornamental camellia plants.

**Resistance to Fungicides**

Since the advent of systemic benzimidazole fungicides such as benomyl (not effective against *Phytophthora*), the idea of therapeutically controlling fungal pathogens after infection has intrigued plant pathologists. Because these fungicides had a specific mode of action (Davidse and Flach, 1977), some populations of the target pathogens became resistant, either by selection of mutants or because resistant individuals existed in the wild population.

Only recently, systemic fungicides active against *Phytophthora* spp. have been discovered. Metalaxyl was the first of several highly selective and effective fungicides (acylalanine family) toxic to *Phytophthora* spp. to be introduced (Urech et al, 1977) (see Schwinn, Chapter 25). Although these fungicides have been highly successful for control of foliar pathogens within the Oomycetes, resistance to the acylalanine fungicides within populations of *P. infestans* has already been found (1981) in potato fields in Ireland (M. D. Coffey, *personal communication*) and in The Netherlands (Davidse et al, 1981). Davidse et al reported that a severe epidemic occurred in The Netherlands in 1980, a year with weather conditions favorable for blight, in fields sprayed with metalaxyl. When isolates from the field were tested on potato leaf disks, they were resistant to metalaxyl and also to the related acylalanine compounds furalaxyl and milfuram. Because resistant isolates were from two different races, it was suggested that resistance arose in the field from at least two different sources. According to Davidse et al (1981), metalaxyl, after being registered in The Netherlands in 1979, was used in 1980 in 50% of the fields. They suggested several factors that could be associated with this rapid development of resistance: 1) evidence that resistant mutants of *P. megasperma* f. sp. *medicaginis* could be developed in the laboratory, which suggested that such mutants could have arisen from a population of *P. infestans* in the field (Davidse, 1981); 2) weather conditions in 1980 that were extremely favorable for late blight; and 3) the extensive use of metalaxyl in 50% of the fields.

Before the observation of resistance of *P. infestans* to metalaxyl, Davidse (1981) demonstrated resistance of some isolates of *P. megasperma* f. sp. *medicaginis*, a root rot pathogen of alfalfa (Kuan and Erwin, 1980) in laboratory experiments. Resistance varied from a low level for some of those single-zoospore isolates from the untreated wild-type culture ($ED_{50}$ value for growth on metalaxyl-amended V-8 juice agar was 5 $\mu$g/ml metalaxyl) to a higher level ($ED_{50} > 100$ $\mu$g/ml) from the wild-type culture treated with the mutagens, ultraviolet light, or N-methyl-N′-nitro-N-nitrosoguanadine. The degree of control of root rot caused by a resistant isolate and by the nonresistant parent was studied by drenching metalaxyl in artificially infested soils. A concentration of 2.5 mg/L controlled the parent isolate, but 20 mg/L failed to control the resistant isolate. A recent paper by Hunger et al (1982) also described different degrees of natural resistance to metalaxyl by various isolates of *P. megasperma*.

These reports suggest that similar resistance may be found in other *Phytophthora* spp. The experimental system Davidse (1981) outlined should be a useful approach before release of new fungicides, not only to the agricultural industry that uses the fungicides, but also to the fungicide manufacturer in assessing risks before developmental costs become too high. If resistance is due to recessive genes, it would be masked in the wild type until selection pressure allowed the expression of resistance. The genetics of resistance should be determined.

The mode of action of the acylalanine fungicides has not yet been elucidated; however, it is likely to be much more specific than in the conventional protectants now in use. If resistance in *Phytophthora* spp. is controlled by a single-gene or two-gene system, resistance will be a problem in the future, especially where metalaxyl is used extensively. Schwinn (Chapter 25) explains how marketing metalaxyl with another fungicide—e.g., one of the dithiocarbamates, which have a broader mode of action—should reduce the threat of development of resistant populations.

## NOMENCLATURE PROBLEMS WITHIN AND AMONG SPECIES

Understanding the extent of normal variability could eventually lead to combining certain species that appear to be closely related morphologically and also to the creation of new species. A few species where such developments are occurring or where particular problems are seen will be discussed in this section.

### *P. cryptogea* and *P. drechsleri*

*P. cryptogea*, described by Pethybridge and Lafferty (1919), is characterized by production of amphigynous antheridia, ellipsoid nonpapillate sporangia, and hyphal swellings in clusters. *P. drechsleri*, described later by Tucker (1931), also produces amphigynous antheridia but has somewhat more elongate nonpapillate sporangia and hyphal swellings in clusters. Tucker (1931) differentiated this fungus from *P. cryptogea* because of the more elongate nonpapillate sporangia and growth at 35°C; *P. cryptogea* did not grow above 30°C. However, when large numbers of isolates of both species have been studied, the temperature threshold difference does not always correlate with sporangial shapes typical of either *P. drechsleri* or *P. cryptogea*. Elongated sporangia and ellipsoid sporangia often occur in the same culture.

Bumbieris (1974) suggested merging both species under *P. cryptogea*, the older name. He compared seven *P. cryptogea* isolates with five *P. drechsleri* isolates, two of which were the mating strain for each species from the Commonwealth Mycological Institute. He concluded that each group could not be classified as a separate species. The width of hyphae, growth rate on unamended cornmeal agar and in lima bean agar amended with malachite green (0.1 $\mu$g/ml), and the appearance and size of the nonpapillate sporangia were similar. Growth at 35°C was less than 1 mm/24 hr for both the *P. cryptogea* isolates and *P. drechsleri* isolates with one exception, CMI 40499 (Bumbieris' P34), which grew at a rate of 3.3 mm/24 hr. Halsall (1976) reported that isolates of *P. cryptogea* and *P. drechsleri* were serologically similar and agreed with Bumbieris that they should be combined. Gerrettson-Cornell (1979) considered both species to be synonymous.

Flowers et al (1973) compared a new isolate from bean with isolates considered to be typical of both *P. drechsleri* and *P. cryptogea*. They also noted the great similarities between the two species and found that growth at 35°C was not a limiting criterion, and they named the bean pathogen *P. cryptogea*.

Duniway (1976) reclassified an isolate from safflower from *P. drechsleri* to *P. cryptogea*. Klisiewicz (1977), working with several isolates pathogenic to safflower, recognized that "morphology of cultures, hyphae, and sporangia of isolates 266 and 473 was judged as similar, more so to that described for *P. drechsleri* . . . than to *P. cryptogea*." Five other isolates were deemed to be closer to *P. cryptogea*. His Figure 1 shows examples of the elongate ellipsoid sporangia, with a swollen base in the sporangiophore, that Tucker (1931) considered to be typical of *P. drechsleri*.

The evidence indicates that many isolates considered to be *P. drechsleri* are morphologically similar to the earlier described *P. cryptogea*. The question arises whether the physiologic character of growth at 35°C is sufficiently important to differentiate *P. drechsleri* from *P. cryptogea*. Although the elongated sporangial shape of *P. drechsleri* differs from the more ellipsoid shape of *P. cryptogea* as shown by Klisiewicz (1977), whether this difference is consistent enough to be a useful criterion seems to be a valid question. An extensive study of isolates collected on a world basis would likely confirm that *P. drechsleri* can be merged with *P. cryptogea*.

### *P. megasperma*

*P. megasperma* includes a broad range of isolates that vary widely in oogonial size. Many of the recognized populations differ culturally. All groups share the characteristics of nonpapillate, obpyriform to ellipsoid, proliferating sporangia; globose oogonia more than 30 μm in diameter; and production of both paragynous (predominant) and amphigynous antheridia. Although at least three groups within the species appear to be specifically pathogenic to single hosts, some of the other groups have wide host ranges (Hansen and Hamm, 1982). Brasier (Chapter 27) also discusses *P. megasperma* as a problem species.

Drechsler (1931) originally described *P. megasperma*, isolated from hollyhock (*Althaea rosea*), because it produced large oogonia averaging 47 μm in diameter. Later, Tompkins et al (1936) amended Tucker's key (1931) to include an isolate from cauliflower with oogonia smaller than 47 μm but larger than 30 μm in diameter. Hildebrand (1959) placed the host-specific soybean root rot pathogen (oogonia about 36 μm in diameter) within the species *P. megasperma*, citing the broader concept of Tompkins et al (1936). However, he differentiated the host specificity of the soybean isolates from other *P. megasperma* isolates by incorrectly using the botanical "variety." He changed *P. sojae* (Kaufmann and Gerdemann, 1958) to *P. megasperma* var. *sojae*. Waterhouse (1963) did not use the botanical term "variety" to designate specific pathogenicity, but, recognizing that both large and small oogonial isolates occurred within *P. megasperma*, she retained Hildebrand's (1959) variety name *sojae*, which included all of the small oogonial isolates (<40 μm in diameter) from soybean, alfalfa, cauliflower, sugarcane, and other hosts, under the name *P. megasperma* var. *sojae* and classified the isolates with oogonia larger than 40 μm under the new variety, *P. megasperma* var. *megasperma*. Although the rationale was correct, this classification has been confusing, because it lumped the host-specific soybean isolates with other isolates. For this reason, Erwin (1965), reclassifying the alfalfa root pathogen; Nagai et al (1978), describing a new disease of rose; and Mircetich and Matheron (1976), describing a root pathogen of cherry trees, did not use the name "*sojae*" but referred to these pathogens as *P. megasperma*.

Kuan and Erwin (1980) compared the oogonial sizes of several isolates of *P. megasperma* from alfalfa, soybean, cherry, walnut, and other hosts. When these isolates were grown on V-8 juice agar, the average diameters of oogonia for all isolates ranged continuously from about 30 to 55 μm, with no gap in the continuous curve by which the morphological var. *megasperma* group could be separated from the var. *sojae* group (Fig. 6) (Kuan and Erwin, 1980). The variability in size of oogonia appeared to be too great to be of diagnostic value.

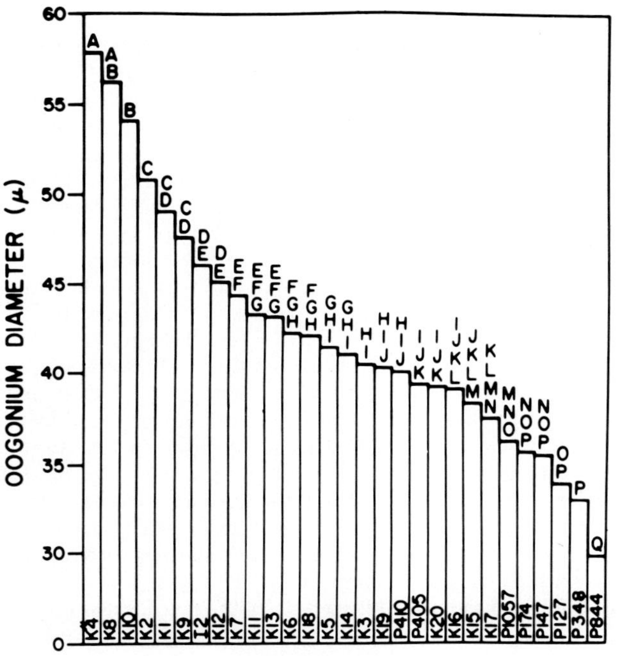


**Fig. 6.** Oogonia diameters of 29 isolates of *Phytophthora megasperma*. Each isolate was grown on V-8 juice agar at 24°C for 1 mo. Values represent the means of 30 measurements of each isolate. Mean values labeled with the same letter are not significantly different ($P = 0.01$), according to Duncan's multiple range test. (Reprinted, by permission, from Kuan and Erwin, 1980)

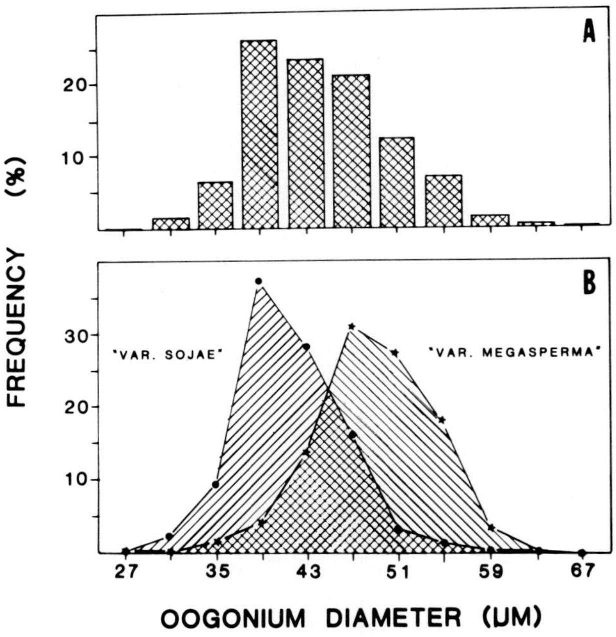


**Fig. 7. A,** Frequency histogram of oogonium diameters from 36 isolates of *Phytophthora megasperma*. **B,** Frequency histogram of oogonium diameters from 12 *P. megasperma* isolates generally considered to be var. *megasperma* (oogonia greater than 45 μm) and 24 isolates considered to be var. *sojae* (oogonia less than 45 μm). Based on measurements of 10 oogonia from each of eight single-zoospore isolates and the parent isolate (90 oogonium measurements) of each of 36 isolate sets. (Reprinted, by permission, from Hamm and Hansen, 1982)

Kuan and Erwin questioned whether oogonial size was a valid criterion for variety status, since there was no clear break between "large" and "small" oogonia.

The study of variability of isolates within *P. megasperma* has been greatly expanded by Hamm and Hansen (1982), who made oogonial and other morphological measurements of eight single-zoospore isolates from each of 36 parental isolates (12 considered to be var. *megasperma* and 24 var. *sojae*) from 14 different hosts. Oogonial sizes of isolates from both varieties overlapped (Fig. 7) (Hamm and Hansen, 1982). Hansen and Hamm (1982) indicated that seven isolates had oogonia with diameter less than 40 $\mu$m, 20 greater than 45 $\mu$m, and 25 were intermediate. Both reports supported the contention by Kuan and Erwin (1980) that size of oogonia was not a dependable characteristic to separate the two varieties.

Hildebrand (1959) based the var. *sojae* concept on the specific pathogenicity of soybean isolates to soybean and not on size of the oogonia. Kuan and Erwin (1980) confirmed the specific pathogenicity of soybean and alfalfa isolates (Hildebrand, 1959) and proposed the use of the forma specialis to recognize the host specificity of these two groups of isolates. The alfalfa pathogen was designated as *P. megasperma* f. sp. *medicaginis* and the soybean pathogen as *P. megasperma* f. sp. *glycinea*. The name "var. *sojae*" was not retained because it implied specific pathogenicity but was based on oogonial size (Newhook et al, 1978; Waterhouse, 1963).

Host specificity of the alfalfa isolates and the soybean isolates has been confirmed by several reports (Hamm and Hansen, 1981; Hildebrand, 1959; Kaufmann and Gerdemann, 1958; Kuan and Erwin 1980; Morgan and Hartwig, 1965; Pratt, 1981). There seems to be little question that no cross-infectivity between these two pathogens occurs in the field, although in greenhouse tests, a small amount of root rot does occur following artificial inoculation in sterile soil (Hamm and Hansen, 1981). However, the alfalfa isolates may have certain minor hosts, because they have been shown to be capable of causing root disease of chickpea (*Cicer arietinum*), a legume, in greenhouse tests (Bhalwa, 1962; Erwin, 1965) and in recent field isolations and pathogenicity tests by Irwin and Dale (1982) in Australia. Alfalfa isolates also caused root rot on *Sesbania* sp., another legume, in greenhouse tests (Erwin, 1965). *P. megasperma* f. sp. *glycinea* isolates from lupine were pathogenic to both soybean and lupine (Jones and Johnson, 1969).

Pratt (1981) reported that a *Phytophthora* isolate from arrowleaf clover (*Trifolium vesiculosum* Savi), which caused a root rot and stand decline, was specifically pathogenic to arrowleaf clover. He confirmed the specific pathogenicity of the arrowleaf clover isolates, soybean isolates, and alfalfa isolates to their respective hosts and named the arrowleaf clover pathogen *P. megasperma* f. sp. *trifolii*.

Brasier (Chapter 27) cautions against too wide an application of forma specialis to designate host specificity before the genetic basis of host specificity is understood. He reinterpreted the numerical root disease data of Hamm and Hansen (1981) to indicate that the system breaks down. Their data indicated that isolates from hosts other than alfalfa caused a small degree of disease on alfalfa in pot tests. Portraying the severity of disease by assigning a number, though useful for comparative research, sacrifices information on the quality of the symptom. Brasier (Chapter 27, Table 2) reinterprets Hamm and Hansen (1981) by changing numerical data to "considerable disease," "moderate disease," and "little or no disease." The values he attributes to moderate disease were in most cases less than two on a zero to four scale. However, Hamm and Hansen (1981) did not indicate whether or not the lesions were on taproots or lateral roots, and they did not consider that these data indicated cross-pathogenicity or invalidated the forma specialis concept. Many fungi will cause perceptible disease on lateral roots in pot tests. However, to prove the actual pathogenicity of a fungus, the quality of the symptom related to what actually occurs in the field is the important factor. Pratt (1981), who proposed *P. megasperma* f. sp. *trifolii* for isolates from arrowleaf clover, stated (emphasis added):

> In each cultivar, *severe disease symptoms* consistently developed only in plants inoculated with isolates originally obtained from that same host. . . . In most plants of arrowleaf clover and susceptible alfalfa inoculated with virulent isolates of *P. megasperma* from the same hosts, roots were *either completely rotted 4 wk after inoculation, or most lateral roots were rotted and large progressive lesions* were present on taproots. In contrast, plants inoculated with isolates from the opposite hosts, or with isolates from soybean, usually had no symptoms or *few lateral roots rotted* and only *small, nonprogressive lesions* on taproots.

However, Brasier's (Chapter 27) caution should be carefully considered, because there is a possibility that a population of isolates could be assigned to a forma specialis based on too few pathogenicity tests and on too few isolates and potential hosts.

Hamm and Hansen (1981) reported that certain isolates of *P. megasperma* from rotted Douglas-fir tree roots were pathogenic following stem-slash inoculations to all cultivars of soybean, even those resistant to certain host-specific races of *P. megasperma* f. sp. *glycinea*. However, a later paper by Hamm and Hansen (1982) reported that the Douglas-fir isolates, though pathogenic by the stem-slash method, were only weakly or not pathogenic in soil to roots of soybean seedlings. The Douglas-fir isolates may be capable of inducing a severe hypersensitive reaction in soybean that may not be related to host specificity. This possibility was alluded to by Hansen and Hamm (1982). The fact that the severe reaction to the Douglas-fir isolates occurred on all cultivars of soybean regardless of their genetic resistance to different races of *P. megasperma* f. sp. *glycinea* also indicates that the reaction was due to hypersensitivity. We have confirmed Hamm and Hansen's (1981) experiments, and the Douglas-fir isolates produced a much more severe symptom response than did *P. cinnamomi*, a nonpathogen of soybean. This phenomenon should be of interest for study of the nature of pathogenesis and hypersensitivity (A. Vaziri, N. T. Keen, and D. C. Erwin, *unpublished*).

Brasier (Chapter 27) also questions whether the evidence by Kuan and Erwin (1980) necessarily invalidates separation of var. *megasperma* and var. *sojae* based on oogonial size. Sansome and Brasier (1974) compared the chromosome number of representatives of some large oogonial (var. *megasperma*) and small oogonial (var. *sojae*) isolates. The two var. *sojae* isolates, both host-specific soybean pathogens, had a basic chromosome number of 10–15, somewhat larger than the 9 or 10 reported for *P. infestans* (Sansome and Brasier, 1973), *P. cactorum* and *P. erythroseptica* (Sansome, 1965, 1966), and *P. cinnamomi* (Brasier and Sansome, 1975). The chromosome number of the three var. *megasperma* isolates was "estimated to be between 22 and 30, the upper limit depending on the assumption that some of the configurations were multivalents." Sansome and Brasier concluded that these var. *sojae* isolates and the var. *megasperma* isolates were different enough in oogonial size, growth characteristics, and chromosome number to be differentiated at least at the varietal level. Although this was true for selected isolates, more cytologic evidence from a wider range of isolates is needed to conclude that each variety can be readily differentiated by numbers of chromosomes. The hypothetical question arises as to what size of oogonia is indicative of polyploidy? Hansen and Hamm (1982) pose the question, "It would be interesting to see whether this difference (chromosome numbers) correlates with spore size, host specificity, or perhaps with the host groups listed here."

Brasier and Sansome (1975) stated, "*P. megasperma* var. *sojae* may have a basic set of nine or ten chromosomes with one or two pairs of extra chromosomes partly homologous with the original set, a polysomic condition." Considering the unique host specificity of the soybean isolates (*P. megasperma* var. *sojae*), it would be interesting to determine whether this polysomic condition is correlated with pathogenicity. Long et al (1975) selfed isolates by producing colonies in successive generations from single germinated oospores and reported that isolate P174 (originally obtained from Hildebrand as highly aggressive) yielded about 40% nonaggressive or weakly aggressive progeny from single-oospore isolates. After being selfed five times, an $S_5$ isolate was more aggressive than the parental P174. Variability was not noted from oospore isolates in the $S_5$ generation. Thus, the question of whether polysomy is correlated with pathogenicity could be approached by observation of both nonaggressive and aggressive isolates of the soybean pathogen.

Much work remains to be done to solve these taxonomic difficulties. The recent work by Hamm and Hansen (1982), showing that single-zoospore isolates can vary from the parental isolates, provides a model for analysis of the potential range of variability within a taxon. The discovery by Sansome and Brasier (1974) that some large oogonial isolates of *P. megasperma* were polyploid should be expanded to determine the extent of this cytogenetic condition on a wide range of isolates. *P. megasperma* encompasses not only at least three groups of fungi with a high degree of host specificity, but also many other isolates with less host specificity. Among these isolates are differences in growth, oogonial size, and sporangial size (Hansen and Hamm, 1982) as well as differences in chromosomal number. I agree with the proposed revision of the description of *P. megasperma* based on a broad morphological concept that includes oogonial diameters of 25–55 $\mu$m (average 40.5 $\mu$m) (Hansen and Hamm, 1982). Arbitrary use of the epithet var. *sojae* for small oogonial isolates confuses the reader, who could assume that all var. *sojae* isolates are host-specific to soybean. For communication purposes, the use of the f. sp. concept designating alfalfa isolates as f. sp. *medicaginis*, the soybean isolates as f. sp. *glycinea* (Kuan and Erwin, 1980), and the clover isolates as f. sp. *trifolii* (Pratt, 1981) is readily understood. All other isolates should simply be designated as *P. megasperma*, and the name var. *sojae* should be discontinued.

### *P. erythroseptica*

*P. erythroseptica* was described by Pethybridge (1913) as the causal agent of pink rot of potato tubers. It produces proliferating nonpapillate sporangia and amphigynous antheridia and is classified in Waterhouse's key (1963) in group VI. In contrast, *P. megasperma* is classified in group V with nonpapillate sporangia but with "all, or the majority of the antheridia paragynous." Savage et al (1968) stated that the descriptions (Waterhouse, 1956) of these two species were very similar except for the type of antheridium produced. A *P. megasperma* isolate from Northern Ireland also causes pink rot of potato tubers (Cairns and Muskett, 1933). The oospore size of *P. erythroseptica* is similar to that of *P. megasperma* var. *sojae* sensu Waterhouse (1963). Savage et al (1968) noted that these similarities, and the work they reported on several isolates, "indicate a strong need for a thorough taxonomic study of these two species."

Recognizing that designation of *P. erythroseptica* is limited to those isolates that produce only amphigynous antheridia (Waterhouse, 1963), Pratt (1981) was concerned about how to classify certain fast-growing isolates from clover that produced mostly paragynous antheridia in agar but mostly amphigynous antheridia in broth. He provisionally identified them as *P. erythroseptica* but also cited the reports of variability of isolates within both *P. erythroseptica* and *P. megasperma* as evidence of need for more research to establish identification criteria for these species. Savage et al (1968) suggested that serologic studies might clarify this relationship. In summary, a large number of isolates should be reexamined.

### *P. nicotianae* var. *parasitica* and *P. nicotianae* var. *nicotianae* sensu Waterhouse (1963)

Waterhouse (1963) differentiated these varieties of *P. nicotianae* based on morphological criteria. Because *P. nicotianae* was described first by Breda de Haan on tobacco in 1896 (Waterhouse, 1963), Waterhouse judged that this name had priority over *P. parasitica*.

The use of *P. nicotianae*, which implies host specificity, for all isolates previously known as *P. parasitica* has confused the pathogenicity character on which Tucker established *P. parasitica* var. *nicotianae* (1931). Tobacco pathologists consider that var. *nicotianae* is based solely on pathogenicity (Johnson and Valleau, 1954). Using the morphological criteria (Waterhouse, 1963), all tobacco isolates (*P. parasitica* var. *nicotianae* sensu Tucker [1931]) are not necessarily synonymous with *P. nicotianae* var. *nicotianae* sensu Waterhouse (1963), and all nontobacco isolates (*P. parasitica* sensu Tucker [1931]) are not necessarily synonymous with *P. nicotianae* var. *parasitica* sensu Waterhouse (1963). This is often overlooked by many authors. Most tobacco pathologists continue to use *P. parasitica* var. *nicotianae* for the tobacco pathogen. It is unfortunate that Tucker (1931) did not use the forma specialis concept rather than variety, because variety should be based on small morphological differences and forma specialis on specific pathogenicity (Staflue et al, 1972). Gallegly (Chapter 12) discusses the anomaly of using "*nicotianae*," which implies host specificity, as the species name.

Morton and Dukes (1967) compared the black shank pathogen, *P. parasitica* var. *nicotianae*, with *P. parasitica* serologically and concluded "that the two varieties of *Ph. parasitica* are serologically very similar." Little other comparative research has been done on differentiation of the var. *nicotianae* from the var. *parasitica*, except for the abstract by Tsao and Sisemore (1978), who reported a comparative study of eight isolates from citrus, one from tomato, and two from tobacco:

> Results showed that all isolates possessed certain diagnostic features which fit either var. *nicotianae* or var. *parasitica* and exhibited a high degree of variability. Some isolates showing characteristics of var. *parasitica* on the basis of oospore size, sporangia shape, and chlamydospore wall thickness also produced characteristics (i.e., sporangium and chlamydospore size, formation of hyphal swellings and radiating hyphae) identifiable as var. *nicotianae*. Several isolates which had previously produced oospores averaging 26–27 $\mu$m (identifiable as var. *nicotianae*) all produced oospores averaging 19–20 $\mu$m (identifiable as var. *parasitica*) in our study. Medium composition also affected oospore size. It appears, therefore, that separation of the species into two varieties is questionable.

This group represents another problem area that needs to be redefined based on a comparative study of a wide range of isolates. There may be enough evidence in the literature to consider f. sp. *nicotianae* with host specificity to tobacco.

### *P. palmivora* and Similar Species with Papillate Sporangia

Since the original description of *P. palmivora* by Butler (1907) as *Pythium* and later as *P. palmivora* (1910), there have been numerous studies on the variability of this species and those closely resembling it. The fungus produces oospores with amphigynous antheridia, chlamydospores, and papillate sporangia. Butler's drawings (Waterhouse, 1956), while apparently accurate, are not very definitive, and the description likewise is not detailed enough to be of much help. Leonian (1934) stated that it "is a widely distributed and cosmopolitan species of omnivorous habit, extremely variable both morphologically and physiologically." He considered the fungus to be one of the three "good species" in the genus *Phytophthora,* although he recognized many others.

The history of the controversies on the many variants of *P. palmivora* is too detailed to cover here. Distribution of mating types of *P. palmivora* was described by Zentmyer (1974) and Zentmyer et al (1973). Waterhouse (1974a) summarized *P. palmivora* and related species on cocoa and also described other species on cocoa (Waterhouse, 1974b). Isolates from black pod of cocoa attributable to *P. citrophthora* have been reported (Kellam and Zentmyer, 1981).

The morphological forms (MF) as defined by Griffin (1977) were based primarily on the varying lengths of the caducous sporangial stalk. MF 1 and MF 2 had short stalks, MF 3 intermediate-length stalks, and MF 4 long stalks. MF 2 differed from MF 1 primarily in the shape of the sporangia (Waterhouse, 1974a).

Brasier and Griffin's (1979) recent summary of much of the pertinent literature shows that many isolates previously considered to be *P. palmivora* were not. After a comprehensive study with 950 isolates identified by various researchers as *P. palmivora* from different regions of the world, they redefined the species *P. palmivora,* described the new species *P. megakarya* (MF 3 of Griffin, 1977), and concluded that another species existed (MF 4), which may be similar to *P. capsici* (Kunimoto et al, 1976). Brasier and Griffin (1979) stated that they no longer supported the existence of MF 2 and that MF 1, MF 2, MF 3, and MF 4 designations should be discontinued to avoid further confusion. They concluded that Griffin's (1977) MF 1 isolates constituted the valid *P. palmivora* species with $n = 9–12$ chromosomes, papillate sporangia varying from near-spherical to ovate-elongate, a short pedicel (2–5 $\mu$m), and worldwide distribution. MF 3 was redescribed as the new species, *P. megakarya,* found only in West Africa, with $n = 5–6$ large chromosomes, papillate sporangia, and a pedicel of medium length (10–30 $\mu$m). The MF 4 type had elongated sporangial stalks (20–150 $\mu$m) and was found in Central and South America and the West Indies. MF 4 has been considered to be closely related to *P. capsici* (Kaosiri et al, 1978; Tsao, 1980; Zentmyer et al, 1977). However, Tsao et al (1981) stated that the sporangial pedicel of typical *P. capsici* isolates was shorter than that of MF 4. The *Piper nigrum* (black pepper) isolates of Tsao and Tummakate (1977) are considered now to be MF 4 (Tsao, 1980).

Chee (1969) studied 260 isolates from rubber with abnormal leaf fall symptoms and divided them into those he considered typical of *P. palmivora* and another group that he described as the new species *P. botryosa. P. botryosa* grew more slowly than *P. palmivora,* had a lower maximum temperature, produced chlamydospores sparsely, produced sporangia in clumps, and produced oospores in both a heterothallic and homothallic manner, whereas *P. palmivora* isolates were all heterothallic. Lengths of pedicels of *P. botryosa* were within the intermediate range (5–20 $\mu$m) (Al-Hedaithy and Tsao, 1979b).

The research on taxonomy of *P. palmivora* and related species has been active in recent years. The trend seems to be toward redescribing the "atypical" types as new species. The recent descriptions are based on studies of large numbers of isolates (Brasier and Griffin, 1979; Chee, 1969), which is in the right direction to recognize variability on a world basis. The designation of *P. megakarya* (Brasier and Griffin, 1979) is the first case in *Phytophthora* in which chromosome size has been invoked for speciation.

## POSSIBLE MECHANISMS OF VARIABILITY

Shaw (Chapter 6) explains in greater detail the known genetic mechanisms of variability and discusses hypotheses for the explanation of variability. Brasier (Chapter 27) also discusses mechanisms of variability.

Heterozygosity, which would naturally be expected to be associated with diploidy of *Phytophthora,* must be a major source of variability that has not been explainable on the basis of haploidy. Basically, it is accepted that over 99% of the zoospores of *Phytophthora* spp. are uninucleate (Long et al, 1975, and many other studies). Because diploid zoospores are heterozygous, they will not necessarily breed true with continued culturing and thus would not be genetically pure. Although this is understandable, it is still not known how the genes, whether recessive or dominant, become rearranged without a sexual or parasexual process.

Parasexuality in nature could be a plausible phenomenon. Hyphal anastomosis, which has been reported by Wilde (1961) in *P. infestans* and by Stephenson et al (1974) in *P. capsici,* could allow transmission of nuclei from one mycelial thallus to another. There are also at least two reports of fusion of zoospores (Gallegly and Eichenmuller, 1959; Hiddema and Kole, 1954). Long and Keen (1977a) constructed heterokaryons of *P. megasperma* var. *sojae* from drug-resistant and auxotrophic strains and recovered monokaryons by isolation of single zoospores. However, because the frequency of heterokaryons was low, despite the use of forcing techniques, they questioned whether heterokaryons could occur in nature unless they could be formed by "mutation, polyploidy, and/or somatic crossing over." Although these processes have not been proved in nature, it seems highly plausible that such mechanisms could be involved.

Boccas (1972, 1980) graphed the variation in growth rate among single-zoospore and single-oospore isolates from the homothallic *P. syringae.* In three generations from a wild-type oospore of *P. syringae,* the range of variability was much greater among single-oospore isolates than from single zoospores. Culture of single oospores derived from single zoospores gave colonies with a wider range of variation than from single oospores. Boccas explained this by genetic recombination in the oospores, following meiosis in gametangia.

Commenting on the role of oospores of heterothallic species that could self, L. Erselius (*personal communication*) points out that *P. cinnamomi* and *P. infestans* are highly successful in the apparent absence of a functional sexual cycle in nature. She states that it is possible that selfed oospores that contain a fusion nucleus will "reveal as homozygotes alleles which were in heterozygote form in the parent. Segregation

Recognition of the breadth of variability within and among

has been redescribed to include only those isolates with

caducous, papillate sporangia with occluded pedicels. Isolates with intermediate-length sporangial stalks and with large chromosomes in the oogonia have been designated *P. megakarya*, a new species. Some isolates with long stalks, designated MF 4, are considered to be similar to *P. capsici*.

The mechanism by which genes are rearranged in diploid mycelium is still not well enough known. The sexual process in which meiosis occurs in the gametangia could be responsible for much variability; however, in nature, the evidence for the role of sexuality in variability is not clear. Fusion of mycelium or zoospores to produce a heterokaryotic mycelial thallus could also be involved. Hypotheses that cytoplasmic inheritance of some factors is responsible for variation have been made.

## ACKNOWLEDGMENTS

I greatly appreciate the comments made by C. M. Brasier, L. Erselius, M. E. Gallegly, P. B. Hamm, E. M. Hansen, J. W. Hendrix, R. C. Pratt, O. K. Ribeiro, P. H. Tsao, and G. A. Zentmyer, who reviewed the manuscript.

## LITERATURE CITED

Al-Hedaithy, S. S. A., and Tsao, P. H. 1979a. The effects of culture media and sporulation methods on caducity and pedicel length of sporangia in selected species of *Phytophthora*. Mycologia 71:392-401.

Al-Hedaithy, S. S. A., and Tsao, P. H. 1979b. Sporangium pedicel length in *Phytophthora* species and the consideration of its uniformity in determining sporangium caducity. Trans. Br. Mycol. Soc. 72:1-13.

Apple, J. L. 1957. Pathogenic, cultural, and physiological variation within *Phytophthora parasitica* var. *nicotianae*. Phytopathology 47:733-740.

Athow, K. L., Laviolette, F. A., and Abney, T. S. 1974. Reaction of soybean germplasm strains to four physiologic races of *Phytophthora megasperma* var. *sojae*. Plant Dis. Rep. 58:789-792.

Athow, K. L., Laviolette, F. A., and Wilcox, J. R. 1979. Genetics of resistance to physiologic races of *Phytophthora megasperma* var. *sojae* in the soybean cultivar Tracy. Phytopathology 69:641-642.

Athow, K. L., Laviolette, F. A., Mueller, E. H., and Wilcox, J. R. 1980. A new major gene for resistance to *Phytophthora megasperma* var. *sojae* in soybean. Phytopathology 70:977-980.

Bhelwa, B. H. 1962. Seed decay, seedling blight, and root rot of *Cicer arietinum* caused by *Phytophthora cryptogea*. M.S. thesis. Ohio State Univ., Columbus. 94 pp.

Black, W., Mastenbroek, C., Mills, W. R., and Peterson, L. C. 1953. A proposal for an international nomenclature of races of *Phytophthora infestans* and of genes controlling immunity in *Solanum demissum* derivatives. Euphytica 2:173-179.

Blackwell, E. 1949. Terminology in *Phytophthora*. Mycol. Pap. 30. Commonw. Mycol. Inst., Kew, Surrey, England. 23 pp.

Boccas, B. 1972. Contribution à l'étude du cycle chez les phytophthora. Analyse du mode de transmission d'un caractère génétique quantitatif chez une espèce homothallique, le *Phytophthora syringae* Kleb. C. R. Hebd. Seances Acad. Sci. Ser. D 275:663-666.

Boccas, B. 1973. Observations préliminaires sur l'hérédité du pouvoir pathogène chez le *P. palmivora* (Butl.) Butl. Cah. ORSTOM Ser. Biol. 20:51-55.

Boccas, B. R. 1980. Genetics of the genus *Phytophthora*. In: Workshop on *Phytophthora* Diseases. Kasaragod, India, September 19–23, 1980. Central Plantation Crops Research Institute, Kerala, India.

Boccas, B. R. 1981. Interspecific crosses between closely related heterothallic *Phytophthora* species. Phytopathology 71:60-65.

Boccas, B., and Zentmyer, G. A. 1976. Genetical studies with interspecific crosses between *Phytophthora cinnamomi* and *Phytophthora parasitica*. Phytopathology 66:477-484.

Brasier, C. M. 1971. Induction of sexual reproduction in single $A^2$ isolates of *Phytophthora* species by *Trichoderma viride*. Nature (London) New Biol. 231:283.

Brasier, C. M. 1975. Stimulation of sex organ formation in *Phytophthora* by antagonistic species of *Trichoderma*. I. The effect in vitro. New Phytol. 74:183-194.

Brasier, C. M. 1978. Stimulation of oospore formation in *Phytophthora* by antagonistic species of *Trichoderma* and its ecological implications. Ann. Appl. Biol. 89:135-139.

Brasier, C. M., and Griffin, M. J. 1979. Taxonomy of '*Phytophthora palmivora*' on cocoa. Trans. Br. Mycol. Soc. 72:111-143.

Brasier, C. M., and Sansome, E. 1975. Diploidy and gametangial meiosis in *Phytophthora cinnamomi, P. infestans,* and *P. drechsleri*. Trans. Br. Mycol. Soc. 65:49-65.

Bruyn, H. L. G. de. 1951. Pathogenic differentiation in *Phytophthora infestans* (Mont.) de Bary. Phytopathol. Z. 18:339-359.

Buddenhagen, I. W. 1958. Induced mutations and variability in *Phytophthora cactorum*. Am. J. Bot. 45:355-365.

Bumbieris, M. 1974. Characteristics of two *Phytophthora* species. Aust. J. Bot. 22:655-660.

Butler, E. J. 1907. An account of the genus *Pythium* and some Chytridiaceae. Mem. Dep. Agric. India Bot. Ser. 1:82-84.

Butler, E. J. 1910. The bud-rot of palms in India. Mem. Dep. Agric. India Bot. Ser. 3:221-280.

Cairns, H., and Muskett, A. E. 1933. Pink rot of the potato. Ann. Appl. Biol. 20:381-403.

Cameron, H. R. 1966. Variability in the genus *Phytophthora*. Phytopathology 56:812-815.

Castro, J., and Zentmyer, G. A. 1969. Mortality of germinated $F_2$ oospores from crosses of $F_1$ single oospore cultures of *Phytophthora infestans*. (Abstr.) Phytopathology 59:10.

Caten, C. E. 1970. Spontaneous variability of single isolates of *Phytophthora infestans*. II. Pathogenic variation. Can. J. Bot. 48:897-905.

Caten, C. E. 1971. Single zoospore variation in *Phytophthora infestans* and attenuation of strains in culture. Trans. Br. Mycol. Soc. 56:1-7.

Caten, C. E. 1974. Intra-racial variation in *Phytophthora infestans* and adaptation to field resistance for potato blight. Ann. Appl. Biol. 77:259-270.

Caten, C. E., and Day, A. W. 1977. Diploidy in plant pathogenic fungi. Annu. Rev. Phytopathol. 15:295-318.

Caten, C. E., and Jinks, J. L. 1968. Spontaneous variability of single isolates of *Phytophthora infestans*. I. Cultural variation. Can J. Bot. 46:329-348.

Chee, K. H. 1969. Variability of *Phytophthora* species from *Hevea brasiliensis*. Trans. Br. Mycol. Soc. 52:425-436.

Chesters, C. G. C. 1968. Morphology as a taxonomic criterion. Pages 517-542 in: The Fungi: An Advanced Treatise. Vol. III. The Fungal Population. G. C. Ainsworth and A. J. Sussman, eds. Academic Press, New York. 737 pp.

Cohen, M. 1950. Direct observation of formation of sexual bodies by combination of two *Phytophthora* isolates. (Abstr.) Phytopathology 40:5-6.

Converse, R. H. 1967. Physiologic races of *Phytophthora fragariae* on strawberry in California, Oregon, and Washington. Phytopathology 57:173-177.

Crandall, B. S., Gravatt, G. F., and Ryan, M. M. 1945. Root disease of *Castanea* species and some coniferous and broadleaf nursery stocks, caused by *Phytophthora cinnamomi*. Phytopathology 35:162-180.

Davidse, L. C. 1981. Resistance to acylalanine fungicides in *Phytophthora megasperma* f. sp. *medicaginis*. Neth. J. Plant Pathol. 87:11-24.

Davidse, L. C., and Flach, W. 1977. Differential binding of methyl benzimidazol-2-yl-carbamate to fungal tubulin as a mechanism of resistance to this antimitotic agent in mutant strains of *Aspergillus nidulans*. J. Cell Biol. 72:174-193.

Davidse, L. C., Looijen, D., Turkensteen, L. J., and Van der Wal, D. 1981. Occurrence of metalaxyl-resistant strains of *Phytophthora infestans* in Dutch potato fields. Neth. J. Plant Pathol. 87:65-68.

Denward, T. 1967. Differentiation in *Phytophthora infestans*. I. A comparative study of eight different biotypes. Hereditas 58:191-220.

Denward, T. 1970. Differentiation in *Phytophthora infestans*. II. Somatic recombination in vegetative mycelium. Hereditas 66:35-48.

Drechsler, C. 1931. A crown rot of hollyhocks caused by *Phytophthora megasperma* n. sp. J. Wash. Acad. Sci. 21:513-526.

Duniway, J. M. 1976. Movement of zoospores of *Phytophthora cryptogea* in soils of various textures and matric potentials. Phytopathology 66:877-882.

Elliott, C. G., and MacIntyre, D. 1973. Genetical evidence on the life-history of *Phytophthora*. Trans. Br. Mycol. Soc. 60:311-316.

Erwin, D. C. 1965. Reclassification of the causal agent of root rot of alfalfa from *Phytophthora cryptogea* to *P. megasperma*. Phytopathology 55:1139-1143.

Erwin, D. C. 1966. Varietal reaction of alfalfa to *Phytophthora megasperma* and variation in virulence of the causal fungus. Phytopathology 56:653-657.

Erwin, D. C., Zentmyer, G. A., Galindo, J., and Niederhauser, J. S. 1963. Variation in the genus *Phytophthora*. Annu. Rev. Phytopathol. 1:375-396.

Eye, L. L., Sneh, B., and Lockwood, J. L. 1978. Inoculation of soybean seedlings with zoospores of *Phytophthora megasperma* var. *sojae* for pathogenicity and race determination. Phytopathology 68:1769-1773.

Flowers, R. A., Erwin, D. C., and Hendrix, J. W. 1973. Isolation of *Phytophthora cryptogea* from bean in Kentucky. Plant Dis. Rep. 57:170-191.

Galindo A., J., and Gallegly, M. E. 1960. The nature of sexuality in *Phytophthora infestans.* Phytopathology 50:123-128.

Galindo A., J., and Zentmyer, G. A. 1964. Mating types in *Phytophthora cinnamomi* Phytopathology 54:238-239.

Gallegly, M. E. 1968. Genetics of pathogenicity of *Phytophthora infestans.* Annu. Rev. Phytopathol. 6:375-396.

Gallegly, M. E., and Eichenmuller, J. J. 1959. The spontaneous appearance of the potato race 4 characters in cultures of *Phytophthora infestans.* Am. Potato J. 36:45-51.

Gerrettson-Cornell, L. 1979. Notes on the morphology of five isolates of *Phytophthora cryptogea* Pethyb. and Laff. from Australia. Phyton (Buenos Aires) 37:105-118.

Graham, K. M. 1954. Nuclear behavior in *Phytophthora infestans.* (Abstr.) Phytopathology 44:490.

Graham, K. M., and Hodgson, W. A. 1965. Effect of major and minor gene interaction on adaptive parasitism in *Phytophthora infestans.* Phytopathology 55:73-75.

Griffin, M. J. 1977. Cocoa *Phytophthora* Workshop, Rothamsted Experimental Station, England, 24–26 May, 1976. PANS 23:107-110.

Haasis, F. A., and Nelson, R. R. 1963. Studies on the biological relationship of species of *Phytophthora* as measured by oospore formation in intra- and interspecific crosses. Plant Dis. Rep. 47:705-709.

Halsall, D. M. 1976. Specificity of cytoplasmic and cell wall antigens from four species of *Phytophthora.* J. Gen. Microbiol. 94:149-158.

Hamm, P. B., and Hansen, E. M. 1981. Host specificity of *Phytophthora megasperma* from Douglas fir, soybean, and alfalfa. Phytopathology 71:65-68.

Hamm, P. B., and Hansen, E. M. 1982. Single-spore isolate variation: The effect on varietal designation in *Phytophthora megasperma.* Can J. Bot. (In press)

Hansen, E. M., and Hamm, P. B. 1982. Morphological differentiation of host-specialized groups of *Phytophthora megasperma.* Phytopathology 72:(In press).

Hendrix, J. W., and Flowers, R. A. 1967. Development of races of *Phytophthora parasitica* var. *nicotianae* in mid-Tennessee in 1966. Tobacco 167:24-25.

Hickman, C. J. 1962. Physiologic races of *Phytophthora fragariae.* Ann. Appl. Biol. 50:95-103.

Hiddema, J., and Kole, A. P. 1954. Enkele waarneming over versmelten ven zoosporen bij *Phytophthora infestans* (Mont.) de Bary. Tijdschr. Plantenziekten 6:138-139.

Hildebrand, A. E. 1959. A root and stalk rot of soybeans caused by *Phytophthora megasperma* var. *sojae* var. nov. Can. J. Bot. 37:927-957.

Hilty, J. W., and Schmitthenner, A. F. 1962. Pathogenic and cultural variability of single zoospore isolates of *Phytophthora megasperma* var. *sojae.* Phytopathology 52:859-862.

Hobe, M. A. 1981. Pathogenic variability of *Phytophthora megasperma* f. sp. *glycinea* isolated from northwest Ohio soils. M.S. thesis. Ohio State Univ., Columbus. 32 pp.

Hunger, R. M., Hamm, P. B., Horner, C. E., and Hansen, E. M. 1982. Tolerance of *Phytophthora megasperma* isolates to metalaxyl. Plant Dis. 66:645-649.

Irwin, J. A. G., and Dale, J. L. 1982. Relationships between *Phytophthora megasperma* isolates from chickpea, lucerne, and soybean. Aust. J. Bot. 30:(In press).

Jeffrey, S. I. B., Jinks, J. L., and Grindle, M. 1962. Intraracial variation in *Phytophthora infestans.* Genetica 32:323-338.

Johnson, E. M., and Valleau, W. D. 1954. Heterothallism in *Phytophthora parasitica* var. *nicotianae.* Phytopathology 44:312-313.

Jones, J. P., and Johnson, H. W. 1969. Lupine, a new host for *Phytophthora megasperma* var. *sojae.* Phytopathology 59:504-507.

Kannaiyan, J., Ribeiro, O. K., Erwin, D. C., and Nene, Y. L. 1980. *Phytophthora* blight of pigeon pea in India. Mycologia 72:169-181.

Kaosiri, T., Zentmyer, G. A., and Erwin, D. C. 1978. Stalk length as a taxonomic criterion for *Phytophthora palmivora* isolates from cacao. Can. J. Bot. 56:1730-1738.

Kaosiri, T., Zentmyer, G. A., and Erwin, D. C. 1980. Oospore morphology and germination in the *Phytophthora palmivora* complex from cacao. Mycologia 72:888-907.

Kaufmann, M. J., and Gerdemann, J. W. 1958. Root and stem rot of soybeans caused by *Phytophthora sojae* n. sp. Phytopathology 48:201-208.

Keeling, B. L. 1980. Research on Phytophthora root rot and stem rot: Isolation, testing procedures, and seven new physiologic races. Pages 367-370 in: World Soybean Conference II, Proceedings. F. T. Corbin, ed. Westview Press, Boulder, CO. 897 pp.

Keeling, B. L. 1982. Four new physiologic races of *Phytophthora megasperma* f. sp. *glycinea.* Plant Dis. 66:334-335.

Kellam, M. K., and Zentmyer, G. A. 1981. Isolation of *Phytophthora citrophthora* from cocoa in Brazil. (Abstr.) Phytopathology 71:230.

Khaki, I. A., and Shaw, D. S. 1974. The inheritance of drug resistance and compatibility type in *Phytophthora drechsleri.* Genet. Res. 23:75-86.

Klisiewicz, J. M. 1977. Identity and relative virulence of some heterothallic *Phytophthora* species associated with root and stem rot of safflower. Phytopathology 67:1174-1177.

Ko, W. H. 1978. Heterothallic *Phytophthora:* Evidence for hormonal regulation of sexual reproduction. J. Gen. Microbiol. 107:15-18.

Ko, W. H. 1980. Hormonal regulation of sexual reproduction in *Phytophthora.* J. Gen. Microbiol. 116:459-463.

Kuan, T.-L., and Erwin, D. C. 1980. Formae speciales differentiation of *Phytophthora megasperma* isolates from soybean and alfalfa. Phytopathology 70:333-338.

Kunimoto, R. K., Aragaki, M., Hunter, J. E., and Ko, W. H. 1976. *Phytophthora capsici,* corrected name for the cause of Phytophthora blight of macadamia racemes. Phytopathology 66:546-548.

Laviola, C. 1968. Studies on the genetics of *Phytophthora infestans.* Diss. Abstr. 29:3573.

Laviolette, F. A., and Athow, K. L. 1977. Three new physiologic races of *Phytophthora megasperma* var. *sojae.* Phytopathology 67:267-268.

Laviolette, F. A., and Athow, K. L. 1981. Physiologic races of *Phytophthora megasperma* f. sp. *glycinea* in Indiana, 1973–1979. Plant Dis. 65:884-885.

Laviolette, F. A., Athow, K. L., Mueller, E. H., and Wilcox, J. R. 1979. Inheritance of resistance in soybeans to physiologic races 5, 6, 7, 8, and 9 of *Phytophthora megasperma* var. *sojae.* Phytopathology 69:270-271.

Leach, S. S., and Rich, A. E. 1969. The possible role of parasexuality and cytoplasmic variation in race differentiation in *Phytophthora infestans.* Phytopathology 59:1360-1365.

Leonian, L. H. 1934. Identification of *Phytophthora* species. W. Va. Agric. Exp. Stn. Bull. 242. 36 pp.

Lockwood, J. L., and Cohen, S. D. 1978. Race determination of *Phytophthora megasperma* var. *sojae,* using differential soybean varieties inoculated with zoospores or incubated on flooded soil samples. Plant Dis. Rep. 62:1019-1021.

Long, M., and Keen, N. T. 1977a. Evidence for heterokaryosis in *Phytophthora megasperma* var. *sojae.* Phytopathology 67:670-674.

Long, M., and Keen, N. T. 1977b. Genetic evidence for diploidy in *Phytophthora megasperma* var. *sojae.* Phytopathology 67:675-677.

Long, M., Keen, N. T., Ribeiro, O. K., Leary, J. V., Erwin, D. C., and Zentmyer, G. A. 1975. *Phytophthora megasperma* var. *sojae:* Development of wild-type strains for genetic research.

Phytopathology 65:592-597.

Maas, J. L. 1972. Growth and reproduction in culture of ten *Phytophthora fragariae* races. Mycopathol. Mycol. Appl. 48:323-334.

Maas, J. L. 1973. *Phytophthora fragariae* race characterization with gas-liquid chromatographic lipid analysis. Mycologia 65:1259-1265.

MacKenzie, D. R. 1980. The problem of variable pests. Pages 183-213 in: Breeding Plants Resistant to Insects. G. Maxwell and P. R. Jennings, eds. John Wiley & Sons, New York. 683 pp.

Malcolmson, J. F. 1969. Races of *Phytophthora infestans* occurring in Great Britain. Trans. Br. Mycol. Soc. 53:417-423.

Malcolmson, J. F. 1970. Vegetative hybridity in *Phytophthora infestans.* Nature (London) 225:971-972.

Malcolmson, J. F. 1976. Assessment of field resistance to blight (*Phytophthora infestans*) in potatoes. Trans. Br. Mycol. Soc. 67:321-335.

Malcolmson, J. F., and Black, W. 1966. New R genes in *Solanum demissum* Lindl. and their complementary races of *Phytophthora infestans* (Mont.) de Bary. Euphytica 15:199-203.

Manning, W. J., and Crossan, D. F. 1966. Evidence for variation in degree of pathogenicity of isolates of *Phytophthora cinnamomi* to broadleaf and coniferous evergreens. Plant Dis. Rep. 50:647-649.

Maronek, D. M., and Hendrix, J. W. 1978. Resistance to race 0 of *Phytophthora parasitica* var. *nicotianae* in tissue cultures of a tobacco breeding line with black shank resistance derived from *Nicotiana longiflora.* Phytopathology 68:233-234.

McIntyre, J. L., and Taylor, G. S. 1978. Race 3 of *Phytophthora parasitica* var. *nicotianae.* Phytopathology 68:35-38.

Mills, W. R. 1940. Adaptive parasitism of *Phytophthora infestans.* (Abstr.) Phytopathology 30:17.

Mills, W. R., and Peterson, L. C. 1952. The development of races of *Phytophthora infestans* on potato hybrids. Phytopathology 42:26.

Mircetich, S. M., and Matheron, M. E. 1976. Phytophthora root rot and crown rot of cherry trees. Phytopathology 66:549-558.

Montgomerie, I. G., and Kennedy, D. W. 1975. Races of *Phytophthora fragariae* from Essex, Hampshire, and Kent. Plant Pathol. 24:191-193.

Morgan, F. L., and Hartwig, E. E. 1965. Physiologic specialization in *Phytophthora megasperma* var. *sojae.* Phytopathology 55:1277-1279.

Mortimer, A. M., and Shaw, D. S. 1975. Cytofluorimetric evidence for meiosis in gametangial nuclei of *Phytophthora drechsleri.* Genet. Res. 25:201-205.

Mortimer, A. M., Shaw, D. S., and Sansome, E. R. 1977. Genetical studies of secondary homothallism in *Phytophthora drechsleri.* Arch. Microbiol. 111:255-259.

Morton, D. J., and Dukes, P. D. 1967. Serological differentiation of *Pythium aphanidermatum* from *Phytophthora parasitica* var. *nicotianae* and *Phytophthora parasitica.* Nature (London) 213:923.

Mueller, E. H., Athow, K. L., and Laviolette, F. A. 1978. Inheritance of resistance to four physiologic races of *Phytophthora megasperma* var. *sojae.* Phytopathology 68:1318-1322.

Nagai, Y., Takeuchi, T., and Watanabe, T. 1978. A stem blight of rose caused by *Phytophthora megasperma.* Phytopathology 68:684-688.

Narasimhan, M. J. 1930. Studies in the genus *Phytophthora* in Mysore. I. Heterothallic strains of *Phytophthora.* Phytopathology 20:201-214.

Newhook, F. J., Waterhouse, G. M., and Stamps, D. J. 1978. Tabular key to the species of *Phytophthora* de Bary. Mycol. Pap. 143. Commonw. Mycol. Inst., Kew, Surrey, England. 20 pp.

Niederhauser, J. S. 1961. Genetic studies of *Phytophthora infestans* and *Solanum* species in relation to late blight resistance in the potato. Pages 491-497 in: Recent Advances in Botany. Univ. Toronto Press, Toronto.

Noon, J. P., and Hickman, C. J. 1974. Oospore production by a single isolate of *Phytophthora capsici* in the presence of chloroneb. Can. J. Bot. 52:1591-1595.

Partridge, J. E., and Keen, N. T. 1976. Association of the phytoalexin kievetone with single-gene resistance of cowpeas to *Phytophthora vignae.* Phytopathology 66:426-429.

Pepin, H. S., and Daubeny, H. A. 1964. The relationships of English and American races of *Phytophthora fragariae.* Phytopathology 54:241.

Pethybridge, G. H. 1913. On the rotting of potato tubers by a new species of *Phytophthora* having a method of sexual reproduction hitherto undescribed. Sci. Proc. R. Dublin Soc. (N.S.) 13:529-564.

Pethybridge, G. H., and Lafferty, H. A. 1919. A disease of tomato and other plants caused by a new species of *Phytophthora.* Sci. Proc. R. Dublin Soc. (N.S.) 15:487-503.

Polach, F. J., and Webster, R. K. 1972. Identification of strains and inheritance of pathogenicity in *Phytophthora capsici.* Phytopathology 62:20-26.

Pratt, R. G. 1981. Morphology, pathogenicity, and host range of *Phytophthora megasperma, P. erythroseptica,* and *P. parasitica* from arrowleaf clover. Phytopathology 71:276-282.

Purss, G. S. 1972. Pathogenic specialization in *Phytophthora vignae.* Aust. J. Agric. Res. 23:453-456.

Rands, R. D. 1922. Streep kanker van kaneel veroorzaakt door *Phytophthora cinnamomi* n. sp. (Stripe canker of cinnamon caused by *Phytophthora cinnamomi* n. sp.). Meded. Inst. Plantenziekten 54:41.

Reddick, D., and Mills, W. 1938. Building up virulence in *Phytophthora infestans.* Am. Potato J. 15:29-34.

Reeves, R. J., and Jackson, R. M. 1974. Stimulation of sexual reproduction in *Phytophthora* by damage. J. Gen. Microbiol. 84:303-310.

Ribeiro, O. K., Erwin, D. C., and Zentmyer, G. A. 1975. An improved synthetic medium for oospore production and germination of several *Phytophthora* species. Mycologia 67:1012-1019.

Robinson, R. A. 1969. Disease resistance terminology. Rev. Appl. Mycol. 48:593-606.

Romero, S., and Erwin, D. C. 1969. Variation in pathogenicity among single-oospore cultures of *Phytophthora infestans.* Phytopathology 59:1310-1317.

Sansome, E. 1965. Meiosis in diploid and polyploid sex organs of *Phytophthora* and *Achlya.* Cytologia 30:103-117.

Sansome, E. 1966. Meiosis in the sex organs of the Oömycetes. Pages 77-83 in: Chromosomes Today. Vol. 1. C. D. Darlington and K. R. Lewis, eds. Oliver & Boyd, Edinburgh.

Sansome, E., and Brasier, C. M. 1973. Diploidy and chromosomal structural hybridity in *Phytophthora infestans.* Nature (London) 241:344-345.

Sansome, E., and Brasier, C. M. 1974. Polyploidy associated with varietal differentiation in the *megasperma* complex of *Phytophthora.* Trans. Br. Mycol. Soc. 63:461-467.

Sansome, E., Brasier, C. M., and Sansome, F. W. 1979. Further cytological studies on the 'L' and 'S' types of *Phytophthora* from cocoa. Trans. Br. Mycol. Soc. 73:293-302.

Satour, M. M., and Butler, E. E. 1968. Comparative morphological and physiological studies of the progenies from intraspecific matings of *Phytophthora capsici.* Phytopathology 58:183-192.

Savage, E. J., Clayton, C. W., Hunter, J. H., Brenneman, J. A., Laviola, C., and Gallegly, M. E. 1968. Homothallism, heterothallism, and interspecific hybridization in the genus *Phytophthora.* Phytopathology 58:1004-1021.

Schick, R. 1932. Über das verhalten von *Solanum demissum, Solanum tuberosum,* und ihren Bastarden gegenuber verschiedenen Herkunften von *Phytophthora infestans.* Zuechter 4:233-237.

Schultz, O. E. 1962. Inter and intra race variation among mass and mono zoospore isolates of *Phytophthora infestans* (Mont.) de Bary. Diss. Abstr. 23:788-789.

Shattock, R. C., Janssen, B. D., Whitbread, R., and Shaw, D. S. 1977. An interpretation of the frequencies of host specific phenotypes of *Phytophthora infestans* in North Wales. Ann. Appl. Biol. 86:249-260.

Shepherd, C. J. 1976. Pigment production from tyrosine by Australian isolates of *Phytophthora* species. Aust. J. Bot. 24:607-617.

Shepherd, C. J., and Pratt, B. H. 1973. Separation of two ecotypes of *Phytophthora drechsleri* Tucker occurring in Australian native forests. Aust. J. Biol. Sci. 26:1095-1107.

Shepherd, C. J., and Pratt, B. H. 1974. Temperature-growth relations and genetic diversity of A2 mating-type isolates of *Phytophthora cinnamomi* in Australia. Aust. J. Bot. 22:231-249.

Shepherd, C. J., Pratt, B. H., and Taylor, P. A. 1974. Comparative morphology and behaviour of A1 and A2 isolates of *Phytophthora cinnamomi.* Aust. J. Bot. 22:461-470.

Simmonds, N. W., and Malcolmson, J. F. 1967. Resistance to late blight in Andigena potatoes. Eur. Potato J. 10:161-166.

Snyder, W. C., Hansen, H. M., and Oswald, J. W. 1957. Cultivars of the fungus, *Fusarium*. J. Madras Univ. Sect. B 27:185-192.

Staflue, F. A., Bonner, C. E. B., McVaugh, R., Meikle, R. D., Rollins, R. C., Ross, R., Schopt, J. M., Schulze, G. M., and DeVilmorin, R. 1972. International code of nomenclature. A. Oosthoek's Vitgeversmaatschappij N.V., Utrecht, The Netherlands. 426 pp.

Stamps, D. J. 1953. Oospore production in paired cultures of *Phytophthora* species. Trans. Br. Mycol. Soc. 36:255-259.

Stephenson, L. W., Erwin, D. C., and Leary, J. V. 1974. Hyphal anastomosis in *Phytophthora capsici*. Phytopathology 64:149-150.

Suhovecky, A. J., and Schmitthenner, A. F. 1955. Soybeans affected by early root rot. Ohio Farm Home Res. 40:85-86.

Timmer, L. W., Castro, J., Erwin, D. C., Belser, W. L., and Zentmyer, G. A. 1970a. Genetic evidence for zygotic meiosis in *Phytophthora capsici*. Am. J. Bot. 57:1211-1218.

Timmer, L. W., Erwin, D. C., and McCormick, W. H. 1970b. Glucose inhibition and inheritance of pigment-producing ability in *Phytophthora*. Mycologia 62:967-977.

Tompkins, C. M., Tucker, C. M., and Gardner, M. W. 1936. Phytophthora root rot of cauliflower. J. Agric. Res. 53:685-692.

Torgeson, D. C. 1954. Root rot of Lawson cypress and other ornamentals caused by *Phytophthora cinnamomi*. Contrib. Boyce Thompson Inst. 17:359-373.

Toxopeus, H. J. 1956. Reflections on the origin of new physiologic races in *Phytophthora infestans* and the breeding for resistance in potatoes. Euphytica 5:221-256.

Tsao, P. H. 1980. Morphology and identity of black pepper *Phytophthora* isolates. Workshop on *Phytophthora* Diseases of Tropical Cultivated Plants, 19–23 September 1980, Kerala, India. Central Plantation Crops Research Institute, Kerala, India.

Tsao, P. H., and Sisemore, D. J. 1978. Morphological variability in *Phytophthora parasitica* (*P. nicotianae*) isolates from citrus, tomato, and tobacco. (Abstr.) Phytopathol. News 12:213.

Tsao, P. H., and Tummakate, A. 1977. The identity of a *Phytophthora* species from black pepper in Thailand. Mycologia 69:631-637.

Tsao, P. H., Ugale, R., Hobbs, S., and Farih, A. 1980. Control of homothallic oospore formation in *Phytophthora parasitica* by culture manipulations. Trans. Br. Mycol. Soc. 75:153-156.

Tsao, P. H., Alizadeh, A., and Tsao, P. W. 1981. Relationship between *Phytophthora capsici* and the black pepper isolates of '*P. palmivora*' morphological form (MF) 4. (Abstr.) Phytopathology 71:262.

Tucker, C. M. 1931. Taxonomy of the genus *Phytophthora* de Bary. Univ. Mo. Agric. Exp Stn. Res. Bull. 153. 208 pp.

Urech, P. A., Schwinn, F. J., and Staub, T. 1977. CGA48988, a novel fungicide for the control of late blight, downy mildews, and related soilborne diseases. Proc. Br. Crop Prot. Conf. 1977:623-631.

van der Plank, J. E. 1963. Plant Diseases: Epidemics and Control. Academic Press, New York. 349 pp.

van der Plank, J. E. 1968. Disease Resistance in Plants. Academic Press, New York. 206 pp.

van der Plank, J. E. 1971. Stability of resistance to *Phytophthora infestans* in cultivars without R genes. Potato Res. 14:263-270.

Waterhouse, G. M. 1956. The genus *Phytophthora*. Misc. Publ. 12. Commonw. Mycol. Inst., Kew, Surrey, England. 120 pp.

Waterhouse, G. M. 1963. Key to the species of *Phytophthora* de Bary. Mycol. Pap. 92. Commonw. Mycol. Inst., Kew, Surrey, England. 22 pp.

Waterhouse, G. M. 1974a. *Phytophthora palmivora* and some related species. Pages 51-70 in: *Phytophthora* Disease of Cocoa. P. H. Gregory, ed. Longman Press, London.

Waterhouse, G. M. 1974b. Other species of *Phytophthora* recorded on cocoa. Pages 71-79 in: *Phytophthora* Disease of Cocoa. P. H. Gregory, ed. Longman Press, London.

Webster, R. K. 1974. Recent advances in the genetics of plant pathogenic fungi. Annu. Rev. Phytopathol. 12:331-353.

Wilcox, J. R. 1976. Breeding for root resistance. Pages 485-490 in: Proc. World Soybean Res. Conf. L. D. Hill, ed. Interstate Printers and Publishers Inc., Danville, IL.

Wilde, P. 1961. Ein beitrag zur kenntnis der variabilitat von *Phytophthora infestans* (Mont.) de Bary. Arch. Mikrobiol. 40:163-195.

Zentmyer, G. A. 1952. A substance stimulating sexual reproduction in *Phytophthora cinnamomi*. (Abstr.) Phytopathology 42:24.

Zentmyer, G. A. 1974. Variation, genetics, and geographical distribution of mating types. Pages 89-102 in: *Phytophthora* Disease of Cocoa. P. H. Gregory, ed. Longman, London.

Zentmyer, G. A. 1979. Stimulation of sexual reproduction in the A2 mating type of *Phytophthora cinnamomi* by a substance in avocado roots. Phytopathology 69:1129-1131.

Zentmyer, G. A. 1980. *Phytophthora cinnamomi* and the diseases it causes. Monogr. 10. Am. Phytopathol. Soc., St. Paul, MN. 96 pp.

Zentmyer, G. A., and Guillemet, F. B. 1981. Evidence for strains of *Phytophthora cinnamomi*. Plant Dis. 65:475-477.

Zentmyer, G. A., Mitchell, D. J., Jefferson, L., Roheim, J., and Carnes, D. 1973. Distribution of mating types of *Phytophthora palmivora*. Phytopathology 63:663-667.

Zentmyer, G. A., Leary, J. V., Klure, L. J., and Grantham, G. L. 1976. Variability in growth of *Phytophthora cinnamomi* in relation to temperature. Phytopathology 66:982-986.

Zentmyer, G. A., Kaosiri, T., and Idosu, G. 1977. Taxonomic variants in the *Phytophthora palmivora* complex. Trans. Br. Mycol. Soc. 69:329-332.

Zentmyer, G. A., Klure, L. J., and Pond, E. C. 1979. The influence of temperature and nutrition on formation of sexual structures by *Phytophthora cinnamomi*. Mycologia 71:55-67.

# 12

# New Criteria for Classifying *Phytophthora* and Critique of Existing Approaches

M. E. Gallegly

CONTENTS



## INTRODUCTION

Tribute must be paid to Grace M. Waterhouse for providing the keys currently used throughout the world for classifying and identifying species of *Phytophthora*. Recognizing that several new species had been described since Tucker's monograph (Tucker, 1931), Waterhouse published her first key in 1963 (Waterhouse, 1963), and a tabular key complementing the earlier one was published in 1978 by her and her associates (Newhook et al, 1978). In 1956, Waterhouse presented a compilation of the original descriptions and figures of all species described up to that time (Waterhouse, 1956), and she updated this in 1970 (Waterhouse, 1970). The history of taxonomic treatments of the genus is presented by Waterhouse (Chapter 10).

The Waterhouse keys have been very useful in creating order at a time when much confusion existed. The task has not been easy, and there is continued need to improve the system to further reduce confusion. Both the 1963 and 1978 keys were put forward as interim documents to be molded by use, and by the integration of new criteria, into a more perfect key. It is in this context that the discussion here is presented.

## THE CURRENT SYSTEM

Morphological characters are the main criteria of the Waterhouse keys. In addition, maximum temperature for growth in culture, colony type, and host-specific pathogenicity are used to sort out some species. The characteristics of sporangia—more specifically, the type of papillae and extent of apical thickening—are the first morphological criteria used. Three groups were recognized: 1) sporangia markedly papillate, with hemispheric apical thickening at least 4 mm deep and a narrow exit pore; 2) sporangia less protuberant, with apical thickening up to 3.5 mm deep and a narrow exit pore; and 3) sporangia nonpapillate, with a very thin apical thickening and a broad exit pore. Each of these three groups was subdivided into those with paragynous or amphigynous antheridia to form six major subdivisions of the genus. Within the six groups, additional morphological characteristics of the sporangia and sporangiophores, the sexual bodies, and other bodies such as chlamydospores and hyphal swellings were used for species identifications along with temperature maxima, colony growth characters, and host range. In the tabular key (Newhook et al, 1978), 80 columns were given, each containing a different character or variation of a character. The keys were presented primarily for identification, without intention of implying that they represented a natural system of classification.

Both the beginning plant pathologist and the experienced mycologist, when first faced with the task of identifying an isolate of *Phytophthora*, quickly discover that the genus is difficult. The beginner must learn to recognize the many morphological characters as well as to develop the art of producing the various stages through cultural manipulations. A seemingly simple task of differentiating between a hyphal swelling and a thin-walled chlamydospore can become confusing. Variation in size and shape of the various morphological characters causes further confusion. Where does one place an isolate when all of the characters fit those of *P. megasperma* except for sizes of the oogonia, which overlap those given for *P. megasperma* var. *megasperma* and *P. megasperma* var. *sojae*?

In time, the beginner learns to use the key and develops a sense of satisfaction and confidence. However, there is a lingering sense that the key is too finely tuned and that some of the characters simply sort out variants rather than separate species. As one includes more and more characters, or splits them into an array of finer components, the number of species that can be named increases accordingly. Thus, one asks where the line should be drawn to differentiate between major characters suitable for species separation and minor characters associated only with variants. The fine tuning of a key with many characters can lead to artificiality, and it could be said that the current key is somewhat artificial.

I believe that the time is now right for the development of a more natural system of classification. Additional studies of interspecific and intraspecific crosses may be necessary to study inheritance of characters. Also, the characters should be examined for phylogenetic relationships. The proper interpretation of the large amount of information already available, and the generation of more information on genetic interrelationships among the currently recognized species, are the major actions needed to mold the Waterhouse key into a more natural system of classification.

## A NATURAL SYSTEM

A natural system of classification should begin with the characteristics of sexuality in the genus. First consideration should be given to sexual patterns and genetic interrelationships

Published with the approval of the Director of the West Virginia Agricultural and Forestry Experiment Station as Scientific Paper 1746.

among the species, followed by differences in morphology of the sexual bodies. The groups based on sexual patterns would be subdivided according to major differences in morphology of the sexual bodies. The sexual groups would be further subdivided according to characters of the asexual stages and then differences in pathogenicity. A system following this order of arrangement might be a more natural system, in that closer relationships would be exhibited. It may be possible, with the information now available, to rearrange the presently recognized species into this system. However, the more natural system may not be practical for identification of species.

## Sexual Patterns

The discovery of mating types and heterothallism in *P. infestans*, the type species (Gough, 1957; Smoot et al, 1958), set the stage for clarification of the sometimes confusing and contradictory observations on the nature of sexual reproduction in the genus. Galindo and Gallegly (1960) found that each isolate of *P. infestans* was self-incompatible but would function as either a male or a female in pairings with isolates of the opposite type. An isolate that formed mostly antheridia in some pairings would form mostly oogonia in pairings with a stronger male of the opposite type. Because relative strengths of maleness or femaleness varied with the isolate, the term mating type was discarded, and the term compatibility type was selected to designate the sexual types in *P. infestans*. All isolates from the United States and Europe were of the same compatibility type and were designated as type A1, whereas isolates from Mexico of the opposite type were designated A2. Fusion of gametangia from an A1 isolate with gametangia of an A2 isolate, resulting in the formation of oospores, was observed by hyphal tracings. Previously, the two compatibility types were found in nature in Mexico in a 1:1 ratio (Gallegly and Galindo, 1958). It was concluded that both A1 and A2 isolates carry both femaleness and maleness, and thus are bisexual and potentially homothallic, but that they function heterothallically because of self-incompatibility (Galindo and Gallegly, 1960).

Both homothallism and heterothallism occur among species of *Phytophthora*. Savage et al (1968) studied 350 isolates, received as 30 species and varieties of the genus, for their ability to form oospores in single culture or in pairings with A1 or A2 compatibility types of *P. infestans*. They recognized three major groups based on patterns of sexuality: 1) homothallic with predominantly paragynous antheridia (seven species), 2) homothallic with predominantly amphigynous antheridia (eight species), and 3) heterothallic with amphigynous antheridia (13 species and one variety). The only isolate received as *P. quininea* failed to form oogonia and could not be placed. Several isolates of some species from both homothallic groups showed no evidence of pairing with either A1 or A2 types of several of the heterothallic species; in most instances, hyphal repulsion occurred with both types. In contrast, all of the heterothallic species paired with either an A1 or an A2 isolate of *P. infestans*. In this way, all of the isolates of the heterothallic species were correctly labeled as to compatibility type. It was later found that oospores formed in almost all intraspecific and interspecific pairings of A1 and A2 isolates. Tracings of gametangial hyphae from *P. infestans* and *P. capsici* crosses revealed that interspecific gametangial fusions occurred to form oospores regardless of which species served as A1 or A2. This indicated that true intraspecific and interspecific hybridization occurred.

Studies of the type described above can be extended to include all of the currently recognized species. Should their patterns of sexual reproduction fall within the three major groups noted above, then these groups would serve as a starting place for a more natural system of classification.

## Genetic Interrelationships

The cytogenetics and genetics of species of *Phytophthora* are presented by Shaw (Chapter 6). Some of the data from cytologic studies may be useful for classification, but genetic studies have not added much to the determination of species relationships because most studies have been of intraspecific selfings and crosses of heterothallic species. Although the offspring of selfings in some homothallic species have been analyzed genetically, there have been no attempts to hybridize homothallic species.

The studies by Savage et al (1968), which standardized the designations of compatibility types among the heterothallic species, revealed that sexual bodies were formed in most interspecific combinations when A1 and A2 types were paired on a proper medium. Numerous oospores were formed in most combinations, indicating no genetic blockage between species. In some species combinations, gametangial fusions were numerous but the oogonial hyphae failed to emerge from the antheridia to form oogonia. Very few mature oospores were produced in pairings where this occurred. This phenomenon of incomplete copulation might indicate genetic blockage and provide evidence that the two species are indeed distinct. This phenomenon was evident in pairings of *P. infestans* with *P. cambivora, P. cinnamomi, P. colocasiae, P. citrophthora, P. meadii*, and *P. palmivora*, as well as in the pairings of *P. cambivora* × *P. citrophthora* and *P. capsici* × *P. cinnamomi*. Oospores failed to form in A1 × A2 combinations of *P. parasitica* var. *nicotianae* × *P. capsici, P. citrophthora* × *P. cambivora, P. infestans* × *P. citrophthora, P. parasitica* var. *nicotianae* × *P. citrophthora, P. citrophthora* × *P. cryptogea, P. infestans* × *P. cryptogea, P. parasitica* var. *nicotianae* × *P. cryptogea, P. infestans* × *P. mexicana*, and *P. parasitica* var. *nicotianae* × *P. mexicana*. Oospores were formed in reciprocal A2 × A1 pairings of the above. Whether or not these observations represent true genetic blockage between species is not known. A thorough search for the prevalence of these phenomena was not made, and the blockages might simply be associated with the particular isolates used.

There can be no question that hybrid oospores are produced in crosses of A1 and A2 types. They have been observed in intraspecific crosses of *P. infestans* (Galindo and Gallegly, 1960) and in interspecific crosses of *P. infestans* × *P. capsici* (Savage et al, 1968). Also, using the same tracing technique, Zimmerman (1973) observed hybrid oospores in intraspecific crosses of *P. capsici, P. parasitica, P. drechsleri*, and *P. infestans*, as well as in an interspecific cross of *P. palmivora* × *P. capsici*. However, selfing in single isolates of both A1 and A2 types has been observed for several species (Savage et al, 1968). Also, single isolates of some heterothallic species can be induced to self with plant extracts (Zentmyer, 1952, 1979), with chloroneb (Noon and Hickman, 1974), by *Trichoderma* species (Brasier, 1975), and by hormonal stimulation (Ko, 1978).

Savage et al (1968) observed that the A1 and A2 designations for opposite compatibility types assumed that the asexual stages of species of *Phytophthora* were haploid. However, they noted that should these stages prove to be diploid, the compatibility type designations could involve dominance and recessiveness, with one type being designated *aa* and the other *Aa*. It is interesting that selfings induced by avocado root extract, chloroneb, and *Trichoderma* species were with the A2 compatibility types. Should the A1 type be the homozygous recessive *aa* and the A2 be the heterozygous

*Aa*, then segregation for compatibility type would be expected in the $F_1$ of selfed A2 isolates. Indeed, this has been observed in *P. parasitica* by W. H. Ko (*unpublished*). Ko obtained 10 A1, 39 A2, and one A1A2 culture from single zoospores picked from an oospore sector induced in the mother A2 culture by chloroneb. This is close to the ratio of 1:3 expected by selfing of an *Aa* type (one *aa* to two *Aa* to one *AA*). However, Ko's statement that the A2 isolate used was a single-zoospore isolate from an aged A1 culture causes one to question the above interpretation, because the homozygous recessive character should not segregate.

Ko's (1978) polycarbonate membrane technique should be useful for testing the above hypothesis, along with a study of the offspring of viable selfed oospores induced by other means. Savage et al (1968) suggested that selfing in single cultures of either compatibility type was induced by stress from low nutrient supply or from the accumulation of some metabolite such as ammonium nitrogen. *Trichoderma* species and chloroneb would provide stress also. Perhaps in some way the self-incompatibility factor is inhibited when the fungus is under stress so that selfing will occur. The hormonal stimulation of selfing may be a different physiologic phenomenon.

A study of interspecific hybridization among species of *Phytophthora* should provide information useful in classification. However, such studies must take into account the selfing phenomenon noted above, and the oogonium germination phenomenon noted by Romero and Gallegly (1963). Techniques should be used to ensure the establishment of single-oospore cultures from true hybrid spores. Otherwise, difficulties in interpretation of results will arise, as noted by Boccas and Zentmyer (1976) with crosses of *P. parasitica* × *P. cinnamomi*. Although there was evidence of recombination for compatibility type, pathogenicity, and temperature response, all recombinant isolates had protein patterns similar to that of *P. parasitica*, indicating that selfing may have occurred. Previously, Zimmerman (1973) established cultures from two germ sporangia from one oospore of *P. palmivora* × *P. capsici* and determined that their characters were similar to the A1 isolate of *P. palmivora* used in the cross. Perhaps a selfed oospore was picked. More recently, Boccas (1981) obtained numerous single-oospore cultures from interspecific crosses of *P. megakarya* × *P. parasitica, P. parasitica* × *P. palmivora, P. parasitica* × *P. capsici, P. capsici* × *P. palmivora*, and *P. cambivora* × *P. cinnamomi*. The progeny varied in optimal and maximum temperatures for mycelial growth, colony type, and pathogenicity, but all but one of the protein patterns of the isolates were of the pattern of one of the parents. Isolate 5 of the *P. capsici* × *P. palmivora* cross had an intermediate protein pattern, intermediate colony type, the A1 compatibility type of the *P. capsici* parent, the maximum temperature of the *P. palmivora* parent, and the pathogenicity of *P. capsici* and was considered to be a product of hybridization.

Boccas (1981) again discussed the problem of induced selfing in both compatibility types from hormonal interchange and the occurrence of selfed and hybrid spores together in paired cultures. He suggested that perhaps only selfed oospores were viable in most crosses. His data also illustrate that recombination for compatibility type may occur among the offspring of selfed A2 types. Among the progeny from *P. megakarya* × *P. parasitica*, two isolates with protein patterns of the A2 *P. parasitica* parent were of compatibility type A1; similarly, one isolate of *P. capsici* × *P. palmivora* had a protein pattern of the A2 *P. palmivora* but was of type A1. The data support the hypothesis that the A2 type is the heterozygous *Aa* compatibility type. Should further genetic work confirm this hypothesis, then perhaps the designations for compatibility types should be changed to *aa* and *Aa* to reflect dominance and recessiveness and diploidy.

I hope that the future will present many more intraspecific and interspecific genetic studies of the currently recognized species to help us to better understand species relationships. Ullrich and Raper (1977) suggest that gene flow within and among populations of fungi probably occurs more easily than in many other organisms and that it is enhanced by heterothallism. They note that gene flow is important in "stabilizing the species with a high level of variability to buffer against the vagaries of the environment." Further, they suggest that interspecific hybridization in the fungi may be of relatively little significance. The results from further studies of intraspecific and interspecific hybridization should help determine the degree of variability for morphological characters now used for classification and whether or not a key character can be easily transferred from one thallus to another. Until this is done, the question of how many natural species are present in the genus *Phytophthora* may remain unanswered.

## Morphology of Sexual Organs

The next obvious step in a natural system of classification would be to use the differences in morphology of the sexual organs. However, certain differences in morphology of these organs may be of no more importance than differences in morphology of certain asexual stages. Obviously, the type of antheridium, whether paragynous or amphigynous, is of major importance and is associated with sexual pattern. Waterhouse (1963) used this in the establishment of her six major groups.

Morphological descriptions of the sexual organs, when known, are included in the original descriptions of the species (Waterhouse, 1970). Variations in the morphology of these stages in subsequent literature are difficult to find, and when such a variation is described, it may be incidental to the major thrust of a paper. More recently, Ho and Zentmyer (1977) and Ho et al (1977) reexamined the morphology of *P. cambivora* and *P. cinnamomi* using scanning electron microscopy. Also, Ho (1979) used this tool to examine the nature of amphigyny in the genus.

Newhook et al (1978) list the morphological variations of antheridia, oogonia, and oospores useful in identifications. They include, in addition to paragyny and amphigyny characteristics of antheridia, whether or not antheridia are subterminal, have tangled hyphae, are elongated, and contain residual oil globules, as well as the number of cells and differences in size. Characters of the oogonia, in addition to variations in size, include whether or not they are ornamented, have tapered bases, have thick walls, are clustered, have eccentric stalks, and become rough with age. Oospore size is used and seems to vary with size of oogonia, except when the oospores are markedly aplerotic. Thickness of the oospore wall is used also.

The question arises whether any of these differences in morphological characters are associated with natural species and which of the key characters now used are simple variants. For instance, *P. cambivora* and *P. cinnamomi* have many characters in common, but the rounded protuberances on the wall of the oogonium of *P. cambivora* serve as a key character separating it from *P. cinnamomi*. Should future genetic studies reveal that this bullate character is easily transferred into isolates now considered to be *P. cinnamomi*, then the character might simply be a morphological variant, and all isolates with the bullate gene would not necessarily be identified as *P. cambivora*.

### Characters of the Asexual Stages

Some characters of the asexual stages may be more important in the establishment of natural species of *Phytophthora* than are some of the sexual stages. Waterhouse (Chapter 10) has meticulously listed those good characters that are stable. The characteristics of the sporangia, such as type of papillation, nature of the exit pore, shape, whether or not they are caducous, and length of the pedicel attached to sporangia of caducous species, probably have evolved as a result of selection pressures applied by the different natural habitats of species. Likewise, the presence or absence and characteristics of other asexual bodies such as chlamydospores and hyphal swellings may have resulted from survival evolution in particular habitats. These characters must be used as major criteria in a key for identification, for they may be the only characters evident in single isolates. Asexual characteristics may or may not have a dominant role in the development of a natural system of classification, but they will play a role.

Once the barrier to oospore germination has been overcome and techniques are developed to ensure that progeny of hybrid oospores are being studied, specific differences in asexual and sexual characters among the current species may be studied genetically. The easy transfer of a particular character by such artificial matings would not necessarily invalidate that character for use in species identifications because, in nature, such associations may never occur.

### Pathogenicity

All species of *Phytophthora* are plant pathogens. Some are pathogenic primarily to foliage and other aboveground parts, and some are considered pathogens of roots and other underground organs. Other species may invade plant parts both above and below ground. Some species, such as *P. infestans, P. fragariae,* and *P. phaseoli*, have narrow host ranges, whereas others, such as *P. cinnamomi* and *P. parasitica*, attack many different host species. Within some species, such as *P. megasperma*, host specificity is designated by forma specialis (f. sp.); e.g., isolates pathogenic to soybean are designated *P. megasperma* f. sp. *glycinea*, whereas those pathogenic to alfalfa are identified as *P. megasperma* f. sp. *medicaginis* (Kuan and Erwin, 1980). Also, physiologic races with differential pathogenicity to differing genotypes of the same host occur within some species, e.g., *P. infestans, P. phaseoli*, and *P. megasperma* f. sp. *glycinea* (see also Erwin, Chapter 11).

The experienced plant pathologist often can identify a *Phytophthora* species by the symptoms produced on a particular plant part. The pathogen of a typical late blight lesion on a potato leaf will be *P. infestans*, that of downy mildew of lima bean will be *P. phaseoli*, and that of red stele of strawberry will be *P. fragariae*. In contrast, some of the root and crown rots as well as fruit and pod rots may be caused by several different species producing similar symptoms, e.g., buckeye rot of tomato. In such cases, the *Phytophthora* species must be identified by other means. Thus, pathogenicity can be used as a taxonomic criterion for some species but is of little value for many others, particularly some of the soilborne species. However, except for new diseases, the host from which an isolate is obtained often serves as a useful guide toward species identification.

## POTENTIAL NONMORPHOLOGICAL CRITERIA

Certain nonmorphological studies are yielding information that may, in time, be useful for separation of species of *Phytophthora*. Hemmes (Chapter 2) presents electron microscopy studies of the ultrastructural organizations of the various organs of *Phytophthora*, but these studies have not yet involved species differences. Hohl (Chapter 3) discusses the nutritional requirements and extracellular enzyme activities among species for possible taxonomic use, and Shaw (Chapter 6) reviews the cytogenetics of the various species. The cytologic studies will not be reviewed here, except to note that differences in chromosome numbers, when known, can be useful in identifications when this technology is available. However, except for the few fastidious species, nutritional differences may not be too helpful. Immunological studies involving serology and protein patterns have provided some helpful information on species relationships and should be even more helpful as they are expanded and as the techniques for doing them are improved.

### Serology

Burrell et al (1966) used immunodiffusion and immunofluorescence techniques to study *P. cactorum, P. cinnamomi*, and *P. erythroseptica* and found distinct species-specific serologic differences among them. They purposely selected these three morphologically distinct species for initial studies in order to develop techniques and to determine whether or not serology could be used to separate species of this genus. In the beginning, the antigen preparations failed to induce antibody formation in the animals, but this was overcome with use of young (5-day-old) mycelial cultures growing in a special liquid medium. After the development of techniques that clearly indicated serologic differences among these three species, five heterothallic species with morphological similarities were studied (Merz et al, 1969).

Savage et al (1968) had considerable difficulty in identifying isolates received as *P. arecae, P. citrophthora, P. meadii, P. mexicana*, and *P. palmivora* and questioned whether or not all were valid species. They found considerable variation in the morphological characters used for separation of these five species and suggested that perhaps the five should be lumped into either one or two species. These same five species, and some of the same isolates, were then studied serologically by Merz et al (1969), along with an isolate of *P. cinnamomi*. On the basis of immunodiffusion and indirect fluorescent antibody tests, *P. cinnamomi* was serologically distinct from the other five species. When tested with antiserum to either *P. meadii* or one isolate of *P. palmivora*, all other species reacted identically. However, when antiserum to *P. arecae* was used, *P. arecae, P. citrophthora*, and one isolate of *P. palmivora* reacted as one group, whereas *P. meadii, P. mexicana*, and another isolate of *P. palmivora* reacted as a second group. At this point, it is not possible to determine whether the new species, *P. megakarya* (Brasier and Griffin, 1979), was among the isolates tested. The results indicated that the antigens in the first group were also present in the second group, but that the isolates in the second group carried additional antigens. Based on oogonial size and pigmentation of the sexual bodies, Savage et al (1968) indicated that *P. citrophthora, P. mexicana*, and one isolate of *P. palmivora* formed one group, whereas *P. arecae, P. meadii*, and another isolate of *P. palmivora* formed another. Thus, the serologic groups differed from the morphological groups. However, the results from both studies indicated that perhaps only one species existed among the isolates used. If so, then *P. citrophthora* would have priority as the species name. If two species are recognized among the five, then one group would be named *P. citrophthora* and the other *P. palmivora*. Based on the results from serologic studies, Merz et al (1969) indicated that the incorporation of the five species into one seemed more advantageous than establishing two serologic groups.

Before the studies noted above, Morton and Dukes (1967) demonstrated distinct serologic differences between the genera *Phytophthora* and *Pythium* using immunodiffusion tests but could not demonstrate differences between *P. parasitica* and *P. parasitica* var. *nicotianae*. Throughout this symposium, differences of opinion have been expressed as to whether there are two separate species, *P. parasitica* and *P. nicotianae*, or whether one is a variety of the other. The results of Morton and Dukes (1967) indicate that they are identical serologically.

It is obvious that much more work will be needed before serologic techniques can be used for species identification or for support of identifications based on morphology. Many more species need to be studied, and the newer serologic techniques need to be examined for use with species of *Phytophthora*. MacDonald and Duniway (1979) used fluorescent antibodies to study the survival of zoospores of *P. megasperma* f. sp. *medicaginis* and *P. cinnamomi* in soil but indicated that their antisera were not species-specific. Genus-specific antiserum was recognized by Merz et al (1969), but cross-absorption techniques were used to obtain species-specific antisera. Although serologic investigations are expensive and time-consuming, the development of a central supply of species-specific antisera would serve a useful need, not only in taxonomy but also for other uses such as the monitoring of inoculum potential in soil. Before this can come about, much more work will be needed.

### Protein Patterns

Perhaps the most promising of the physiologic methods for use in classification and identification of species of *Phytophthora* is the determination of protein patterns by gel electrophoresis. The rationale used is that valid species should differ in their protein makeup and that such differences can be detected by the rate at which the proteins move in a gel aided by an electric current. Electrophoretic measurement of protein differences can be much less difficult and time-consuming than measurement by serologic methods. Although the reliability of protein patterns as a taxonomic tool for other fungi is controversial, the data thus far obtained with species of *Phytophthora* appear to offer reliability. Starch and acrylamide gels have been used in slab and tubular electrophoresis to produce these patterns.

The first evidence that protein patterns may be useful in identification of pythiaceous fungi was provided by Clare (1963), who used starch gel electrophoresis to produce patterns for six species of *Pythium*. Each species produced a distinctly different pattern, with one band common to all species. Three isolates of one species had identical patterns, as did two isolates of another species. Using the same method, Clare and Zentmyer (1966) found that the protein patterns of *P. cinnamomi, P. citrophthora*, and *P. palmivora* were distinctly different and that the patterns of different isolates of the same species were identical or similar, regardless of geographic or host source. Using polyacrylamide gel electrophoresis, Hall et al (1969) obtained similar results with *P. cactorum, P. cinnamomi*, and *P. palmivora*, but the *P. palmivora* isolates showed two bands, which were different quantitatively and identical qualitatively. Esterase patterns were also examined, but 13 different patterns were found among the isolates of the three species. In retrospect, one wonders whether the two different bands among the isolates of *P. palmivora* are associated with any of the different morphological forms of the species.

Gill and Powell (1968b), using polyacrylamide gel (disc) electrophoresis, found distinct protein patterns for *P. cactorum, P. fragariae*, and *P. sojae*. However, this technique did not differentiate the eight physiologic races of *P. fragariae* (Gill and Powell, 1968a); all isolates of *P. fragariae* had nearly identical protein patterns regardless of race. Later, Gill and Zentmyer (1978) found that the protein patterns from polyacrylamide gel (disc) electrophoresis of 30 isolates of *P. cinnamomi* were identical except for a minor variation by one isolate. Likewise, five isolates of *P. cactorum* had identical patterns. The *P. cinnamomi* isolates were from 17 different hosts and from widely separated geographic areas. Further, neither the length of time an isolate had been in culture nor the compatibility types A1 and A2 affected the protein patterns. Boccas and Zentmyer (1976) and Boccas (1981) determined the protein patterns of the parents used in their genetic studies of interspecific hybrids and compared the protein patterns of cultures established from single germinating oospores. They concluded that all but one of the cultures were from selfed oospores. Previously, Zentmyer et al (1974) used protein patterns to help identify *P. citricola* obtained from avocado roots.

Thus, distinctly different protein patterns have been obtained for seven species of *Phytophthora*. Except for *P. palmivora* and *P. citrophthora*, all of these species are easily separated by morphological criteria. It is now time to study the protein patterns of the more difficult taxonomic groups, and closely related species, to determine whether this technology will help clarify taxonomic controversy in this genus and become an identification tool for use along with morphological criteria. Perhaps this technology will provide guidance in the *P. megasperma* complex, in the varieties of *P. parasitica*, and in groups such as the one containing *P. cinnamomi, P. cambivora, P. cryptogea*, and *P. drechsleri*. Also, the technique should be helpful in studies of the morphological forms of *P. palmivora* and in determining the relationship of this species to *P. arecae, P. citrophthora, P. meadii, P. megakarya*, and *P. mexicana*. An ideal study would use all species and would be conducted in one laboratory under the same experimental conditions. Although protein patterns from gel electrophoresis appear to be suitable for identifications at the species level, they may not differentiate further divisions within the species, such as physiologic races. Perhaps newer refinements, such as isoelectric focusing, will bring out these smaller differences.

### The Computer

The increasing availability of microcomputers to research laboratories is creating a need for a computer program for identification of *Phytophthora* species. The criteria now available from Newhook et al (1978) should be quite suitable for developing an initial program. The synoptic key of Ho (1981) is a beginning. Should future data from electrophoretically produced protein patterns confirm that each species has its own unique pattern, perhaps data such as the $E_f$ values can be programmed for use. A computerized key for identification may, or may not, be different from a key for classification.

## SUMMARY

Tribute is paid to Grace M. Waterhouse for providing the current key used throughout the world for identifying species of *Phytophthora*. It has created order where there was confusion. The key works, once one learns to produce and recognize the many morphological criteria, but to many plant pathologists, the key seems to be too finely tuned, and there is a lingering sense that some of the morphological characters sort out variants rather than species.

A more natural system of classification is proposed, beginning with the characteristics of sexuality rather than with the characteristics of sporangial papillae. The three major groups of homothallic paragynous, homothallic amphigynous, and heterothallic amphigynous species would be further subdivided by genetic relationship, sexual morphology, asexual morphology, and host-specific pathogenicity. Differences in chromosomal characters, serologic relationships, and protein patterns should play a strong role in the rearrangement of species into more closely related groups and in determining the validity of species.

It is recognized that much more information on genetic interrelationships, chromosomes, serology, and protein patterns will be needed to mold the Waterhouse key into a more natural system of classification. Moreover, it is recognized that the more natural system of classification may not be practical for identification, particularly in the field where the technology provided by sophisticated laboratory equipment may not be available, and that a separate or companion identification key may be needed.

## LITERATURE CITED

Boccas, B. R. 1981. Interspecific crosses between closely related heterothallic *Phytophthora* species. Phytopathology 71:60-65.

Boccas, B., and Zentmyer, G. A. 1976. Genetical studies with interspecific crosses between *Phytophthora cinnamomi* and *Phytophthora parasitica*. Phytopathology 66:477-484.

Brasier, C. M. 1975. Stimulation of sex organ formation in *Phytophthora* by antagonistic species of *Trichoderma*. I. The effect in vitro. New Phytol. 74:183-194.

Brasier, C. M., and Griffin, M. J. 1979. The taxonomy of *Phytophthora palmivora* on cocoa. Trans. Br. Mycol. Soc. 72:111-144.

Burrell, R. G., Clayton, C. W., Gallegly, M. E., and Lilly, V. G. 1966. Factors affecting the antigenicity of the mycelium of three species of *Phytophthora*. Phytopathology 56:422-426.

Clare, B. G. 1963. Starch-gel electrophoresis of proteins as an aid in identifying fungi. Nature (London) 200:803-804.

Clare, B. G., and Zentmyer, G. A. 1966. Starch gel electrophoresis of proteins from species of *Phytophthora*. Phytopathology 56:1334-1335.

Galindo A., J., and Gallegly, M. E. 1960. The nature of sexuality in *Phytophthora infestans*. Phytopathology 50:123-128.

Gallegly, M. E., and Galindo, J. 1958. Mating types and oospores of *Phytophthora infestans* in nature in Mexico. Phytopathology 48:274-277.

Gill, H. S., and Powell, D. 1968a. Polyacrylamide gel (disc) electrophoresis of physiologic races A-1 to A-8 of *Phytophthora fragariae*. Phytopathology 58:722-723.

Gill, H. S., and Powell, D. 1968b. The use of polyacrylamide gel disc electrophoresis in delimiting three species of *Phytophthora*. Phytopathol. Z. 63:23-29.

Gill, H. S., and Zentmyer, G. A. 1978. Identification of *Phytophthora* species by disc electrophoresis. Phytopathology 68:163-167.

Gough, F. J. 1957. The sexual stage of *Phytophthora infestans*, the cause of potato and tomato late blight. Ph.D. dissertation. W. Va. Univ., Morgantown. 47 pp.

Hall, R., Zentmyer, G. A., and Erwin, D. C. 1969. Approach to taxonomy of *Phytophthora* through acrylamide gel-electrophoresis of proteins. Phytopathology 59:770-774.

Ho, H. H. 1979. The nature of amphigyny in *Phytophthora*. Mycologia 71:1057-1064.

Ho, H. H. 1981. Synoptic keys to the species of *Phytophthora*. Mycologia 73:705-714.

Ho, H. H., and Zentmyer, G. A. 1977. Morphology of *Phytophthora cinnamomi*. Mycologia 69:701-713.

Ho, H. H., Zentmyer, G. A., and Erwin, D. C. 1977. Morphology of sex organs of *Phytophthora cambivora*. Mycologia 69:641-646.

Ko, W. H. 1978. Heterothallic *Phytophthora*: Evidence for hormonal regulation of sexual reproduction. J. Gen. Microbiol. 107:15-18.

Kuan, T.-L., and Erwin, D. C. 1980. Formae speciales differentiation of *Phytophthora megasperma* isolates from soybean and alfalfa. Phytopathology 70:333-338.

MacDonald, J. D., and Duniway, J. M. 1979. Use of fluorescent antibodies to study the survival of *Phytophthora megasperma* and *P. cinnamomi* zoospores in soil. Phytopathology 69:436-441.

Merz, W. G., Burrell, R. G., and Gallegly, M. E. 1969. A serological comparison of six heterothallic species of *Phytophthora*. Phytopathology 59:367-370.

Morton, D. J., and Dukes, P. D. 1967. Serological differentiation of *Pythium aphanidermatum* from *Phytophthora parasitica* var. *nicotianae* and *Ph. parasitica*. Nature (London) 213:923.

Newhook, F. J., Waterhouse, G. M., and Stamps, D. J. 1978. Tabular key to the species of *Phytophthora* de Bary. Mycol. Pap. 143. Commonw. Mycol. Inst., Kew, Surrey, England. 20 pp.

Noon, J. P., and Hickman, C. J. 1974. Oospore production by a single isolate of *Phytophthora capsici* in the presence of chloroneb. Can. J. Bot. 52:1591-1595.

Romero, S., and Gallegly, M. E. 1963. Oogonium germination in *Phytophthora infestans*. Phytopathology 53:899-903.

Savage, E. J., Clayton, C. W., Hunter, J. H., Brenneman, J. A., Laviola, C., and Gallegly, M. E. 1968. Homothallism, heterothallism, and interspecific hybridization in the genus *Phytophthora*. Phytopathology 58:1004-1021.

Smoot, J. J., Gough, F. J., Lamey, H. A., Eichenmuller, J. J., and Gallegly, M. E. 1958. Production and germination of oospores of *Phytophthora infestans*. Phytopathology 48:165-171.

Tucker, C. M. 1931. Taxonomy of the genus *Phytophthora* de Bary. Mo. Agric. Exp. Stn. Res. Bull. 153. 208 pp.

Ullrich, R. C., and Raper, J. R. 1977. Evolution of genetic mechanisms in fungi. Taxon 26:169-179.

Waterhouse, G. M. 1956. The genus *Phytophthora*, diagnoses (or descriptions) and figures from the original papers. Misc. Publ. 12. Commonw. Mycol. Inst., Kew, Surrey, England. 120 pp.

Waterhouse, G. M. 1963. Key to the species of *Phytophthora* de Bary. Mycol. Pap. 92. Commonw. Mycol. Inst., Kew, Surrey, England. 22 pp.

Waterhouse, G. M. 1970. The genus *Phytophthora* de Bary. Mycol. Pap. 122. Commonw. Mycol. Inst., Kew, Surrey, England. 104 pp.

Zentmyer, G. A. 1952. A substance stimulating sexual reproduction in *Phytophthora cinnamomi*. (Abstr.) Phytopathology 42:24

Zentmyer, G. A. 1979. Stimulation of sexual reproduction in the $A^2$ mating type of *Phytophthora cinnamomi* by a substance in avocado roots. Phytopathology 69:1129-1131.

Zentmyer, G. A., Jefferson, L., Hickman, C. J., and Chang-Ho, Y. 1974. Studies of *Phytophthora citricola*, isolated from *Persea americana*. Mycologia 66:830-845.

Zimmerman, R. K. 1973. Problems in the establishment of interspecific hybrids in the genus *Phytophthora* de Bary. Ph.D. dissertation. W. Va. Univ., Morgantown. 78 pp.

# 13

# Summary of the Open Discussion Session on Taxonomy of *Phytophthora*

M. E. Gallegly

CONTENTS



## INTRODUCTION

More than a century has passed since Anton de Bary established the genus *Phytophthora* in 1876 by describing the nature of *P. infestans*. During these years, more than 60 species and subspecies have been described, and today about 50 species, varieties, morphological forms, and formae speciales remain as separate entities. Only six taxonomic keys have been presented, beginning with Rosenbaum's in 1917. Currently, the 1963 key of Waterhouse and the updated (in 1978) tabular key of Newhook, Waterhouse, and Stamps are the main ones in use.

The genus *Phytophthora* is difficult, particularly when one is trying to identify an isolate. Both the beginning plant pathologist and the experienced mycologist have difficulties. Extreme variability within and among species is the rule rather than the exception. Paragyny, amphigyny, homothallism, and heterothallism exist in the genus. Among the heterothallic species, pairing of A1 and A2 compatibility types is required for sexual spore formation by some. Natural media, or synthetic media containing 3-$\beta$-hydroxy sterols, are required for formation of oospores, viable sporangia, and chlamydospores. Species differ in their ability to sporulate on different natural media. Light and darkness affect both sporulation and spore germination. Some species are considered fastidious in nutrient requirements for growth. Certain species have a narrow host range, whereas others are pathogenic to many plants in different families. These are just a few examples of the complexity in the genus, and because of this complexity, it is fitting that a session updating the taxonomy of the genus *Phytophthora* be a part of this symposium and the resulting book.

An open discussion was held following the presentations of formal papers by Waterhouse (Chapter 10), Erwin (Chapter 11), Gallegly (Chapter 12), and Brasier (part of Chapter 27). Brasier and Gallegly jointly presented orally the paper on new criteria for classifying species of *Phytophthora* and critique of existing approaches because of the sudden and unexpected illness of C. J. Shepherd, who was originally scheduled to cover this subject.

The highlights of the open discussion are presented here. A symbol (•) is used to separate contributions to the discussion.

## OPEN DISCUSSION

The format for the open discussion was for members of a panel consisting of Waterhouse (Richmond, Surrey, England), Newhook (University of Auckland, New Zealand), Stamps (Commonwealth Mycological Institute, Kew, Surrey, England), Erwin (University of California, Riverside), and Brasier (Forest Research Station, Wrecclesham, Farnham, England) to respond to questions and comments from the floor. Gallegly (West Virginia University) served as moderator for the session.

*The discussion began with disagreement over Brasier's statement that the botanical rules of nomenclature may not be adequate for classification of species of* Phytophthora *and that physiologic criteria may be needed.* Physiology has not helped much so far, and we should look at what fungi are and not at what they do. We need a way to obtain high germination percentages of oospores so that heritability studies can be made to better our understanding of ploidy, and we need a medium that will maintain isolates as they appear in nature (Edward Butler, University of California, Davis). • The point made on the use of physiology in classification was that morphological characters should be redefined in physiologic terms, and as for a uniform medium, there may never be one (Brasier). • Much is known about the nutritional requirements for vegetative growth, but physiologic requirements for spore formation and oospore germination are not well understood (H. R. Hohl, University of Zurich, Switzerland). • Natural and synthetic media that support oospore production by many species are available, and good oospore germination can be obtained with *P. megasperma* and some other homothallic species, but good germination of heterothallic species is still blocked (O. K. Ribeiro, University of California, Riverside). • Physiologic characters, e.g. nutritional requirements, are much too variable to be used reliably for taxonomic purposes; in contrast, biochemical characters, such as the structure of macromolecules, are remarkably constant and someday might be the decisive key to settle taxonomic controversies (S. Bartnicki-Garcia, University of California, Riverside).

One may feel uneasy about the use of physiologic characters, particularly nutritional differences, for taxonomic purposes. There are too many simple mutations for nutritional deficiencies, which may not show up until the isolate is in artificial culture. If a nutritional deficiency is linked with a gene required for pathogenicity, that deficiency could quickly become widespread and be identified as a new species if nutritional differences were to play a major role in classification (C. H. Canaday, Ohio State University, Columbus). • Morphological characters should not be played against biochemical ones, because there are good characters of both types. What is important is whether or not the character is stable. Theoretically, characters of multigenic control might be so stable that there would be no differences between species; likewise, characters of single genes with multiple copies in the genome that show qualitative differences would be stable because a mutation by one copy would not change the stability of that character (Hohl). • The recognition and use of stable characters for taxonomy is ideal, but even these will vary, and the limits of variability of each character must be defined. Lack of understanding and recognition of such variability may explain the disagreements over so many species with certain characters that fall between the limits (Erwin).

We hope that physiologic studies will help identify the morphological structures that have some meaning; the plant pathologist in the field can then make better identification, with isolates being sent to the backup authority, who may use physiologic characters (Newhook). • Thus, classification and identification are two different procedures; classifiers need extra tools to support moves to split or lump, whereas identifiers do not want to use difficult things such as chromosome numbers and certain physiologic characters that may require the use of sophisticated equipment (Waterhouse). • Identifiers want to base their identifications on something they can look at and link with basic things (Stamps). • Taxonomy is order but also communication (Bartnicki-Garcia).

*Discussion turned to the question of the use of trinomial designations.* They should be abandoned because of redundancy; if enough differences exist, different species should be named (Bartnicki-Garcia). • The trinomial system is not necessary, particularly in regard to designations of varieties, because differences between varieties within a species are so small in comparison with the large differences between the species containing varieties. • *The panel was asked*: Are varieties needed and valid, specifically in *P. nicotianae* (P. H. Tsao, University of California, Riverside)? • If anyone had suggested that *P. nicotianae* var. *nicotianae* and *P. nicotianae* var. *parasitica* should be set up as two separate species, this would have been done, because they are morphologically different, but by no means should they be lumped together (Waterhouse). • There is an anomaly with this species as the varieties are now named, because "*nicotianae*" implies host specificity and "*parasitica*" implies a broad host range. Although variety designations are not supposed to be used in the sense of formae speciales to designate host-specific variants, the nontaxonomist is left with the impression that *P. nicotianae* var. *parasitica* is a variety of a host-specific species that has a broad host range. In spite of the rules of nomenclature, it is more logical that the species name should imply the broad host range and the variety name the more narrow one. Naming two separate species would resolve this anomaly (Gallegly).

*The trinomial system that uses forma specialis (f. sp.) to indicate host specificity was supported because of its usefulness to plant pathologists.* The f. sp. designation carries a meaning similar to that of cultivar designations for economically important higher plants (Erwin). • The concept of forma specialis is that various formae are indistinguishable morphologically but can be distinguished pathogenetically. The formae of *P. megasperma* show some morphological differences, however, and do not conform to this concept. Until it is accepted that f. sp. has more meaning mycologically, this designation should be reserved for pathogenic characters (Newhook). • This concept was used to designate an isolate of *P. drechsleri*, with specific pathogenicity to pigeon pea and closely related species, as *P. drechsleri* f. sp. *cajani* (J. Kannaiyan, ICRISAT, Andhra Pradesh, India). • A system similar to the forma specialis is used for the *Pseudomonas* leaf-spotting organisms; more than 50 have been lumped into *Pseudomonas syringae*, and pathovars (pv.) are used to indicate host specificity; for example, *Pseudomonas syringae* pv. *phaseolicola* (N. T. Keen, University of California, Riverside). • Although plant pathologists are using the pathovar system, the International Committee on Systematic Bacteriology has not accepted it (P. W. Tooley, University of Wisconsin, Madison). • It is not as important to consider trinomials from the standpoint of the identifier as it is to have some means of easily identifying a species to communicate with others (J. E. Dimitman, California State Polytechnic University, Pomona).

*The questions asked by Brasier about the confusion existing in* P. megasperma *stimulated considerable discussion.* The current key lists two varieties, the large-oogonium var. *megasperma* and the small-oogonium var. *sojae*. More recently, Kuan and Erwin proposed that the morphologically different subgroups designated by variety be discarded and that forma specialis subgroups be erected to designate host-specific pathogenicity differences between alfalfa (f. sp. *medicaginis*) and soybean (f. sp. *glycinea*) isolates. • *In response to the question of whether or not all large-spored isolates came from woody plants and all small-spored ones from herbaceous plants*: Results from research in Oregon indicate no correlation between host source and oogonium size (P. B. Hamm, Oregon State University, Corvallis). Large-oogonium and small-oogonium isolates were obtained from both the herbaceous alfalfa and the woody Douglas-fir. Although the data indicate a slight correlation between host type and oogonium size, there are many exceptions. However, the data clearly showed that oogonium size among isolates of *P. megasperma* forms a continuum from large to small, as previously shown by Kuan and Erwin. Support was given to the proposal that morphological varieties in *P. megasperma* should not be recognized and that a new description of the species recognizing such variability should be accepted. A large number of isolates should be examined for similarities rather than for differences (Hamm). • *The question was asked*: Since Hildebrand sorted out variety *sojae* not only on the basis of small oogonia but also because of its growth at a higher temperature, is there a correlation among the temperature optimum, host specificity, and oogonial size of the many isolates of *P. megasperma* now available? Within the forma specialis concept, recent work shows that isolates from a particular host are pathogenic to that host only in the proper environment; some isolates of *P. megasperma* are pathogenic at 20° C but not at 25° C when flooding is done. This suggests that a thorough investigation of environmental effects on disease development is needed before designation of host specificity (Wayne Wilcox, University of California, Davis). *Many of the arguments presented here briefly on* P. megasperma *are covered in more detail and are documented in Chapter 27 by Brasier and Chapter 11 by Erwin.*

Perhaps the situation for *P. megasperma* is one in which morphological characters do not give the types of clear-cut differences that are needed for identification. Again, more backup information is needed and could be provided by physiologic, genetic, and cytologic studies. The basic morphological criteria must not be discarded, but other types of criteria are needed to use hand-in-hand with the morphological criteria. The potential new criteria must be sorted into good ones and bad ones. We must get through the phase of experimental taxonomy and come back to redefine morphological criteria within the new parameters (Brasier).

# 14

# Role of Physical Factors in the Development of *Phytophthora* Diseases

J. M. Duniway

CONTENTS



## INTRODUCTION

Climate influences the development of all diseases caused by *Phytophthora* spp., and the climatic constraints on certain species of *Phytophthora* determine the geographic distribution of the diseases they incite. For example, *P. cinnamomi* is not a pathogen of consequence in a field where winter temperatures are repeatedly or consistently below freezing (Zentmyer, 1980). It is also not an important pathogen of Douglas-fir in the Pacific Northwest, where Roth and Kuhlman (1966) found that there was little coincidence between sufficiently warm and sufficiently moist soil conditions for disease development. In contrast, the Mediterranean climates of Australia have periods, especially in the fall and spring, when both soil moisture and temperature are highly suitable for *P. cinnamomi* to be active (Shea, 1975; Zentmyer, 1980), and it has become a destructive pathogen in the jarrah forests of Western Australia.

Unfortunately, *Phytophthora* is a large and diverse genus, and where one species may fail to be a pathogen, another may succeed. The cold climates that restrict pathogenic activities by *P. cinnamomi* in soil, for example, may not restrict the activities of *P. syringae, P. cactorum, P. megasperma, P. lateralis,* or several other species. The climate in most agricultural regions of the world will permit a number of *Phytophthora* spp. to be active pathogens, at least during some part of the year. To a large extent, the seasonal activities of the host and the influence of physical factors of the environment determine the time course of epidemics caused by *Phytophthora* spp.

The microenvironment around aerial parts of plants routinely cycles between highly permissive and very restrictive conditions for growth, inoculum production, dispersal, and infection by *Phytophthora* spp. In fact, the early associations of late blight epidemics with certain weather conditions in Ireland and England were so strong that the disease was first attributed to the effects of bad weather on potato. It soon became evident, however, that *P. infestans* and late blight epidemics are highly responsive to weekly or even daily changes in the weather. In many potato-growing regions where the weather is sufficiently variable during crop growth, physical parameters such as rainfall, dew period, humidity, temperature, and solar radiation are used to forecast the onset or even the rate of development of late blight epidemics (e.g., DeWeille, 1964; Krause and Massie, 1975). A great deal of epidemiologic research has been done on late blight, partly because of its predictability, but for the most part, the role of physical factors in the epidemiology of other *Phytophthora* diseases is considerably less well known.

A majority of *Phytophthora* spp. are soilborne. Compared to the aerial environment, the soil has relatively stable temperatures, a relatively high and stable water status, and very little or no solar radiation. Therefore, almost none of the physical parameters that influence *Phytophthora* spp. in soil change back and forth between permissive and restrictive states as rapidly as do the physical parameters of the aerial environment. Considerations of temperature effects on late blight of foliage and tubers illustrate some of the differences between the soil and aerial environments. Among other things, Sato (1979) found that soil temperatures during and after rainfall had to be below 18° C for *P. infestans* to infect tubers, largely because the release and motility of zoospores were enhanced at cooler temperatures. Soil temperatures at tuber depth are fairly constant and tend to approach the mean ambient temperature of the preceding few weeks. Therefore, although mean temperatures above 18° C will limit tuber infection, they may still allow the development of considerable foliage blight because diurnal fluctuations in ambient temperature may frequently include night temperatures that are highly suitable. The moisture requirements for *P. infestans* to infect foliage and growing tubers also differ greatly; a few hours of rain or dew at suitable temperatures suffice for foliage infection (e.g., DeWeille, 1964), whereas a much greater accumulation of water in the soil from rain or irrigation is required for tuber infection (Lapwood, 1977; Sato, 1979).

Although the soil environment is less variable over time and is generally less apparent to the agriculturalist or plant pathologist than is the aerial environment, some of the effects of physical factors in the soil on Phytophthora root rots, particularly the effects of very wet soil conditions, are as striking as the effects of wetness on *Phytophthora* diseases of aerial plant parts. In fact, root rots incited by many *Phytophthora* spp. on a wide variety of plants are aggravated when the soil is saturated for prolonged periods (e.g., Wilcox and Mircetich, 1979; other citations in Duniway, 1979; Zentmyer, 1980). It is impossible to discuss the roles of all the physical factors that influence *Phytophthora* diseases in one chapter. Therefore, the effects of water status on the development of *Phytophthora* diseases, especially the soilborne examples, will be emphasized.

Frequently, the behavior of an epidemic is determined largely by the direct effects of physical factors on the pathogen.

This is thought to be the case for many *Phytophthora* diseases, and the bulk of research addressing the role of physical factors in their epidemiology has focused on the direct effects of temperature and moisture on the behavior of *Phytophthora* spp. For the most part, the asexual reproductive cycle—the repeating cycle from parasitic mycelium to sporangia, zoospores, and more parasitic mycelium—prevails during the epidemic development of *Phytophthora* diseases. Therefore, some of the direct influences of water status and, to a lesser extent, temperature and solar radiation on the biology of sporangia, zoospores, and mycelium are reviewed here. However, the effects of physical factors on disease development can be influenced to some extent by a response of the host or by the activities of other microorganisms, which may stimulate or suppress the pathogen. Much is said elsewhere in this volume about the role of competitive microorganisms (Malajczuk, Chapter 16; Shea and Broadbent, Chapter 26; Weste, Chapter 18) and about the physiology of resistance and disease (Keen and Yoshikawa, Chapter 21); some of the influences of physical factors on host responses to *Phytophthora* spp. are also reviewed here. More general discussions of the epidemiology of *Phytophthora* diseases and more detailed reviews of the effects of physical factors on the biology of *Phytophthora* spp. can be found in other chapters of this volume (Gisi, Chapter 8; MacKenzie et al, Chapter 23; Mitchell and Kannwischer-Mitchell, Chapter 19; Ribeiro, Chapter 4).

The status of water in fungi and the substrates on which they grow can best be described using the concept of water potential. Water potential is the chemical potential of water per unit volume and has the dimensions of pressure. The units used here are bars and millibars (mb). One bar equals 100 kPa and nearly equals 1 atm. The water potential of pure water at atmospheric pressure is defined arbitrarily as zero. At any given temperature, the potential of liquid water is a logarithmic function of its activity ($a_w$) and, therefore, its equilibrium relative humidity (RH). For example, at a water potential of −15 bars, a value at which crop plants frequently wilt, the equilibrium RH of air is 98.9%; at a water potential of −100 bars, which is much too dry for *Phytophthora* spp. to be active (Duniway, 1979), the RH is 92.8%. Useful tables relating water potential to humidity and showing the conversion of units can be found in Slavík (1974).

The major forces acting on water can be expressed as components of water potential in the equation $\Psi = \psi_s + \psi_p + \psi_m$, where $\Psi$ is the total water potential and $\psi_s$, $\psi_p$, and $\psi_m$ are the solute, pressure, and matric components, respectively. Solute potential, $\psi_s$, represents the influence of dissolved substances on water and is equivalent to but opposite in sign from osmotic pressure. As it is used here, the pressure potential, $\psi_p$, represents a hydrostatic pressure, such as the positive $\psi_p$ identified with turgor or the negative $\psi_p$ in the xylem elements of transpiring plants. Matric potential, $\psi_m$, represents a reduction in $\Psi$ due to absorption and capillary forces that are of primary importance in soil. Older expressions of soil moisture suction, sometimes given as centimeters of water, or pF values (Griffin, 1972), and negative pressure terms used in soil hydrology can usually be equated to the negative $\psi_m$ component of soil $\Psi$. The soil contains a system of interconnecting capillary pore spaces; although soil pores are heterogeneous in shape, there are fairly definite relationships between the size of soil pores that can retain liquid water and $\psi_m$ (Griffin, 1972; Slavík, 1974). Therefore, as a fully saturated soil begins to dry or drain, $\psi_m$ is the first component of $\Psi$ to change appreciably, and as it decreases, the water in progressively smaller pore spaces is replaced by air. By the time a soil dries to $\psi_m = -0.3$ bar, a $\psi_m$ value at which soil water content is frequently considered to represent field capacity, most pore spaces and pore necks larger than 10 $\mu$m in diameter will have drained. As the soil dries further, both $\psi_m$ and $\psi_s$ change, and partitioning soil $\Psi$ into its component potentials becomes increasingly difficult. Because the configuration of water in very wet soils is clearly linked to $\psi_m$, zoospore dispersal, solute diffusion, and aeration in soil all interact strongly with $\psi_m$, and it is one of the more useful parameters in the epidemiology of diseases caused by soilborne species of *Phytophthora.*

## DIRECT EFFECTS OF PHYSICAL FACTORS ON BEHAVIOR OF *PHYTOPHTHORA*

### Sporangium Formation and Dispersal in Aerial Environments

Much research has been done on the effects of ambient humidity and temperature on sporangium formation by *P. infestans* on potato tissue. In atmospheres saturated to 100% RH, sporangia appear within 8 hr at temperatures within the optimal range of 18–22° C, but their appearance requires progressively longer saturated periods as temperatures shift away from the optimal range and approach the lower (3° C) or upper (26° C) limit for sporangia to form (Crosier, 1934; DeWeille, 1964). Generally, *P. infestans* does not form sporangia in an atmosphere of less than 91% RH and forms them only sparingly at 97% RH (Crosier, 1934; DeWeille, 1964). Although their temperature requirements vary, most other *Phytophthora* spp. that form sporangia aerially on host surfaces also require ambient humidities higher than 95–97% RH to do so, and frequently their sporangium formation is most prolific when host surfaces are wetted to some extent with liquid water (Waterhouse, 1931). There are some important exceptions. For example, *P. palmivora* is reported to form sporangia on cocoa pods over the entire 70–90% RH range but none at 100% RH (Rocha and Machado, 1973). There are also reports that the presence of abundant liquid water on potato foliage suppresses sporulation by *P. infestans* (Rotem et al, 1978). Thus, the optimal conditions for sporangia to form aerially may represent a compromise between a requirement for water at very high $\Psi$ values and for adequate gas exchange with the air. It is impossible, however, to infer the actual $\Psi$ requirements for sporangia to form aerially from the existing data on humidity and wetness requirements, because the water status of the substrates was not measured and the substrates were not at equilibrium with the ambient humidity. Fortunately, many of the existing data describing humidity and wetness effects on sporulation will suffice for epidemiologic purposes (Rotem et al, 1978). In fact, there is probably more to be gained from a micrometerological analysis to show the distribution of permissive humidities in space and time within plant canopies (e.g., Hirst and Stedman, 1960) than there is to be gained from more precise knowledge of the $\Psi$ requirements for *Phytophthora* spp. to sporulate aerially.

Once sporangia are formed, their effective dispersal to initiate infections at new sites is limited mainly by the adverse effects of low humidity, solar radiation, and extreme temperatures on their viability. Caducous sporangia that are formed above ground are readily dispersed by splashing or running water, and dispersal in water very much reduces the threat of desiccation. Water may account for much local dispersal within plant canopies and, especially when aided by wind, splash dispersal mechanisms may move sporangia over horizontal distances of several meters or more (e.g., Hunter and Kunimoto, 1974). The sporangia of only two species, *P.*

*infestans* and *P. phaseoli*, are readily shed into the air without the aid of moving water (Al-Hedaithy and Tsao, 1979; Waterhouse, 1974), and the risk of drying is greatest for the sporangia of these species that are dispersed by the wind. In fact, a slight drying associated with a decrease in ambient humidity may actually stimulate the release of sporangia into the air (Leach, 1975; Rotem and Cohen, 1974). The survival of detached sporangia during periods of low humidity is conditioned by several factors, including temperature (Rotem and Cohen, 1974), their previous exposure to wetness (Cohen et al, 1974), and the manner in which they are rewetted (DeWeille, 1964). Sporangia may also need to mature at high humidities before they can germinate (DeWeille, 1964; Rotem et al, 1978), and their subsequent germination may shift from indirect to direct with desiccation (Crosier, 1934).

Unfortunately, because most experiments on sporangial viability have been done on surfaces that may not have been equilibrated fully with the reported environmental conditions, the reliability of many of the data on the effects of humidity on sporangial viability is doubtful. The study by Warren and Colhoun (1975) illustrates one of the complexities. They found that detached sporangia of *P. infestans* carried a film of water, and when glass slides bearing detached sporangia were kept at 90 and 95% RH, sporangia died within a few minutes after the adhering water film dried. Similarly, sporangia of *P. palmivora* died rapidly when they were removed from papaya fruit without adhering water and held at 80% RH (Hunter and Kunimoto, 1974). There is good evidence that sporangia of both *P. infestans* and *P. palmivora* can effectively persist for at least several hours at moderate humidities while attached to sporangiophores (Hunter and Kunimoto, 1974; Rotem and Cohen, 1974; Zan, 1962), and detached sporangia of *P. infestans* are known to persist for several hours at low ambient humidities after they are deposited on leaf surfaces (Rotem and Cohen, 1974). Therefore, attachment or deposition on substrates of high water status and the very high humidities at plant surfaces probably account for more of the sporangial viability in the field than is generally recognized. Wallin (1953) illustrated this possibility nicely when he found viable sporangia of *P. infestans* on tomato and potato leaflets under what were seemingly very adverse ambient conditions for sporangia to survive in the field.

Although light has many effects on the behavior of *Phytophthora* spp. (Ribeiro, Chapter 4), the adverse effects of solar radiation on the viability of exposed inoculum affect disease development directly. Observations of cloudy and sunny periods have aided late blight forecasts for a long time, but with few exceptions (e.g., Rotem et al, 1970), solar radiation effects on disease development have not been quantified. Only DeWeille (1964) appears to have systematically evaluated the effects of solar radiation on the sporangia of *P. infestans*. He found that the ultraviolet component of solar radiation influenced both the germination and the viability of sporangia. Small doses of ultraviolet radiation stimulated subsequent germination, but larger doses reversed the stimulatory effect and were lethal. In the field, the dose of solar radiation that is stimulatory is readily attained even on cloudy days, and the detrimental effects of solar radiation are more conspicuous. For example, a 2-hr exposure to diffuse sunlight was enough to reduce the germinability of sporangia from 65 to 40% (DeWeille, 1964). As with humidity and temperature (e.g., DeWeille, 1964; Rotem and Cohen, 1974), we really have very little precise information about the effects of solar radiation on the viability of sporangia. It would be helpful to know much more, because to a large extent the times and distances of effective dispersal of sporangial inoculum in aerial environments are governed by the adverse effects of the sun and other physical factors on sporangial viability.

### Sporangium Formation in Soil

The water requirements for *Phytophthora* spp. to form sporangia in soil are reviewed elsewhere (Duniway, 1979; Gisi, Chapter 8; Gisi et al, 1980). Both the minimum and maximum $\psi_m$ values for sporangia to form in soil depend on a number of factors including the species, its depth in the soil, and soil texture. For example, when mycelium of *P. cactorum*, *P. cinnamomi*, or *P. cryptogea* was buried in soil at constant $\psi_m$ values, optimal sporangium formation occurred in the −40 to −200 mb $\psi_m$ range; the lowest $\psi_m$ or $\Psi$ value at which sporangia formed was −2.5 to −4 bars (Duniway, 1975a, 1975b; Gisi et al, 1980). In fully saturated soil at $\psi_m = 0$, *P. cryptogea* and *P. cinnamomi* formed few if any sporangia on buried mycelium but formed many on mycelium placed on the soil surface. As may be the case for sporangium formation aerially, optimal $\psi_m$ values for sporangium formation in soil by these two species probably represent a compromise between their requirements for liquid water and aeration (Duniway, 1975a; Gisi et al, 1980). The responses of *P. cactorum* and *P. palmivora* to high $\psi_m$ values differed from those of the above species; although they formed more sporangia in soils drained to $\psi_m \leqslant -10$ mb, they also formed significant numbers of sporangia on buried mycelium in fully saturated soil at $\psi_m = 0$ (Gisi et al, 1980). In fact, the research to date suggests that species such as *P. cactorum*, *P. cinnamomi*, *P. cryptogea*, and *P. palmivora* are probably capable of forming sporangia over most of the range of $\psi_m$ values that predominates in agricultural soils during crop growth.

The effects of $\psi_m$ on sporangium formation by *Phytophthora* spp. in soil may be influenced by the nature of the substrate. The formation of sporangia by *P. cambivora* and *P. megasperma* on colonized pieces of leaf or root tissue was largely confined to $\psi_m$ values higher than −10 mb (Pfender et al, 1977; Sugar, 1977), but when washed mycelium of *P. megasperma* was buried in soil, $\psi_m$ values less than −10 mb were optimal for sporangium formation (MacDonald, 1977). Because transfer to a dilute medium containing few nutrients generally enhances sporangium formation (Waterhouse, 1931), a more rapid loss of solutes from excised tissues in saturated soil than in somewhat drained soil might explain this phenomenon.

To investigate the effects of a more realistic substrate on sporangium formation by *P. cryptogea*, I uniformly inoculated roots of young safflower seedlings with a suspension of zoospores just before they were transplanted into soil at constant $\psi_m$ values. After incubation for 48 hr, seedling roots were gently washed from the soil, and their entire length was surveyed for sporangia with a microscope. As expected from the behavior of mycelial disks in soil (Duniway, 1975a), many sporangia formed at depths between 1 and 3.8 cm (the maximum depth of root penetration) at $\psi_m = -150$ mb, whereas sporangium formation at $\psi_m = 0$ was confined to depths of less than 1 cm (Fig. 1). Although more sporangia were formed at −150 mb than at 0 mb if all depths are considered, a significant amount of sporulation occurred at both $\psi_m$ values. In this regard, small $\psi_m$ deficits appear to have their most important influence on the depth at which sporangium formation by *P. cryptogea* can occur in soil.

Some of the cyclic changes in soil moisture that occur routinely in agriculture can influence sporangium formation. For example, once sporangium formation by *P. cambivora* on buried, colonized leaf disks was initiated at $\psi_m = 0$, sporangia continued to form when soil was drained to $\psi_m = -50$ mb, a $\psi_m$

value that was evidently too dry for them to be initiated (Sugar, 1977). Under constant conditions, *P. parasitica* formed significant numbers of sporangia at negative $\psi_m$ values ranging down to −300 mb but formed few at $\psi_m = 0$ (Ioannou and Grogan, 1977; Sterne and McCarver, 1980). If, however, mycelial disks of *P. parasitica* were first held in soil for a few days at $\psi_m$ values of −200 or −300 mb, large numbers of new sporangia were formed within a few hours when the soil was saturated to $\psi_m = 0$ (Bernhardt and Grogan, 1982). Sporangial formation by *P. capsici, P. cambivora,* and *P. megasperma* can also be stimulated when an unsaturated soil becomes saturated (Bernhardt and Grogan, 1982; Sugar, 1977). *P. capsici* responds in this way to saturation more slowly than *P. parasitica,* but *P. cambivora* and *P. megasperma* may respond more quickly. For example, sporangia of *P. megasperma* failed to form when colonized leaf disks were held in soil at −25 mb $\psi_m$, but when the disks were subsequently flooded and maintained in soil at $\psi_m = 0$, numerous sporangial initials formed within 30–60 min, and these matured to release zoospores within 3–6 hr (W. F. Wilcox and S. M. Mircetich, *personal communication*). In contrast, cyclic variations in soil $\psi_m$ did not stimulate sporangium production by *P. cinnamomi* (Gisi et al, 1980) or *P. cryptogea* (Bernhardt and Grogan, 1982).

Diversity among *Phytophthora* spp. in the capacity to form sporangia aerially, on agar media, when submersed in water (Waterhouse, 1931), or at various oxygen and carbon dioxide concentrations (Mitchell and Zentmyer, 1971b) suggests that many species will respond differently to small $\psi_m$ deficits in soil. Furthermore, the amount of reported diversity in sporangium formation that is inherent among *Phytophthora* spp., or even among isolates of one species, and the amount that can be attributed to experimental conditions remain to be determined. In all probability, enhanced sporangium formation by some, but not all, *Phytophthora* spp. is one of the reasons certain Phytophthora root rots are aggravated by saturated soil conditions.

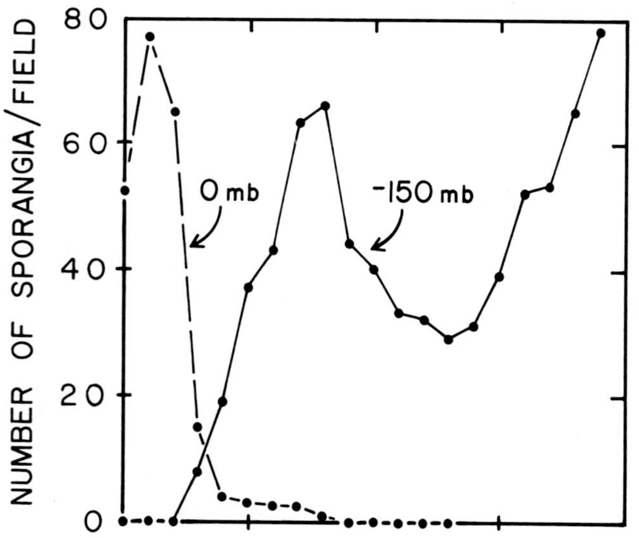


**Fig. 1.** Sporangium formation by *P. cryptogea* on roots of living safflower seedlings at various depths in soil maintained at $\psi_m$ values of 0 and −150 mb for 48 hr after inoculation. A microscope was used to count the sporangia in each field as one edge of a root was scanned from the level of the soil surface to the root tip. Representative data from one of 10 seedlings in each treatment are shown.

In addition to water and its interactions with aeration, temperature influences sporangium formation in soil. In the example of *P. cinnamomi,* the temperature range over which sporangia can form appears to be more narrow than the range over which mycelium will grow, and there is circumstantial evidence that the effects of soil temperature on sporangium formation may sometimes confine the pathogenic activities of *P. cinnamomi* to certain seasons of the year (Nesbitt et al, 1979; Shea, 1975; Zentmyer, 1980). Several soilborne *Phytophthora* spp. form sporangia over a broader temperature range than does *P. cinnamomi,* including the following species and ranges: *P. drechsleri,* 9–33° C; *P. cambivora,* 9–30° C; and *P. megasperma,* 6–24 or 27° C, depending on the isolate (Pfender et al, 1977; Sugar, 1977). Species such as these may be less limited by soil temperature than is *P. cinnamomi.* Nevertheless, the rate at which all *Phytophthora* spp. form sporangia is influenced by temperature, and there may be interactions between temperature and the soil water status at which sporangia will form (Sneh and McIntosh, 1974). The time course and abundance of sporangium formation need to be researched further under the physical conditions found in natural soils.

### Sporangium Dispersal and Persistence in Soil

Sporangia are not very likely to be dispersed from their point of formation within soil. One reason is that sporangia of most soilborne *Phytophthora* spp. are not readily shed; i.e., they are not caducous (Al-Hedaithy and Tsao, 1979; Waterhouse, 1974). A second reason is the physical structure of soil; most natural soils do not contain many pores large enough for the unimpeded movement of sporangia with water flow. In fact, soil constraints on sporangium dispersal are probably much greater than are the constraints on zoospore dispersal (Duniway, 1976; Zan, 1962). For those species that do not form sporangia quickly (e.g., within a few hours after soil becomes saturated), the capacity of sporangia to effectively persist in the place where they are formed influences the speed and abundance with which inoculum in the form of zoospores becomes available when the soil is subsequently wetted by rain or irrigation.

There are many general accounts of persistence by *Phytophthora* spp. in soil, but only a few have considered the effects of soil conditions on sporangial viability. Although some of the literature on soilborne *Phytophthora* spp. gives an impression that sporangia are ephemeral, those of *P. palmivora* and *P. infestans* may remain viable for weeks or even months at intermediate levels of soil moisture (Turner, 1965; Zan, 1962). Sporangia of *P. cactorum* remained viable for up to 12 wk in soil at −0.2 to −5 bars $\psi_m$ and 10° C, and their persistence was only somewhat less in more moist soils and at 24° C (Sneh and McIntosh, 1974). However, sporangial viability in *P. cactorum* was lost rapidly in both air-dried and fully saturated soils (Gisi, 1975; Sneh and McIntosh, 1974). Furthermore, sporangia of *P. cactorum* that were exposed to progressively drier conditions lost the capacity to germinate indirectly before they lost the capacity to germinate directly (Meyer and Schönbeck, 1975). In contrast, sporangia of *P. cryptogea* and *P. megasperma* did not shift from one mode of germination to the other when they were dried in soil over a period of 2 days to $\Psi$ values between 0 and −85 bars (MacDonald and Duniway, 1978b). Instead, sporangia of *P. cryptogea* were unaffected by drying to −10 bars, and 50% remained viable when they were dried to −45 bars; sporangia of *P. megasperma* were somewhat more sensitive, since 50% were not viable after drying to −15 bars. Evidently, many sporangia of *P. cryptogea, P. megasperma,* and *P. cactorum* could remain viable for several days, at least under physical conditions normally encountered in the root zones of crop plants. The activities of other microorganisms, particularly in warm and wet soils, probably offer the greatest hazard to sporangial survival. More information is needed on the capacity of many *Phytophthora* spp. to persist as sporangia in natural soils.

### Sporangium Germination

Wetness and temperature are, without doubt, the main physical factors governing the germination behavior of sporangia. Indeed, most of the literature on the genus *Phytophthora* suggests that sporangia must be in contact with liquid water to germinate either directly by the growth of germ tubes or indirectly by the release of zoospores (Duniway, 1979). Once sporangia on aerial plant surfaces are wetted, temperature influences the relative rates of direct and indirect germination. For example, when sporangia of *P. infestans* were placed in water at 9–15° C, 40–70% germinated indirectly within 3 hr and at least 20% did so within 3 hr at 3,6, and 18° C; fewer than 2% germinated directly within 16 hr at 18° C, and only 4% did so after 12 hr at 24° C, the optimal temperature for direct germination (Crosier, 1934). The speed and abundance with which sporangia germinate indirectly suggest that indirect germination by sporangia of *P. infestans* contributes more to infection than does direct germination, especially when dew is the main source of surface wetness. In fact, the temperatures and durations of wetness periods that are most strongly correlated with late blight increases in the field are those that are most suitable for indirect germination (e.g., DeWeille, 1964; Krause and Massie, 1975). The same correlations and the requirements for infection under more controlled conditions (e.g., Rotem et al, 1971) also suggest that indirect germination is a rate-limiting step in the infection process. Therefore, the steps in the infection process that follow indirect germination appear to have little influence on the speed with which *P. infestans* infects potato foliage. Perhaps steps such as zoospore migration, germination, and penetration are much more rapid than is indirect germination under the same conditions.

In soil, where moisture and temperature change less rapidly than on foliage, one might expect more direct germination to occur. However, sporangia of *P. infestans* on or in soil may also germinate indirectly, because a soil temperature of 18° C or less is required for it to infect potato tubers in the field (Sato, 1979). Many soilborne *Phytophthora* spp. germinate indirectly at higher temperatures than does *P. infestans* and are capable of producing zoospores rapidly in water at temperatures that are common in agricultural soils. When sporangia of various species were placed in water or dilute soil extract, indirect germination predominated over direct germination at the following temperatures: *P. cactorum*, 15–27° C (Sneh, 1974); *P. capsici*, 10–24° C (Katsura, 1971); *P. cambivora*, 9–27° C (Sugar, 1977); *P. drechsleri*, 9–30° C (Sugar, 1977); *P. cryptogea*, 8–30° C (MacDonald and Duniway, 1978b), and *P. megasperma*, 12–28 or 8–21° C, depending on the isolate (MacDonald and Duniway, 1978b; Sugar, 1977). In addition, Gisi (Chapter 8) found that zoospore release began 12–55 min after sporangia of *P. cactorum, P. cinnamomi, P. cryptogea,* and *P. palmivora* were placed in water at room temperature, which was presumably in the 20–25° C range. It is also noteworthy that sporangia of *P. parasitica* may be initially committed to germinate indirectly, for direct germination occurs only after early stages of zoosporogenesis are reversed (Hemmes and Hohl, 1975). Furthermore, the intense competition among soil microorganisms for nutrients may suppress direct germination in natural soils. Although the underlying mechanisms were not determined, Marx and Bryan (1969) found that sporangia of *P. cinnamomi* germinated only indirectly in the presence of soil bacteria.

Even though the temperature and nutrient conditions in soil may tend to favor indirect germination, and the release of zoospores multiplies the number of inoculum units several times, the capacity of sporangia to germinate directly in soil needs to be researched more carefully. For one thing, there is limited evidence that sporangia may be able to germinate directly at $\Psi$ values that are too low to permit indirect germination (Duniway, 1979), and it is important to know whether infection can occur by means of direct germination at soil moisture levels lower than those usually associated with the release of zoospores. Experiments on *Pythium* spp. are interesting in this regard, because periodic saturation of soil increased root infection by *Pythium vexans*, a species that germinates indirectly, but increased infection by *Pythium irregulare*, a species that germinates directly, to a much lesser extent (Biesbrock and Hendrix, 1970). Sporangia of *Pythium ultimum* germinate only directly and can be stimulated to do so abundantly in soil at $\psi_m$ values down to −50 or −100 mb (Stanghellini and Hancock, 1971). In the case of *Phytophthora* spp., sporangia of *P. palmivora* have been observed to germinate directly to infect nearby roots in soil (Ko, 1971), and in a saturated natural soil, germ tubes from sporangia of *P. cactorum* grew up to 380 μm in length (Gisi, 1975). When stimulated by glucose, a large percentage of the *P. cactorum* sporangia in soil at $\psi_m$ values of −0.16 and −0.3 bar germinated directly, but new sporangia formed terminally on those germ tubes that grew within the rhizosphere of apple seedlings at −0.3 bar (Sneh and McIntosh, 1974). Although direct germination by sporangia of other *Phytophthora* spp. has been observed at $\psi_m$ values between −50 and −300 mb (Duniway, 1975a; Pfender et al, 1977; Sugar, 1977) and even at $\Psi$ values down to −3 bars (Duniway, 1975b), no clear data show the minimum $\psi_m$ values at which sporangia can germinate directly in soil. In natural soils, however, the direct germination by sporangia of *Phytophthora* spp. may generally be confined to the relatively wet soil conditions that enhance diffusion of nutrients from roots or other substrates.

Recent experiments have clearly shown that the indirect germination of sporangia is stimulated by and is largely confined to the same levels of soil saturation at which almost all Phytophthora root rot diseases are most severe. Such a correlation certainly suggests that indirect germination contributes more to root infection than does direct germination. In one recent study, when the $\psi_m$ component of soil $\Psi$ was increased from −150 mb to test the influence of various higher $\psi_m$ values on the final differentiation and release of zoospores by sporangia of *P. cryptogea* and *P. megasperma*, 77–99% of the sporangia released zoospores within 4 hr at 0 (fully saturated) and −1 mb $\psi_m$; but at final $\psi_m$ values of −5, −10, and −25 mb, the percentages of sporangia releasing zoospores within 24 hr were reduced to about 30, 10, and 0, respectively (MacDonald and Duniway, 1978a). Furthermore, zoospore release by these two species was related more to $\psi_m$ than to soil water content (MacDonald and Duniway, 1978a), and intermittent wetting to $\psi_m = 0$ was not detrimental to zoospore release; the total amount of zoospore release was determined by the total time at $\psi_m = 0$ (MacDonald and Duniway, 1978b). Several studies have shown that various solutes can retard or even inhibit zoospore release, and the effects of solutes and their implications for possible mechanisms of zoospore release are reviewed in detail by Gisi (Chapter 8). The effects of $\psi_s$ on zoospore release that

more about the persistence and germination behavior of zoospore cysts.

### Inoculum Other than Sporangia and Zoospores

Oospores, chlamydospores, and even parts of hyphae probably survive longer under relatively harsh conditions than do the sporangia or zoospores of *Phytophthora* spp.; these more persistent structures can be important forms of primary inoculum (e.g., Gerlach et al, 1976; Zentmyer, 1980). If the initial infections by primary inoculum can develop extensively enough to incite disease, then the disease is in effect a single-cycle disease and the behavior of primary inoculum is of utmost importance. For example, safflower crops in Australia have exhibited advanced symptoms of Phytophthora root rot under dry conditions, and it has been suggested that chlamydospores of *P. drechsleri* can germinate directly to initiate the disease under such conditions (Cother and Griffin, 1974). Although the impact of an infection depends on its location on a host, and although the extent to which individual infections by *Phytophthora* spp. can develop in their hosts has not been researched adequately, the highly stimulatory effects of prolonged or repeated periods of dew, rain, or saturation of soil on *Phytophthora* diseases suggest that single infections by most *Phytophthora* spp. do not spread rapidly enough in established hosts to incite discernible disease and that several infection foci or secondary infections are usually required. Therefore, most *Phytophthora* diseases can be multiple-cycle, compound-interest diseases, and the behavior of secondary inoculum, which usually takes the form of sporangia and zoospores, has more of an impact on the rate of disease development than does the behavior of primary inoculum, given that minimum amounts of primary inoculum are present and function (van der Plank, 1963; see also MacKenzie et al, Chapter 23). Furthermore, the capacity of persistent structures of *Phytophthora* spp. to germinate and infect directly has not been widely demonstrated, and many of the oospores, chlamydospores, and hyphal fragments that do persist in soil may form sporangia and zoospores upon germination (e.g., Marx and Bryan, 1969; Zentmyer, 1980). For these reasons, I have emphasized the influence of physical factors on sporangia and zoospores.

Some literature indicates the extent to which soil water status influences chlamydospores and oospores. Although contradictory data have been published (Gisi, 1975; Meyer and Schönbeck, 1975), some *Phytophthora* spp. appear to form chlamydospores and/or oospores under relatively dry conditions (Malajczuk, 1975; Reeves, 1975; Sneh and McIntosh, 1974) where other forms of inoculum are produced less abundantly. Limited experimentation leaves little doubt that water status influences the survival of chlamydospores in soil (e.g., Chee, 1973), and rather exciting research on several genera has shown the probable importance of water status to oospore survival and germination in soil. In relatively wet soils, parasitism of oospores can be extensive (Sneh et al, 1977; see also Malajczuk, Chapter 16), and *Pythium ultimum* oospores can be converted to a state from which they germinate in nonsterile soils at $\Psi$ values higher than −7 bars (Lumsden and Ayers, 1975). Very wet soil conditions are generally most unfavorable for oospore survival, and although oospores evidently tolerate considerable drying, intermediate levels of soil moisture favor survival (Meyer and Schönbeck, 1975; Sneh and McIntosh, 1974). Oospores of *P. cactorum* (Sneh and McIntosh, 1974) and chlamydospores of *P. parasitica* (Ioannou and Grogan, 1977) appear to have the same water requirements for sporangial formation upon germination in soil as do mycelia, but chlamydospores of *P. drechsleri* and *P. cinnamomi* may germinate directly at fairly low $\Psi$ values (Cother and Griffin, 1974; Malajczuk, 1975). As was noted earlier for sporangia, the mode of germination and infection by propagules of *Phytophthora* spp. in soil may depend on interactions between $\psi_m$ and the diffusion of nutrients from roots as well as the inherent water requirements of the fungus. For example, when chlamydospores of *P. cinnamomi* were used as inoculum in a sandy soil maintained at constant $\psi_m$ values, root rot of *Persea indica* seedlings increased with each increase in $\psi_m$ above −0.25 bar and was most severe at $\psi_m = 0$ (Sterne et al, 1977a). The addition of nutrients or the use of a clay soil, however, permitted severe root rot to develop over a greater part of the 0 to −0.25 bar $\psi_m$ range, probably because the added nutrients or a greater diffusion of root exudates in the clay soil promoted direct germination and the growth of germ tubes (Sterne et al, 1977b). Although they are not reviewed here, the effects of temperature on all facets of oospore and chlamydospore behavior may equal or exceed those of moisture (e.g., Gerlach et al, 1976; Zentmyer, 1980).

### Mycelial Growth

A number of recent studies examining the effects of water status on the growth of *Phytophthora* spp. are reviewed by Gisi (Chapter 8) and Duniway (1979). With very few exceptions, the effects of water status on growth have been examined in pure culture, where the results are influenced greatly by the nature of the media and solutes used to vary $\psi_s$. Nevertheless, although growth by *Phytophthora* spp. is confined to higher $\psi_s$ values than is growth by many plant-pathogenic members of the Ascomycetes and Fungi Imperfecti (e.g., Cook and Duniway, 1981; Duniway, 1979; Griffin, 1972), it is probably safe to conclude that *Phytophthora* spp. grow well at the $\Psi$ values they are most likely to encounter in growing host plants. The older water relations literature referred to a soil $\Psi$ value of −15 bars as the permanent wilting point, and considerable research has now shown that the productive growth of crop plants is generally confined to tissue $\Psi$ values much higher than −15 bars (e.g., citations in Slavík, 1974). Those *Phytophthora* spp. that have been tested maintained at least one-quarter to one-half their optimal growth rate at $\psi_s$ or $\Psi$ values as low as −15 bars (Duniway, 1979). Therefore, the data obtained in pure culture suggest that under most agricultural conditions, the direct effects of $\Psi$ on mycelial growth by *Phytophthora* spp. are not likely to limit disease development. This hypothesis obviously needs further testing, but it is also supported by the fact that other stages in the life cycles of *Phytophthora* spp., such as the sporangia and zoospore stages already discussed, appear to be much more sensitive to changes in $\Psi$ than is mycelium.

Unfortunately, no research has been done to show the effects of tissue $\Psi$ on the growth of a *Phytophthora* sp. in a host. The only study that has come close to doing so measured the development of symptoms of root rot in safflower plants after inoculation with *P. cryptogea*. With the exception of completely saturated soil, a variety of water regimes under which susceptible safflower plants were maintained after inoculation were found to be almost equally suitable for the development of root rot (Duniway, 1977b). The suitable water regimes included soil $\Psi$ values that gradually declined from −0.4 bar to −4, −8, or −12 bars. Although pathogen growth was not measured directly, the drier $\Psi$ values used would prevent secondary infections by sporangia or zoospores, and the disease that developed in safflower probably reflected the growth of the pathogen from the initial points of infection (Duniway, 1977b). The aerial parts of host plants can develop

Ψ values that are considerably more negative than the Ψ values of roots or stems. It would be interesting to know more about the effects of low plant Ψ on the growth and sporulation capacity of mycelium of *Phytophthora* spp. that attack aerial plant parts. In the case of soilborne species, extremely wet conditions that lower the availability of oxygen and raise carbon dioxide concentrations in soil may also be expected to limit growth. However, as will be discussed later, the impacts of the extreme oxygen and carbon dioxide concentrations found in most agricultural soils on host susceptibility probably exceed their direct effects on the growth rate of *Phytophthora* spp. (Mitchell and Zentmyer, 1971a).

Temperature influences the growth rate of mycelium and thereby influences, at least to some extent, the rate of disease development. This fact has been recognized for a long time, and a considerable amount of data in the literature describes the relative growth rates of *Phytophthora* spp. at various permissive temperatures as well as the uppermost and lowermost temperatures that permit growth (e.g., Tucker, 1931). Although temperature effects on growth cannot by themselves provide an accurate forecast of disease development, the temperature in a given climate must permit some growth if the disease is to develop at all. Furthermore, the effects of temperature on root diseases incited by soilborne *Phytophthora* spp. sometimes parallel closely the effects of temperature on the growth rate of mycelium, even though other aspects of their disease cycles may be influenced somewhat differently by temperature. For example, the extent to which *P. cinnamomi* suppresses the growth of avocado seedlings in soils at various temperatures is closely correlated with the effects of temperature on the growth of mycelium (Zentmyer, 1980). On the other hand, the optimal temperature for mycelial growth of *P. parasitica* is not the optimal temperature for causation of root rot of parsley or fuchsia. One isolate of *P. parasitica* that grew best at 28–31° C caused the most rapid development of crown rot in parsley at 36° C (Hine and Aragaki, 1963). Another isolate of *P. parasitica* caused severe root rot of fuchsia at 30° C but caused none at 24° C (Patil and Young, 1960), and most isolates of this species grow almost as well at 24° C as they do at 30° C (Tucker, 1931). Therefore, it appears that the higher temperatures may have reduced the ability of parsley and fuchsia to withstand or tolerate pathogenic attack by *P. parasitica*. The host may also mediate some of the effects of temperature on other *Phytophthora* diseases but in comparison to the data on growth of *Phytophthora* spp. in pure culture, very little is known about the effects of temperature on pathogen growth or disease development in host plants. In fact, many of the effects of temperature, especially changes in temperature, on *Phytophthora* diseases need more study. For example, the extent to which mycelium of *P. infestans* survives in potato at high temperatures determines whether or not late blight epidemics will resume after hot weather has arrested epidemics in the field (Kable and MacKenzie, 1980). Similar effects of temperature may apply to other *Phytophthora* diseases.

## HOST SUSCEPTIBILITY AND PREDISPOSITION

Although the physical requirements of the pathogen establish some of the climatic constraints on disease development, in the final analysis, pathogen behavior and disease epidemiology represent an integration of pathogen and host physiology. As was suggested above by the discussion of temperature effects on Phytophthora root rots of parsley and fuchsia, there is little doubt that the influence of physical factors on *Phytophthora* diseases is determined in part by the behavior of the host. Furthermore, the inherent resistance of a host to pathogenic invasion will modify the extent to which physical conditions that influence inoculum behavior also influence disease development. If, for example, a pathogen spreads rapidly within a host from a few initial points of infection, disease development in that host will not be highly dependent on external climatic conditions that favor inoculum formation, dispersal, and numerous new infections. If, on the other hand, the pathogen does not spread far from individual infection foci, disease development will depend on the cumulative effects of many infections and will depend greatly on the factors that influence inoculum behavior outside the host.

Even though it was argued earlier in this paper that most *Phytophthora* diseases tend to fall into the latter of these two categories and require large numbers of infection foci or secondary infections, the extent to which infection foci can spread within a host does appear to vary greatly among many of the host-*Phytophthora* spp. combinations that have been examined. *P. cinnamomi* generally invades only the small feeder roots of avocado, but it readily invades the large roots and the major part of the root system of *Banksia* spp. from Western Australia (Zentmyer, 1980). The disease incited by *P. cinnamomi* in avocado is more likely to require those conditions that allow many new or repeated infections than is the disease incited by the same fungus in *Banksia* spp. There is considerable interest in the horizontal resistance of potatoes to *P. infestans,* and late blight development on cultivars with horizontal resistance is likely to require more and longer periods of time that are physically suitable for infection than is blight development on more susceptible cultivars. For example, the hours of surface wetness required for equivalent amounts of inoculum of *P. infestans* to incite equivalent numbers of lesions on different selections of *Solanum phurija* is inversely correlated with the rate of lesion enlargement on the same selections (Guzmán-N., 1964). Furthermore, the number of sporangia deposited per unit of uninfected leaf area influences the length and temperature of wet periods required for *P. infestans* to infect potato (Rotem et al, 1971). The efficiency with which inoculum functions apparently depends on both the physical conditions and susceptibility of the host. The interactions between the level of genetic resistance in a host and the physical factors influencing disease development are certainly large enough to warrant more consideration in epidemiologic research.

It is becoming more apparent that the resistance of plants to attack by *Phytophthora* spp. is not constant and that it changes with the physiologic status of the host. Predisposition phenomena are widely known in plant pathology, and laboratory investigations into mechanisms of plant resistance to several *Phytophthora* spp. have shown that resistance can be modified by extreme physical conditions (e.g., Johnson and Klisiewicz, 1969; Ward et al, 1981). However, only recently is it becoming recognized that some of the physical conditions that occur routinely in agriculture can significantly predispose or otherwise modify the resistance of plants to attack by *Phytophthora* spp. Unlike many of the quantitative facets of epidemiology, predisposition has been more clearly demonstrated for diseases caused by soilborne species of *Phytophthora* than for diseases caused by the airborne species. Tobacco, however, can be predisposed to downy mildew by water stress (Rotem et al, 1968), and it may be suspected that aerial parts of plants may also be predisposed by environmental conditions to infection by *Phytophthora* spp.

One relatively well-researched example of predisposition to a Phytophthora root rot, which may seem somewhat incongruous after the preceding discussions of inoculum behavior, is the predisposition of safflower by water stress. In experiments under controlled conditions, the induction of water stress in safflower plants before inoculation with zoospores increased their susceptibility to root rot caused by *P. cryptogea* (Duniway, 1977b). Well-watered safflower plants had leaf Ψ values higher than −6 bars, and significant predisposition of the susceptible cultivar Nebraska 10 occurred if water was withheld and leaf Ψ values were depressed to −13 bars. Comparable depressions in leaf Ψ predisposed the normally resistant cultivar Biggs to a greater extent. In addition, the combined effects of a mild depression in leaf Ψ from −4.0 to −8.8 bars and inoculation of the Biggs cultivar were significant, whereas neither treatment alone had a significant effect on the growth of Biggs plants.

Two separate field experiments have confirmed the importance of water stress to predisposition in safflower by showing that the occurrence of water stress before irrigation increased the incidence of Phytophthora root rot in several safflower cultivars (Duniway, 1978; Zimmer and Urie, 1967). The interactive effects of irrigation and water stress on Phytophthora root rot in the susceptible cultivar Nebraska 10, as demonstrated in the more recent field experiment (Duniway, 1978), are shown in Figure 3. Nebraska 10 plants grown on available soil moisture without irrigation did not develop symptoms of root rot, but the application of one or four irrigations increased root rot incidence significantly. The disease links with brief surface irrigations suggest that the water requirements for zoospore release and dispersal are, indeed, important factors in disease development. The occurrence of stress before one irrigation, however, increased disease incidence compared to that induced by one irrigation without stress, and the disease that developed after the stress was relieved finally equaled the level of disease that developed after four irrigations in the absence of stress (Fig. 3). In more resistant safflower cultivars, irrigation had a less stimulatory effect on root rot unless an irrigation was preceded by stress. When stressed and irrigated, the cultivars Gila and C2517, for example, developed as much disease as did Nebraska 10 (J. M. Duniway, *unpublished*). Therefore, in at least some safflower cultivars under irrigated conditions, water stress was the major physical factor determining the development of Phytophthora root rot.

Stresses caused by drought or salinity have recently been found to predispose at least two kinds of ornamental plants to Phytophthora root rot. While water stress did not increase the incidence or severity of root and crown rot in a susceptible rhododendron cultivar, Purple Splendor, an induction of water stress before inoculation rendered the normally resistant cultivar Caroline susceptible to extensive attack by *P. cinnamomi* (Blaker and MacDonald, 1981). The rhododendrons that were watered to avoid stress maintained leaf Ψ values between −2 and −6 bars, whereas the rhododendrons that were stressed for several days had minimum leaf Ψ values between −16 and −20 bars; the stress was relieved shortly before the soil was saturated and inoculated with motile zoospores. In somewhat analogous experiments, the susceptibility of chrysanthemum roots to *P. cryptogea* was found to increase with the degree of salinity stress to which they were exposed before inoculation (MacDonald, 1982). Only 20% of the unstressed chrysanthemum roots developed lesions 72 hr after inoculation, whereas 70 and 88% of roots previously exposed to 0.1 and 0.2 *M* NaCl, respectively, developed lesions. Furthermore, microscopic examination showed that more zoospores encysted on the surfaces of roots that had been stressed by salt; the lesions that developed on the salt-stressed roots also spread rapidly, and large portions of the root system became necrotic (MacDonald, 1982). Evidently, stress and/or the relief of stress may stimulate zoospore attraction or attachment to roots and may decrease the resistance of root tissue to pathogenic invasion.

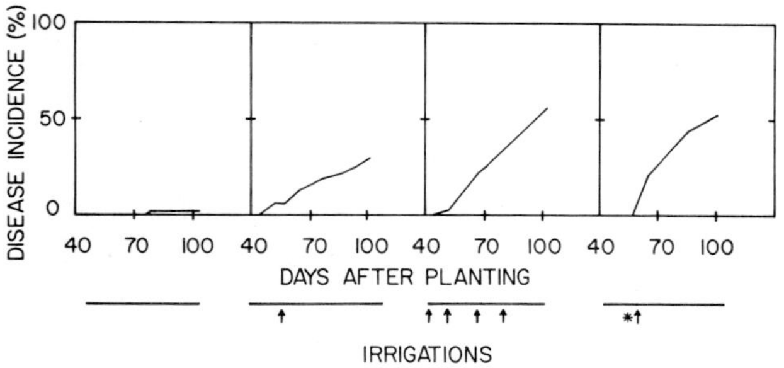


**Fig. 3.** The development of Phytophthora root rot in the susceptible safflower cultivar Nebraska 10 subjected to various irrigation schedules in the field. Surface irrigations were applied at the times indicated by arrows, and the asterisk denotes the occurrence of stress before irrigation. No rain fell during crop growth.

Oxygen has a very low solubility and diffusion rate in water, and when a soil is flooded or saturated, oxygen may be depleted in sites of high metabolic activity (e.g., Drew and Lynch, 1980; Griffin, 1972; Stolzy, 1974). As a result, roots experience varying degrees of oxygen deficiency in saturated soils, and physiologic damage resulting from both the direct and indirect effects of low oxygen concentrations in soils may predispose roots to infection (Drew and Lynch, 1980; Stolzy, 1974). Actually, Phytophthora root rots are among the most interesting examples of flood predisposition to infectious diseases, because the same wet soil conditions that enhance zoospore formation and dispersal can predispose some hosts to infection. Predisposition of alfalfa was demonstrated by flooding plants in sterile soil before inoculating them with zoospores of *P. megasperma* f. sp. *medicaginis* (Kuan and Erwin, 1980). The percentage of alfalfa seedlings killed after inoculation progressively increased from 50 to 90 as the duration of the flooding treatment before inoculation increased from 0 to 5 days. The mechanisms by which flooding predisposes alfalfa are not clearly known, but low oxygen levels may make roots more leaky and thus more attractive to zoospores (Kuan and Erwin, 1980). The predisposing effect of flooding on the susceptibility of the rhododendron cultivar Caroline to root and crown rot is even more dramatic. Caroline, which is normally resistant, became severely diseased when it was flooded for 48 hr before being inoculated with zoospores of *P. cinnamomi* (Blaker and MacDonald, 1981).

The impact of low oxygen concentrations in soil on the severity of Phytophthora root rot was apparently first examined by Stolzy et al (1965). They found root rot of *Citrus sinensis* in soil infested with *P. citrophthora* and *P. parasitica* to be greater in a nitrogen atmosphere than in air and greater under prolonged periods of soil saturation. A lack of compensating growth by uninfected roots was one of the reasons that low oxygen treatments increased the damage caused to *C. sinensis* by *Phytophthora* spp. (Stolzy et al, 1965) and that high levels of soil moisture increased the severity of Phytophthora root rot in soybean (Kittle and Gray, 1979). Because the growth of healthy roots is sensitive to oxygen concentration (Drew and Lynch, 1980; Stolzy, 1974), a lack of root regeneration is likely to aggravate any Phytophthora root rot problem when soils are saturated for prolonged periods. It is also noteworthy that postinoculation treatments of cocoa pods with 2 or 4% oxygen greatly increased the rate of pathogenic attack by *P. palmivora* (Spence, 1961). Such results suggest that the effects of low oxygen concentrations on the resistance mechanisms that are manifested in roots after infection warrant further investigation.

*Phytophthora* spp. cause root and crown rots in a very large number of diverse plant species, and predisposition to Phytophthora root rot by flooding, drought, or salinity stresses has not been researched in hosts other than alfalfa, safflower, rhododendron, and chrysanthemum. Therefore, although the diversity of the few known examples suggests that they are not unique, we do not really know whether stress predisposition phenomena are widespread among the multitude of other *Phytophthora* diseases. Note also that although periods of drought or salinity stress are not conducive to inoculum formation or infection by soilborne *Phytophthora* spp., the inoculum of those few species that have been examined under stress can, nevertheless, effectively persist through the stress cycles that cause predisposition (Duniway, 1979). Of course, the flooded or saturated soil conditions that usually relieve a drought or salinity stress, and which may also predispose a host, are very suitable for zoospore release and dispersal.

Epidemiologic research on *Phytophthora* diseases has tended to emphasize the influence of physical factors on the pathogen on the assumption that the effects of climate on pathogen behavior to a large extent govern the rate of disease development. The research demonstrating predisposition has shown the fallacy of this assumption and the importance of research on the host. If, for example, a host is predisposed under flooded or saturated soil conditions, research on sporangia and zoospores alone, though very necessary, cannot fully demonstrate the effects of such conditions on disease development. Studies on the host were obviously necessary to demonstrate predisposition by drought and salinity. Although space and time do not permit a review here, the limited studies that have been done on mechanisms by which infection of roots by *Phytophthora* spp. damages plants physiologically (Duniway, 1977a; Kittle and Gray, 1980; Sterne et al, 1978) do suggest that the manifestations of that damage as symptoms and yield losses, like infection processes, are conditioned by physical factors in the environment.

## CONCLUDING REMARKS

Most of the data reviewed here come from laboratory studies, but more precise research on the epidemiology of *Phytophthora* diseases is also needed in the field. For example, in late blight, empirical correlations between weather conditions and disease development that were established in the field have been used more than laboratory data to forecast epidemics (e.g., DeWeille, 1964; Krause and Massie, 1975). Experiments under controlled conditions are needed to isolate variables and to establish many of the causal relationships that underlie disease epidemiology. Careful field experimentation and observation, however, are just as necessary. Research on the influence of physical factors on disease epidemiology ought to at least include accurate measurements of the physical conditions actually encountered by the pathogen and host in the field. Furthermore, for many of the diseases caused by soilborne *Phytophthora* spp., the net effects of physical factors on the pathogen, the host, and the other microorganisms that may stimulate or suppress disease development must be considered.

*Phytophthora* diseases are very diverse, and no one course of epidemiologic research is satisfactory for research on many of them. In fact, many of the advancements that have been made come from the diversity of the research approaches taken. When designing research on the effects of physical factors on *Phytophthora* diseases, however, consideration of the following goals may be helpful: 1) define the parameters and dimensions of infection periods, i.e. those periods of time during which physical conditions are conducive to infection; 2) examine host and pathogen responses that underlie infection periods; 3) determine the physical requirements and limits of effective dispersal; 4) determine the conditions under which inoculum and infections persist between infection periods; 5) examine possible predisposition effects of conditions between infection periods; 6) determine the extent and conditions under which infection foci grow to induce disease and crop loss; and 7) monitor or manipulate the physical environment appropriately.

The last goal is the ultimate one if physical conditions can be manipulated sufficiently to control disease. The potentials for this are probably the greatest in irrigated agriculture where soil moisture and, to a lesser degree, surface wetness (e.g., Rotem et al, 1970) can be manipulated through variations in irrigation schedules, volumes applied, means of delivery, and features of surface and soil drainage. In greenhouses, temperatures may also be controlled. However, knowledge of physical factors that influence *Phytophthora* diseases has many applications in less controlled circumstances, especially in predicting the threat of disease development. More precise knowledge of the role of physical factors in disease development may enhance the detection of *Phytophthora* spp. in soil and optimize potentials for chemical control and has universal application in selecting conditions for pathogenicity tests to determine host range or find genetic resistance. Furthermore, knowledge of predisposing factors is essential to realistically screen for and effectively deploy genetic resistance.

## SUMMARY

The published literature on physical factors that influence the epidemiology of *Phytophthora* diseases is much too voluminous and diverse for complete review here. The topics chosen for review here really represent a meshing of only some of the literature with the author's own research interests in the effects of soil water on the epidemiology of Phytophthora root rots. Nevertheless, water status and temperature have large impacts on all *Phytophthora* diseases and I hope that this review has given the reader some new insights into the mechanisms by which these major physical factors of the environment influence disease development.

### ACKNOWLEDGMENTS

I thank E. A. Bernhardt, N. S. Blaker, J. D. MacDonald, S. M. Mircetich, F. J. Newhook, and W. F. Wilcox for providing data that were unpublished or manuscripts that were in press. I thank Carol Lindquist for assistance in the literature search and proofreading. Preparation of the manuscript was supported in part by USDA grant 59-2063-0-1-426-0.

### LITERATURE CITED

Al-Hedaithy, S. S. A., and Tsao, P. H. 1979. Sporangium pedicel length in *Phytophthora* species and the consideration of its uniformity in determining sporangium caducity. Trans. Br. Mycol. Soc. 72:1-13.

Benjamin, M., and Newhook, F. J. 1982. Effect of glass microbeads on *Phytophthora* zoospore motility. Trans. Br. Mycol. Soc. 78:43-46.

Bernhardt, E. A., and Grogan, R. G. 1982. Effect of soil matric potential on formation and indirect germination of sporangia of *Phytophthora parasitica, P. capsici,* and *P. cryptogea.* Phytopathology 72:507-511.

Biesbrock, J. A., and Hendrix, F. F. 1970. Influence of soil water and temperature on root necrosis of peach caused by *Pythium* spp. Phytopathology 60:880-882.

Bimpong, C. E., and Clerk, G. C. 1970. Motility and chemotaxis in zoospores of *Phytophthora palmivora* (Butl.) Butl. Ann. Bot. (London) 34:617-624.

Blaker, N. S., and MacDonald, J. D. 1981. Predisposing effects of soil moisture extremes on the susceptibility of rhododendron to Phytophthora root and crown rot. Phytopathology 71:831-834.

Chee, K. H. 1973. Production, germination and survival of chlamydospores of *Phytophthora palmivora* from *Hevea brasiliensis*. Trans. Br. Mycol. Soc. 61:21-26.

Cohen, Y., Perl, M., Rotem, J., Eyal, H., and Cohen, J. 1974. Ultrastructural and physiological changes in sporangia of *Pseudoperonospora cubensis* and *Phytophthora infestans* exposed to water stress. Can. J. Bot. 52:447-450.

Cook, R. J., and Duniway, J. M. 1981. Water relations in the life-cycles of soilborne plant pathogens. Pages 119-139 in: Water Potential Relations in Soil Microbiology. Soil Sci. Soc. Am. Spec. Publ. 9. 151 pp.

Cother, E. J., and Griffin, D. M. 1974. Chlamydospore germination in *Phytophthora drechsleri*. Trans. Br. Mycol. Soc. 63:273-279.

Crosier, W. 1934. Studies in the biology of *Phytophthora infestans* (Mont.) DeBary. Cornell Univ. Agric. Exp. Stn. Mem. 155. 40 pp.

DeWeille, G. A. 1964. Forecasting crop infection by the potato blight fungus. K. Ned. Meteorologisch Inst., Meded. Verh. 82. 144 pp.

Drew, M. C., and Lynch, J. M. 1980. Soil anaerobiosis, microorganisms, and root function. Annu. Rev. Phytopathol. 18:37-66.

Duniway, J. M. 1975a. Formation of sporangia by *Phytophthora drechsleri* in soil at high matric potentials. Can. J. Bot. 53:1270-1275.

Duniway, J. M. 1975b. Limiting influence of low water potential on the formation of sporangia by *Phytophthora drechsleri* in soil. Phytopathology 65:1089-1093.

Duniway, J. M. 1976. Movement of zoospores of *Phytophthora cryptogea* in soils of various textures and matric potentials. Phytopathology 66:877-882.

Duniway, J. M. 1977a. Changes in resistance to water transport in safflower during the development of Phytophthora root rot. Phytopathology 67:331-337.

Duniway, J. M. 1977b. Predisposing effect of water stress on the severity of Phytophthora root rot in safflower. Phytopathology 67:884-889.

Duniway, J. M. 1978. Irrigation, plant water stress and varietal interactions affecting the development of Phytophthora root rot of safflower in the field. (Abstr.) Phytopathol. News 12:148.

Duniway, J. M. 1979. Water relations of water molds. Annu. Rev. Phytopathol. 17:431-460.

Gerlach, W. W. P., Hoitink, H. A. J., and Schmitthenner, A. F. 1976. Survival and host range of *Phytophthora citrophthora* in Ohio nurseries. Phytopathology 66:309-311.

Gisi, U. 1975. Investigations on the soil phase of *Phytophthora cactorum* (Leb. and Cohn) Schroet. with fluorescent optical direct observation. Z. Pflanzenkr. Pflanzenschutz 82:355-377. (In German)

Gisi, U., Zentmyer, G. A., and Klure, L. J. 1980. Production of sporangia by *Phytophthora cinnamomi* and *P. palmivora* in soils at different matric potentials. Phytopathology 70:301-306.

Griffin, D. M. 1972. Ecology of Soil Fungi. Syracuse Univ. Press, Syracuse, NY. 193 pp.

Guzmán-N., J. 1964. Nature of partial resistance of certain clones of three *Solanum* species to *Phytophthora infestans*. Phytopathology 54:1398-1404.

Hemmes, D. E., and Hohl, H. R. 1975. Ultrastructural changes and a second mode of flagellar degeneration during ageing of *Phytophthora palmivora* sporangia. J. Cell Sci. 19:563-577.

Hine, R. B., and Aragaki, M. 1963. Influence of soil temperature on a crown rot disease of parsley caused by *Phytophthora parasitica*. Phytopathology 53:1113-1114.

Hirst, J. M., and Stedman, O. J. 1960. The epidemiology of *Phytophthora infestans*. I. Climate, ecoclimate and the phenology of disease outbreak. Ann. Appl. Biol. 48:471-488.

Hunter, J. E., and Kunimoto, R. K. 1974. Dispersal of *Phytophthora palmivora* sporangia by wind-blown rain. Phytopathology 64:202-206.

Ioannou, N., and Grogan, R. G. 1977. The influence of soil matric potential on the production of sporangia by *Phytophthora parasitica* in relation to its isolation from soil by baiting techniques. (Abstr.) Proc. Am. Phytopathol. Soc. 4:173.

Johnson, L. B., and Klisiewicz, J. M. 1969. Environmental effects on safflower reaction to *Phytophthora drechsleri*. Phytopathology 59:469-472.

Kable, P. F., and MacKenzie, D. R. 1980. Survival of *Phytophthora infestans* in potato stem lesions at high temperatures and implications for disease forecasting. Plant Dis. 64:165-167.

Katsura, K. 1971. Some ecological studies on zoospore of *Phytophthora capsici* Leonian. Rev. Plant Prot. Res. 4:58-70.

Kittle, D. R., and Gray, L. E. 1979. The influence of soil temperature, moisture, porosity, and bulk density on the pathogenicity of *Phytophthora megasperma* var. *sojae*. Plant Dis. Rep. 63:231-234.

Kittle, D. R., and Gray, L. E. 1980. Effects of infection by *Phytophthora megasperma* var. *sojae* on the water relations of soybean. Crop Sci. 20:504-507.

Kliejunas, J. T., and Ko, W. H. 1974. Effect of motility of *Phytophthora palmivora* zoospores on disease severity in papaya seedlings and substrate colonization in soil. Phytopathology 64:426-428.

Ko, W.-H. 1971. Direct observation of fungal activities on soil. Phytopathology 61:437-438.

Krause, R. A., and Massie, L. B. 1975. Predictive systems: Modern approaches to disease control. Annu. Rev. Phytopathol. 13:31-47.

Kuan, T. L., and Erwin, D. C. 1980. Predisposition effect of water saturation of soil on Phytophthora root rot of alfalfa. Phytopathology 70:981-986.

Lapwood, D. H. 1977. Factors affecting the field infection of potato tubers of different cultivars by blight (*Phytophthora infestans*). Ann. Appl. Biol. 85:23-42.

Leach, C. M. 1975. Influence of relative humidity and red-infrared radiation on violent spore release by *Drechslera turcica* and other fungi. Phytopathology 65:1303-1312.

Lumsden, R. D., and Ayers, W. A. 1975. Influence of soil environment on the germinability of constitutively dormant oospores of *Pythium ultimum*. Phytopathology 65:1101-1107.

MacDonald, J. D. 1977. Influence of soil water on the sporangia and zoospores of three root-infecting species of *Phytophthora*. Ph.D. thesis, Univ. Calif., Davis. 109 pp.

MacDonald, J. D. 1982. Effect of salinity stress on the development of Phytophthora root rot of chrysanthemum. Phytopathology 72:214-219.

MacDonald, J. D., and Duniway, J. M. 1978a. Influence of the matric and osmotic components of water potential on zoospore discharge in *Phytophthora*. Phytopathology 68:751-757.

MacDonald, J. D., and Duniway, J. M. 1978b. Temperature and water stress effects on sporangium viability and zoospore discharge in *Phytophthora cryptogea* and *P. megasperma*. Phytopathology 68:1449-1455.

MacDonald, J. D., and Duniway, J. M. 1978c. Influence of soil texture and temperature on the motility of *Phytophthora cryptogea* and *P. megasperma* zoospores. Phytopathology 68:1627-1630.

MacDonald, J. D., and Duniway, J. M. 1979. Use of fluorescent antibodies to study the survival of *Phytophthora megasperma* and *P. cinnamomi* zoospores in soil. Phytopathology 69:436-441.

Malajczuk, N. 1975. Interactions between *Phytophthora cinnamomi* Rands and roots of *Eucalyptus calophylla* R. Br. and *Eucalyptus marginata* Donn. ex. Sm. Ph.D. thesis, Univ. West. Aust., Nedlands. 371 pp.

Marx, D. H., and Bryan, W. C. 1969. Effect of soil bacteria on the mode of infection of pine roots by *Phytophthora cinnamomi*. Phytopathology 59:614-619.

McIntosh, D. L. 1972. Effects of soil water suction, soil temperature, carbon and nitrogen amendments, and host rootlets on survival in soil of zoospores of *Phytophthora cactorum*. Can. J. Bot. 50:269-272.

Meyer, D., and Schönbeck, F. 1975. Investigations on the development of *Phytophthora cactorum* (Leb. et Cohn) Schroet. in soil. Z. Pflanzenkr. Pflanzenschutz 82:337-354. (In German)

Mitchell, D. J., and Zentmyer, G. A. 1971a. Effects of oxygen and carbon dioxide tensions on growth of several species of *Phytophthora*. Phytopathology 61:787-791.

Mitchell, D. J., and Zentmyer, G. A. 1971b. Effects of oxygen and carbon dioxide tensions on sporangium and oospore formation by *Phytophthora* spp. Phytopathology 61:807-811.

Nesbitt, H. J., Malajczuk, N., and Glenn, A. R. 1979. Effect of soil moisture and temperature on the survival of *Phytophthora cinnamomi* Rands in soil. Soil Biol. Biochem. 11:137-140.

Newhook, F. J., Young, B. R., Allen, S. D., and Allen, R. N. 1981. Zoospore motility of *Phytophthora cinnamomi* in particulate substrates. Phytopathol. Z. 101:202-209.

Patil, S. S., and Young, R. A. 1960. The influence of temperature on development of *Phytophthora parasitica* root rot of fuchsia. Phytopathology 50:386-388.

Pfender, W. F., Hine, R. B., and Stanghellini, M. E. 1977. Production of sporangia and release of zoospores by *Phytophthora megasperma* in soil. Phytopathology 67:657-663.

Reeves, R. J. 1975. Behaviour of *Phytophthora cinnamomi* Rands in different soils and water regimes. Soil Biol. Biochem. 7:19-24.

Rocha, H. M., and Machado, A. D. 1973. The effect of light, temperature and relative humidity on sporulation of *Phytophthora palmivora* (Butl.) Butl. in cacao pods. Rev. Theobroma 3:22-25. (In Portuguese)

Rotem, J., and Cohen, Y. 1974. Epidemiological patterns of *Phytophthora infestans* under semi-arid conditions. Phytopathology 64:711-714.

Rotem, J., Cohen, Y., and Spiegel, S. 1968. Effect of soil moisture on the predisposition of tobacco to *Peronospora tabacina.* Plant Dis. Rep. 52:310-313.

Rotem, J., Palti, J., and Lomas, J. 1970. Effects of sprinkler irrigation at various times of the day on development of potato late blight. Phytopathology 60:839-843.

Rotem, J., Cohen, Y., and Putter, J. 1971. Relativity of limiting and optimum inoculum loads wetting durations, and temperatures for infection by *Phytophthora infestans.* Phytopathology 61:275-278.

Rotem, J., Cohen, Y., and Bashi, E. 1978. Host and environmental influences on sporulation in vivo. Annu. Rev. Phytopathol. 16:83-101.

Roth, L. F., and Kuhlman, E. G. 1966. *Phytophthora cinnamomi,* an unlikely threat to Douglas-fir forestry. For. Sci. 12:147-159.

Ruddick, S. M., and Williams, S. T. 1972. Studies on the ecology of actinomycetes in soil. V. Some factors influencing the dispersal and adsorption of spores in soil. Soil Biol. Biochem. 4:93-103.

Sato, N. 1979. Effect of soil temperature on the field infection of potato tubers by *Phytophthora infestans.* Phytopathology 69:989-993.

Shea, S. R. 1975. Environmental factors of the northern jarrah forest in relation to pathogenicity and survival of *Phytophthora cinnamomi.* Bull. 85. For. Dep., West. Australia, Perth. 83 pp.

Sing, V. O., and Bartnicki-Garcia, S. 1975. Adhesion of *Phytophthora palmivora* zoospores: Electron microscopy of cell attachment and cyst wall fibril formation. J. Cell Sci. 18:123-132.

Slavík, B., ed. 1974. Methods of Studying Plant Water Relations. Springer-Verlag, New York. 449 pp.

Sneh, B. 1974. Recovery of sporangia of *Phytophthora cactorum* from soil. Can. J. Bot. 52:1777-1778.

Sneh, B., and McIntosh, D. L. 1974. Studies on the behavior and survival of *Phytophthora cactorum* in soil. Can. J. Bot. 52:795-802.

Sneh, B., Humble, S. J., and Lockwood, J. L. 1977. Parasitism of oospores of *Phytophthora megasperma* var. *sojae, P. cactorum, Pythium* sp., and *Aphanomyces euteiches* in soil by oomycetes, chytridiomycetes, hyphomycetes, actinomycetes, and bacteria. Phytopathology 67:622-628.

Spence, J. A. 1961. Probable mechanism of resistance of varieties of cocoa to black pod disease caused by *Phytophthora palmivora* (Butl.) Butl. Nature (London) 192:278.

Stanghellini, M. E., and Hancock, J. G. 1971. The sporangium of *Pythium ultimum* as a survival structure in soil. Phytopathology 61:157-164.

Sterne, R. E., and McCarver, T. H. 1980. Formation of sporangia by *Phytophthora cryptogea* and *P. parasitica* in artificial and natural soils. Soil Biol. Biochem. 12:441-442.

Sterne, R. E., Zentmyer, G. A., and Kaufmann, M. R. 1977a. The effect of matric and osmotic potential of soil on Phytophthora root disease of *Persea indica.* Phytopathology 67:1491-1494.

Sterne, R. E , Zentmyer, G. A., and Kaufmann, M. R. 1977b. The influence of matric potential, soil texture, and soil amendment on root disease caused by *Phytophthora cinnamomi.* Phytopathology 67:1495-1500.

Sterne, R. E., Kaufmann, M. R., and Zentmyer, G. A. 1978. Effect of Phytophthora root rot on water relations of avocado: Interpretation with a water transport model. Phytopathology 68:595-602.

Stolzy, L. H. 1974. Soil atmosphere. Pages 335-361 in: The Plant Root and Its Environment. E. W. Carson, ed. Univ. Press Va., Charlottesville. 691 pp.

Stolzy, L. H., Letey, J., Klotz, L. J., and Labanauskas, C. K. 1965. Water and aeration as factors in root decay of *Citrus sinensis.* Phytopathology 55:270-275.

Sugar, D. 1977. The development of sporangia of *Phytophthora cambivora, P. megasperma,* and *P. drechsleri* and severity of root and crown rot in *Prunus mahaleb* as influenced by soil matric potential. M.S. thesis, Univ. Calif., Davis. 56 pp.

Tucker, C. M. 1931. Taxonomy of the genus *Phytophthora* de Bary. Mo. Agric. Exp. Stn. Res. Bull. 153. 208 pp.

Turner, P. D. 1965. Behavior of *Phytophthora palmivora* in soil. Plant Dis. Rep. 49:135-137.

Uppal, B. N. 1926. Relation of oxygen to spore germination in some species of the Peronosporales. Phytopathology 16:285-292.

van der Plank, J. E. 1963. Plant Diseases: Epidemics and Control. Academic Press, New York. 349 pp.

Wallin, J. R. 1953. The production and survival of sporangia of *Phytophthora infestans* on tomato and potato plants in the field. Phytopathology 43:505-508.

Ward, E. W. B., Stössel, P., and Lazarovits, G. 1981. Similarities between age-related and race-specific resistance of soybean hypocotyls to *Phytophthora megasperma* var. *sojae.* Phytopathology 71:504-508.

Warren, R. C., and Colhoun, J. 1975. Viability of sporangia of *Phytophthora infestans* in relation to drying. Trans. Br. Mycol. Soc. 64:73-78.

Waterhouse, G. M. 1931. The production of conidia in the genus *Phytophthora.* Trans. Br. Mycol. Soc. 15:311-321.

Waterhouse, G. M. 1974. Other *Phytophthora* species recorded on cacao. Pages 71-79 in: *Phytophthora* Disease of Cocoa. P. H. Gregory, ed. Longman, London. 348 pp.

Wilcox, W., and Mircetich, S. M. 1979. The influence of different levels of soil moisture on Phytophthora root rot and crown rot of Mahaleb cherry rootstock. (Abstr.) Phytopathology 69:1049.

Young, B. R., Newhook, F. J., and Allen, R. N. 1979. Motility and chemotactic response of *Phytophthora cinnamomi* zoospores in 'ideal soils.' Trans. Br. Mycol. Soc. 72:395-401.

Zan, K. 1962. Activity of *Phytophthora infestans* in soil in relation to tuber infection. Trans. Br. Mycol. Soc. 45:205-221.

Zentmyer, G. A. 1980. *Phytophthora cinnamomi* and the diseases it causes. Monogr. 10. Am. Phytopathol. Soc., St. Paul, MN. 96 pp.

Zimmer, D. E., and Urie, A. L. 1967. Influence of irrigation and soil infestation with strains of *Phytophthora drechsleri* on root rot resistance of safflower. Phytopathology 57:1056-1059.

# 15

# Role of Chemical Factors in Development of *Phytophthora* Diseases

A. F. Schmitthenner and C. H. Canaday

## CONTENTS



## INTRODUCTION

The chemical factors considered in this review are high fertility and salinity, nitrogen (N), potassium (K), phosphorus (P), calcium (Ca), magnesium (Mg), sulfur (S), hydrogen ion concentration ($H^+$), and micronutrient elements. The effects of these factors have been compiled many times in the past, but little attention has been given to their effects on the diseases caused by *Phytophthora* spp. The one notable exception is the recent monograph on *P. cinnamomi* by Zentmyer (1980). Good general reviews on certain aspects of chemical effects on disease development have been presented by Chapman (1965), Newhook and Podger (1972), Broadbent and Baker (1974), and Huber (1980).

Most investigations have been concerned with late blight of potato caused by *P. infestans*, root rots caused by *P. cinnamomi*, and black shank of tobacco caused by *P. parasitica* var. *nicotianae* (hereinafter referred to as *P. parasitica*). There are some studies on *Phytophthora* diseases of soybean (*P. megasperma* f. sp. *glycinea*), sweet pepper (*P. capsici*), citrus (*P. parasitica* and *P. citrophthora*), tomato and roselle (*P. parasitica*), betelvine (*P. palmivora*), chrysanthemum (*P. cryptogea*), and pigeon pea (*P. drechsleri*). There are few or no reports for several other important diseases such as cocoa black pod (*P. palmivora*), alfalfa root rot (*P. megasperma* f. sp. *medicaginis*), safflower root rot (*P. cryptogea*), apple collar rot (*P. cactorum*), red stele of strawberry (*P. fragariae*), downy mildew of bean (*P. phaseoli*), or the many *Phytophthora* problems of woody plants and herbaceous ornamentals. This review discusses first the effects of high fertility and salinity, followed by individual chemical factors, their modes of action, and their potential for enhancing or controlling disease.

Approved for publication as Journal Article 55-81 of the Ohio Agricultural Research and Development Center, Wooster.

## EFFECTS OF CHEMICAL FACTORS

### High Fertility and Salinity

Moderate to high balanced fertility has been reported to increase the severity of a number of diseases caused by *Phytophthora* spp. (Table 1). Multigenic resistance of potato to late blight may not be effective under conditions of high fertility (Gallegly and Niederhauser, 1959; Main and Gallegly, 1964). In solution culture, multigenic resistance to late blight in tomato (Wilson and Gallegly, 1960) and potato (Main and Gallegly, 1964) can be detected using tenth-strength Hoagland's solution but not in Hoagland's solution three times normal strength. Marks et al (1972) reported more dieback caused by *P. cinnamomi* in eucalyptus trees treated with 17-9-7 fertilizer than in unfertilized trees. Complete fertilizer has been reported to reverse ohia decline in Hawaii in which *P. cinnamomi* may be involved (Kliejunas and Ko, 1976). The response was improved by the addition of fungicide. Soybean rot caused by *P. megasperma* f. sp. *glycinea* was more severe under high fertility (Dirks et al, 1980). High fertility effects may be due predominantly to high amounts of nitrogen.

Bingham et al (1958) commented on the possible confounding of the effects of high fertility and high salinity in solution culture experiments. They found that root rot of avocado seedlings inoculated with *P. cinnamomi* developed more slowly with high rates of N, P, and K; but these same solutions also had very high total salt concentrations. Anderson (1951) reported that high salt concentrations decreased development of *P. cinnamomi* on pineapple. In contrast, black shank of tobacco was increased with high concentrations of salts (Apple, 1961; McCarter, 1965). Recently, MacDonald (1982) showed that Phytophthora root rot of chrysanthemum, caused by *P. cryptogea*, can be increased by subjecting hydroponically grown plants to a brief period of salinity stress before inoculating them with zoospores. Insufficient information is available at this time to draw any general conclusions concerning the effects of high salinity on *Phytophthora* diseases.

### Nitrogen

There are as many examples of N increasing the severity of disease as there are of N decreasing disease severity (Table 2). High N has been reported to both increase (Awan and Struchtemeyer, 1957; Gallegly and Niederhauser, 1959; Weindlmayr, 1965) and decrease (Langbein and Pehl, 1962; Lowings and Acha, 1959; Schaffnit and Volk, 1927) severity of potato late blight. These apparently contradictory data have been reconciled by Langbein and Pehl (1962), Herlihy and Carroll (1969), and Herlihy (1970). They pointed out that high N reduced the size of late blight lesions (less disease) but at the same time increased the size of the canopy, which creates an environment conducive to sporulation and spread of *P. infestans*. The end result is increased severity of disease. Walerych et al (1970) reported that $NO_3^-$-N increased the severity of late blight, but $NH_4^+$-N and urea had no effect. Pepper blight is another foliar disease enhanced by high N (Elenkov and Bakharieva, 1975).

Soybean root rot caused by *P. megasperma* f. sp. *glycinea* (Canaday and Schmitthenner, 1979) and pigeon pea blight caused by *P. drechsleri* (Pal and Grewal, 1976) are increased by high N. Effects of N on diseases caused by *P. parasitica* vary. Tobacco black shank (Apple, 1961; McCarter, 1965) and roselle stem rot (Olunloyo and Adeniji, 1976) were more severe under high N, whereas tomato stem lesion formation (Weststeijn, 1973), betelvine blight (Thyagarajan et al, 1972), and philodendron leaf spot (Harkness and Reynolds, 1964) were less severe. Klotz et al (1958) reported that $NO_3^-$-N decreased citrus root rot, whereas $NH_4^+$-N and urea had the opposite effect.

Some very interesting effects of high N have been explored with root rots caused by *P. cinnamomi*. Bingham et al (1958) demonstrated that high $NO_3^-$-N reduced avocado root rot in solution culture, but they concluded that the high salinity of the solution was responsible for the effects. Zentmyer and Bingham (1956) reported that $NO_2^-$ was more toxic to *P. cinnamomi* than to avocado in solution culture. This response

TABLE 1. Effects of balanced fertility (Fert) and salinity (Sal) on *Phytophthora* diseases

| Pathogen | Host and disease | Factor | Effect[a] | References |
|---|---|---|---|---|
| *P. infestans* | Potato late blight | Fert | + | Main and Gallegly (1964) |
| | Tomato late blight | Fert | + | Wilson and Gallegly (1960) |
| *P. parasitica* | Tobacco black shank | Fert | + | McCarter (1965) |
| | | Sal | + | Apple (1961) |
| *P. megasperma* f. sp. *glycinea* | Soybean rot | Fert | + | Dirks et al (1980) |
| *P. cinnamomi* | Eucalyptus root rot | Fert | + | Marks et al (1972) |
| | Ohia decline | Fert | − | Kliejunas and Ko (1976) |
| | Avocado root rot | Sal | − | Bingham et al (1958) |
| | Pineapple root rot | Sal | − | Anderson (1951) |
| *P. cryptogea* | Chrysanthemum root rot | Sal | + | MacDonald (1982) |

[a] Disease increased (+) or decreased (−) with high levels of fertility or salinity.

TABLE 2. Effects of nitrogen (N) on *Phytophthora* diseases

| Pathogen | Host and disease | Factor[a] | Effect[b] | References |
|---|---|---|---|---|
| *P. infestans* | Potato late blight | N | + | Awan and Struchtemeyer (1957)<br>Gallegly and Niederhauser (1959)<br>Weindlmayr (1965) |
| | | $NO_3^-$ | + | Walerych et al (1970) |
| | | $NH_4^+$, urea | 0 | Walerych et al (1970) |
| | | N | − | Langbein and Pehl (1962)<br>Lowings and Acha (1959)<br>Schaffnit and Volk (1927) |
| | Potato tuber rot | N | + | Herlihy (1970)<br>Herlihy and Carroll (1969) |
| *P. capsici* | Pepper blight | N | + | Elenkov and Bakharieva (1975) |
| *P. parasitica* | Tobacco blank shank | N | + | Apple (1961)<br>McCarter (1965) |
| | Roselle stem rot | N | + | Olunloyo and Adeniji (1976) |
| | Tomato stem lesions | N | − | Weststeijn (1973) |
| | Betelvine wilt | N | − | Thyagarajan et al (1972) |
| | Philodendron leaf spot | N | − | Harkness and Reynolds (1964) |
| | Citrus root rot | $NO_3^-$ | − | Klotz et al (1958) |
| | | $NH_4^+$, urea | + | Klotz et al (1958) |
| *P. megasperma* f. sp. *glycinea* | Soybean rot | N | + | Canaday and Schmitthenner (1979) |
| *P. drechsleri* | Pigeon pea blight | N | + | Pal and Grewal (1976) |
| *P. cinnamomi* | Avocado root rot | $NH_4^+$ | − | Broadbent and Baker (1974) |
| | | $NO_2^-$ | − | Zentmyer and Bingham (1956) |
| | | $NO_3^-$ | − | Bingham et al (1958) |
| | | $NH_3$ | − | Gilpatrick (1969) |
| | Pineapple root rot | N | 0 | Anderson (1951) |
| | Pine little leaf | N | + | Newhook and Podger (1972) |
| | | | − | Campbell and Copeland (1954)<br>Hepting et al (1945)<br>Roth et al (1948) |
| *P. cinnamomi* and *P. parasitica* | In vitro | $NH_3$, pH 8<br>$HNO_2$, pH 6 | − | Tsao and Oster (1981)<br>Tsao and Zentmyer (1979) |

[a] N = any form of nitrogen. Where nitrogen forms were compared, the forms are indicated.
[b] Disease increased (+), decreased (−), or did not change (0) with high levels of N.

to N depended on pH; 2 ppm was effective at pH 4.5 and 40 ppm at pH 6.5. Zentmyer (1963) reported on the efficacy of a high-N organic residue, alfalfa meal, for control of avocado root rot, and Gilpatrick (1969) demonstrated that levels of $NH_3$ toxic to *P. cinnamomi* were present in soil amended with alfalfa meal. Tsao and Zentmyer (1979) and Tsao and Oster (1981) proposed that $HNO_2$ and $NH_3$ produced at pH 6.0 and 8.0, respectively, are responsible for the inhibition of *P. cinnamomi* in soils amended with high rates of urea or other organic N. It should also be noted that high levels of $NH_4^+$-N are characteristic of *P. cinnamomi*-suppressive soil in Australia (Broadbent and Baker, 1974).

Effects on little leaf of pine, also incited by *P. cinnamomi*, varied with geographic location. Heavy N fertilization increased little leaf in New Zealand (Newhook and Podger, 1972). In the United States, little leaf was first treated as a nitrogen deficiency, and application of nitrogen fertilizer greatly improved the condition of infected trees and halted the spread of the disease (Campbell and Copeland, 1954; Hepting et al, 1945; Roth et al, 1948).

In general, the effects of nitrogen on *Phytophthora* diseases vary with different host-pathogen combinations. Additional work is needed on the effect of different forms of N on the severity of diseases caused by *Phytophthora* spp.

### Potassium

High rates of K generally are thought to decrease disease severity. However, results obtained with diseases caused by *P. infestans* have not been consistent (Table 3). In two cases, high K was associated with a reduction in late blight of potato (Alten and Orth, 1940; Awan and Struchtemeyer, 1957), but in three cases, high K had either no effects or variable effects (Herlihy and Carroll, 1969; Szczotka et al, 1973; Weindlmayr, 1965). High K reduced pepper blight (Elenkov and Bakharieva, 1975) and pigeon pea blight (Pal and Grewal, 1976). Response to K may depend on the rate of N used (Pal and Grewal, 1976; Weindlmayr, 1965).

Most studies on K have been with diseases caused by *P. parasitica*. In one study, high K increased betelvine wilt (Thyagarajan et al, 1972); in another, high K in combination with low Ca increased disease caused by *Phytophthora* on citrus (Chapman and Brown, 1942). Kincaid et al (1970) found a moderate quadratic relationship between K and tobacco black shank. Potassium at 560 kg/ha was associated with the least disease.

The decrease in avocado (Bingham et al, 1958) and pineapple (Anderson, 1951) root rot with high K was thought to involve high salinity as well. More work is needed with K, particularly on the relationship of high N fertilizer to levels of K in plant tissues.

### Phosphorus

The effects of P on *Phytophthora* diseases also have been inconsistent (Table 4). In general, high P is associated with less late blight (Awan and Struchtemeyer, 1957; Borys, 1966; Wojciechowski et al, 1964) and less tuber rot (Herlihy, 1970; Herlihy and Carroll, 1969) of potatoes. In general, these effects are the opposite of those of N. Results with diseases caused by *P. parasitica* have ranged from less disease (Thyagarajan et al, 1972) to no response (Dukes and Apple, 1968; McCarter, 1965) to increased disease (Apple, 1961; Olunloyo and Adeniji, 1976).

The most interesting response to P has been with little leaf of pine caused by *P. cinnamomi*. In New Zealand, correction of P deficiency has halted the spread of this disease (Newhook, 1970; Newhook and Podger, 1972; Weston, 1956) and resulted in recovery of diseased plants, similar to the effect of N (Hepting et al, 1945) in the United States.

Overall, P appears to have fewer effects on *Phytophthora* diseases (except on late blight of potato) than other major chemical factors, although this may simply reflect fewer studies with this factor.

### Calcium and Magnesium

There are some reports that Ca may have direct effects on *Phytophthora* diseases (Table 5). No response to Ca was reported with late blight of potato (Borys, 1964; El Fahl and Calvert, 1976). The addition of small quantities of Ca as $CaCl_2$ increased the percentage of plants killed by pepper blight without raising the pH (Muchovej et al, 1980). In extensive soil studies with tobacco black shank, exchangeable Ca and Mg were positively correlated with increases in disease (Dukes and

TABLE 3. Effects of potassium (K) on *Phytophthora* diseases

| Pathogen | Host and disease | Factors | Effect[a] | References |
|---|---|---|---|---|
| *P. infestans* | Potato late blight | K | − | Alten and Orth (1940) |
| | | | | Awan and Struchtemeyer (1957) |
| | | K | 0 | Herlihy and Carroll (1969) |
| | | K | ± | Szczotka et al (1973) |
| | | High K, high N | + | Weindlmayr (1965) |
| *P. capsici* | Pepper blight | K | − | Elenkov and Bakharieva (1975) |
| *P. drechsleri* | Pigeon pea blight | High K, low N | − | Pal and Grewal (1976) |
| *P. parasitica* | Tobacco black shank | K | 0 | Apple (1961) |
| | | | | Dukes and Apple (1968) |
| | | | | McCarter (1965) |
| | | | − | Kincaid et al (1970) |
| | Tomato stem lesions | K | 0 | Weststeijn (1973) |
| | Roselle stem rot | K | − | Olunloyo and Adeniji (1976) |
| | Philodendron leaf spot | K | 0 | Harkness and Reynolds (1964) |
| | Betelvine wilt | K | + | Thyagarajan et al (1972) |
| | Citrus gummosis | High K, low Ca | + | Chapman and Brown (1942) |
| *P. cinnamomi* | Avocado root rot | K | − | Bingham et al (1958) |
| | Pine little leaf | K | 0 | Roth et al (1948) |
| | Pineapple root rot | K | − | Anderson (1951) |

[a]Disease increased (+), decreased (−), was highly variable (±), or did not change (0) with high levels of K.

Apple, 1968; Kincaid et al, 1972). In solution culture tests (Moore and Wills, 1967), low Ca retarded development of black shank without directly affecting the fungus. In another study, Ca increased disease in susceptible and moderately resistant but not highly resistant cultivars (Wills and Moore, 1969), and Mg was equally effective in increasing disease. Other studies have indicated that Mg is more effective than Ca in increasing black shank (Kincaid et al, 1970; McCarter, 1965).

Results reported with eucalyptus root rot are also of interest. Boughton et al (1978) reported suppression of jarrah root rot by application of calcium carbonate. Halsall (1980) found that high calcium had no effect on susceptible eucalyptus species but increased root rot of tolerant species. High Ca is also one of the characteristics of *P. cinnamomi*-suppressive soil (Broadbent and Baker, 1974). In general, however, high Ca increases the severity of *Phytophthora* diseases.

## pH and Sulfur

In general, high pH results in more severe diseases caused by *Phytophthora* spp. Low pH and application of sulfur to reduce the pH of soil or growth media have been quite effective in controlling a number of *Phytophthora* diseases (Table 6). Low pH reduces both late blight (Barnes and Chestnutt, 1966; El Fahl and Calvert, 1976) and tuber rot (Hutchinson and Mulligan, 1971; McKee, 1968) of potatoes. Tobacco black shank can be controlled by lowering the pH. Critical pH values reported are 4.5 (Kincaid and Gammon, 1954), 4.6 (McCarter, 1965), and 4.8 (Troutman and LaPrade, 1962). However, controlling black shank by low pH may not be a practical solution, because liming is important for increased leaf quality (Kincaid et al, 1970). The increase in pepper blight caused by addition of $CaCO_3$ and $Ca(OH)_2$ has been explained by an increase in pH with a resultant decrease in soluble aluminum (Muchovej et al, 1980).

Diseases caused by *P. cinnamomi* also respond to pH. Avocado root rot was controlled in nutrient solutions at pH 3.0 (Bingham and Zentmyer, 1954). At higher pH, the rate at which roots became diseased increased, but above pH 8.0, disease development was retarded. Growth of avocado is also adversely affected at pH 3.0. Pineapple root rot was controlled when sulfur was added to reduce the pH below 3.8, which is satisfactory for pineapple growth (Pegg, 1977). This low pH could not be attained with aluminum or ammonium sulfate. Other diseases reduced by low pH include citrus root rot (Tsao, 1959), rhododendron wilt (White, 1937), and betelvine wilt (Turner, 1969). Lowering the pH with sulfur may be a practical method for controlling *Phytophthora* diseases of

TABLE 4. Effects of phosphorus (P) on *Phytophthora* diseases

| Pathogen | Host and disease | Factors | Effect[a] | References |
|---|---|---|---|---|
| *P. infestans* | Potato late blight | P | − | Awan and Struchtemeyer (1957) |
| | | | | Borys (1966) |
| | | | | Wojciechowski et al (1964) |
| | | | 0 | Weindlmayr (1965) |
| | Potato tuber rot | P | − | Herlihy (1970) |
| | | | | Herlihy and Carroll (1969) |
| *P. parasitica* | Tobacco black shank | P | + | Apple (1961) |
| | | | 0 | Dukes and Apple (1968) |
| | | | | McCarter (1965) |
| | Roselle stem rot | P | + | Olunloyo and Adeniji (1976) |
| | Betelvine wilt | P | − | Thyagarajan et al (1972) |
| *P. cinnamomi* | Pine litte leaf | P | − | Newhook (1970) |
| | | | | Newhook and Podger (1972) |
| | | | | Weston (1956) |
| | | | + | Roth et al (1948) |
| *P. drechsleri* | Pigeon pea blight | P | 0 | Pal and Grewal (1976) |

[a] Disease increased (+), decreased (−), or did not change (0) with high levels of P.

TABLE 5. Effects of calcium (Ca) and magnesium (Mg) on *Phytophthora* diseases

| Pathogen | Host and disease | Factor | Effect[a] | References |
|---|---|---|---|---|
| *P. infestans* | Potato late blight | Ca | 0 | Borys (1964) |
| | | | | El Fahl and Calvert (1976) |
| *P. capsici* | Pepper blight | Ca | + | Muchovej et al (1980) |
| *P. parasitica* | Citrus gummosis | High Ca, low K | − | Chapman and Brown (1942) |
| | Tobacco black shank | Ca | + | Kincaid et al (1972) |
| | | | | Moore and Wills (1967) |
| | | | | Dukes and Apple (1968) |
| | | Ca, Mg | + | Wills and Moore (1969) |
| | | Ca | 0 | Kincaid et al (1970) |
| | | | | McCarter (1965) |
| | | Mg | + | Kincaid et al (1970) |
| | | | | McCarter (1965) |
| *P. cinnamomi* | Eucalyptus root rot | Ca | − | Boughton et al (1978) |
| | | Ca-solution | 0 | Halsall (1980) |
| | | Ca-soil | + | Halsall (1980) |
| | Avocado root rot | Ca-soil | − | Broadbent and Baker (1974) |
| | Pine little leaf | Ca, Mg | + | Roth et al (1948) |

[a] Disease increased (+), decreased (−), or did not change (0) with high levels of factors.

acid-tolerant plants. However, *Phytophthora* diseases are not always severe at near-neutral pH (Broadbent and Baker, 1974).

### Other Chemical Factors

There are sporadic reports of the effects of some other chemical factors on *Phytophthora* diseases (Table 7). An attempt was made to control potato late blight by foliage sprays with 5% NaCl (Bollen, 1965), but the potatoes were more adversely affected than the disease. Variable results were obtained with chloride and sulfate ions (Borys, 1964).

The effects of micronutrients on potato tuber rot (Mudich, 1967) and little leaf of pine (Roth et al, 1948) have also been reported. Molybdenum and manganese in phosphate fertilizer reduced potato tuber blight, but copper and zinc had no effect. Addition of boron to a complete fertilizer reduced little leaf of pine, but other heavy metals were ineffective. Skaptason et al (1940) reported copper buildup in soil, after a number of years of continuous spray application on potatoes, to levels high enough to significantly lower germination of *P. infestans* sporangia.

Of recent interest among minor factors is the report of aluminum toxicity to *P. capsici* (Muchovej et al, 1980). The possibilities of using aluminum in soil to inhibit other *Phytophthora* spp. on aluminum-tolerant hosts need to be explored.

## MODES OF ACTION

### Direct Inhibition of *Phytophthora*

The direct inhibition mode of action is exemplified by the effect of low pH on *P. cinnamomi* and *P. parasitica*. Bingham and Zentmyer (1954) concluded that *P. cinnamomi* could not tolerate a pH of 3.0 in nutrient solution. In experiments with avocado fruit in contact with saturated soil at pH 3.0, Zentmyer (1980) did not observe infection and suggested that this was because of the lack of formation of sporangia and zoospores at this pH. Pegg (1977) was unable to recover *P. cinnamomi* from pineapple roots at a pH less than 3.9 and found marked reduction in sporangium production in the laboratory at pH 3.8. Troutman and LaPrade (1962) reported little sporulation of *P. parasitica* at pH 4.0, at which no black shank of tobacco occurred.

A second example is the work on nitrogen effects. Zentmyer and Bingham (1956) showed that $NO_2^-$ was toxic to mycelial growth and zoospore germination of *P. cinnamomi*. Rates required for inhibition changed with the pH of the solution. Gilpatrick (1969) reported toxic levels of $NH_3$ in soil amended with alfalfa meal in which *P. cinnamomi* was suppressed (Zentmyer, 1963). Tsao and Zentmyer (1979) demonstrated that 0.1% urea controlled avocado root rot when it was mixed in soil and incubated before planting. Tsao and Oster (1981) presented evidence that $NH_3$ and $HNO_2$ formed in urea-amended soils at levels high enough to account for inhibition of *P. cinnamomi* and *P. parasitica*. Ammonia was present at high pH during ammonification and $HNO_2$ at low pH during subsequent nitrification. Tsao and Oster (1981) proposed that $NH_3$ and $HNO_2$ were responsible for inhibition of *Phytophthora* wherever high rates of organic N were used.

A third example of direct inhibition is the accumulation of copper in soil from repeated fungicide applications (Skaptason et al, 1940).

### Indirect Inhibition of *Phytophthora*

Examples of the indirect inhibition mode of action include

TABLE 6. Effects of hydrogen ion ($H^+$) and sulfur (S) on *Phytophthora* diseases

| Pathogen | Host and disease | Factor | Effect[a] | References |
|---|---|---|---|---|
| *P. infestans* | Potato late blight | S | – | Barnes and Chestnutt (1966) |
| | | | | El Fahl and Calvert (1976) |
| | Potato tuber rot | S | – | Hutchinson and Mulligan (1971) |
| *P. erythroseptica* | Potato pink rot | S | – | McKee (1968) |
| *P. parasitica* | Tobacco black shank | $H^+$ | – | Dukes and Apple (1968) |
| | | | | Kincaid and Gammon (1957) |
| | | | | Kincaid et al (1970) |
| | | | | McCarter (1965) |
| | | | | Troutman and LaPrade (1962) |
| | | S | – | Kincaid and Gammon (1954) |
| | Citrus root rot | $H^+$ | – | Tsao (1959) |
| *P. capsici* | Pepper blight | $H^+$ | – | Muchovej et al (1980) |
| *P. cinnamomi* | Avocado root rot | $H^+$ | – | Bingham and Zentmyer (1954) |
| | Pineapple root rot | S | – | Pegg (1977) |
| | Rhododendron wilt | $H^+$ | – | White (1937) |
| *P. palmivora* | Betelvine wilt | $H^+$ | – | Turner (1969) |

[a] Disease increased (+) or decreased (–) with high levels of $H^+$ or S.

TABLE 7. Effects of other chemical factors on *Phytophthora* diseases

| Pathogen | Host and disease | Factor | Effect[a] | References |
|---|---|---|---|---|
| *P. infestans* | Potato late blight | NaCl | 0 | Bollen (1965) |
| | | $Cl^-$, $SO_4^-$ | ± | Borys (1964) |
| | Potato tuber blight | Mo, Mn | – | Mudich (1967) |
| | | Cu, Zn | 0 | Mudich (1967) |
| | | Cu | – | Skaptason et al (1940) |
| *P. capsici* | In vitro | $Al^{3+}$ | – | Muchovej et al (1980) |
| *P. cinnamomi* | Pine little leaf | B | – | Roth et al (1948) |
| | | Zn, Mn, Cu, Mo, Co, Ni, and Fe | 0 | Roth et al (1948) |

[a] Disease increased (+), decreased (–), was highly variable (±), or did not change (0) with high levels of factors.

salinity, aluminum toxicity, root exudates triggered by chemical factors, and the development of suppressive microflora. Effects of salinity on *Phytophthora* due to excessive amounts of N, P, and K have been reported for solution culture only. Reports by Anderson (1951) and Bingham et al (1958) involving *P. cinnamomi* suggested that the high salinity affected growth of the fungus. In reports on tobacco black shank (Apple, 1961) and chrysanthemum root rot (MacDonald, 1982), high salinity apparently predisposed plants to infection.

Possible involvement of aluminum toxicity in pH and calcium effects on *Phytophthora* is another indirect inhibitory mechanism. Muchovej et al (1980) presented evidence that $Al^{3+}$ and not pH per se was inhibitory to *P. capsici* at pH 4.8. They proposed that alkaline Ca salts raised the pH, immobilized exchangeable $Al^{3+}$, detoxified the soil, and increased disease significantly. Nonalkaline Ca salts affected pH and disease only slightly. This mechanism may explain some of the reports that Ca increased *Phytophthora* diseases (Table 5) at pH values too high for direct inhibition by $H^+$. Additional testing of this hypothesis is needed, especially with different *Phytophthora* spp. and in soils with a range of aluminum levels. Also, the effect of $Al^{3+}$ on sporangium development and germination needs to be determined.

Moore and Wills (1967) suggested another indirect mechanism. They found that low Ca in the soil at the time of inoculation retarded the development of black shank of tobacco but that growth of the pathogen was not affected by low Ca. They suggested that low Ca could have altered the composition of the root exudates, making them inhibitory to *P. parasitica*.

Chemical factors might also favor inhibitory soil organisms. For example, *P. cinnamomi*-suppressive soils in Australia are characterized by high calcium and nitrogen (Broadbent and Baker, 1974), conditions often conducive to severe *Phytophthora* damage in other areas. However, suppressive soils are also characterized by high organic matter and larger bacterial and actinomycete populations than conducive soils. High Ca and N may be involved in the development of suppressive microflora.

### Alleviation of Mineral Deficiencies

Little leaf of pine is a good example of the mechanism of alleviation of mineral deficiencies. In the southern United States, little leaf disease was controlled by the application of nitrogen fertilizer (Campbell and Copeland, 1954; Hepting et al, 1945; Roth et al, 1948). Apparently, little leaf symptoms are the result of N deficiency caused by extensive root damage from *P. cinnamomi* and are reversed by excessive nitrogen application. In New Zealand, the disease was controlled by application of phosphorus to P-deficient soils. Newhook (1970) attributed this marked decrease in disease to increased top growth, which intercepted rain, and increased root growth, which removed water more rapidly from the soil, thus preventing waterlogging and creating an environment less favorable for *P. cinnamomi*. Boughton et al (1978) explained the decrease of jarrah dieback caused by *P. cinnamomi* by correction of Ca deficiency in acid sands, which increased the number of fine roots. They also suggested that increased mycorrhizal development might be involved.

### Increasing Leaf Growth and Succulence

The mechanism of increased leaf growth and succulence is invoked frequently to explain increased severity of late blight of potato under conditions of high fertility (Awan and Struchtemeyer, 1957; Gallegly and Niederhauser, 1959; Main and Gallegly, 1964; Wilson and Gallegly, 1960) and high N (Weindlmayr, 1965). Langbein and Pehl (1962) and Herlihy (1970) have suggested that increased foliage (higher relative humidity) provides a more favorable environment for sporangium formation and germination. Apple (1961) noted that N-deficient tobacco plants were less susceptible to *P. parasitica* and that high N increased succulence and predisposed resistant tobacco plants to black shank. Other diseases in this category may include roselle stem rot, soybean root rot, pigeon pea blight, and pepper blight (Table 2). The nature of this mechanism has not been elucidated except for suggestions that succulent tissues that are high in N are colonized more easily by *Phytophthora* spp. than are nonsucculent tissues.

### Altered C/N Ratio in Tissue and Accelerated Maturity

*P. infestans* on potato and tomato may differ significantly from other diseases caused by *Phytophthora* in terms of the effects of N on disease development. There is evidence that N-deficient potato leaves are more susceptible than high-N leaves (Langbein and Pehl, 1962; Lowings and Acha, 1959; Schaffnit and Volk, 1927). Chlorotic leaves are more susceptible than fresh green leaves, and susceptibility has been associated with the onset of protein degradation in leaves (Grummer, 1955). Factors that induce maturity would make leaves more susceptible (Schaffnit and Volk, 1927). Langbein and Pehl (1962) reported that high N increased N levels in leaves and lowered susceptibility. Lowings and Acha (1959) noted that *P. infestans* hyphae grew more slowly in high-N leaves but that growth was stimulated by the onset of senescence. Walerych et al (1970) reported that $NO_3^-$-N increased the C/N ratio of tomato leaves, resulting in increased organic acids and greater susceptibility. Other N forms reduced the C/N ratio and resulted in resistant, high-N leaves. Alten and Orth (1940) have associated high K in potato with a decrease in foliage N, accompanied by buildup of arginine to levels toxic to *P. infestans*. The opposite effects have been reported with tuber blight caused by *P. infestans* (Herlihy, 1970; Herlihy and Carroll, 1969). Here, high N increased the number of blighted tubers by delaying tuber maturity. Apparently, *P. infestans* does not readily penetrate mature tubers. High P accelerates tuber maturity and decreases tuber blight. Thus, tissue maturity and tissue C/N composition are involved in the effects of chemical factors on late blight.

## INTEGRATION OF CHEMICAL FACTORS FOR CONTROL

Very few chemical factors considered in this review are useful by themselves for control of diseases caused by *Phytophthora* spp. Nonetheless, five chemical factors may have direct application in control programs: 1) incorporation of high rates of organic N before planting, to use the toxicity to *Phytophthora* propagules in soil of $NH_3$ and $HNO_2$ where economically and culturally feasible; 2) reduction of pH to below 4.0 for acid-tolerant plants; 3) reduction of pH to below 5.0 in high-aluminum soil for aluminum-tolerant plants; 4) judicious application of deficient minerals to optimize uptake by root systems reduced by infection by *Phytophthora* spp.; and 5) avoidance of indiscriminate overfertilization in attempts to maximize productivity.

It should be very obvious from this review that no principles apply to all *Phytophthora* spp. Each *Phytophthora*-host combination may present special problems. More information on effects of different forms of N and of N/K, N/P, K/P, K/Ca, Ca/Mg, and other ratios is needed to help control diseases and maximize crop yield. Manipulation of chemical factors also might be useful for evaluating resistance and tolerance in the field. An intriguing prospect is that high fertility accompanied by high organic matter may be involved in suppressive soils. Manures, composted sludge, plant residues, or agricultural waste would be good sources of organic matter provided they did not contain available chemical factors that might enhance disease before a stable suppressive environment has been established.

The potential of chemical factors for controlling *Phytophthora* diseases has not been fully exploited. Many combinations of chemical, pathogen, host, and cultivar have not been studied and should receive serious consideration in development of integrated disease control systems.

## SUMMARY

Chemical factors vary considerably in their effects on *Phytophthora* diseases. High fertility increases severity of potato and tomato late blight, tobacco black shank, soybean root rot, eucalyptus dieback, and ohia decline. High salinity decreases avocado and pineapple root rot but increases tobacco black shank and chrysanthemum root rot. High N both increases and decreases severity of potato late blight. High N increases severity of pepper blight, tobacco black shank, roselle stem rot, soybean root rot, pigeon pea blight, and little leaf of pine in New Zealand; but it decreases the severity of tomato stem lesions, betelvine wilt, philodendron leaf spot, and little leaf of pine in the United States. Nitrate-N increases potato late blight but decreases citrus root rot compared to other N forms. Nitrite-N and ammonia-N are toxic to *Phytophthora*, and toxic levels may occur in soils amended with high rates of organic N.

Potassium may increase disease (potato late blight, betelvine wilt, and citrus gummosis), decrease disease (potato late blight, pepper blight, pigeon pea blight, and roselle stem rot), or have no effect (potato late blight, tobacco black shank, tomato stem lesions, philodendron leaf spot, and little leaf of pine). Phosphorus decreased little leaf of pine in New Zealand, potato late blight, and betelvine wilt and either had no effect or increased tobacco black shank, roselle stem rot, pigeon pea blight, and little leaf of pine in the United States. Calcium increased pepper blight, tobacco black shank, and little leaf of pine but decreased eucalyptus root rot and citrus gummosis.

Addition of sulfur and low pH reduced the severity of all diseases caused by *Phytophthora* spp. that have been studied (potato late blight, tobacco black shank, citrus root rot, avocado root rot, pineapple root rot, rhododendron wilt, and betelvine wilt). At very low pH ($<$3.8), $H^+$ is toxic, and at slightly higher pH (3.8–5.0), aluminum toxicity may be a factor. There are few reports on the effects of heavy metals.

With the exceptions of $H^+$, $NH_3$, and $HNO_2$, effects of chemical factors may vary with different *Phytophthora*-host combinations. More information is needed on interactions between N forms and other nutrients before their role in *Phytophthora* diseases can be totally understood. Optimal adjustment of soil fertility and related chemical factors should be a part of any integrated control program.

## LITERATURE CITED

Alten, F., and Orth, H. 1940. Untersuchen über den Aminosäurengehalt und die Anfälligkeit der Kartoffell gegen die Kraut- und Knollenfäule (*Phytophthora infestans* de By.). Phytopathol. Z. 13:243-271.

Anderson, E. J. 1951. Effect of nutrient salt concentrations on infection of pineapple roots by *Phytophthora cinnamomi.* (Abstr.) Phytopathology 41:2.

Apple, J. L. 1961. The development of black shank in tobacco as influenced by host nutrition. Phytopathology 51:386-389.

Awan, A. B., and Struchtemeyer, R. A. 1957. The effect of fertilization on the susceptibility of potatoes to late blight. Am. Potato J. 34:315-319.

Barnes, E. D., and Chestnutt, D. M. B. 1966. Some aspects of the control of common scab in potatoes. Rec. Agric. Res. Minist. Agric. North. Irel. 15:35-42.

Bingham, F. T., and Zentmyer, G. A. 1954. Relation of hydrogen-ion concentration of nutrient solution to Phytophthora root rot of avocado seedlings. Phytopathology 44:611-614.

Bingham, F. T., Zentmyer, G. A., and Martin, J. P. 1958. Host nutrition in relation to Phytophthora root rot of avocado seedlings. Phytopathology 48:144-148.

Bollen, G. J. 1965. Some experiments on the influence of sodium chloride on germination of sporangia and on infection by zoospores of *Phytophthora infestans.* (Abstr.) Eur. Potato J. 8:182.

Borys, M. W. 1964. Influence of calcium, magnesium, chloride, and sulphate nutrition on resistance of potato leaves to *Phytophthora infestans* deBary. Acta Microbiol. Pol. 13:221-226.

Borys, M. W. 1966. Influence of $H_2PO_4^-$ nutrition of potato plants on resistance of their leaves to *Phytophthora infestans* de By. Phytopathol. Z. 57:301-309.

Boughton, T. J., Malajczuk, N., and Robson, A. D. 1978. Suppression of infection of jarrah roots by *Phytophthora cinnamomi* with application of calcium carbonate. Aust. J. Bot. 26:611-615.

Broadbent, P., and Baker, K. F. 1974. Behaviour of *Phytophthora cinnamomi* in soils suppressive and conducive to root rot. Aust. J. Agric. Res. 25:121-137.

Campbell, W. A., and Copeland, O. L., Jr. 1954. Littleleaf disease of shortleaf and loblolly pine. U.S. Dep. Agric. Circ. 940. 41 pp.

Canaday, C. H., and Schmitthenner, A. F. 1979. The effect of nitrogen on Phytophthora root rot of soybean. (Abstr.) Phytopathology 69:539.

Chapman, H. D. 1965. Chemical factors of the soil as they affect microorganisms. Pages 120-141 in: Ecology of Soil-Borne Pathogens. K. F. Baker and W. C. Snyder, eds. Univ. Calif. Press, Berkeley. 571 pp.

Chapman, H. D., and Brown, S. M. 1942. Some fungal infections of citrus in relation to nutrition. Soil Sci. 54:303-313.

Dirks, V. A., Anderson, T. R., and Bolton, E. F. 1980. Effect of fertilizer and drain location on incidence of Phytophthora root rot in soybeans. Can. J. Plant Pathol. 2:179-183.

Dukes, P. D., and Apple, J. L. 1968. Inoculum potential of *Phytophthora parasitica* var. *nicotianae* as related to factors of the soil. Tobacco Sci. 12:200-207.

Elenkov, E., and Bakharieva, V. 1975. [Mildew, *Phytophthora capsici*, a dangerous disease of pepper in Bulgaria.] Rastit. Zash. 23:31-35.

El Fahl, A. M., and Calvert, E. L. 1976. The effect of soil treatment with sulphur and lime on the incidence of potato diseases with special reference to blight. Rec. Agric. Res. Minist. Agric. North. Irel. 24:7-12.

Gallegly, M. E., and Niederhauser, J. S. 1959. Genetic controls of host-parasite interactions in the Phytophthora late blight disease. Pages 168-182 in: Plant Pathology: Problems and Progress 1908–1958. C. S. Holton, G. W. Fischer, R. W. Fulton, H. Hart, and S. E. A. McCallan, eds. Univ. Wis. Press, Madison. 588 pp.

Gilpatrick, J. D. 1969. Role of ammonia in the control of avocado root rot with alfalfa meal soil amendment. Phytopathology 59:973-978.

Grummer, G. 1955. Die Bezeihung zwischen dem Eiweiszstoffwechsel von Kulturpflanzen und ihrer anfälligkeit gegen parasitische Pilze. Phytopathol. Z. 24:1-42.

Halsall, D. M. 1980. Calcium nutrition and the infection of eucalypt seedlings by *Phytophthora cinnamomi.* Aust. J. Bot. 28:19-25.

Harkness, R. W., and Reynolds, J. E. 1964. Effect of nitrogen and potassium nutrition on the Phytophthora leaf spot of *Philodendron.* Proc. Fla. State Hortic. Soc. 77:475-481.

Hepting, G. H., Buchanan, T. S., and Jackson, L. W. R. 1945. Little leaf disease of pine. U.S. Dep. Agric. Circ. 716. 15 pp.

Herlihy, M. 1970. Contrasting effects of nitrogen and phosphorus on potato tuber blight. Plant Pathol. 19:65-68.

Herlihy, M., and Carroll, P. J. 1969. Effects of N, P and K and their interactions on yield, tuber blight and quality of potatoes. J. Sci. Food Agric. 20:513-517.

Huber, D. M. 1980. The role of mineral nutrition in defense. Pages 381-406 in: Plant Disease, An Advanced Treatise. Vol. V. J. G. Horsfall and E. B. Cowling, eds. Academic Press, New York. 534 pp.

Hutchinson, R. W., and Mulligan, T. E. 1971. The control of tuber blight in Kerr's pink potatoes by application of sulfur to the soil. Rec. Agric. Res. Minist. Agric. North. Irel. 19:13-17.

Kincaid, R. R., and Gammon, N., Jr. 1954. Incidence of tobacco black shank directly related to soil pH. Plant Dis. Rep. 38:852-853.

Kincaid, R. R., and Gammon, N., Jr. 1957. Effect of soil pH on the incidence of three soil-borne diseases of tobacco. Plant Dis. Rep. 41:177-179.

Kincaid, R. R., Martin, F. G., Gammon, N., Jr., Breland, H. L., and Pritchett, W. L. 1970. Multiple regression of tobacco black shank, root knot, and coarse root indexes on soil pH, potassium, calcium, and magnesium. Phytopathology 60:1513-1516.

Kincaid, R. R., Martin, F. G., and Rhoads, F. M. 1972. Regression of tobacco black-shank index on soil calcium. Phytopathology 62:302.

Kliejunas, J. T., and Ko, W. H. 1976. Association of *Phytophthora cinnamomi* with small ohia decline on the island of Hawaii. Phytopathology 66:116-121.

Klotz, L. J., DeWolfe, T. A., and Wong, P.-P. 1958. Decay of fibrous roots of citrus. Phytopathology 48:616-622.

Langbein, H., and Pehl, P. 1962. Einfluss der Stickstoffversorgung auf den *Phytophthora*-befall der Kartoffeln. Meded. Landbouwhogesch. Opzoekingsstn. Staat Gent 27:1186-1198.

Lowings, P. H., and Acha, I. G. 1959. Some factors affecting growth of *Phytophthora infestans* (Mont.) de Bary. I. *P. infestans* on living potato leaves. Trans. Br. Mycol. Soc. 42:491-501.

MacDonald, J. D. 1982. Effect of salinity stress on the development of Phytophthora root rot of chrysanthemum. Phytopathology 72:214-219.

Main, C. E., and Gallegly, M. E. 1964. The disease cycle in relation to multigenic resistance of potato to leaf blight. Am. Potato J. 41:387-400.

Marks, G. C., Kassaby, F. Y., and Fagg, P. C. 1972. Die-back tolerance of eucalyptus species in relation to fertilization and soil populations of *Phytophthora cinnamomi*. Aust. J. Bot. 21:53-65.

McCarter, S. M. 1965. Effect of certain soil environmental factors and host nutrition on black shank disease development in tobacco. Diss. Abstr. 26:6955.

McKee, R. K. 1968. Plant pathology research division. Rep. Res. Tech. Work, Minist. Agric. North. Irel. 1967:83-93.

Moore, L. D., and Wills, W. H. 1967. Calcium nutrition in relation to black shank disease of tobacco. Plant Dis. Rep. 51:641-644.

Muchovej, J. J., Maffia, L. A., and Muchovej, R. M. C. 1980. Effect of exchangeable soil aluminum and alkaline calcium salts on the pathogenicity and growth of *Phytophthora capsici* from green pepper. Phytopathology 70:1212-1214.

Mudich, A. 1967. Effect of trace-elements bound to superphosphate on the resistance of potato tubers to *Phytophthora infestans* (Mont.) de Bary. Acta Phytopathol. Acad. Sci. Hung. 2:295-302.

Newhook, F. J. 1970. *Phytophthora cinnamomi* in New Zealand. Pages 173-176 in: Root Diseases and Soil-Borne Pathogens. T. A. Toussoun, R. V. Bega, and P. E. Nelson, eds. Univ. Calif. Press, Berkeley. 252 pp.

Newhook, F. J., and Podger, F. D. 1972. The role of *Phytophthora cinnamomi* in Australian and New Zealand forests. Annu. Rev. Phytopathol. 10:299-326.

Olunloyo, O. A., and Adeniji, M. O. 1976. The influence of NPK fertilizers on the root and stem rot disease of roselle caused by *Phytophthora parasitica* var. *nicotianae*. Niger. J. Plant Prot. 2:84-87.

Pal, M., and Grewal, J. S. 1976. Effect of NPK fertilizers on Phytophthora blight of pigeon-pea. Indian J. Agric. Sci. 46:32-35.

Pegg, K. G. 1977. Soil application of elemental sulphur as a control of *Phytophthora cinnamomi* root and heart rot of pineapple. Aust. J. Exp. Agric. Anim. Husb. 17:859-865.

Roth, E. R., Toole, E. R., and Hepting, G. H. 1948. Nutrition aspects of the littleleaf disease of pine. J. For. 46:578-587.

Schaffnit, E., and Volk, A. 1927. Uber den Einfluss der Ernährung auf die Empfänglichkeit der Pflanzen fur Parasiten (I. Teil). Forsch. Geb. Pflanzenkr. Immun. Pflanzenreich 3:1-45.

Skaptason, J. B., Peterson, L. C., and Blodgett, F. M. 1940. The copper content of Long Island soils in relation to tuber rot of potatoes caused by *Phytophthora infestans* (Mont.) de Bary. Am. Potato J. 17:88-92.

Szczotka, Z., Borys, M. W., and Wojciechowski, J. 1973. Relation between $K^+$-nutrition of potatoes and their leaflets' resistance to *Phytophthora infestans* de By. Phytopathol. Z. 76:57-66.

Thyagarajan, P., Rao, A. V., Varadarajan, S., and Sundararajan, R. 1972. Studies on betelvine wilt disease—Influence of nitrogen and phosphorus in the control of betelvine wilt disease. Madras Agric. J. 59:187-189.

Troutman, J. L., and LaPrade, J. L. 1962. Effect of pH on the black shank disease of tobacco. Va. Agric. Exp. Stn. Tech. Bull. 158. 12 pp.

Tsao, P. H. 1959. Phytophthora fibrous root rot of citrus affected by soil factors. (Abstr.) Phytopathology 49:553.

Tsao, P. H., and Oster, J. J. 1981. Relation of ammonia and nitrous acid to suppression of *Phytophthora* in soils amended with nitrogenous organic substances. Phytopathology 71:53-59.

Tsao, P. H., and Zentmyer, G. A. 1979. Suppression of *Phytophthora cinnamomi* and *Ph. parasitica* in urea-amended soils. Pages 191-199 in: Soil-Borne Plant Pathogens. B. Schippers and W. Gams, eds. Academic Press, London. 686 pp.

Turner, G. J. 1969. Effects of hydrogen ion concentration on *Phytophthora palmivora* from *Piper nigrum*. Trans. Br. Mycol. Soc. 52:419-423.

Walerych, H., Wojciechowski, J., and Borys, M. W. 1970. Resistance of tomato leaves to *Phytophthora infestans* de By as affected by N-$NO_3$, N-$NH_4NO_3$ or N-$(NH_2)_2CO$. Phytopathol. Z. 68:244-257.

Weindlmayr, J. 1965. Untersuchungen über den Einflusz gesteigerter Stickstoff-, Kali- und Phosphorgaben auf die *Phytophthora*-Anfälligkeit von Kartoffelpflanzen in Nährlösungskultur. Bodenkultur 16:144-168.

Weston, G. C. 1956. Fertilizer trials in unthrifty pine plantations at Riverhead Forest. N.Z. J. For. 7:35-46.

Weststeijn, G. 1973. *Phytophthora nicotianae* var. *nicotianae* on tomatoes. Neth. J. Plant Pathol. 79, Suppl. 1. 86 pp.

White, R. P. 1937. Rhododendron wilt and root rot. N.J. Agric. Exp. Stn. Bull. 615. 32 pp.

Wills, W. H., and Moore, L. D. 1969. Calcium nutrition and black shank of tobacco. Phytopathology 59:346-351.

Wilson, J. B., and Gallegly, M. E. 1960. Nutrition and environment in relation to development of late blight on resistant and susceptible varieties of tomatoes. (Abstr.) Phytopathology 50:574.

Wojciechowski, J., Dzięciolowska, M., Borys, M., and Krzywański, Z. 1964. Influence of $NaH_2PO_4$-nutrition of potato plants on the resistance of their leaves to *Phytophthora infestans* de By. Phytopathol. Z. 51:198-203.

Zentmyer, G. A. 1963. Biological control of Phytophthora root rot of avocado with alfalfa meal. Phytopathology 53:1383-1387.

Zentmyer, G. A. 1980. *Phytophthora cinnamomi* and the diseases it causes. Monogr. 10. Am. Phytopathol. Soc., St. Paul, MN. 96 pp.

Zentmyer, G. A., and Bingham, F. T. 1956. The influence of nitrite on the development of Phytophthora root rot of avocado. Phytopathology 46:121-124.

# 16

# Microbial Antagonism to *Phytophthora*

N. Malajczuk

CONTENTS



## INTRODUCTION

The genus *Phytophthora* is unique among pathogenic fungi known to man in that the species within this genus colonize and initiate disease on nearly all parts of the host plant. Also, species in *Phytophthora* have both narrow and broad host ranges (Ribeiro, 1978). This fact alone is a problem when discussing microbial antagonism to *Phytophthora* spp. However, our knowledge related to this topic to date has been restricted to the *Phytophthora* species that cause root rot or those, such as *P. infestans,* whose life cycle includes a period associated with the below-ground parts of the plant. An exciting stage of work now emerging concerns antagonism toward foliar plant pathogens on aerial plant surfaces (the phyllosphere). A number of extensive reviews recently published (Fokkema, 1978; Skidmore, 1976) suggest that the principles governing antagonism to foliar pathogens are generally the same as those reported for antagonism to root pathogens.

The *Phytophthora* species that are found in soil and that are involved in aspects of root and stem collar damage include *P. boehmeriae, P. cactorum, P. cambivora, P. capsici, P. cinnamomi, P. erythroseptica, P. fragariae, P. heveae, P. infestans, P. lateralis, P. megasperma* var. *megasperma, P. megasperma* var. *sojae, P. nicotianae* var. *nicotianae, P. nicotianae* var. *parasitica, P. palmivora, P. primulae, P. quininea, P. richardiae, P. verrucosa,* and *P. vignae.* Their activity in most part is governed by the predisposition of host tissue, as influenced by physical, chemical, and biological properties of the soil. Their modes of penetration are similar, involving either mycelial or zoospore inoculum. To date most research, including my own, has been on *P. cinnamomi,* primarily because it has the greatest impact on both agricultural and forest crops (Zentmyer, 1980). Therefore, this chapter on aspects of microbial antagonism concentrates on this fungus.

## PRINCIPLES OF MICROBIAL INTERACTIONS

Microorganisms interact as a consequence of their growth and development in the environment. These interactions can be either beneficial or detrimental to the microorganism. Beneficial interactions stimulate an increase in the inoculum potential or biomass to enable the microorganism to dominate the particular substrate. When interactions are detrimental, the microorganism fails to become established and may or may not be eliminated from the environment. *P. cinnamomi* is a good example because in natural soil, it is dependent on metabolites from other bacteria and fungi to stimulate production of either sporangia or oospores (Ayers, 1971; Brasier, 1971; Zentmyer, 1965), but it also is severely antagonized by many biological agents (Malajczuk, 1979b; Malajczuk et al, 1977b).

The term antagonism encompasses a number of mechanisms that result in the suppression of *Phytophthora* species in the soil, rhizosphere, or infection court. Broadly, these processes are amensalism, parasitism, predation, and competition, all of which may operate together or independently to reduce inoculum.

### Amensalism

Amensalism is the production of antibiotics or toxic metabolic by-products, either volatile or not volatile, that cause lysis. Amensalism has been the subject of a number of reviews (Baker, 1968; Baker and Cook, 1974; Brian, 1957, 1960; Garrett, 1970; Jackson, 1965; Katznelson, 1965). The evidence that antibiosis is important in the ecology of soil microorganisms is largely circumstantial, primarily because antibiotics in soil or the rhizosphere are difficult to extract and detect, which in turn is probably because of their adsorption onto clay and humus particles or their unstable nature (Brian, 1957). Park (1967) agrees that the evidence for presence of antibiotics in soil is inconclusive but argues that antibiosis probably does occur in microhabitats with adequate nutrients, such as in the rhizosphere-rhizoplane. Additions of nutrients as organic amendments to soil have contributed to the increased lysis and reduction of *Phytophthora* spp. (Nesbitt et al, 1979; Zentmyer, 1963). A range of soil microorganisms that produce antibiotics against pathogenic fungi in vitro have been reported, including bacteria, actinomycetes, and fungi (see reviews by Baker [1968], Baker and Snyder [1965], and Baker and Cook [1974]). Some of these reports simply assume that a correlation between disease suppression and antibiotic production on agar is an adequate indication of antibiosis in the soil or rhizosphere; however, care must be taken in this assumption.

### Predation and Parasitism

Predation and parasitism involve active contact among microorganisms resulting in either the hyphal degradation of cells or the mycophagy of whole propagules (Alexander, 1971; Barnett and Binder, 1973; Boosalis, 1964; Boosalis and Mankau, 1965). Viruses, bacteria, and some fungi have been recorded to colonize both vegetative and reproductive fungal structures. The mode of penetration and subsequent lysis of the structures differ according to the active agent concerned. Extracellular secretion of toxins or degrading enzymes are typical of both fungal and bacterial hyperparasites. A number of fungi, such as *Trichoderma* spp., show typical coiling around the host hyphae (Dennis and Webster, 1971a, 1971b, 1971c).

Mycophagous animals have received little attention in the literature but possibly play a major role in causing a decline of fungal propagules in soil and the rhizosphere-rhizoplane. Amoebas in particular are known to cause perforation in spores and hyphae (Old, 1978). Little information is available on the roles of the Collembola and mites, but they too may be involved in the decline of fungal populations in soil.

### Competition

Competition may be used broadly to denote factors favoring one species over another, but in a narrower sense it can mean demand by two or more microorganisms for the same resource in excess of immediate supply. Clark (1965) defined competition as "the injurious effect of one organism on another because of the utilization or removal of some resources of the environment." Resources include nutrients as well as oxygen and space. Space and oxygen are probably important variables that can change in the rhizosphere-rhizoplane to the detriment of pathogen establishment. Space and nutrient competition appear to go hand in hand as well, because occupation of a site by microorganisms must diminish the limited supply of nutrients either in soil or the rhizosphere-rhizoplane. Carbon, nitrogen, and vitamins are important in this respect because they determine the growth and infection of soilborne pathogens in competition with other microorganisms (Baker, 1968).

## ANTAGONISM TO *PHYTOPHTHORA* SPECIES IN THE SOIL

### Survival of Propagules

Survival of plant pathogens in soil in the absence of host tissue, as mycelium on organic substances and/or as resting propagules, has been discussed by Garrett (1970). Survival is markedly influenced by a combination of environmental conditions, including physical, chemical, and biological variables acting upon the plant pathogens. Physical and chemical variables are reviewed by Duniway (Chapter 14) and Schmitthenner and Canaday (Chapter 15). Biological parameters are discussed below.

Survival periods reported for a range of *Phytophthora* spp. are shown in Table 1. Undoubtedly, absolute values are meaningless in that the survival studies were performed under different environmental conditions; however, the data do reflect the relative survival periods of propagules. It is apparent that all propagules naturally decline in viability with time and presumably would eventually become extinct in the absence of host tissue. Mycelia of all *Phytophthora* species represent a stage in the life cycle that is least capable of long-term survival, followed by zoospores and sporangia, while chlamydospores and oospores are more resistant to an unfavorable environment.

Studies on the influence of soil microorganisms on *Phytophthora* species are few. In general, the competitive saprophytic ability of *Phytophthora* spp. is extremely low, and rapid lysis of mycelium occurs in natural soil (Hine and Trujillo, 1966; Kuhlman, 1964; Lacey, 1965; Legge, 1953; Tsao, 1969b; Turner, 1965; Vujičić and Park, 1964). These studies require the use of sterile soils with appropriate comparisons with unsterile soil. Although the obvious dangers of sterilization must be realized (Van Bragt et al, 1971), the studies do provide an insight into the influence of microbial antagonism to *Phytophthora* species.

Tsao and Bricker (1963) showed that diffusates prepared from natural soils inhibited mycelial growth of *P. parasitica*, whereas those prepared from autoclaved soils stimulated

TABLE 1. Survival periods of *Phytophthora* propagules in soil[a]

| Species | Propagule | | | |
|---|---|---|---|---|
| | Mycelium | Zoospores/Sporangia | Chlamydospores | Oospores |
| *P. cactorum* | 3–14 (A,B) | 7–35 (A) | > 105 (B) | > 365 (C) |
| *P. cinnamomi* | 1–60 (D–F) | 3–42 (D,G) | 84–365 (D,H,I) | |
| *P. cryptogea* | | 14 (J) | | |
| *P. drechsleri* | 10–12[b] (K); > 30[c] (K) | 15 (L)/35 (M) | | |
| *P. infestans* | poor growth[b] (N); good growth[c] (N) | > 63/77 (O) | | |
| *P. megasperma* | 7–8 (K) | 5–14 (G,M,P) | > 210 (Q) | |
| *P. palmivora* | | 10–180 (R) | | |
| *P. parasitica* | < 7[b] (S,T); > 14[c] (S,T) | | | |

[a]Approximate survival time in days. Key literature citations are given in parentheses: A = Sneh and McIntosh (1974), B = Gisi and Meyer (1973), C = Legge (1953), D = Hwang and Ko (1978), E = Reeves (1975), F = Malajczuk and Theodorou (1979), G = MacDonald and Duniway (1979), H = Kassaby et al (1977), I = Weste and Vithanage (1979), J = Bumbieris (1979), K = Mehrotra (1972), L = Hickman (1970), M = Duniway (1975), N = Lacey (1965), O = Zan (1962), P = Ho (1969), Q = Basu (1980), R = Turner (1965), S = Tsao (1969b), T = Tsao (1969a).

[b]Unsterile soil.

[c]Sterile soil.

mycelial growth. Further, in natural soil amended with a supply of carbon and nitrogen, rapid lysis followed an initial burst of mycelial growth (Fig. 1) (Tsao, 1969b). Similar series of experiments have been performed with *P. drechsleri* (Mehrotra, 1972), *P. palmivora* (Turner, 1965), and *P. cinnamomi* (Malajczuk and Theodorou, 1979). In all these examples, the lysis of the mycelium was associated with intense bacterial colonization of hyphae and the breakdown of the cytoplasmic content. This suggests a general phenomenon of microbial antagonism.

Lysis of mycelium is invariably associated with the formation of zoosporangia and chlamydospores (Tsao, 1969b). In fact, we have found a direct relationship between lysis of mycelium of *P. cinnamomi* and the number of zoosporangia produced in a range of different soil types and incubation temperatures (Fig. 2). If one regards zoosporangia and chlamydospores as reproductive structures, this phenomenon is consistent with Klebs' principle (discussed by Kauffman [1926]) in that vegetative growth precedes reproduction and reproduction does not occur as long as conditions favor growth. When severe antagonism of hyphae occurs, vegetative growth ceases and reproduction accelerates. In contrast, Broadbent and Baker (1974a) and Malajczuk and Parker (*unpublished*) reported soils that are suppressive to *P. cinnamomi* where the pathogen failed to produce sporangia even though hyphal lysis was extensive. These soils may simply be examples where sporangia-stimulating bacteria are themselves suppressed.

Sporangium formation and survival in soils are poorly understood. Information may be lacking because sporangia for most *Phytophthora* species represent a transient stage before zoospore formation or direct germination. Sneh and McIntosh (1974) found that *P. cactorum* sporangia survived for up to 35 days in soil at optimal soil temperatures, although rapid lysis of these structures occurred within 15 days.

*Phytophthora* zoospores and cysts are effective inocula once they are brought into contact with host tissue (Hickman, 1970). However, their biology in soil has received little attention. Reported survival periods range from a matter of days to a few weeks (Hickman and Ho, 1966; Zan, 1962). An exception to this general situation was found by Turner (1965) for *P. palmivora,* which remained viable for up to 6 mo.

The role of microorganisms influencing the survival of the motile zoospores and cysts of *P. cinnamomi* has recently been demonstrated (Fig. 3). Zoospores introduced as cysts into soil leachates showed little difference in survival in either the sterile or unsterile treatments. However, when zoospores were introduced as motile zoospores, viability during the initial sampling period was significantly less in the unsterile treatment than in the sterile soil leachate. The vulnerability of these zoospores to antagonism by soil microorganisms could be expected, considering their thin-walled nature and the gaps for microbial penetration during loss of flagella before encystment (Fig. 4F).

Chlamydospores and oospores are thick-walled structures and can survive for considerable periods in soil (Table 1). These are the propagules that initiate disease in host plants when conditions become favorable for their germination and growth. The enforced dormancy in soils is markedly affected

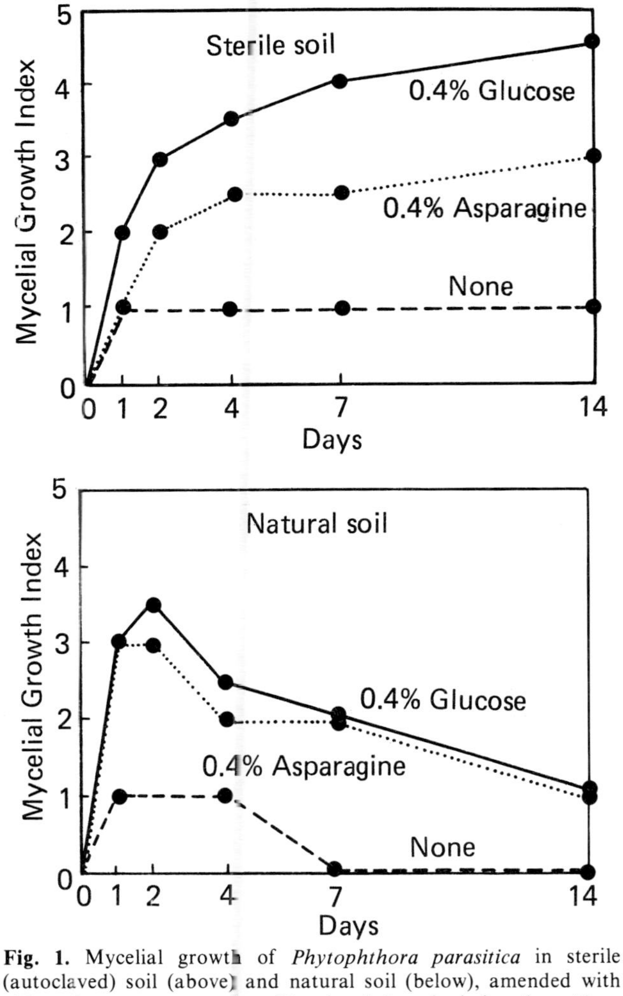


**Fig. 1.** Mycelial growth of *Phytophthora parasitica* in sterile (autoclaved) soil (above) and natural soil (below), amended with either glucose or asparagine. (Reprinted, by permission, from Tsao, 1969b)

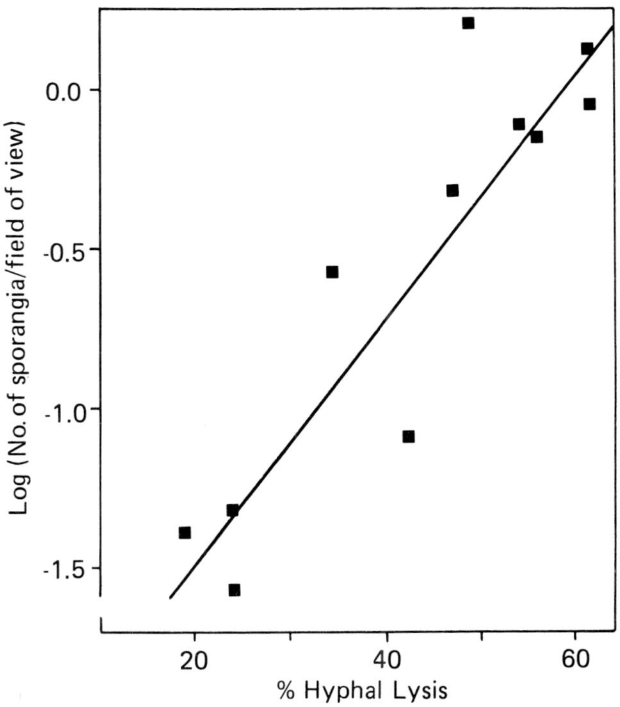


**Fig. 2.** Correlation between the percentage of hyphal lysis (HL) of *Phytophthora cinnamomi* and the logarithm of sporangium production (log S). Data collected from hyphae incubated in three soil types at four temperatures. Log S = −2.28 + 0.039HL; $r = 0.785$ ($P < 0.01$).

by a phenomenon termed fungistasis (Lockwood, 1977). Although microorganisms have been implicated, the exact mechanism is still questionable. However, the lack of nutrients required for the germination of chlamydospores and the presence of inhibitory compounds produced by microorganisms and/or soil appear to be the main factors involved. This is supported by studies on *P. cinnamomi* (Mircetich et al, 1968; Tsao and Bricker, 1968) showing that exogenous nutrients are required for germination and that soil sterilization suppresses soil fungistasis. The possible consequences of fungistasis are twofold: first, the promotion of *Phytophthora* survival by preventing its germination, which in turn does not allow infection; and second, the eventual elimination of propagules by microbial attrition following their prolonged incubation in soil.

### Suppressive and Conducive Soils

In considering microbial antagonism to *Phytophthora* spp. in soil, it is becoming increasingly obvious that soils can influence disease development. Because natural soils differ in microbial composition as well as in nutrient content, and because *Phytophthora* growth is largely governed by nutrient availability and the intensity of inhibitory substances, differential growth on different soils may be expected. Soils that favor the expression of disease have been termed conducive. Soils in which disease is minimal or diminishing even in the presence of susceptible host tissue, with or without reduction of the pathogen population, are termed suppressive.

Investigations of suppressiveness suggest that the principal soil component involved is microbial, because sterilization by autoclaving, irradiation, or steam pasteurization renders soils conducive (Broadbent and Baker, 1974a; Halsall, 1978; Malajczuk et al, 1977a). This natural biological control is relevant in terms of explaining the establishment of diseases caused by *Phytophthora* in both agricultural and forest areas and may be the key to understanding and developing potential control measures for the eradication of these diseases.

Of particular relevance are studies of soils suppressive and conducive to *P. cinnamomi* in Australia. The pathogen is widely distributed (Newhook and Podger, 1972) and extremely destructive in many of the indigenous communities (Brown, 1976; Malajczuk and Glenn, 1981; Shea, 1975; Weste, 1974). Its spread and development suggest that the fungus has only recently been introduced (Newhook and Podger, 1972). However, in a number of locations throughout Australia, the impact of the disease in natural communities is minimal, which suggests that these soils may be suppressive to the pathogen. Malajczuk and McComb (1979), Malajczuk (1979a, 1979b), and Weste and Vithanage (1977, 1978) have reported microbial populations in soil types of natural eucalypt communities in eastern and western Australia. Invariably, the populations differed qualitatively and quantitatively in both soil and the rhizospheres of plant species, with greater numbers being found in suppressive soils (Table 2). Broadbent and Baker (1974a) and Malajczuk and Parker (*unpublished*) showed that there was a corresponding increase in the percentage of antagonistic bacteria and actinomycetes. This increase was correlated with reduced sporulation of the fungus and increased hyphal lysis. Although specific isolates of *Bacillus, Streptomyces,* and fluorescent pseudomonads have been isolated that may be implicated in the suppression, more probably a range of microbial components work in concert to suppress the pathogen.

In these preliminary investigations of soil types influencing disease development, many of the suppressive soils tested were krasnozems originating from basaltic rock. These soils are typically high in clay content, exchangeable cations, organic matter content, and microbial activity. Subsequent investigations by Broadbent, Baker, and Malajczuk (*unpublished*) have found other suppressive soils not of basaltic origin. One particular soil occurred under an undisturbed rain forest where the original parent material was a beach sand. Suppression of the pathogen was associated with the humus layer of the soil profile showing intense microbial activity. Nesbitt et al (1979) subsequently

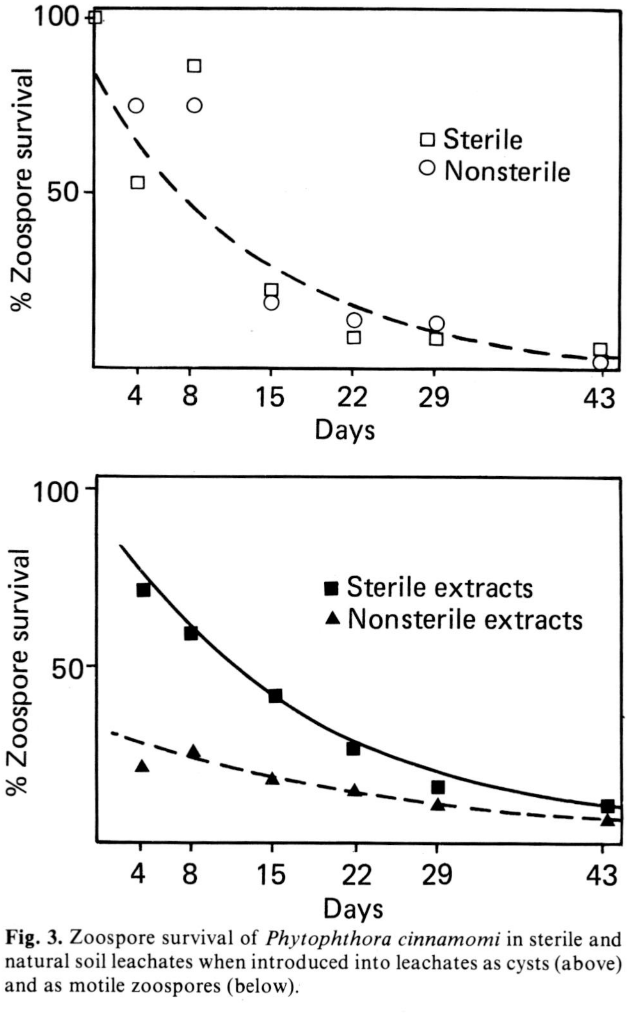


**Fig. 3.** Zoospore survival of *Phytophthora cinnamomi* in sterile and natural soil leachates when introduced into leachates as cysts (above) and as motile zoospores (below).

TABLE 2. Estimate of total populations of bacteria and actinomycetes in suppressive loam and conducive lateritic soils and in the rhizosphere of *Eucalyptus calophylla* (resistant) and *E. marginata* (susceptible) seedlings growing in these soils

| Soil type | | Rhizosphere | |
|---|---|---|---|
| Microorganisms | Soil | *E. calophylla* | *E. marginata* |
| Loam | | | |
| Bacteria | $32.8 \times 10^5$ | $18.8 \times 10^7$ | $14.5 \times 10^6$ |
| Actinomycetes | $11.2 \times 10^6$ | $22.2 \times 10^6$ | $69.8 \times 10^6$ |
| % Antagonists | 4.2 | 10.68 | 15.92 |
| Laterite | | | |
| Bacteria | $52.4 \times 10^5$ | $72.5 \times 10^6$ | $36.6 \times 10^6$ |
| Actinomycetes | $13.6 \times 10^5$ | $32.3 \times 10^6$ | $16.3 \times 10^6$ |
| % Antagonists | 0.75 | 11.68 | 2.00 |

*Source*: Malajczuk (1979b).

demonstrated that adding increasing amounts of decomposing litter to conducive soil increased the suppression of *P. cinnamomi* as indicated by hyphal lysis and formation of abortive sporangia. Thus, in natural ecosystems, which are either undisturbed or have intense microbial activity resulting in a high and rapid accumulation and incorporation of litter into the soil profile, pathogens like *P. cinnamomi* presumably would find it difficult to become established and initiate disease.

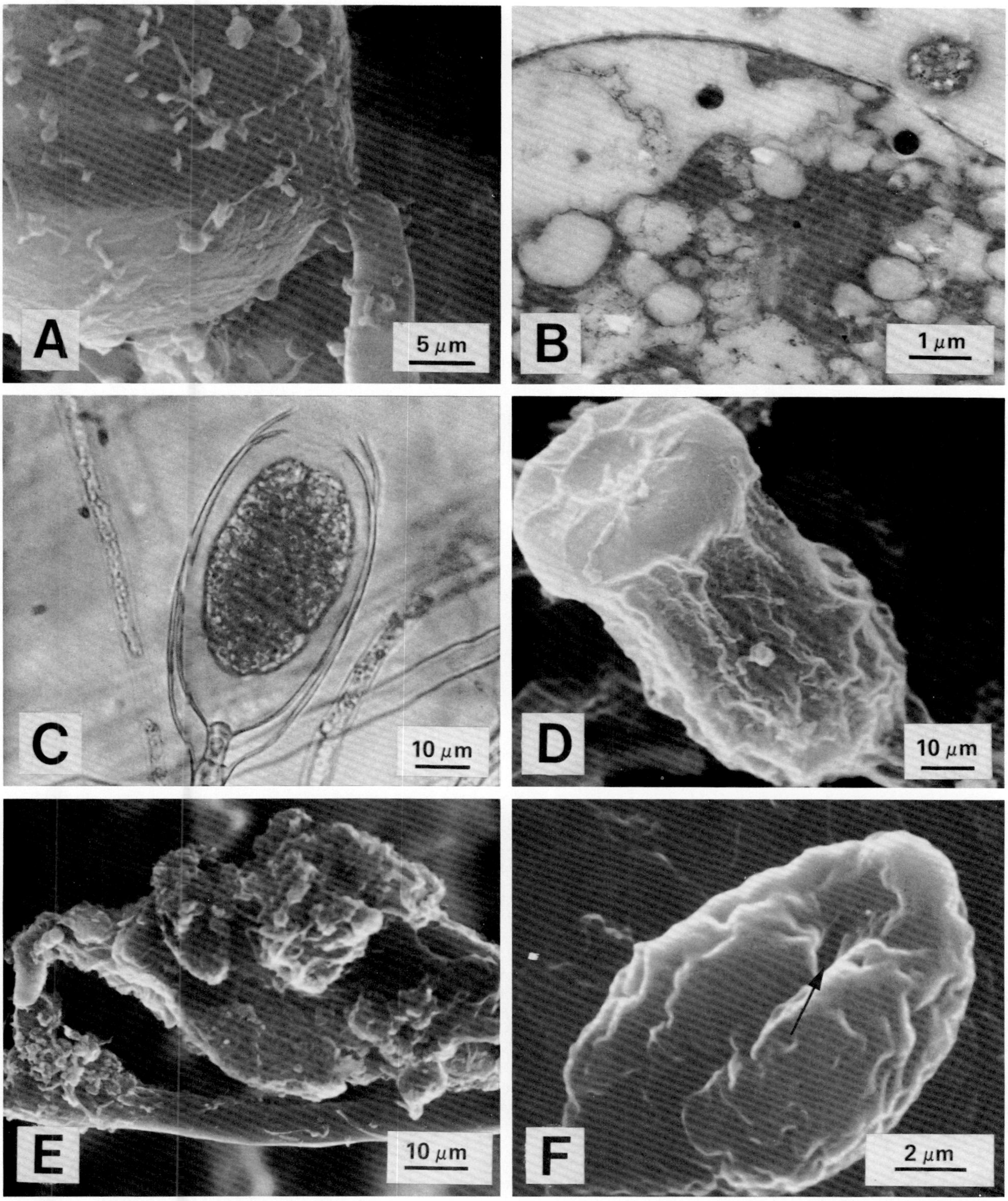


**Fig. 4.** *Phytophthora cinnamomi* sporangia and zoospore in unsterile leachates: **A,** Scanning electron micrograph (SEM) of sporangia, showing intense colonization of the surface by bacteria. **B,** Transmission electron micrograph of sporangia, showing lysis of cytoplasm immediately beneath the colonizing bacteria. **C,** Abortive sporangia, showing collapsed cytoplasm. **D,** SEM of abortive sporangia, showing collapse of side walls. **E,** SEM of abortive sporangia, showing bacterial colonization. **F,** SEM of zoospore before encystment, showing gaps created immediately following loss of flagella (arrow).

## Microbial Antagonists

### Bacteria and Actinomycetes

By virtue of their large populations in soil, bacteria constitute a significant microbial component influencing the behavior of soilborne *Phytophthora* propagules (Baker and Cook, 1982). Both groups of microorganisms depend on adequate soil moisture to grow and spread in soil (Cook and Papendick, 1972; Duniway, 1979). Many researchers have observed bacteria intensely colonizing *Phytophthora* hyphae. Recently, Malajczuk et al (1977b) conducted a light and electron microscopic study of the *P. cinnamomi* hyphae-soil interface, showing the relationship of bacteria on hyphae in unsterile soil or soil leachates. Bacteria that accumulated on or near fungal hyphae (Fig. 5) included a range of morphologically distinct types occurring as individuals or in microcolonies of 5–40 cells. Bacteria were attached either with the long axes directly against the hyphae or on the hyphae by their polar tips (Fig. 5A–D). Accumulation of bacteria was associated with slime development as single strands or intertwined within sheets enveloping groups of hyphae (Fig. 5E and F). Samples of hyphae colonized by bacteria prepared for transmission electron microscopy and stained with ruthenium red demonstrated the presence of acid mucopolysaccharides that were obviously involved in the attachment process binding bacteria to surfaces (Fig. 6B) (Marshall and Cruickshank, 1973).

The attraction of bacteria to hyphae appears to be a chemotactic response, because inert glass fibers failed to attract bacteria in the way that hyphae typically do (Nesbitt et al, 1981b). This observation raises the possibility that materials are exuded from the hyphae of *Phytophthora* spp., thus stimulating a hyphosphere population in much the same way that plant root exudates attract a rhizosphere population (Bowen and Rovira, 1976). Recently, Nesbitt et al (1981a) demonstrated the translocation and exudation of metabolites of phenylalanine and glucose from hyphae of *P. cinnamomi* that may serve as chemotactic attractants and/or a metabolic substrate for bacteria. More work is needed, especially on the chemical analysis of exudates, to determine the mechanism of attraction, because this may help to elucidate the significance of hyphosphere populations. In the case of *P. cinnamomi*, attraction of a sporangia-stimulating bacterial population may be postulated to be advantageous to the pathogen for survival in soils.

In most experiments, we have observed a positive correlation between hyphal lysis and soil bacterial populations. In one experiment, we added increasing amounts of soil organic matter (= soil bacteria) (Gray et al, 1968) to a microbiologically depauperate lateritic soil and determined hyphal lysis of *P. cinnamomi* incubated as mats on cellophane (Figs. 6A and 7). Hyphal lysis occurred within 24 hr in soils containing more than 50% organic matter and was maximal within 3–5 days, whereas in lateritic soil, little lysis (<2%) was observed over this period. In control treatments (sterile soil), little or no hyphal lysis occurred after 6 days. The number of bacteria firmly adhering to hyphae also increased with time in the soils with high organic matter. Examination of bacterial numbers on lysed and viable hyphae compared with dead (autoclaved) hyphae and inert fiber controls showed increased numbers on lysed hyphae (Fig. 8). Although it is tempting to suggest that this increased bacterial population initiated lysis, it can be argued that the adhering bacteria are simply utilizing the products of hyphal lysis. Tu (1978) reported a large increase in the number of bacteria initially growing on active hyphal tips of *P. megasperma* var. *sojae*, resulting in hyphal lysis. He interpreted this as evidence for suppression of the pathogen. Nesbitt et al (1978) noted that hyphal tips (terminal 50 μm of hyphae) of *P. cinnamomi* lysed more rapidly than the rest of the hyphae (Fig. 6C). It is therefore possible that colonizers of hyphae represent hypholytic bacteria. Further experimentation is needed to confirm this suggestion.

Antibiotic production in vitro by bacteria antagonistic to *Phytophthora* species is well recognized (Table 3, Fig. 6D). These include *Bacillus, Rhizobium, Flavobacterium,* and *Pseudomonas* spp. The fact that many of these species have been isolated from the hyphosphere of *Phytophthora* spp. suggests that these microorganisms may be prime candidates involved in secreting extracellular metabolites that initiate lysis. This interpretation is consistent with the lack of direct penetration of the hyphal walls during the early stages of hyphal lysis. Bacteria have been observed within the hyphae at a later stage of lysis, presumably having entered via breaks in the walls initiated by other microbial agents (Fig. 6E).

Antagonism of sexual and asexual reproductive structures of *Phytophthora* spp. has also been observed. Broadbent and

TABLE 3. Antagonism of bacteria and actinomycete toward *Phytophthora* species

| Species | Antagonistic bacteria | Reaction recorded | Reference |
|---|---|---|---|
| *P. cactorum* | *Rhizobium* | Antibiotic production | Drapeau et al (1973) |
| *P. cinnamomi* | *Bacillus* | Antibiotic production | Malajczuk et al (1977b) |
| | Fluorescent *Pseudomonas* | Hyphal lysis | Broadbent et al (1971) |
| | *Streptomyces* | Sporangium abortion | Broadbent and Baker (1974a,1974b) |
| | *S. griseoloalbus* | Antibiotic production | Rose et al (1980) |
| *P. citrophthora* | *Bacillus* | Antibiotic production | Broadbent et al (1971) |
| | Fluorescent *Pseudomonas* | Hyphal lysis | Broadbent and Baker (1974a) |
| | *Streptomyces* | Sporangium abortion | Broadbent and Baker (1974a) |
| *P. cryptogea* | *B. subtilis* | Antibiotic production | El-Goorani et al (1976) |
| | *Arthrobacter* | Antibiotic production | Erwin and Katznelson (1961) |
| *P. erythroseptica* | Bacterial isolate | Oospore parasitism (?) | Wynn and Epton (1979) |
| *P. infestans* | *Streptomyces* | Antibiotic production | Lacey (1965) |
| *P. megasperma* | *Rhizobium* | Hyphal lysis | Tu (1978,1979) |
| var. *sojae* | Bacterial isolates | Antibiotic production | Vaartaja et al (1979) |
| | *Pseudomonas* | Oospore parasitism (?) | Sneh et al (1977) |
| | *Actinoplanes missouriensis* | Oospore parasitism (?) | Sneh et al (1977) |
| *P. nicotianae* | *Bacillus* | Antibiotic production | Broadbent et al (1971) |
| var. *parasitica* | *Streptomyces* | Antibiotic production | Broadbent et al (1971) |
| *P. parasitica* | *Streptomyces* | Parasitism of oospores | Honour and Tsao (1973) |
| | *Pseudomonas* | Antibiotic production | Sarbini and Kommedahl (1977) |
| | *Bacillus* | Antibiotic production | Sarbini and Kommedahl (1977) |

Baker (1974a, 1974b) and Malajczuk et al (1977b) demonstrated inhibition of zoospore differentiation by *P. cinnamomi* sporangia in suppressive soil in the presence of soil bacteria. The cytoplasm came to one of two possible fates—either it was expelled in an undifferentiated state (Figs. 4A–E and 6F) or it lysed within the sporangia after the operculum failed to open. Transmission electron micrographs of these abortive sporangia showed bacteria on the outer surface immediately above the area of collapsing cytoplasm. This observation lends further support to the hypothesis that

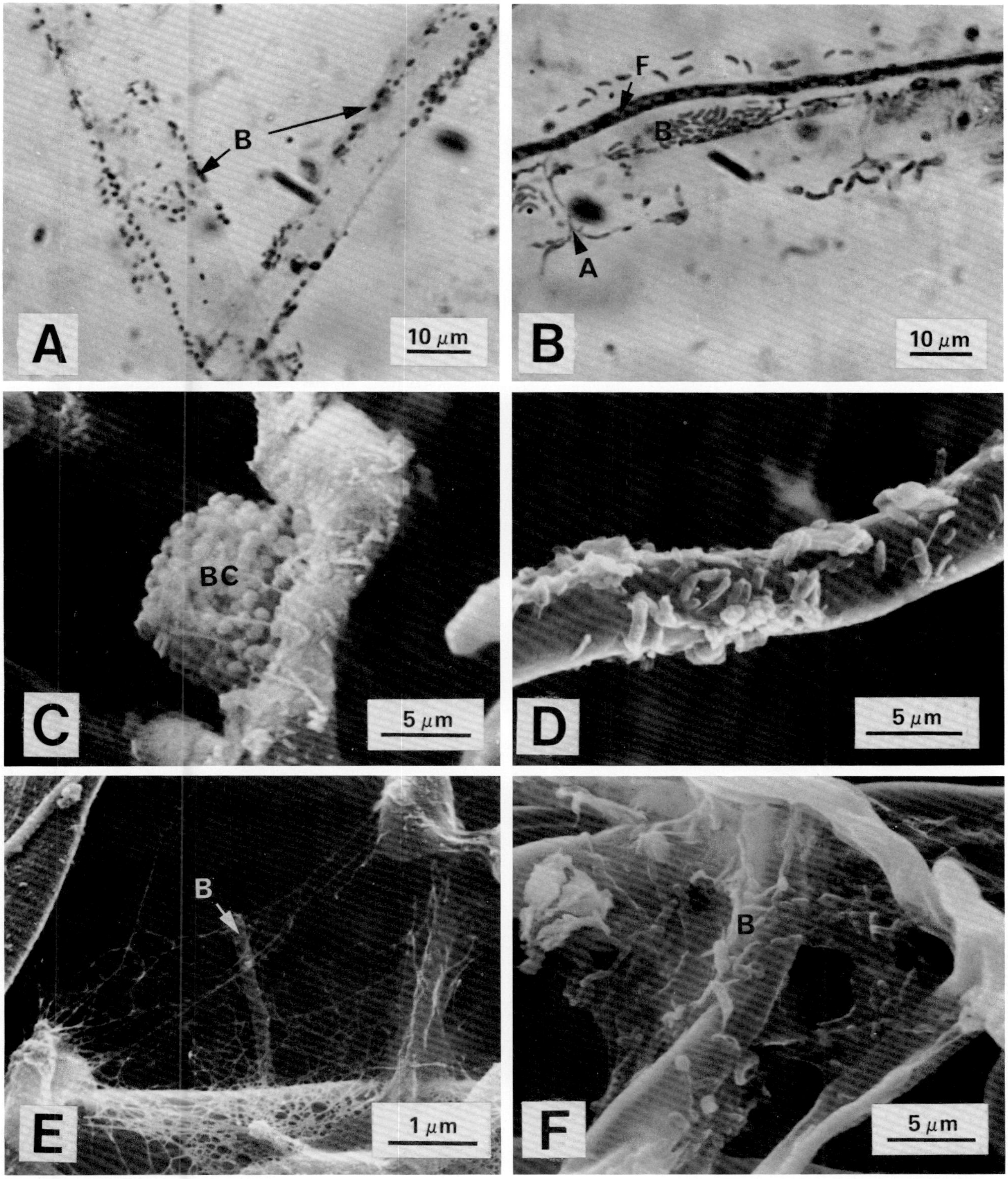


**Fig. 5.** *Phytophthora cinnamomi* hyphae at soil interface in unsterile soil: **A,** Lysed hyphae with bacteria (B) colonizing the hyphal surface. **B,** Lysed hypha with bacterial colonies (B), actinomycete filaments (A), and fungal hyphae (F) colonizing the hyphal surface. **C,** Scanning electron micrograph (SEM) of hypha with bacterial colony (BC) on its surface. **D,** SEM of hypha with bacterial colonies appressed to its surface. **E,** SEM of hyphae, showing end-on attachment of bacteria (B) and associated slime filaments. **F,** SEM of bacterial colonies (B) associated with slime enveloping hyphae.

some bacteria are actively involved in the production of extracellular compounds capable of initiating lysis. It may be profitable to pursue further research in the identification of these compounds for use as potentially specific fungicidal compounds.

Chlamydospores and oospores, being thick walled, are more resistant to antagonism by bacteria, presumably because they do not exude compounds to attract soil bacteria while in a dormant state. Chlamydospores of *P. cinnamomi* incubated in soil showed little or no colonization by bacteria after 10 wk

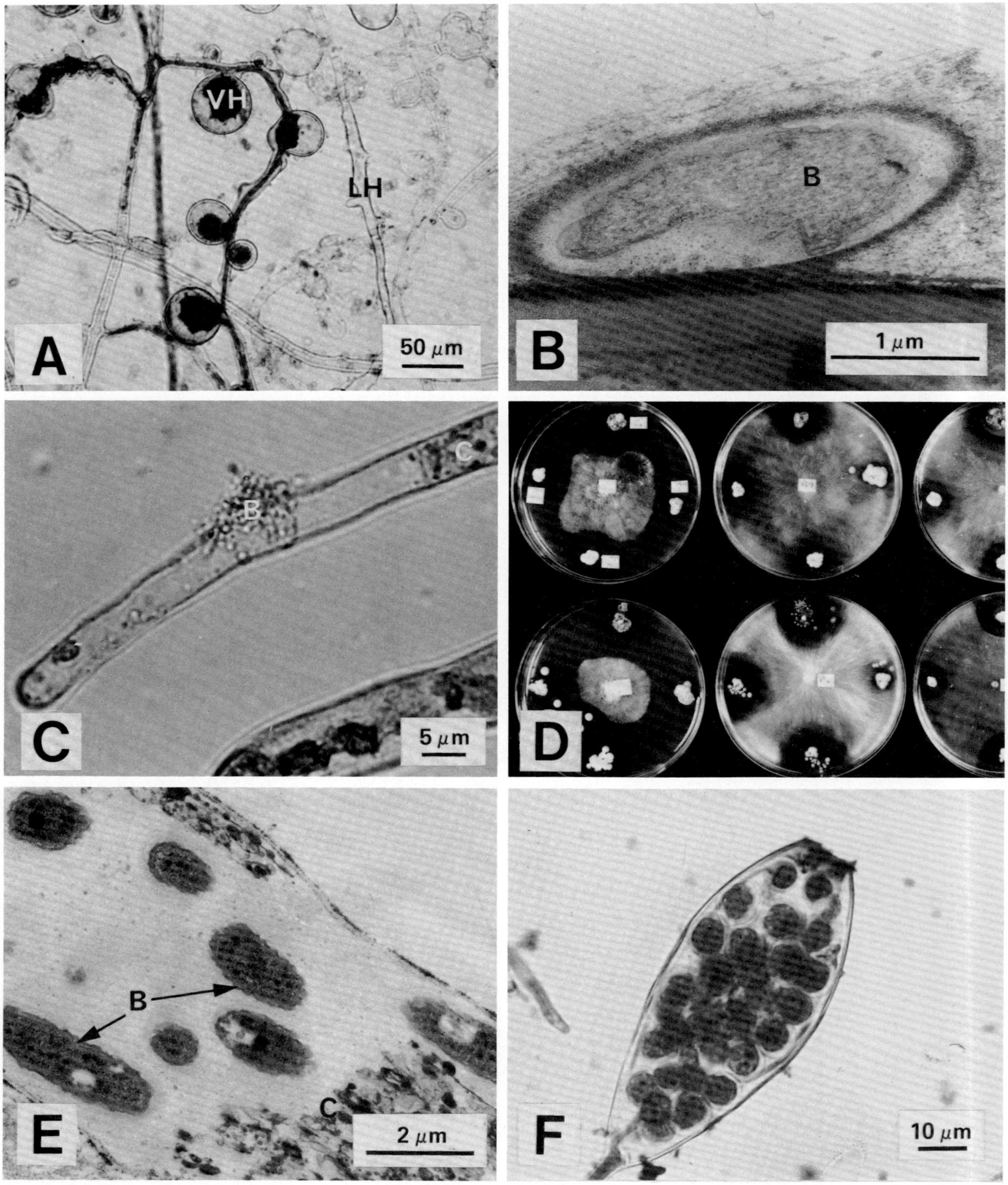


**Fig. 6. A,** Lysed hyphae (LH) and viable hyphae (VH) of *Phytophthora cinnamomi* when the mycelium is incubated in soil leachates. **B,** Transmission electron micrograph (TEM) showing bacteria (B) and mucopolysaccharide envelope attached to hypha of *P. cinnamomi.* **C,** Bacterial colony (B) on *P. cinnamomi* hyphal tip, showing absence of cytoplasm (C) immediately beneath the colony. **D,** Antibiotic production by *Streptomyces* sp. isolates from soil, showing inhibition zones on agar against *P. cinnamomi.* **E,** TEM of *P. cinnamomi* hypha, showing bacteria (B) invading and initiating lysis of cytoplasm (C). **F,** Viable sporangia of *P. cinnamomi,* showing zoospores before their release.

even though most of the accompanying mycelium was lysed and buried in slime. Laboratory experiments (N. Malajczuk, *unpublished*) using selected antibiotic-producing bacteria at $10^6$ cells per gram of soil showed some colonization of *P. cinnamomi* chlamydospores, although their viability was unaffected. To date, no predatory bacterium capable of secreting extracellular enzymes to break down fungal cell walls has been isolated from *Phytophthora* propagules. Although both Wynn and Epton (1979) and Sneh et al (1977) found bacteria associated with lysed oospores, it may be assumed that in these instances the bacteria were not the primary parasites, because parasitic chytrids and actinomycetes were also found attached to the spores.

Actinomycetes are also readily isolated from *Phytophthora* propagules and have been shown to be active parasitic agents of *P. megasperma* var. *sojae* and *P. cactorum* oospores (Sneh et al, 1977). Actinomycetes are a major group of microorganisms that produce an antibiotic in vitro against many of the *Phytophthora* species found in agricultural and forest soils in Australia (Broadbent et al, 1971; N. Malajczuk and C. A. Parker, *unpublished*). They are a dominant microorganism in *Phytophthora*-suppressive soils (Malajczuk and McComb, 1979; Weste and Vithanage, 1977, 1978). Because antinomycetes are more prevalent in dry soils (Chen and Griffin, 1966; Kouyeas, 1964) and produce more antibiotics at low water potential (Wong and Griffin, 1974), their activity on *Phytophthora* species may be restricted to the resting structures that are formed under adverse soil conditions where zoospore movement and hyphal growth are minimal. Such conditions would prevail in well-drained soils typical of those suppressive to the pathogen.

**Fungi**

Although numerous studies have reported antagonism among a great diversity of fungi (Barnett and Binder, 1973; Boosalis, 1964; Boosalis and Mankau, 1965), the extent of these interactions between *Phytophthora* and other groups under in vitro and in vivo conditions and their possible importance in the mechanism of fungal propagule reduction remain largely unexplored (Table 4). There have been a limited number of observations of antibiotic production by soil fungi such as *Penicillium* and *Trichoderma* spp., with the main emphasis on basidiomycetes because of their importance as possible biological deterrents against *Phytophthora* spp. when in symbiotic association with plant roots.

Species of *Trichoderma* have attracted special interest because of their ability to induce development of sex organs in some normally sterile isolates of *Phytophthora* spp. (Brasier, 1971, 1972, 1975a, 1975b; Pratt et al, 1972; Reeves and Jackson, 1972). Invariably, this effect is associated with the antagonism of vegetative growth by either volatile or soluble metabolites (Brasier, 1975a). In these in vitro studies, *Trichoderma* volatiles reduced the growth of *Phytophthora,* followed by a vacuolation of cell contents that eventually resulted in lysis of hyphal tips. A similar reaction occurred when both *Phytophthora* and *Trichoderma* spp. were grown on agar; however, the *Trichoderma* eventually overgrew the *Phytophthora* culture and parasitized the hyphae. Antibiotics subsequently described by Dennis and Webster (1971a, 1971b) may account for much of the antagonistic activity of *Trichoderma* spp.

Studies of the behavior of *P. cinnamomi* in soil reaffirmed the role of *Trichoderma* spp. as common antagonists involved in the stimulation of oospore formation, lysis of hyphae (Malajczuk and Theodorou, 1979; Reeves, 1975), and chlamydospore production (Kelley, 1977). Clearly, the formation of resting structures in the presence of potential fungal antagonists parallels the role of antagonistic bacteria and is an important mechanism involved in the survival of *Phytophthora.* The fact that *Trichoderma* spp. are commonly associated with decaying plant material in soil (Danielson and Davey, 1973; Malajczuk and McComb, 1979) highlights how *Phytophthora* spp. have developed survival mechanisms for their persistence in soils. Whether the metabolites are directly involved in the stimulation of the resting structures or whether this is a secondary response to breakdown of the cytoplasm within *Phytophthora* hyphae remains to be determined. Some evidence favoring the latter suggestion was provided by Mukerjee and Roy (1962), who showed the induction of *P. parasitica* oospores by a soil bacterium, *Xanthomonas* sp. In this context, it would be interesting to determine the range of

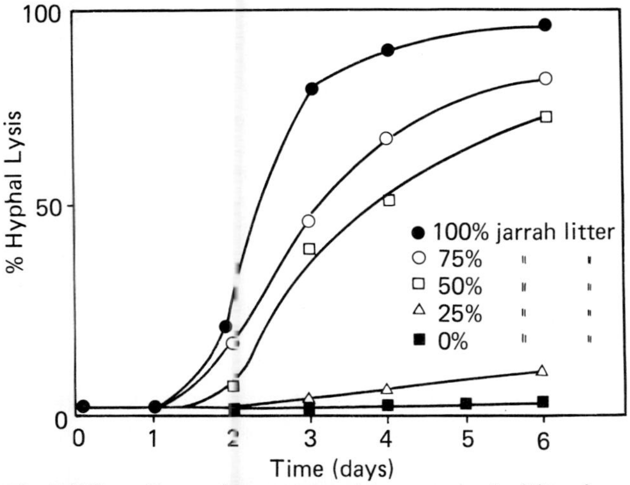


**Fig. 7.** Effect of increasing amounts of decomposing leaf litter from *Eucalyptus marginata* (organic matter = microbial populations) added to lateritic soil on lysis of *Phytophthora cinnamomi* hyphae. (Reprinted, by permission, from Nesbitt et al, 1979)

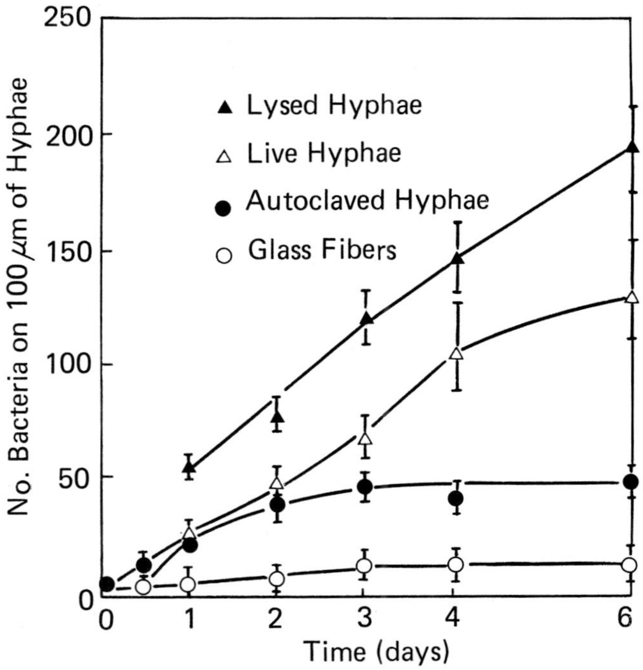


**Fig. 8.** Bacterial colonization of lysed and viable hyphae of *Phytophthora cinnamomi* compared with that of autoclaved hyphae and inert glass fiber controls incubated in jarrah leaf litter leachates. (Reprinted, by permission, from Nesbitt et al, 1981b)

soil fungi that may also initiate this reaction, considering the large number of species involved in the decomposition of plant tissue (Swift et al, 1979).

Hyphal lysis of *Phytophthora* initiated by soil fungi is rapid and, as observed with *Trichoderma* spp., involves the contact and coiling of the parasite around the hyphae before penetration (Fig. 9A). Presumably, the attraction of these antagonistic fungi to *Phytophthora* hyphae is mediated by exudation of hyphal metabolites, although information is lacking and research is warranted in this area.

Mycoparasitism of the resting structures of oomycetes has been reported as a natural phenomenon occurring in soil (Drechsler, 1938). However, the significance of this antagonism in reducing the population levels of *Phytophthora* is little understood. A number of parasitic fungi have been identified that are capable of penetrating the thick walls of both chlamydospores and oospores (Table 4). Sneh et al (1977) and Wynn and Epton (1979) identified oomycetes, hyphomycetes, and chytridiomycetes as parasitic agents of oospores of *P. megasperma* var. *sojae, P. cactorum,* and *P. erythroseptica.* Soil water status appears to govern the succession of fungi invading oospores; oomycetes and chytrids predominate in moist soils, while hyphomycetes have been isolated in relatively dry soils. We have observed chytrids parasitizing *P. cinnamomi* sporangia (Fig. 9B) in soil leachates, but they are rarely found in soil per se.

There is some suggestion that fungal parasites may show a degree of host specificity to oospores of the different *Phytophthora* species, which suggests that selected fungi might be exploited for manipulation as biological controlling agents against the pathogens.

TABLE 4. Antagonism of fungi to *Phytophthora* species

| Antagonistic fungi | *Phytophthora* species | Reaction[a] | Reference[b] |
|---|---|---|---|
| Chytridiales | | | |
| *Canteriomyces stigeoclonii* | *P. megasperma* | 5 | C |
| *Hyphochytrium catenoides* | *P. cactorum* | 5 | A |
| | *P. erythroseptica* | 5 | J |
| | *P. megasperma* | 5 | C |
| *Pleolpidium inflatum* | *P. erythroseptica* | 5 | H |
| | *P. megasperma* | 3 | H |
| *Rhizidiomycopsis* | *P. erythroseptica* | 5 | H |
| *japonicus* | *P. megasperma* | 5 | C |
| Oomycetales | | | |
| *Leptolegnia* sp. | *P. megasperma* | 5 | C |
| *Pythium* sp. | *P. megasperma* | 5 | C |
| Mucorales | | | |
| *Mucor spinosus* | *P. infestans* | 1 | K |
| Moniliales (= Hyphomycetes) | | | |
| *Alternaria alternata* | *P. erythroseptica* | 5 | H |
| | *P. megasperma* | 5 | C |
| *Cephalosporium* sp. | *P. megasperma* | 5 | C |
| *Cladosporium herbarum* | *P. infestans* | 1 | K |
| *Dactylella spermatophaga* | *P. cactorum* | 2,5 | C |
| | *P. cinnamomi* | 2,5 | C |
| | *P. megasperma* | 2,5[c] | A,C[c] |
| *Diheterospora chlamydosporia* | *P. megasperma* | 5 | C |
| *Fusarium oxysporum* | *P. erythroseptica* | 5 | H |
| | *P. megasperma* | 5 | C |
| *Gliocladium roseum* | *P. erythroseptica* | 5 | H |
| | *P. megasperma* | 1 | L |
| *Humicola fuscoatra* | *P. erythroseptica* | 5 | H |
| | *P. megasperma* | 5 | C |
| *Monosporium sylvaticum* | *P. infestans* | 1 | K |
| *Penicillium patulum* | *P. cryptogea* | 1 | I |
| *Rhizoctonia* | *P. infestans* | 1 | K |
| *Trichoderma harzianum* | *P. cinnamomi* | 1 | D |
| *T. lignorum* | *P. megasperma* | 2 | M |
| *T. polysporum* | *P. cinnamomi* | 6 | E |
| *T. viride* | *P. cactorum* | 2 | B |
| | *P. cinnamomi* | 1,2[c] | F,G[c] |
| | *P. erythroseptica* | 2 | B |

[a] 1 = hyphal lysis from antibiotic production; 2 = hyphal parasitism; 3 = zoospore/sporangia parasitism; 4 = chlamydospore parasitism; 5 = oospore parasitism; 6 = stimulation of the fungus.

[b] A = Drechsler (1938), B = Dennis and Webster (1971c), C = Sneh et al (1977), D = Kelley and Rodriguez-Kabana (1976), E = Kelley (1977), F = Brasier (1975a), G = Reeves (1975), H = Waterhouse (1940), I = El-Goorani et al (1976), J = Wynn and Epton (1979), K = Lacey (1965), L = Vaartaja et al (1979), M = Durrell (1968).

[c] Respectively.

**Soil Microfauna**

A large component of soil microbiota that has been largely ignored by pathologists is the microfauna. Only recently, Old and Darbyshire (1978) and Old and Patrick (1979) demonstrated the mycophagous habit of vampyrellid and testate amoebas on a range of morphologically different fungal species. This involved the elaborate physical and possibly chemical production of holes in the fungal walls and the digestion of the cellular contents (Old, 1977). Old and Darbyshire (1978) and Old and Oros (1980) showed lysis of oospores of *P. fragariae* and of hyphae and chlamydospores of *P. cinnamomi,* respectively. In the latter study, large holes were observed that were typical of spore destruction initiated by the vampyrellid amoeba *Arachnula impatiens* Cienk. (Old, 1978). Malajczuk (1979b) observed similar holes in *P. cinnamomi* hyphae incubated in soils suppressive to the pathogen (Fig. 9C). These soils were characterized as totally "alive" with a wide range of morphologically different types of testate and naked amoebas (Fig. 9D–F) as well as bacteria. Ciliated protozoa were invariably associated with discharging sporangia. Similar groups belonging to the Hartmannellidae group have been reported ingesting and lysing *P. cinnamomi* zoospores (Palzer, 1976). Conducive soil also contained many of these amoebas but at much lower numbers. Mycophagous amoebas appear to be widespread, and their action appears to be nonspecific. They probably exert a general antagonism toward most fungal propagules in both forest and agricultural soils such that opportunity for manipulating them for biological control of *Phytophthora* may be limited. However, studies should be initiated to examine their specific role (if any) in disease suppression before such conclusions are reached.

Predaceous nematodes and mites are groups of soil fauna that have not received any attention with regard to investigations of their antagonism to *Phytophthora* propagules. As with amoebas, their food preference may be nonspecific. Reports investigating the role of soil nematodes and mites in root disease development are somewhat discouraging. Meagher et al (1978) reported that nematodes increased the severity of root disease, as did soil mites (Shew and Beute, 1979).

## ANTAGONISM TO *PHYTOPHTHORA* SPECIES IN THE RHIZOSPHERE

### Root Exudates and Rhizosphere Phenomena

As a plant root grows through soil, the abrasive action of soil particles, together with the expansion of root cells, results in the leakage and loss of organic compounds termed root exudates (Hale and Moore, 1979; Hale et al, 1971, 1978; Rovira, 1969; Smith, 1977). Among the substances exuded are sugars, amino acids, organic acids, vitamins, nucleotides, flavones, and enzymes which vary quantitatively and qualitatively among plant species (Rovira, 1965).

Considerable evidence shows that root exudates are of major importance in stimulating germination of *Phytophthora* chlamydospores, growth of mycelium, and the attraction of zoospores in the rhizosphere (Hickman, 1970; Zentmyer,

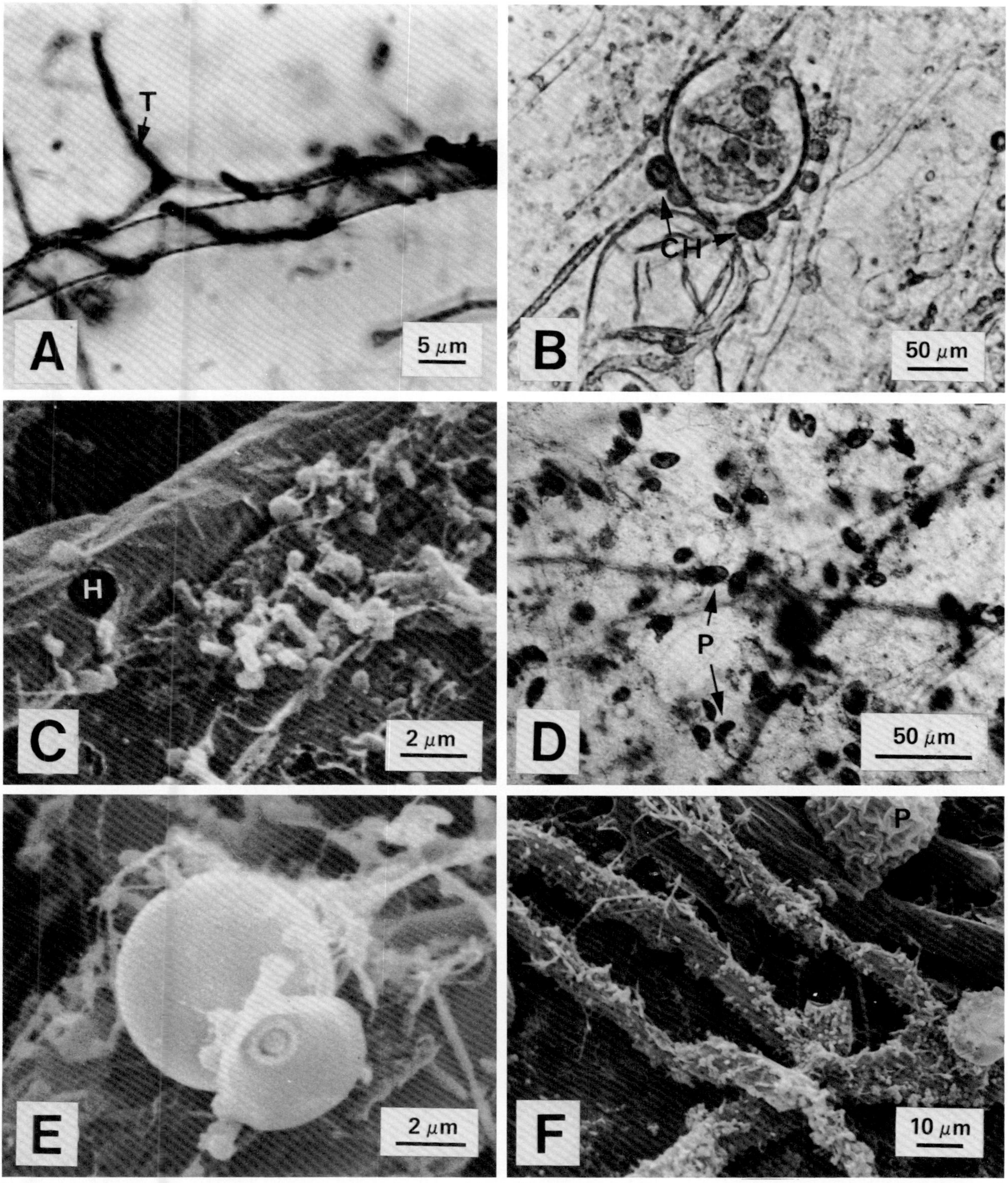


**Fig. 9.** Colonization of *Phytophthora cinnamomi* by soil antagonists: **A,** Hypha colonized by antagonistic *Trichoderma* sp. (T). **B,** Sporangia colonized by chytrids (CH). **C,** Scanning electron micrograph (SEM) of hypha incubated in suppressive soil, showing holes (H) typical of perforations initiated by amoebas. **D,** Intense development of protozoa colonies (P) associated with lysed mycelium incubated in suppressive soil. **E,** SEM of testate amoebas, which are common in suppressive soil, colonizing *P. cinnamomi* hypha; **F,** SEM of hyphae incubated in suppressive soil, showing intense colonization by bacteria and presence of parasitic protozoa (P).

1980). This response appears to be nonspecific and mediated by susceptible and resistant hosts alike (Chi and Sabo, 1978; Ho and Zentmyer, 1977; Malajczuk and McComb, 1977). Stimulation of the pathogen may, however, be negated by the presence of an antagonistic microflora in the rhizosphere (Bowen and Rovira, 1976; Schroth and Hildebrand, 1964). In fact, Garrett (1970) described the rhizosphere as the outmost defense of a plant against root attack with implications for the susceptibility or resistance of plant species to infection. Selective stimulation of antagonistic microbial populations to certain pathogens is known to be partly determined by the host (Neal et al, 1970) and is clearly exhibited in symbiotic associations between the root nodule bacterium (*Rhizobium*) and its legume host (Vincent, 1965) and between mycorrhizal fungi and tree roots (Bowen and Theodorou, 1973; Harley, 1959).

### Rhizosphere Microorganisms

A number of reports indicate that rhizosphere populations per se can limit disease caused by *Phytophthora*. Palzer (1976) conducted a number of precise inoculation experiments using zoospores of *P. cinnamomi* on lupine (*Lupinus angustifolius* L.) roots with and without an established rhizosphere. Although the evidence clearly indicated that rhizosphere microorganisms significantly reduced the infection of lupines by zoospores (Table 5), the mechanism for this suppression was not determined.

Malajczuk et al (1977a) provided a more detailed study of this nature when they examined the differential susceptibility of eucalypts to infection by *P. cinnamomi*. In conducive, lateritic soil, *Eucalyptus marginata* was susceptible, while *E. calophylla* showed resistance to the pathogen. When grown aseptically on sterile sand, seedlings of both eucalypts succumbed readily to infection. When small quantities of lateritic soil were added to the sterile sand, the differential susceptibility became apparent. Because the quantities of natural soil added were too insignificant to alter the physical or chemical nature of the sterile sand, the basis of the resistance of *E. calophylla* was ascribed to the differential selection of rhizosphere populations (Malajczuk et al, 1977a). Subsequent studies indicated that *E. calophylla* seedlings harbored a more intense rhizosphere population that was significantly more antagonistic to *P. cinnamomi* than that found in the *E. marginata* rhizosphere (Table 2). Although *E. marginata* seedlings succumbed to infection by the pathogen, the amount of diseased root was invariably less in natural soil than in sterile sand.

TABLE 5. Probability of zoospore infection of lupine (*Lupinus angustifolius*) roots with and without rhizosphere population

| Experiment | Rhizosphere micro-organisms | Dose (average no. of zoospores) | No. of lupines Inoculated | No. of lupines Infected | Probability of single spore infecting[a] |
|---|---|---|---|---|---|
| 1 | Absent | 3.1 | 11 | 1 | 0.030 |
| | Present | 15.5 | 11 | 0 | <0.006 |
| 2 | Absent | 8.6 | 40 | 21 | 0.087 |
| | Present | 71.6 | 26 | 16 | 0.013 |
| 3 | Absent | 12.3 | 34 | 19 | 0.066 |
| | Present | 111.8 | 40 | 33 | 0.016 |

[a]Based on an exponential model detailed by Bald (1937) and Peto (1953).

*Source:* Palzer (1976).

Examination of eucalypt roots under light and scanning electron microscopy revealed microcolonies of bacterial and actinomycete cells in cell junctions where infection by *P. cinnamomi* normally takes place (Fig. 10A) (Malajczuk and McComb, 1977). Encysted zoospores on root surfaces have also been shown to be intensely colonized by both bacterial and actinomycete cells (Fig. 10B), suggesting active antagonism to the fungus. Although the rhizosphere bacteria and actinomycetes can significantly affect *Phytophthora* propagules when attracted to the root surface, their low surface occupancy—between 1 and 20% (Bowen and Rovira, 1976)—may not offer total protection against *Phytophthora* spp., which are rapid in their movement to and infection of root tissue. In this respect, symbiotic systems of plant roots may be more significant.

### Root Symbiotic Microorganisms

Symbiotic microorganisms associated with plant roots have the advantage over other rhizosphere colonizers that the association is specific, allowing a rapid buildup of these microorganisms in the rhizosphere in preference to other rhizosphere inhabitants. Besides colonizing the rhizosphere, symbiotic microorganisms invade the host tissue, eliciting both morphological and biochemical changes in host tissue that may incite additional resistance mechanisms to infection by *Phytophthora* species.

#### Rhizobial Associations

The legumes form symbiotic associations with members of the genus *Rhizobium* that lead to the formation of nodules that participate in the acquisition of nitrogen (Vincent, 1965). The population of rhizobia can be as high as $10^8$ cells per gram in soil and the rhizosphere (Chatel and Parker, 1973a, 1973b). Such high population levels undoubtedly reflect the importance of rhizobia in influencing the establishment of alien microorganisms such as plant pathogens in the rhizosphere. Trinick (1970) demonstrated the various ways in which rhizobia interact with other microorganisms in the rhizosphere of subterranean clover (*Trifolium subterraneum*), including antibiosis and competition. More specifically, when grown on agar, rhizobia have been shown to produce antibiotics against a range of root rot pathogens including *P. cactorum* (Drapeau et al, 1973) and to colonize hyphae of *P. megasperma*, initiating lysis (Tu, 1978, 1979).

The evidence for plant protection by rhizobia against *Phytophthora* spp. is somewhat contradictory. Chou and Schmitthenner (1974) found that inoculation of soybean roots with *Rhizobium* had no effect on reducing infection by *P.*

TABLE 6. Severity of root rot[a] on soybean (*Glycine max*) caused by *Phytophthora megasperma* var. *sojae* at different concentrations of fungal inoculum and different populations of rhizobia (*Rhizobium japonicum*) in soil

| Concentration of fungal inoculum in soil (v/v) | Population of *Rhizobium* per cm³ of soil: $10^4$ | $10^3$ | $10^2$ | $10^1$ | 0 |
|---|---|---|---|---|---|
| 0 | 0 | 0 | 0 | 0 | 0 |
| $10^4$ | 0.2 | 0.5 | 1.5 | 2.2 | 2.5 |
| $10^3$ | 2.0 | 3.8 | 4.2 | 6.3 | 7.2 |
| $10^2$ | 5.5 | 6.5 | 7.2 | 8.0 | 8.5 |
| $10^1$ | 6.1 | 7.5 | 8.0 | 8.3 | 9.0 |

[a]Average root rot rating from 0 to 9; 0 = free from root rot, 9 = extensive root rot.

*Source*: Tu (1978).

*megasperma* var. *sojae*, whereas Tu (1978) observed in glasshouse trials that root rot caused by this pathogen was substantially reduced when the concentration of rhizobia in soil was increased (Table 6). Tu (1980) confirmed these findings in a more recent study showing reduced severity of root rot caused by *P. megasperma* in alfalfa (*Medicago sativa* L.) with increased concentration of rhizobia application. In both of Tu's studies, plants were not completely protected. The

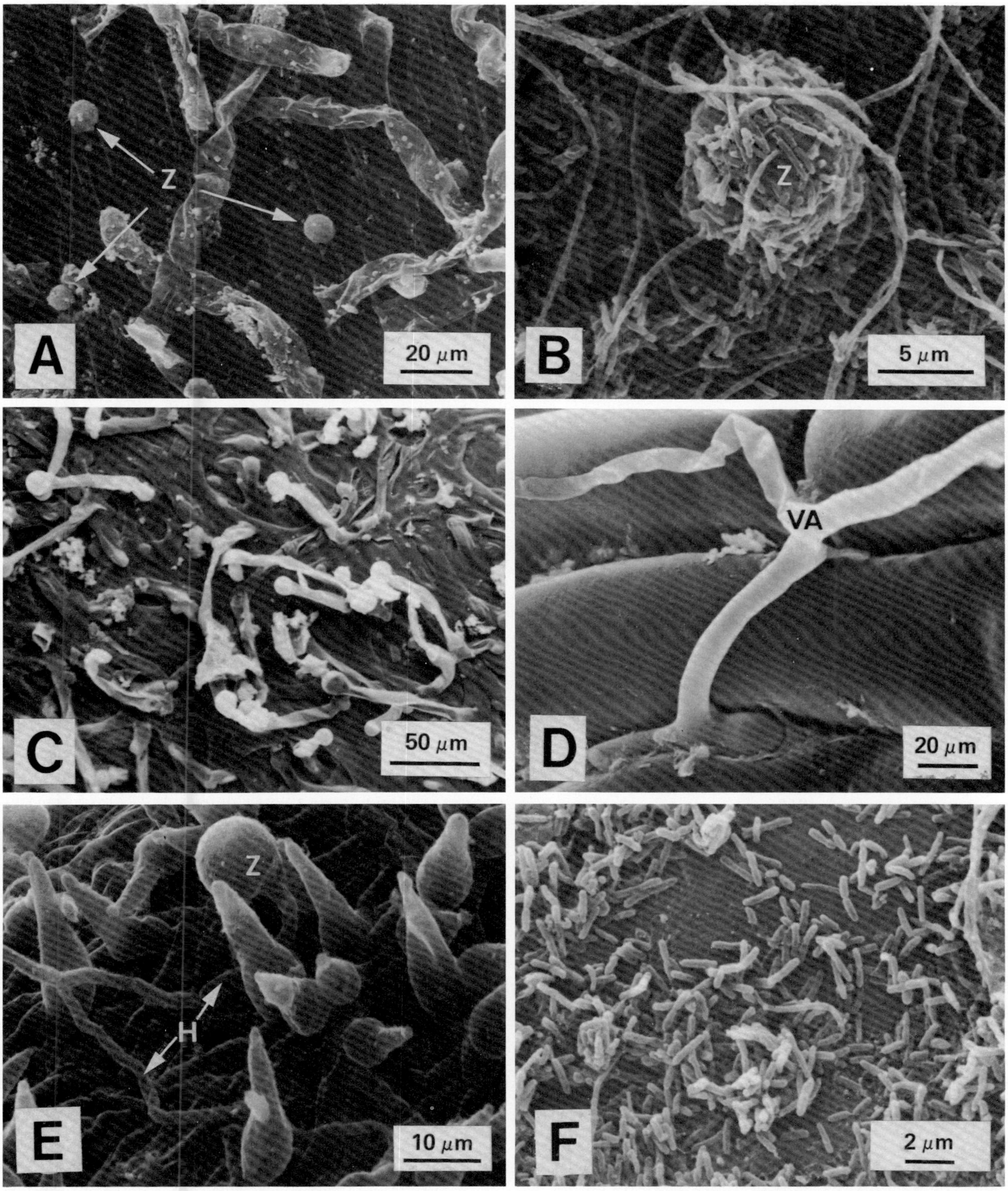


**Fig. 10. A,** Scanning electron micrograph (SEM) of *Eucalyptus* root, showing zoospores (Z) of *Phytophthora cinnamomi* encysted on and penetrating epidermal cells. **B,** SEM of *P. cinnamomi* zoospore (Z) intensely colonized by bacteria and actinomycetes in the rhizosphere of a *Eucalyptus* root. **C,** SEM of *Eucalyptus* ectomycorrhizal root surface, showing intense covering of the epidermal cells by a fungal mantle. **D,** SEM of *Trifolium subterraneum* root surface, showing typical vesicular-arbuscular mycorrhizal fungal mycelium (VA) penetrating between epidermal cells. **E,** SEM of *Eucalyptus* ectomycorrhizal root surface showing a germinating zoospore (Z) with hyphae (H) growing at random on the mantle. **F,** SEM of bacterial colonies in the rhizosphere of a *Eucalyptus* root infected with *P. cinnamomi.*

failure to obtain disease-free plants may reflect the speed at which the *Phytophthora* zoospores colonized and infected roots, preventing any interaction with rhizobia. Obviously, the ecology of rhizobia in natural soils and the rhizosphere must be understood before conclusions of success or failure of *Rhizobium* inoculation are reached.

**Mycorrhizal Associations**

Mycorrhizae are widely distributed among the higher plant forms, being recognized in 1,000 genera of vascular plants (Gerdemann, 1968). Most agricultural crop species possess endomycorrhizae of the vesicular-arbuscular type, whereas forest species tend to form ectomycorrhizae (Malajczuk and Lamont, 1981). These associations with plant roots have evolved with the exploitation of different habitats, and they function principally in nutrient uptake (Bowen, 1973). Their role as biological deterrents to root infection by pathogenic fungi has only recently been recognized (Marx, 1972; Schönbeck, 1979; Zak, 1964). This latter function does seem to be a logical development, because mycorrhizal associations, like root rot initiated by *Phytophthora* spp., are intimately associated with the fine, unsuberized root systems of plant species. Possession of this ecological niche by mycorrhizae would stimulate plant growth; possession by *Phytophthora* spp. would depress plant growth.

An examination of the anatomy of ectomycorrhizal and endomycorrhizal roots reveals a basic difference between them, which may be reflected in the different mechanisms for resistance to *Phytophthora* species. Ectomycorrhizae are characterized by the fungus producing an intricate mesh of hyphae on the root surface, forming tightly woven tissue termed the mantle (Fig. 10C). This sheath varies in thickness from a few microns to more than 100 μm. In transverse section, the ectomycorrhizal fungus can be seen to penetrate intercellularly in a network pattern, termed the Hartig net. Marked changes have been noted in host tissue associated with this penetration (Chilvers and Pryor, 1965; Marks and Foster, 1973). These changes include thickening of the intertangential and radial walls of the cortex, elongation of epidermal cells, and development of tanniferous cells containing phenol compounds. Mycorrhizal roots are typically thicker and more branched than nonmycorrhizal roots and lack root hairs. The fungal symbionts involved are mainly basidiomycetes and ascomycetes (Trappe, 1962). The vesicular-arbuscular mycorrhizae, on the other hand, lack a fungal mantle and Hartig net formation. The fungus penetrates root epidermal cells and proliferates between and within cortical cells (Fig. 10D). Inside the cells, it usually branches to form the characteristic multibranched hyphae termed arbuscules. The fungal symbionts in this instance are members of the Zygomycete family Endogonaceae.

The role of ectomycorrhizae in limiting infection by *Phytophthora* (mainly *P. cinnamomi*) has been reported for a wide range of tree hosts and fungal symbionts (Malajczuk, 1979b; Marais and Kotzé, 1976; Marx, 1973; Ross and Marx, 1972). However, vesicular-arbuscular mycorrhizae appear to have little or no effect on disease development (Table 7). One positive case of reduced disease severity was reported by Davis and Menge (1980), who attributed tolerance of citrus seedlings to *P. parasitica* to the fact that mycorrhizal seedlings had the ability to compensate for the loss of potential growth nutrients (mainly phosphorus) in decaying roots by absorbing more nutrients from healthy roots. Although this may not appear to be an active mechanism for antagonism to the pathogen, it may be the most important means by which all mycorrhizal plants survive in infested soil.

Active mechanisms against *Phytophthora* invasion that contribute to a reduction in the *Phytophthora* inoculum have been attributed to one or a combination of the following factors: First, altering the root exudation pattern may reduce the chemotactic response of *Phytophthora* zoospores to roots and reduce the stimulus for germination of chlamydospores and oospores. Marx and Davey (1969b) observed this phenomenon when they examined the chemotactic behavior of *P. cinnamomi* zoospores toward ectomycorrhizae and nonmycorrhizal roots of shortleaf and loblolly pine seedlings. Zoospores invariably germinated faster and more vigorously at the growing tip of nonmycorrhizal roots, while encysted zoospores on the ectomycorrhizal roots germinated slowly and grew poorly. Marx (1972) speculated that mycorrhizal roots may exude compounds somewhat different from those exuded by nonmycorrhizal roots, which might account for these differences. No work of this nature has been done for vesicular-arbuscular mycorrhizae, although it is suspected that increased root exudation may result from points of mycorrhizal penetration of root tissue, which in turn may further stimulate *Phytophthora* growth.

Second, antibiotics may be secreted (certain ectomycorrhizal fungi have been shown to produce antibiotics on agar culture that are active against mycelium of *Phytophthora* spp.) (Table 8). Marx (1969a, 1969b) also reported that a diatretyne nitrite compound produced by the ectomycorrhizal fungus *Leucopaxillus cerealis* var. *piceina* (Pick) inhibited germination of zoospores of *P. cinnamomi* at 50–70 parts per billion and killed zoospores at two parts per million. This and related compounds were extracted from ectomycorrhizal roots formed with the fungal symbiont and from the adjacent substrate on which seedlings were growing axenically. These compounds were functional components in the resistance system of feeder root infection by the pathogen. Because fungal symbionts of vesicular-arbuscular mycorrhizae have not been cultured to date, similar conclusions about antibiotic production have not been determined.

Third, provision of a mechanical barrier to the pathogen in the form of a fungal mantle may be important. Considering its elaborate structure, an ectomycorrhizal mantle (Fig. 10C) might be expected to pose a formidable barrier to penetration by *Phytophthora* zoospores and mycelium. Histologic evidence for the prevention of entry by the pathogen has been presented for ectomycorrhizae of pines (Marx and Davey, 1969a, 1969b) and eucalypts (Malajczuk, 1979b) formed by different fungal symbionts, including fungi that do not produce antibiotics. *P. cinnamomi* zoospores could be seen

TABLE 7. Effect of vesicular-arbuscular mycorrhizae on disease caused by *Phytophthora* species

| Species | Host | Resistance response by mycorrhizal plant | Reference |
|---|---|---|---|
| *P. cinnamomi* | Avocado | No difference | Mataré and Hattingh (1978) |
| *P. cinnamomi* | Avocado | Decreased | Davis et al (1978) |
| *P. megasperma* | Soybean | Decreased | Ross (1972) |
| *P. megasperma* | Soybean | Slight increase | Chou and Schmitthenner (1974) |
| *P. megasperma* | Soybean | No difference | Davis et al (1978) |
| *P. palmivora* | Papaya | No response | Ramirez (1974) |
| *P. parasitica* | Citrus | Increased | Davis et al (1978) |
| *P. parasitica* | Citrus | Increased (influenced by nutrients) | Davis and Menge (1980) |

germinating, and hyphae could be seen growing randomly over the mantle surface (Fig. 10E). This is obviously atypical of *Phytophthora* zoospores, which normally encyst rapidly, germinate and penetrate plant tissue directly. Where the mantle was incomplete or damaged by nematodes (Barham et al, 1974), the cortical cells were invaded. The lack of a similar mantle in vesicular-arbuscular mycorrhizae is possibly the single main factor that significantly reduces their role as potential deterrents to invasion by *Phytophthora* (Fig. 10D).

Fourth, antagonistic mycorrhizosphere microorganisms may affect the microbial population in the vicinity of mycorrhizal roots. This antagonism may be distinct from that in a nonmycorrhizal rhizosphere. This microbial population is not only unique (Rambelli, 1973), but the colonizing intensity is greater than that in the rhizosphere (Katznelson et al, 1962; Malajczuk, 1979a). Inoculation of eucalypt mycorrhizae with zoospores or mycelium of *P. cinnamomi* showed that within hours the hyphae became colonized by bacteria and lysis of mycelium was evident (N. Malajczuk, *unpublished*). Presumably, the failure of zoospores or mycelium to rapidly penetrate the mantle allowed antagonism of *P. cinnamomi* by the mycorrhizosphere microorganisms.

## MICROBIAL ANTAGONISM WITHIN PLANT TISSUE

The nature of antagonism has so far been considered external to the root tissue. *Phytophthora* species adapted to this environment and not totally inhibited by antagonistic rhizosphere microorganisms usually penetrate tissue of both resistant and susceptible species and cultivars (Hohl and Stössel, 1976; Hohl and Suter, 1976; Klisiewicz and Johnson, 1968; Malajczuk et al, 1977a; Tippett et al, 1976). Further development of the disease depends on the pathogen's ability to overcome resistance mechanisms within the root tissue. Host resistance mechanisms to *Phytophthora* species include mechanical barriers such as thickness of wall, suberization, and lignification and physiologic-biochemical barriers such as phytoalexin production, hypersensitive reactions, or presence of phytotoxic phenols (Hohl and Suter, 1976; Keen et al, 1971, Shimony and Friend, 1975; Tippett and Malajczuk, 1979; Tomiyama, 1971). (See also Keen and Yoshikawa, Chapter 21.) In response to these mechanisms, the fungus is either lysed or it produces resting structures.

TABLE 8. Antagonism to *Phytophthora* species by basidiomycetes

| Basidiomycete | *Phytophthora* species and references[a] |
|---|---|
| *Agaricus langei* | *P. cinnamomi* (D) |
| *Suillus luteus* | *P. cactorum* (D), *P. cambivora* (D), *P. cinnamomi* (A,D), *P. cryptogea* (D), *P. palmivora* (D), *P. parasitica* (D) |
| *Bovista brunnea* | *P. cinnamomi* (A) |
| *Clavulina amethystina* | *P. cinnamomi* (A) |
| *Clitocybe eucalyptorum* | *P. cinnamomi* (A) |
| *Clitocybe infundibuliformis* | *P. cinnamomi* (A) |
| *Collybia butyracea* | *P. cinnamomi* (A) |
| *Collybia* sp. | *P. cinnamomi* (A) |
| *Cortinarius austro-venetus* | *P. cinnamomi* (A) |
| *Cortinarius walkeri* | *P. cinnamomi* (A) |
| *Geastrum* | *P. cinnamomi* (A) |
| *Gymnopilus* | *P. cinnamomi* (A) |
| *Hyprophoropsis aurantiaca* | *P. cinnamomi* (A) |
| *Hygrophorus niveus* | *P. cinnamomi* (A) |
| *Laccaria laccata* | *P. heveae* (D) |
| *Lactarius deliciosus* | *P. cactorum* (D), *P. cambivora* (D), *P. cinnamomi* (A,D), *P. palmivora* (D), *P. parasitica* (D) |
| *Leucopaxillus cerealis* var. *piceina* | *P. cactorum* (D), *P. cambivora* (D), *P. cinnamomi* (D), *P. citrophthora* (D), *P. cryptogea* (D), *P. drechsleri* (D), *P. heveae* (D), *P. palmivora* (D), *P. parasitica* (D) |
| *Naematoloma fasciculare* | *P. cinnamomi* (A) |
| *Omphalina* sp. | *P. cinnamomi* (A) |
| *Rhizopogon vinicolor* | *P. cinnamomi* (B) |
| *Scleroderma bovista* | *P. cactorum* (C), *P. cinnamomi* (C), *P. citricola* (C), *P. drechsleri* (D), *P. heveae* (C) |

[a] References: A = Pratt (1971), B = Zak (1971), C = Marx and Bryan (1969), D = Marx (1969a).

The role of secondary invaders, either saprophytic or mildly parasitic, and their interaction with *Phytophthora* propagules in the infection court have been totally neglected. Obviously, the penetration of root tissue by *Phytophthora* mycelium and zoospores increases the leakage of root exudates, which in turn stimulates an increase in the microbial population in the rhizosphere and within plant tissue. In a separate experiment, we showed that diseased roots of eucalypt species both susceptible and resistant to *P. cinnamomi* had a preponderance of antibiotic-producing *Penicillium, Aspergillus,* and *Trichoderma* spp. as well as numerous bacterial colonies in the infection area (Fig. 10F). Because it was extremely difficult to isolate *P. cinnamomi* from diseased feeder roots, we presumed that microbial antagonism resulted in elimination of the fungus. This conclusion was supported by the observation of root regeneration occurring immediately behind the region of root damage in both susceptible and resistant species. In this case, the rate of growth of the *Phytophthora* mycelium in relation to the multiplication and growth rate of the secondary invaders again may determine the success or failure of the pathogen. More precise investigations into this aspect of interaction within plant tissue of both susceptible and resistant species of cultivars are clearly needed to determine the contribution (if any) of these antagonists to resistance of the host.

The use of avirulent races of related fungi for cross-protection against a pathogen, which is a variation on this theme, may have some practical importance. Much of the early work in this area has been in the field of plant virology. Matthews (1970) postulated that two identical virus isolates should react similarly but that some degree of cross-reaction will be observed between two related isolates. Further, infection of a plant by one strain of a virus causing only mild disease symptoms may protect the plant from a more severe strain by either inciting a host reaction or affecting the pathogenic strain directly.

Evidence of cross-protection by *Phytophthora* spp. is limited. However, studies by Klarman and Gerdemann (1963a, 1963b) showed that phytoalexin production was stimulated in soybean tissue of both susceptible and resistant cultivars following infection by both pathogenic and nonpathogenic strains of *P. megasperma* var. *sojae*. Subsequent work on a susceptible soybean cultivar by Paxton and Chamberlain (1967) demonstrated a marked degree of local cross-protection to the pathogen when plants were initially inoculated with a nonpathogenic fungus, *P. cactorum*. In a recent study, Borstel (1979) demonstrated in pot culture the cross-protection of two Western Australia forest species, *Banksia grandis* Willd. and *Casuarina fraseriana* Miq., against infection by *P. cinnamomi* when the tree seedlings were preinoculated with *Pythium irregulare* Buism. The mechanism of pathogen suppression was not determined, although there were indications that a hypersensitive reaction in roots occurred after inoculation with *Pythium irregulare* that reduced the attractiveness of roots to *P. cinnamomi*. Further, many *Pythium* species are readily isolated from diseased roots in field samples from forest areas where *P. cinnamomi* occurs but cannot be isolated (N. Malajczuk, *unpublished*). This suggests that the phenomenon of cross-protection may be widespread, involving many other *Pythium* spp.

Host resistance incited by mildly parasitic pathogens such as *Pythium irregulare* (Robertson, 1973) may offer a potential method for biological control or an explanation of it against many pathogenic *Phytophthora* species. The costs of reduced crop yield from *Pythium* damage need to be weighed against the benefits of protection from more severe losses from *Phytophthora*.

## MANIPULATION OF ANTAGONISM AGAINST *PHYTOPHTHORA* SPECIES

### Organic Matter

The augmentation of soil with organic matter to promote biological control of soilborne pathogens has gained considerable notice in the last few years because of some spectacular results obtained by a number of researchers in both glasshouse and field experiments (Baker and Cook, 1974). The role of these amendments appears to be related to the suppression of the pathogen in both soil and rhizosphere, although little experimentation appears to have been done to clarify the precise mechanisms involved (Cook, 1977). The quality of the organic amendments has a strong influence on whether or not pathogens are suppressed. This has led to investigations of a range of different organic substrates for disease control. Control of *Phytophthora* by this means has had an interesting history and has resulted in cases of effective control of the pathogen, in particular of *P. cinnamomi*.

Early investigations by Zentmyer (1963) showed that amending soils with alfalfa meal reduced the severity of root rot of avocado seedlings caused by *P. cinnamomi*. Increased microbial activity and presumably microbial antagonism to the pathogen were considered the main factors associated with disease suppression. Subsequent research by Gilpatrick (1969a, 1969b) indicated that ammonia production associated with the microbial decomposition of alfalfa meal also had a major effect on pathogen survival. Although glasshouse trials and in vitro studies of the pathogen showed this marked response, field experimentation failed to give total control of the pathogen. In fact, alfalfa meal proved to be phytotoxic to root growth (Gilpatrick, 1969b). Differences in pot and field results are often difficult to explain; in this instance, the rate of alfalfa meal application as well as the rate at which it decomposes and the rate of buildup of microbial antagonists may be possible explanations.

In this context, composting of hardwood bark before soil mixes are prepared for nursery-grown plants has been successful in controlling *P. cinnamomi* and other pathogens in the ornamental plant industry (Hoitink et al, 1975, 1976). The composting process involved the natural, aerobic, thermophilic breakdown of bark with nitrogen amendments. This allowed for rapid buildup of various mycophagous and antagonistic microorganisms, including actinomycetes, fluorescent pseudomonads, and *Trichoderma* spp. (Hoitink, 1980), that contribute to lysis of *P. cinnamomi* hyphae, zoospores, and germinating cysts (Hoitink et al, 1977).

Perhaps the best example of field control of *Phytophthora* with organic amendments is in Australia, where a number of avocado growers have adopted "organic farming" to maintain high levels of organic matter in soil by green cropping beneath the trees with legumes and forage crops and incorporating these into soil with minimal disturbance. Additional organic matter is added as mulches using straw. Fowl manure and dolomite are also added to the surface to encourage rapid breakdown of the organic matter (Pegg, 1977). The buildup in the nutrient status of soils enhances microbial activity, favoring an antagonistic soil microflora and an orchard free of root rot. Similar studies by Tsao (1977) showed promising in vitro control of *P. cinnamomi* and *P. parasitica* in avocado and citrus soils using nitrogenous organic amendments. Chicken manure, alfalfa meal, and processed sewage sludge rendered the soil suppressive (increased hyphal lysis and sporangial abortion) as a result of increased microbial activity associated with the decomposing amendments in soil.

It follows logically that depletion of the organic status of soils may be the all-encompassing factor that has led to severe outbreaks of disease caused by *P. cinnamomi* in Australia and perhaps diseases caused by other *Phytophthora* species throughout agricultural and forest soils. Interestingly, in Australian forest areas, litter is removed regularly by a fire program conducted on a cycling basis. The overall aim of this program is to reduce the fire hazard in forests during the dry summer months. However, this practice may inadvertently be reducing the microbial buffering capacity of soils to withstand invasion by *P. cinnamomi,* thereby contributing to severe outbreaks of disease in the indigenous eucalypt forests. Reduced microbial activity (Malajczuk et al, 1977a; Weste and Vithanage, 1977) and development of few mycorrhizal roots (Malajczuk and Hingston, 1981) have been shown to be characteristics of Australian soils low in organic matter.

The regular burning of the valuable *E. marginata* (jarrah) forest in Western Australia is being questioned in light of observations and experimentation. Regular, low-intensity fires are creating favorable conditions for the development of *P. cinnamomi* in these soils by reducing the organic matter status and favoring development of a susceptible proteaceous understory (Shea, 1975; Shea and Malajczuk, 1977). Field experiments have been established to evaluate the effect of moderate- to high-intensity fires with longer intervals between burns (Shea et al, 1979). This approach would allow litter to accumulate and to decompose over a longer period of time, encouraging the development of antagonistic microorganisms (Nesbitt et al, 1979) and ectomycorrhizae (Malajczuk and Hingston, 1981). In addition, high-intensity fires favor the development of a leguminous understory in preference to the susceptible proteaceous understory. The common legume *Acacia pulchella* R.Br. has been shown to be resistant to infection by the pathogen (Tippett and Malajczuk, 1979) and appears to be able to reduce the infection of susceptible jarrah seedlings grown with it in pots (Shea and Malajczuk, 1977). There is evidence of hyphal lysis and sporangial abortion of *P. cinnamomi* mediated by microorganisms that develop in the rhizosphere and soil under legumes, which may be responsible for reducing the inoculum of the pathogen and decreasing the susceptibility of *E. marginata* seedlings (Shea and Malajczuk, 1977).

### Fertilizers

The application of fertilizers to the environment to improve the nutrient supply for plant growth can also increase the resistance of plants by influencing an array of changes in both soil and rhizosphere microflora. These changes may be mediated via host physiology, e.g. root exudates, or via the chemical status of soil, e.g. pH (Baker and Cook, 1974; Cook, 1977; Henis and Katan, 1975).

Nitrogen fertilizers have been shown to have a wide range of effects on *Phytophthora* disease severity as well as on propagules. In vitro studies report that $NO_3^-$ and $NH_4^+$ fertilizers are highly toxic to *Phytophthora* (McIntosh, 1972; Tsao et al, 1975; Zentmyer and Bingham, 1956) on the one hand, while stimulating sporulation (Halsall, 1978) and increasing disease severity (Apple, 1961) on the other. In fact, Tsao and Zentmyer (1979) showed that $NH_4^+$ and $NO_3^-$ applied to soil were toxic to *P. cinnamomi* sporangia but had little or no toxicity to *P. parasitica*. This differential tolerance may explain some of the differences in the results of many studies. More probably, however, metabolic by-products produced by the utilization of these nitrogen compounds and their effect on *Phytophthora* are the keys to these discrepancies in the literature. In a recent study by Tsao and Zentmyer (1979), suppression of *P. cinnamomi* and *P. parasitica* occurred in pots when the soil was amended with urea. Aqueous extracts of amended soil inhibited germination and formation of *Phytophthora* sporangia, whereas the filter-sterilized urea was nontoxic to the pathogens. This type of suppression is an example of broadly defined biological control (Baker and Cook, 1974).

Phosphorus is a nutrient element that is second only to nitrogen in its importance in plant and microbial growth. Phosphorus has been shown to be important for the development of mycorrhizae in soils where the levels are suboptimal for plant growth (Bowen, 1973). It should follow, therefore, that increased mycorrhizal root development would influence the development of diseases caused by *Phytophthora*. According to Newhook (1970), the little-leaf disease caused by *P. cinnamomi* in stands of *Pinus radiata* on clay soils in New Zealand is indirectly a phosphorus deficiency problem, because application of phosphorus fertilizers to some of these stands not only improved tree vigor directly but also increased the development of ectomycorrhizal root growth. The resultant denser crowns intercepted more rainfall and removed more water through transpiration. Treated sites were thus drier and less favorable for *P. cinnamomi*.

Boughton et al (1978) reported suppression of infection of *E. marginata* roots by *P. cinnamomi* after calcium carbonate was applied. The decrease in frequency of *P. cinnamomi* isolated from treated seedlings was correlated with an increase in the development of ectomycorrhizal roots.

Application of sulfur to pineapple crops infested with *P. cinnamomi* was found by Pegg (1977) to control the pathogen. The suppression appeared to be related to an explosion in the population of *Trichoderma viride* in sulfur-treated soils. *T. viride* is known to be a potential antagonist of *P. cinnamomi* (Table 4).

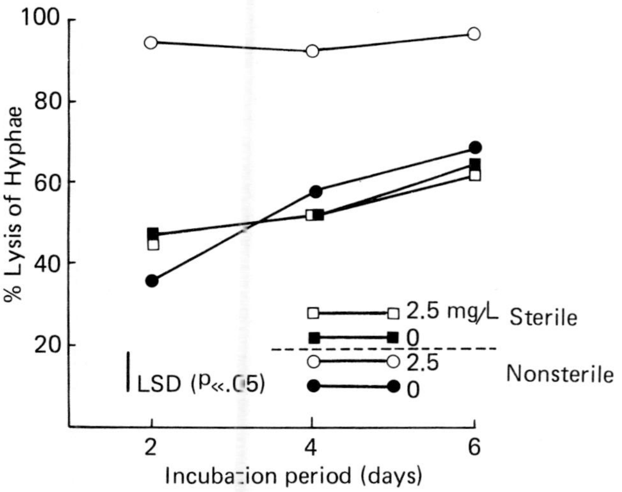


**Fig. 11.** Lysis of *Phytophthora cinnamomi* hyphae inoculated in sterile and natural soil leachates amended with different rates of metalaxyl (Ridomil). LSD = least significant difference.

## Pesticides

As with fertilizers, pesticides can act indirectly in controlling plant pathogens by stimulating microbial antagonists (Rodriguez-Kabana and Curl, 1980). Examples of this integrated approach to controlling *Phytophthora* are few, possibly because most researchers are preoccupied with the direct action of pesticides. Recently, Boughton and Malajczuk (*unpublished*) found a marked interaction between soil microorganisms and the fungicide metalaxyl (Ridomil, Ciba-Geigy) (Fig. 11). Hyphal lysis of *P. cinnamomi* was accelerated in unsterile leachates containing metalaxyl compared with sterile treatments with equivalent levels of fungicide.

Similar types of studies may be necessary to determine the mechanisms of action and interaction of other potential pesticides with soil and rhizosphere microorganisms, because these studies may help us to develop prudent application rates of pesticides in the field to minimize possible side effects associated with an overkill situation.

## Introduced Antagonists

The introduction of specific antagonists into soil for biological control of plant pathogens aims to emulate chemical control without the possible side effects. However, the poor success of this approach to date reflects the lack of appreciation by pathologists of the complex interactions that exist in nature. Often, pathologists tend to neglect preliminary work of finding out how the antagonist survives, grows, and functions in soil, instead embarking immediately on investigations into the success or failure of these antagonists in limiting disease. Successful examples of biological control of *Phytophthora* by this approach are limited to situations where the antagonists are introduced into pretreated soil; e.g., fumigated or steamed soil that contains a much simpler microflora and fauna with reduced interactive complexity.

Malajczuk (*unpublished*) introduced a range of bacterial and actinomycete isolates into sterile soil and examined their effect on root rot caused by *P. cinnamomi* on two eucalypts. These selected isolates of microorganisms had different effects on disease development in both eucalypts that varied with time of sampling (Table 9). *Streptomyces* isolates were most effective in limiting disease on both eucalypts, whereas bacterial isolate 8 was effective only with *E. marginata* seedlings. No single microorganism totally suppressed disease development; however, all isolates were antibiotic to *P. cinnamomi* on agar. Failure to protect eucalypt seedlings with these microorganisms may have been caused by either the failure of the isolates to colonize the soil substrate or their failure to grow and function in the rhizosphere. Sarbini and Kommedahl (1977) reported biological control of *P. megasperma* var. *sojae* by inoculation of soybean pellets with an antagonistic *Pseudomonas* sp. isolate; the antagonists colonized and persisted in the rhizosphere for considerable lengths of time.

The few fungi that have been tested for suppression of disease caused by *Phytophthora* have also given mixed results. El-Goorani et al (1976) found that *Penicillium patulum* prevented *P. cryptogea* from attacking summer squash seedlings in autoclaved soil, whereas Vaartaja et al (1979) reported that the antagonist *Gliocladium virens* did not significantly protect soybean plants from infection by *P. megasperma* var. *sojae* in either greenhouse or field experiments, even though *G. virens* was readily isolated from the soil.

## CONCLUSIONS

Microbial antagonism to *Phytophthora* species may be expected in many situations involving a wide range of microflora and fauna. The ability of the pathogens to counter these antagonistic interactions by producing propagules that enable them to survive under these adverse conditions has evolved over long periods of time. The challenge for pathologists is to anticipate any possible changes in the life stage of *Phytophthora* in which control might be brought about. Control by use of antagonistic microorganisms is considered the ultimate goal and in many cases may be the only means available. However, to date the state of the art is severely deficient, not only in the number of examples of biological control achieved by these means, but also in our understanding of it. We need to develop an understanding of the antagonists as well as of the behavior of the pathogen in soil, in the rhizosphere, and within plant tissue. Bowen's approach (1980) of developing theoretical models to interpret the changes of the pathogen in relationship to the antagonist under different environmental conditions will help to establish principles for predicting epidemics or control of the pathogen. Using guides provided in situations where *Phytophthora* species are naturally suppressed, e.g. in suppressive soil, will help not only to formulate model systems but also to test models. Selective stains, fluorescent brighteners (Tsao, 1969a), and immunofluorescent techniques (Malajczuk et al, 1975, 1978) would be powerful tools to help visualize the activity of both *Phytophthora* and antagonist(s) in situ. Scanning and transmission electron microscopy may then indicate the type and nature of interaction. Only with more of this type of knowledge will we gain an understanding of and subsequent realization of biocontrol of *Phytophthora*.

TABLE 9. Effect of root inoculation with bacteria on total dry weight of plants and disease severity of *Eucalyptus marginata* and *E. calophylla* seedlings grown in sterile soil infested with *Phytophthora cinnamomi*

| | | *E. calophylla* | | *E. marginata* | |
|---|---|---|---|---|---|
| Isolate | Plant age (wk) | Mean plant dry weight (mg) | Mean root rot index[a] | Mean plant dry weight (mg) | Mean root rot index[a] |
| *Streptomyces* 1 | 4 | 7.00 | 0.5 | 2.22 | 2.5 |
| | 12 | 16.75 | 0.5 | 5.43 | 3.5 |
| *Streptomyces* 3 | 4 | 4.03 | 0.5 | 2.82 | 2.0 |
| | 12 | 9.00 | 3.5 | 3.55 | 4.0 |
| *Bacillus* 5 | 4 | 4.09 | 1.5 | 1.29 | 5.0 |
| | 12 | 17.35 | 2.0 | 3.16 | 4.5 |
| *Pseudomonas* 7 | 4 | 2.24 | 4.5 | 1.78 | 3.5 |
| | 12 | 6.78 | 4.5 | 4.21 | 4.5 |
| Bacterial isolate 8 | 4 | 2.44 | 4.5 | 3.80 | 1.5 |
| | 12 | 4.73 | 5.0 | 14.02 | 1.0 |
| Controls | | | | | |
| No *P. cinnamomi* inoculum | 4 | 7.54 | 0.0 | 5.08 | 0.0 |
| | 12 | 19.65 | 0.0 | 19.39 | 0.0 |
| *P. cinnamomi*-inoculated | 4 | 2.03 | 5.0 | 0.16 | 5.0 |
| | 12 | 1.47 | 5.0 | 3.92 | 5.0 |

[a] Mean of five replicates based on a scale of 0–5; 0 = uninfected, 5 = 81–100% root rot.

## SUMMARY

Behavior of *Phytophthora* species is markedly affected by soil microorganisms, resulting in either stimulation of or antagonism to fungal propagules. Forms of antagonism include predation, parasitism, amensalism, and competition and occur both in soil and in the rhizosphere of plant species. The degree of suppression of disease caused by *Phytophthora* varies with soil type and cultural practices.

Hyphal lysis is rapid and is invariably associated with spore formation. Sporangia and zoospores are themselves subjected to breakdown rapidly following their initiation and release in soil. Chlamydospores and oospores survive for considerably longer periods in soil before microbial degradation.

Stimulation of microbial activity in the hyphosphere is the result of organic compounds exuded from growing hyphae and the autolysis of older portions of hyphae. Bacteria commonly associated with hyphae include *Pseudomonas, Bacillus,* and *Streptomyces* spp. Many of the species isolated from lysed hyphae have been shown to be antagonistic in vitro; however, the precise roles of these specific bacteria in hyphal and sporangial lysis are still uncertain. Chytrids and *Trichoderma* spp. are active parasites of *Phytophthora* hyphae that contribute to their breakdown and decay in soil. Protozoa and fungal mites are the common mycophagous biota that attack hyphae and chlamydospores. Small amoebas also ingest and lyse zoospores.

The rhizosphere poses a formidable microbial barrier and may contribute to the resistance of some plant species to root disease. Symbiotic microorganisms (mycorrhizal fungi and rhizobia) have been implicated in antagonism to *Phytophthora* species.

The fact that microorganisms are involved in the suppression of diseases caused by *Phytophthora* spp. strongly suggests that the manipulation of these microorganisms by various management techniques may be possible to biologically control these diseases in both agriculture and forestry. More work is needed to develop a better understanding of the interaction of soil and rhizosphere microorganisms with *Phytophthora*.

## LITERATURE CITED

Alexander, M. 1971. Microbial Ecology. John Wiley & Sons, New York. 511 pp.

Apple, J. L. 1961. The development of black shank in tobacco as influenced by host nutrition. Phytopathology 51:386-389.

Ayers, W. A. 1971. Induction of sporangia in *Phytophthora cinnamomi* by a substance from bacteria and soil. Can. J. Microbiol. 17:1517-1523.

Baker, K. F., and Cook, R. J. 1974. Biological Control of Plant Pathogens. W. H. Freeman and Company, San Francisco. 433 pp.

Baker, K. F., and Snyder, W. C., eds. 1965. Ecology of Soil-Borne Plant Pathogens: Prelude to Biological Control. Univ. Calif. Press, Berkeley. 571 pp.

Baker, R. 1968. Mechanisms of biological control of soil-borne pathogens. Annu. Rev. Phytopathol. 6:263-294.

Bald, J. G. 1937. The use of numbers of infections for comparing the concentration of plant virus suspensions. I. Dilution experiments with purified suspensions. Ann. Appl. Biol. 24:33-55.

Barham, R. O., Marx, D. H., and Ruehle, J. L. 1974. Infection of ectomycorrhizal and nonmycorrhizal roots of shortleaf pine by nematodes and *Phytophthora cinnamomi.* Phytopathology 64:1260-1264.

Barnett, H. L., and Binder, F. L. 1973. The fungal host-parasite relationship. Annu. Rev. Phytopathol. 11:273-292.

Basu, P. K. 1980. Production of chlamydospores of *Phytophthora megasperma* and their possible role in primary infection and survival in soil. Can. J. Plant Pathol. 2:70-75.

Boosalis, M. G. 1964. Hyperparasitism. Annu. Rev. Phytopathol. 2:363-376.

Boosalis, M. G., and Mankau, R. 1965. Parasitism and predation of soil microorganisms. Pages 374-391 in: Ecology of Soil-Borne Plant Pathogens: Prelude to Biological Control. K. F. Baker and W. C. Snyder, eds. Univ. Calif. Press, Berkeley.

Borstel, P. J. 1979. Cross protecting several West Australian native plants against *Phytophthora cinnamomi* Rands using *Pythium irregulare* Buism. Honours dissertation. Univ. West. Aust., Perth. 130 pp.

Boughton, T. J., Malajczuk, N., and Robson, A. D. 1978. Suppression of the infection of jarrah roots by *Phytophthora cinnamomi* with application of calcium carbonate. Aust. J. Bot. 26:611-615.

Bowen, G. D. 1973. Mineral nutrition of ectomycorrhizae. Pages 151-205 in: Ectomycorrhizae: Their Ecology and Physiology. G. C. Marks and T. T. Kozlowski, eds. Academic Press, New York.

Bowen, G. D. 1980. Misconceptions, concepts and approaches in rhizosphere biology. Pages 283-304 in: Contemporary Microbial Ecology. D. C. Ellwood, J. N. Hedger, M. J. Latham, J. M. Lynch, and J. H. Slater, eds. Academic Press, London.

Bowen, G. D., and Rovira, A. D. 1976. Microbial colonization of plant roots. Annu. Rev. Phytopathol. 14:121-144.

Bowen, G. D., and Theodorou, C. 1973. Growth of ectomycorrhizal fungi around seeds and roots. Pages 107-150 in: Ectomycorrhizae: Their Ecology and Physiology. G. C. Marks and T. T. Kozlowski, eds. Academic Press, New York.

Brasier, C. M. 1971. Induction of sexual reproduction in single $A^2$ isolates of *Phytophthora* species by *Trichoderma viride.* Nature (London) New Biol. 231:283.

Brasier, C. M. 1972. Observations on the sexual mechanism in *Phytophthora palmivora* and related species. Trans. Br. Mycol. Soc. 58:237-251.

Brasier, C. M. 1975a. Stimulation of sex organ formation of *Phytophthora* by antagonistic species of *Trichoderma.* I. The effect in vitro. New Phytol. 74:183-194.

Brasier, C. M. 1975b. Stimulation of sex organ formation in *Phytophthora* by antagonistic species of *Trichoderma.* II. Ecological implications. New Phytol. 74:195-198.

Brian, P. W. 1957. The ecological significance of antibiotic production. Pages 168-188 in: Microbial Ecology. R. E. O. Williams and C. C. Spicer, eds. Cambridge Univ. Press, Cambridge.

Brian, P. W. 1960. Antagonistic and competitive mechanisms limiting survival and activity of fungi in soil. Pages 115-129 in: The Ecology of Soil Fungi; An International Symposium. D. Parkinson and J. S. Waid, eds. Liverpool Univ. Press, Liverpool.

Broadbent, P., and Baker, K. F. 1974a. Behaviour of *Phytophthora cinnamomi* in soils suppressive and conducive to root rot. Aust. J. Agric. Res. 25:121-137.

Broadbent, P., and Baker, K. F. 1974b. Association of bacteria with sporangium formation and breakdown of sporangia in *Phytophthora* spp. Aust. J. Agric. Res. 25:139-145.

Broadbent, P., Baker, K. F., and Waterworth, Y. 1971. Bacteria and actinomycetes antagonistic to fungal root pathogens in Australian soils. Aust. J. Biol. Sci. 24:925-944.

Brown, B. N. 1976. *Phytophthora cinnamomi* associated with patch death in tropical rain forests in Queensland. Aust. Plant Pathol. Soc. Newsl. 5:1–4.

Bumbieris, M. 1979. Aspects of the biology of *Phytophthora cryptogea.* Aust. J. Bot. 27:11-16.

Chatel, D. L., and Parker, C. A. 1973a. Survival of field-grown rhizobia over the dry summer period in western Australia. Soil Biol. Biochem. 5:415-423.

Chatel, D. L., and Parker, C. A. 1973b. The colonization of host-root and soil by rhizobia. I. Species and strain differences in the field. Soil Biol. Biochem. 5:425-432.

Chen, A. W.-C., and Griffin, D. M. 1966. Soil physical factors and the ecology of fungi. V. Further studies in relatively dry soils. Trans. Br. Mycol. Soc. 49:419-426.

Chi, C. C., and Sabo, F. E. 1978. Chemotaxis of zoospores of *Phytophthora megasperma* to primary roots of alfalfa seedlings. Can. J. Bot. 56:795-800.

Chilvers, G. A., and Pryor, L. D. 1965. The structure of eucalypt mycorrhizas. Aust. J. Bot. 13:245-259.

Chou, L. G., and Schmitthenner, A. F. 1974. Effect of *Rhizobium japonicum* and *Endogone mosseae* on soybean root rot caused by *Pythium ultimum* and *Phytophthora megasperma* var. *sojae.* Plant Dis. Rep. 58:221-225.

Clark, F. E. 1965. The concept of competition in microbial ecology. Pages 339-345 in: Ecology of Soil-Borne Plant Pathogens: Prelude to Biological Control. K. F. Baker and W. C. Snyder, eds. Univ. Calif. Press, Berkeley.

Cook, R. J. 1977. Management of the associated microbiota. Pages 145-166 in: Plant Disease: An Advanced Treatise. Vol. I. J. G. Horsfall and E. B. Cowling, eds. Academic Press, New York.

Cook, R. J., and Papendick, R. I. 1972. Influence of water potential of soils and plants on root disease. Annu. Rev. Phytopathol. 10:349-374.

Danielson, R. M., and Davey, C. B. 1973. The abundance of *Trichoderma* propagules and the distribution of species in forest soils. Soil Biol. Biochem. 5:485-494.

Davis, R. M., and Menge, J. A. 1980. Influence of *Glomus fasciculatus* and soil phosphorus on *Phytophthora* root rot of citrus. Phytopathology 70:447-452.

Davis, R. M., Menge, J. A., and Zentmeyer, G. A. 1978. Influence of vesicular-arbuscular mycorrhizae on Phytophthora root rot of three crop plants. Phytopathology 68:1614-1617.

Dennis, C., and Webster, J. 1971a. Antagonistic properties of species-groups of *Trichoderma.* I. Production of non-volatile antibiotics. Trans. Br. Mycol. Soc. 57:25-39.

Dennis, C., and Webster, J. 1971b. Antagonistic properties of species-groups of *Trichoderma.* II. Production of volatile antibiotics. Trans. Br. Mycol. Soc. 57:41-48.

Dennis, C., and Webster, J. 1971c. Antagonistic properties of species-groups of *Trichoderma.* III. Hyphal interaction. Trans. Br. Mycol. Soc. 57:363-369.

Drapeau, R., Fortin, J. A., and Gagnon, C. 1973. Antifungal activity of *Rhizobium.* Can. J. Bot. 51:681-682.

Drechsler, C. 1938. Two hyphomycetes parasitic on oospores of root-rotting Oomycetes. Phytopathology 28:81-103.

Duniway, J. M. 1975. Formation of sporangia by *Phytophthora drechsleri* in soil at high matric potentials. Can. J. Bot. 53:1270-1275.

Duniway, J. M. 1979. Water relations of water molds. Annu. Rev. Phytopathol. 17:431-460.

Durrell, L. W. 1968. Hyphal invasion by *Trichoderma viride.* Mycopathol. Mycol. Appl. 35:138-144.

El-Goorani, M. A., Farag, S. A., and Shehata, M. R. A. 1976. The effect of *Bacillus subtilis* and *Penicillium patulum* on in vitro growth and pathogenicity of *Rhizoctonia solani* and *Phytophthora cryptogea.* Phytopathol. Z. 85:345-352.

Erwin, D. C., and Katznelson, H. 1961. Suppression and stimulation of mycelial growth of *Phytophthora cryptogea* by certain thiamine-requiring and thiamine-synthesizing bacteria. Can. J. Microbiol. 7:945-950.

Fokkema, N. J. 1978. Fungal antagonisms in the phyllosphere. Ann. Appl. Biol. 89:115-119.

Garrett, S. D. 1970. Pathogenic Root-Infecting Fungi. Cambridge Univ. Press, Cambridge. 294 pp.

Gerdemann, J. W. 1968. Vesicular-arbuscular mycorrhiza and plant growth. Annu. Rev. Phytopathol. 6:397-418.

Gilpatrick, J. D. 1969a. Role of ammonia in the control of avocado root rot with alfalfa meal soil amendment. Phytopathology 59:973-978.

Gilpatrick, J. D. 1969b. Effect of soil amendments upon inoculum survival and function in *Phytophthora* root rot of avocado. Phytopathology 59:979-985.

Gisi, U., and Meyer, D. 1973. Ökologische Untersuchungen an *Phytophthora cactorum* (Leb. et Cohn) Schroet. im Boden mit direkten Beobachtungsmethoden. Phytopathol. Z. 76:276-279.

Gray, T. R. G., Baxby, P., Hill, I. R., and Goodfellow, M. 1968. Direct observation of bacteria in soil. Pages 171-192 in: The Ecology of Soil Bacteria. T. R. G. Gray and D. Parkinson, eds. Liverpool Univ. Press, Liverpool.

Hale, M. G., and Moore, L. D. 1979. Factors affecting root exudation. II. 1970–1978. Adv. Agron. 31:93-100.

Hale, M. G., Foy, C. L., and Shay, F. J. 1971. Factors affecting root exudation. Adv. Agron. 23:89-109.

Hale, M. G., Moore, L. D., and Griffin, G. J. 1978. Root exudates and exudation. Pages 163-203 in: Interactions Between Non-Pathogenic Soil Microorganisms and Plants. Y. R. Dommergues and S. V. Krupa, eds. Elsevier Scientific Publishing Co., Amsterdam.

Halsall, D. M. 1978. Examination of a forest soil suppressive to *Phytophthora cinnamoni.* Pages 360-363 in: Microbial Ecology. M. W. Loutit and J. A. R. Miles, eds. Springer-Verlag, Berlin.

Harley, J. L. 1959. The Biology of Mycorrhiza. Leonard Hill, London. 234 pp.

Henis, Y., and Katan, J. 1975. Effect of inorganic amendments and soil reaction on soil-borne plant diseases. Pages 100-106 in: Biology and Control of Soil-Borne Plant Pathogens. G. W. Bruehl, ed. Am. Phytopathol. Soc., St. Paul, MN.

Hickman, C. J. 1970. Biology of *Phytophthora* zoospores. Phytopathology 60:1128-1135.

Hickman, C. J., and Ho, H. H. 1966. Behaviour of zoospores in plant-pathogenic phycomycetes. Annu. Rev. Phytopathol. 4:195-220.

Hine, R. B., and Trujillo, E. E. 1966. Manometric studies on residue colonization in soil by *Pythium aphanidermatum* and *Phytophthora parasitica.* Phytopathology 56:334-336.

Ho, H. H. 1969. Notes on the behavior of *Phytophthora megasperma* var. *sojae* in soil. Mycologia 61:835-838.

Ho, H. H., and Zentmyer, G. A. 1977. Infection of avocado and other species of *Persea* by *Phytophthora cinnamomi.* Phytopathology 67:1085-1089.

Hohl, H. R., and Stössel, P. 1976. Host-parasite interfaces in a resistant and a susceptible cultivar of *Solanum tuberosum* inoculated with *Phytophthora infestans:* Tuber tissue. Can. J. Bot. 54:900-912.

Hohl, H. R., and Suter, E. 1976. Host-parasite interfaces in a resistant and a susceptible cultivar of *Solanum tuberosum* inoculated with *Phytophthora infestans:* Leaf tissue. Can. J. Bot. 54:1956-1970.

Hoitink, H. A. J. 1980. Composted bark, a lightweight growth medium with fungicidal properties. Plant Dis. 64:142-147.

Hoitink, H. A. J., Schmitthenner, A. F., and Herr, L. J. 1975. Composted bark for control of root rot in ornamentals. Ohio Rep. 60:25-26.

Hoitink, H. A. J., Herr, L. J., and Schmitthenner, A. F. 1976. Survival of some plant pathogens during composting of hardwood tree bark. Phytopathology 66:1369-1372.

Hoitink, H. A. J., Van Doren, D. M., Jr., and Schmitthenner, A. F. 1977. Suppression of *Phytophthora cinnamomi* in a composted hardwood bark potting medium. Phytopathology 67:561-565.

Honour, R. C., and Tsao, P. H. 1973. Lysis of *Phytophthora parasitica* oospores in soil. No. 36 in: Abstracts of Papers. 2nd Int. Congr. Plant Pathol., September 5–12, 1973, Minneapolis, MN. Int. Soc. Plant Pathol./Am. Phytopathol. Soc., St. Paul, MN.

Hwang, S. C., and Ko, W. H. 1978. Biology of chlamydospores, sporangia, and zoospores of *Phytophthora cinnamomi* in soil. Phytopathology 68:726-731.

Jackson, R. M. 1965. Antibiosis and fungistasis of soil microorganisms. Pages 363-373 in: Ecology of Soil-Borne Plant Pathogens: Prelude to Biological Control. K. F. Baker and W. C. Snyder, eds. Univ. Calif. Press, Berkeley.

Kassaby, F. Y., Fagg, P. C., and Marks, G. C. 1977. Survival in, and colonisation of mountain forest soil by *Phytophthora cinnamomi* Rands in eastern Victoria. Aust. For. Res. 7:173-183.

Katznelson, H. 1965. Nature and importance of the rhizosphere. Pages 187-207 in: Ecology of Soil-Borne Plant Pathogens: Prelude to Biological Control. K. F. Baker and W. C. Snyder, eds. Univ. Calif. Press, Berkeley.

Katznelson, H., Rovatt, J. W., and Peterson, E. A. 1962. The rhizosphere effect of mycorrhizal and non-mycorrhizal roots of yellow birch seedlings. Can. J. Bot. 40:377-382.

Kauffman, C. H. 1926. Klebs' theory of the control of developmental processes in organisms, and its application to fungi. Proc. 4th Int. Bot. Congr. 2:1603-1611.

Keen, N. T., Sims, J. J., Erwin, D. C., Rice, E., and Partridge, J. E. 1971. 6a-Hydroxyphaseollin: An antifungal chemical induced in soybean hypocotyls by *Phytophthora megasperma* var. *sojae.* Phytopathology 61:1084-1089.

Kelley, W. D. 1977. Interactions of *Phytophthora cinnamomi* and *Trichoderma* spp. in relation to propagule production in soil cultures at 26°C. Can. J. Microbiol. 23:288-294.

Kelley, W. D., and Rodriguez-Kabana, R. 1976. Competition between *Phytophthora cinnamomi* and *Trichoderma* spp. in autoclaved soil. Can. J. Microbiol. 22:1120-1127.

Klarman, W. L., and Gerdemann, J. W. 1963a. Induced susceptibility in soybean plants genetically resistant to *Phytophthora sojae.* Phytopathology 53:863-864.

Klarman, W. L., and Gerdemann, J. W. 1963b. Resistance of soybeans to three *Phytophthora* species due to the production of a phytoalexin. Phytopathology 53:1317-1320.

Klisiewicz, J. M., and Johnson, L. B. 1968. Host-parasite relationship in safflower resistant and susceptible to *Phytophthora* root rot. Phytopathology 58:1022-1025.

Kouyeas, V. 1964. An approach to the study of moisture relations of soil fungi. Plant Soil 20:351-363.

Kuhlman, E. G. 1964. Survival and pathogenicity of *Phytophthora cinnamomi* in several western Oregon soils. For. Sci. 10:151-158.

Lacey, J. 1965. The infectivity of soils containing *Phytophthora infestans.* Ann. Appl Biol. 56:363-380.

Legge, B. J. 1953. Studies of species of *Phytophthora.* Ph.D. thesis, Univ. Birmingham. 136 pp.

Lockwood, J. L. 1977. Fungistasis in soils. Biol. Rev. Cambridge Philos. Soc. 52:1-43.

MacDonald, J. D., and Duniway, J. M. 1979. Use of fluorescent antibodies to study the survival of *Phytophthora megasperma* and *P. cinnamomi* zoospores in soil. Phytopathology 69:436-441.

Malajczuk, N. 1979a. The microflora of unsuberized roots of *Eucalyptus calophylla* R.Br. and *Eucalyptus marginata* Donn. ex Sm. seedlings grown in soils suppressive and conducive to *Phytophthora cinnamomi* Rands. II. Mycorrhizal roots and associated microflora. Aust. J. Bot. 27:255-272.

Malajczuk, N. 1979b. Biological suppression of *Phytophthora cinnamomi* in eucalypts and avocados in Australia. Pages 635-652 in: Soil-Borne Plant Pathogens. B. Schippers and W. Gams, eds. Academic Press, London.

Malajczuk, N., and Glenn, A. R. 1981. *Phytophthora cinnamomi*—A threat to the heathlands. Pages 241-247 in: Ecosystems of the World (D. W. Goodall, ser. ed.), Vol. 9: Heathlands and Related Shrublands of the World. B. Analytical Studies. R. L. Specht, ed. Elsevier, Amsterdam.

Malajczuk, N., and Hingston, F. J. 1981. Ectomycorrhizae associated with jarrah. Aust. J. Bot. 29:453-462.

Malajczuk, N., and Lamont, B. B. 1981. Specialized roots of symbiotic origin in heathlands. Pages 165-182 in: Ecosystems of the World (D. W. Goodall, ser. ed.), Vol. 9: Heathlands and Related Shrublands of the World. B. Analytical Studies. R. L. Specht, ed. Elsevier, Amsterdam.

Malajczuk, N., and McComb, A. J. 1977. Root exudates from *Eucalyptus calophylla* R.Br. and *Eucalyptus marginata* Donn. ex Sm. seedlings and their effect on *Phytophthora cinnamomi* Rands. Aust. J. Bot. 25:501-514.

Malajczuk, N., and McComb, A. J. 1979. The microflora of unsuberized roots of *Eucalyptus calophylla* R.Br. and *Eucalyptus marginata* Donn ex Sm. seedlings grown in soils suppressive and conducive to *Phytophthora cinnamomi* Rands. I. Rhizosphere bacteria, actinomycetes and fungi. Aust. J. Bot. 27:235-254.

Malajczuk, N., and Theodorou, C. 1979. Influence of water potential on growth and cultural characteristics of *Phytophthora cinnamomi.* Trans. Br. Mycol. Soc. 72:15-18.

Malajczuk, N., McComb, A. J., and Parker, C. A. 1975. An immunofluorescence technique for detecting *Phytophthora cinnamomi* Rands. Aust. J. Bot. 23:289-309.

Malajczuk, N., McComb, A. J., and Parker, C. A. 1977a. Infection by *Phytophthora cinnamomi* Rands of roots of *Eucalyptus calophylla* R.Br. and *Eucalyptus marginata* Donn. ex Sm. Aust. J. Bot. 25:483-500.

Malajczuk, N., Nesbitt, H. J., and Glenn, A. R. 1977b. A light and electron microscope study of the interaction of soil bacteria with *Phytophthora cinnamomi* Rands. Can. J. Microbiol. 23:1518-1525.

Malajczuk, N., Bowen, G. D., and Greenhalgh, F. C. 1978. A combined fluorescent antibody and soil sieving technique to count chlamydospores of *Phytophthora cinnamomi* in soil. Soil Biol. Biochem. 10:437-438.

Marais, L. J., and Kotzé, J. M. 1976. Ectomycorrhizae of *Pinus patula* as biological deterrents to *Phytophthora cinnamomi.* S. Afr. For. J. 99:35-39.

Marks, G. C., and Foster, R. C. 1973. Structure, morphogenesis, and ultrastructure of ectomycorrhizae. Pages 1-4 in: Ectomycorrhizae: Their Ecology and Physiology. G. C. Marks and T. T. Kozlowski, eds. Academic Press, New York.

Marshall, K. C., and Cruickshank, R. H. 1973. Cell surface hydrophobicity and the orientation of certain bacteria at interfaces. Arch. Microbiol. 91:29-40.

Marx, D. H. 1969a. The influence of ectotrophic mycorrhizal fungi on the resistance of pine roots to pathogenic infections. I. Antagonism of mycorrhizal fungi to root pathogenic fungi and soil bacteria. Phytopathology 59:153-163.

Marx, D. H. 1969b. The influence of ectotrophic mycorrhizal fungi on the resistance of pine roots to pathogenic infections. II. Production, identification, and biological activity of antibiotics produced by *Leucopaxillus cerealis* var. *piceina.* Phytopathology 59:411-417.

Marx, D. H. 1972. Ectomycorrhizae as biological deterrents to pathogenic root infections. Annu. Rev. Phytopathol. 10:429-454.

Marx, D. H. 1973. Growth of ectomycorrhizal and nonmycorrhizal shortleaf pine seedlings in soil with *Phytophthora cinnamomi.* Phytopathology 63:18-23.

Marx, D. H., and Bryan, W. C. 1969. *Scleroderma bovista,* an ectotrophic mycorrhizal fungus of pecan. Phytopathology 59:1128-1132.

Marx, D. H., and Davey, C. B. 1969a. The influence of ectotrophic mycorrhizal fungi on the resistance of pine roots to pathogenic infections. III. Resistance of aseptically formed mycorrhizae to infection by *Phytophthora cinnamomi.* Phytopathology 59:549-558.

Marx, D. H., and Davey, C. B. 1969b. The influence of ectotrophic mycorrhizal fungi on the resistance of pine roots to pathogenic infections. IV. Resistance of naturally occurring mycorrhizae to infections by *Phytophthora cinnamomi.* Phytopathology 59:559-565.

Mataré, R., and Hattingh, M. J. 1978. Effect of mycorrhizal status of avocado seedlings on root rot caused by *Phytophthora cinnamomi.* Plant Soil 49:433-435.

Matthews, R. E. F. 1970. Plant Virology. Academic Press, New York. 778 pp.

McIntosh, D. L. 1972. Effects of soil water suction, soil temperature, carbon and nitrogen amendments, and host rootlets on survival in soil of zoospores of *Phytophthora cactorum.* Can. J. Bot. 50:269-272.

Meagher, J. W., Brown, R. H., and Rovira, A. D. 1978. The effects of cereal cyst nematode (*Heterodera avenae*) and *Rhizoctonia solani* on growth and yield of wheat. Aust. J. Agric. Res. 29:1127-1137.

Mehrotra, R. S. 1972. Behavior of zoospores of *Phytophthora megasperma* var. *sojae* and *P. drechsleri* in soil. Can. J. Bot. 50:2125-2130.

Mircetich, S. M., Zentmyer, G. A., and Kendrick, J. B., Jr. 1968. Physiology of germination of chlamydospores of *Phytophthora cinnamomi.* Phytopathology 58:666-671.

Mukerjee, N., and Roy, A. B. 1962. Microbial influence on the formation of oospores in culture by *Phytophthora parasitica* var. *sabdariffae.* Phytopathology 52:583-584.

Neal, J. L., Jr., Atkinson, T. G., and Larson, R. I. 1970. Changes in the rhizosphere microflora of spring wheat induced by disomic substitution of a chromosome. Can. J. Microbiol. 16:153-158.

Nesbitt, H. J., Malajczuk, N., and Glenn, A. R. 1978. Bacterial colonization of *Phytophthora cinnamomi* Rands. Pages 371-375 in: Microbial Ecology. M. W. Loutit and J. A. R. Miles, eds. Springer-Verlag, Berlin.

Nesbitt, H. J., Malajczuk, N., and Glenn, A. R. 1979. Effect of organic matter on the survival of *Phytophthora cinnamomi* Rands in soil. Soil Biol. Biochem. 11:133-136.

Nesbitt, H. J., Malajczuk, N., and Glenn, A. R. 1981a. Translocation and exudation of metabolites in *Phytophthora cinnamomi.* Trans. Br. Mycol. Soc. 76:503-505.

Nesbitt, H. J., Malajczuk, N., and Glenn, A. R. 1981b. Bacterial colonization and lysis of *Phytophthora cinnamomi.* Trans. Br. Mycol. Soc. 77:47-54.

Newhook, F. J. 1970. *Phytophthora cinnamomi* in New Zealand. Pages 173-176 in: Root Diseases and Soil-Borne Pathogens. T. A. Toussoun, R. V. Bega and P. E. Nelson, eds. Univ. Calif. Press, Berkeley.

Newhook, F. J., and Podger, F. D. 1972. The role of *Phytophthora cinnamomi* in Australian and New Zealand forests. Annu. Rev. Phytopathol. 10:299-326.

Old, K. M. 1977. Giant soil amoebae cause perforation of conidia of *Cochliobolus sativus.* Trans. Br. Mycol. Soc. 68:277-281.

Old, K. M. 1978. Perforation and lysis of fungal spores by soil amoebae. Ann. Appl. Biol. 89:128-131.

Old, K. M., and Darbyshire, J. F. 1978. Soil fungi as food for giant amoebae. Soil Biol. Biochem. 10:93-100.

Old, K. M., and Oros, J. M. 1980. Mycophagous amoebae in Australian forest soils. Soil Biol. Biochem. 12:169-175.

Old, K. M., and Patrick, Z. A. 1979. Giant soil amoebae, potential biocontrol agents. Pages 617-628 in: Soil-Borne Plant Pathogens. B. Schippers and W. Gams, eds. Academic Press, London.

Palzer, C. 1976. Zoospore inoculum potential of *Phytophthora cinnamomi.* Ph.D. thesis. Univ. West. Aust., Perth. 222 pp.

Park, D. 1967. The importance of antibiotics and inhibiting substances. Pages 435-447 in: Soil Biology. A. Burges and F. Raw, eds. Academic Press, London.

Paxton, J. D., and Chamberlain, D. W. 1967. Acquired local resistance of soybean plants to *Phytophthora* spp. Phytopathology 57:352-353.

Pegg, K. G. 1977. Soil application of elemental sulphur as a control of *Phytophthora cinnamomi* root and heart rot of pineapple Aust. J. Exp. Agric. Anim. Husb. 17:859-865.

Peto, S. 1953. A dose-response equation for the invasion of microorganisms. Biometrics 9:320-335.

Pratt, B. H. 1971. Isolation of basidiomycetes from Australian eucalypt forest and assessment of their antagonism to *Phytophthora cinnamomi.* Trans. Br. Mycol. Soc. 56:243-250.

Pratt, B. H., Sedgley, J. H., Heather, W. A., and Shepherd, C. J. 1972. Oospore production in *Phytophthora cinnamomi* in the presence of *Trichoderma koningii.* Aust. J. Biol. Sci. 25:861-863.

Rambelli, A. 1973. The rhizosphere of mycorrhizae. Pages 299-349 in: Ectomycorrhizae: Their Ecology and Physiology. G. C. Marks and T. T. Kozlowski, eds. Academic Press, New York.

Ramirez, B. N. 1974. Influence of endomycorrhizae on the relationship of inoculum density of *Phytophthora palmivora* in soil to infection of papaya roots. M.S. thesis, Univ. Florida, Gainesville. 45 pp.

Reeves, R. J. 1975. Behaviour of *Phytophthora cinnanomi* Rands in different soils and water regimes. Soil Biol. Biochem. 7:19-24.

Reeves, R. J., and Jackson, R. M. 1972. Induction of *Phytophthora cinnamomi* oospores in soil by *Trichoderma viride.* Trans. Br. Mycol. Soc. 59:156-159.

Ribeiro, O. K. 1978. A Source Book of the Genus *Phytophthora.* J. Cramer, Vaduz, Liechtenstein. 417 pp.

Robertson, G. I. 1973. Occurrence of *Pythium* spp. in New Zealand soils, sands, pumices, and peat, and on roots of container-grown plants. N.Z. J. Agric. Res. 16:357-365.

Rodriguez-Kabana, R., and Curl, E. A. 1980. Nontarget effects of pesticides on soil-borne pathogens and disease. Annu. Rev. Phytopathol. 18:311-332.

Rose, S. L., Li, C.-Y., and Hutchins, A. S. 1980. A streptomycete antagonistic to *Phellinus weirii, Fomes annosus,* and *Phytophthora cinnamomi.* Can. J. Microbiol. 26:583-587.

Ross, E. W., and Marx, D. H. 1972. Susceptibility of sand pine to *Phytophthora cinnamomi.* Phytopathology 62:1197-1200.

Ross, J. P. 1972. Influence of *Endogone* mycorrhiza on Phytophthora rot of soybean. Phytopathology 62:896-897.

Rovira, A. D. 1965. Plant root exudates and their influence upon soil microorganisms. Pages 170-186 in: Ecology of Soil-Borne Plant Pathogens: Prelude to Biological Control. K. F. Baker and W. C. Snyder, eds. Univ. Calif. Press, Berkeley.

Rovira, A. D. 1969. Plant root exudates. Bot. Rev. 35:35-57.

Sarbini, G., and Kommedahl, T. 1977. Effect of bacterial seed treatment on stand and yield of soybeans in *Phytophthora*-infested soil. (Abstr.) Proc. Am. Phytopathol. Soc. 4:146-147.

Schönbeck, F. 1979. Endomycorrhiza in relation to plant diseases. Pages 271-280 in: Soil-Borne Plant Pathogens. B. Schippers and W. Gams, eds. Academic Press, London.

Schroth, M. N., and Hildebrand, D. C. 1964. Influence of plant exudates on root-infecting fungi. Annu. Rev. Phytopathol. 2:101-132.

Shea, S. R. 1975. Environmental factors of the northern jarrah forest in relation to pathogenicity and survival of *Phytophthora cinnamomi.* Bull. For. Dep. West. Aust. No. 85. 83 pp.

Shea, S. R., and Malajczuk, N. 1977. Potential for control of eucalypt dieback in Western Australia. Pages 13-19 in: Seminar Papers, Annu. Conf. Aust. Nurserymen's Assoc. Ltd., Hobart.

Shea, S. R., McCormick, J., and Portlock, C. C. 1979. The effect of fires on regeneration of leguminous species in the northern jarrah (*Eucalyptus marginata* Sm.) forest of Western Australia. Aust. J. Ecol. 4:195-205.

Shew, H. D., and Beute, M. K. 1979. Evidence for the involvement of soilborne mites in *Pythium* pod rot of peanut. Phytopathology 69:204-207.

Shimony, C., and Friend, J. 1975. Ultrastructure of the interaction between *Phytophthora infestans* and leaves of two cultivars of potato (*Solanum tuberosum* L.) Orion and Majestic. New Phytol. 74:59-65.

Skidmore, A. M. 1976. Interactions in relation to biological control of plant pathogens. Pages 507-528 in: Microbiology of Aerial Plant Surfaces. C. H. Dickinson and T. F. Preece, eds. Academic Press, London.

Smith, W. H. 1977. Tree root exudates and the forest soil ecosystem: Exudate chemistry, biological significance and alteration by stress. Pages 289-302 in: The Belowground Ecosystem: A Synthesis of Plant-Associated Processes. J. K. Marshall, ed. Colorado State Univ., Fort Collins.

Sneh, B., and McIntosh, D. L. 1974. Studies on the behavior and survival of *Phytophthora cactorum* in soil. Can. J. Bot. 52:795-802.

Sneh, B., Humble, S. J., and Lockwood, J. L. 1977. Parasitism of oospores of *Phytophthora megasperma* var. *sojae, P. cactorum, Pythium* sp., and *Aphanomyces euteiches* in soil by oomycetes, chytridiomycetes, hyphomycetes, actinomycetes, and bacteria. Phytopathology 67:622-628.

Swift, M. J., Heal, O. W., and Anderson, J. M. 1979. Studies in Ecology, Vol. 5: Decomposition in Terrestrial Ecosystems. Univ. Calif. Press, Los Angeles. 372 pp.

Tippett, J., and Malajczuk, N. 1979. Interaction of *Phytophthora cinnamomi* and a resistant host, *Acacia pulchella.* Phytopathology 69:764-772.

Tippett, J. T., Holland, A. A., Marks, G. C., and O'Brien, T. P. 1976. Penetration of *Phytophthora cinnamomi* into disease tolerant and susceptible eucalypts. Arch. Microbiol. 108:231-242.

Tomiyama, K. 1971. Cytological and biochemical studies of the hypersensitive reaction of potato cells to *Phytophthora infestans.* Pages 387-401 in: Morphological and Biochemical Events in Plant-Parasite Interactions. S. Akai and S. Ouchi, eds. Phytopathol. Soc. Jpn., Tokyo.

Trappe, J. M. 1962. Fungus associates of ectotrophic mycorrhizae. Bot. Rev. 28:538-606.

Trinick, M. J. 1970. The ecology of *Rhizobium*—Interactions between *Rhizobium* strains and other microorganisms. Ph.D. thesis. Univ. West. Aust., Perth. 353 pp.

Tsao, P. H. 1969a. Vital fluorescent labeling as applied to soil ecological studies of *Phytophthora parasitica.* (Abstr.) Phytopathology 59:16.

Tsao, P. H. 1969b. Studies on the saprophytic behavior of *Phytophthora parasitica* in soil. Proc. 1st Int. Citrus Symp. 3:1221-1230.

Tsao, P. H. 1970. Selective media for isolation of pathogenic fungi. Annu. Rev. Phytopathol. 8:157-186.

Tsao, P. H. 1977. Prospects of biological control of citrus root disease fungi. Proc. Int. Soc. Citriculture 3:857-863.

Tsao, P. H., and Bricker, J. L. 1963. Growth inhibition of *Phytophthora cinnamomi* by soil diffusates on agar medium (Abstr.) Phytopathology 53:892.

Tsao, P. H., and Bricker, J. L. 1968. Germination of chlamydospores of *Phytophthora parasitica* in soil. (Abstr.) Phytopathology 58:1070.

Tsao, P. H., and Zentmyer, G. A. 1979. Suppression of *Phytophthora cinnamomi* and *Ph. parasitica* in urea-amended soils. Pages 191-199 in: Soil-Borne Plant Pathogens. B. Schippers and W. Gams, eds. Academic Press, London.

Tsao, P. H., Bhalla, H. S., and Zentmyer, G. A. 1975. Effects of certain organic amendments on survival and activity of *Phytophthora cinnamomi* in soil. (Abstr.) Proc. Am. Phytopathol. Soc. 2:40.

Tu, J. C. 1978. Protection of soybean from severe Phytophthora root rot by *Rhizobium.* Physiol. Plant Pathol. 12:233-240.

Tu, J. C. 1979. Evidence of differential tolerance among some root rot fungi to rhizobial parasitism in vitro. Physiol. Plant Pathol. 14:171-177.

Tu, J. C. 1980. Incidence of root rot and overwintering of alfalfa as influenced by rhizobia. Phytopathol. Z. 97:97-108.

Turner, P. D. 1965. Behavior of *Phytophthora palmivora* in soil. Plant Dis. Rep. 49:135-137.

Vaartaja, O., Pitblado, R. E., Buzzell, R. I., and Crawford, L. G. 1979. Chemical and biological control of Phytophthora root and stalk rot of soybean. Can. J. Plant Sci. 59:307-311.

Van Bragt, J., Mossel, D. A. A., Pierik, R. L. M., and Veldstra, H. 1971. Effect of sterilization on components in nutrient media. Misc. Papers 9. Landbouwhogeschool, Wageningen, The Netherlands. 146 pp.

Vincent, J. M. 1965. Environmental factors in the fixation of nitrogen by the legume. Pages 384-435 in: Soil Nitrogen. W. B. Bartholomew and F. E. Clark, eds. Am. Soc. Agron., Madison, WI.

Vujičić, R., and Park, D. 1964. Behaviour of *Phytophthora erythroseptica* in soil. Trans. Br. Mycol. Soc. 47:455-458.

Waterhouse, G. M. 1940. A chytrid allied to *Pleolpidium inflatum* Butler. Trans. Br. Mycol. Soc. 24:7-19.

Weste, G. 1974. *Phytophthora cinnamomi*—The cause of severe disease in certain native communities in Victoria. Aust. J. Bot. 22:1-8.

Weste, G., and Vithanage, K. 1977. Microbial populations of three forest soils: Seasonal variations and changes associated with *Phytophthora cinnamomi.* Aust. J. Bot. 25:377-383.

Weste, G., and Vithanage, K. 1978. Effect of *Phytophthora cinnamomi* on microbial populations associated with the roots of forest flora. Aust. J. Bot. 26:153-167.

Weste, G., and Vithanage, K. 1979. Survival of chlamydospores of *Phytophthora cinnamomi* in several non-sterile, host-free forest soils and gravels at different soil water potentials. Aust. J. Bot. 27:1-9.

Wong, P. T. W., and Griffin, D. M. 1974. Effect of osmotic potential on streptomycete growth, antibiotic production and antagonism to fungi. Soil Biol. Biochem. 6:319-325.

Wynn, A. R., and Epton, H. A. S. 1979. Parasitism of oospores of *Phytophthora erythroseptica* in soil. Trans. Br. Mycol. Soc. 73:255-259.

Zak, B. 1964. Role of mycorrhizae in root disease. Annu. Rev. Phytopathol. 2:377-392.

Zak, B. 1971. Characterization and classification of mycorrhizae of Douglas fir. II. *Pseudotsuga menziesii* + *Rhizopogon vinicolor.* Can. J. Bot. 49:1079-1084.

Zan, K. 1962. Activity of *Phytophthora infestans* in soil in relation to tuber infection. Trans. Br. Mycol. Soc. 45:205-221.

Zentmyer, G. A. 1963. Biological control of Phytophthora root rot of avocado with alfalfa meal. Phytopathology 53:1383-1387.

Zentmyer, G. A. 1965. Bacterial stimulation of sporangium production in *Phytophthora cinnamomi.* Science 150:1178-1179.

Zentmyer, G. A. 1980. *Phytophthora cinnamomi* and the Diseases It Causes. Monogr. 10. Am. Phytopathol. Soc., St. Paul, MN. 96 pp.

Zentmyer, G. A., and Bingham, F. T. 1956. The influence of nitrite on the development of Phytophthora root rot of avocado. Phytopathology 46:121-124.

# 17

# Factors Affecting Isolation and Quantitation of *Phytophthora* from Soil

Peter H. Tsao

CONTENTS



## INTRODUCTION

Detection and/or isolation of *Phytophthora* fungi from newly infected, fresh plant tissues that contain viable and actively growing mycelia is generally quite easy, especially for those *Phytophthora* spp. that are relatively fast-growing. Incubation of tissues in water, soil extract, or a suitable mineral salts solution to induce sporangium formation allows quick detection, and placing of tissues from advancing margins of fresh lesions on water agar or other ordinary agar media usually results in isolation of the target fungus. However, isolation of *Phytophthora* from old, badly decayed or dried, infected tissues in an advanced stage of disease or from naturally infested soils, especially those containing low populations, is extremely difficult or impossible unless special techniques are used. The difficulty is usually due to antagonism and interference from fast-growing associated secondary microflora and to the slow emergence of the surviving, dormant *Phytophthora* propagules (e.g., chlamydospores) in the decayed tissue or soil. Special isolation techniques such as baiting and selective media, properly used, usually overcome these difficulties. When coupled with a quantitative manipulation of the soil sample, these isolation techniques can also be effectively used for quantitation of *Phytophthora* in soil.

This review discusses the procedures and problems in the detection, isolation, and quantitation of *Phytophthora* propagules from soil and the factors affecting their recovery and accurate enumeration. Factors affecting recovery from infected plant tissue are essentially the same as those affecting recovery from infested soil and will also be discussed parallelly but only briefly. This chapter is not meant to be a complete compilation and review of all the methods and factors reported in the literature but serves as a comprehensive and up-to-date discussion and analysis of the subject. Only information exemplary of the individual topics will be cited and discussed, along with my personal perspective on the subject based on more than 20 yr of experience in the diagnosis, detection, isolation, and quantitation of *Phytophthora* diseases or fungi in temperate, subtropical, and tropical regions. The literature review for this chapter was terminated in March 1981. Portions of the subject have been previously reviewed, in varying extent, by Menzies (1963), Domsch and Schwinn (1965), Tuite (1969), Waterhouse and Stamps (1969), Tsao (1970), Katsura (1971), Johnson and Curl (1972), Schmitthenner (1973), Ribeiro (1978), and Zentmyer (1980).

## DETECTION AND ISOLATION OF *PHYTOPHTHORA* WITH BAITING TECHNIQUES

Proper choice and use of plant parts (e.g., fruits, seeds, seedlings, cotyledons, leaves, and leaf disks of susceptible host plants) as baits, or traps, can usually bring about selective isolation of target *Phytophthora* spp. from soil. Plant materials that have been used successfully for baiting various *Phytophthora* spp. from soil are listed in Table 1.

Various methods employing the baiting (or trapping, or host-infection) techniques have been reported (Table 1), but they can be grouped into three general categories: 1) Inserting soil (or infected tissue) into a wound or hole made on a fleshy fruit (e.g., apple, cocoa pod, or watermelon). This method requires a relatively large fruit to provide a thick substrate and can involve only a small sample of soil or tissue per replicate. 2) Planting seeds, seedlings, rooted cuttings, etc. to soil in pots or in the field, followed by heavy watering. This method can sample a larger volume of soil but is generally of longer duration and more time-consuming. 3) Floating, partially immersing, or otherwise incubating baits of various types in a water and soil mixture, oversaturated with water or with a high water/soil ratio, in containers of various sizes. This is perhaps the most widely used baiting method and is considered effective for isolating most *Phytophthora* spp. Isolation can be attempted also from diseased plant tissues by mixing them with water and performed with baiting, or a small amount of diseased tissues can be added to the soil sample to increase the chance of recovery.

Success of baiting technique depends on the susceptibility of the bait and on incubation conditions.

### Susceptibility of Bait

Baiting materials are generally selected from the susceptible hosts to which the target *Phytophthora* sp. to be isolated is pathogenic. Among susceptible hosts or plant parts, some have been found to be more sensitive as baits than others for a given *Phytophthora* sp. For example, for isolating *P. cinnamomi* from azalea and Fraser fir soils, leaf disks of *Rhododendron* spp. were more successful than blue lupine radicles and some other bait materials (Shew et al, 1979a). Lupine radicles are excellent for isolating *P. megasperma* but are unsuitable for isolating *P. cactorum* (Dance et al, 1975; Hargreaves and Duncan, 1978) and are less suitable than pine needles as bait for the isolation of *P. cinnamomi* and some other *Phytophthora* spp. from soil (Dance et al, 1975). Interestingly, for isolating *P. cactorum* from apple soils, safflower seedlings were more effective than apple fruit, the natural host (Banihashemi and Mitchell, 1975).

Pathogenicity of the isolates recovered from soil is often the concern of the investigator; therefore, preference for a natural host often dictates the choice of a bait. Hence the choice of *Eucalyptus* sp. (Marks and Kassaby, 1974) and *Persea* spp. (Zentmyer, 1980) for *P. cinnamomi* and of citrus (Klotz and DeWolfe, 1958) for *P. parasitica.*

Comparisons of baiting materials for their relative effectiveness in recovering other *Phytophthora* spp. from soil have also been reported (Canaday and Schmitthenner, 1978; Chee and Foong, 1968; Chee and Newhook, 1965; Hansen et al, 1979; Hargreaves and Duncan, 1978; Marks and Kassaby, 1974; McIntosh, 1964). Dance et al (1975) listed several criteria for an ideal bait.

TABLE 1. Baiting techniques for the detection and isolation of *Phytophthora* spp.

| Species<br>Authors | Bait material | Conditions of baiting and remarks |
|---|---|---|
| *P. cactorum* | | |
| Schwinn (1961) | Apple fruit | Buried partly in wet soil in large flat; 24° C, 90–100% RH, 5–6 days. |
| McIntosh (1964) | Pear fruit | Immersed partly in 50 cc of soil plus 100 cc of water; 6–7 days; pear better than apple, peach, and apricot fruits as bait. |
| Van der Scheer (1971) | Apple fruit, pear fruit | Buried partly in wet soil; 20° C, until rot occurs; isolation from apple and strawberry soils; pear better than apple as bait. |
| Sewell et al (1974) | Apple fruit | Immersed partly in 20-mm soil layer with 5 mm of water above soil; optimum 20° C, 14 days; *P. citricola* and *P. syringae* also recovered from apple soils at lower incubation temperatures. |
| Banihashemi and Mitchell (1975) | Safflower seedlings | Planted in steamed soil before adding 15-mm layer of test soil in contact with hypocotyl, or directly to test soil; optimum 28° C, 3–4 wk; better than apple fruit; used also in quantitative assays of apple soils. |
| Molot and Beyries (1976) | Strawberry fruit | Immature fruit inoculated with soil in small hole; 22° C, 3–6 days; used for isolation from strawberry soils; also susceptible to five other *Phytophthora* spp. |
| *P. cambivora* | | (See Dance et al [1975] under *P. cinnamomi* and Mircetich and Matheron [1976] under *P. megasperma.*) |
| *P. capsici* | | (See Lee and Varghese [1974] and Ponchet et al [1972] under *P. parasitica.*) |
| *P. cinnamomi* | | |
| Campbell (1949) | Apple fruit | Holes made in fruit, filled with soil; 15–27° C, 5–10 days; suitable also for many other *Phytophthora* spp., but also colonized by *Pythium* and other soil fungi. |
| Anderson (1951) | Pineapple leaves and rooted crowns | Immersed in 400–800 cc of water added to about 10–20 cc of soil; 21–24° C, 4 days. |
| Zentmyer et al (1960) | Avocado fruit, *Persea indica* seedlings | Fruit embedded partly in flooded soil; 27°C, 2–4 days; seedlings planted in wet soil, 2–3 days. |
| Klemmer and Nakano (1962) | Pineapple leaves and rooted crowns | Leaf base or root tips immersed in water-soil mixture at least 5 cm above the container bottom; 18–27° C. |
| Chee and Newhook (1965) | Lupine radicles | Root tips immersed in large volume of water over 20-mm soil layer; 2 days; isolation from pine soils; better than apple fruit; also susceptible to four other *Phytophthora* spp. |
| Marks et al (1972) | Lupine radicles, excised | Pieces 1–2 cm long floated on water 2–3 mm over 30 cc of soil in Petri plate; 2–3 days; used for quantitative assays of eucalyptus soils. |
| Pratt and Heather (1972) | Lupine radicles | Immersed in 150 cc of water added to 26 g of soil; 17–24° C, 5 days; lesions distinguishable from those caused by eight other *Phytophthora* spp. except *P. drechsleri.* |
| Gerrettson-Cornell (1974) | Apple slices, lupine radicles | Immersed in 200 cc of water added to 25 g of soil; 4–10 days; both baits were satisfactory. |
| Marks and Kassaby (1974) | Eucalyptus cotyledons | Detached cotyledons floated on water 5–10 mm over 50–60 cc of soil; 22–24° C, within 60 hr; 2–4 times more sensitive than lupine radicles; used for quantitative assays of eucalyptus soils. |
| Dance et al (1975) | Cedar needles, pine needles, lupine radicles | Floated on or immersed partly in water added to soil; both 16 and 22° C, 3 days; conifer needles also good for *P. cambivora, P. citricola, P. cryptogea, P. drechsleri, P. hibernalis,* and a few other *Phytophthora* spp. |

[*continued on next page*]

TABLE 1—Continued

| Species<br>Authors | Bait material | Conditions of baiting and remarks |
|---|---|---|
| Linderman and Zeitoun (1977) | Eucalyptus leaf disks | Floated on water added to soil or roots; 20° C, 24 hr; *Pythium* spp. interfere with results. |
| Pegg (1977) | Avocado leaf pieces | Floated on water added to soil; 4 days; recovery up to 89%. |
| Greenhalgh (1978) | Eucalyptus cotyledons, lupine seedlings, pear fruit | Floated on water added to soil at various depths; 25° C under continuous light, 3–7 days; all baits successful, but more frequently with pear; used in quantitative assays. |
| Zentmyer and Ohr (1978) | *Persea indica* seedlings | Immersed partly in large volume of water added to 1.2–2.5 cm soil layer; 5–7 days; isolation from avocado soils. |
| Shew et al (1979a) | Azalea leaf disks, cedar needles, fir seedlings, lupine radicles | Floated on or immersed in 100 ml of water added to 5–50 g of soil; 20° C, 1–2 days; recovery was best with azalea leaf disks and least with lupine radicles. |
| *P. citricola* | | (See Dance et al [1975] under *P. cinnamomi* and Sewell et al [1974] under *P. cactorum.*) |
| *P. citrophthora* | | (See Klotz and DeWolfe [1958] and Tsao [1960] under *P. parasitica.*) |
| *P. cryptogea* | | (See Dance et al [1975] under *P. cinnamomi.*) |
| *P. drechsleri* | | |
| Alavi and Strange (1979) | Cantaloupe seed and seedlings | Planted in wet soil mixed with diseased tissues of *Cucumis* spp.; 20–35° C, 7 days. |
| *P. fragariae* | | |
| Duncan (1976, 1979) | Strawberry rooted runners and seedlings | Planted to wet soil in pots or in the field; 5 wk or longer; used in quantitative assays of strawberry soils. |
| *P. heveae* | | (See Lee and Varghese [1974] under *P. parasitica.*) |
| *P. hibernalis* | | (See Dance et al [1975] under *P. cinnamomi.*) |
| *P. infestans* | | |
| Lacey (1965) | Potato tuber slices | Soil placed on wet surface of tuber slice; 20° C, 6 days. |
| *P. lateralis* | | |
| Ostrofsky et al (1977) | Port Orford cedar tissue sections | Branchlet or root sections floated on water added to soil or organic matter in petri plate, or inserted in deeper soil; 18° C, 6 days; better than seedlings. |
| *P. meadii* | | (See Chee and Foong [1968] under *P. palmivora.*) |
| *P. megasperma* (all f. sp. or var.) | | |
| Marks and Mitchell (1970) | Alfalfa seedlings, injured | Floated on water 2–3 mm over 30 cc of soil in petri plate; 20–22° C, 6 days; used for quantitative assays of alfalfa soils. |
| Eye et al (1976) | Soybean seedlings | Placed in 30–40 cc of water added to 20–40 g of soil in petri plate; 3 days; for isolation from soybean soils. |
| Mircetich and Matheron (1976) | Pear fruit | Immersed partly in 500 cc of soil with 1 cm of water above soil; 20° C, 3 days; for isolation from cherry soils; also good for *P. cambivora.* |
| Paxton (1976) | Soybean seed | Buried in wet soil; 20–30° C, 7 days; for isolation from soybean soils. |
| Canaday and Schmitthenner (1978) | Soybean leaf disks | Floated on 50 cc of water added to 12 g of soil; 3 hr; for isolation from soybean soils. |
| Hargreaves and Duncan (1978) | Pine needles, lupine seedlings, apple fruit, pear fruit | Needles and seedlings placed in water 5 mm over 100 g of soil; 16 and 22° C, 3 days; fruit buried in larger amounts of soil; 22° C, 10 days; for isolation from strawberry and raspberry soils. |
| Stovold and Carratt (1978) | Alfalfa seedlings, injured | Same as in Marks and Mitchell (1970) except using nylon mesh placed below seedlings to prevent roots from becoming enmeshed in the soil. |
| *P. melonis* | | |
| Katsura (1971) | Cucumber fruit | Immersed partly in water 1 cm over 20–30 g of soil; 20 and 28° C, within 5 days. |
| *P. palmivora* (all morphological forms) | | |
| Orellana (1954) | Cocoa pods | Placed on or in soil, or inoculated with a loopful of soil suspension on the pod surface and incubated in moist box; 1–4 days; for isolation from cocoa soils. |
| Holliday and Mowat (1963) | Apple fruit, black pepper leaves | Soil inserted into apple the same as in Campbell's (1949) method. Black pepper leaves immersed partly in water-soil mixture. For isolation (of MF 4?) from black pepper soils. Black pepper leaf disks used successfully by Tsao (*unpublished*). |

[*continued on next page*]

TABLE 1—Continued

| Species<br>Authors | Bait material | Conditions of baiting and remarks |
|---|---|---|
| Turner (1965) | Cocoa pods | Soil inserted beneath flaps of endocarp tissue of unripe pods; for isolation from cocoa soils. |
| Chee and Foong (1968) | Cocoa pods | Soil or diseased *Hevea* tissues inserted into unripe green pod as in Campbell's (1949) apple method; 26–30° C, 4–5 days; also good for *P. meadii* and *P. parasitica*. |
| Okaisabor (1971) | Cocoa pod tissues | Small blocks placed in wet soil; 25° C, 4 days; premoistened for 5 days if soil samples were dry. |
| Newhook and Jackson (1977) | Cocoa pods and tissues | Four methods: pod on soil, soil in pod, pod tissue in flooded soil (the best), and pod tissue on soil; used in quantitative assays of cocoa soils. |
| *P. parasitica* (*P. nicotianae*) (both var.) | | |
| Klotz and DeWolfe (1958) | Apple fruit, lemon fruit, orange fruit | Soil or citrus tissues inserted into apple the same as in Campbell's (1949) method. Lemon or orange placed on surface of saturated soil; 20° C, 4 or more days; also good for *P. citrophthora* and *P. syringae* from citrus soils. |
| Tsao (1960) | Lemon fruit | Immersed partly in 150 cc of water added to 25 cc of soil; 25° C, 6 days; also good for *P. citrophthora*; used for quantitative assays of citrus soils. |
| Jenkins (1962) | Tobacco leaves | Petiole end immersed in water-soil mixture; as in Tsao (1960). Used for quantitative assays of black shank fungus in tobacco soils. |
| Narasimhan and Ramakrishnan (1969) | Castorbean seed | Buried in soil; 63 hr; for isolation from betelvine soils or irrigation water. |
| Ponchet et al (1972), Ricci (1972) | Carnation petals | Immature petals from buds floated on soil suspension in petri plate; 24–35° C, 2–4 days; used for quantitative assays of carnation soils; also invaded by *P. capsici*. |
| Grimm and Alexander (1973) | Citrus leaf pieces | Small leaf pieces, from various *Citrus* spp., floated on water 1–2 cm above 100 cc of soil; 22–28° C, 3–4 days; calamondin fruit equally effective as bait. |
| Lee and Varghese (1974) | Apple fruit, eggplant fruit | Methods not given. Both fruits also good for *P. capsici*. Cocoa pod allows isolation of *P. palmivora* and *P. heveae*. |
| Ioannou and Grogan (1977) | Tomato fruit | Green fruit placed in flooded soil from tomato fields, after premoistening for 1 wk. |
| *P. syringae* | | (See Klotz and DeWolfe [1958] under *P. parasitica* and Sewell et al [1974] under *P. cactorum*.) |
| Unspecified *Phytophthora* spp. | | |
| Waterhouse and Stamps (1969) | 19 different plants | Seeds, seedlings, leaves, or fruits floated on or immersed in water-soil mixture; for isolation of various species of *Phytophthora* and *Pythium*. |
| Banihashemi (1970) | Cucumber fruit | Immersed partly in water 1 cm over 40–60 g of soil; 20° C, 4–7 days; for isolation of *Phytophthora* spp. causing cucurbit root rot. Cantaloupe, muskmelon, and watermelon fruits also are satisfactory. |
| Fatemi (1974) | Hemp and safflower seeds | Boiled seeds placed on nylon screen over 10–15 mm layer of saturated soil; 20° C, 2–3 days; good for various species of *Phytophthora* and *Pythium*. |

Age or maturity of the bait materials often determines their effectiveness as well. For example, an immature lemon fruit while still green is not susceptible to infection by *P. citrophthora* and *P. parasitica*, and overripe lemon fruit is subject to colonization by *Geotrichum, Trichoderma,* and other soil fungi; thus, neither type of fruit would be suitable for use as bait for isolating the citrus phytophthoras (Tsao, 1960). Apple fruit, on the other hand, is more suitable as bait while green; mature apples, especially when overripe, are often colonized by mucoralous fungi, *Trichoderma,* and *Fusarium* spp. that interfere with or prevent *Phytophthora* isolation. McIntosh (1964, 1977) also used green pear fruit, and Molot and Beyries (1976) used immature strawberry fruit for baiting *P. cactorum*.

### Incubation Conditions

Although infection of bait by mycelium may frequently be involved in certain baiting techniques, oospores, chlamydospores, and perhaps sporangia are the surviving propagules in soil and, along with zoospores, are primarily responsible for bait colonization. In baiting systems involving soils that are saturated or have high water/soil ratios, zoospores are likely to be the main infective propagules. The conditions for bait incubation, therefore, should enhance spore germination, sporangium formation, and zoospore release in the water/soil mixture to ensure maximal infection of baits.

#### Temperature

Many investigators use fluctuating day and night temperatures for bait incubation; this allows a range of temperatures suitable for the germination and sporulation of various spore stages for any given *Phytophthora* sp. It is important to include a cool period (perhaps 22°C or below), which is conducive to zoospore formation in most *Phytophthora* spp. Bait incubation in such a "natural" environment is usually feasible during "mild" seasons, but in extreme climates (e.g., in the hot tropics or desert areas or in very hot and cold seasons), incubating baits in enclosed quarters with suitable temperature regulations is necessary. Isolation of *Phytophthora* spp. that have low growth-temperature requirements (such as *P. fragariae, P. hibernalis,*

*P. infestans, P. lateralis* and *P. syringae*) necessitates low-temperature (below 20° C) incubation of the baits (Dance et al, 1975; Hargreaves and Duncan, 1978; Lacey, 1965; Ostrofsky et al, 1977; Sewell et al, 1974). Bait incubation at low temperature (22° C or below) was found to be suitable also for *Phytophthora* spp. requiring higher growth temperatures, such as *P. cactorum, P. cambivora, P. cinnamomi, P. citricola,* and *P. megasperma* (Dance et al, 1975; Hargreaves and Duncan, 1978; Linderman and Zeitoun, 1977; Marks and Mitchell 1970; Mircetich and Matheron, 1976; Pratt and Mitchell 1975; Sewell et al, 1974; Shew et al, 1979a). Conditions favorable for spore germination and sporangium and zoospore formation are likely reasons for successful recoveries. Banihashemi and Mitchell (1975), however, found 28° C more favorable than 24, 20, and 16° C for isolating *P. cactorum* with safflower seedlings as bait in their greenhouse pot system, in which bait colonization by mycelium rather than zoospores was likely involved. Okaisabor (1971) reported a much higher recovery of *P. palmivora* at 25° C than at 20 or 30° C.

**Light**

Because light stimulates oospore germination and sporangium formation in most *Phytophthora* spp. studied (see Ribeiro, Chapter 4), bait should be exposed to some light period during incubation. When feasible or when climate permits, bait incubation in a proper "natural" environment is suitable, because it provides not only fluctuating day and night temperatures but also alternating light and dark periods. Gerrettson-Cornell and Townsend (1974) and Banihashemi and Mitchell (1975, 1976) showed experimentally higher recoveries of *P. cinnamomi* and *P. cactorum,* respectively, when baits were incubated under proper lighting conditions. Success in isolation of other *Phytophthora* spp. with baiting techniques might also be affected by light in the same way. Even though incubation temperature is an important factor, bait incubation in a temperature-controlled dark room or an incubator without artificial lighting is not recommended.

**Aeration**

Perhaps because of the aerophilic or negative geotactic nature of motile zoospores, *Phytophthora* infections on partially immersed bait materials (e.g., avocado, lemon, or cucurbit fruits or whole black pepper leaves), in a system involving high water/soil ratios, usually occur at or just below the water line. Care should therefore be taken to expose the bait or portions of large bait to the water-air interface by not allowing the entire bait to submerge or, in the case of leaf disks or pieces by floating them on the water surface (Grimm and Alexander, 1973). To ensure floating, melted paraffin can be brushed onto the leaf surface before disks are cut (J. L. Troutman, *personal communication*). Sporangia can form and later germinate to release zoospores in saturated soil unless aeration is inadequate (see Gisi, Chapter 8). Theoretically, therefore, recovery of *Phytophthora* should be enhanced by occasional stirring of the water-soil mixture during bait incubation.

**Water Quality**

*Phytophthora* zoospores are sensitive to toxic ions present in unpurified water. Tap water originating from copper pipes or pipes with copper fixtures should be avoided for use in baiting. Although the scale of baiting experiments is usually large, perhaps too large to use glass-distilled water, relatively inexpensive deionized water of good quality should definitely be considered. Grimm and Alexander (1973) used rainwater, which would also be suitable. The report by Gerrettson-Cornell (1976) clearly illustrated the importance of water quality in baiting in which blue lupine seedlings were used to isolate *P. cinnamomi. Phytophthora* recovery was 94, 32, and 0%, respectively, when glass-distilled water, deionized water, and distilled water from a metal still were used.

**Water/Soil Ratio**

High water/soil ratios in baiting techniques enhance *Phytophthora* recovery, because soils may contain chemical substances that inhibit sporangium formation and germination, which would result in negative *Phytophthora* recovery if soil is not adequately diluted. In soils treated with organic nitrogenous amendments (Tsao and Oster, 1981; Tsao and Zentmyer, 1979) or with fungicides, a small *Phytophthora* population might remain if original treatment dosages are sublethal. The viable *Phytophthora* propagules might be prevented from germinating and from subsequently infecting the bait if the soil sample is only saturated or flooded but not adequately diluted with water to reach a high water/soil ratio. An erroneous negative recovery could result under these circumstances.

Many reports on baiting techniques have described adequate dilution of the soil by water, which would prevent such a fungistatic inhibition. Examples are 200 cc of water to 25 g of soil (Gerrettson-Cornell, 1974), 150 cc of water to 26 g of soil (Pratt and Heather, 1972), and 150 cc of water to 25 cc of soil (Tsao, 1960), resulting in high water/soil ratios of about 8:1 or 6:1. Similar high dilutions were used by Chee and Newhook (1965), McIntosh (1977), and Zentmyer and Ohr (1978). Most of the reported procedures on baiting in the literature, however, employ a relatively low water/soil ratio (about 1:1 or lower); a relatively large amount (50–500 cc) of soil is placed in a container and enough water is added to flood the soil and to reach about 5–10 mm above the soil surface (e.g. Banihashemi, 1970; Grimm and Alexander, 1973; Marks and Kassaby, 1974; Mircetich and Matheron, 1976; Sewell et al, 1974). Insufficient dilution (or low water/soil ratio) also results when petri plates are used as containers, because they are too shallow to hold enough water to cover the soil more than a few millimeters above the surface (Marks and Mitchell, 1970; Marks et al, 1972; Onesirosan, 1971; Pratt and Mitchell, 1973, 1975; Stovold and Carratt, 1978). As a rule, when the inoculum densities of *Phytophthora* in the soil samples are high, positive recovery will be achieved regardless of the water/soil ratios used if other incubation requirements are met. However, with soils of extremely low inoculum densities, *Phytophthora* might be recovered only if sufficiently high (about 4:1 or higher) water/soil ratios are used. This has been shown experimentally in a series of tests (P. H. Tsao and L. A. Bower, *unpublished*) employing the serial dilution end point method (Tsao, 1960) and azalea leaf disks as bait for the isolation of *P. cinnamomi* from soils amended with organic nitrogenous substances. Recovery was negative from a large sample tested under a low (1:2) water/soil ratio, whereas an identical sample but only one-eighth as large (and therefore containing only one-eighth as much inoculum) yielded *Phytophthora* when tested under a higher (6:1) water/soil ratio.

If a small container is used for baiting, the rationale of using the maximum allowable amount of soil per container, in order to "increase the chance of *Phytophthora* recovery," would not be sound. Too large a sample per replicate is not necessarily conducive to positive recovery because, aside from the difficulty of achieving adequate dilution, the *Phytophthora* propagules residing at the bottom of a deep layer (say, 3 cm) of soil would not have a chance to infect the bait even if germination is allowed to occur. The maximum allowable amount of water should be used per container, and the amount of soil used should be dictated by the dilution chosen and reduced accordingly. "Increasing the chance of *Phytophthora* recovery" can be accomplished by using more replicates instead of more soil per replicate.

### Reliability of Positive Detection

Infection or colonization of bait materials generally produces characteristic, recognizable *Phytophthora* lesions or symptoms, but they are not always a reliable indication of *Phytophthora* infections because of similarities to the lesions resulting from infections by some other pathogens. Detection of *Phytophthora* recovery by means of sporangium formation on the bait, as determined by microscopic examination, is deemed more reliable and has been used by various workers (Banihashemi and Mitchell, 1975; Eye et al, 1976; Grimm and Alexander, 1973; Jenkins, 1962; Lacey, 1965; Marks and Kassaby, 1974; Marks et al, 1972; Onesirosan, 1971; Ostrofsky et al, 1977; Pratt and Heather, 1972; Pratt and Mitchell, 1973, 1975; Stovold and Carratt, 1978) for detecting *P. cactorum, P. cinnamomi, P. infestans, P. lateralis, P. megasperma, P. palmivora,* and *P. parasitica.* However, because incubation conditions chosen are not always favorable for *Phytophthora* sporulation, absence of sporangium formation does not necessarily signify a negative recovery. Special procedures involving transfer of colonized bait tissues to soil extract, Petri solution, or other mineral salts solutions and incubation under lights to induce sporulation have been found useful (Holliday and Mowat, 1963; Pratt and Heather, 1972; P H. Tsao, *unpublished*). Marks et al (1972) reported that soil bacteria and *Fusarium* spp. also infected the bait (excised lupine radicles), suppressing sporulation and making it difficult to observe the sporangia of the target *Phytophthora* (*P. cinnamomi*). Dance et al (1975) also pointed out the dangers inherent in identification by sporangia only. Most workers, therefore, prefer the ultimately reliable method of positive *Phytophthora* identification—by plating infected bait tissues on a selective medium to allow emergence of *Phytophthora* from the tissues, followed by identification (Canaday and Schmitthenner, 1978; Dance et al, 1975; Gerrettson-Cornell, 1974, 1976; Greenhalgh, 1978; Marks and Kassaby, 1974; Marks and Mitchell, 1970; Mircetich and Matheron, 1976; Pegg, 1977; Ponchet et al, 1972; Pratt and Heather, 1972; Pratt and Mitchell, 1973; Ricci, 1972; Shew et al, 1979a, Zentmyer and Ohr, 1978).

## DETECTION AND ISOLATION OF *PHYTOPHTHORA* WITH SELECTIVE AGAR MEDIA

The development and use of selective agar media in plant pathology began only in the early 1960s, and *Phytophthora* was among the first plant-pathogenic fungi for which selective isolation media were devised and reported (the review by Tsao [1970] gives the historical development). Since then, a great number of good selective media have been reported and widely used for isolation and quantitation of many *Phytophthora* spp. Table 2 contains a chronologic list of such media and some of their salient features.

Selective media are perhaps still more often used for isolation from plant tissue than from soil. Small pieces of infected tissues, even some from old, dried, or badly decayed lesions (Tsao et al, 1976), can be directly plated, with or without prior surface-sterilization, onto a suitable selective medium and properly incubated. If the medium is effective, surface-sterilization is not necessary. In fact, improper use of a disinfectant often results in undesired killing of the target *Phytophthora* sp. near the surface of the tissue sample being disinfected. Therefore, with thoroughly washed plant tissue samples, except for the purpose of fulfilling Koch's Postulates, surface-sterilization is not advisable when an effective selective agar medium is used for *Phytophthora* isolation. To minimize bacterial growth, blotting the wet tissue pieces dry on absorbent paper before plating is deemed a very important step, however.

For soil isolation, selective agar media are usually employed in quantitative assays of *Phytophthora* populations (see later section on quantitation). However, if the inoculum density of *Phytophthora* in soil is high, qualitative isolation can also be accomplished on effective selective media. It can be performed by placing a dry or wet soil mass, or by sprinkling small crumbs of soil, on the medium surface (direct inoculation method), by pouring cool molten agar medium over a small quantity of heavy soil suspension (Warcup's soil plate method), or by using a less precisely controlled soil dilution plate method normally used for quantitative studies.

### Composition of Selective Media

An effective selective agar medium contains appropriate concentrations of several properly chosen selective antimicrobial agents, each inhibitory to a wide spectrum of undesired, nontarget fungi or bacteria (including actinomycetes) but nontoxic to *Phytophthora.* It is based on the principle of selective inhibition, the most effective and commonly used approach in developing selective isolation media for pathogenic fungi (Tsao, 1970). Another common principle, selective enhancement (i.e., to provide in the medium an energy source or other nutrients that are assimilable only by a limited number of microorganisms including the target fungus or fungi, thus reducing or excluding undesired organisms in the isolation plate [Tsao, 1970]), should also be an effective approach for isolating *Phytophthora* but has not yet been employed in the development of any *Phytophthora* isolation media reported in the literature. A suitable basal agar medium in which to incorporate the selective agents also contributes to the success of the isolation medium.

#### Basal Medium

An efficient selective medium for *Phytophthora* isolation employs a relatively "weak" agar medium as the basal medium, which minimizes bacterial growth yet allows adequate, though not luxuriant, *Phytophthora* growth. Besides the resistant bacteria, certain fungi (e.g., species of *Pythium* and *Mortierella*) also appear, as they are not inhibited by the antifungal agents used in most of the *Phytophthora* isolation media. When these nontarget fungi emerge on a weak basal medium, their growth is also sufficiently sparse so that *Phytophthora* colonies in their midst are still visible.

Cornmeal agar is one such weak agar medium and is among the most commonly employed basal media for *Phytophthora* isolation; it is used in 14 of the 35 selective media listed in Table 2. The basal media for a number of other *Phytophthora* isolation media are V-8 juice agar (Ko et al, 1978; McCain et al, 1967; Pratt and Mitchell, 1973), potato-dextrose agar (Fujisawa and Masago, 1975; Green and Pratt, 1970; Masago et al, 1977; Pratt and Mitchell, 1973), malt agar (Molot and Nourrisseau, 1972; Ponchet et al, 1972; Ricci, 1972), and lima bean agar (Follin, 1971). These richer media produce more visible and denser *Phytophthora* colonies than the weak cornmeal agar but also support luxuriant bacterial growth and thick, dense colonies of other fungi that obscure *Phytophthora*

TABLE 2. Selective agar media for the detection and isolation of *Phytophthora* spp.[a]

| Authors | Antimicrobial agents[b] | Basal medium[c] | Nature, usage, and remarks |
|---|---|---|---|
| Eckert and Tsao (1960, 1962) | Pimaricin, 100 ppm ($\mu$g/ml)<br>Penicillin, 50 ppm<br>Polymyxin, 50 ppm | CMA | Unsuitable for isolation from soil because of high pimaricin concentration inhibitory to spore germination. |
| Haas (1964) | Pimaricin, 2 ppm<br>Penicillin, 80 units/ml<br>Polymyxin, 370 units/ml<br>PCNB, 10 ppm | CMA | For *P. megasperma* var. *sojae*. Direct isolation from soil successful because of pimaricin at 2 ppm. |
| Hendrix and Kuhlman (1965) | Nystatin, 100 units/ml<br>PCNB, 100 ppm<br>Streptomycin, 50 ppm<br>Rose bengal, 60 ppm | SA plus yeast extract | For *P. cinnamomi*. Combination of streptomycin and rose bengal inhibitory to spore germination. |
| Hollomon (1965) | Pimaricin, 2,000 ppm<br>Vancomycin, 1,000 ppm<br>Chloramphenicol, 50 ppm | PA | For *P. infestans* from potato leaves. Antibiotic concentrations exceedingly high, inhibitory to spore germination. |
| Tsao and Menyonga (1966) | Pimaricin, 100 ppm<br>Vancomycin, 200 ppm | CMA | Unsuitable for isolation from soil because of high pimaricin concentration. |
| Ocana and Tsao (1966), Tsao and Ocana (1969) | Pimaricin, 10 ppm<br>Vancomycin, 200 ppm<br>PCNB, 100 ppm | CMA | For *P. cinnamomi, P. citrophthora, P. palmivora, P. parasitica,* and other *Phytophthora* spp. Low pimaricin concentration allows spore germination and isolation from soil. |
| Frosheiser (1967) | Pimaricin, 50 ppm<br>or nystatin, 100 ppm<br>Penicillin, 100 ppm<br>Polymyxin, 100 ppm | CMA | For *P. megasperma* from alfalfa roots. A modified Eckert-Tsao medium. |
| McCain et al (1967) | Nystatin, 50 ppm<br>Vancomycin, 100 ppm<br>PCNB, 10 ppm | V8A | For *P. cinnamomi* and other *Phytophthora* spp. |
| Flowers and Hendrix (1969) | Gallic acid, 425 ppm<br>Nystatin, 100 units/ml<br>PCNB, 25 ppm<br>Penicillin, 80 units/ml<br>Rose bengal, 0.5 ppm | SA plus yeast extract and thiamin | For *P. parasitica* var. *nicotianae* and other *Phytophthora* spp. A modified Hendrix-Kuhlman medium. |
| Green and Pratt (1970) | Pimaricin, 100 ppm<br>Penicillin, 50 ppm<br>Neomycin, 35 ppm | PDA | For *P. citricola* from walnut roots and baits. |
| Follin (1971) | Benomyl, 25 ppm<br>Penicillin, 50 ppm<br>Polymyxin, 50 ppm | LBA | For *P. parasitica* from *Hibiscus* spp. Benomyl was used as a commonly available substitute for pimaricin. |
| Wheeler and Boyle (1971) | Pimaricin, 100 ppm<br>Chloramphenicol, 125 ppm | WA | For *P. parasitica* from plant tissues. |
| Alconero et al (1972) | Nystatin, 100 ppm<br>Penicillin, 50 ppm<br>Polymyxin, 50 ppm | CMA | For *P. palmivora* (MF 4?) from black pepper tissues and soil near roots. |
| Molot and Nourrisseau (1972) | Benomyl, 15 ppm<br>PCNB, 100 ppm<br>Penicillin, 500 ppm | MA | For *P. cactorum* from strawberry soils. |
| Ponchet et al (1972) | Benomyl, 15 ppm<br>PCNB, 100 ppm<br>Penicillin, 250 ppm<br>Polymyxin, 250 ppm | MA | For *P. parasitica* from carnation tissues and soils. Polymyxin concentration likely too high for spore germination. |
| Sneh (1972) | Pimaricin, 2 ppm<br>Nystatin, 10 ppm<br>PCNB, 100 ppm<br>Penicillin, 60 ppm<br>Polymyxin, 60 ppm<br>Chloramphenicol, 60 ppm | SA plus corn oil | For *P. cactorum* from soil. Also good for *P. cinnamomi, P. drechsleri,* and *P. parasitica.* Rose bengal and additional pimaricin were added to the plates after soil was washed off. |
| Pratt and Mitchell (1973) | Pimaricin, 50 ppm<br>Vancomycin, 100 ppm | PDA or V8A | For *P. megasperma* from alfalfa seedlings and baits. |
| Schmitthenner (1973) | Benomyl, 10 ppm<br>PCNB, 20 ppm<br>Neomycin, 100 ppm<br>Chloramphenicol, 10 ppm | Dilute V8A | For various *Phytophthora* spp. primarily from tissues. |
| Broadbent and Baker (1974) | Pimaricin, 10 ppm<br>Penicillin, 50 ppm | 1/4 PDA | For *P. cinnamomi* from roots and soil. |
| Castillo and Fernandez-Northcote (1974) | Pimaricin, 10 ppm<br>Cefalotin, 400 ppm<br>Bromthymol blue, 50 ppm<br>Na azide, 1 ppm | CMA | For various *Phytophthora* spp. from plant tissues, but not successful from soil. |
| Banihashemi and Mitchell (1975) | Pimaricin, 10 ppm<br>Vancomycin, 200 ppm<br>Polymyxin, 175 units/ml<br>Tergitol NPX, 100 ppm | SA plus corn oil | For *P. cactorum* from safflower and apple tissues and soil. |

[*continued on next page*]

TABLE 2—Continued

| Authors | Antimicrobial agents[b] | Basal medium[c] | Nature, usage, and remarks |
|---|---|---|---|
| Fujisawa and Masago (1975) | Benomyl, 10 ppm<br>Nystatin, 50 ppm<br>PCNB, 25 ppm<br>Rifampicin, 10 ppm<br>Ampicillin, 500 ppm | PDA | The BNPRA medium, forerunner of the medium of Masago et al (1977). |
| McIntosh (1975) | Benomyl, 20 ppm<br>Na taurocholate, 200 ppm | SA | Na taurocholate did not sufficiently control bacteria but increased hyphal growth of *P. cactorum*. |
| Mircetich and Matheron (1976) | Pimaricin, 5 ppm<br>Vancomycin, 300 ppm<br>PCNB, 25 ppm | CMA | A modified Tsao-Ocana medium. For *P. cambivora, P. drechsleri, P. megasperma* from cherry tissues and soil. |
| Newhook and Jackson (1977) | Pimaricin, 10 ppm<br>Penicillin, 50 ppm<br>PCNB, 100 ppm | CMA | For *P. palmivora* from leaf debris in cocoa plantations. |
| Shepherd and Forrester (1977) | Pimaricin, 20 ppm<br>Vancomycin, 75 ppm<br>Ampicillin, 150 ppm<br>PCNB, 50 ppm | Dilute V8A | For *P. cinnamomi* from root tissues and baits. |
| Masago et al (1977) | Benomyl, 10 ppm<br>Nystatin, 25 ppm<br>PCNB, 25 ppm<br>Rifampicin, 10 ppm<br>Ampicillin, 500 ppm<br>Hymexazol, 25–50 ppm | PDA | Addition of hymexazol selectively suppresses *Pythium* spp. For various *Phytophthora* spp. from tissues and soil. |
| Tsao and Guy (1977) | Pimaricin, 10 ppm<br>Vancomycin, 200 ppm<br>PCNB, 100 ppm<br>Hymexazol, 50 ppm | CMA | A combination of media of Tsao and Ocana and Masago et al. Hymexazol selectively suppresses *Pythium* and *Mortierella* spp. but is not suitable for certain *Phytophthora* spp. sensitive to it. |
| Hargreaves and Duncan (1978) | Carbendazim, 10 ppm<br>Penicillin, 100 ppm<br>Polymyxin, 20 ppm<br>Streptomycin, 10 ppm | SA plus sterol | For *P. megasperma* and other *Phytophthora* spp. from several baits. Not yet attempted for soil isolation. |
| Kannwischer and Mitchell (1978) | Pimaricin, 10 ppm<br>Ampicillin, 250 ppm<br>Rifampicin, 10 ppm<br>PCNB, 100 ppm | CMA | For *P. parasitica* var. *nicotianae* and other *Phytophthora* spp. Hymexazol at 50 ppm added in recent years (*personal communication*). |
| Ko et al (1978) | Nystatin, 50 ppm<br>Ampicillin, 100 ppm<br>PCNB, 10 ppm | V8A | For *P. cinnamomi* from roots and baits. |
| Hansen et al (1979) | Pimaricin, 20 ppm<br>Vancomycin, 200 ppm<br>Penicillin, 200 ppm | CMA | For several *Phytophthora* spp. from apple bait used in isolation from forest tree nurseries. |
| Julis et al (1979) | Pimaricin, 10–25 ppm<br>Vancomycin, 125 ppm<br>Penicillin, 100 ppm | CMA | For *P. cactorum* and several other *Phytophthora* spp. from apple bait. |
| Shew et al (1979b) | Pimaricin, 5 ppm<br>PCNB, 35 ppm<br>Chloramphenicol, 10 ppm<br>Hymexazol, 50 ppm | SA plus yeast extract and thiamin | For *P. cinnamomi* from azalea and Fraser fir soils. |
| Papavizas et al (1981) | Pimaricin, 5 ppm<br>Vancomycin, 200 ppm<br>Penicillin, 100 ppm<br>PCNB, 100 ppm<br>Benomyl, 2.5 ppm<br>Hymexazol, 20 ppm | CMA | For *P. capsici* from tissues and soils of pepper, squash, and eggplant fields; pH adjusted to 3.8–4. |

[a]Unless otherwise indicated, the media are reportedly or presumably applicable to isolation from both plant tissue and soil. Only the selective media containing both antifungal and antibacterial agents are listed in this table.

[b]Only generic names of antibiotics and chemical or common names for other antimicrobial agents are used in this table. PCNB = pentachloronitrobenzene.

[c]CMA = cornmeal agar, SA = synthetic agar, PA = pea agar, V8A = V-8 juice agar, PDA = potato-dextrose agar, LBA = lima bean agar, WA = water agar, MA = malt agar.

emergence and detection. However, when properly diluted, these media might be more suitable for use; examples are the one-fourth-strength potato-dextrose agar, 1% V-8 juice agar, and dilute V-8 juice agar chosen by Broadbent and Baker (1974); Greenhalgh (1978); Shepherd and Forrester (1977) and Schmitthenner (1973), respectively. A weak agar medium such as cornmeal agar cannot be used for isolating certain fastidious *Phytophthora* spp. such as *P. infestans;* for these species, more nutritious basal media such as pea agar (Holloman, 1965), dilute oatmeal or V-8 juice agar (C. Y. Shen and P. H. Tsao, *unpublished*), or some others must be used.

Several selective *Phytophthora* isolation media employed various formulations of synthetic agar media as the basal media (Banihashemi and Mitchell, 1975; Flowers and

Hendrix, 1969; Hargreaves and Duncan, 1978; Hendrix and Kuhlman, 1965; McIntosh, 1975; Shew et al, 1979b; Sneh, 1972). Most but not all of these contain thiamin, sterols, or vegetable oils to meet the growth and sporulation requirements of *Phytophthora* spp. Preparation of synthetic media is time-consuming and prone to error, and a given formulation does not usually meet the growth requirements of all *Phytophthora* spp. I believe that the many commonly available natural media can easily satisfy the requirement of a good basal medium without resorting to complex synthetic formulations.

The natural pH of most selective media falls within the acidic range. Some authors (e.g., Hendrix and Kuhlman, 1965; Papavizas et al, 1981) further lower the medium pH to control excessive bacterial growth. Most *Phytophthora* spp. tolerate low pH for mycelial growth quite well but do not sporulate or germinate well in highly acidic media (e.g., below pH 4.5) (see Ribeiro, Chapter 4). The value of low pH in *Phytophthora* isolation media has not been thoroughly investigated.

**Antifungal Agents**

Because of the unique properties of the Oomycetes, especially the family Pythiaceae (see Bartnicki-Garcia and Wang, Chapter 9), *Phytophthora* and some other related fungi respond to certain toxicants very differently from the vast majority of fungi. Excellent selective chemicals are, therefore, available for use in the development of *Phytophthora* isolation media.

Polyene antibiotics are known to be strongly antifungal except to a small number of fungi, including those in the Pythiaceae. Eckert and Tsao (1960, 1962) were among the first to report the noninhibitory nature of nystatin and pimaricin to *Phytophthora* and *Pythium* spp. and the use of these polyene antibiotics in selective isolation of pythiaceous fungi from infected roots. Each of these two antibiotics has since been widely used; of the 35 selective media listed in Table 2, 20 contain pimaricin, seven contain nystatin, one contains both, and one contains either one of these two polyenes. Pimaricin is more potent and has a wider spectrum than nystatin and is generally the more effective of the two. Nystatin was more available in the past, however, and became the choice of some investigators, especially in some countries where pimaricin was not obtainable locally.

Benomyl and pentachloronitrobenzene (PCNB), two synthetic antifungal compounds, are also selectively nontoxic to *Phytophthora,* Pythiaceae in general, and a large number of other fungi. Twenty-one of the 35 media in Table 2 employ either or both of these fungicides. They have much narrower antifungal spectra than the polyene antibiotics and are used primarily as supplements to pimaricin or nystatin or as substitutes, especially when the polyene antibiotics are not available. When used alone as antifungal agent, neither benomyl nor PCNB adequately suppresses many undesired fungi from diseased tissues or soils. Gallic acid is another such supplement that, when combined with nystatin and PCNB in a selective medium, effectively reduced secondary fungal growth (Flowers and Hendrix, 1969).

All of the above five selective antifungal agents are generally noninhibitory to *Pythium* as well as *Phytophthora.* In fact, almost all the *Phytophthora* isolation media reported before 1977 (and not containing hymexazol) are suitable for isolating *Pythium* spp. (Tsao, 1970). These media, interestingly, are also noninhibitory to most species of *Mortierella* (Mircetich and Matheron, 1976; Ponchet et al, 1972; Tsao, 1970; Tsao and Guy, 1977), a genus in the Mucorales within the Zygomycetes but physiologically and toxicologically behaving very similarly to members of the Pythiaceae within the Oomycetes. When soil populations of *Pythium* and *Mortierella* are high, isolation and quantitation of *Phytophthora* can become difficult and problematic.

In this connection, the discovery by Masago et al (1977) of the unique selectivity of the fungicide hymexazol (or HMI, the acronym used by these authors) against *Pythium* and its use in their BNPRA isolation medium was a major breakthrough in the field of selective *Phytophthora* isolation. They reported that hymexazol at 25 or 50 ppm ($\mu$g/ ml) completely or greatly inhibited the growth of 12 *Pythium* spp. but was virtually noninhibitory to 12 *Phytophthora* spp. tested (Masago et al, 1977). Tsao and Guy (1977) partially confirmed the findings of Masago et al and reported, in addition, the inhibition of *Mortierella* spp. by hymexazol. Their PVPH medium containing 50 ppm hymexazol (Tsao and Guy, 1977) worked well for the isolation and quantitation of *P. cinnamomi* and *P. parasitica* when combined with the soil dilution plate or soil sprinkle method, and *Phytophthora* detection on isolation plates was generally free of interfering *Pythium* and *Mortierella* colonies.

Not all *Pythium* spp. are sensitive to hymexazol, however; at 50 ppm, the chemical allowed almost normal growth of *Pythium vexans* (Tsao and Guy, 1977), which I have frequently encountered in root or soil samples of avocado, black pepper, and citrus. Hansen et al (1979) also reported the insensitivity to hymexazol of six *Pythium* isolates present in forest tree nurseries. Furthermore, not all *Phytophthora* spp. are insensitive to hymexazol; at 25 or 50 ppm, the chemical partially inhibited the growth of certain isolates of *P. cactorum* (Hansen et al, 1979; Tsao and Guy, 1977), *P. lateralis* (Hansen et al, 1979), and *P. palmivora* (Tsao and Guy, 1977) and completely inhibited that of *P. infestans* (C. Y. Shen and P. H. Tsao, *unpublished*).

In spite of these anomalies, hymexazol at 50 ppm is an effective supplement in a medium containing other selective antifungal agents when used in *Phytophthora* isolation from soils, roots, or other underground plant parts in which *Pythium* and *Mortierella* are likely inhabitants. Hymexazol should be excluded from an otherwise already effective selective medium if the medium is to be used for the isolation of *P. cactorum, P. infestans, P. palmivora,* or other sensitive species from infected leaves, fruits, trunk barks, or other aboveground plant parts that might not contain *Pythium* or *Mortierella* spp.

**Antibacterial Agents**

*Phytophthora* spp. are generally very sensitive to antagonism by bacteria (see Malajczuk, Chapter 16), which are mainly responsible for the difficulties encountered in *Phytophthora* isolation. Interference from bacteria occurs even on selective media, because of bacterial populations insensitive to the antibacterial agents used. Hasty and improper choice of potent antibacterial agents, used in some media, which are toxic to *Phytophthora* may also result in failure of recovery from tissues or soils.

Among the antimicrobial chemicals, antibiotics are the most effective and widely used for bacterial inhibition in modern fungal isolation media (Tsao, 1970). Although most fungi are insensitive to antibacterial antibiotics, members of the Peronosporales (including Pythiaceae) behave like prokaryotic organisms in that they are sensitive to many antibacterial antibiotics (Tsao, 1970; Bartnicki-Garcia and Wang, Chapter 9), especially to those with anti-gram-negative or broad spectra. These antibiotics, especially at high concentrations, inhibit the growth and/or spore germination of many, though not all, *Phytophthora* spp. Some notable examples of the so-called antibacterial antibiotics known to

transferring agar blocks containing *Phytophthora* hyphae from advancing margins of the colony to a fresh selective medium plate is a necessary step in obtaining and later maintaining a pure culture in a suitable growth medium in test tube slants as stock cultures. Most antimicrobial agents used in selective media are fungistatic or bacteriostatic and will allow the inhibited propagules of the contaminants in seemingly "clean" agar areas of the original isolation plate to later germinate and grow in an ordinary growth medium, unless this purification process is first performed. Purification of sporangia and other propagules from contaminated soil extract or mineral salts solution can be similarly performed by picking off these reproductive structures and blotting them dry on absorbent paper before plating them onto a selective medium.

## QUANTITATION OF *PHYTOPHTHORA* PROPAGULES IN SOIL

The baiting techniques and selective agar media discussed above are primarily qualitative isolation tools, but each can be adapted to quantitation of *Phytophthora* propagules in soil when combined with a quantitative manipulation of the soil sample before isolation. The success of any quantitative method depends, of course, on the proper choice and use of the qualitative isolation techniques; the previous discussions on factors affecting detection and isolation of *Phytophthora* with the baiting techniques and selective media are, therefore, applicable also to this section on quantitation. Only additional important aspects related to quantitative procedures will be discussed here.

### With Baiting Techniques

#### Serial Dilution End Point Method

The serial dilution end point (SDEP) method can be effectively used in conjunction with appropriate baits to obtain quantitative data on the *Phytophthora* disease potential (or inoculum potential) in soil. The method was devised in the late 1950s by Tsao (1960) at a time when no quantitative methods of any kind were available for studying the ecology of *P. citrophthora* and *P. parasitica* in citrus soils. Briefly, the method consists of using lemon fruit as bait to detect the citrus phytophthoras in soil samples serially diluted (in twofold dilutions of 1/2, 1/4,... 1/64) with a sterile soil and incubated under standardized conditions. Soils with different degrees of *Phytophthora* infestation produce different end points of fruit infection in the dilution series. From these end points, levels of disease potential can be estimated by the disease potential index (DPI), defined as the reciprocal of the highest dilution that produces infection on the bait under the test conditions. A soil with a DPI of 16 is considered to have a disease potential about eight times as high as that of a soil with a DPI of 2. The method is similar in principle to that of Nusbaum et al (1952).

The SDEP method has since been used or adapted by other workers in conjunction with various appropriate baits to obtain quantitative data on a number of other *Phytophthora* spp. in soil: *P. cactorum* (Banihashemi and Mitchell, 1975), *P. cinnamomi* (Greenhalgh, 1978; Marks et al, 1972, 1975; Weste and Marks, 1974; Weste and Ruppin, 1975), *P. megasperma* (Marks and Mitchell, 1970; Pratt and Mitchell, 1973), *P. palmivora* (Newhook and Jackson, 1977; Onesirosan, 1971), *P. parasitica* var. *nicotianae* (Jenkins, 1962), and *P. parasitica* (Grimm and Alexander, 1973). The dilutions were extended upward to 1/256 or even higher for some species and with some soils. Certain other modifications in the procedures or terms were reported by some of these workers, such as: 1) the use of uneven (1/3, 1/5, 1/9, etc.) instead of serial dilutions by Mitchell and his associates (Banihashemi and Mitchell, 1975; Pratt and Mitchell, 1973); 2) weighing instead of equally dividing the diluted soil samples by Grimm and Alexander (1973); 3) mixing in a rotating container (e.g., Greenhalgh, 1978); 4) use of an inoculum potential index (IPI) by Onesirosan (1971); and 5) use of a population density index (PDI) by Marks et al (1972, 1975) and by Weste and Ruppin (1975).

The success of the SDEP method for quantitation, as in qualitative isolation, depends greatly on the incubation conditions, which should favor sporulation, germination, and bait infection by the target *Phytophthora* to be quantified. However, as accuracy and reproducibility of a quantitative method necessitate the use of a standardized, but not necessarily optimal, set of conditions (i.e., the same unvarying incubation temperature, diluent soil, termination time, etc., for each sample and in every test), *Phytophthora* propagules in a given sample container may not culminate in bait infection and thus may not be positively scored (Tsao, 1960). This is especially so if the populations in the samples are extremely low. Therefore, lack of bait infection in a dilution does not necessarily indicate absence of the fungus in the soil of that dilution. Likewise, a DPI of zero for a soil sample does not always indicate absence of the fungus in that soil sample.

Chemical factors in soil and susceptibility of the bait play roles as important as physical incubation conditions in *Phytophthora* recovery from soil with baiting techniques, and the importance of a proper diluent soil and a suitable bait in the SDEP method cannot be overemphasized. Unless the ecological study requires the use of the same soil (as the soil sample to be studied, except sterilized) as the diluent soil, routine quantitations of a large number of diverse soil samples should involve a diluent soil of near-neutral pH, low in organic content, and an easy-to-mix physical property. That the disease potential of a *Phytophthora* sp. in soil is strongly influenced by the diluent soil can be well illustrated by the following examples. The five soils (numbers 59–63) in Table 1 of Tsao (1960) and Table 2 of Tsao and Garber (1960) had been adjusted to about pH 4.5, 5, 6, 7, and 7.5, respectively, before they were infested with identical amounts of *P. parasitica* inoculum. The SDEP test was performed using the five sterilized soils, each with its respective pH value, as diluent soils—one for each of the five soil samples, so as to maintain the original soil pH of the samples. Although the inoculum density of *P. parasitica* in the samples was identical, the DPIs produced were 0, 1.5, 8, 8, and 8, respectively, in the five soils of different pH (Tsao, 1960). The DPIs of 0 and 1.5 were estimates of the disease-producing ability, or the disease potential (or inoculum potential), of the fungus under the given environmental conditions and on a given susceptible host and were not at all indications of inoculum density. In a subsequent test (P. H. Tsao, *unpublished*), the disease potential of soil 59 (pH 4.5) was again tested using two sterile soils of different pH (4.5 and 7) as the diluent soils; the difference in soil chemical environments generated by the two different diluent soils produced DPIs of 0 and 8, respectively. Similarly, in a study by Marks and Kassaby (1974), the eucalyptus cotyledons and lupine roots, used as baits to detect *P. cinnamomi* from two identical soils, produced DPIs of 16 and 2, respectively. These data convincingly underscore the need for using the proper term—disease potential index or inoculum potential index (Onesirosan, 1971)—when performing the SDEP method and for avoiding the use of this method to assess the inoculum density or population density of *Phytophthora* spp. in soil. Under identical conditions, however, the DPI should be directly correlated with the inoculum density of the pathogen in soil.

**Most Probable Number Method**

The most probable number (MPN) method is also known as the dilution frequency or maximum likelihood method. Widely used in determining bacterial populations in soil and water, it has been reported in earlier years as also applicable in quantitation of certain soilborne plant-pathogenic fungi such as species of *Fusarium, Thielaviopsis,* and *Gaeumannomyces* when combined with a suitable bait. Briefly, the method consists of preparing a series of (usually 10-fold) aqueous dilutions and assaying each in large numbers of replicates with an appropriate bait. The "code number" or "significant number" is chosen from the positive-reading replicates of consecutive dilutions, and the MPN is obtained from a precalculated table. Multiplying this number by the dilution factor gives the MPN of fungus propagules per unit weight of soil in the original sample. Menzies (1963) reviewed the earlier papers on the MPN method as applied to soilborne fungi.

Because of the time-consuming nature of the method and, perhaps, the recent availability of other more efficient quantitative methods such as the soil dilution plate method with selective media, the adaptation of the MPN method to *Phytophthora* quantitation has been infrequent. It has been reported for use on *P. parasitica* with carnation petals as bait (Ricci, 1972) and on *P. fragariae* with rooted strawberry runners as bait (Duncan, 1976, 1977). (Papavizas et al [1981] also attempted to use the MPN method to quantify *P. capsici* in soil but performed the tests in combination with a selective agar medium [see next section] instead of a bait.) The MPN analysis is generally deemed not sufficiently sensitive to accurately quantify *Phytophthora* in soil, especially in low populations. Ricci (1972) reported that, under his test conditions, only 30% accuracy could be obtained, but Duncan (1976) found it more useful and more informative than the SDEP test, which produces DPI data. Ricci (1974) discussed further the factors affecting accuracy of the MPN method along with a mathematical treatment of the analysis.

## With Selective Agar Media

Unlike the baiting techniques, the success of which relies on bait susceptibility and incubation conditions favorable for *Phytophthora* sporulation, germination, and infection on the bait, selective agar media allow direct germination of "germinable" propagules of any form (resistant mycelial fragment, oospore, chlamydospore, sporangium, and zoospore) and establishment of a colony on the medium. Therefore, unlike the SDEP and the MPN methods, both of which employ baiting techniques and are designed to measure disease potential or inoculum potential of *Phytophthora* in soil, quantitative methods employing selective agar media are capable of measuring inoculum densities. In addition, selective media allow quantitation when the large quantities of soil samples usually needed in baiting techniques are not available.

**Soil Dilution Plate Method**

Johnson and Curl (1972) discussed the basic principles, various procedures, and factors influencing the performance of the soil dilution plate (SDP) method for quantitation of a number of soilborne fungal pathogens, most of which are found in soil at much higher populations than *Phytophthora* spp. Specific points on the SDP method as related to *Phytophthora* are discussed below.

Generally considered the most accurate method and by far the most common for quantitation of soil populations of *Phytophthora,* the SDP method is effective when *Phytophthora* inoculum densities in the soil samples are relatively high (e.g., greater than 10 propagules per gram [ppg] of soil on a dry-weight basis). It is especially useful in quantitation of artificially infested experimental soils containing high *Phytophthora* populations. Naturally infested field soils usually contain low *Phytophthora* populations, often less than 1 ppg, and therefore require extremely low dilutions (e.g., 1/10 or lower) to increase the chance of *Phytophthora* detection on each dilution plate. Unless the antimicrobial agents in the selective medium are extremely effective against the myriad undesired microflora in the soil suspension of low dilution (i.e., of high soil concentration), the SDP method is not deemed an efficient method for *Phytophthora* quantitation in most naturally infested soils.

Because *Phytophthora* propagule distribution in natural field soils is generally uneven, the soil sample used for making the initial dilution should not be less than 10 g for each replicate dilution series. Most workers used 10 g or more for a number of diverse *Phytophthora* spp. (e.g., Kliejunas and Nagata, 1979; McIntosh, 1977; Ponchet et al, 1972; Shea et al, 1980). For uniformly mixed, artificially infested soils with high *Phytophthora* populations, 2–3 g may be considered sufficient. Most workers used sterile water as diluent, but some chose to use weak agar solution to minimize propagule settling in the soil suspension (e.g., Flowers and Hendrix, 1972; Otrosina and Marx, 1975). Although dilutions of 1/25 or higher have been used successfully with naturally infested soils with high *Phytophthora* populations (e.g., Flowers and Hendrix, 1969, 1972; Medeiros, 1976; Tsao and Ocana, 1969), 1/10 or lower dilutions were necessary for *Phytophthora* quantitation in most natural soils (e.g., Hendrix and Kuhlman, 1965; Kliejunas and Nagata, 1979; McIntosh, 1977; Otrosina and Marx, 1975; Papavizas et al, 1981; Ponchet et al, 1972; Shea et al, 1980; Sneh, 1972). As a rule, lower soil dilutions result in greater interference from undesired microflora developing on dilution plates and demand a highly efficient selective medium.

An overwhelming majority of workers apply the soil suspension, usually 1 ml per plate but sometimes more, onto the surface of solidified selective agar medium, then spread the suspension with the aid of a "suitable" tool such as a soup spoon, "hockey stick," or "rubber policeman." A uniform distribution of propagules on the medium surface is difficult to achieve, however, especially in a large experiment when speed is essential and an extra hand to spread the suspension is not available. Furthermore, unless the medium in the plate has been "predried" at a high temperature, often the day before, the water in the soil suspension added to the plate might not be quickly absorbed by the agar medium and thus might accumulate at one side of the plate for a considerable length of time. This might allow the formation and release of zoospores from the preexisting sporangia in the suspension, which would adversely affect the quantitation results. This surface application procedure, however, has the advantage of allowing the heavy soil suspension to be washed off with water one to several days after plating, to facilitate the detection and counting of *Phytophthora* colonies on the medium after further incubation (e.g., Flowers and Hendrix, 1969; Hendrix and Kuhlman, 1965; Papavizas et al, 1981; Shea et al, 1980; Shew et al, 1979b). A few workers (e.g., Masago et al, 1977; Ponchet et al, 1972; Tsao and Ocana, 1969) applied the soil suspension before pouring the cooled (43–45°C) molten agar medium to the plates, which were then rotated back and forth for uniform mixing. This procedure allows a better separation and more uniform distribution of *Phytophthora* propagules in the plate but runs the risk of the high agar temperature killing the sensitive propagules. In an experiment comparing these two procedures in quantitation of *P. parasitica* populations in soil, colony counts obtained with the surface application procedure were less than half those obtained with the procedure that involved mixing with cooled molten agar. This result was achieved if and when the temperature of the molten agar medium was precisely controlled, by the use of a water bath, at 43°C at the time of mixing with stock solution of antimicrobial agents and pouring the plates (P. H. Tsao, *unpublished*). Unless precise control of agar temperature is feasible, this latter procedure of applying the soil suspension is not recommended.

**Other Quantitative or Semiquantitative Methods**

Warcup's soil plate method (see Johnson and Curl, 1972) and the soil sprinkle (or direct inoculation) method (Tsao and Guy, 1977; Tsao and Ocana, 1969; Zentmyer, 1980), which are used primarily for qualitative isolation, can be adapted for quantitation of *Phytophthora* with selective media. The results obtained therefrom are understandably only semiquantitative in nature because of imprecise control of sample size and nonuniform distribution of propagules. Soil moisture also greatly affects the ease of spreading soil particles evenly on the plate with the soil sprinkle method. Papavizas et al (1981) performed the MPN test in combination with a selective medium but found the method too insensitive to detect populations of *P. capsici* less than 46 ppg.

## SELECTION, COLLECTION, TRANSPORT, STORAGE, AND PREISOLATION TREATMENTS OF SAMPLES

Negative recovery of *Phytophthora* from tissues or soils might not always be due to improper choice of isolation techniques or faulty isolation procedures in the laboratory. It could be simply due to the absence of *Phytophthora* in the selected sample at the time of collection or the loss of viable propagules during transport or storage if the samples are improperly handled.

Aside from certain ecological studies, e.g. on population distribution, in which a rigid or random scheme, depending on the experiment, must be followed, soil samples in a diagnostic study or disease survey can be discriminately selected to maximize *Phytophthora* recovery. Soils near the root zone or next to other underground plant parts are expected to harbor higher populations. Because *Phytophthora* succumbs easily to dryness and because soil near the ground surface is more subject to high temperature from solar radiation and to drying, soil samples for *Phytophthora* isolation should be chosen from at least 3–5 cm below the soil surface. For similar reasons, soil at the north or northeast corner (in the northern hemisphere) of a tree under foliage canopy would be the most desirable.

Although outside the realm of the topic of soil isolation, a brief discussion of selection and collection of infected plant tissues is pertinent here. *Phytophthora* infections on stems and roots result in discoloration and are usually restricted to the bark or cortical regions, with hyphae extending internally to the cambium but seldom (at most a few cells deep) into the wood. Dark lesions in the wood are largely from secondary invaders. *Phytophthora* lesions on fruits, leaves, and petioles are almost always brown, purplish brown, or almost black and are firm in texture. Soft, water-soaked lesions are probably not those produced by *Phytophthora* or may already have been invaded by secondary microorganisms. Proper recognition of *Phytophthora* symptoms of other kinds also improves the chances of isolation.

As with any fungus, *Phytophthora* would lose its viability if the collected tissue or soil samples were exposed to drying or high temperature during transit before reaching the laboratory. This often happens, especially in an enclosed vehicle in hot summer or in the tropics. Samples should be collected in plastic bags or similar closed containers. Loss of viability during sample transport can be avoided by the use of a portable, insulated "ice chest" or similar container. Most *Phytophthora* spp. are sensitive to extreme cold as well as heat, and excessively low temperatures during sample transport are also to be avoided Therefore, samples should not be in direct contact with the ice and should be placed on a raised rack to prevent contact with the icy water from melted ice accumulated at the bottom. Volumes of ice in the container need not be large, because the purpose of the insulated container is only to maintain a cool, nonlethal temperature during transit.

For sample storage, before isolation experiments, both excessively high and low temperatures should be avoided. The general practice of storing samples in a "refrigerator" (4–6°C) to reduce growth of secondary bacteria and fungi also reduces the chance of *Phytophthora* recovery because of its sensitivity to extreme cold. A cool incubator of 10–15°C is generally more desirable than a cold refrigerator. Needless to say, lengthy storage at any temperature is not desirable, and chance of recovery is best if isolation is performed immediately after samples are collected.

Under certain circumstances, such as with extremely dry soil samples, a preisolation moistening of the sample might enhance *Phytophthora* recovery. Chee and Newhook (1965) premoistened the dry soils for 28 hr, and Okaisabor (1971) incubated his soils at 20% water content for 5 days before testing. Ioannou and Grogan (1977) and Canaday and Schmitthenner (1978) both incubated dry soil samples at field capacity for 7 days before flooding the soil for baiting tests. The reasons for this increase in *Phytophthora* recovery are not well known but are presumably the result of activation of the germination process of the dormant propagules (chlamydospores or oospores) and/or formation of sporangia in the wet soil. Premoistening is a useful treatment for diagnostic isolation but would not be desirable in quantitative soil assays because premature population changes might be induced in the sample. Because germination of a single chlamydospore or oospore in wet soil could result in formation of more than one "germ-sporangium" and perhaps numerous zoospores, premoistening increases the inoculum density of *Phytophthora* appreciably before the quantitative assay.

Wet-sieving, a preisolation treatment to facilitate recovery of chlamydospores, mainly of *P. cinnamomi*, from large quantities of soil, was first used by McCain et al (1967) and later adapted by a number of other workers (e.g., Brodrick et al, 1976; Greenhalgh, 1978; Kliejunas and Nagata, 1980; Shew et al, 1979b). Overreliance on the 38-$\mu$m pore-size sieve, used by almost all workers, invariably results in loss of *Phytophthora* propagules smaller than 38 $\mu$m during the wet-sieving process and adversely affects quantitation. Kliejunas and Nagata (1980) reported that the recovery of *P. cinnamomi* from wet-sieved soil samples was less than 20% of that from samples not wet-sieved. Wet-sieving is not a desirable or recommendable procedure for quantitative assays of *Phytophthora* populations in soil.

## CONCLUSIONS

Historically one of the most difficult fungal groups to be isolated from old diseased tissues and soils, *Phytophthora* fungi are now among the easiest to isolate because of the recent availability of excellent techniques. However, even with the varied baiting techniques now at our disposal and more than 20 yr since the development of the first group of selective media, many pathologists in different parts of the world still use such media as potato-dextrose agar or water agar for the isolation of *Phytophthora.* Although positive recovery with these archaic techniques may occur with newly infected tissues, these workers often do not realize the meaninglessness of the negative recoveries they encounter in their endeavors. Failure to recover *Phytophthora* because of inadequate isolation techniques may result in an incorrect diagnosis of a disease problem. A falsely low recovery in a quantitative soil study caused by inadequate enumeration techniques may result in an improper conclusion in a fungicidal treatment, or biological control, soil amendment, or other control treatment of the soil or in a faulty conclusion in soil ecological studies. Difficulties and obstacles notwithstanding, proper techniques accompanied by appropriately chosen procedures and test conditions, therefore, are imperative for any well-planned study on diagnosis or disease control.

Many excellent baiting techniques for isolating *Phytophthora* have been discussed in this chapter, and the number of selective *Phytophthora* isolation media in the literature has increased fourfold since the last review on the topic (Tsao, 1970). Other methods of detection and counting of *Phytophthora* propagules, utilizing an immunofluorescence technique, have also been devised (Malajczuk et al, 1975, 1978). However, they have been used successfully, or sometimes only adequately, for less than half of the 40-odd species in this notorious genus (Tables 1 and 2). Success on other *Phytophthora* spp. awaits further testing or development of new media or techniques. Some of the existing selective media were so hastily and nonscientifically devised that their lack of success for isolation from soil is not surprising. Future development of new media will, I hope, benefit from past mistakes and from discussions in this review. It also must be realized that no single bait or selective medium is perfect or even adequate for all species of *Phytophthora.* Blind adaptation to the target species of a technique used for other species, without preliminary testing and modifications, often leads to less than satisfactory results. It is hoped that further development and judicious use of selective isolation and quantitation techniques for *Phytophthora* will advance our understanding of the ecology of these important plant pathogens, especially in the areas of inoculum density, population dynamics, inoculum potential, saprophytic behavior and survival in soil, and the intricate relationships of these to biological control.

## SUMMARY

Detection or isolation of *Phytophthora* fungi from old, badly decayed or dried, infected plant tissues or naturally infested soils is extremely difficult or impossible unless special techniques are used. The difficulty is usually due to fast growth of the associated microflora and to the slow emergence of surviving, resistant *Phytophthora* propagules (e.g., chlamydospores) in decayed tissue or soil. Proper choice and use of plant parts (e.g., fruits, seedlings, leaves, or leaf disks of the host plants) as baits can usually succeed in selectively isolating target *Phytophthora* from soil in a water mixture. Success depends on susceptibility of the baiting material and on incubation conditions that maximally enhance the formation of zoospores, the primary infective propagules colonizing the baits. High water/soil ratios (4:1 or greater) cause a dilution of inhibitory chemical factors in soil, thus enhancing the formation of sporangia and zoospores of *Phytophthora* in the soil-water mixture and subsequent infection of the bait.

Selective media have become widely used for isolation and quantitation of many *Phytophthora* spp. An efficient selective medium employs a relatively weak agar medium as the basal medium (which minimizes bacterial growth yet allows adequate but not luxuriant *Phytophthora* growth) and contains appropriate concentrations of several properly chosen, selective antimicrobial agents, each inhibitory to a wide spectrum of undesired nontarget fungi or bacteria, but nontoxic to *Phytophthora.* The polyene antibiotic, pimaricin, is an ideal selective antifungal antibiotic; other antifungal agents such as benomyl, hymexazol, nystatin, and pentachloronitrobenzene are useful substitutes and/or supplements. Suitable antibacterial antibiotics for *Phytophthora* isolation media are ampicillin, penicillin, rifampicin, and vancomycin or their combinations, but not chloramphenicol, chlortetracycline, novobiocin, oxytetracycline, polymyxin, streptomycin, or tetracycline. These latter antibacterial agents are also inhibitory to *Phytophthora* growth and/or spore germination and, therefore, should not be used for *Phytophthora* isolation.

Selective chemicals should not be used at excessively high concentrations, which, though not inhibitory to mycelial growth, might inhibit germination of dormant *Phytophthora* propagules responsible for survival in soil or in decayed tissue. Incubation of isolation plates at relatively low temperature (e.g., 18° C) generally enhances *Phytophthora* recovery, due to growth suppression of undesired associated bacteria.

Both the baiting method and selective medium can be adapted to quantitation of *Phytophthora* in soil by combining each with a quantitative manipulation of the soil sample. The serial dilution end point method and the most probable number method have been used in conjunction with appropriate baits to obtain quantitative data on *Phytophthora* disease potential (or inoculum potential, but not inoculum density) in soil. Inoculum density data are best obtained by the use of a selective medium in conjunction with the soil dilution plate method.

## ACKNOWLEDGMENTS

I thank Leslie Bower Batick for her excellent assistance in every aspect of the manuscript preparation and Pamela W. Tsao for her patience and encouragement as well as her assistance during the preparation of this chapter.

## LITERATURE CITED

Alavi, A., and Strange, R. N. 1979. A baiting technique for isolating *Phytophthora drechsleri,* causal agent of crown rot of *Cucumis* species in Iran. Plant Dis. Rep. 63:1084-1086.

Alconero, R., Albuquerque, F., Almeyda, N., and Santiago, A. G. 1972. Phytophthora root rot of black pepper in Brazil and Puerto Rico. Phytopathology 62:144-148.

Anderson, E. J. 1951. A simple method for detecting the presence of *Phytophthora cinnamomi* Rands in soil. Phytopathology 41:187-189.

Banihashemi, Z. 1970. A new technique for isolation of *Phytophthora* and *Pythium* species from soil. Plant Dis. Rep. 54:261-262.

Banihashemi, Z., and Mitchell, J. E. 1975. Use of safflower seedlings for the detection and isolation of *Phytophthora cactorum* from soil and its application to population studies. Phytopathology 65:1424-1430.

Banihashemi, Z., and Mitchell, J. E. 1976. Factors affecting oospore germination in *Phytophthora cactorum,* the incitant of apple collar rot. Phytopathology 66:443-448.

Broadbent, P., and Baker, K. F. 1974. Behaviour of *Phytophthora cinnamomi* in soils suppressive and conducive to root rot. Aust. J. Agric. Res. 25:121-137.

Brodrick, H. T., Zentmyer, G. A., and Wood, R. 1976. Comparison of various methods for the isolation of *Phytophthora cinnamomi* from avocado soils. Calif. Avocado Soc. Yearb. 59:87-91.

Butler, E. E., and Hine, R. B. 1958. Use of novobiocin for isolation of fungi from soil. Soil Sci. 85:250-254.

Campbell, W. A. 1949. A method of isolating *Phytophthora cinnamomi* directly from soil. Plant Dis. Rep. 33:134-135.

Canaday, C. H., and Schmitthenner, A. F. 1978. A simple baiting technique for isolation of *Phytophthora megasperma* var. *sojae* from soil. (Abstr.) Phytopathol. News 12:148.

Carnes, D. W. 1976. The sensitivity of *Phytophthora palmivora* to chloramphenicol. Ph.D. dissertation. Univ. Calif., Riverside. 75 pp.

Castillo G., E., and Fernandez-Northcote, E. N. 1974. Nuevo medio selectivo para el aislamiento de especies de *Phytophthora* y *Pythium*. Fitopatologia 9:1-7.

Chee, K. H., and Foong, K. M. 1968. Use of cacao pod for recovering *Phytophthora* species pathogenic to *Hevea brasiliensis*. Plant Dis. Rep. 52:5.

Chee, K. H., and Newhook, F. J. 1965. Improved methods for use in studies on *Phytophthora cinnamomi* Rands and other *Phytophthora* species. N.Z. J. Agric. Res 8:88-95.

Dance, M. H., Newhook, F. J., and Cole, J. S. 1975. Bioassay of *Phytophthora* spp. in soil. Plant. Dis. Rep. 59:523-527.

Domsch, K. H., and Schwinn, F. J. 1965. Nachweis und isolierung von Pflanzenpathogenen bodenpilzen mit selektiven verfahren. Zentralbl. Bakteriol., Suppl. 1:461-485.

Duncan, J. M. 1976. The use of bait plants to detect *Phytophthora fragariae* in soil. Trans. Br. Mycol. Soc. 66:85-89.

Duncan, J. M. 1977. On the use of maximum likelihood analyses in strawberry red core detection. Trans. Br. Mycol. Soc. 68:127.

Duncan, J. M. 1979. In situ bait plants as a method for detecting red core in strawberry fields. Ann. Appl. Biol. 92:307-312.

Eckert, J. W., and Tsao, P. H. 1960. A preliminary report on the use of pimaricin in the isolation of *Phytophthora* spp. from root tissues. Plant Dis. Rep. 44:660-661.

Eckert, J. W., and Tsao, P. H. 1962. A selective antibiotic medium for isolation of *Phytophthora* and *Pythium* from plant roots. Phytopathology 52:771-777.

El-Goorani, M. A., and Tarabeih, A. M. 1973. A note on the effect of certain antibacterial antibiotics on certain pythiaceous fungi. Phytopathol. Mediterr. 12:92-93.

Ersek, T. 1975. The sensitivity of *Phytophthora infestans* to several antibiotics. Z. Pflanzenkr. Pflanzenschutz 82:614-617.

Eye, L. L., Sneh, B., and Lockwood, J. L. 1976. Isolation and identification of *Phytophthora megasperma* var. *sojae* from soil using soybean seedlings as baits. (Abstr.) Proc. Am. Phytopathol. Soc. 3:239.

Fatemi, J. 1974. A simple technique for isolation of *Phytophthora* and *Pythium* species from soil. Phytopathol. Mediterr. 13:120-121.

Flowers, R. A., and Hendrix, J. W. 1969. Gallic acid in a procedure for isolation of *Phytophthora parasitica* var. *nicotianae* and *Pythium* spp. from soil. Phytopathology 59:725-731.

Flowers, R. A., and Hendrix, J. W. 1972. Population density of *Phytophthora parasitica* var. *nicotianae* in relation to pathogenesis and season. Phytopathology 62:474-477.

Follin, J. C. 1971. L'utilisation du benomyl pour l'isolement sélectif des Pythiacées. Coton Fibres Trop. (Fr. Ed.) 26:467.

Frosheiser, F. I. 1967. Phytophthora root rot of alfalfa in Minnesota. Plant Dis. Rep. 51:679-681.

Fujisawa, Y., and Masago, H. 1975. Investigation on the selective medium for isolation of *Phytophthora* spp. (Abstr.) Ann. Phytopathol. Soc. Jpn. 41:267. (In Japanese)

Gerrettson-Cornell, L. 1974. A comparative test of isolation of *Phytophthora cinnamomi* Rands between the lupin baiting and a newly devised apple trap. Phyton 32:35-36.

Gerrettson-Cornell, L. 1976. Effect of different methods of water demineralization on the recovery of *Phytophthora cinnamomi* Rands by lupin baiting. Phyton 34:85-88.

Gerrettson-Cornell, L., and Townsend, S. R. 1974. A study of a lupin baiting technique for isolation of *Phytophthora cinnamomi* Rands under different lighting conditions. Phyton 32:39-43.

Green, R. J., and Pratt, R. G. 1970. Root rot of black walnut seedlings caused by *Phytophthora citricola*. Plant Dis. Rep. 54:583-585.

Greenhalgh, F. C. 1978. Evaluation of techniques for quantitative detection of *Phytophthora cinnamomi*. Soil Biol. Biochem. 10:257-259.

Grimm, G. R., and Alexander, A. F. 1973. Citrus leaf pieces as traps for *Phytophthora parasitica* from soil slurries. Phytopathology 63:540-541.

Haas, J. H. 1964. Isolation of *Phytophthora megasperma* var. *sojae* in soil dilution plates. (Abstr.) Phytopathology 54:894.

Hansen, E. M., Hamm, P. B., Julis, A. J., and Roth, L. F. 1979. Isolation, incidence and management of *Phytophthora* in forest tree nurseries in the Pacific Northwest. Plant Dis. Rep. 63:607-611.

Hargreaves, A. J., and Duncan, J. M. 1978. Detection of *Phytophthora* species in field soils by simple baiting procedures. Soil Biol. Biochem. 10:343-345.

Hendrix, F. F., and Kuhlman, E. G. 1965. Factors affecting direct recovery of *Phytophthora cinnamomi* from soil. Phytopathology 55:1183-1187.

Holliday, P., and Mowat, W. P. 1963. Foot rot of *Piper nigrum* L. (*Phytophthora palmivora*). Phytopathol. Pap. 5. Commonw. Mycol. Inst., Kew, Surrey, England. 62 pp.

Hollomon, D. W. 1965. A medium for the direct isolation of *Phytophthora infestans*. Plant Pathol. 14:34-35.

Ioannou, N., and Grogan, R. G. 1977. The influence of soil matric potential on the production of sporangia by *Phytophthora parasitica* in relation to its isolation from soil by baiting techniques. (Abstr.) Proc. Am. Phytopathol. Soc. 4:173.

Jenkins, S. F. 1962. Preliminary studies for estimating the disease potential of *Phytophthora parasitica* var. *nicotianae* in infested tobacco soils. Plant Dis. Rep. 46:825-826.

Johnson, L. F., and Curl, E. A. 1972. Methods for Research on the Ecology of Soil-Borne Plant Pathogens. Burgess Publishing Co., Minneapolis, MN. 247 pp.

Julis, A. J., Clayton, C. N., and Sutton, T. B. 1979. Occurrence of *Phytophthora* sp. in North Carolina apple orchard soils. Plant Dis. Rep. 63:147-150.

Kannwischer, M. E., and Mitchell, D. J. 1978. The influence of a fungicide on the epidemiology of black shank of tobacco. Phytopathology 68:1760-1765.

Katsura, K. 1971. Phytophthora Diseases of Plants—Theory and Practice. Seibundo Shinkosha, Tokyo. 128 pp.

Klemmer, H. W., and Nakano, R. Y. 1962. Techniques in isolation pythiaceous fungi from soil and diseased pineapple tissue. Phytopathology 52:955-956.

Kliejunas, J. T., and Nagata, J. T. 1979. *Phytophthora cinnamomi* in Hawaiian forest soils: Seasonal variations in population levels. Phytopathology 69:1268-1272.

Kliejunas, J. T., and Nagata, J. T. 1980. *Phytophthora cinnamomi* in Hawaiian forest soils: Technique for enumeration and types of propagules recovered. Soil Biol. Biochem. 12:89-91.

Klotz, L. J., and DeWolfe, T. A. 1958. Techniques for isolating *Phytophthora* spp. which attack citrus. Plant Dis. Rep. 42:675-676.

Ko, W. H., Chang, H. S., and Su, H. J. 1978. Isolates of *Phytophthora cinnamomi* from Taiwan as evidence for an Asian origin of the species. Trans. Br. Mycol. Soc. 71:496-499.

Lacey, J. 1965. The infectivity of soils containing *Phytophthora infestans*. Ann. Appl. Biol. 56:363-380.

Lee, B. S., and Varghese, G. 1974. Studies on the genus *Phytophthora* in Malaysia. I. Isolation techniques, comparative morphology and physiology and reaction to antibiotics. Malays. Agric. Res. 3:13-21.

Linderman, R. G., and Zeitoun, F. 1977. *Phytophthora cinnamomi* causing root rot and wilt of nursery-grown native western azalea and salal. Plant Dis. Rep. 61:1045-1048.

Malajczuk, N., McComb, A. J., and Parker, C. A. 1975. An immunofluorescence technique for detecting *Phytophthora cinnamomi* Rands. Aust. J. Bot. 23:289-309.

Malajczuk, N., Bowen, G. D., and Greenhalgh, F. C. 1978. A combined fluorescent antibody and soil sieving technique to count chlamydospores of *Phytophthora cinnamomi* in soil. Soil Biol. Biochem. 10:437-438.

Marks, G. C., and Kassaby, F. Y. 1974. Detection of *Phytophthora cinnamomi* in soils. Aust. For. 36:198-203.

Marks, G. C., and Mitchell, J. E. 1970. Detection, isolation, and pathogenicity of *Phytophthora megasperma* from soils and estimation of inoculum levels. Phytopathology 60:1687-1690.

Marks, G. C., Kassaby, F. Y., and Reynolds, S. T. 1972. Die-back in mixed hardwood forests of eastern Victoria: A preliminary report. Aust. J. Bot. 20:141-154.

Marks, G. C., Kassaby, F. Y., and Fagg, P. C. 1975. Variation in population levels of *Phytophthora cinnamomi* in eucalyptus forest soils of eastern Victoria. Aust. J. Bot. 23:435-449.

Masago, H., Yoshikawa, M., Fukada, M., and Nakanishi, N. 1977. Selective inhibition of *Pythium* spp. on a medium for direct isolation of *Phytophthora* spp. from soils and plants. Phytopathology 67:425-428.

McCain, A. H., Holtzmann, O. V., and Trujillo, E. E. 1967. Concentration of *Phytophthora cinnamomi* chlamydospores by soil sieving. Phytopathology 57:1134-1135.

McIntosh, D. L. 1964. *Phytophthora* spp. in soils of the Okanagan and Similkameen Valleys of British Columbia. Can. J. Bot. 42:1411-1415.

McIntosh, D. L. 1975. An improved agar medium for isolating *Phytophthora cactorum* from soil. Can. J. Bot. 53:1444-1445.

McIntosh, D. L. 1977. *Phytophthora cactorum* propagule density levels in orchard soils. Plant Dis. Rep. 61:528-532.

Medeiros, A. G. 1976. Sporulation of *Phytophthora palmivora* (Butl.) Butl. in relation to epidemiology and chemical control of cacao black pod disease. Ph.D. dissertation. Univ. Calif., Riverside. 220 pp.

Menzies, J. D. 1963. The direct assay of plant pathogen populations in soil. Annu. Rev. Phytopathol. 1:127-142.

Mircetich, S. M. 1970. Inhibition of germination of chlamydospores of *Phytophthora cinnamomi* by some antimicrobial agents in *Phytophthora* selective media. Can. J. Microbiol. 16:1227-1230.

Mircetich, S. M., and Matheron, M. E. 1976. Phytophthora root and crown rot of cherry trees. Phytopathology 66:549-558.

Molot, P. M., and Beyries, A. 1976. Contamination in vitro de la fraise et de la tomate immature par diverses espèces de *Phytophthora.* Ann. Phytopathol. 8:91-94.

Molot, P. M., and Nourrisseau, J. G. 1972. Mise en évidence et étude de la conservation du *Phytophthora cactorum* dans le sol (travaux réalisés sur le fraisier). Pages 87-92 in: Actas Congr. Unione Fitopatol. Mediterr., 3rd, Oeiras.

Narasimhan, V., and Ramakrishnan, K. 1969. Studies on wilt disease of betelvine. Madras Agric. J. 56:68-73.

Newhook, F. J., and Jackson, G. V. H. 1977. *Phytophthora palmivora* in cocoa plantation soils in the Solomon Islands. Trans. Br. Mycol. Soc. 69:31-38.

Nusbaum, C. J., Lucas, G. B., and Chaplin, J. F. 1952. Estimating the inoculum potential of *Phytophthora parasitica* var. *nicotianae* in the soil. (Abstr.) Phytopathology 42:286.

Ocana, G., and Tsao, P. H. 1965. Origin of colonies of *Phytophthora parasitica* in selective pimaricin media in soil dilution plates. (Abstr.) Phytopathology 55:1070.

Ocana, G., and Tsao, P. H. 1966. A selective agar medium for the direct isolation and enumeration of *Phytophthora* in soil. (Abstr.) Phytopathology 56:893.

Okaisabor, E. K. 1971. Studies on some aspects of the soil phase of *Phytophthora palmivora* (Butl.) Butl. Pages 484-489 in: Proc. Int. Cocoa Conf., 3rd, Accra, Ghana.

Onesirosan, P. T. 1971. The survival of *Phytophthora palmivora* in a cacao plantation during the dry season. Phytopathology 61:975-977.

Orellana, R. G. 1954. Contribution to the study of survival, dissemination and control of *Phytophthora* of cacao. Proc. Inter-Am. Cacao Conf., 5th, Turrialba, Costa Rica. Sec. Fitopatol. 1:1-10.

Ostrofsky, W. D., Pratt, R. G., and Roth, L. F. 1977. Detection of *Phytophthora lateralis* in soil organic matter and factors that affect its survival. Phytopathology 67:79-84.

Otrosina, W. J., and Marx, D. H. 1975. Populations of *Phytophthora cinnamomi* and *Pythium* spp. under shortleaf and loblolly pines in littleleaf disease sites. Phytopathology 65:1224-1229.

Papavizas, G. C., Bowers, J. H., and Johnston, S. A. 1981. Selective isolation of *Phytophthora capsici* from soils. Phytopathology 71:129-133.

Paxton, J. D. 1976. A new technique for isolating *Phytophthora megasperma* var. *sojae* from field soil. (Abstr.) Proc. Am. Phytopathol. Soc. 3:238.

Pegg, K. G. 1977. Soil application of elemental sulphur as a control of *Phytophthora cinnamomi* root and heart rot of pineapple. Aust. J. Exp. Agric. Anim. Husb. 17:859-865.

Perches, S., and Galindo, J. 1967. Supervivencia del *Phytophthora infestans* (Mont.) de Bary, causante del tizon tardio de la papa y jitomate. Agrociencia 1:92-98.

Ponchet, J., Ricci, P., Andreoli, C., and Auge, G. 1972. Méthodes sélectives d'isolement du *Phytophthora nicotianae* f. sp. *parasitica* (Dastur) Waterh. à partir du sol. Ann. Phytopathol. 4:97-108.

Pratt, B. H., and Heather, W. A. 1972. Method for rapid differentiation of *Phytophthora cinnamomi* from other *Phytophthora* species isolated from soil by lupin baiting. Trans. Br. Mycol. Soc. 59:87-96.

Pratt, R. G., and Mitchell, J. E. 1973. Conditions affecting the detection of *Phytophthora megasperma* in soils of Wisconsin alfalfa fields. Phytopathology 63:1374-1379.

Pratt, R. G., and Mitchell, J. E. 1975. The survival and activity of *Phytophthora megasperma* in naturally infested soils. Phytopathology 65:1267-1272.

Ribeiro, O. K. 1978. A Source Book of the Genus *Phytophthora.* Cramer, Vaduz, Liechtenstein. 417 pp.

Ricci, P. 1972. Moyens d'étude de l'inoculum du *Phytophthora nicotianae* f. sp. *parasitica* (Dastur) Waterh., parasite de l'oeillet dans le sol. Ann. Phytopathol. 4:257-276.

Ricci, P. 1974. Mesure de la densité d'inoculum d'un agent pathogène dans le sol à l'aide d'une technique d'isolement par "tout ou rien." Ann. Phytopathol. 6:441-453.

Schmitthenner, A. F. 1973. Isolation and identification methods for *Phytophthora* and *Pythium.* Pages 94-110 in: Proc. Woody Ornamental Dis. Workshop, 1st, 24–25 January 1973. Univ. Mo., Columbia. 128 pp.

Schwinn, F. J. 1961. Der Nachweis von *Phytophthora cactorum* (Leb. et Cohn) Schroet. im Boden. Mitt. Biol. Bundesanst. Land Forstwirtsch. Berlin-Dahlem 104:42-44.

Sewell, G. W. F., Wilson, J. F., and Dakwa, J. T. 1974. Seasonal variations in the activity in soil of *Phytophthora cactorum, P. syringae,* and *P. citricola* in relation to collar rot disease of apple. Ann. Appl. Biol. 76:179-186.

Shea, S. R., Gillen, K. J., and Leppard, W. I. 1980. Seasonal variation in population levels of *Phytophthora cinnamomi* Rands in soil in diseased, freely-drained *Eucalyptus marginata* Sm. sites in the northern jarrah forest of south-western Australia. Prot. Ecol. 2:135-156.

Shepherd, C. J., and Forrester, R. I. 1977. Influence of isolation method on growth rate characteristics of populations of *Phytophthora cinnamomi.* Aust. J. Bot. 25:477-482.

Shew, H. D., Benson, D. M., and Grand, L. F. 1979a. A comparison of baits for isolating *Phytophthora cinnamomi* from soil. (Abstr.) Phytopathology 69:532.

Shew, H. D., Benson, D. M., and Grand, L. F. 1979b. A quantitative soil assay for *Phytophthora cinnamomi.* (Abstr.) Phytopathology 69:1045.

Siradhana, B. S. 1972. Inhibition of zoosporangial germination in *Phytophthora nicotianae* var. *parasitica* by antibiotics. Indian Phytopathol. 25:146-147.

Sneh, B. 1972. An agar medium for the isolation and macroscopic recognition of *Phytophthora* spp. from soil on dilution plates. Can. J. Microbiol. 18:1389-1392.

Stovold, G. E., and Carratt, D. E. 1978. Occurrence of *Phytophthora megasperma* in alfalfa growing soils of New South Wales. Plant Dis. Rep. 62:742-744.

Tsao, P. H. 1960. A serial dilution end-point method for estimating disease potentials of citrus phytophthoras in soil. Phytopathology 50:717-724.

Tsao, P. H. 1970. Selective media for isolation of pathogenic fungi. Annu. Rev. Phytopathol. 8:157-186.

Tsao, P. H., and Garber, M. J. 1960. Methods of soil infestation, watering, and assessing the degree of root infection for greenhouse in situ ecological studies with citrus phytophthoras. Plant Dis. Rep. 44:710-715.

Tsao, P. H., and Guy, S. O. 1977. Inhibition of *Mortierella* and *Pythium* in a *Phytophthora*-isolation medium containing hymexazol. Phytopathology 67:796-801.

Tsao, P. H., and Menyonga, J. M. 1966. Response of *Phytophthora* spp. and soil microflora to antibiotics in the pimaricin-vancomycin

*Phytophthora*. The soil is diluted with sterilized soil in a series. Soil samples are saturated with distilled water or copper-free tap water, and a susceptible host is incubated at a temperature that is conducive to disease expression. In the case of *P. parasitica* or *P. citrophthora* in citrus soil, a lemon fruit in the "silver" stage is incubated in a container so that the fruit is about half submerged. After incubation, a visible lesion, characteristic of the disease, is evidence for expression of the "disease potential" as the reciprocal of the smallest fraction of soil on which the disease is detected.

Determination of a disease potential index and inoculum density for *Phytophthora* depends on the methods used. Because methodology is important for the interpretation of some of our experimental results that will be covered here, these methods will be described.

To determine the incidence of *P. cinnamomi* in a plant community, four samples (0.5 kg) of soil and fine roots were collected from each of four scattered diseased, but not dead, plants. Sampling on a grid basis is not recommended because *P. cinnamomi* is concentrated on roots and is not uniformly distributed. Soil samples were not refrigerated, because low temperatures usually favor other microorganisms. Each sample was spread evenly in the shape of a circle; subsamples were taken from alternate quadrants, and the process was repeated until the residue was of the required size (50 g). The mixed particle size in soil necessitates such a sampling method as described by Leeper (1967). Assays were made within a few hours of sample collection.

The serial dilution end point method of Tsao (1960) has been adapted to determine the disease potential index; the reciprocal of the smallest fraction (in our work, not diluted with sterile soil) of a 50-g sample of soil and roots in 100 ml of deionized distilled water from which *Phytophthora* could be isolated was used as the index. Three replicates (plastic cups) were therefore set up for a 50-g sample of soil and each fraction of it in a series up to 1/256. Three germinated seeds of either *Lupinus angustifolius* (Chee and Newhook, 1965) or *Eucalyptus sieberi* (Marks and Kassaby, 1974) with radicles 1 cm long were placed in each cup as bait (Weste and Ruppin, 1975). The seedling baits were incubated at 20–25° C for 48–72 hr in the saturated sample of soil and plated on selective agar. The lupine seeds were supported on perforated polystyrene foam floats with the radicles projecting through the perforations. The eucalypt seedlings were floated on the water.

Both baits gave identical results when tested for *P. cinnamomi*. The eucalypt seedlings had several advantages: they belong to the native flora, they floated without support, and the disappearance of the red pigment in cotyledons indicated infection; however, 4 wk were required for the seedlings to reach a suitable size. Lupine seedlings were ready for assay in only 3 days and were thus more frequently used.

After 48–72 hr, depending on the temperature, the eucalypt cotyledons or lupine radicles were excised, washed, dried carefully on sterile filter paper, and plated on selective antibiotic agar (Tsao et al, 1976). The index value was the reciprocal of the fraction; for example, if all samples were positive for isolation of *P. cinnamomi*, the index value was 256; if only those samples containing 50 g of soil were positive, the index value was 1.

This technique provides a measure of the disease potential of that particular sample for sporangial and zoospore formation under the field conditions of collection. It does not measure the potential due to propagules contained in large infected roots. The result may be distorted by the presence of antagonistic bacteria in the soil. The method gives a zero population in hot, dry weather and when soil temperatures fall below 10° C; however, this zero value cannot be interpreted as absence of the pathogen, because soil samples taken from the same source when soil temperatures are above 10° C or when the soil matric potential is adequate may have a high disease potential. In fact, on several occasions, disease potential indexes changed within 13 days of changes in soil temperature and within 2 days of changes in soil matric potential.

The most important procedures are 1) collection of soil samples from living hosts, 2) inclusion of fine roots in the samples, 3) drying of washed roots or cotyledons of the bait before plating to reduce numbers of bacteria (bacterial slime completely suppresses *Phytophthora*), and 4) culturing of the bait on selective antibiotic $P_{10}VP$ agar (Tsao, 1970) for positive identification of *P. cinnamomi*. The identification of *Phytophthora* species only by microscopic examination of roots or cotyledons of the bait is unsatisfactory because sporangial characters are variable and because sporangia may be suppressed or aborted by various antagonists in the soil.

Greenhalgh (1978) demonstrated a direct relationship between the number of chlamydospores of *P. cinnamomi* added to soils and the frequency of detection by baiting. More research is needed on calibration of pathogen populations by using several methods for a range of soils infested with a series of concentrations of inoculum for a defined range of environmental and seasonal fluctuations. Changes in pathogen populations with time, host, race of the pathogen, and environment may then be compared with the standard series.

Inoculum density may be measured directly by soil sieving. Samples of soil and roots (50 g, oven-dry basis) collected and subsampled as described were passed through a nest of sieves with different pore sizes (McCain et al, 1967). The contents of each sieve were plated separately in appropriate dilutions with water (1 ml per plate) onto $P_{10}VP$ agar (Tsao, 1970). After 48 hr, the soil was washed from each plate. The total incubation time (at 22° C) was 6 days. The number of colonies that developed on the plates from each 50-g sample was recorded. Colonies were subcultured, if required, to separate them from each other and from any contaminants. Colonies were examined frequently to detect coralloid growth, vesicles, and chlamydospores, which aid in identification of *Phytophthora* species.

## RELATION BETWEEN DISEASE POTENTIAL AND PATHOGEN POPULATIONS

The increase in the incidence of disease with an increase of pathogen population for several diseases is shown in Table 1. However, infection and disease severity may fluctuate independently of population levels. Saturation of soil and the associated reduction in oxygen levels may make the host more susceptible. Thus, disease potential may increase independently of the pathogen population.

The converse situation, in which a high population of *Phytophthora* is associated with a low disease potential, has also been recorded. For example, in the Fynbos of the Western Cape Province of South Africa in a remote area at high elevation, *P. cinnamomi* can be recovered from all catchment waters and from the roots of indigenous plants, but there is little evidence of plant damage, whereas at lower elevations, the same fungus causes disease and death in the same native species grown commercially (von Broembsen, 1979). The difference may be associated with increasing host resistance, with altitude, or with race or mating type of the pathogen. In the Fynbos, *P. cinnamomi* is apparently in ecological balance with the host species.

TABLE 1. Examples of increase in disease associated with increase in pathogen populations of several *Phytophthora* species

| Species | Initial infection or inoculum density | Final disease or pathogen population | Dispersal unit | Reference |
|---|---|---|---|---|
| *P. infestans* | One infection in $10^5$ potatoes | Epidemic disease | Sporangia, zoospores | Newhook (1980) |
| *P. palmivora* | 0.5 chlamydospores per gram of soil | 50% infection | Chlamydospores | Ramirez and Mitchell (1975) |
| | 10.0 chlamydospores per gram of soil | 95% infection | Chlamydospores | Ramirez and Mitchell (1975) |
| | $10^3$ zoospores per plant | 75% infection | Zoospores | Ramirez and Mitchell (1975) |
| | One infection per cocoa pod | 5–10 × $10^6$ sporangia per pod | Sporangia | Turner (1974) |
| | | (5–10 × $10^6$ × 25 zoospores per pod) | Zoospores | Turner (1974) |
| *P. parasitica* var. *nicotianae* | $10^5$ spores per tobacco plant | 100% infection of plant | Sporangia | Gooding and Lucas (1959) |
| *P. cinnamomi* | One sporangium per seedling | 100 sporangia per seedling | Sporangia | Byrt and Holland (1978) |

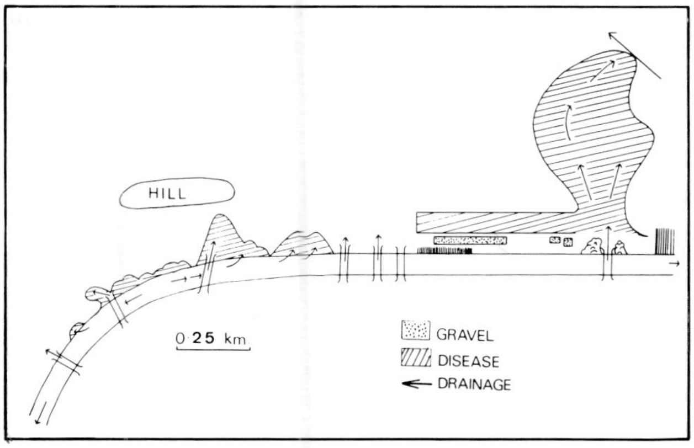


**Fig. 2.** Disease spread from gravel infested with *Phytophthora cinnamomi* from a road on which the gravel had been piled at Wilson's Promontory National Park, Victoria, Australia. (Reprinted, by permission, from Weste, 1975a)

## CHANGES IN THE POPULATION OF *PHYTOPHTHORA*

Increases in population density usually precede an increase in disease, which is reflected by both extension or incidence and severity of the disease. At any one time, the size of the *Phytophthora* population depends on the amount and form of the original inoculum and its rate of change, which in turn depends on characteristics of the total host-pathogen system in its ecological community. Attributes of the systems that play a role in population increase from infection to epidemic may be classified into those due to the form and amount of initial inoculum, those due to characteristics of *Phytophthora* species, those due to the hosts and the ecological community in which the susceptible species are growing, and those due to the environment.

### The Form and Amount of Initial Inoculum

If the initial inoculum is in infected nursery stock, the pathogen in the tissue of susceptible hosts is in an ideal situation. The subsequent disease will escalate at the maximum rate depending on the percentage of infected nursery stock and the favorableness of the environment to the pathogen.

If the inoculum arrives as windblown sporangia or water-carried zoospores, any increase in population will depend on whether this inoculum reaches suitable points of entry on the susceptible hosts. When the initial inoculum occurs in infested quarries or gravel pits or is in contaminated soil or gravel used in road building or piled at roadsides, disease development and extension will depend on dispersal of the inoculum into a crop or forest, as has been demonstrated for *P. cinnamomi* in Australian forests (Weste, 1975a) (Fig. 2). Initial inoculum may also be carried on equipment, such as spray vehicles and bulldozers, particularly if these are used off roads in field and forest. A bulldozer used in bushfire work was taken from infested forests (East Gippsland in eastern Victoria) in 1962 and removed from its carrier in an old gravel pit at Wilson's Promontory National Park in southern Victoria. The gravel from this pit became infested and later was piled beside the roads in various parts of the national park. Later, tongues of diseased plants radiating from these sites could be seen in the forest.

TABLE 2. The influence of initial inoculum on the development of black pod of cacao caused by *Phytophthora palmivora*

| Initial inoculum | *P. palmivora* infection (% of pods) after 15 days | 30 days | 45 days | 60 days |
|---|---|---|---|---|
| Trees with five rotted pods around base | 35.7 | 77.6 | 91.7 | 94.2 |
| Trees with no rotted pods near base | 11.3 | 26.7 | 57.1 | 75.8 |

*Source*: Muller (1974).

The initial inoculum of *P. palmivora* may also be dispersed by agents such as ants and rodents from infested litter to cocoa trees and from tree to tree (Table 2) (Okaisabor, 1974). Attendant ants build protective nests of detritus (which may be infested) on clusters of mealybugs on stalks of cocoa pods. *P. cinnamomi* was spread by feral pigs wallowing in the mud of Queensland rain forests (Brown, 1976).

### Characteristics of *Phytophthora* Responsible for Population Changes

#### Mycelial Growth

Mycelium can grow within the host, from host to host through plant contacts, and through the soil. Dieback disease caused by *P. cinnamomi* has been observed to extend uphill from infected plants at various rates depending on the site. It is difficult to conceive how this could occur except by mycelial growth. In the Brisbane Ranges, Victoria, disease extended 6.6 m/yr up a slope of 4° (Weste et al, 1976). Podger (1972) reported mycelial spread at 3.6 m/yr in Western Australia jarrah forests. Contact between infected potatoes or aerial parts of potato plants allows transmission of *P. infestans* by mycelial spread. Reeves (1975) demonstrated that mycelium of *P. cinnamomi* invaded roots and root pieces in nonsterile soil.

**Latent Period**

The latent period is the time that elapses between penetration and sporulation of the pathogen. Naturally, the shorter the latent period, the more rapidly pathogen populations increase. A short latent period is a characteristic of the *Phytophthora* life cycle, but small variations occur among the species. The susceptibility or resistance of the host also influences the latent period (see MacKenzie et al, Chapter 23). The latent period is only 24 hr for *P. cinnamomi* on eucalypt seedlings (Byrt and Holland, 1978). It is also brief for *P. infestans* but longer for *P. palmivora*.

**Rate of Reproduction**

Each sporangium of *P. cinnamomi* formed 24 hr after eucalypt seedling inoculation releases 10–30 zoospores within an hour (Byrt and Holland, 1978; Zentmyer, 1980). In Australian forests, zoospore production is confined to periods when soil moisture is sufficient, namely in spring and early summer. In Victorian and Western Australian forests, no sporangia are formed during the dry summer period. Some sporangia production occurs during winter and autumn, but it is limited by low soil temperature (Shea et al, 1980; Weste and Vithanage, 1979b) (Fig. 3).

Reproduction is rapid in *P. infestans*, with the result that one blighted seed tuber in 100,000 provides adequate initial inoculum for a major late blight epidemic (Robinson, 1976). One cocoa pod infected with *P. palmivora* can bear 10 lesions, each with $5 \times 10^6$ sporangia releasing 20–25 zoospores each (Turner, 1974). Rapid production of sporangia for most species of *Phytophthora* depends on suitable temperatures and moisture levels and results in the release of zoospores, which are the prime infective agents. Chlamydospores can also increase in numbers in soil and gravel under suitable conditions (Weste, 1982).

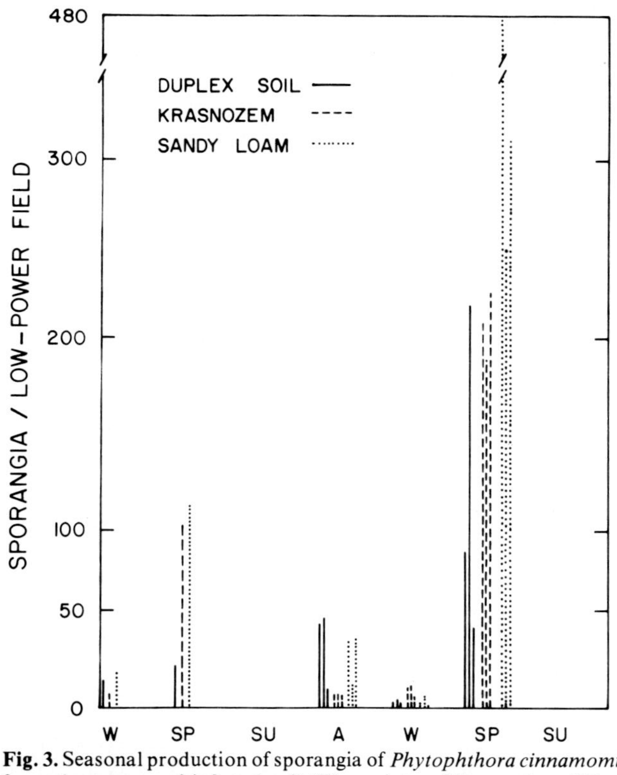


**Fig. 3.** Seasonal production of sporangia of *Phytophthora cinnamomi* from three types of infested soil (W = winter, SP = spring, SU = summer, and A = autumn). Numbers represent numbers of sporangia from 15 random counts per low-power field (3 mm$^2$). (Reprinted, by permission, from Weste and Vithanage, 1979b)

**Dispersal**

Whole sporangia (caducous) are dispersed by some *Phytophthora* species, such as *P. infestans, P. palmivora, P. phaseoli*, and *P. cactorum*. The first two species shed sporangia over a range of conditions varying from dry air to water. The flight range for *P. infestans* is greater than 60 km, and the population can spread hundreds of kilometers in the course of one season (Robinson, 1976). About 21 other species shed whole sporangia, but they do so only in water or at relative humidity values between 98 and 100%. Whole sporangia may be carried in wind and rain (Trione and Roth, 1957), in seed potatoes, on spray vehicles, etc.

About 26 species do not normally shed sporangia (Waterhouse, 1974), but zoospores are released and are the principal agents of dispersal and infection, especially under wet conditions. Zoospores are biflagellate and swim or are transported by irrigation, in surface drainage water, or via water-filled pores in wet soils. Duniway (1976) recorded that zoospores of *P. cryptogea* swam 25–35 mm in surface water over flooded soil or through a coarse-textured soil mix at matric potentials ($\psi_m$) of −0.001 bar, but swam only 5 mm at −0.05 bar and not at all at −0.1 bar. In silt and sandy loams, zoospore movement was limited to 5 mm at $\psi_m$ of −0.001 bar and was absent at −0.01 bar. Downward movement of zoospores was only 8–12 mm in loams and 40 mm in the coarse soil mix. Duniway concluded that the helical movement of zoospores required large continuous pores that are filled with water. In field, garden, and forest, these pores are supplemented by cracks and holes that occur naturally in soil and become filled with water during rain. Allen and Newhook (1973) demonstrated that the helical motion of zoospores required the water-filled pores to be about 190 μm in diameter.

In drier conditions and/or higher temperatures, sporangia germinate directly to produce a germ tube instead of zoospores; for example, *P. infestans* sporangia produce germ tubes at 20°C but zoospores at 12°C (Crosier, 1934). *P. palmivora* sporangia produce germ tubes at 30–34°C and zoospores at 10–34°C (Clerk, 1972). Mycelium, chlamydospores, and oospores may be dispersed passively in transplanted infected plants; by the movement of soil, gravel, and dead plant material by various agricultural, horticultural, or forestry practices; or by adhering to equipment and machinery, off-road vehicles, shoes, tires, or animals (Kliejunas and Ko, 1976; Podger, 1972).

**Orientation**

Zoospores of most or all species of *Phytophthora* swim toward nutrient sources, which are root exudates in soilborne species. This process, called chemotaxis, affects populations by increasing zoospore vitality and by decreasing the wastage of inoculum. Chemotaxis also increases the disease potential apart from actual population numbers. Maximum exudate is extruded from the region of root elongation, which is also the region of zoospore penetration.

Chemotaxis was first observed by Goode (1956) and later by Zentmyer (1960, 1961). Zentmyer reported a differential attraction to susceptible compared with resistant hosts. Turner (1963), working with *P. palmivora*, and Chi and Sabo (1978), working with *P. megasperma*, also obtained differential responses to susceptible hosts. However, Hinch and Weste (1979) found that zoospores of *P. cinnamomi* swam toward roots of all 23 Australian species investigated, whether or not they were susceptible. The chemotactic response was stimulated by amino acids, sugars, growth factors, and ethanol.

Troutman and Wills (1964) reported electrotaxis, Katsura and Miyata (1966) observed zoospores swimming against a current (rheotaxis), and Cameron and Carlile (1977) reported that the accumulation of zoospores of several *Phytophthora* species near a liquid surface was the result of negative geotaxis. All these kinds of directed movements or taxes assist in increasing the effectiveness of the zoospore population and reducing wastage of inoculum. Carlile (Chapter 7) treats this subject in greater detail.

**Zoospore Encystment**

Zoospores of root-infecting species encyst at the root surface, and zoospores of species that cause disease of aerial plant tissues encyst on the leaf or stem surface. They lose their flagella and produce a protective wall. Zoospores may encyst before reaching the host surface in response to continuous contact with solid particles or in response to cations at concentrations that are likely to occur in certain soil solutions. In this respect, the relative strength of response is in the following order: $Fe^{3+} > Mn^{2-} > Sr^{2+} > K^{+} > Ca^{2+} > Na^{+}$. Ferric iron triggered encystment at 20 $\mu M$ and potassium ions at 2 m$M$ after exposure for 20 min. Sodium and potassium ions caused partial loss of viability, but the cysts that formed in response to $Fe^{3+}$ were completely viable (Byrt et al, 1982a, 1982b).

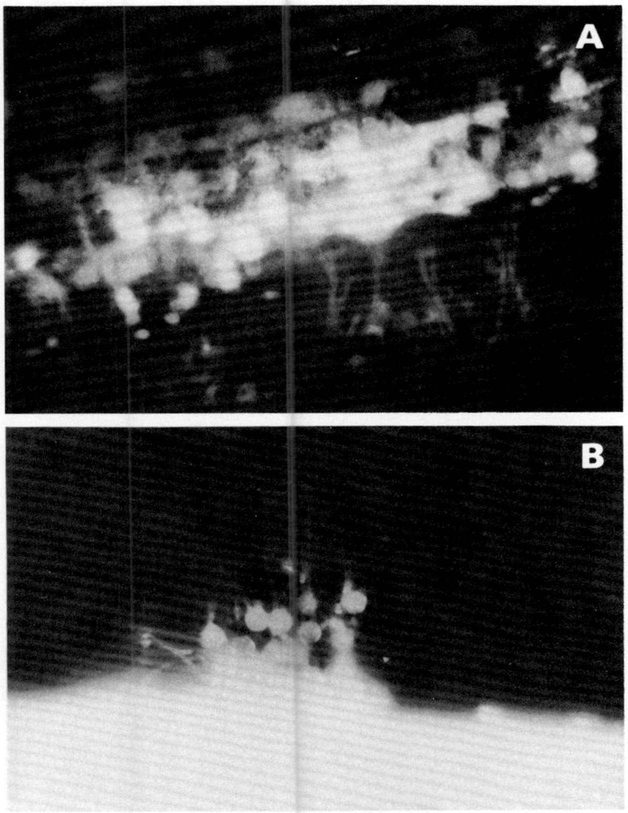


**Fig. 4. A,** Zoospore cysts of *Phytophthora cinnamomi* germinating on a root of corn (*Zea mays*). Germ tubes are chemotropically oriented toward the zone of root elongation. **B,** Zoospore cysts of *P. cinnamomi* on the root of a grass plant, *Themeda australis*. Germ tubes are oriented away from the root surface, indicating interaction, competition, and antagonism among cysts. (Reprinted, by permission, from Hinch and Weste, 1979)

Within 1–6 hr of cyst formation, a germ tube is produced that is chemotropically oriented toward the zone of elongation on roots and penetrates the host tissue. However, when several zoospores encyst at the same site, interference may occur and the germ tubes often become oriented away from the root (Fig. 4) (Hinch and Weste, 1979). If penetration is not successful, the cyst produces a papilla or microsporangium, from which a secondary zoospore emerges. The population density of infective propagules is thereby maintained.

**Chlamydospore Production**

Sporangia and zoospores are essentially adapted to wet conditions. When soil $\psi_m$ is reduced to between −0.3 bar and −35 bars, chlamydospores of *P. cinnamomi* are differentiated in plant tissue, in soil, or in organic matter. Hence, when samples of sieved soil from three types of forest soil were plated by the method of McCain et al (1967), the maximum number of chlamydospores was obtained during the dry autumn period (Fig. 5) (Weste and Vithanage, 1978b).

Chlamydospore populations in soils are dynamic; in contrast to the usual picture of resistant spores that simply survive in a static population, chlamydospores have been observed by use of fluorescent dyes to increase in number in nonsterile, host-free soils, producing mycelium and more chlamydospores or sporangia and zoospores, depending on $\psi_m$ (Fig. 6). The cycle depends on a minimum organic content in the substrate. Chlamydospores do not survive without organic matter in soil, but when content of organic matter in soil is high, microbial competition and antagonism are important negative factors (Weste and Vithanage, 1979a).

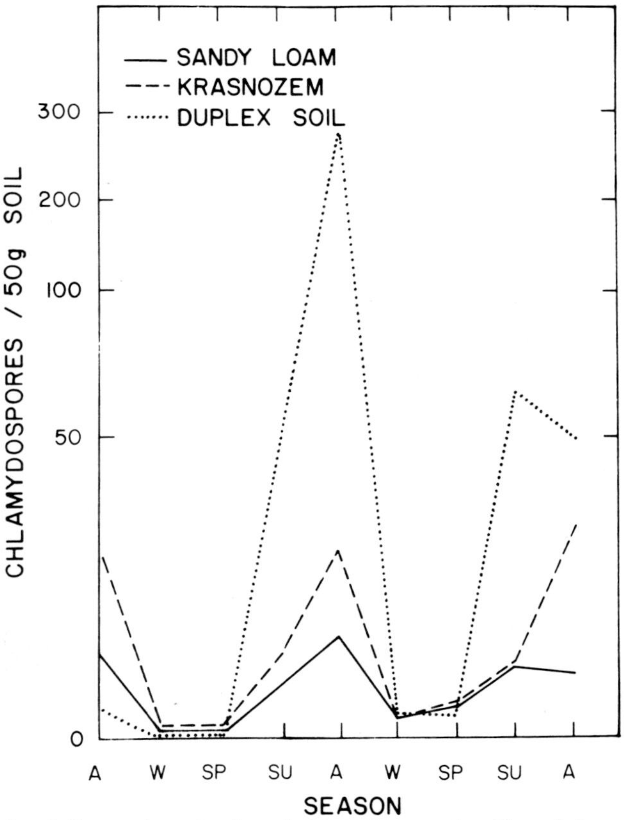


**Fig. 5.** Seasonal production of chlamydospores by *Phytophthora cinnamomi* in three types of infested soil as measured by soil sieving (W = winter, SP = spring, SU = summer, and A = autumn). (Reprinted, by permission, from Weste and Vithanage, 1978b)

**Oospore Production**

The physiology of oospore production seems to vary among species of *Phytophthora*. In some homothallic species such as *P. cactorum*, oospores are common under various circumstances, but in many heterothallic species they are extremely rare. Only one mating type of *P. infestans* occurs in Europe. Oospores are not found except in Mexico, where both mating types coexist. Both mating types occur in *P. cinnamomi*, but the A2 mating type is most common; the A1 is seldom isolated except in mountains of New Guinea and South Africa. Oospores form occasionally in old cultures and in response to antagonistic fungi like the common soil species of *Trichoderma* (Brasier, 1971; Reeves and Jackson, 1972) and in response to avocado root extract (Zentmyer, 1979). Oospores of *P. parasitica*, ordinarily heterothallic, formed in cultures originated from aged inocula (Tsao et al, 1980). Oospores are probably the most resistant structures produced and serve to maintain the population when the pathogen has almost ceased to grow because of adverse environmental factors.

**Competitive Saprophytic Ability**

A pathogen that can maintain its population as a saprophyte in the soil in the absence of a host has an advantage over obligate parasites. The genus *Phytophthora* encompasses about 50 species. At one end of the scale are the biotrophs, like *P. infestans* and *P. megasperma* f. sp. *glycinea*, which are host-specific and have little competitive saprophytic ability. Species-specific biotrophic parasites often achieve ecological balance with their hosts and do not usually kill, so that survival is less important. Species like *P. cinnamomi* and *P. cactorum* have both a parasitic and a saprophytic life cycle, have a wide host range, cause necrosis and death of the host, can both grow and reproduce in the absence of a host and, under suitable conditions of temperature and moisture, can compete to a limited extent with other soil microorganisms for organic substrate. At the other end of this scale are the little-known species of *Phytophthora* that are entirely saprophytic and cause no disease. Some of these are marine and have been isolated from mangrove roots (Fell and Master, 1975; Pegg et al, 1980; Weste et al, 1982). *Phytophthora* species with competitive saprophytic ability can live on organic particles and on litter, which acts as a reservoir of inoculum or a population in reserve, which is very difficult to remove or control in a soil that is conducive to disease.

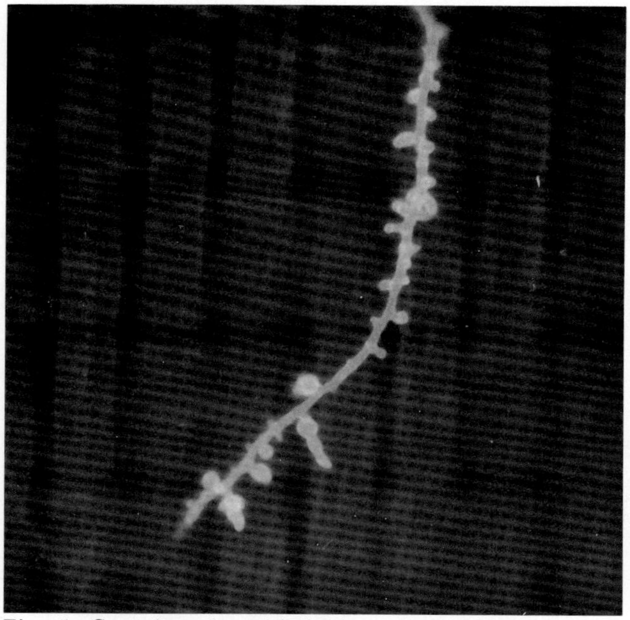

**Fig. 6.** Growth and multiplication of chlamydospores of *Phytophthora cinnamomi* in infested gravel in the absence of a host over time. Fluorescent chlamydospores were obtained by growing them on a medium containing 200 μg/ml fluorescent dye. The photograph was taken 20 days after burial in soil of nylon mesh inoculated with chlamydospores.

## Characteristics of the Hosts and Their Ecological Communities

Natural forest communities are very different from agricultural fields or orchards. In the natural forest, tree species are genetically diverse. The essential feature of agriculture is high productivity aimed at maximum production per unit of labor. The large fields, the gangs of combine harvesters, and bulk storage, marketing, and processing all demand a uniform product and hence a uniform crop. Because epidemics require frequent contacts between the pathogen population and the host population, the likelihood of a severe epidemic is increased if both populations are large and uniform.

**Susceptibility of Forest Species**

The pathogen in a natural forest community depends on the susceptibility of the various forest species. In a diverse forest, these range from susceptible through tolerant to resistant (Fig. 7). The pathogen infects susceptible hosts and rapidly reproduces on them; it also infects tolerant hosts, but no symptoms appear and reproduction may be curtailed; and it does not infect resistant hosts. The percentage of susceptible hosts and their closeness to the source of inoculum influence both pathogen and host populations. *P. cinnamomi* reduced the chestnut population in North America, and *P. lateralis* reduced the population of the Port Orford cedars of Oregon.

In Australian forests, the susceptibility of the tree stratum is overwhelmed by the far greater susceptibility of the understory species. Australian dry sclerophyll forests have a mixed woody shrub understory, many species of which are highly susceptible to *P. cinnamomi* and die out rapidly on a large scale when invaded by the pathogen. Some of the dominant understory species are so susceptible that they serve as indicators of *P. cinnamomi* (Fig. 1) (Weste, 1981). The plants die a few weeks

**Fig. 7.** *Eucalyptus macrorhyncha* trees killed by *Phytophthora cinnamomi* along the road in a native forest in the Brisbane Ranges, Victoria, Australia. (Reprinted, by permission, from Weste and Marks, 1974)

after infection if the soil is conducive and the forest is under environmental stress. Infection of such species results in a huge increase in pathogen population, which in turn affects the tree stratum. This usually leads to a true epidemic situation, with a characteristic disease front marking the limit of the epidemic.

*Predisposition to disease.* Factors such as waterlogging, drought, low nutrient status, and shortage of light predispose trees to disease. Little-leaf disease, caused by *P. cinnamomi* in the southeastern United States, developed on soils deficient in nutrients (Zak, 1957). The more susceptible species *Pinus echinata* suffered a higher percentage of mortality than *Pinus taeda* (Campbell and Copeland, 1954). Susceptibility and disease proneness both acted to increase the population of *P. cinnamomi*. Waterlogging and drought also affect the health of the tree root system and hence its disease proneness. This will be discussed with environmental factors.

*Age of stand.* The susceptibility of species of forest trees varies with age. When *P. cinnamomi* caused disease in a krasnozem soil (a red-brown clay loam with little differentiation in the profile), young seedlings of *Pinus radiata* up to the age of 5 yr were killed, but trees became resistant on the well-drained soil after 5 yr (Weste, 1980).

*Wounding and damage from forest practices.* Thinning, selective felling, and road and mill construction cause wounding and damage to the forest flora. *P. cinnamomi* and *P. lateralis* penetrate wounded stem tissue and cause a necrotic canker just above ground level that can kill the host (Shea et al, 1980; Weste, 1975b).

*Root invasion.* When the pathogen is present in the soil, roots of the forest flora can grow into the inoculum and become infected. The pathogen may reproduce on the diseased roots and so increase in population.

**Fig. 8.** Stem canker of peach caused by *Phytophthora cactorum* and *P. cinnamomi* in Victoria, Australia. Note the lesion and gum exudates. (Courtesy P. Taylor)

**Susceptibility of Crops**

Soil cultivation influences the initial distribution of a pathogen population in a field or orchard soil. The susceptibility of the species in monoculture influences subsequent changes in pathogen population. A susceptible crop undiluted by resistant species is a feast for the pathogen. A period of fallow between crops reduces the pathogen population to some extent, depending on the length of the fallow period and the saprophytic ability of the pathogen species. If infected residues from the previous crop are allowed to remain in the soil during the fallow period, the species of *Phytophthora* can survive and grow on them. For example, *P. infestans* survives in the remains of old infected tubers and increases rapidly in a new crop of potatoes, *P. fragariae* survives in strawberry plant residue, and *P. palmivora* survives in diseased pods of the cocoa plant. In most agricultural crops, *Phytophthora* species are present in the residues of the preceding crop. Volunteer plants between crops may also nullify the effect of fallow. For example, *P. infestans* will persist on volunteer plants after the potato crop has been removed.

Close planting increases the opportunities for host-parasite contact, results in higher pathogen populations, and therefore increases the likelihood of an epidemic. Such epidemics are characteristic of infested nurseries where seedlings die from damping-off disease.

Penetration through wounds and grafts may increase the pathogen population and subsequent disease in orchards. If the graft of fruit trees is too near the soil level, infection rapidly leads to collar rot and death, as, for example, with *P. cactorum* and *P. cinnamomi* in Victorian peach orchards (Fig. 8) (Taylor, 1977). Wounds made by pruning and machinery also allow the pathogen to penetrate more easily.

Irrigation of peach trees by sprinkling them with water containing zoospores of *P. cactorum* and/or *P. cinnamomi* resulted in rapid death from crown rot at the collar region. Irrigation increased the propagule population of *P. cinnamomi* in soil (Shea et al, 1980) (Fig. 9).

The site chosen for agricultural development may favor population increase of the pathogen as well as predispose the

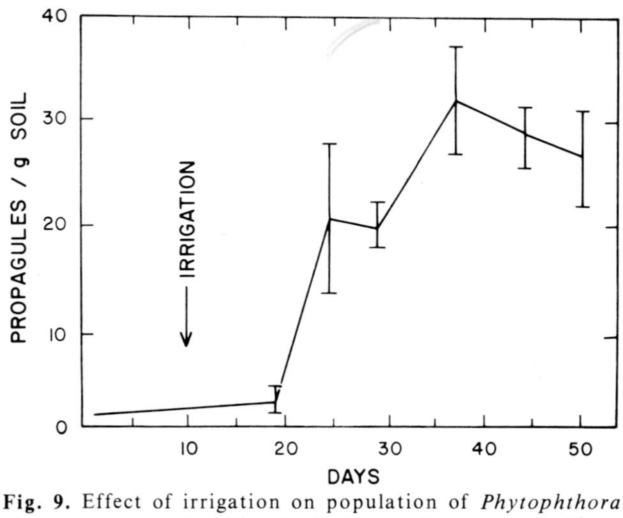


**Fig. 9.** Effect of irrigation on population of *Phytophthora cinnamomi* propagules in forest soils. Propagules were counted from colonies obtained by wet-sieving onto agar plates. (Reprinted, by permission, from Shea et al, 1980)

crop to disease. For example, the crop may be grown on land where the soil remains water-saturated for a long time, which favors zoospore production, dispersal, and infection while reducing the health of the host root system. The roots of the host may grow into inoculum and become infected. These roots will be a nutrient source that may support a rapid increase in the population. The pathogen may be transported on machinery used for cultivation or harvesting and on seed (Zentmyer et al, 1958).

### Environmental Factors and Populations of *Phytophthora*

Although the pathogen inside the susceptible host is only indirectly influenced by the environment, reproduction on the surface of the host and in all soilborne populations is strongly influenced by environmental changes such as temperature and moisture. About half the variations in the progress of epidemics caused by *P. infestans* were attributable to temperature and rainfall (Schröter and Ullrich, 1965).

The contributions of independent variables such as temperature and moisture may be assessed by multiple regression analysis: $Y = a + b_1x_1 + b_2x_2 + b_3x_3 + \ldots$, where $Y$ is an appropriate disease or population measure; $a$ is the intercept; $x_1, x_2, x_3, \ldots$ are the independent variables; and $b_1, b_2, b_3, \ldots$ are their partial regression coefficients. The number of independent variables is unlimited; they usually include temperature, leaf wetness, etc. The coefficients can be determined with a computer (van der Plank, 1975).

#### Temperature

Growth and population of the pathogen are related more to the extent and frequency of favorable and unfavorable temperatures than to the mean daily temperature. Schröter (1965) found that data on growth and population increase in relation to temperature could be fitted by the following equation: $Y = \sin^2(a_1x + a_2x^2 + a_3x^3)$, where $Y$ is the biological activity, e.g. rate of growth or population increase, and $x = 100\,(t - t_n)/(t_x - t_n)$, where $t$ is the actual temperature, $t_n$ is the minimum temperature, and $t_x$ is the maximum temperature for the particular activity of that organism.

Temperatures below the minimum or above the maximum may completely suppress the pathogen; for example, *P. cinnamomi* is pathogenic to avocado between 15 and 27°C but not at 33°C (Zentmyer, 1981). The fungus is rarely found in soils at temperatures below 10°C in Victorian native forests (Weste and Ruppin, 1977). Hosts such as eucalypts also vary in susceptibility at different temperatures (B. Grant, *personal communication*). Low temperatures in winter explain in part why Douglas-fir forests on the Pacific coast of the United States are not seriously jeopardized by *P. cinnamomi* (Roth and Kuhlman, 1966) and also explain the difference in the impact of the same pathogen on montane (hilly to mountainous) and lowland forests in East Gippsland (Marks et al, 1975). *P. phaseoli* does not cause disease when the air temperature exceeds 29.5°C.

*Phytophthora* populations are unlikely to overwinter in frozen surface soil but may survive at different depths beneath the surface or in host tissue. Disease potential depends on the proportion of time when the temperature is suitable for reproduction and infection. Table 3 presents temperature ranges for a number of *Phytophthora* species, but because of the interaction between temperature and other factors, critical temperatures are relative rather than absolute. Rotem et al (1971) conducted a factorial experiment with 120 combinations of temperature, concentration of inoculum of *P. infestans*, and duration of wetting. The minimum, maximum, and optimal levels of each factor depended on the levels of the other two, so limiting levels were relative rather than absolute. For example, 6 hr of wetness was insufficient for infection at 5°C but optimal at 15°C. Therefore, the effect of temperature change on the pathogen, on the host, and on their interaction in relation to other factors such as moisture and duration of wetting must be studied.

A good correlation was obtained between the soil pathogen populations of *P. cinnamomi* measured as disease potentials and concurrent soil temperatures registered by a recording thermometer with leads buried permanently 3–6 cm deep in soil near the sampling site. The correlation coefficients demonstrated a highly significant relationship ($P < 0.001$) between disease potential and soil temperatures from autumn to early summer. Because $\psi_m$ greater than −1 bar was not limiting, the pathogen population was primarily controlled by temperature (Weste and Ruppin, 1977). Results were highly significant for individual plots and for the total site area for dry shrubby sclerophyll forests on duplex soils in the Brisbane Ranges (Fig. 10), for sclerophyll shrubby woodlands on deep sands at Wilson's Promontory (Fig. 11), and for wetter

TABLE 3. Approximate temperature ranges for growth and reproduction of some *Phytophthora* species[a]

| | Temperature (°C) | | | | |
|---|---|---|---|---|---|
| Species | Growth | Sporangia | Zoospores | Chlamydospores | Oospores |
| *P. botryosa* | 9–32 | 9–32 (abundant) | | 9–32 (rare) | 9–32 (abundant) |
| *P. cactorum* | 4–29 | 10–20 | | | |
| *P. cinnamomi* | 5–36 | 12–30 | 15–28 | 9–28 | 12–30 |
| *P. cryptogea* | 3–25 | 5–25 | | | |
| *P. drechsleri* | 5–37 | | | | |
| *P. fragariae* | 1–30 | | | | |
| *P. heveae* | 11.5–32.5 | 27–30 | 21 | rare | 12–32 (abundant) |
| *P. infestans* | <5–28 | | 5–21 | | 5–28 (rare) |
| *P. lateralis* | 3–25 | 5–25 | 5–25 | 15–25 | 5 (rare) |
| *P. megasperma* | | 8–24 | 12–16 | | |
| *P. nicotianae* var. *nicotianae* | <0–38 | 15–32 | 17–27 | | |
| *P. nicotianae* var. *parasitica* | 5–38 | 15–32 | 17–27 | | |
| *P. palmivora* | 11–35 | 15–32.5 | 20–24 | 32 | 15–20 |
| *P. parasitica* | | 27–30 | | 12–25 | |

[a] No references are given because the table is an approximate composite from many authors and ignores many factors such as media, moisture, etc.

sclerophyll forests on krasnozem soil at Narbethong (foothills of the Great Dividing Range, Victoria, Australia). Changes in disease potential in response to temperature were often quite dramatic. When the soil temperature fell below 10° C, disease potential declined from near maximum to zero, and *P. cinnamomi* was not detected by any method in 13 days in completely different forests growing on different types of soil. Marks et al (1973) recorded a similar population decline in cultivated soils of eastern Victoria during the winter. When soil temperatures rose, disease potential increased, but more gradually. The decrease in pathogen activity observed each winter, the increase observed in spring, and the further increase in activity during the warmer months may be partly explained by the influence of temperature on populations of the pathogen.

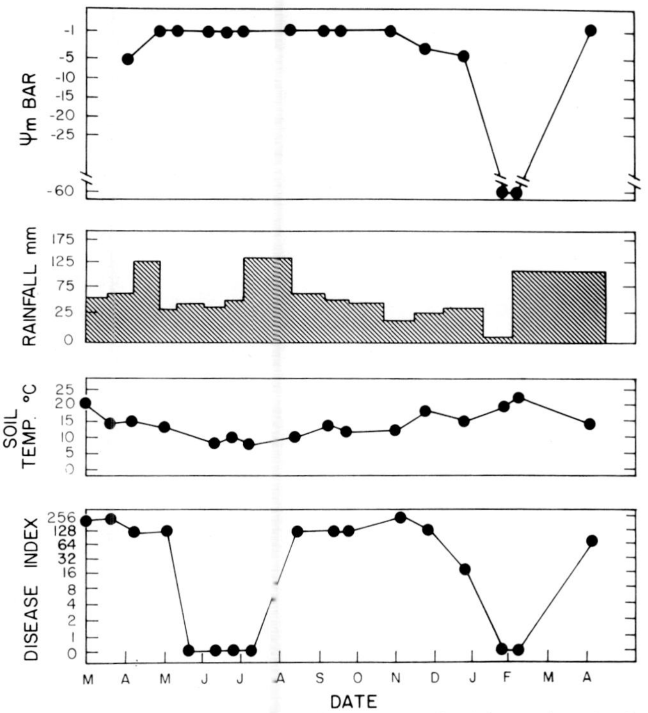


**Fig. 10.** Concurrent records of disease potential (disease index) of *Phytophthora cinnamomi*, soil temperatures, rainfall, and soil water matric potentials at 5-cm depth in a diseased forest, Brisbane Ranges, Victoria, Australia. The disease index was calculated as the highest dilution of soil at which *P. cinnamomi* could be detected by baiting. (Reprinted, by permission, from Weste and Ruppin, 1977)

## Moisture

The moisture level has an important influence on the growth and reproduction of *Phytophthora* species. The high moisture requirement is satisfied by high relative humidity and free water on the plant surface for species such as *P. infestans* and *P. palmivora* that are pathogens of leaves or pods. A film of water on the leaf for 6–8 hr, depending on the temperature, allows infection and penetration by *P. infestans* (Crosier, 1934). Sporangial formation and dispersal of *P. infestans* (Lapwood, 1968) and *P. palmivora* (Okaisabor, 1974) occur at high relative humidities.

Soilborne *Phytophthora* species require high soil moisture, and many species require free water for production and dispersal of zoospores. The approximate moisture requirements for population changes in some *Phytophthora* species are presented in Table 4. Soil moisture is expressed as $\psi_m$, which recognizes the varying availability of water that is characteristic of different soils and allows comparison between them. This is discussed fully by Duniway (Chapter 14). As an illustration of the difference in soil moisture characteristics, Reeves (1975) concluded that a shower of rain may suffice for sporangial formation and zoospore release on sandy but not on clay soil.

Species of *Phytophthora* vary in water requirements; for example, maximum zoospore formation and release occur on flooded soils for *P. megasperma* (Pfender et al, 1977) but under slightly drier conditions for *P. cinnamomi* (Table 4). Both pathogen population and disease severity increased with increasing $\psi_m$, with the greatest at zero for *P. megasperma* f.

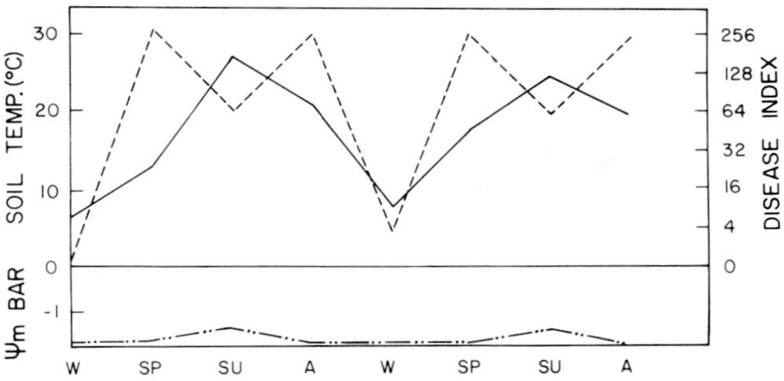


**Fig. 11.** Concurrent records of disease potential (disease index) of *Phytophthora cinnamomi*, soil temperatures, and soil water matric potentials at 5-cm depth for a diseased woodland, Wilson's Promontory, Victoria, Australia. (W = winter, SP = spring, SU = summer, and A = autumn.) The disease index was calculated as the highest dilution of soil at which *P. cinnamomi* could be detected by baiting. (Reprinted, by permission, from Weste and Ruppin, 1977)

TABLE 4. Moisture requirements for growth and reproduction of some *Phytophthora* species

| | Atmospheric relative humidity (%) or soil matric water potential (bars)[a] | | | | |
|---|---|---|---|---|---|
| Species | Sporangia | Zoospores | Chlamydospores | Mycelium | References |
| *P. cactorum* | +0.001—−0.3 bar | | | | Sneh and McIntosh (1974) |
| *P. cinnamomi* | −0.01→−2.5 bars | >−0.01 bar | >−0.03→−100 bars | −25→−35 bars | Sommers et al (1970) |
| *P. cryptogea* | −0.025—−0.3 bar | −0.001→−0.025 bar | | 0→−0.4 bar | MacDonald and Duniway (1978) |
| *P. drechsleri* | −0.3→−3.5 bars | 0→−0.3 bar | | 0→−40 bars | Duniway (1975a) |
| *P. megasperma* | 0→−2.3 bars | −0.001→−0.025 bar | | | MacDonald and Duniway (1978), Sommers et al (1970) |
| *P. palmivora* | +0.001—−15 bars | 0→−0.05 bar | | | Rocha and Machado (1973) |
| *P. parasitica* | −0.025—−0.3 bar | | | 0→−26 bars | Sommers et al (1970) |
| *P. infestans* | 95–97% | 97–100% | | | Rotem et al (1971) |

[a]Indicates range of moisture requirement.

sp. *medicaginis* on alfalfa (Kuan and Erwin, 1980). Zoospore dispersal in free water runoff after heavy rain or in streams was reported in the Fynbos, Western Cape, South Africa (von Broembsen, 1979). Irrigation channels spread *P. cinnamomi* in California avocado orchards (Zentmyer and Richards, 1952) and both *P. cactorum* and *P. cinnamomi* in Victorian peach orchards (Taylor, 1977). Shea et al (1980) provided evidence that irrigation increased the number of propagules of *P. cinnamomi* in soil (Fig. 9).

Seasonal changes in populations of *P. cinnamomi*, measured as disease potentials in three ecologically different Victorian forests, were not correlated with the moisture content of soils but were significantly correlated ($P < 0.001$) with $\psi_m$ during summer and autumn (Figs. 10 and 11) (Weste and Ruppin, 1977). Pathogen populations were reduced by low temperatures, even though water was never limiting during winter and spring. During summer and autumn in the dry shrubby sclerophyll forest on duplex soils in the Brisbane Ranges, in the sclerophyll shrubby woodland on sands at Wilson's Promontory, and in the wetter sclerophyll forest on krasnozem soil at Narbethong, $\psi_m$ fell below −9 bars, which limited growth and reproduction of the pathogen and subsequently the disease potential. In the Brisbane Ranges and at Narbethong, $\psi_m$ was lower than −60 bars in February–March 1974, and the disease potential was zero. Two days after a rain, the disease potential rose to near its maximum level. The rapid increase was probably caused by the production of sporangia and zoospores in the wet soils (Weste and Ruppin, 1975, 1977). Similar population declines and increases after rain were recorded in 1974 and 1975 in the same forests. Although temperature and moisture are suitable for growth and infection by *P. cinnamomi* for only brief periods in the Western Australia jarrah forests (Shea, 1975), disease is severe. These results indicate problems that require further investigation for soilborne *Phytophthora* species.

Better calibration of methods for determining disease potential by infesting a range of soils with a series of inoculum concentrations for a defined range of environmental conditions is needed. This would enable the establishment of the relationship between propagule density and disease severity for a specified crop or forest in a particular environment. Changes in disease potential with time, host, pathogen, and environment might then be compared with the standard series and related to changes in pathogen populations. We need to know why disease potential, as determined by host plant assay, falls rapidly to zero at limiting soil temperature or $\psi_m$ and then climbs rapidly to its maximum at optimal soil moisture and temperature. Is the reserve population of the pathogen within host tissue? Too little is known about pathogen distribution, sporangial production, and zoospore release in infested soil. Investigation may be practicable with the aid of mycelium grown in a fluorescent medium (Gisi, 1975).

Gisi et al (1977, 1980) obtained a strong correlation between $\psi_m$ and the number of sporangia produced by *P. cinnamomi, P. cactorum*, and *P. palmivora*. The numbers of sporangia produced by each species were similar on both sandy loam and clay soils at the same $\psi_m$, but they were not correlated with soil moisture content.

Most species of *Phytophthora* are extremely sensitive to variations in $\psi_m$ but are not as sensitive to changes in the osmotic component of soil water potential. Some species grow well on agar prepared with seawater, others will tolerate only diluted seawater. The solute concentration tolerated varied with the salt used (Duniway, 1979; Fell and Master, 1975). Increasing the salt concentration reduced growth of *P. cinnamomi*, depending on the osmotica (Sterne et al, 1976). The salt concentration that normally occurs in forest or agricultural soil is unlikely to be high enough to affect or suppress pathogen populations but may or may not affect the colonization of mangrove in salt marshes (Pegg et al, 1980; Weste and Cahill, 1982; Weste et al, 1982). Sommers et al (1970) investigated the effects of increased salt concentration on *P. cinnamomi, P. parasitica*, and *P. megasperma*. MacDonald and Duniway (1978) investigated the effects of matric and osmotic components of water potential on zoospore discharge of *P. cryptogea* and *P. megasperma*. The results demonstrated extreme sensitivity to $\psi_m$.

Increasing $\psi_m$ toward zero results in increasing populations of soil microorganisms, some of which are antagonistic to *Phytophthora*. This may explain the vulnerability of many species to bacterial lysis when growing in moist soils.

## Light

Light is apparently not an important population determinant for *Phytophthora* species, but different species vary in light requirement. Sporangial formation for *P. cinnamomi* does not depend on light but is greater in the ultraviolet region of the spectrum at 100 $\mu W/cm^2$ but not at 200 $\mu W/cm^2$ (Zentmyer and Ribeiro, 1977). Light inhibits

TABLE 5. Changes in numbers and viability of chlamydospores of *Phytophthora cinnamomi* incubated on nylon mesh squares[a] after burial in different nonsterile, hostfree substrates

| Medium | Matric potential (bars) | Chlamydospores (no.)[b] after burial in soil for: | | | Viability after 28 days | |
|---|---|---|---|---|---|---|
| | | 10 days | 20 days | 30 days | In mesh | In substrate |
| Glass beads | −1.4 | 218 | 162 | 127 | − | − |
| (2.5–3.5 mm in | −1.1 | 93 | 162 | 104 | + | + |
| diameter) | −1.0 | 144 | 144 | 87 | + | + |
| Gravel | −0.9 | 118 | 119 | 95 | − | + |
| | −0.7 | 294 | 90 | 62 | − | + |
| | −0.6 | 173 | 140 | 87 | − | − |
| Sand | −0.7 | 46 | 57 | 69 | − | + |
| | −0.7 | 137 | 79 | 63 | − | + |
| | −0.7 | 75 | 55 | 78 | − | − |
| Soil | −32.6 | 102 | 173 | 207 | + | − |
| | −19.6 | 328 | 142 | 201 | + | − |
| | − 6.9 | 287 | 118 | 249 | + | − |

[a]About 38 per square.

[b]Data are the means of six counts, comprising three replicates from each of two isolates. Viability was determined by plating on selective agar medium (Tsao et al, 1976).

[c]+ = Viable, − = not viable.

sexual reproduction of most species (Brasier, 1969). The role of light on reproduction of *Phytophthora* is discussed in more detail by Ribeiro (Chapter 4).

### Soil pH

*Phytophthora* species tolerate a relatively wide pH range from 3.8 to 9; however, *P. cinnamomi* has been controlled in pineapple plantations by the addition of elemental sulfur, which reduces the soil pH below 3.8 but does not affect the growth of pineapple plants (Pegg, 1977). A soil pH of 6–6.5 was optimal for sporangial production by *P. cinnamomi* (Zentmyer, 1980).

### Soil Characteristics

The soil plays an important role in the population dynamics of soilborne species, since most of the life cycle takes place there. Zoospores are dispersed in water over soil and infect roots growing in soil. In some species, mycelium and chlamydospores grow and reproduce in soil (Table 5) (Weste, 1982). The existence of a *Phytophthora* population in the soil depends first on its introduction into the soil and then on whether the soil is conducive or suppressive to the pathogen. Suppressive soils have high populations of bacteria and actinomycetes, which lyse the hyphae and suppress sporangial formation of species such as *P. citrophthora* and *P. cinnamomi* (Broadbent and Baker, 1974; Nesbitt et al, 1979a). Suppressive soils contain relatively high exchangeable calcium, magnesium, and organic nitrogen. However, because suppressiveness in soil may be negated by steam sterilization and reintroduced by the addition of field soil, the cause is probably microbial in origin.

Bouhot (1979) measured the biological resistance of both agricultural and undisturbed forest soils to fungi. He defined the biological resistance of soils as the ability to hinder the pathogenic activity of a virulent inoculum. He measured the disease severity that developed after the addition of natural inoculum in a given time and compared the results with a sterile soil and a suppressive soil to provide a scale of disease risk. His technique may provide data useful in assessing disease risk from soilborne *Phytophthora* spp. Biological control is also discussed by Shea and Broadbent (Chapter 26).

Once the pathogen has been introduced into a conducive soil population, the increase in inoculum and subsequent disease development depend on physical factors such as soil depth, slope, drainage, and structure; chemical factors such as aeration, nutrient concentration, and organic content; and biological factors such as the populations of microorganisms inhabiting rhizosphere, rhizoplane, and soil. These factors influence pathogen populations individually and also interact to produce various combinations more or less favorable to disease.

Shallow soils may contain a higher concentration of plant roots that become colonized more rapidly, resulting in development of a high pathogen population. Soil and roots also dry out more quickly in shallow soils in dry weather. Water stress in hosts with fine roots, which may be decayed, often causes death. Shallow soils may be underlaid with an impervious clay layer or rock base that impedes drainage. Thus, the small volume of soil quickly becomes saturated by a small amount of rain in which zoospores may disperse. The anaerobic root environment may predispose roots to disease. Shallow soils are classified as high risks for the dieback disease caused by *P. cinnamomi* in Victorian dry sclerophyll forests. Deeper soil is associated with reduced disease severity (Table 6) (Weste et al, 1976).

Low-lying or poorly drained land remains wet for long periods, which allows greater opportunity for sporangial production and zoospore release. The runoff water from nearby infected plants may also carry zoospores. Inoculum spreads in a fan-shaped zone from drainage channels or road culverts (Fig. 12) (Weste and Law, 1973; Weste and Taylor, 1971). Progress of disease on downhill slopes with drainage water has been measured at 400 m/yr in both a sclerophyll shrubby woodland and a dry sclerophyll forest. Along the contour, however, it varied from 171 m/yr at the height of an epidemic on shallow soils (25-cm depth) to 18 m/yr on deep sands (more than 85 cm deep) and 12 m/yr on krasnozem soil (Weste et al, 1973) (Fig. 13). The shallow soils of the Brisbane Ranges are characterized by a particularly steep soil water potential gradient. A reduction of only 2.5% in moisture content of soil changes the soil from saturation, which promotes zoospore formation, release, and dispersal, to water stress, which endangers survival of the plant.

Aeration in the soil affects both the population of the pathogen and the health of the roots. In anaerobic soil, plants may be predisposed to disease. Mycelial growth in vivo of *P. cinnamomi* has a wide tolerance for various levels of oxygen and carbon dioxide (Zentmyer, 1980). Other species, such as *P. cactorum*, are more sensitive (Durbin, 1959). Sporangial formation by several species of *Phytophthora* is reduced by a decrease in oxygen or an increase in carbon dioxide (Mitchell and Zentmyer, 1971; Nesbitt et al, 1979b). *P. drechsleri* formed sporangia on, but not below, the surface of saturated soil (Duniway, 1975b). Zoospores of *P. citrophthora* germinated more rapidly in an atmosphere containing 40 ppm oxygen than in one with less oxygen (Klotz et al, 1966). *P. cinnamomi* grew satisfactorily at a low oxygen concentration that damaged higher plants, retarded growth, and caused a

TABLE 6. Correlation of the related factors of soil depth, organic matter content, and microbial population density with disease development caused by *Phytophthora cinnamomi* in transect samples

| Transect sample[a] | Soil depth (cm) | Organic content (%) | Microorganisms ($\times 10^5$) per gram of soil | Disease potential index[b] | Symptoms |
|---|---|---|---|---|---|
| 0 | 25 | 2.29 | 45 | 256 | Severe |
| 1 | 30 | | 112 | 256 | Severe |
| 2 | 25 | | 147 | 64 | Severe |
| 3 | 40 | | 224 | 0 | Nil |
| 4 | 50 | 8.03 | 659 | 0 | Nil |

[a] Samples were taken at 1-m intervals from a transect from a diseased to a healthy area in the Brisbane Ranges, Victoria.

[b] The disease potential index is the reciprocal of the smallest fraction of a 50-g soil sample that provides a positive indication of the pathogen by bait isolation.

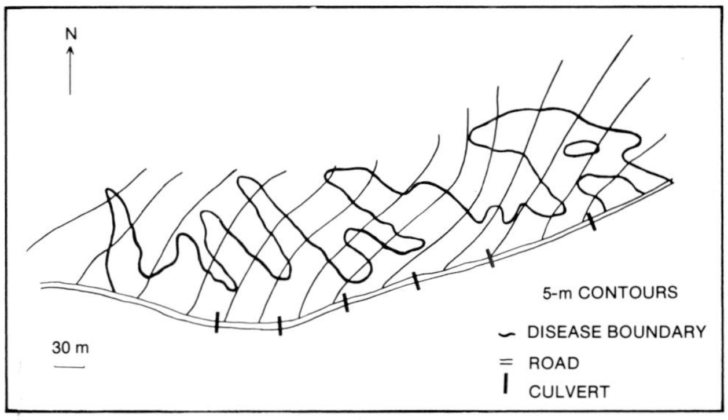


**Fig. 12.** Extension of disease caused by *Phytophthora cinnamomi* from infested gravel via road culverts, Wilson's Promontory, Victoria, Australia. (Reprinted, by permission, from Weste and Law, 1973)

proliferation of highly susceptible adventitious roots. High moisture and low oxygen together increased root exudation, which in turn stimulated propagule germination and growth (Drew and Lynch, 1980). Host susceptibility is also enhanced. Drew and Lynch (1980) ascribe decreased production of phenol oxidases and phytoalexins in roots, suppression of mycorrhizae, reduction of nitrogen fixation, and an increase in antagonistic microorganisms in soil to waterlogging and anaerobiosis.

Movement of zoospores to host roots depends on pore size, which is related to soil texture and structure and soil water status (Griffin, 1972, 1978). Free water allows greater zoospore dispersal, because in addition to surface water, all large pores are water-filled.

The effect of nutrients in soil on *Phytophthora* populations is complex. Nickel and cobalt in soil increased resistance of potato to *P. infestans* (Isaeva, 1969). Ammonia from decomposing alfalfa residue inhibited *P. cinnamomi* (Gilpatrick, 1969). In vitro nutrient requirements for *P. cinnamomi* have been reviewed by Zentmyer (1980). The sterol requirement may have an important effect on the population of *Phytophthora* propagules (see Hohl, Chapter 3). Nutrient requirements are more likely to limit in vitro than in vivo growth and populations, except as a result of competition with soil microbes.

Soil populations of *Phytophthora* species, as indicated by measurements of disease potential, vary with the organic matter content of the soil. Saprophytic growth of fungi depends on organic substrate for nutrients, but on the other hand, soils with high organic matter content contain more bacteria and actinomycetes, many of which are competitive, antagonistic, or predatory to fungi. Figure 14 demonstrates the association between the organic content of soil and the numbers of microorganisms for a range of soils and gravels.

Disease potential varies inversely with the number of soil microorganisms (Weste and Vithanage, 1977, 1978a) (Table 6). Marx and Haasis (1965) showed that microorganisms, or their diffusates, stimulated sporangial formation in *P. cinnamomi*. Some of the bacteria involved have since been identified (Chee and Newhook, 1966; Zentmyer, 1965). Common species of *Trichoderma*, a soil saprophyte, stimulate sexual reproduction in the A2 mating type of some heterothallic *Phytophthora* species, such as *P. cambivora, P. palmivora*, and *P. cinnamomi* (Pratt et al, 1972; Reeves and Jackson, 1972). Brasier (1975a, 1975b) speculates that this response evolved as a defense mechanism. Malajczuk et al (1977) studied the association between bacteria and *P. cinnamomi* and demonstrated that the bacteria involved in sporangial stimulation, sporangial breakdown, zoospore inhibition, and hyphal lysis were also antagonistic in vitro and affected the population density. Species of *Pseudomonas, Bacillus*, and *Streptomyces* that were identified may be the agents of biological control in the future.

Certain ectomycorrhizal fungi are antagonistic to *P. cinnamomi*; for example, *Leucopaxillus cerealis* var. *piceina* on *Pinus echinata* inhibited mycelial and zoospore infection (Marx and Bryan, 1969). Malajczuk (1979) showed that the different species of mycorrhizal fungi that grew on the susceptible jarrah (*Eucalyptus marginata*) differed from those on the tolerant marri (*E. calophylla*). Also, the rhizosphere of *E. calophylla* contained more actinomycetes than that of *E. marginata* (Malajczuk and McComb, 1979). On the basis of this paper, it appears likely that different species of forest trees have different symbionts and unique populations of bacteria and/or actinomycetes in the mantle of the mycorrhiza and that these differences, which are both qualitative and quantitative, probably exert differing effects on the population of *Phytophthora*. We need more information on the association between populations of soilborne *Phytophthora* species and specific soil microorganisms. Malajczuk (Chapter 16) covers this area in more detail.

In Australian forests, the most severe disease occurs on soils with an extremely low microbial population of actinomycetes and aerobic bacteria (Table 7) (Podger, 1972; Weste and Vithanage, 1978a). A seasonal study in Australia demonstrated that pathogen (*P. cinnamomi*) and microbial

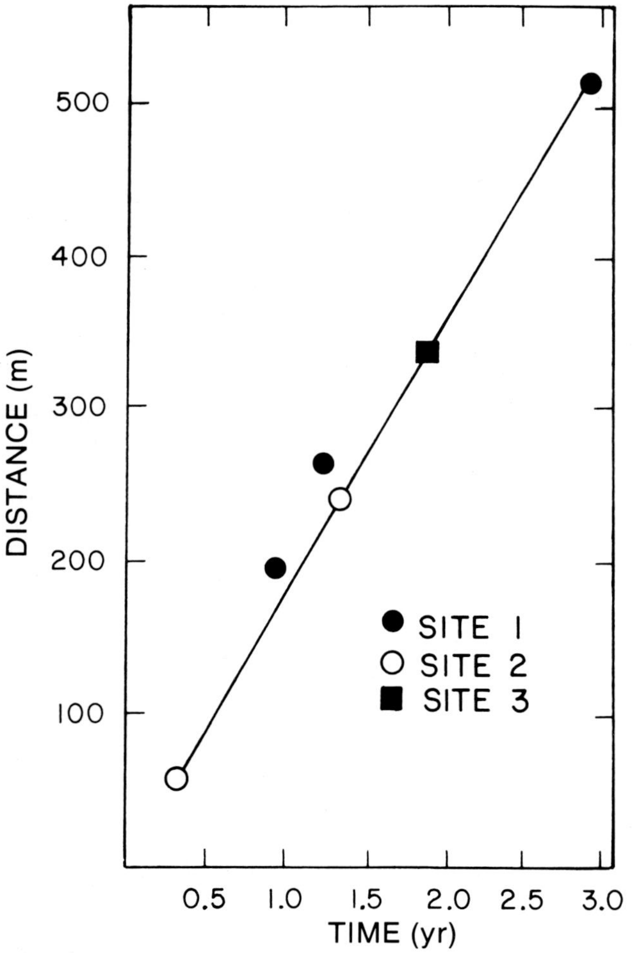


**Fig. 13.** Rates of extension of disease caused by *Phytophthora cinnamomi* on level land at three different sites, Brisbane Ranges, Victoria, Australia. (Reprinted, by permission, from Weste et al, 1973)

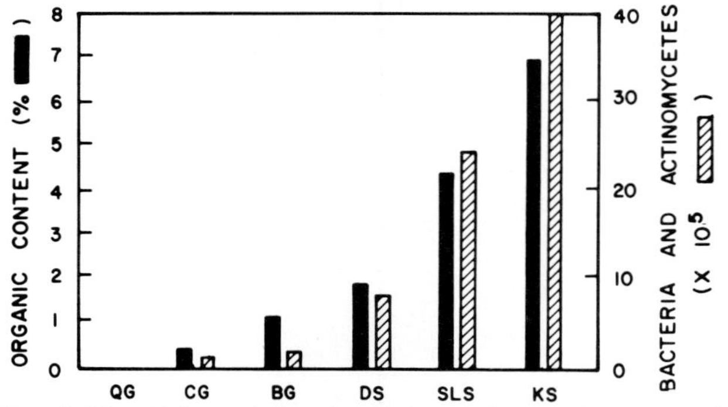


**Fig. 14.** Microbial populations in relation to the organic content of various soils. QG = quarried gravel; CG = colluvial gravel; BG = buckshot gravel; DS = duplex soil; SLS = sandy loam soil; KS = krasnozem soil.

populations were out of phase. Microbial populations were maximum during autumn and winter, whereas pathogen populations were maximum during spring and early summer. Thus, in a natural situation, biological control did not take place (Weste and Vithanage, 1978a).

### Interaction Among Factors Affecting Population

Environmental factors vary seasonally, particularly temperature, rainfall, and day length. They interact with each other and with the pathogen to produce seasonal fluctuations in *Phytophthora* populations, changes in host population and growth, and variations in site factors. Susceptible crops may be sown consecutively or planted into soil containing *Phytophthora* inoculum in the form of infected crop residue. In "permanent" forest stands, infected root residues are also continually being added to the soil. Root growth into this inoculum is also seasonal. Environmental factors such as moisture and temperature are continually changing both daily and seasonally. These directly influence the population of the pathogen, which continually changes in response to these variables. Seasonal cyclic variations in disease potential are likely to reflect these changes.

As an illustration of seasonal fluctuation in a *Phytophthora* population, consider the production of sporangia or chlamydospores by *P. cinnamomi* in nonsterile soils from three entirely different forest communities (Weste and Vithanage, 1978b, 1979a) (Figs. 3 and 5). Few or none formed during the winter when soil was cold; large numbers formed during spring when the soil was both warm and wet; and none developed during the summer when the soil was warm but dry. Despite continued dryness, large numbers of chlamydospores formed in autumn. The figures for both sporangia and chlamydospores represent a mean of actual counts made microscopically. For the same three forest soils, disease potential as measured by host-plant assay was also seasonal, reaching a maximum in the spring and early summer and falling to zero in the cold and dry periods. Based on the conclusions made by Rotem et al (1971) that environmental factors are relative and not absolute, Colhoun (1973) commented that reports on the influence of individual factors on field conditions should be regarded with extreme caution.

### Rate of Population Change

The rate of population change is of vital importance to the pathogen because the season during which temperature and moisture are suitable for infection of susceptible host material may be limited. Many characteristics of *Phytophthora*, such as its relatively short life cycle in which motile zoospores are formed, combine to make reproduction and dispersal rapid and efficient. The rate of sporangial production varies not only with the species of *Phytophthora* but also with the host. Shea et al (1978) reported that numbers of sporangia were lower on more resistant legumes than on the susceptible *Banksia grandis* in the jarrah forest understory.

Resistance to *P. cinnamomi*, which has a wide host range, differs fundamentally from resistance to the species-specific *P. megasperma* f. sp. *glycinea*. Resistance to the species-specific pathogen may rely on the genetic incompatibility of interacting surface molecules such as glycoproteins and lectins, whereas resistance to a pathogen with a broad host range may depend entirely on the rate of host response to the invading pathogen. The response may consist of rapid and continued root regeneration, as in grasses and sedges, or reduction of transpiration and water loss or suppression of sporangia, as in legumes.

Rate of sporangial formation also varies with the species. Gregory (1974) commented on the slower rate of reproduction of *P. palmivora* compared to *P. infestans*. He concluded that the danger of the black pod disease becoming an epidemic was due to the prolonged time ($t$) over which the multiplication rate ($r$) operated. In terms of the multiple-cycle disease or the compound interest equation, $x = x_0 e^{rt}$, where $x$ is the amount of diseased tissue, $x_0$ is the amount of initial inoculum, and e is the natural logarithm value, time is an important exponential factor. Infected pods that remain on the tree or on nearby ground are the source of initial inoculum ($x_0$) for the main or later crop. The prolonged flowering and cropping period of cocoa allows a dangerously long time ($t$) for the black pod epidemic to develop. MacKenzie et al (Chapter 23) discuss factors that influence the rate of population change.

## SURVIVAL OF *PHYTOPHTHORA*

The chances of continued existence of a pathogen are increased if the pathogen has the capacity not only to parasitize the host but also to survive in its absence. Many biologically successful pathogens survive as dormant or resting propagules and are also saprophytic, at first in dead host tissue and later in soil or water. Different species of *Phytophthora* vary in their capacity to persist either as dormant propagules or as saprophytic colonizers of dead organic material. However, accurate data on survival of different *Phytophthora* species are scarce. More precise information might be obtained from the use of inoculum that is isotopically labeled, for example with $^{14}C$ in hyphal walls, to locate the pathogen in soil organic particles or in plant residue.

### Survival in Host Tissue

*P. infestans* and *P. erythroseptica* survive in stored potato tubers, which provides enough inoculum to start epidemics in susceptible host populations if environmental conditions are favorable. Survival lasts at least from one season to the next, usually 6–8 mo. The inoculum is in the form of mycelium, which is protected and kept moist within host tissue and survives at temperatures below zero. Oospores survive for longer periods. Although oospores are common for a homothallic species such as *P. erythroseptica*, they are rare for heterothallic species such as *P. infestans*. Except in Mexico, where both A1 and A2 compatibility types occur, only the A1 is known. *P. palmivora* survives 2–4 wk on infected cocoa pods. It overwinters on rubber trees but not on cocoa (Turner, 1974) or on ground litter.

TABLE 7. Correlation of density of microorganisms and organic matter content with disease risk for three Victorian soils

| Soil type | Organic matter content of soil (%) | Number of microorganisms per gram of soil | | | | Relative disease risk |
|---|---|---|---|---|---|---|
| | | Fungi | Bacteria | Actinomycetes | Total | |
| Shallow duplex | 2.29 | $20 \times 10^4$ | $2.5 \times 10^4$ | $16 \times 10^4$ | $4 \times 10^5$ | Severe |
| Krasnozem | 2.75 | $40 \times 10^4$ | $15 \times 10^5$ | $12 \times 10^5$ | $3 \times 10^6$ | Moderate |
| Garden loam | 7.15 | $11 \times 10^4$ | $75 \times 10^5$ | $12 \times 10^6$ | $2 \times 10^7$ | Low |

The survival of soilborne *Phytophthora* species in infected host tissue depends primarily on the $\psi_m$. Shea et al (1980) recovered *P. cinnamomi* from dead *B. grandis* roots 2 yr after the plant died. The pathogen was recovered from suberized and partially suberized perennial roots of jarrah and from the collar region but not the fine roots of *Banksia* during the dry summer. Although both fine roots and proteoid roots were invaded during the warm wet spring or when irrigated, *P. cinnamomi* was not recovered. Shea et al (1980) concluded that the stump, lower stem, and root system of *B. grandis* provided a large, protected reservoir of inoculum from which large populations of zoospores invaded jarrah trees. Victorian forests have dry periods during late summer and autumn, but the rains are sufficient to maintain a large population of *P. cinnamomi* in the soil during spring and early summer.

Survival in fine roots, as detected by plating on agar, depends partly on the $\psi_m$ of the soil in which the roots are buried (Weste, 1982). In a series of experiments conducted in 1980, *P. cinnamomi* survived 30 days in fragile roots of 10-day-old seedlings of lupine (*Lupinus angustifolius* L.) buried in a nonsterile loam soil at three different $\psi_m$. However, when roots were buried in glass beads, gravel, or sand at the same $\psi_m$, the roots dried quickly and shriveled, and the pathogen failed to survive (Table 8). By contrast, chlamydospores on nylon mesh and buried in the same soils produced mycelium and more chlamydospores (Fig. 6), and about one-third survived for 30 days (Table 5).

Garrett (1970) pointed out that pathogens have an initial advantage over saprophytic soil microorganisms in that they already occupy the infected root before it dies and is decomposed. Saprophytic survival can then be interpreted in terms of relative growth rate, temperature, and affinity for the substrate (Gray, 1976). The $\psi_m$ of the substrate is important. In peach orchards in which *P. cinnamomi* causes stem cankers, the canker tissue is rapidly colonized by soil saprophytes. Consequently, the pathogen is difficult to isolate except from the advancing margin of necrotic tissue (P. Taylor, *personal communication*). In soils with less organic matter and fewer microorganisms, such as the Western Australian laterites, *P. cinnamomi* survives for years in dead host tissue if moisture is sufficient.

### Survival in the Absence of a Host

A pathogen must be able to survive between active parasitic phases, for example, on a crop, or it may become extinct. The surviving inoculum forms the springboard for epidemics. The soilborne species like *P. cinnamomi* and *P. cactorum*, which not only survive between hosts but also increase in population and distribution in a saprophytic phase in the soil, possess a decided competitive advantage. Most epidemics initiated by *Phytophthora* species originate from soilborne inoculum; for example, *P. palmivora*, cause of black pod disease (Okaisabor, 1974), and *P. nicotianae* var. *parasitica*, cause of fruit rot of guava (Mehrotra and Prasad, 1980).

How does *Phytophthora* multiply and survive in the absence of a host? Can we explain the fluctuations of disease potential? We need more quantitative information on the competitive saprophytic ability of soilborne *Phytophthora* species. Where is the pathogen during cold or dry periods when the disease potential is zero? What are the modes and rates of the increase of population when rain falls and soil temperatures rise? We need more information on the numbers of propagules per gram of soil in relation to organic particles or dead roots to calculate and compare the competitive saprophytic ability of different *Phytophthora* species in the same soil or of the same *Phytophthora* species in different soils. Competitive saprophytic ability can be measured by placing uniform fragments of dead plant tissue in the soil to trap the pathogen and then measuring the percentage of bait colonized by the *Phytophthora* species.

*Phytophthora* may survive in soil in the form of mycelium, sporangia, zoospores, cysts, chlamydospores, and oospores. Drying, cold, and antagonistic or competitive microorganisms are limiting factors. Hornby (1975) demonstrated that the take-all fungus, *Gaeumannomyces graminis* var. *tritici*, survived much longer in large soil particles than in small ones. The pathogen survived 14 mo in coarse particles (bigger than 4,000 $\mu$m) but only 5 mo in fine particles (300–700 $\mu$m). *P. cinnamomi* survived longer in gravels than in fine soils (Weste and Vithanage, 1979a) maintained at the same $\psi_m$ and temperature if there was an organic substrate in the gravel. Shea et al (1980) demonstrated that *P. cinnamomi* survived on larger organic particles in irrigated soils from jarrah forests of Western Australia. Survival was in the form of mycelium but not chlamydospores (Table 9).

Zoospores, which lack cell walls and have a short life, may be the most vulnerable propagules; mycelium, which is lysed by bacteria, may be the next most vulnerable, followed by sporangia, then cysts. The least vulnerable propagules may be chlamydospores, which appear to be adapted for survival because of their thick walls and tolerance for minimum water

TABLE 8. Formation and viability of chlamydospores of *Phytophthora cinnamomi* in roots of inoculated lupine seedlings at different times after being buried in different nonsterile substrates

| Medium | Matric potential (bars) | Chlamydospores (no.) per root[a] and viability[b] in roots | | | | | | Viability[b] in substrate after 30 days |
|---|---|---|---|---|---|---|---|---|
| | | 10 days | | 20 days | | 30 days | | |
| | | No. | Viability | No. | Viability | No. | Viability | |
| Glass beads | −1.4 | 0 | + | 30 | + | 0 | − | − |
| (2.5–3.5 mm in diameter) | −1.1 | 161 | + | 35 | − | 3 | − | − |
| | −1.0 | 514 | + | 281 | + | 5 | − | − |
| Gravel | −0.9 | 360 | + | 0 | − | 0 | − | − |
| | −0.7 | 485 | + | 4 | − | 0 | − | − |
| | −0.6 | 393 | + | 0 | − | 3 | − | − |
| Sand | −0.7 | 55 | + | 0 | − | 0 | − | − |
| | −0.7 | 150 | + | 185 | + | 0 | − | − |
| | −0.7 | 87 | + | 264 | + | 0 | − | − |
| Soil | −32.6 | 1 | + | 26 | + | 36 | + | + |
| | −19.6 | 4 | + | 136 | + | 41 | + | + |
| | −6.9 | 10 | + | 30 | − | 45 | + | + |

[a] Each result is the mean of six replicates, three from each of two isolates.

[b] + = Viable, − = not viable. In some cases, viability of *P. cinnamomi* was due to mycelium in the root.

potential, and finally oospores, which may be constitutively dormant.

**Mycelium**

Sneh and McIntosh (1974) found that the rate of lysis of mycelium of *P. cactorum* depended on temperature. Lysis required 3 days at 29°C but 45 days at 4°C. Most sporangia and oospores survived freezing but not drying. Only a few oospores germinated after drying. *P. parasitica* was recovered after 180 days from moist soil at 12°C (Holdaway and Tsao, 1971). *P. parasitica* var. *nicotianae* overwinters in high populations in the upper soil (Flowers and Hendrix, 1972), and *P. fragariae* may persist in soil in the absence of strawberry for many years (Bain and Demaree, 1945). *P. cinnamomi* survived 6 yr in moist soil at 20°C in the absence of a host (Zentmyer and Mircetich, 1966). Persistence by mycelium varies with density of the microbial population. Washed mycelium of *P. cinnamomi* survived and infected forest species when buried in undisturbed forest soils (Podger, 1972; Weste, 1974).

**Sporangia and Zoospores**

Survival of sporangia and zoospores is based on the maxim "populate or perish." In suitable conditions of temperature and moisture, *Phytophthora* sporulates rapidly, and zoospores become widely dispersed away from micro-environmental hazards by swimming toward nutrients exuded by host roots. Exudates overcome soil fungistasis and exert a chemotactic effect on the zoospores. This gives the motile zoospores a survival advantage over competitive microbes. Germinating zoospore cysts may also have a competitive advantage because of the chemotropic orientation of the germ tubes toward the host root.

Tests of the survival of zoospore cysts of *P. megasperma* and *P. cinnamomi* in soil at various $\psi_m$ values (MacDonald and Duniway, 1978) demonstrated that *P. megasperma* cysts survived only 2 wk at $\psi_m$ values ranging from 0 to −15 bars, whereas *P. cinnamomi* cysts survived 3 wk at −1 bar, 6 wk at between −5 and −15 bars, and 1 wk at −75 bars (Fig. 15). These results are similar to those obtained for chlamydospores of *P. cinnamomi* (Weste and Vithanage, 1979a). MacDonald and Duniway (1978) showed that cysts survived and germinated without change of form. Both zoospore cysts and chlamydospores of *P. cinnamomi* actually increased in numbers following infestation of a nonsterile, hostfree soil. By contrast, zoospore cysts of *P. megasperma* var. *sojae* lost infectivity in soil within 24 hr, and those of *P. drechsleri* survived only 15 days in moist soil (Mehrotra, 1972). Zoospores of *P. palmivora* remained infective for 6–18 mo when stored in soil at 50% moisture-holding capacity but for only 10–32 days in saturated or air-dried soil (Turner, 1965).

TABLE 9. Number and type of propagules of *Phytophthora cinnamomi* detected in fractions of soil collected from three irrigated plots

| | Sieve pore size ($\mu$m)[a] | | | | |
|---|---|---|---|---|---|
| Origin of colony | 150 | 63 | 38 | 8 | Residue |
| Soil organic matter | 304 | 102 | 35 | 23 | 12 |
| Sporangia | 0 | 5 | 2 | 3 | 1 |
| Chlamydospores | 1 | 2 | 4 | 0 | 1 |
| Zoospores | 0 | 0 | 0 | 0 | 0 |
| Total | 305 | 109 | 41 | 26 | 14 |

[a] Fractions were determined by sieving a 90-g (wet-weight) sample on sieves of different pore sizes.

*Source*: Shea et al (1980).

*P. cactorum* zoospores survived only a few weeks in soil at −0.1 to −0.2 bar but longer at matric water potential between −2 and −5 bars.

**Chlamydospores**

Survival of chlamydospores in soil is apparently similar to that of zoospore cysts. In addition, chlamydospores can survive in the gastrointestinal tracts of birds and termites, which have dispersed the fungus in Western Australian jarrah forests (Keast and Walsh, 1979).

Chlamydospores are delineated from hyphae by a septum and a resistant wall. Survival may depend on thickness of the wall, which is variable in many species. We do not know what conditions promote the formation of a thick wall in the chlamydospore or whether the thick wall surrounds some chlamydospores from inception or results from subsequent modification. Microchlamydospores have been recorded for some *Phytophthora* species (Cother and Griffin, 1973). Their production by uncontaminated isolates requires verification, possibly with immunofluorescent techniques as suggested by Tippett (1978).

In experiments measuring survival of chlamydospores in a range of six different nonsterile, hostfree soils, each at three different $\psi_m$ values, numbers of chlamydospores increased in hostfree, nonsterile gravels and soils (Fig. 16). Survival was a maximum in gravel, 52% of the particles of which were greater than 2 mm in diameter and which was low in organic matter and microorganisms. Survival was a minimum in krasnozem soil, 58% of the particles of which were less than 0.004 mm in diameter and which had a higher organic matter content than the gravel and a large population of microorganisms. Survival of chlamydospores was maximum at $\psi_m$ of −5 bars.

The relation between survival of chlamydospores and $\psi_m$ is illustrated in Figure 17 (Weste and Vithanage, 1979a). Survival of chlamydospores of *P. cinnamomi* depends on $\psi_m$. Zentmyer and Mircetich (1966) demonstrated that chlamydospores survived 6 yr if the soil moisture was greater than 3%. Chlamydospores were not detected as survival propagules in Western Australian jarrah soils, probably because of their requirements for moisture (Shea et al, 1980).

Chlamydospores of *P. cinnamomi* form in soil, gravel, or plant tissue during dry periods (Weste, 1982). They germinate under favorable conditions and may grow to form mycelia and sporangia or more chlamydospores. The latter in turn may remain dormant until conditions become suitable, then germinate to produce infective mycelia, sporangia, and zoospores, or more chlamydospores. The cycle may

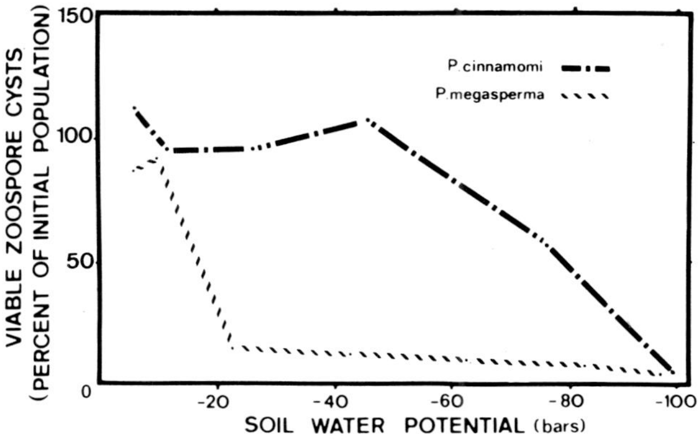


**Fig. 15.** Viability of zoospore cysts of *Phytophthora cinnamomi* and *P. megasperma* as a function of the total water potential to which they were dried in soil. (Reprinted, by permission, from MacDonald and Duniway, 1979)

apparently continue for at least 5 yr (Weste, 1975a), provided there is a nutrient source (organic material) and a microbial population that is not highly competitive.

### Oospores

Experimental evidence indicates that oospores of *Phytophthora* have great survival value. In some species such as *P. erythroseptica* and *P. palmivora*, oospores have been observed frequently, but in others, particularly the heterothallic species such as *P. cinnamomi*, they occur very rarely. However, oospore formation may be stimulated in *P. cinnamomi* by volatiles from *Trichoderma* spp. and in avocado roots or by root extract (Zentmyer, 1979). Oospore formation requires sterols (Elliott et al, 1964); oospores are apparently constitutively dormant and germination is slow, with each species requiring different conditions. Table 10 presents some pregermination temperatures at which oospore germination for different species was stimulated. Oospores formed at one season may germinate after differing periods of

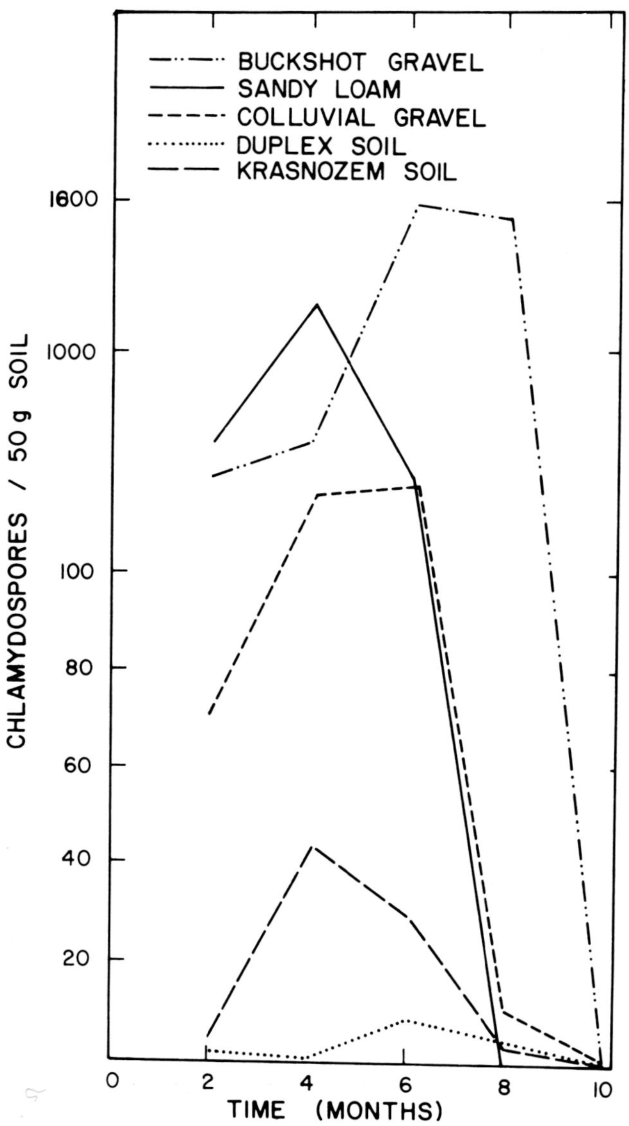


**Fig. 16.** Changes in the population of chlamydospores of *Phytophthora cinnamomi* over time on nylon mesh buried in nonsterile, hostfree soils and gravels at soil water matric potentials of −5 bars. Chlamydospores were counted after wet-sieving onto agar plates. (Reprinted, by permission, from Weste and Vithanage, 1979a)

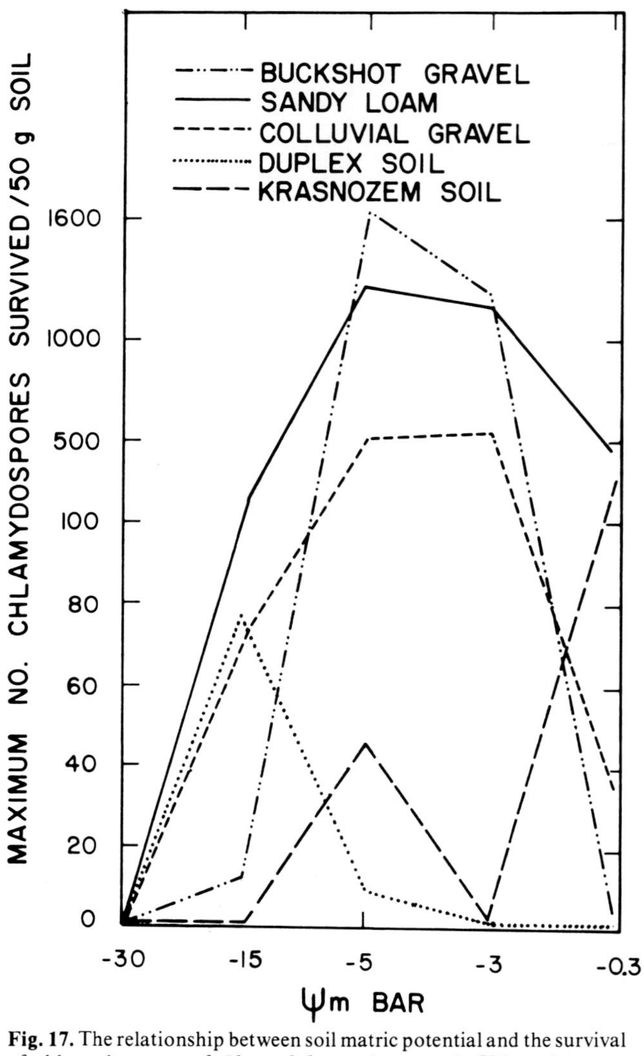


**Fig. 17.** The relationship between soil matric potential and the survival of chlamydospores of *Phytophthora cinnamomi*. Chlamydospores were counted after wet-sieving onto agar plates. (Reprinted, by permission, from Weste and Vithanage, 1979a)

TABLE 10. Examples of pregermination temperatures that stimulated oospore germination for six *Phytophthora* species

| Species | Temperature (°C) | Period (days) | Germination (%) | Reference |
|---|---|---|---|---|
| *P. infestans* | 4 | 45 | 64 | Romero and Erwin (1969) |
| *P. drechsleri* | 3 | 56 | Maximum | Klisiewicz (1970) |
| *P. palmivora* | 12 | 7 | Maximum | Kaosiri et al (1980) |
| *P. megasperma* var. *sojae* | 36 | 48 | 50 | Long et al (1975) |
| *P. cactorum* | 1 | | Maximum | Blackwell (1943) |
| *P. cinnamomi* | 24 | | Maximum | Zentmyer (1980) |

dormancy, thus maintaining a low but continuing population. Persistence of *P. fragariae* in soil is probably due to oospores within plant tissue (Duncan, 1980).

**Fail-Safe Devices**

Propagules of *Phytophthora* show alternative germination behaviors, depending on limiting environmental factors. This is surely an extraordinary adaptation for survival. Sporangia germinate directly by means of a germ tube or form motile zoospores that can swim toward a more suitable environment. Zoospores encyst and from the cyst, a germ tube may penetrate the root, or the cyst may produce a papilla or microsporangium, from which another zoospore may emerge. Chlamydospores germinate to produce mycelium and more chlamydospores, some of which may be microchlamydospores, or chlamydospores may germinate to form sporangia and zoospores. The production of a sporangium by a chlamydospore enables it to survive in fallow soils or gravel deposits, or it may provide an effective device for bridging hosts in space (Mircetich and Zentmyer, 1966). These alternative survival and germination processes are fail-safe devices that have evolved over time. The behavior that the propagule follows depends on certain limiting conditions such as moisture level, temperature, nutrient concentration, and levels of fungistatic factors in soil.

What is the prime risk for survival of species of *Phytophthora* in hostfree, nonsterile soil? Drying of soil is perhaps the greatest hazard to survival of water molds that require sufficient water. Experiments by Shea et al (1980) demonstrated that irrigation markedly increased the number of propagules isolated from soil (Fig. 18). Other factors are soil temperature and the occurrence and population of other microorganisms in soil. Although growth of *Phytophthora* is strongly temperature-dependent, survival of propagules is less sensitive to extremes of temperature unless these are also accompanied by water saturation or dryness. The sensitivity to temperature varies with the species. Temperatures ranging from below zero to 35–45°C, depending on duration, are lethal for various species (Zentmyer, 1980).

**Microbial Hazards**

The microorganisms responsible for suppressiveness in soil adversely affect survival of *Phytophthora* species in soil. These microorganisms cause hyphal lysis and abortive sporangia, which reduce the pathogen population in the soil. Some of these microorganisms have been identified as species of *Pseudomonas, Bacillus*, and *Streptomyces*, which are most active at 25–27°C and at moisture content approximating field capacity (Broadbent and Baker, 1974; Malajczuk et al, 1977; Nesbitt et al, 1979b). *Trichoderma* species, which are common soil fungi, are also antagonistic to *Phytophthora* species such as *P. cactorum, P. cinnamomi, P. erythroseptica, P. parasitica*, and *P. infestans* (Brasier, 1971; Reeves and Jackson, 1972). The reaction of *Phytophthora* species to the presence of antagonistic microorganisms is usually to reproduce. In the presence of antagonistic bacteria, sporangia and zoospores are produced; and in the presence of *Trichoderma*, oospores form; both responses are survival mechanisms.

Although specific microorganisms are antagonistic to *Phytophthora* species, others are effective in biological control simply because they compete for a substrate. Microbial populations exist in spatially discrete communities on certain favored sites of the rhizoplane, such as the longitudinal junctions of cells, and occupy only 4–10% of the root surface (Bowen and Rovira, 1976). Competition for substrate may be for root exudates, secretions, mucilages, mucigel, or lysates. Microorganisms that act locally to increase the permeability of the root cell have a competitive advantage (Cahill and Weste, 1982). The populations of microorganisms in the rhizosphere are directly affected by rooting intensity, growth of roots, and propagule germination. Root exudate may be doubled when host plants are stressed for water and is also increased by deficiencies in phosphate. Both survival of *Phytophthora* and disease potential in infested soil varied with the numbers of microorganisms (Weste and Vithanage, 1978a). One factor responsible for the longer survival of *P. cinnamomi* in gravels than in krasnozem soil was the small microbial population of the gravel. Garden soil contained $10^7$ microorganisms per gram of oven-dried soil, whereas soil from the Brisbane Ranges contained $10^5$ and gravel $10^2$ or $10^3$ microorganisms per gram of soil (Table 7). The relation between microbial numbers and active populations of *Phytophthora* has already been discussed and applies to a small extent to survival in soil. After an epidemic of Phytophthora root rot has passed through a susceptible forest, the disease potential is reduced and isolation of *P. cinnamomi* becomes difficult. Are the dead plants rapidly invaded by saprophytic microorganisms? Does the soil become suppressive? These problems need to be studied in greater detail in nonsterile soil. When Tippett (1978) added microbially rich soil to Western Australian laterite soils, most of the *P. cinnamomi* propagules in plant roots disappeared. When access to this soil was impeded by membranes with small pores (0.2 $\mu$m), *P. cinnamomi* survived in root cells and on the root surface and sporulated.

**Fire**

Bushfires are frequent in Australian dry sclerophyll forests susceptible to *P. cinnamomi*. The effects of fire on the population of *P. cinnamomi*, which have been evaluated, demonstrated that control was not significant because fires heated soil only to a depth of a few centimeters (Marks et al, 1975; Weste, 1974). However, fire is being used to control *Phytophthora* by replacing the fire-sensitive *B. grandis* with a less susceptible understory to reduce disease potential and remove a long-standing reservoir of inoculum in *Banksia* stumps and roots (Shea, 1975).

**Depth**

Where and at what depth in the soil do soilborne *Phytophthora* species survive? The evidence favors large

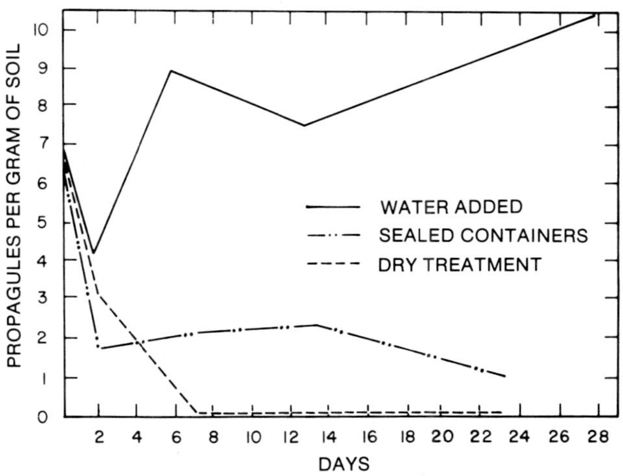


**Fig. 18.** Survival of *Phytophthora cinnamomi* in the absence of a host under different watering regimes. Propagules were counted after wet-sieving onto agar plates. (Reprinted, by permission, from Shea et al, 1980)

organic particles and dead root tissue. *Phytophthora* species can be isolated from soil at all depths at which roots grow; for instance, to rock or subsoil base at depths of about 85 or 91–105 cm (Weste and Ruppin, 1977; Zentmyer, 1980). It therefore follows that despite absence of *Phytophthora* from surface soils, residual inoculum at greater depths may be sufficient to start an epidemic. Consequently, samples for measurements of population must be collected from all depths. Otherwise, even though populations in the upper soil levels correlate well with disease potential, the data are incomplete and liable to be misinterpreted.

## CONCLUSION

The evidence suggests that some species of *Phytophthora* colonize dead organic matter in nonsterile soils without a host and therefore possess limited competitive saprophytic ability if the soil is moist. Competitive saprophytic ability ranges from little or none in the case of species-specific biotrophs, such as *P. infestans* and *P. megasperma* f. sp. *glycinea*, to considerable in the case of *P. cinnamomi*. A few nonpathogenic species, such as the marine species, are entirely saprophytic. Survival in the soil provides a reservoir of initial inoculum that is difficult to eradicate and that is usually sufficient to start an epidemic.

## SUMMARY

Plant diseases caused by species of *Phytophthora*, like those caused by other pathogens, depend on the population density. The relation is linear when numbers are low but becomes sigmoid when numbers of propagules are high. Populations of *Phytophthora* are continually changing; these changes are of great importance in disease development and must therefore be assessed in soil, water, and plant tissue. Most studies illustrate the normal process of increasing disease with increasing population, but at times, disease potential fluctuates independently of the population because of changes in the environment or in host susceptibility without concurrent changes in the population density of the pathogen. Increased *Phytophthora* populations usually precede disease increase; hence the factors responsible for changes in the population are important for disease forecasts and management. These factors are related to the pathogen, the initial inoculum, host susceptibility, and each particular type of environment. The rates of growth, reproduction, and dispersal of the various species of *Phytophthora*; their competitive saprophytic ability; host susceptibility; and the ecology and cultivation practices used for the agricultural crop, forest, or nursery all influence the population of *Phytophthora* in soil, water, or host tissue. The population density changes with seasonal changes in temperature and moisture. *Phytophthora* species grow, reproduce, and infect over a temperature range that varies with the species. Within that temperature range, numbers depend on relative humidity or $\psi_m$, because the density of the population is largely due to production of spores, which depend on high humidity or the presence of free water. Populations of soilborne *Phytophthora* species may be effectively reduced in a soil that is suppressive because of high numbers of antagonistic or competitive microorganisms. Conversely, *Phytophthora* populations may increase rapidly on shallow conducive soils invaded during the saprophytic phase of the *Phytophthora* life cycle. Shallow, poorly drained soils may contain large numbers of host roots that are subject to waterlogging. This provides ideal conditions for zoospore formation and dispersal, infection, and predisposition of roots to infection. The risk of high disease potential is great.

The rate of change in population may be important for crop and forest diseases, because seasonal changes in temperature and moisture may severely limit the time during which the population of *Phytophthora* can increase. Hence, the short life cycle, brief latent period, rapid reproduction, and dispersal of some species of *Phytophthora* enable them to increase sufficiently in a conducive environment to produce an epidemic. In some instances, a change of host species may reduce the rate of increase of the population and thus prevent an epidemic. The results discussed in this chapter confirm conclusions by MacKenzie et al (Chapter 23) that in general, diseases caused by *Phytophthora* species are multiple-cycle diseases that potentially are capable of increasing according to a compound interest model.

*Phytophthora* species are protected by continued occupancy of the colonized host tissue. Survival depends on the maintenance of a sufficient $\psi_m$ in diseased tissue. In the absence of a host, the chances of survival are increased by the capacity for independent saprophytic activity. Saprophytism is absent in biotrophic species, which grow and reproduce only in living host tissue, but important in species that have a saprophytic phase during which growth and reproduction occur independently of the living host. Survival may be in the form of mycelium in organic particles. Survival is greater in larger particles and in gravels with small microbial populations. Chlamydospores and zoospore cysts survive for long periods if the $\psi_m$ value is favorable (near zero). Oospores are the most resistant survival unit. They are considered to be constitutively dormant because they require particular conditions for germination. Survival of *Phytophthora* propagules is enhanced by a number of alternative germination procedures, or fail-safe devices, which may adapt to limiting environmental factors. Thus, most propagules germinate to produce either a germ tube or sporangia, which in turn may either germinate directly or produce zoospores. The greatest hazards to survival are dryness and high populations of antagonistic or competitive soil microorganisms.

## LITERATURE CITED

Allen, R. N., and Newhook, F. J. 1973. Chemotaxis of zoospores of *Phytophthora cinnamomi* to ethanol in capillaries of soil pore dimensions. Trans. Br. Mycol. Soc. 61:287-302.

Bain, H. F., and Demaree, J. B. 1945. Red stele root disease of the strawberry caused by *Phytophthora fragariae*. J. Agric. Res. 70:11-30.

Baker, R. 1971. Analyses involving inoculum density of soil-borne plant pathogens in epidemiology. Phytopathology 61:1280-1292.

Baker, R. 1978. Inoculum potential. Pages 137-157 in: Plant Disease. An Advanced Treatise. Vol. II. J. G. Horsfall and E. B. Cowling, eds. Academic Press, New York.

Blackwell, E. 1943. On germinating the oospores of *Phytophthora cactorum*. Trans. Br. Mycol. Soc. 26:93-103.

Bouhot, D. 1979. Estimation of inoculum density and inoculum potential: Techniques and their value for disease prediction. Pages 21-34 in: Soil-Borne Plant Pathogens. B. Schippers and W. Gams, eds. Academic Press, London.

Bowen, G. D., and Rovira, A. D. 1976. Microbial colonization of plant roots. Annu. Rev. Phytopathol. 14:121-144.

Brasier, C. M. 1969. The effect of light and temperature on reproduction in vitro in two tropical species of *Phytophthora*. Trans. Br. Mycol. Soc. 52:105-113.

Brasier, C. M. 1971. Induction of sexual reproduction in single $A_2$ isolates of *Phytophthora* species by *Trichoderma viride*. Nature (London) New Biol. 231:283.

Brasier, C. M. 1975a. Stimulation of sex organ formation in *Phytophthora* by antagonistic species of *Trichoderma*. I. The effect in vitro. New Phytol. 74:183-194.

Brasier, C. M. 1975b. Stimulation of sex organ formation in *Phytophthora* by antagonistic species of *Trichoderma*. II.

Ecological implications. New Phytol. 74:195-198.

Broadbent, P., and Baker, K. F. 1974. Behaviour of *Phytophthora cinnamomi* in soils suppressive and conducive to root rot. Aust. J. Agric. Res. 25:121-137.

Brown, B. N. 1976. *Phytophthora cinnamomi* associated with patch death in tropical rain forests in Queensland. Aust. Plant Pathol. Soc. Newsl. 5:1-4.

Byrt, P. N., and Holland, A. A. 1978. Infection of axenic *Eucalyptus* seedlings with *Phytophthora cinnamomi* zoospores. Aust. J. Bot. 26:169-176.

Byrt, P. N., Irving, H. R., and Grant, B. R. 1982a. The effect of cations on zoospores of the fungus *Phytophthora cinnamomi*. J. Gen. Microbiol. 128:1189-1198.

Byrt, P. N., Irving, H. R., and Grant, B. R. 1982b. The effect of organic compounds on the encystment and germination of zoospores of the cinnamon fungus, *Phytophthora cinnamomi*. J. Gen. Microbiol. (In press)

Cahill, D., and Weste, G. 1982. The formation of callose deposits as a response to infection with *Phytophthora cinnamomi*. Trans. Br. Mycol. Soc. (In press)

Cameron, J. N., and Carlile, M. J. 1977. Negative geotaxis of zoospores of the fungus *Phytophthora*. J. Gen. Microbiol. 98:599-602.

Campbell, W. A., and Copeland, O. L., Jr. 1954. Littleleaf disease of shortleaf and loblolly pines. U.S. Dep. Agric. Circ. 940. 41 pp.

Chee, K.-H., and Newhook, F. J. 1965. Improved methods for use in studies on *Phytophthora cinnamomi* Rands and other *Phytophthora* species. N.Z. J. Agric. Res. 8:88-95.

Chee, K.-H., and Newhook, F. J. 1966. Relationship of microorganisms to sporulation of *Phytophthora cinnamomi* Rands. N.Z. J. Agric. Res. 9:32-43.

Chi, C. C., and Sabo, F. E. 1978. Chemotaxis of zoospores of *Phytophthora megasperma* to primary roots of alfalfa seedlings. Can. J. Bot. 56:795-800.

Clerk, G. C. 1972. Germination of sporangia of *Phytophthora palmivora* (Butl.) Butl. Ann. Bot. 36:801-807.

Colhoun, J. 1973. Effects of environmental factors on plant disease. Annu. Rev. Phytopathol. 1 :343-364.

Commonwealth Mycological Institute. 1973. A guide to the use of terms in plant pathology. Pap. No. 17, Commonw. Mycol. Inst., Kew, Surrey, England.

Cother, E. J., and Griffin, D. M. 1973. Formation of chlamydospores by *Phytophthora drechsleri*. Trans. Br. Mycol. Soc. 61:379-383.

Crosier, W. 1934. Studies in the biology of *Phytophthora* infestans (Mont.) de Bary. Cornell Univ. Agric. Exp. Stn. Memo 155.

Drew, M. C., and Lynch, J. M. 1980. Soil anaerobiosis, microorganisms, and root function. Annu. Rev. Phytopathol. 18:37-67.

Duncan, J. M. 1980. Persistence of mycelium of *Phytophthora fragariae* in soil. Trans. Br. Mycol. Soc. 75:383-387.

Duniway, J. M. 1975a. Formation of sporangia by *Phytophthora drechsleri* in soil at high matric potentials. Can. J. Bot. 53:1270-1275.

Duniway, J. M. 1975b. Limiting influence of low water potential on the formation of sporangia by *Phytophthora drechsleri* in soil. Phytopathology 65:1089-1093.

Duniway, J. M. 1976. Movement of zoospores of *Phytophthora cryptogea* in soils of various textures and matric potentials. Phytopathology 66:877-882.

Duniway, J. M. 1979. Water relations of water molds. Annu. Rev. Phytopathol. 17:431-460.

Durbin, R. D. 1959. Factors affecting the vertical distribution of *Rhizoctonia solani* with special reference to $CO_2$ concentration. Am. J. Bot. 46:22-25.

Elliott, C. G., Hendrie, M. E., Knights, B. A., and Parker, W. 1964. A steroid growth factor requirement in a fungus. Nature (London) 203:427-428.

Fell, J. W., and Master, I. M. 1975. Phycomycetes (*Phytophthora* spp. nov. and *Pythium* sp. nov.) associated with degrading mangrove (*Rhizophora mangle*) leaves. Can. J. Bot. 53:2908-2922.

Flowers, R. A., and Hendrix, J. W. 1972. Population density of *Phytophthora parasitica* var. *nicotianae* in relation to pathogenesis and season. Phytopathology 62:474-477.

Garrett, S. D. 1970. Pathogenic Root-Infecting Fungi. Cambridge Univ. Press, London. 294 pp.

Gilpatrick, J. D. 1969. Role of ammonia in the control of avocado root rot with alfalfa meal soil amendment. Phytopathology 59:973-978.

Gisi, U. 1975. Investigations on the soil phase of *Phytophthora cactorum* (Leb. et Cohn) Schroet. with fluorescent optical direct observation. Z. Pflanzenkr. Pflanzenschutz 82:355-377.

Gisi, U., Oertli, J. J., and Schwinn, F. J. 1977. Water and salt relations of sporangia of *Phytophthora cactorum* (Leb. et Cohn) Schroet. in vitro. Phytopathol. Z. 89:261-284.

Gisi, U., Zentmyer, G. A., and Klure, L. J. 1980. Production of sporangia by *Phytophthora cinnamomi* and *P. palmivora* in soils at different matric potentials. Phytopathology 70:301-306.

Goode, P. M. 1956. Infection of strawberry roots by zoospores of *Phytophthora fragariae*. Trans. Br. Mycol. Soc. 39:367-377.

Gooding, G. V., and Lucas, G. B. 1959. Effect of inoculum level on the severity of tobacco black shank. Phytopathology 49:274-276.

Gray, T. R. G. 1976. The Survival of Vegetative Microbes. Soc. Gen. Microbiol. Symp. 26. T. R. G. Gray and J. R. Postgate, eds. Cambridge Univ. Press, Cambridge. 432 pp.

Greenhalgh, F. C. 1978. Evaluation of techniques for quantitative detection of *Phytophthora cinnamomi*. Soil Biol. Biochem. 10:257-259.

Gregory, P. H. 1974. *Phytophthora* Disease of Cocoa. Longman, London. 348 pp.

Griffin, D. M. 1972. Ecology of Soil Fungi. Syracuse Univ. Press, Syracuse. 193 pp.

Griffin, D. M. 1978. Effect of soil moisture on survival and spread of pathogens. Pages 175-197 in: Water Deficits and Plant Growth. Vol. 5. T. T. Kozlowski, ed. Academic Press, New York.

Hinch, J., and Weste, G. 1979. Behaviour of *Phytophthora cinnamomi* zoospores on roots of Australian forest species. Aust. J. Bot. 27:679-691.

Holdaway, B. F., and Tsao, P. H. 1971. Survival of *Phytophthora parasitica* in soils. (Abstr.) Phytopathology 61:1321.

Hornby, D. 1975. Inoculum of the take-all fungus: Nature, measurement, distribution and survival. Bull. OEPP 5(4):319-333.

Isaeva, G. Y. 1969. Influence of nickel and cobalt on the resistance of potato to diseases. Nauk. Pr. Zhitomir. Sil's'kogospod. Inst. 16:110-112. (Rev. Plant Pathol. 49:507)

Kaosiri, T., Zentmyer, G. A., and Erwin, D. C. 1980. Oospore morphology and germination in the *Phytophthora palmivora* complex from cacao. Mycologia 72:888-907.

Katsura, K., and Miyata, Y. 1966. Movement of zoospores of *Phytophthora capsici*. III. Rheotaxis. Sci. Rep. Kyoto Prefect. Univ. Agric. 18:81-88.

Keast, D., and Walsh, L. G. 1979. Passage and survival of chlamydospores of *Phytophthora cinnamomi* Rands, the causal agent of forest dieback disease, through the gastrointestinal tracts of termites and wild birds. Appl. Environ. Microbiol. 37:661-664.

Kliejunas, J. T., and Ko, W. H. 1976. Dispersal of *Phytophthora cinnamomi* on the island of Hawaii. Phytopathology 66:457-460.

Klisiewicz, J. M. 1970. Factors affecting production and germination of oospores of *Phytophthora drechsleri*. Phytopathology 60:1738-1742.

Klotz, L. J., Stolzy, L. H., De Wolfe, T. A., and Szuszkiewicz, T. E. 1966. Distribution of root-rotting fungi in citrus orchards as affected by soil oxygen supply. Calif. Agric. 20(6):15.

Kuan, T. L., and Erwin, D. C. 1980. Predisposition of water saturation of soil on Phytophthora root rot of alfalfa. Phytopathology 70:981-986.

Lapwood, D. H. 1968. Observations on the infection of potato leaves by *Phytophthora infestans*. Trans. Br. Mycol. Soc. 51:233-240.

Leeper, G. W. 1967. Introduction to Soil Science. Melbourne Univ. Press, Parkville, Victoria. 253 pp.

Long, M., Keen, N. T., Ribeiro, O. K., Leary, J. V., Erwin, D. C., and Zentmyer, G. A. 1975. *Phytophthora megasperma* var. *sojae*: Development of wild-type strains for genetic research. Phytopathology 65:592-597.

MacDonald, J. D., and Duniway, J. M. 1978. Influence of the matric and osmotic components of water potential on zoospore discharge in *Phytophthora*. Phytopathology 68:751-757.

MacDonald, J. D., and Duniway, J. M. 1979. Use of fluorescent antibodies to study the survival of *Phytophthora megasperma* and *P. cinnamomi* zoospores in soil. Phytopathology 69:436-441.

Malajczuk, N. 1979. The microflora of unsuberized roots of

# 19

# Relationship of Inoculum Density of *Phytophthora* Species to Disease Incidence in Various Hosts

D. J. Mitchell and M. E. Kannwischer-Mitchell

CONTENTS



## INTRODUCTION

A considerable amount of information on the relationships of inoculum densities of *Phytophthora* spp. to disease incidences in various hosts has accumulated since Schmitthenner (1970) reviewed the significance of populations of these fungi in soil only 12 yr ago but presented no quantitative information. Most of the current information, however, is from experiments conducted under defined conditions in the laboratory and greenhouse. Little information has been generated on the roles of initial inoculum of any soilborne phytopathogenic fungus or of time on the development of epidemics of soilborne diseases under field conditions. As Vanderplank (1975) contends, initial inoculum should be studied in as much detail as subsequent infection rates in the quantitative analyses of epidemics. Conceptualization of the quantitative relationships of inoculum of *Phytophthora* spp. to disease incidences in various hosts is an important introduction to eventual development of programs for forecasting disease. The objectives of this chapter are to summarize current knowledge on the relationships of inoculum densities of *Phytophthora* spp. to incidences of soilborne diseases, to enumerate some of the problems slowing progress in quantitation of these diseases, and to propose new approaches to quantitation of soilborne diseases caused by these fungi.

## INOCULUM-DISEASE RELATIONSHIPS IN THE FIELD

Recent advances in the development of media selective for the isolation of *Phytophthora* spp. from soil as reviewed by Tsao (1970 and Chapter 17) provide a foundation for new techniques that allow detection and quantitation of very low populations of these fungi in soil. Inoculum densities of *Phytophthora* spp. detected in soils associated with roots of perennial hosts ranged from less than one to approximately 30 propagules per gram of soil. Soil samples from root zones of apple and *Citrus* spp. contained 1–19 propagules of *P. cactorum* and 0.2–32 propagules of *P. parasitica* per gram of soil, respectively (McIntosh, 1977; M. E. Kannwischer-Mitchell and D. J. Mitchell, *unpublished*). In several studies with *P. cinnamomi*, the following inoculum levels were detected in root zones: less than one propagule per gram of soil from pines with little-leaf disease (Otrosina and Marx, 1975), less than five propagules per gram of soil with several species of plants native to Australian forests (Weste and Vithanage, 1978), 5–12 propagules per gram of soil from dying *Xanthorrhoea australis* (Malajczuk et al, 1978), 14 propagules per gram of soil from diseased avocado (Hwang and Ko, 1978), and 1–30 propagules per gram of soil in nurseries with dying Fraser fir seedlings (Hendrix and Kuhlman, 1965). Shew (1980) reported populations in Fraser fir nurseries similar to those found in the latter study but detected less than five propagules of *P. cinnamomi* per gram of soil from Fraser fir plantations.

When populations of *Phytophthora* spp. associated with perennial plants are determined through the seasons of the year, higher inoculum levels may occur at times when environmental conditions are optimum for the formation of zoospores. Although *P. cinnamomi* was not demonstrated to cause ohia decline, Kliejunas and Nagata (1979) followed seasonal variations in populations of the fungus and detected up to 330 propagules (66% of which were zoospores) per gram of soil in ohia forests. Isolation techniques must be selected with consideration of these variations; for example, sieving to retain spores may be useful when populations are very low but may lead to low population estimates if zoospores constitute all or some of the inoculum.

Inoculum densities of *Phytophthora* spp. detected in fields with histories of diseases caused by these pathogens are generally low or undetectable when soil is sampled before planting. Papavizas et al (1981) did not detect *P. capsici* when soil from eight fields, thought to be naturally infested, was sampled before planting and assayed on a selective medium, but up to 24 propagules per gram of soil were recovered from soil in plant rows of several crops in six of the eight fields during the growing season. Only 1–2.4 chlamydospores of *P. cinnamomi* per gram of soil were reported from soil that had been in pineapple cultivation for about 20 yr (McCain et al, 1967). Kannwischer and Mitchell (1978) and Flowers and Hendrix (1972) detected 0.2–0.75 and less than 10 propagules of *P. parasitica* var. *nicotianae* per gram of soil, respectively, in tobacco fields before planting.

The relationships of inoculum densities of *Phytophthora* spp. to disease incidences in annual or perennial crops under natural field conditions have been characterized only in a few studies and quantified only partially. Hendrix and Kuhlman (1965) noted that in a nursery where 50% of a stand of Fraser fir was killed, 1–30 propagules of *P. cinnamomi* per gram of soil were recovered. It should be emphasized that in the above zoospores; 98% of the papaya seedlings treated with $2.4 \times 10^3$ sporangia per container were killed.

## INFLUENCES ON INOCULUM-DISEASE RELATIONSHIPS

Many of the biological, chemical, and physical factors that influence plant diseases may influence critically the relationship of inoculum density to disease incidence. Although a great volume of literature is available on such influences on disease, only information on the influence of various factors over time on the quantitative associations of inoculum of various soilborne plant pathogens to disease will be addressed.

Of the physical factors in soil that influence soilborne diseases caused by *Phytophthora* spp., only the influences of soil water and temperature on the relationship of inoculum to disease have been examined quantitatively. Diseases caused by these fungi are favored by high soil water content. Sterne et al (1977a, 1977b) found that at 15 chlamydospores of *P. cinnamomi* per gram of sandy loam soil, 80–90, 50–90, 10–50, and 4–8% of *Persea indica* roots were diseased at matric potentials of 0, −0.05, −0.10, and −0.25 bar, respectively; in clay soil, significant differences in disease were not observed at matric potentials of less than −0.25 bar. The authors suggested that nutrient availability rather than matric potential per se limited disease development at a matric potential of −0.25 bar in the sandy loam soil; chlamydospore germination and germ tube growth were reduced significantly at a matric potential of −0.25 bar compared to potentials approaching zero, but the effects were reversed by the addition of glucose and asparagine to the soil. In contrast to the influence of matric potential, osmotic potential in the range found in avocado groves (−0.37 to −2.37 bars) did not influence disease. The influence of soil water on formation of sporangia and chlamydospores, on zoospore release from sporangia, and on zoospore motility and dispersal in soil has been addressed by Duniway (1979; Chapter 14). Preinoculation water stress of hosts can result in increased disease severity at defined inoculum levels (Duniway, 1977; Kuan and Erwin, 1980). Kuan and Erwin (1980) found that at $2 \times 10^4$ zoospores of *P. megasperma* f. sp. *medicaginis* per container of alfalfa seedlings, mortality ranged from 90% with plants predisposed by 5 days of growth in flooded soil to 50% with plants not predisposed. Water saturation of soil apparently predisposed roots to root rot by increasing root damage and exudation of nutrients, which increased the chemotactic attraction of zoospores to roots.

Extremes of temperature limit disease development by directly affecting pythiaceous fungi in soil (Zentmyer and Erwin, 1970) or by reducing infection or subsequent disease at defined inoculum levels (Banihashemi and Mitchell, 1975; Gray et al, 1973; Mitchell, 1978; Shew, 1980). Banihashemi and Mitchell (1975) reported 85% infection of safflower seedlings exposed to $10^3$ oospores of *P. cactorum* per gram of soil at 28°C but only 35% infection at 24°C. Shew (1980) observed 32–46% infection and similar levels of subsequent mortality of Fraser fir seedlings at 10 chlamydospores of *P. cinnamomi* per gram of soil maintained at 19, 22, or 25°C; however, at 16°C only 16% of the trees died although 41% were infected.

Other physical factors that may influence the quantitative association of inoculum of *Phytophthora* spp. and disease include soil hydrogen ion concentration, aeration, and structure. Tsao and Martin (1963) reported 26, 67, and 51% reductions in numbers of healthy citrus root tips in soils at pH 4.2–5.5, 6.0–6.5, and 7.0–8.4, respectively, with a standard inoculum density of *P. parasitica*. Little is known about the influence of oxygen, carbon dioxide, or the various volatiles that are commonly present in soil on the infection process (Stolzy et al, 1965, 1967), but generally growth and reproduction of *Phytophthora* spp. are limited less than those of many other soilborne pathogens by low oxygen or high carbon dioxide concentrations (Dukes and Apple, 1965; Klotz et al, 1963; Mitchell and Mitchell, 1973; Mitchell and Zentmyer, 1971a, 1971b; Uppal, 1926; Vujičić and Colhoun, 1966; Zentmyer and Erwin, 1970). Since *Phytophthora* spp. are tolerant of aeration extremes, aeration may be an important influence on survival, interactions with other organisms, the infection process, and subsequent disease development.

Many biological factors influence disease, but little quantitative information is available on interactions of soilborne organisms with *Phytophthora* spp. Microbial antagonism to *Phytophthora* spp. and the role of suppressive soil in disease development are reviewed by Malajczuk (Chapter 16). Some soilborne organisms may interact additively or synergistically with *Phytophthora* spp. in causing disease. Interactions between nematodes and *Phytophthora* spp. may be the most common form of such interactions (Powell, 1971). For example, Schlang and Sikora (1978) reported that disease occurred 10 days earlier and disease intensity was 40% greater on *Gerbera jamesonii* with a combination of *Meloidogyne arenaria* and *P. cryptogea* than with the fungus alone.

Some host factors that relate to tolerance or resistance to disease and thus influence disease incidence at any defined level of inoculum include host age, plant succulence, and root exudation. Disease incidence at defined inoculum levels may decrease with increasing plant age (Baumer, 1980; Bonnet et al, 1978; Gray et al, 1973; Ko, 1971; Ramirez and Mitchell, 1975). For example, Baumer (1980) found that susceptibility of alfalfa seedlings decreased logarithmically with age of plants at several concentrations of zoospore inoculum of *P. megasperma* f. sp. *medicaginis*. The influence of host succulence on disease development was noted by Gooding and Lucas (1959); at $10^5$ zoospores of *P. parasitica* var. *nicotianae* per plant, all of the tobacco plants died when treated daily with a nutrient solution, but only 20% of the plants died when treated on alternate days. The influence of variation in host root exudates on the quantitative relationship of inoculum to disease has not been addressed in the literature, but manipulations of root exudates may allow reductions in disease incidence at defined inoculum densities.

Several factors in experimental design may influence greatly the significance of inoculum density-disease incidence relationships. The evaluation of a combination of a susceptible host with an inoculum form under environmental conditions that simulate field conditions as closely as possible is essential for fundamental work. Selection of the inoculum source is particularly important with residual inoculum. Thick-walled chlamydospores produced in some natural media or chlamydospores recovered from soil or infested plant debris may differ greatly from thin-walled chlamydospores produced under laboratory conditions in respect to germination, survival, and numbers required for infection (Kadooka and Ko, 1973; Shew, 1980). Shew (1980), for example, observed higher levels of infection of Fraser fir with chlamydospores of *P. cinnamomi* recovered from naturally infested soil than with the same density of chlamydospores produced in culture. Production and treatment of inoculum, such as aging in soil or exposure to enzymes, before inoculation may be particularly important when oospores are employed. The amount of organic matter associated with inoculum also may be influential in supporting limited growth through soil or proliferation of inoculum. In artificial systems simulating field conditions in which inoculum most probably exists as free resting spores in soil (Holdaway and Tsao, 1971; Kannwischer and Mitchell, 1978, 1981; Mitchell, 1978; Shew, 1980), defined initial inoculum levels may be established by infesting soil with chlamydospores or oospores exposed to treatments such as sonication (Kannwischer and Mitchell, 1981; Mitchell, 1978), freezing (Kittle and Gray, 1979b), or desiccation (Gray et al, 1973) to eliminate other viable forms of the pathogen. Future work hopefully will provide information on residual inoculum units consisting of propagules of the fungus, host remains, and soil components.

Mechanisms of exposure of host to inoculum in experiments should be based on natural associations of inoculum and host. With host-pathogen combinations in which the host is seeded or transplanted into soil infested with residual inoculum, resting spores should be mixed into the planting medium. Lower levels of chlamydospores, for example, were required for $ID_{50}$ values when inocula were mixed into soil (Kannwischer and Mitchell, 1981; Mitchell, 1978; Shew, 1980) than when inoculum suspensions were poured over the soil (Hwang and Ko, 1978; Kliejunas and Ko, 1974; Ko and Chan, 1974). Because dispersal of zoospores or their secondary structures, such as encysted zoospores or microsporangia, over distances greater than a few centimeters generally depends on moving water (Duniway, 1979; Hickman, 1970; Hickman and Ho, 1966; Katsura and Miyata, 1966; Mitchell et al, 1978), introduction of zoospore inoculum to hosts may be best organized to expose stems or crowns to infested water on the surface of the soil with subsequent drainage through roots growing in soil. When such methods have been employed (Baumer, 1980; Kannwischer and Mitchell, 1981; Mitchell, 1978; Mitchell et al, 1978), lower $ID_{50}$ values were observed than with methods in which zoospore suspensions were applied directly to soil (Gooding and Lucas, 1959; Hwang and Ko, 1978; Kliejunas and Ko, 1974; Ko and Chan, 1974; Kuan and Erwin, 1980). In the latter method, a high proportion of zoospores may be immobilized. Zoospores are induced to encyst by such factors as collisions with other zoospores or solid surfaces, changes in hydrogen ion concentration or temperature, variations in water currents (rheotaxis), and variations in nutrients or ions (Bimpong and Clerk, 1970; Halsall, 1977; Hickman, 1970; Hickman and Ho, 1966; Ho and Hickman, 1967; Katsura and Miyata, 1966). Thus, extreme care must be exercised in attempting to determine the numbers of motile zoospores required for specific levels of infection, because some or all of the above factors may be involved in inoculation of plants and because fewer motile than encysted zoospores of several fungi are required for the same incidence of infection (Bhalla, 1968; Hwang and Ko, 1978; Kliejunas and Ko, 1974). When data from various studies are compared, it is helpful to have information on the proportion of zoospores that were motile at the time of inoculation. Whenever possible, inoculum levels established by counting and diluting any type of spore should be monitored in the soil or other medium in which the host is exposed to the inoculum as the experiment progresses.

Variation in pathogenicity may be measured quantitatively by comparing $ID_{50}$ values of isolates of *Phytophthora* spp. or by comparing disease incidences with various isolates at the same inoculum density. Although considerations of variations in pathogenicity among isolates have been included in several studies (Gray et al, 1973; Kittle and Gray, 1979b), no reports are available on the influence of pathogen variability on carefully defined relationships of inoculum density to disease incidence.

## APPLICATIONS OF INOCULUM DENSITY-DISEASE MODELS

### Pragmatic Applications

Quantitatively defined associations of inoculum with disease provide valuable models for examining the influences of many factors in disease. At present, the most important information derived from these models quantitatively depicts pragmatic influences, such as host resistance, pathogen variability, biological control agents, fungicides, or physical factors, on disease. Host resistance or tolerance to several *Phytophthora* spp. has been evaluated by quantifying disease at defined inoculum levels of zoospores (Baumer, 1980; Byrt and Holland, 1978; Eye et al, 1978; Gooding and Lucas, 1959; Pratt et al, 1975) or oospores (Baumer, 1980; Gray et al, 1973; Kittle and Gray, 1979b). For example, Baumer (1980) demonstrated that resistance of alfalfa lines to *P. megasperma* f. sp. *medicaginis* could be quantitatively defined in terms of decreased disease incidence and decreased rate of disease development compared to a susceptible line at corresponding concentrations of inoculum. The author concluded that either oospores or zoospores could be used to screen for disease resistance in seedlings because $ID_{50}$ values obtained for several alfalfa lines with varying levels of root-rot resistance indicated levels of resistance that agreed with known resistance reactions of adult plants of these lines. Resistance was detected in alfalfa seedlings only when levels of inoculum that resulted in less than 100% disease incidence were used.

Although little quantitative information is available on the control of *Phytophthora* spp. with biological agents or soil suppressive to the pathogen, models of the influence of such biological factors on the relationships of inoculum densities of soilborne plant pathogens to disease incidences have provided information on other host-pathogen combinations (Marois and Mitchell, 1981; Rouse and Baker, 1978; Wijetunga and Baker, 1979). Rouse and Baker (1978) developed a model that allowed precise differentiation of effectiveness among biological control measures, such as the addition of cellulose or chitin to soil, by determining how efficiency of inoculum of *Rhizoctonia solani* was influenced by these treatments in the relationship of inoculum density to disease incidence in radish seedlings. Marois and Mitchell (1981) selected microbial antagonists to *F. oxysporum* f. sp. *radicis lycopersici* based on the ability of the antagonists to increase the ratio of inoculum density of the pathogen to disease incidence in tomato seedlings. A composite of five fungal antagonists significantly reduced pathogen inoculum efficiency and provided control of tomato crown rot under growth-chamber and field conditions (Marois et al, 1981). Similar models should be instrumental in the selection and study of biological agents for the control of *Phytophthora* spp. The effectiveness of *Phytophthora* spp. for the control of weeds can be evaluated by quantifying the association of inoculum to disease that allows the selection of minimum inoculum concentrations for efficient control. Mitchell and Ridings (1975) determined an $ID_{50}$ with milkweed vine of approximately 0.5 chlamydospore of *P. palmivora* (reported as *P. citrophthora*) per gram of soil in laboratory studies. Subsequent field applications of only 1.5 chlamydospores per square centimeter of soil under citrus trees provided greater than 95% mortality of the weeds and minimized selection pressure on the pathogen to form variants that might have affected citrus (Ridings et al, 1976).

Quantitatively defined host-pathogen systems also may be used to critically evaluate fungicides or other control measures for *Phytophthora* spp. Kannwischer et al (1976), for example, reported that at 0.5 chlamydospore of *P. nicotianae* var.

1979; Kannwischer and Mitchell, 1978; Papavizas et al, 1981, Weste and Vithanage, 1978), no quantitative information is available on the dissemination or efficiency of that inoculum in secondary cycles of disease. Approaches such as those recently reported with foliage diseases may be useful in studies with soilborne diseases. For example, parameters representing both sporulation potential and infection efficiency have been included in a stochastic model for a foliage disease (Rouse et al, 1980).

Recently, several reports have contained information on the efficiency of primary (but not of secondary) inoculum in soilborne diseases. The efficiency of oospores of *P. megasperma* f. sp. *medicaginis* and sclerotia of *Rhizoctonia solani* in inducing infections was calculated to be 4.5% (Baumer, 1980) and less than 10% (Baker and Drury, 1981), respectively. Tomimatsu and Griffin (1982) detected an efficiency of germinating microsclerotia of *Cylindrocladium crotalariae* in infection of nearly 100%, but an efficiency of microsclerotia in disease (percentages of infections that develop into necrosis) of only 0.27–0.28%. Thus, inoculum density, infection incidence, and subsequent disease incidence must all be quantitated carefully in attempts to determine inoculum efficiency or effectiveness. Studies on the efficiency of inoculum in secondary cycles, however, will have to be conducted with zoospores or forms of inoculum that would be moved naturally by splashing rain or mechanical means. It will be of interest to determine numbers of such propagules produced per plant or unit area of soil by assaying with selective media or baiting systems, but experiments to determine how far and how efficiently inoculum is disseminated may depend on following the progression of both infection and subsequent disease over time and space in susceptible hosts arranged around one initially infected plant.

The rate of spread of diseases will depend on many factors, such as slope, drainage, soil depth, and physical, chemical, and biological factors in the soil. Weste and Ruppin (1977) observed progression of disease caused by *P. cinnamomi* of 18–171 m/yr in various relatively level soils and of about 400 m/yr in areas with mountain range runoff.

Simulation models of polycyclic soilborne diseases caused by *Phytophthora* spp. in the field are not immediately foreseeable, but significant amounts of information are available for the initial quantitative conceptualization of these diseases. Although present theoretical models on the relationship of inoculum density to disease incidence are not suitable practically to predict disease under field conditions (Bouhot, 1979; Geypens, 1974), inoculum density will be an important factor in the development of models. Bloomberg (1979a, 1979b) developed a simulation model of root rot of Douglas-fir seedlings caused by *Fusarium oxysporum* that was based partially on the relationship of inoculum distribution to soil depth. The model was used to predict seedling root growth, infection, and mortality over many parameters, including concentrations and distributions of inoculum. Hopefully, the work of Bloomberg (1979a, 1979b) will provide the impetus for modeling soilborne diseases that the early work by Baker and his co-workers (Baker, 1971; Baker et al, 1967) did to stimulate investigations of the nature of the inoculum density-disease incidence relationship. However, as difficult as the future seems in reference to the collection of biological data suitable for the determination of the relationship of inoculum density to disease incidence in the field and to the subsequent development of simulation models of diseases caused by soilborne species of *Phytophthora*, the ultimate goal of developing models of crop losses in total agroecosystems must be envisioned. Thus, diseases caused by *Phytophthora* spp. and the role of inocula in the development of diseases will be but parts of models on crop production. Other components involved in crop production that are closely associated with the development of soilborne diseases and that have been predicted by simulation models include plant root growth (Bartos and Jameson, 1974; Lambert et al, 1975), activity of microorganisms in the rhizosphere (Balis, 1975; Newman and Watson, 1977), and crop losses (Ferriss, 1981; Stynes et al, 1979). Thus, important information and approaches are already available for the future development of simulation models that will predict disease and crop losses caused by soilborne *Phytophthora* spp. within a holistic framework of crop production.

## SUMMARY

Inoculum densities of *Phytophthora* spp. of less than one to approximately 30 propagules per gram of soil have been detected in soils associated with roots of perennial hosts. Higher inoculum levels may be observed when environmental conditions are optimum for the formation of zoospores if appropriate techniques for their detection are employed. Inoculum densities in infested fields sampled before planting of annual crops generally have been undetectable or less than a few propagules per gram of soil; however, significant amounts of disease can occur at these low levels. Subsequently, populations of up to several hundred propagules per gram of soil may be detected in the root zones of plants during the growing season

Under artificial conditions with *P. cinnamomi, P. parasitica* var. *nicotianae*, or *P. palmivora* mixed into soil, low percentages of infection occur at only 0.05–0.1 chlamydospore per gram of soil; 50% infection occurs at 0.1–0.9 chlamydospore per gram of soil; and 100% infection of various hosts occurs at 5–10 chlamydospores per gram of soil. Much higher inoculum concentrations of chlamydospores may be necessary for comparable levels of disease when inoculum is drenched onto the soil surface instead of uniformly mixed into soil.

Depending on the host-pathogen system, time of host exposure to inoculum, and treatment of inoculum, less than 1.0 to more than 100 oospores of homothallic *Phytophthora* spp. per gram of soil are required for 50% infection of various hosts under growth chamber or greenhouse conditions. Quantitation of root infection with oospores of heterothallic species has not been reported.

Although low percentages of plants were infected when single or a few zoospores were added directly to plant parts or to water in which plant parts were floating, 50–300 zoospores per plant were required for 50% infection of several rooted hosts under temporarily flooded conditions. Much higher levels of zoospores ($10^4$–$10^5$ zoospores per plant) were required for 50% infection or mortality of various hosts by different *Phytophthora* spp. High $ID_{50}$ values may often be the result of experimental designs (e.g., drenching zoospore suspensions onto soil) that reduce the efficiency of zoospores in dispersal or host infection.

The ultimate usefulness of information on quantitative relationships of inoculum levels of *Phytophthora* spp. to incidences of disease depends on the recognition of many factors, including variation in fitness of the pathogen, variation in host tolerance or resistance at different growth stages, limitation or variation in time of host exposure to the pathogen, and limitation of infection or disease expression due to suppressive physical, chemical, and biological components of the environment. Information on these relationships has been useful for the evaluation of disease control with chemicals, host resistance, or biological agents; for the evaluation of biological interactions; and for the determination of environmental influences on disease.

Many factors must be elucidated before predictive systems can be developed for disease incidence or severity in the field based even in part on inoculum densities of *Phytophthora* spp. Some progress has been made in the development of sensitive assays for the detection of extremely low levels of inoculum, but their utility will depend on the elucidation and compensation made for distribution patterns of inoculum in the field. If populations are generally distributed in clumped patterns, evaluations of the degree of clumping plus the mean population may be important in determining the probability that a plant root system will contact inoculum. The relative distribution patterns of inoculum should be considered in determining number and size of soil samples required to achieve statistical accuracy. Disease progression rates must be determined and can be ascertained fairly easily, but correlating disease progression rates to initial inoculum densities will be much more difficult. Finally, because soilborne diseases caused by *Phytophthora* spp. are probably polycyclic, it is important to elucidate quantitatively the production, dissemination, and efficiency of secondary inoculum in the spread of disease. Ecological and epidemiologic concepts must be integrated to determine the dynamic interactions of inoculum and disease incidence within polycyclic and polyetic (disease cycles that extend over seasons) processes of epidemics.

## LITERATURE CITED


Baker, R. 1971. Analyses involving inoculum density of soil-borne plant pathogens in epidemiology. Phytopathology 61:1280-1292.

Baker, R. 1978. Inoculum potential. Pages 137-157 in: Plant Disease, An Advanced Treatise. Vol. II. How Disease Develops in Populations. J. G. Horsfall and E. B. Cowling, eds. Academic Press, New York. 436 pp.

Baker, R., and Drury, R. 1981. Inoculum potential and soilborne pathogens: The essence of every model is within the frame. Phytopathology 71:363-372.

Baker, R., Maurer, C. L., and Maurer, R. A. 1967. Ecology of plant pathogens in soil. VII. Mathematical models and inoculum density. Phytopathology 57:662-666.

Balis, C. 1975. A theoretical approach to competitive saprophytic colonization of substrates as a function of inoculum density. Ann. Inst. Phytopathol. Benaki (N.S.) 11:73-93.

Banihashemi, Z., and Mitchell J. E. 1975. Use of safflower seedlings for the detection and isolation of *Phytophthora cactorum* from soil and its application to population studies. Phytopathology 65:1424-1430.

Bartos, D. L., and Jameson, D A. 1974. A dynamic root model. Am. Midl. Nat. 91:499-504.

Baumer, J. S. 1980. Effect of sitosterol on the plasma membrane composition of *Phytophthora megasperma* f. sp. *glycinea* and inoculum density-disease incidence relationships of Phytophthora root rot of alfalfa. Ph.D. thesis. Univ. Calif., Riverside. 87 pp.

Bhalla, H. S. 1968. Number of zoospores of *Aphanomyces euteiches* required for infection of peas. (Abstr.) Phytopathology 58:1043.

Bimpong, C. E., and Clerk, G. C. 1970. Motility and chemotaxis in zoospores of *Phytophthora palmivora* (Butl.) Butl. Ann. Bot. (London) 34:617-624.

Bloomberg, W. J. 1979a. A model of damping-off and root rot of Douglas-fir seedlings caused by *Fusarium oxysporum*. Phytopathology 69:74-81.

Bloomberg, W. J. 1979b. Model simulations of infection of Douglas-fir seedlings by *Fusarium oxysporum*. Phytopathology 69:1072-1077.

Bonnet, P., Maia, N., Tello-Marquina, J., and Venard, P. 1978. Pouvoir pathogène du *Phytophthora parasitica* (Dastur): Facteurs de variabilité et notion de spécialisation parasitaire. Ann. Phytopathol. 10:15-29.

Bouhot, D. 1979. Estimation of inoculum density and inoculum potential: Techniques and their value for disease prediction. Pages 21-34 in: Soil-Borne Plant Pathogens. B. Schippers and W. Gams, eds. Academic Press, New York. 686 pp.

Byrt, P. N., and Holland, A. A. 1978. Infection of axenic eucalyptus seedlings with *Phytophthora cinnamomi* zoospores. Aust. J. Bot. 26:169-176.

Campbell, C. L., and Pennypacker, S. P. 1980. Distribution of hypocotyl rot caused in snapbean by *Rhizoctonia solani*. Phytopathology 70:521-525.

Campbell, C. L., and Powell, N. T. 1980. Progression of disease induced by soilborne pathogens: Tobacco black shank. Prot. Ecol. 2:177-182.

Dukes, P. D., and Apple, J. L. 1965. Effect of oxygen and carbon dioxide tensions on growth and inoculum potential of *Phytophthora parasitica* var. *nicotianae*. Phytopathology 55:666-669.

Duniway, J. M. 1977. Predisposing effect of water stress on the severity of *Phytophthora* root rot in safflower. Phytopathology 67:884-889.

Duniway, J. M. 1979. Water relations of water molds. Annu. Rev. Phytopathol. 17:431-460.

Eye, L. L., Sneh, B., and Lockwood, J. L. 1978. Inoculation of soybean seedlings with zoospores of *Phytophthora megasperma* var. *sojae* for pathogenicity and race determination. Phytopathology 68:1769-1773.

Ferris, H. 1981. Mathematical approaches to the assessment of crop damage. Pages 405-420 in: Plant Parasitic Nematodes, Vol. III. B. M. Zuckerman and R. A. Rohde, eds. Academic Press, New York. 508 pp.

Ferriss, R. S. 1981. Calculating rhizosphere size. Phytopathology 71:1229-1231.

Ferriss, R. S. 1982. Relationship of infection and damping-off of soybean to inoculum density of *Pythium ultimum*. Phytopathology 72:1397-1403.

Flowers, R. A., and Hendrix, J. W. 1972. Population density of *Phytophthora parasitica* var. *nicotianae* in relation to pathogenesis and season. Phytopathology 62:474-477.

Geypens, M. 1974. Inoculum potential of soil-borne plant pathogenic fungi: Problems encountered in analysis and significance in epidemiology. Agro-Ecosystems 1:177-192.

Gilligan, C. A. 1979. Modeling rhizosphere infection. Phytopathology 69:782-784.

Goodell, P., and Ferris, H. 1980. Plant-parasitic nematode distributions in an alfalfa field. J. Nematol. 12:136-141.

Gooding, G. V., and Lucas, G. B. 1959. Effect of inoculum level on the severity of tobacco black shank. Phytopathology 49:274-276.

Gray, F. A., Hine, R. B., Schonhorst, M. H., and Naik, J. D. 1973. A screening technique useful in selecting for resistance in alfalfa to *Phytophthora megasperma*. Phytopathology 63:1185-1188.

Grogan, R G., Sall, M. A., and Punja, Z. K. 1980. Concepts for modeling root infection by soilborne fungi. Phytopathology 70:361-363.

Halsall, D. M. 1977. Effects of certain cations on the formation and infectivity of *Phytophthora* zoospores. II. Effects of copper, boron, cobalt, manganese, molybdenum, and zinc ions. Can. J. Microbiol. 23:1002-1010.

Hendrix, F. F., Jr., and Kuhlman, E. G. 1965. Factors affecting direct recovery of *Phytophthora cinnamomi* from soil. Phytopathology 55:1183-1187.

Hickman, C. J. 1970. Biology of *Phytophthora* zoospores. Phytopathology 60:1128-1135.

Hickman, C. J., and Ho, H. H. 1966. Behaviour of zoospores in plant-pathogenic Phycomycetes. Annu. Rev. Phytopathol. 4:195-220.

Ho, H. H. 1969. Notes on the behavior of *Phytophthora megasperma* var. *sojae* in soil. Mycologia 61:835-838.

Ho, H. H., and Hickman, C. J. 1967. Factors governing zoospore responses of *Phytophthora megasperma* var. *sojae* to plant roots. Can. J. Bot 45:1983-1994.

Holdaway, B. F., and Tsao, P. H. 1971. Survival of *Phytophthora parasitica* in soils. (Abstr.) Phytopathology 61:1321.

Hwang, S. C., and Ko, W. H. 1978. Biology of chlamydospores, sporangia, and zoospores of *Phytophthora cinnamomi* in soil. Phytopathology 68:726-731.

Kadooka, J. Y., and Ko, W. H. 1973. Production of chlamydospores by *Phytophthora palmivora* in culture media. Phytopathology 63:559-562.

Kannwischer, M. E. 1979. Quantitative studies on the development and control of black shank of tobacco, and on the survival of *Phytophthora parasitica* var. *nicotianae* in soil. Ph.D. thesis, Univ. Fla., Gainesville. 111 pp.

Kannwischer, M. E., and Mitchell, D. J. 1978. The influence of a fungicide on the epidemiology of black shank of tobacco. Phytopathology 68:1760-1765.

Kannwischer, M. E., and Mitchell, D. J. 1981. Relationships of numbers of spores of *Phytophthora parasitica* var. *nicotianae* to infection and mortality of tobacco. Phytopathology 71:69-73.

Kannwischer, M. E., Mitchell, D. J., and Sanden, G. E., 1976. Apparent infection rates of black shank of tobacco. (Abstr.) Proc. Am. Phytopathol. Soc. 3:338.

Katsura, K., and Miyata, Y. 1966. Movement of zoospores of *Phytophthora capsici*. III. Rheotaxis. Sci. Rep. Kyoto Prefect. Univ. Agric. 18:51-56.

Kim, S. H., Kantzes, J. G., and Weaver, L. O. 1974. Infection of aboveground parts of bean by *Pythium aphanidermatum*. Phytopathology 64:373-380.

Kittle, D. R., and Gray, L. E. 1979a. Storage and use of *Phytophthora megasperma* var. *sojae* oospores as inoculum. Phytopathology 69:821-823.

Kittle, D. R., and Gray, L. E. 1979b. Production and pathogenicity of oospores of *Phytophthora megasperma* var. *sojae*. Plant Dis. Rep. 63:628-631.

Kliejunas, J. T., and Ko, W. H. 1974. Effect of motility of *Phytophthora palmivora* zoospores on disease severity in papaya seedlings and substrate colonization in soil. Phytopathology 64:426-428.

Kliejunas, J. T., and Nagata, J. T. 1979. *Phytophthora cinnamomi* in Hawaiian forest soils: Seasonal variations in population levels. Phytopathology 69:1268-1272.

Klotz, L. J., Stolzy, L. H., and DeWolfe, T. A. 1963. Oxygen requirement of three root-rotting fungi in a liquid medium. Phytopathology 53:302-305.

Ko, W.-H. 1971. Direct observation of fungal activities on soil. Phytopathology 61:437-438.

Ko, W. H., and Chan, M. J. 1974. Infection and colonization potential of sporangia, zoospores, and chlamydospores of *Phytophthora palmivora* in soil. Phytopathology 64:1307-1309.

Kraft, J. M., and Erwin, D. C. 1968. Effects of inoculum substrate and density on the virulence of *Pythium aphanidermatum* to mung bean seedlings. Phytopathology 58:1427-1428.

Kuan, T. L., and Erwin, D. C. 1980. Predisposition effect of water saturation of soil on Phytophthora root rot of alfalfa. Phytopathology 70:981-986.

Lambert, J. R., Baker, D. N., and Phene, C. J. 1975. Simulation of soil processes under growing row crops. Pap. 75-2580. Am. Soc. Agric. Eng., St. Joseph, MI.

Legge, B. J. 1953. Studies on species of *Phytophthora*. Ph.D. thesis, Univ. Birmingham. 136 pp.

Leonard, K. J. 1980. A reinterpretation of the mathematical analysis of rhizoplane and rhizosphere effects. Phytopathology 70:695-696.

Lucas, G. B. 1975. Diseases of Tobacco. 3rd ed. Howard E. Parker & Sons, Fuquay-Varina, NC. 621 pp.

Madden, L. V. 1980. Quantification of disease progression. Prot. Ecol. 2:159-176.

Malajczuk, N., Bowen, G. D., and Greenhalgh, F. C. 1978. A combined fluorescent antibody and soil sieving technique to count chlamydospores of *Phytophthora cinnamomi* in soil. Soil Biol. Biochem. 10:437-438.

Marois, J. J., and Mitchell, D. J. 1981. Effects of fumigation and fungal antagonists on the relationships of inoculum density to infection incidence and disease severity in Fusarium crown rot of tomato. Phytopathology 71:167-170.

Marois, J. J., Mitchell, D. J., and Sonoda, R. M. 1981. Biological control of Fusarium crown rot of tomato under field conditions. Phytopathology 71:1257-1260.

McCain, A. H., Holtzmann, O. V., and Trujillo, E. E. 1967. Concentration of *Phytophthora cinnamomi* chlamydospores by soil sieving. Phytopathology 57:1134-1135.

McIntosh, D. L. 1972. Effects of soil water suction, soil temperature, carbon and nitrogen amendments, and host rootlets on survival in soil of zoospores of *Phytophthora cactorum*. Can. J. Bot. 50:269-272.

McIntosh, D. L. 1977. *Phytophthora cactorum* propagule density levels in orchard soil. Plant Dis. Rep. 61:528-532.

McIntyre, J. L., and Taylor, G. S. 1976. Screening tobacco seedlings for resistance to *Phytophthora parasitica* var. *nicotianae*. Phytopathology 66:70-73.

McKee, R. K. 1969. Effects of ultraviolet irradiation on zoospores of *Phytophthora infestans*. Trans. Br. Mycol. Soc. 52:281-291.

Meyer, D., and Schönbeck, F. 1975. Investigations on the development of *Phytophthora cactorum* (Leb. et Cohn) Schroet. in soil. Z. Pflanzenkr. Pflanzenschutz 82:337-354.

Mircetich, S. M., and Zentmyer, G. A. 1966. Production of oospores and chlamydospores of *Phytophthora cinnamomi* in roots and soil. Phytopathology 56:1076-1078.

Mitchell, D. J. 1978. Relationships of inoculum levels of several soilborne species of *Phytophthora* and *Pythium* to infection of several hosts. Phytopathology 68:1754-1759.

Mitchell, D. J., and Mitchell, J. E. 1973. Oxygen and carbon dioxide concentration effects on the growth and reproduction of *Aphanomyces euteiches* and certain other soil-borne plant pathogens. Phytopathology 63:1053-1059.

Mitchell, D. J., and Ridings, W. H. 1975. The infection of seedlings of *Morrenia odorata* with chlamydospores of *Phytophthora citrophthora*. (Abstr.) Proc. Am. Phytopathol. Soc. 2:77.

Mitchell, D. J., and Zentmyer, G. A. 1971a. Effect of oxygen and carbon dioxide tensions on growth of several species of *Phytophthora*. Phytopathology 61:787-791.

Mitchell, D. J., and Zentmyer, G. A. 1971b. Effect of oxygen and carbon dioxide tensions on sporangium and oospore formation by *Phytophthora* spp. Phytopathology 61:807-812.

Mitchell, D. J., Kannwischer, M. E., and Moore, E. S. 1978. Relationship of numbers of zoospores of *Phytophthora cryptogea* to infection and mortality of watercress. Phytopathology 68:1446-1448.

Mitchell, J. E. 1979. The dynamics of the inoculum potential of populations of soil-borne plant pathogens in the soil ecosystem. Pages 3-20 in: Soil-Borne Plant Pathogens. B. Schippers and W. Gams, eds. Academic Press, New York. 686 pp.

Newman, E. I., and Watson, A. 1977. Microbial abundance in the rhizosphere: A computer model. Plant Soil 48:17-56.

Otrosina, W. J., and Marx, D. H. 1975. Populations of *Phytophthora cinnamomi* and *Pythium* spp. under shortleaf and loblolly pines in littleleaf disease sites. Phytopathology 65:1224-1229.

Papavizas, G. C., Bowers, J. H., and Johnston, S. A. 1981. Selective isolation of *Phytophthora capsici* from soils. Phytopathology 71:129-133.

Pfender, W. F., Hine, R. B., and Stanghellini, M. E. 1977. Production of sporangia and release of zoospores by *Phytophthora megasperma* in soil. Phytopathology 67:657-663.

Powell, N. T. 1971. Interactions between nematodes and fungi in disease complexes. Annu. Rev. Phytopathol. 9:253-274.

Pratt, R. G., Mitchell, J. E., and Wills, D. A. 1975. Resistance and susceptibility to *Phytophthora megasperma* expressed in alfalfa cotyledons. Phytopathology 65:365-369.

Ramirez, B. N., and Mitchell, D. J. 1975. Relationship of density of chlamydospores and zoospores of *Phytophthora palmivora* in soil to infection of papaya. Phytopathology 65:780-785.

Ridings, W. H., Mitchell, D. J., and Schoulties, C. L. 1976. Biological control of milkweed vine in citrus with a pathotype of *Phytophthora citrophthora*. (Abstr.) Proc. Am. Phytopathol. Soc. 3:273-274.

Roth, D. A., and Griffin, G. J. 1981. Cylindrocladium root rot of black walnut seedlings and inoculum pattern in nursery soil. Can. J. Plant Pathol. 3:1-5.

Rouse, D. I., and Baker, R. 1978. Modeling and quantitative analyses of biological control mechanisms. Phytopathology 68:1297-1302.

Rouse, D. I., Nelson, R. R., and MacKenzie, D. R. 1980. A stochastic model of horizontal resistance based on frequency distributions. Phytopathology 70:951-954.

Sauve, R. J., and Mitchell, D. J. 1975. The influence of oospore density of four *Pythium* spp. in soil on infection and disease of several hosts. (Abstr.) Proc. Am. Phytopathol. Soc. 2:140.

Schlang, J., and Sikora, R. A. 1978. Interactions between *Meloidogyne arenaria, Verticillium dahliae* and *Phytophthora cryptogea* on *Gerbera jamesonii*. Z. Pflanzenkr. Pflanzenschutz

85:203-209.

Schmitthenner, A. F. 1970. Significance of populations of *Pythium* and *Phytophthora* in soil. Pages 25-27 in: Root Diseases and Soil-Borne Pathogens. T. A. Toussoun, R. V. Bega, and P. E. Nelson, eds. Univ. Calif. Press, Berkeley. 252 pp.

Shew, H. D. 1980. The biology and epidemiology of Fraser fir root rot caused by *Phytophthora cinnamomi*. Ph.D. thesis. N.C. State Univ., Raleigh. 68 pp.

Sneh, B., and McIntosh, D. L. 1974. Studies on the behavior and survival of *Phytophthora cactorum* in soil. Can. J. Bot. 52:795-801.

Spencer, J. A., and Cooper, W. E. 1967. Pathogenesis of cotton (*Gossypium hirsutum*) by *Pythium* species: Zoospore and mycelium attraction and infectivity. Phytopathology 57:1332-1338.

Stanghellini, M. E., von Bretzel, P., Kronland, W. C., and Jenkins, A. D. 1982. Inoculum densities of *Pythium aphanidermatum* in soils of irrigated sugar beet fields in Arizona. Phytopathology 72:1481-1485.

Stasz, T. E., and Harman, G. E. 1980. Interactions of *Pythium ultimum* with germinating resistant or susceptible pea seeds. Phytopathology 70:27-31.

Sterne, R. E., Zentmyer, G. A., and Kaufmann, M. R. 1977a. The effect of matric and osmotic potential of soil on Phytophthora root disease of *Persea indica*. Phytopathology 67:1491-1494.

Sterne, R. E., Zentmyer, G. A., and Kaufmann, M. R. 1977b. The influence of matric potential, soil texture, and soil amendment on root disease caused by *Phytophthora cinnamomi*. Phytopathology 67:1495-1500.

Stolzy, L. H., Letey, J., Klotz, L. J., and Labanauskas, C. K. 1965. Water and aeration as factors in root decay of *Citrus sinensis*. Phytopathology 55:270-275.

Stolzy, L. H., Zentmyer, G. A., Klotz, L. J., and Labanauskas, C. K. 1967. Oxygen diffusion, water, and *Phytophthora cinnamomi* in root decay and nutrition of avocados. Proc. Am. Soc. Hortic. Sci. 90:67-76.

Stynes, B. A., Wallace, H. R., and Veitch, L. G. 1979. A synoptic approach for crop loss assessment used to study wheat. I. An appraisal of the physical and chemical soil properties in the study area. Aust. J. Soil Res. 17:2 7-225.

Taylor, J. D., Griffin, G. J., and Garren, K. H. 1981. Inoculum pattern, inoculum density-disease incidence relationships, and population fluctuations of *Cylindrocladium crotalariae* microsclerotia in peanut field soil. Phytopathology 71:1297-1302.

Thomson, S. V., and Allen, R. M. 1976. Mechanisms of survival of zoospores of *Phytophthora parasitica* in irrigation water. Phytopathology 66:1198-1202.

Tomimatsu, G. S., and Griffin, G. J. 1982. Inoculum potential of *Cylindrocladium crotalariae*: Infection rates and microsclerotial density-root infection relationships on peanut. Phytopathology 72:511-517.

Tsao, P. H. 1970. Selective media for isolation of pathogenic fungi. Annu. Rev. Phytopathol. 8:157-186.

Tsao, P. H., and Martin, J. P. 1963. Effect of soil exchangeable hydrogen and calcium ratio and the resulting pH on infection of citrus fibrous roots by *Phytophthora parasitica*. (Abstr.) Phytopathology 53:1143.

Uppal, B. N. 1926. Relation of oxygen to spore germination in some species of the Peronosporales. Phytopathology 16:285-292.

van der Plank, J. E. 1963. Plant Diseases: Epidemics and Control. Academic Press, New York. 349 pp.

van der Plank, J. E. 1975. Principles of Plant Infection. Academic Press, New York. 216 pp.

Vujičić, R., and Colhoun, J. 1966. Asexual reproduction in *Phytophthora erythroseptica*. Trans. Br. Mycol. Soc. 49:245-254.

Weste, G., and Ruppin, P. 1977. *Phytophthora cinnamomi*: Population densities in forest soils. Aust. J. Bot. 25:461-475.

Weste, G., and Vithanage, K. 1978. Seasonal variation in numbers of chlamydospores in Victorian forest soils infected with *Phytophthora cinnamomi*. Aust. J. Bot. 26:657-662.

Wijetunga, C., and Baker, R. 1979. Modeling of phenomena associated with soil suppressive to *Rhizoctonia solani*. Phytopathology 69:1287-1293.

Zadoks, J. C., and Schein, R. D. 1980. Epidemiology and plant-disease management, the known and the needed. Pages 1-17 in: Comparative Epidemiology, A Tool for Better Disease Management. J. Palti and J. Kranz, eds. Centre for Agric. Publ. and Documentation, Wageningen, The Netherlands. 122 pp.

Zentmyer, G. A. 1980. *Phytophthora cinnamomi* and the diseases it causes. Monogr. 10. Am. Phytopathol. Soc., St. Paul, MN. 96 pp.

Zentmyer, G. A., and Erwin, D. C. 1970. Development and reproduction of *Phytophthora*. Phytopathology 60:1120-1143.

# 20

# Some Major Epidemics Caused by *Phytophthora*

P. H. Gregory

CONTENTS



## INTRODUCTION

The importance of the genus *Phytophthora*, both to humanity and to the development of the science of plant pathology, has been obvious ever since *P. infestans* devastated the potato crop in Western Europe in 1845. Its greatest impact was the potato blight epidemic in Ireland (Woodham-Smith, 1962). Austin Bourke's graph (Fig. 1) shows the steady growth of population in England and Wales during the 19th century. By contrast, Ireland's population suffered a catastrophic fall, due first to starvation and later to fevers and emigration (Bourke, 1965). The decrease in population continued long after the famine years, doubtless because migrants wrote home from the New World, saying, "Come and join us; it's better here." The effect on both sides of the Atlantic was profound.

A recent survey of *Phytophthora* lists 43 species plus four varieties (Newhook et al, 1978), and a search in abstracting journals indicates disease in more than a thousand combinations of host plant and *Phytophthora*, many of only minor import but some attaining the status of major epidemics.

An epidemic has been defined as "a widespread temporary increase of an infectious disease." Such an increase may occur because the pathogen has invaded a new area, because of changes in agronomic practices, or because of seasonal fluctuations in the crop, the weather, or other factors.

*Phytophthora* spp. are remarkably plastic fungi. They exist as a variety of structures that are able to change freely from one to the other according to external conditions, although not all possible transitions have been observed in all species. The mycelium, the absorbing metabolizing structure, may produce sexual organs (antheridia and oogonia followed by oospores), hyphal swellings and chlamydospores, or sporangia capable of germinating as mycelia or emitting zoospores. After encystment, these zoospores, in turn, may grow as mycelia, repeat as zoospores, or even produce microsporangia (Blackwell, 1949; Drechsler, 1930; Ho and Zentmyer, 1977).

How do these structures, which have been described in detail by taxonomists, fit into the annual disease cycle? The basic assumption that they are functional is adopted here because it generates testable hypotheses and so is valuable in planning research. The contrary assumption, that they are functionless, does not generate hypotheses and is impossible to prove. In this chapter, stress is laid on that remarkable microbe, the *Phytophthora* zoospore (Waterhouse, 1962). This survey is not an attempt to prove anything; rather it is a search for ideas to explore.

On the basis of function, I have distinguished two broad classes of spores in the fungi (Gregory, 1966): memnospores or "stay-put" spores, exemplified by oospores and chlamydospores, and xenospores or "go-places" spores, exemplified by zoospores. The *Phytophthora* sporangium is borderline. In *P. infestans*, the sporangium is clearly a xenospore, shed into dry air and apparently able to travel to distances measured in kilometers. Some other species have sporangia that are caducous in water (Waterhouse, 1974), where they can travel by flow or splash. Still other species have noncaducous (persistent) sporangia that germinate in situ.

Obviously, the zoospore migrates by swimming. In moist soil or on wet leaves, it is probably self-propelled to distances measured in millimeters, but it is capable of traveling much farther in splash droplets.

Cameron and Carlile (1977) found that zoospores of *P. cactorum*, *P. palmivora*, and *P. nicotianae* tend to swim upward against gravity (negative geotaxis), resulting in high concentration of zoospores just below a surface film. This phenomenon could well be an important factor in epidemiology, because at the surface of wet soil, the zoospores would be in an ideal position for dissemination through the air by the rain-splash mechanism, activated by falling raindrops

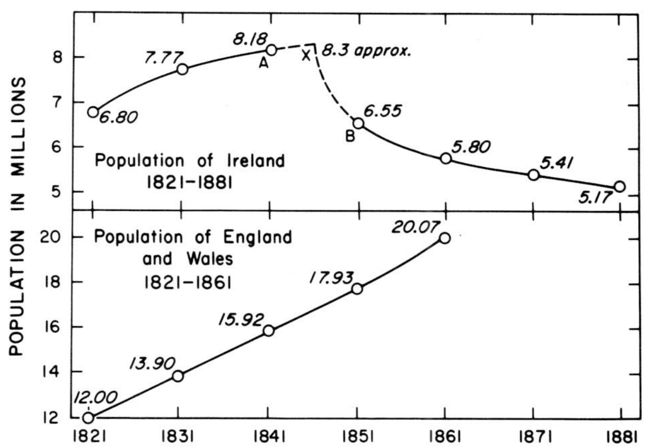


**Fig. 1.** Population changes in Ireland and in England and Wales from 1821 to 1881. The X marks the date of the potato blight epidemic in Ireland. (Courtesy Austin Bourke)

or drip from vegetation (Gregory, 1973). The fall of 100 mm of rain on 1 $m^2$ of ground could splash into the air over $10^9$ droplets large enough to carry a zoospore. In still air, these could be propelled by the energy of the splash to heights of about 1 m, and in wind they could travel horizontally to tens of meters before the zoospore died as the droplet evaporated. One wonders whether the shearing forces in the splash would strip the zoospore of its flagella. Research is needed on splashing of zoospores and its effects on viability. Meanwhile, splash of zoospores from surface concentrations appears highly significant in epidemiology.

This comparative survey attempts a synthetic view of *Phytophthora* as a going concern, a genus of living organisms with an array of potential morphological states adapted to parasitize plants in a highly competitive environment. The survey of the functions of structures in the epidemic cycle is mainly confined to five examples: *P. infestans* on potato, *P. colocasiae* on taro, *P. syringae* on apple, *Phytophthora* spp. on cocoa, and *P. cinnamomi* in Australian forests.

## *P. INFESTANS* ON POTATO

*P. infestans* is one of the two *Phytophthora* species with sporangia that are shed easily into dry air, the other species being *P. phaseoli* (Waterhouse, 1974), about which we know comparatively little.

In the 1840s, *P. infestans* arrived in Europe and, to judge from a map (Fig. 2) compiled by Bourke (1965), it spread at a speed typical of those other well-documented, wind-dispersed fungi, the pathogens of coffee rust, maize rust, and tobacco blue mold, which invaded virgin territory at about 500 km per year.

A century later, the scene was reenacted on a less spectacular scale in Africa. More potatoes were needed in East Africa to feed Allied troops during World War II, and seed potatoes were sent to Kenya from the United Kingdom. The fact that the small area devoted to growing potatoes in Kenya was free from *P. infestans* was overlooked in the emergency. Blight was first reported from Kenya in 1941; data on its subsequent spread were assembled by Cox and Large (1960).

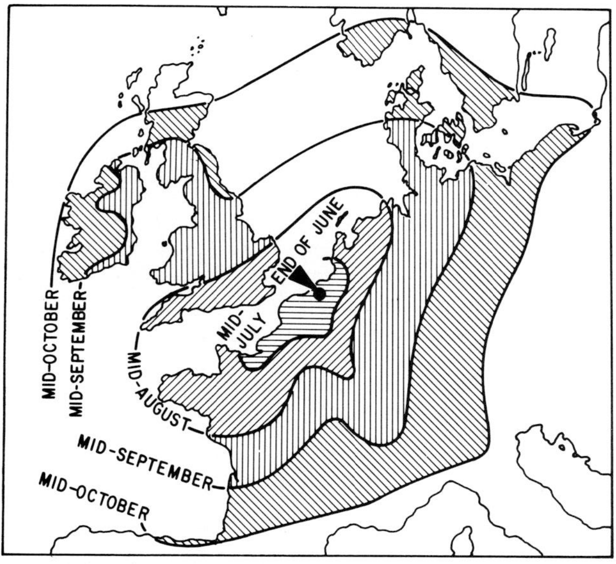


**Fig. 2.** Approximate dates of first reports of potato blight in Europe in 1845, commencing in Belgium on 24 June. (Courtesy Austin Bourke)

### Annual Cycle of *P. infestans*

In the north temperate region, the fungus overwinters in the form of slight or inapparent lesions on "seed" tubers, from which it very occasionally invades sprouts, producing sporangia when the sprouts emerge. In the isles of Scilly, in the extreme southwest of Britain, a search would probably reveal sporulating lesions any day of the year. In Holland, van der Zaag (1956) estimated that the annual epidemic arises from an infected sprout from approximately one seed tuber per square kilometer (Boyd, 1980).

### Persistence in Soil

According to Cox and Large (1960), economic loss of the harvested crop from blight depends on how early 70% of the foliage is killed. But the most serious loss occurs when the harvested tubers rot in storage from infection acquired at harvest from "infectivity" in moist soil. Nowadays in the United Kingdom, as Cox and Large point out, *P. infestans* mainly serves to adjust the market price—blight is slight in dry summers, when yields are smaller, and severe in wet summers, when the yield per hectare is heavy.

Zan (1962) and Lacey (1965) agree that soil infectivity persists for up to 11 wk; this is not long enough to bridge the 6-mo gap between successive potato crops but is ample time for heavy tuber infection if the crop is left in the ground after the foliage has been killed by blight.

The form in which this soil infectivity persists is a mystery. In temperate regions, neither chlamydospores nor oospores of *P. infestans* are normally found (only the A1 compatibility type occurs outside Mexico). Roots are apparently not attacked, although in vitro experiments by Lacey (1962) showed sporangia forming on root tips where zoospore suspensions had concentrated, and brown spots were seen on the epidermis. Zan (1962) attributed soil infectivity to small mycelia arising from germinated sporangia or zoospores, but such mycelia were not demonstrated microscopically. However, such dwarf mycelia of *P. parasitica* were found in irrigation water by Thomson and Allen (1976).

In the north temperate region, the *P. infestans* cycle is maintained by: 1) mycelium overwintering in "seed" tubers; 2) sporangia and zoospores spreading above ground; and 3) unidentified structures persisting for a limited time in soil.

## *P. COLOCASIAE* ON TARO

Taro (*Colocasia esculenta*), a tropical aroid with a thousand cultivars, is a staple food for millions of people in developing countries. Most of the produce is consumed locally and never impinges on international markets; perhaps for this reason its problems have been relatively neglected, especially taro leaf blight, which is often followed by rhizome rot during storage. This neglect is being redressed by studies by Gollifer et al (1980) and Putter (1980).

*P. colocasiae* has survival devices that are less simple than those of *P. infestans*. Sporangia are shed in water but not in wind. Spread is by sporangia and zoospores carried by splash between plants and plantings. Butler and Kulkarni (1913) found that the fungus survives in rhizomes. Because chlamydospores and oospores are produced, persistence of soil infectivity should not be a mystery. Geographic distribution is probably restricted to Southeast Asia and the Pacific areas (Holliday, 1980), and quarantine precautions are therefore necessary to prevent invasion of Africa and tropical America.

### *P. SYRINGAE* ON APPLE FRUITS

In the United Kingdom, *P. syringae* has been found occasionally since 1922 on rotting apples. Recently, severe losses during cold storage have been associated with changes in orchard practices (Upstone, 1978).

*P. syringae* is a homothallic, relatively low-temperature fungus; optimal temperature for growth is 20° C. The fungus perennates as oospores (Harris, 1979), which abound in fallen leaves on the orchard floor. Sporangia from germinating oospores produce zoospores available for rain splash. My theory is that the oospores are held in fragments of leaf and not easily splashed, whereas the zoospores are of appropriate size to be splashed.

During the relatively warm and dry months of June, July, and August in England, the fungus is quiescent. It is active from September to May, and most active from September to November, the time when fruits mature. Fruit is contaminated by splash from infective soil during rain at harvesttime (Upstone and Gunn, 1978). The fungus is considered to colonize apple leaves after leaf fall in late October and November, a season with favorable temperature and adequate water. This would rank as true saprophytic growth, but why the leaves are not infected on the tree on the same occasions as the fruit is not clear. The oospores evidently have variable dormancy, because the soil remains infective for many years.

The factors that have promoted increased incidence of disease in recent years are: 1) shorter trees, achieved by dwarfing rootstocks and the method of pruning, which causes fruits to form in a zone with a greater risk of splash from soil; and 2) the practice of maintaining bare soil around the trees by use of herbicides, thereby replacing the older grass floor, which had a mulching effect. The removal of the normal leaf mulch layer around the base of cocoa trees greatly increases the incidence of splashborne spores of *P. megakarya*.

The annual cycle presumably depends on zoospores for infecting the leaves and fruit; it is unusual in depending on oospores for perennation. Counts in infected orchards indicated a potential annual increment of 100 million oospores per square meter (Harris, 1979).

### *PHYTOPHTHORA* SPECIES ON COCOA

The average annual loss from black pod of cocoa is estimated by Padwick (1956) to be 10% of the world crop; this would be worth over $300 million (in U.S. dollars) at current prices. The pathogen, *P. palmivora*, seems to have migrated with its host from the American tropics to new plantations overseas. The cocoa tree also encountered additional *Phytophthora* spp. in its travels.

*P. palmivora*, in the strict sense, is widespread in the tropics and varies in severity from place to place. We now know that another species, *P. megakarya*, which was described recently by Brasier and Griffin (1979) and was once considered to be *P. palmivora*, is the cause of the pod disease in Nigeria and Cameroon. Still another species—morphological form (MF) 4, which may be a form of *P. capsici* (Zentmyer et al, 1981)—occurs commonly in Brazil and Cameroon and is also found in Mexico, Guatemala, El Salvador, and Costa Rica.

Unlike both *P. infestans* and *P. phaseoli*, sporangia of the cocoa *Phytophthora* spp. are seldom airborne although they are caducous; they are not caught by Hirst spore traps in cocoa plantations. Dispersal of sporangia in dry air is negligible in the annual disease cycle. However, cocoa is a perennial tree crop, and it is possible that windblown sporangia could occasionally bring inoculum to a new plantation, although inoculum is likely to have arrived already on planting material such as infected seed pods, rooted cuttings, or seedlings from the nursery. Within the plantation, inoculum is distributed by the rain-splash mechanism and by living vectors, especially ants.

If it were not for the usual dry season of the tropics, *Phytophthora* would cycle unrestrictedly in the cocoa plantation. The trees carry susceptible fruits for about 10 mo of each year. Each pod remains on the tree at risk of being infected for 5 mo, and the overlapping generations of pods give opportunities for infection to spread by splashing sporangia or zoospores in a rainy climate.

#### *P. palmivora*

In Ghana, *P. palmivora* survives the dry season partly in the soil but partly as dormant mycelium in flower cushions on the trunk and branches. From there, as the pod develops, the mycelium travels down the peduncle and produces sporangia on the developing pod (a development similar to the origin of *P. infestans* foci from sprouts on potato "seed" tubers). In the early part of the season, 70% of initial pod infections originate from infected flower cushions. Transfer of inoculum by ants also plays a role. Inoculum from soil accounts for many early-season infections in the zone at the base of the tree. Chlamydospores are formed on diseased pods (Asare-Nyako, 1971; Dade, 1927; Evans, 1973).

#### *P. megakarya*

The epidemiology of *P. megakarya* in Nigeria is very different from that of *P. palmivora*. This was discovered by the International Cocoa Black Pod Research Team, whose reports were compiled by Gregory and Maddison (1981). I will try to simplify a complicated story.

Pods are very seldom infected from dormant mycelia in flower cushions. Although *P. megakarya* forms chlamydospores sparingly in culture in the dark, my associates tell me that they have never found chlamydospores in a cocoa plantation. Nevertheless, a massive reservoir of *P. megakarya* evidently exists in the soil of cocoa plantations.

After the dry season, the first pod infections come mainly, though not exclusively, from the ground in one of two ways. They may be carried up the tree by species of ants that build soil tents to protect coccids and aphids that they have pastured on peduncles in order to tap the rich flow of nutrients into developing pods (Taylor and Griffin, 1981). Alternatively, they may be brought up from the ground by the splash mechanism from falling raindrops or drips from leaf tips. In the terminology of Maddison (1981), these "initiators" infect nearby pods by splash or contact, giving rise to localized infection "sequences."

Through much of the annual disease cycle, the plantation floor's store of inoculum may be replenished by sporangia and zoospores but apparently not by chlamydospores, unless we have inadvertently overlooked their presence. Additional sources of soil infection besides sporangia and zoospores washed down in rain are abscised flowers, husks of partly infected pods abandoned after the farmer has removed the beans, and old mummified pods remaining on the tree.

What happens in the soil is still a mystery. One would expect that careful hygiene over a period of years, by decreasing the annual increment to near zero, would deplete the soil reservoir, but this is not so. When Griffin et al (1981) applied what they refer to as "ultimate hygiene" in Nigeria, the results were surprising. On a strip of 500 trees across a plantation (Fig. 3), all *Phytophthora*-infected pods were counted and removed daily before they could sporulate—hard work for both assistants and senior scientists except during the dry season. For comparison, on two strips of 250 trees on each side of the hygiene treatment, the infected pods were left in position to sporulate. At the triweekly harvest, ripe pods and infected pods were counted and removed from all three strips. The experiment lasted from 1975 until 1979. In the first 3 yr, the number of infected pods recorded in the daily hygiene strip decreased somewhat, but so did the numbers in the adjacent triweekly harvested strips and in neighboring plantations. Then in 1978, the number of infections on the ultimate hygiene strip rose again, almost to the 1975 level (and again in unison with nearby plantations).

Isolations from plantation soil by baiting techniques help complete the picture. The amount of *Phytophthora* in the soil varied from year to year and with the time of year. In general, more inoculum was found in 1978 than in 1979. In both the daily (ultimate hygiene) and triweekly removal areas, more propagules were found in samples collected in September than in other months. So even in the absence of replenishment from overhead, the soil population still showed the usual seasonal fluctuation. After 3 yr, the mean number of infective propagules per gram of dry soil (excluding negative samples) was 21 for the triweekly removal area and 15 for the ultimate hygiene area.

The conclusion seems inescapable that, whatever the agent responsible, *P. megakarya* inoculum in the soil is not slowly dying out but is self-sustaining under cocoa trees and may even increase without augmentation from the aboveground parts of the trees. Several explanations seem possible.

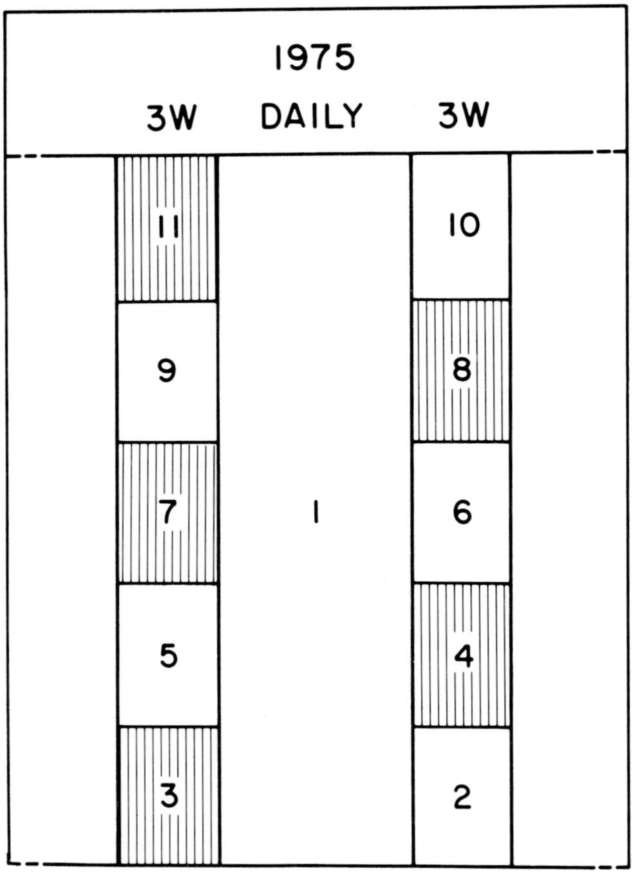


**Fig. 3.** Layout of experiment on "ultimate hygiene" in cocoa plantation at Cocoa Research Institute of Nigeria (Gambari), showing areas where black pods were removed every 3 wk (3W) and areas receiving "ultimate hygiene" treatment (Daily). (Hatched areas refer to an experiment on basal pod suppression.) (Reprinted, by permission, from Griffin et al, 1981)

One, *P. megakarya* may be surviving in the soil as small inactive mycelia derived from germinating sporangia or zoospores (the explanation favored by Zan [1962] for *P. infestans*).

Two, *P. megakarya* may be growing saprophytically as a soil invader. This seems unlikely because *Phytophthora* is difficult to isolate from plantation soil except by baiting with cocoa pod tissue, and even this is soon overrun by vigorous molds.

Three, a partial explanation may be found if research confirms that *P. megakarya* is an unrecognized root pathogen of cocoa, as suggested by Gregory (1981), comparable to the nonlethal root infections of *Pinus* described by Newhook (1961). Turner and Asomaning (1962) developed a seedling root inoculation test for resistance. Turner (1963) remarked that "further investigation is required to determine whether root infection of cacao is of similar importance to the fibrous root infection of citrus by *P. citrophthora*." Onesirosan (1971) isolated *Phytophthora* from roots of mature cocoa trees. Ward and Griffin (1981) isolated both *P. palmivora* and *P. megakarya* regularly over a period of 2 yr from surface-sterilized mature cocoa roots that had no apparent lesions. This must surely imply that the fungus was growing inside the root tissue in inapparent lesions. If not, then it must be living outside the root, and surface-sterilization must be assumed to have been successful against all the vigorous soil saprophytes on the outside of the roots but not against *Phytophthora*; this seems improbable.

Meanwhile, soil infectivity by *P. megakarya* is evidently mediated by some structure still undetermined.

## *P. CINNAMOMI* IN AUSTRALIAN FORESTS

How long *P. cinnamomi* has inhabited Australia is controversial (Newhook and Podger, 1972; Pratt and Heather, 1973; Weste, 1974; Zentmyer, 1980). What is disturbing is its relentless spread during the last 40 yr through the jarrah (*Eucalyptus marginata*) forests of Western Australia (Podger et al, 1965) and during the last 10 yr into coastal forests of Victoria, where entire plant ecosystems are being obliterated (Weste, 1974; Weste and Ruppin, 1975, 1977; Weste and Vithanage, 1978, 1979a, 1979b; Weste et al, 1976).

*P. cinnamomi* attacks roots, and almost the entire action takes place below ground. Local spread from a focus is clearly by zoospores, whereas survival is considered to be by chlamydospores. With only the A2 compatibility type present, oospores are not found, although the Brasier *Trichoderma* effect is not excluded (Brasier, 1975).

### The Role of Road Gravel

In both Western Australia and Victoria, contaminated gravel used in making forest roads has been incriminated in spreading the pathogen. In the jarrah forests, before the role of *P. cinnamomi* was understood, road gravel was being taken from areas already devastated and moved into areas still free from disease. In Victoria, where most outbreaks originated along the roads, one large gravel pit belonging to the Country Roads Board was an important source of inoculum.

Weste et al (1976) report two distinct patterns of spread from road gravel: 1) downslope by zoospores in drainage water after rain, to distances of up to 400 m/yr; and 2) uphill and along the contour, to distances of up to 15 or 20 m/yr (exceptionally more), which was attributed to zoospores swimming, to contact between roots, or to mycelium growing through the soil. One may argue that 15–20 m/yr is unexpectedly rapid for spread by any of the modes suggested and that wind transport of rain-splashed zoospores from surface concentrations as a result of negative geotaxis is a more plausible mechanism. This mechanism could also explain why *P. cinnamomi* may be found in the soil at distances of a few tens of meters beyond the edge of infected patches (Zentmyer and Richards, 1952).

### Population in Soil

That rather obscure property, "soil infectivity," varies with the season. Weste and Ruppin (1975, 1977) took samples of soil plus roots from affected patches at short intervals over 2 yr. Using the *Lupinus angustifolius* technique of Chee and Newhook (1965), they determined the serial dilution end point to obtain the population density index (PDI), as defined by Tsao (1960). The PDI tends to be maximal in spring and autumn when soil conditions are most favorable; the soil is too cold in winter and too dry in summer. Increase can be rapid when conditions become favorable. At one site, the PDI jumped from zero to 128 after 2 days of rain, possibly suggesting that chlamydospores were germinating to produce sporangia and thereby liberating zoospores (no dormant propagule would be included in the PDI). Infectivity increased rapidly in spring when temperatures rose in conjunction with adequate soil moisture. Survival during winter was considered to be in infected roots. Although the precise morphological state remains uncertain, chlamydospores are a possible candidate for the mechanism of soil infectivity.

### Chlamydospores

Chlamydospores were observed frequently inside root tissues, and they germinated readily, producing sporangia and zoospores. However, data on their seasonal occurrence raise some doubts as to their role.

Weste and Vithanage (1978) counted chlamydospores in soil samples from infected sites at intervals throughout 2 yr, using the ingenious sieving method of McCain et al (1967). After chlamydospores were concentrated in the sample by wet-sieving through screens of appropriate mesh, the residues were plated on media selective for *Phytophthora*. After a short incubation, during which *Phytophthora* mycelium penetrated the agar, the soil was washed off the agar layer with tap water, leaving germinated chlamydospores anchored in position for being counted under the microscope. The number of chlamydospores increased to a maximum in summer and autumn, not in winter as might be expected. Very few chlamydospores were detected in winter. One explanation might be that chlamydospores enter a dormant phase in winter, and dormant ungerminated chlamydospores would be washed off the isolation plates by this technique. We need to explore the possibility of chlamydospores being held dormant in soil.

The behavior of road gravel remains a puzzle. One naturally infected pile of this gravel was remarkable. It was a small pile, with a volume of less than 2 $m^3$, located beside a forest road. Although visible plant material was lacking, the pile remained infective for 5 yr, despite a severe drought in 1972–1973 (Weste, 1974). *P. cinnamomi* was easily isolated from any small random sample taken from the base of the heap.

Yet when Weste and Vithanage (1979a) tested washed chlamydospores grown in vitro for survival after they were stored in several different nonsterile soils at various moisture contents, they found that chlamydospores generally survived in moderately dry soils and in infertile soils such as gravel soils, for up to 8–10 mo. An exception was this anomalous "Country Road Board" gravel, in which organic matter content, bacteria, and actinomycetes all registered zero. Here, chlamydospore survival at 2 mo was nil at all moisture levels.

Gravelly soils may be particularly favorable for growth of *P. cinnamomi*. Weste and Vithanage (1979b) compared the stimulation exerted by various soil extracts on the production of sporangia from preformed, culture-grown mycelial mats. The greatest stimulation came from extracts of a gravelly soil from the Brisbane Ranges.

These facts about road gravel seem contradictory. We want to know whether the gravel contained a massive dormant population of *P. cinnamomi* that was slowly awakening. Or, because the particular heap had infected a gully area of forest, could it have been repeatedly recontaminated by rain-splashed zoospores blown back from the diseased patch, zoospores that perhaps encysted in the gravel and reemerged later?

As with *P. infestans* and *P. megakarya*, we are faced with perennation by a structure of the fungus not positively identified.

## SPECULATIONS ON SOIL INFECTIVITY

The epidemic cycles of some phytophthoras give evidence of "soil infectivity"—a disembodied ghost or poltergeist. The problem is to connect the ghost with a body. What morphological state or states mediate "soil infectivity"? The answer could be different in different species. The following speculations represent a personal attempt at synthesis, using the comparative method in a search for ideas.

With *P. colocasiae*, the problem seems not to arise. Splashed sporangia or zoospores, windborne from plant to plant, and the species' ability to produce oospores and chlamydospores suffice to explain the facts.

With *P. syringae* on apple fruit in Kent orchards, the answer is clear: long-term survival, possibly without saprophytic growth, of dormant oospores on fallen leaves renews the epidemic each year.

With *P. infestans*, spread is by windborne sporangia and zoospores. But, in addition, soil can remain infective for 11 wk after harvest, and Zan (1962) considered that persistence is mediated by small mycelia from germinated spores.

With *P. palmivora* on cocoa, persistence is largely by infection of flower cushions on the trunk and branches, aided by a reservoir in the soil.

With *P. megakarya* on cocoa, inocula for sequence initiators are carried from the soil reservoir by several possible agents, but the nature of the perennating structure in the soil is obscure.

With *P. cinnamomi*, saprophytic growth of mycelium in nonsterile soil has been deduced from infectivity detected at points up to 3 cm from the inoculum (Zentmyer and Mircetich, 1966). Evidence from persistence in road gravel in Australia is puzzling.

### Saprophytic Mycelium in Soil?

The hypothesis of saprophytic growth in the soil is attractive but presents some difficulties. Most *Phytophthora* spp. grow well as saprophytes in pure culture, but their competitive ability is slight. Isolating a *Phytophthora* sp. in the presence of its soil competitors has always proved difficult. Complex media laced with antibiotics and fungicides can often be used (although they have not been suitable for routine isolation from the soil of Nigerian cocoa plantations). The general formula for success is to bait the soil with portions of living tissue of a susceptible host, thus giving the pathogenic *Phytophthora* sp. a head start over more sturdy mold competitors. This scenario is repeated so often in the literature that the hypothesis of saprophytic growth through natural soil still raises doubts, and alternative explanations may be sought.

Generally, *P. cinnamomi* hyphae are lysed within 17 days by the action of soil bacteria and actinomycetes; even resistant chlamydospores are attacked (Reeves, 1975). Tsao (1969) found that mycelium of *P. parasitica* did not grow through natural soil unless extra nutrients were added, and he concluded that survival in soil is by resting spores, which, however, were not positively identified. Gisi (1975) found that, even with nutrients added, growth of *P. cactorum* from mycelium through soil was less than 1 mm.

The most probable number of infective propagules of *P. megakarya* in cocoa plantation soils during the rainy season was estimated by the dilution technique to be 15–20 per gram of dry soil near Ibadan and 160/g in the higher rainfall area near Ikom in eastern Nigeria (Ward and Griffin, 1981). These numbers do not suggest survival as mycelia but rather suggest the presence of dormant spores, some of which germinate when appropriately triggered, perhaps by exudate from the approaching tip of a growing rootlet (Bimpong and Clerk, 1970); this is an example of the small-spore principle enunciated by Garrett (1973).

When Hickman and English (see Hickman and Ho, 1966) placed mycelial inocula of *P. fragariae* in nonsterile soil, mycelium failed to grow out from a food base; instead, sporangia were produced around the edge of the inoculum. (Sporangia production no doubt was stimulated by soil microbes, an example of the phenomenon discovered by Zentmyer [1965].) Is this a clue? Sporangia may produce free-swimming zoospores.

### Repeated Emergence of Zoospores

The phenomenon of diplanetism of zoospores in *Saprolegnia* has puzzled me for over half a century. From my student days it seemed to me that there was no adequate explanation for the encystment of a free-swimming zoospore, followed almost immediately by emergence as a zoospore again—a second chance perhaps, but why encyst? It is of course possible that diplanetism serves to replace worn flagella. Repeated emergence of zoospores in *Phytophthora* is well known (Blackwell, 1949; Drechsler, 1930). What if repeated emergence of encysted *Phytophthora* zoospores can be delayed, perhaps by soil fungistasis? If it provided a second chance in the time dimension as well as in space, encystment would become intelligible.

A hint of such a possibility comes from an intriguing paper on *P. palmivora* by Turner (1963). Washed roots of cocoa and of 20 other mostly tropical species of plants were placed in dialysis tubes (for convenience) in deionized water, left in air for 15 hr, then placed, tube and all, in a sporangial suspension for 2.5 hr before being scored for germination of sporangia and zoospores. Sporangial germination was fairly high in root extracts of cocoa and one or two other species (and in the water control) but was inhibited in extracts of a dozen other species. Turner commented:

> although sporangial germination occurs in all soil extracts, the resulting zoospores are almost certainly inhibited in the soil by a fungistatic factor. . . . This fungistasis can easily be countered by exudates of cocoa leaves, pods and roots.

However, Tsao (1969) found that encysted zoospores of *P. parasitica* need amino acids for germination.

Later, in a crucial experiment, Turner (1965) infested soil with a suspension of *P. palmivora* zoospores that had been passed through filter paper and showed that the inoculum remained viable in soils of intermediate water content for at least 18 mo. However, survival was very short in soil at either 0 or 100% water-holding capacity. Turner also observed that *P. palmivora* zoospores placed in dialyzed root extracts became encysted within a relatively thick wall.

If this proves that zoospore encystment can be induced, maintained, and canceled by events and conditions in field soil, then some of the puzzles that have emerged in this survey of disease cycles can now be explained: 1) the persistence of *P. infestans* in soil (Lacey, 1962; Zan, 1962); 2) the persistence of *P. cinnamomi* in gravel (Weste and Vithanage, 1979a); 3) the persistence of infectivity in soil inoculated with filtered zoospores for the 18 mo recorded by Turner (1965); and 4) the survival of *P. megakarya* in cocoa plantation soils (Ward and Griffin, 1981).

## DISCUSSION

Knowledge about fungi in such a complex and opaque medium as soil is gained with difficulty. Clearly, more information is needed about the repeated emergence of zoospores and about possible dormancy in all *Phytophthora* structures: mycelium, chlamydospores, sporangia, zoospores, and even oospores, which have already been investigated extensively. Experiments are needed not only with soil in the laboratory, but also in field soil, which is liable to flood in tropical rainstorms and to drain equally rapidly, and which has cracks and earthworm burrows that help zoospore migration.

Meanwhile, from examples discussed above, we can synthesize a picture of the potentialities of the idealized *Phytophthora* zoospore; these potentialities must be reckoned with, even though not all are expressed in every species of *Phytophthora*. The zoospore can swim, encyst, reemerge as a swimming zoospore again, or produce a hypha that either penetrates the host or produces a minisporangium. Negative geotaxis tends to bring concentrations of zoospores to the surface of wet soil, where they are in a position to be dispersed by rain splash, drip splash, or surface runoff.

The examples in this survey suggest a testable hypothesis: Encysted zoospores (and chlamydospores) in the soil may enter a dormant phase, perhaps induced by soil fungistasis, and this dormancy may be broken by changed conditions, such as exudate from a growing root tip.

Perhaps in the past we have concentrated too much on the ephemeral nature of the zoospore and written it off too soon. In my speculations, a zoospore liberated from a sporangium infects immediately if a susceptible host is nearby. If unsuccessful at the first try, it lies dormant, waiting for a growing root tip to approach and trigger it.

Whatever the outcome of these speculations, the genus *Phytophthora* clearly possesses some of the most sophisticated weaponry known among plant pathogens. The mycelium is an active undercover communications network. Oospores, and perhaps chlamydospores, may be likened to time bombs, programmed for activation over widely different periods. The sporangium is a multipurpose weapon capable of emitting a flotilla of 20 or so minisubmarines, each oriented by gravity and able to home in on a chemical target against which it is equipped for biological warfare. No wonder, then, that the plant pathologists at Riverside decided to convene this International Symposium on *Phytophthora*!

## SUMMARY

Selected diseases caused by the socially important genus *Phytophthora* are compared in an attempt to relate the diverse structural elements of *Phytophthora* spp. to their survival in the disease cycle and their spread (often by splash from surface concentrations of zoospores). *P. infestans* on potato, *P. colocasiae* on taro, *P. syringae* on apple fruits, *P. palmivora* and *P. megakarya* on cocoa, and *P. cinnamomi* in Australian forests are discussed. Some of the examples illustrate perennation as mycelium and oospores. In other examples, the structure or structures that mediate soil infectivity are unknown, or the evidence is ambiguous. Research is needed on dormancy of chlamydospores and zoospores in soil and on repeated emergence of zoospores as part of the disease cycle.

### LITERATURE CITED

Asare-Nyako, A. 1971. Sources of primary field inoculum of *Phytophthora palmivora* in cocoa fields in Ghana. Pages 18-23 in: Première Réunion du Sous-Groupe de travail Afrique sur *Phytophthora palmivora*, 18–22 Octobre 1971. Communication No. 16. Institut Français du Café et du Cacao (IFCC), Paris.

Bimpong, C. E., and Clerk, G. C. 1970. Motility and chemotaxis in zoospores of *Phytophthora palmivora* (Butl.) Butl. Ann. Bot. 34:617-624.

Blackwell, E. 1949. Terminology in *Phytophthora*. Mycol. Pap. 30. Commonw. Mycol. Inst., Kew, Surrey, England. 24 pp.

Bourke, P. M. A. 1965. The potato, blight, weather—And the Irish famine. Ph.D. thesis, National University of Ireland, Dublin. 610 pp.

Boyd, A. E. W. 1980. Development of potato blight (*Phytophthora infestans*) after planting infected seed tubers. Ann. Appl. Biol. 95:301-309.

Brasier, C. M. 1975. Stimulation of sex organ formation in *Phytophthora* by antagonistic species of *Trichoderma*. I. The effect *in vitro*. New Phytol. 74:183-194.

Brasier, C. M., and Griffin, M. J. 1979. Taxonomy of '*Phytophthora palmivora*' on cocoa. Trans. Br. Mycol. Soc. 72:111-143.

Butler, E. J., and Kulkarni, G. S. 1913. Colocasia blight caused by *Phytophthora colocasiae* Rac. Mem. Dep. Agric. India 5:233-261.

Cameron, J. N., and Carlile, M. J. 1977. Negative geotaxis of zoospores of the fungus *Phytophthora*. J. Gen. Microbiol. 98:599-602.

Chee, K. H., and Newhook, F. J. 1965. Improved methods for use in studies on *Phytophthora cinnamomi* Rands and other *Phytophthora* species. N.Z. J. Agric. Res. 8:88-95.

Cox, A. E., and Large, E. C. 1960. Potato blight epidemics throughout the world. U.S. Dep. Agric. Agric. Handb. 174. 230 pp.

Dade, H. A. 1927. Economic significance of cacao pod diseases and factors determining their incidence and control. Gold Coast Dep. Agric. Bull. 6. 58 pp.

Drechsler, C. 1930. Repetitional diplanetism in the genus *Phytophthora*. J. Agric. Res. 40:557-573.

Evans, H. C. 1973. Invertebrate vectors of *Phytophthora palmivora*, causing black pod disease of cocoa in Ghana. Ann. Appl. Biol. 75:331-345.

Garrett, S. D. 1973. Deployment of reproductive resources by plant-pathogenic fungi: An application of E. J. Salisbury's generalization for flowering plants. Acta Bot. Indica 1:1-9.

Gisi, U. 1975. Untersuchungen über die Bodenphase von *Phytophthora cactorum* (Leb. et Cohn) Schroet. mit fluoreszenzoptischer Direktbeobachtung. Z. Pflanzenkr. Pflanzenschutz 82:355-377.

Gollifer, D. E., Jackson, G. V. H., and Newhook, F. J. 1980. Survival of inoculum of the leaf blight fungus *Phytophthora colocasiae* infecting taro, *Colocasia esculenta* in the Solomon Islands. Ann. Appl. Biol. 94:379-390.

Gregory, P. H. 1966. The fungus spore: What it is and what it does. Pages 1–13 in: The Fungus Spore. M. F. Madelin, ed. Butterworth, London.

Gregory, P. H. 1973. The Microbiology of the Atmosphere, 2nd ed. Leonard Hill (Intertext), Aylesbury, England. 377 pp.

Gregory, P. H. 1981. Speculations on *Phytophthora* as a cocoa root pathogen. Pages 267-269 in: Proc. 7th Int. Cocoa Res. Conf., Douala, Cameroon, 1979. J. de Lafforest, ed. Transla-Inter Ltd., London.

Gregory, P. H., and Maddison, A. C., eds. 1981. Epidemiology of *Phytophthora* on Cocoa in Nigeria. Phytopathol. Pap. 25. Commonw. Mycol. Inst., Kew, Surrey, England. 188 pp.

Griffin, M. J., Idowu, O. L., Maddison, A. C., Taylor, B., and Ward, M. R. 1981. Source of infection. Pages 75-95 in: Epidemiology of *Phytophthora* on Cocoa in Nigeria. P. H. Gregory and A. C. Maddison, eds. Phytopathol. Pap. 25. Commonw. Mycol. Inst., Kew, Surrey, England. 188 pp.

Harris, D. C. 1979. The occurrence of *Phytophthora syringae* in fallen apple leaves. Ann. Appl. Biol. 91:309-312.

Hickman, C. J., and Ho, H. H. 1966. Behaviour of zoospores in plant pathogenic Phycomycetes. Annu. Rev. Phytopathol. 4:195-220.

Ho, H. H., and Zentmyer, G. A. 1977. Morphology of *Phytophthora cinnamomi*. Mycologia 69:701-713.

Holliday, P. 1980. Fungus diseases of tropical crops. Cambridge University Press, Cambridge. 607 pp.

Lacey, J. 1962. The activity of *Phytophthora infestans* (Mont.) deBary in soil. Ph.D. thesis, University of Reading, England. 226 pp.

Lacey, J. 1965. The infectivity of soils containing *Phytophthora infestans*. Ann. Appl. Biol. 56:363-380.

Maddison, A. C. 1981. Terminology and methods of analysis of *P. megakarya* infection sequences. Appendix 1 in: Epidemiology of *Phytophthora* on Cocoa in Nigeria. P. H. Gregory and A. C. Maddison, eds. Phytopathol. Pap. 25. Commonw. Mycol. Inst., Kew, Surrey, England. 188 pp.

McCain, A. H., Holtzmann, O. V., and Trujillo, E. E. 1967. Concentration of *Phytophthora cinnamomi* chlamydospores by soil sieving. Phytopathology 57:1134-1135.

Newhook, F. J. 1961. Non-lethal pathological infection of roots. Nature (London) 191:615-616.

Newhook, F. J., and Podger, F. D. 1972. The role of *Phytophthora cinnamomi* in Australian and New Zealand forests. Annu. Rev. Phytopathol. 10:299-326.

Newhook, F. J., Waterhouse, G. M., and Stamps, D. J. 1978. Tabular key to the species of *Phytophthora* de Bary. Mycol. Pap. 143. Commonw. Mycol. Inst., Kew, Surrey, England. 20 pp.

Onesirosan, P. T. 1971. The survival of *Phytophthora palmivora* in a cocoa plantation during the dry season. Phytopathology 61:975-977.

Padwick, G. W. 1956. Losses caused by plant diseases in the colonies. Phytopathol. Pap. 1. Commonw. Mycol. Inst., Kew, Surrey, England. 60 pp.

Podger, F. D., Doepel, R. F., and Zentmyer, G. A. 1965. Association of *Phytophthora cinnamomi* with a disease of *Eucalyptus marginata* forest in Western Australia. Plant Dis. Rep. 49:943-947.

Pratt, B. H., and Heather, W. A. 1973. The origin and distribution of *Phytophthora cinnamomi* Rands in Australian native plant communities and the significance of its association with particular plant species. Aust. J. Biol. Sci. 26:559-573.

Putter, C. A. J. 1980. The management of epidemic levels of epidemic diseases under tropical subsistence farming conditions. Pages 93-103 in: Comparative Epidemiology: A Tool for Better Disease Management. J. Palti and J. Kranz, eds. Centre for Agric. Pub. and Documentation, Wageningen, The Netherlands.

Reeves, R. J. 1975. Behaviour of *Phytophthora cinnamomi* Rands in different soils and water regimes. Soil Biol. Biochem. 7:19-24.

Taylor, B., and Griffin, M. J. 1981. The role and relative importance of different ant species in the dissemination of black pod disease of cocoa. Pages 114-129 in: Epidemiology of *Phytophthora* on Cocoa in Nigeria. P. H. Gregory and A. C. Maddison, eds. Phytopathol. Pap. 25. Commonw. Mycol. Inst., Kew, Surrey, England. 188 pp.

Thomson, S. V., and Allen, R. M. 1976. Mechanisms of survival of zoospores of *Phytophthora parasitica* in irrigation water. Phytopathology 66:1198-1202.

Tsao, P. H. 1960. A serial dilution end-point method for estimating disease potentials of citrus phytophthoras in soil. Phytopathology 50:717-724.

Tsao, P. H. 1969. Studies on the saprophytic behavior of *Phytophthora parasitica* in soil. Proc. 1st Int. Citrus Symp. 3:1221-1230.

infection in this system were perhaps a consequence of the previously discussed macerating factor produced by *P. capsici* (Yoshikawa et al, 1977).

Yoshikawa et al (1978b) discovered that the translation inhibitor blasticidin S blocked phytoalexin accumulation in soybean hypocotyls inoculated with incompatible races of *P. megasperma* f. sp. *glycinea*; the fungus grew in these tissues as if it were a genetically compatible race. Significantly, the same results were obtained when inhibitor-treated plants were inoculated with normally nonpathogenic (to soybean) isolates of *P. capsici, P. cinnamomi, P. melonis, P. palmivora,* and *P. parasitica* var. *parasitica*. These observations are similar to those of Chamberlain and Gerdemann (1966), who showed that preinoculation heat treatment of soybean hypocotyls destroyed both resistance expression and phytoalexin production in response to the normally nonpathogenic fungi *P. cactorum* and *Helminthosporium sativum*. However, the heat-treated plants remained resistant to inoculation with three saprophytic fungi, which suggests that phytoalexin production is the normal mechanism rendering soybean resistant to the nonpathogenic fungal species but not to the saprophytes. Even more important, the results suggest that *Phytophthora* spp. generally have the potential to colonize soybean tissues but that this capability is not normally expressed by nonpathogenic species because the plant invokes an active defense mechanism to restrict their colonization. Therefore, the various *Phytophthora* spp. probably do not possess unique, specialized machinery for colonizing their respective host tissues; instead, they may be equipped with common, relatively unsophisticated mechanisms.

## PHYSIOLOGY OF PLANTS INFECTED BY *PHYTOPHTHORA* SPP.

Plants infected by *Phytophthora* spp. show several visible and metabolic disease symptoms (Table 1) that in general resemble diseases caused by other facultative plant pathogens. However, *Phytophthora* spp. may have more specialized relationships with their hosts than other pathogens do (for example, the widespread production of haustorial structures in host cells [Hohl and Suter, 1976, and others]). In addition, Garraway and Pelletier (1966) noted that $^{14}CO_2$ assimilates accumulate in infection sites of *P. infestans* on potato leaflets and concluded that this resulted in part from fungus-induced "abnormal metabolic sinks" at the infection sites that caused preferential long-distance transport of metabolites to the infection site. Because long-distance mobilization of nutrients to infection sites has been well documented in diseases incited by obligate plant parasites (Daly, 1976), the results may further indicate that *P. infestans* has features in common with these pathogens. Farrell (1971), however, concluded that the accumulation of $^{14}CO_2$ metabolites at infection sites of *P. infestans* was solely the result of increased photosynthesis caused by stomatal dysfunction, which allowed greater uptake of carbon dioxide, and was therefore not analogous to the accumulation of carbon assimilates that occurs in diseases incited by obligate parasites.

Water shortage in aboveground plant parts, a common symptom in many *Phytophthora* diseases (Duniway, 1977), may indirectly cause many of the metabolic alterations noted in Table 1. Water shortage appears to result from loss of roots rotted by the fungus but may also arise from stomatal dysfunction resulting in increased transpiration rates (Farrell et al, 1969) and possible translocation of the fungus polysaccharides discussed in the last section to the aboveground plant parts, where they may cause vascular dysfunction, wilting, and necrosis (Woodward et al, 1980a).

## MECHANISMS OF HOST RESISTANCE

Considerable information is available on natural mechanisms of resistance to *Phytophthora* spp. Escape mechanisms, such as the failure of zoospore attraction and encystment to occur or failure of the fungi to penetrate and infect properly, exemplify such mechanisms. Epidemiologic factors and predispositioning are also very important in determining plant infection. We consider here only those mechanisms that operate after successful initial infection by *Phytophthora* spp. These may be divided into mechanisms conferring general resistance—that is, resistance of an entire plant species to an entire *Phytophthora* species (nonhost resistance)—and mechanisms conferring specific resistance—that is, resistance that is effective within a plant species to only certain races of a single *Phytophthora* species and that is usually controlled by single host resistance genes (Keen, 1982b). General resistance mechanisms against *Phytophthora* spp. include structural features of the host; preformed chemical inhibitors; induced structural barriers; and hypersensitive reactions and phytoalexins. The only conferral mechanism for specific resistance that is supported by substantial evidence is the hypersensitive reaction and phytoalexins.

### General Resistance

Although structural features of the host function as resistance factors to potential pathogens in plants (Royle, 1976), little evidence indicates that such mechanisms are involved with resistance to *Phytophthora* spp. Hohl and Stössel (1976), however, noted that a resistant potato cultivar produced encasements around haustoria of *P. infestans* and speculated that the encasements were associated with

TABLE 1. Some metabolic changes observed in plants infected by *Phytophthora* spp. and their possible causes

| Symptom | Possible cause | References |
|---|---|---|
| Water shortage and necrosis in aboveground parts | Loss of roots; stomatal dysfunction; effect of pathogen metabolites on water transport | Farrell et al (1969), Duniway (1977), Keen et al (1975), Woodward et al (1980b) |
| Altered mineral contents | Altered water supply? | Weste et al (1980) |
| Increased auxin levels | Stimulated host production? | Fehrmann and Dimond (1967b, 1967c) |
| Respiratory increases | Damage to host cells? | Brenneman and Black (1979), Held et al (1965), Rubin and Artsikhovskaya (1964), Tomiyama et al (1959) |
| Increases in polyphenol oxidase and other enzymes | Damage to host cells? | Held et al (1965), Rubin and Artsikhovskaya (1964), Pitt (1976) |
| Altered accumulation of photosynthates | Directed translocation or stomates remain open | Garraway and Pelletier (1966), Farrell et al (1968) |
| Premature senescence | Vascular dysfunction? | ... |

resistance. Friend et al (1973) also suggested that induced lignification was associated with resistance to this fungus.

Preformed chemical inhibitors in plant tissues may be important in resistance to *Phytophthora* spp. Schwinn (1965) suggested that mustard oils function in this manner in cruciferous plants, and Zaki et al (1980) presented evidence that borbonols—alkyl-substituted lactones present in certain resistant species of *Persea*—function as resistance factors against *P. cinnamomi*. However, involvement of these factors has not yet been proven. Similarly, preformed levels of simple phenolic compounds (Clarke, 1969; Friend et al, 1973) and enzymes converting them to more toxic products (Fehrmann and Dimond, 1967a) have been suggested but not proven to play a role as resistance factors.

Hypersensitive reactions appear to be involved with the general resistance of several plants to nonpathogenic *Phytophthora* spp. (Friend and Shimony, 1978; Jones et al, 1975a; Tippett and Malajczuk, 1979). Although the major observable feature of this type of defense is necrosis of host cells at the pathogen infection site, the role of cell death per se in arresting pathogen development is controversial. One view is that the inducible production of phytoalexins is a major factor inhibiting pathogen development in hypersensitive reactions.

Phytoalexins are antibiotic products of secondary plant biosynthesis. They are generally not present in significant amounts in normal plant tissue but are inducibly formed in response to inoculation with incompatible pathogens or after treatment of plant tissue with certain chemicals called elicitors, some of which are obtained from pathogens (Keen, 1982b; Keen and Bruegger, 1977). Several plants produce phytoalexins in response to inoculation with nonpathogenic races or species of *Phytophthora* (Table 2), and a few phytoalexins may confer general resistance to nonpathogenic *Phytophthora* spp. (Ingham et al, 1977; Ward, 1976; Yoshikawa et al, 1978b).

Ward and coworkers critically studied the role of the phytoalexin capsidiol in conferring general resistance in pepper leaves and fruits to *P. infestans*. The compatible fungus, *P. capsici*, rapidly colonized pepper tissues, but penetration and initial colonization by *P. infestans* resulted in a necrotic hypersensitive reaction in host cells after a few hours (Jones et al, 1975a). Capsidiol accumulated to fungitoxic levels in these tissues at about 12 hr, coincident with a decrease in fungal growth (Jones et al, 1975b). The compatible *P. capsici* strain, on the other hand, did not elicit significant production of capsidiol in the tissues. In addition, *P. capsici* was much less sensitive to inhibition by capsidiol in vitro. Capsidiol might therefore account for inhibition of *P. infestans* growth in pepper.

In summary, the evidence to date indicates that phytoalexins may be involved in general (nonhost) resistance of certain plants to *Phytophthora* spp., but conclusive proof for this role is lacking. Genetic tests such as those elegantly employed by Tegtmeier and VanEtten (1982) with pathogenic and nonpathogenic strains of *Fusarium solani* f. sp. *pisi* would be useful to more critically test the association of phytoalexin production with general resistance to *Phytophthora* spp.

## Specific Resistance

Several plants contain single genes for resistance to incompatible races of certain *Phytophthora* spp., and some of these plants undergo hypersensitive reactions with an accompanying accumulation of phytoalexins (Table 2). In all of these cases, accumulation of the phytoalexins is much greater in response to inoculation with incompatible fungus races than compatible, which suggests a priori that phytoalexins may be causally related to resistance expression. This possibility has been critically investigated in two gene-for-gene host-parasite systems involving *Phytophthora* spp.

### The Potato-*P. infestans* Interaction

Bills (1977) and Kuć et al (1979) have reviewed research on the potato-*P. infestans* system; we will emphasize more recent findings and assess the available evidence. The expression of single-gene host resistance in potato to *P. infestans* involves a classic hypersensitive reaction that has been extensively studied (see Tomiyama et al, 1976). Based primarily on the observation that hypersensitive host cell death in response to an incompatible fungus race occurs well before fungus growth is inhibited, it appears that cell necrosis is not the cause of fungal inhibition per se (Sato and Tomiyama, 1976). Tomiyama et al (1968) first isolated and identified the sesquiterpene phytoalexin, rishitin, from slices of potato tuber that had been inoculated with an incompatible race of *P. infestans*. Because little rishitin accumulated in slices inoculated with a compatible race, the phytoalexin appeared to be involved with resistance expression. Several related terpenoids subsequently isolated from hypersensitive potato tissues accumulated more or less coordinately with rishitin (Bills, 1977; Gross, 1979).

TABLE 2. Host-parasite systems involving *Phytophthora* spp. in which the production of phytoalexins is associated with resistance expression

| Host-parasite system | Type of resistance | Phytoalexins | References |
|---|---|---|---|
| Pepper-*P. infestans* | General | Capsidiol | Jones et al (1975b), Ward (1976) |
| Lupine and *Neonotonia wightii*-*P. megasperma* f. sp. *glycinea* | General | Wighteone | N. T. Keen (*unpublished*), Ingham et al (1977) |
| Peperomia-*P. parasitica* var. *nicotianae* | General | Unidentified | Siradhana et al (1969) |
| Soybean-*Phytophthora* spp. | General | Glyceollin | Yoshikawa et al (1978b) |
| Strawberry-*P. fragariae* | Specific? | Unidentified | Mussell and Staples (1971) |
| Carnation-*P. parasitica* | Specific? | Unidentified | P. Ricci (*unpublished*) |
| Cucumber-*P. drechsleri* | Specific? | Unidentified | A. Alavi and R. N. Strange (*unpublished*) |
| Tobacco-*P. parasitica* var. *nicotianae* | Specific | Rishitin and two others | Budde and Helgeson (1981) |
| Alfalfa-*P. megasperma* f. sp. *medicaginis* | Specific? | Medicarpin | Vaziri et al (1981) |
| Soybean-*P. megasperma* f. sp. *glycinea* | Specific | Glyceollin | Keen (1971), Yoshikawa et al (1978a) |
| Potato-*P. infestans* | Specific | Rishitin and related compounds | Sato et al (1971), Varns et al (1971) |
| Cowpea-*P. vignae* | Specific | Kievitone | Partridge and Keen (1976) |
| Safflower-*P. drechsleri* | Specific | Safynol | Allen and Thomas (1971) |

The available evidence suggests but does not prove a role for the potato terpenoid compounds in the expression of single-gene resistance in this system. The data summarized in Table 3 constitute correlative evidence supporting a causal role, but several uncertainties exist with the potato-*P. infestans* system. First, most of the biochemical work has been done with potato tuber slices, and the reaction of tuber slices to the fungus is not always the same as the reaction of leaves or sprouts of the same cultivar (Lisker and Kuć, 1978; Zacharius et al, 1976). The observation that the response of tuber slices to incompatible fungus races depends on the time after slicing is also disturbing. For instance, slices aged for 72 hr before inoculation were highly resistant to *P. infestans*, but rishitin did not accumulate; some other mechanism, possibly phenolic compounds, was therefore responsible for resistance (Kuć et al, 1979). Sato et al (1971), Nakajima et al (1975), and Sato and Tomiyama (1976) concluded that the terpenoid levels were sufficiently high in the localized infection court of an incompatible *P. infestans* race to explain fungus inhibition, but the data are not as convincing as the corresponding work done with the soybean-*P. megasperma* f. sp. *glycinea* system discussed in the next section.

In view of the foregoing considerations, rishitin and the related compounds cannot be definitely concluded to confer resistance to incompatible *P. infestans* races. Neither do the data exclude this possibility. It is surprising that more efforts have not been made to explore the interactions between the fungus and potato leaves, the natural plant organ primarily attacked by the fungus in the field. Perhaps some of the uncertainties are the result of artifacts associated with the widespread use of tuber slices, an unnatural host tissue.

Fractions have been prepared from *P. infestans* that elicit production of the potato terpenoids in potato tuber slices (Henfling et al, 1980a; Lisker and Kuć, 1977). However, approximately equal quantities of the compounds were elicited irrespective of the host genotype or parasite race used. Perhaps for this reason, Doke and Tomiyama (1980b) and Doke et al (1980) directed attention to the presence of low-molecular-weight cytoplasmic β-glucans, similar if not identical to the mycolaminarans isolated from several *Phytophthora* spp. by Wang and Bartnicki-Garcia (1973,1980). The glucans from several races of *P. infestans* specifically suppressed production of terpenoid phytoalexins induced by the nonspecific fungus elicitors (Doke and Tomiyama, 1980b; Doke et al, 1980). Therefore, the suppressors may be associated with the expression of race-specific resistance, but several questions remain to be answered before this mechanism can be considered established (see Keen, 1982a).

### The Soybean-*P. megasperma* f. sp. *glycinea* Interaction

The expression of single-gene resistance in soybean to *P. megasperma* Drechs. f. sp. *glycinea* Kuan and Erwin (formerly *P. megasperma* var. *sojae*) has been extensively studied (for earlier references, see Keen and Bruegger, 1977). The evidence strongly indicates a functional association between phytoalexin accumulation at the infection site and the expression of disease resistance (Table 3). Several isoflavonoid compounds have been observed to accumulate in soybean tissues after challenge with incompatible pathogens. The only important one, both quantitatively and from the standpoint of antibiotic activity, is glyceollin, a 6a-hydroxypterocarpan isolated from *Glycine max* as three isomers with approximately equal biological activity (Ingham et al, 1981; Lyne et al, 1976).

As in the potato-*P. infestans* system, several resistance genes have been defined in soybeans, and at least 16 pathogenic races of the fungus have been isolated from nature. Therefore, the soybean-*P. megasperma* f. sp. *glycinea* system behaves as a gene-for-gene interaction (Keen, 1982b). As is general in these systems, resistance genes in the host are inherited as dominant characters. Virulence genes have not been defined in the parasite by formal genetic studies, but incompatible races are physiologically dominant, consistent with genetic observations in other gene-for-gene systems indicating that avirulence is generally the dominant character. This is important because it indicates that specificity is determined by the interaction of resistance gene products in the host and avirulence gene products in the pathogen, and a compatible reaction occurs only in the absence of this specific interaction (see Ellingboe, 1976; Keen, 1982b). Consistent with this proposition, there is no evidence for the production of specific pathogen suppressors like those discussed in the last section from *P. infestans*. Coupled with the biochemical data in Table 3, these factors led to the formulation of a pathogen elicitor-host receptor model for determination of single gene-mediated resistance in the soybean-*P. megasperma* f. sp. *glycinea* host-parasite system (Keen, 1982b).

TABLE 3. Evidence for the association of phytoalexin accumulation with the expression of resistance in the potato-*Phytophthora infestans* and soybean-*P. megasperma* f. sp. *glycinea* host-parasite systems

| Evidence | References | |
|---|---|---|
| | Potato-*P. infestans* system | Soybean-*P. megasperma* f. sp. *glycinea* system |
| High phytoalexin accumulation occurs only in incompatible host-parasite combinations | Varns et al (1971) | Keen (1971), Paxton and Chamberlain (1967) |
| The expression of resistance and phytoalexin production can be simultaneously reversed by heat treatments, chemical inhibitors, or elution of the phytoalexin | Tomiyama (1973), Henfling et al (1980b), Doke et al (1976) | Klarman and Gerdemann (1963), Chamberlain (1972) |
| Phytoalexins reach an inhibitory concentration in the pathogen infection court at the time pathogen development is inhibited | Sato et al (1971), Sato and Tomiyama (1976) | Yoshikawa et al (1978a, 1978b) |
| Pathogen elicitors and/or suppressors reproduce the same relative specificity for phytoalexin production as living fungus races | Doke and Tomiyama (1980b) | Keen and Legrand (1980), M. Yoshikawa and N. T. Keen (*unpublished*) |
| Prior induction of or application of phytoalexin at the inoculation site of a normally compatible fungus race leads to a resistant reaction | ... | Bridge and Klarman (1973), Chamberlain and Paxton (1968) |
| Incompatible but not compatible fungus races exhibit ultrastructural aberrations in the host similar to those produced by the phytoalexin in vitro | ... | M. Yoshikawa (*unpublished*) |

As summarized in Table 3, incompatible host-parasite combinations lead to much greater accumulation of glyceollin than compatible ones. Similarly, preinoculation heat treatments or the use of inhibitors of DNA transcription or mRNA translation blocks phytoalexin production and converts normally incompatible interactions to completely compatible ones. Yoshikawa et al (1978a, 1978b) observed that an incompatible fungus race stopped growing shortly after an $ED_{90}$ concentration of glyceollin was reached in host cells at the margin of the advancing fungus colony. When inhibitors of glyceollin accumulation were administered to the same tissues, glyceollin did not accumulate and the fungus continued to grow, just as if the tissue had been inoculated with a genetically compatible fungus race. These studies demonstrate a cause and effect relationship between glyceollin accumulation and the observed expression of host resistance.

Converse experiments, in which glyceollin accumulation was induced in genetically compatible plants by ultraviolet light before inoculation (Bridge and Klarman, 1973) or the phytoalexin was exogenously applied to the inoculation site before inoculation (Chamberlain and Paxton, 1968) resulted in resistant reactions. These experiments also suggest that glyceollin inhibits fungus development in soybean tissues.

Recent unpublished work by Yoshikawa constitutes an independent confirmation of the role of glyceollin in disease resistance. Glyceollin caused ultrastructural plasma membrane abnormalities in *P. megasperma* f. sp. *glycinea* hyphae in vitro. Significantly, the same defects were observed in fungus hyphae growing in an incompatible soybean genotype shortly after their growth stopped. When a compatible soybean genotype was inoculated, however, the membrane defects were not observed. These results also strongly indicate that glyceollin accumulation is the functional basis of resistance expression.

Murch and Paxton (1977) reported that low light intensities led to diminished phytoalexin accumulation and to susceptibility in genetically resistant plants. Murch and Paxton (1979) also observed that reduced aeration of roots decreased phytoalexin production and increased disease susceptibility in hypocotyls of a resistant soybean genotype. They further noted (Murch and Paxton, 1980) that applying high levels of salts such as $CaCl_2$ to inoculated resistant soybean plants decreased glyceollin production but did not increase susceptibility. This evidence may argue against the role of glyceollin in resistance expression; however, $CaCl_2$-treated plants of the resistant genotype were not inoculated with a compatible fungus race, so the observed failure of the incompatible fungus to grow may have been caused by an inhibitory effect of the high salt concentration on the fungus.

Using zoospore inoculation of intact but etiolated soybean hypocotyls, Ward et al (1979) confirmed the association of glyceollin production with resistance expression. Ward et al (1980) also reported that the systemic fungicide metalaxyl induced significant production of glyceollin as well as hypersensitive necrosis at the inoculation point of soybean hypocotyls inoculated with a normally compatible fungus race. Accordingly, disease control by the fungicide could be explained at least in part by phytoalexin production. If this is correct, then the fungicide would be acting as a sensitizer (Keen and Bruegger, 1977).

Ward et al (1981) reported differences in the susceptibility or resistance of etiolated soybean hypocotyls when the tops or lower portions of the hypocotyls were inoculated with zoospores. They further noted age-related differences in the resistance of the hypocotyls and did not always obtain a close correlation between observed resistance and glyceollin production (Lazarovits et al, 1981). However, the fact that etiolated hypocotyls were used in this work may account for the position differences, which have not been noted by other groups using nonetiolated plants and mycelial inoculation. Also, no attention was given to the localized production of glyceollin at the infection sites, a factor of great importance in assessing the role of phytoalexins in resistance expression (Yoshikawa et al, 1978a, 1978b).

Metabolic inhibitors have been shown to block glyceollin production and convert genetically resistant soybean hypocotyls to complete susceptibility (Table 3). The objection may be raised, however, that such inhibitors of primary metabolism may simultaneously block other unknown defense mechanisms and may also have secondary effects on plant metabolism besides the primary target. In recent unpublished work, we observed that the herbicide glyphosate is a potent inhibitor of glyceollin production in soybean hypocotyls inoculated with *P. megasperma* f. sp. *glycinea* and also blocks the expression of disease resistance when applied to inoculated seedling cuttings of a resistant genotype at a minimum concentration of about 5 $\mu g/ml$. Amrhein et al (1980a, 1980b) reported that glyphosate is a specific inhibitor of the higher plant enzyme converting shikimate to chorismate. Because glyceollin production requires function of the shikimate pathway, depletion of biosynthetic precursors is thought to cause the inhibitory effect of glyphosate. The herbicide did not cause visible plant toxicity at the low concentrations and exposure times used, and feeding plants phenylalanine and tyrosine reversed the inhibitory effects of glyphosate on both phytoalexin accumulation and resistance expression. Accordingly, glyphosate appears to be a relatively specific inhibitor of resistance expression, thus circumventing objections that may be raised against the use of inhibitors of primary metabolic processes such as protein biosynthesis. As such, the observations with glyphosate strongly support the role of glyceollin in resistance expression.

Zähringer et al (1979) reported that particulate fractions from elicitor-treated soybean cotyledons contained an enzyme(s) catalyzing a specific prenylation of 3,6,9a-trihydroxypterocarpan at the 2 and 4 positions but at no other carbons. Extracts from cotyledons not exposed to an elicitor were devoid of transferase activity. Because the trihydroxypterocarpan appears to be a biosynthetic precursor of glyceollin (Ingham et al, 1981), the results indicate that the transferase(s) was involved in glyceollin biosynthesis. These results are significant because they demonstrate that the transferase is a strictly inducible enzyme formed only after elicitor challenge of soybean tissue. The observations therefore offer direct experimental support for the hypothesis (Keen and Bruegger, 1977) that one or more enzymes in the terminal biosynthetic pathway to glyceollin are synthesized de novo after elicitor challenge or inoculation with an incompatible fungus race.

Zähringer et al (1978) also noted that elicitor-treated soybean cotyledons, but not those exposed to water, exhibited an activation of phenylalanine ammonia-lyase, an enzyme thought to play an important regulatory role in the biosynthesis of phenolic compounds in plants.

Taken collectively, these recent observations support earlier indications that glyceollin accumulation is the mechanism that inhibits fungus development in genetically resistant soybean cultivars. Accordingly, it appears well established that phytoalexin production accounts for resistance expression in the soybean-*P. megasperma* f. sp. *glycinea* host-parasite system. A diagrammatic summation of the known sequence of events taking place in an incompatible reaction to the fungus is presented in Table 4.

The specific recognition of incompatible *P. megasperma* f.

sp. *glycinea* races that predicates phytoalexin accumulation in soybean occurs soon after inoculation and may accordingly be mediated by cell surface molecules (Table 4). Supporting this interpretation, surface glycoproteins extracted from cell walls of several *P. megasperma* f. sp. *glycinea* races gave the same relative specificity for glyceollin production in near-isogenic resistant and susceptible soybean cultivars as the living fungus races (Keen and Legrand, 1980). The carbohydrate portions were the elicitor-active moieties and contained glucose and mannose as neutral sugars. We have also recently observed that Pronase releases glycopeptides from fungus cell walls that originate from the glycoproteins.

Yoshikawa et al (1981) performed experiments that disclosed a second class of mannose containing surface molecules in *P. megasperma* f. sp. *glycinea*. Elicitor-active carbohydrates were released from fungus cell walls by an enzyme preparation obtained from soybean tissues. Similar to the surface glycoproteins, these carbohydrates contained glucose and mannose and appeared to function as race-specific glyceollin elicitors. Yoshikawa and Keen (*unpublished*) have purified two closely related proteins of about 35 kd and isoelectric points at pH 8.7 and 10.5 to homogeneity from the crude soybean extracts. Both enzymes release the carbohydrate elicitor from fungus walls and also possess endoglucanase activity against mycolaminaran. The fungus carbohydrates released from cell walls by the Yoshikawa enzyme therefore appear to be covalently linked to the branched $\beta$-1,3-glucan backbone of the *P. megasperma* f. sp. *glycinea* cell wall. We are now approaching elucidation of the chemical structures of the surface carbohydrates from several fungus races, because they appear to be the race-specific elements recognized by resistant genotypes of soybeans to invoke phytoalexin production.

Yoshikawa, Keen, and Wang (*unpublished*) recently demonstrated that specific receptors exist on soybean membranes for the phytoalexin elicitor, mycolaminaran. Preparations of mycolaminaran from *Phytophthora* spp. elicit glyceollin production in soybean cotyledons but do not exhibit race specificity and are only about 2% as active on a concentration basis as the surface glucomannans discussed previously. However, because they are relatively homogeneous, are produced in large amounts by *Phytophthora* spp., and can be made radioactive by growing the fungi on a medium supplemented with [$^{14}$C]glucose, mycolaminarans were chosen in the initial search for plant receptors that recognize fungal carbohydrate elicitors. The 10,000–30,000 *g* centrifuge fraction of soybean cotyledon membranes bound [$^{14}$C]mycolaminaran with an affinity constant of about 10 $\mu M$. Competition studies with unlabeled mycolaminaran, various mycolaminaran derivatives, or other carbohydrates indicated that a single class of specific receptors exists on soybean membranes for mycolaminaran. The elicitor activity of mycolaminaran in soybean appears, therefore, to be mediated by its initial binding to membrane receptors. We have not yet addressed the possibility that analogous receptors exist for the more active glucomannan elicitors discussed above, but these latter carbohydrates are not recognized by the mycolaminaran receptor.

TABLE 4. Approximate sequence of events in soybean plants during an incompatible reaction to *Phytophthora megasperma* f. sp. *glycinea*

| Approximate time (hours) | Event |
|---|---|
| 0 | Inoculation or elicitor treatment |
| 0–4 | Host recognition |
| 2 | Increased poly adenosine mRNA synthesis |
| 4 | DNA transcription required to here |
| 5 | mRNA translation required to here |
| 6 | Increased phenylalanine ammonia-lyase (PAL) activity |
| ? | Inducible prenyl transferase activity appears |
| 6.5 | Glyceollin accumulation starts |
| 8 | Glyceollin reaches an in vivo $ED_{90}$ concentration against the fungus |
| 9 | Fungus growth stops |

## SUMMARY

The mechanisms whereby *Phytophthora* spp. colonize host tissues have not been conclusively elucidated but may involve mechanical penetration of cell walls and possibly the activity of macerating enzymes. Experiments with inhibitors of host defense mechanisms indicate that the colonization mechanism(s) is similar in all *Phytophthora* spp. and that the important determinant of the host range of various species may be whether the host invokes an active defense mechanism to inhibit fungus growth. Phytotoxic metabolites are produced by *Phytophthora* spp. but do not appear to be important in host colonization; they may, however, contribute to disease symptoms such as water stress in infected plants. The many metabolic alterations observed in plants infected by *Phytophthora* spp. resemble those caused by other facultative plant pathogens. Preformed chemical inhibitors may contribute to host defenses against certain *Phytophthora* spp., but the evidence for this role is as yet inconclusive.

Hypersensitive defense reactions are involved in the expression of both general and specific (single gene-mediated) resistance of certain plants to *Phytophthora* spp. Phytoalexins appear to be directly responsible for the inhibition of fungal development in the hypersensitive reactions of potato, pepper, and soybean. A large body of evidence strongly implicates the phytoalexin glyceollin with expression of resistance in soybean to nonpathogenic *Phytophthora* spp. (general resistance) and to incompatible races of *P. megasperma* f. sp. *glycinea* (specific resistance). Surface carbohydrates of incompatible races of the latter fungus may be the factors that are specifically recognized by incompatible soybean genotypes to invoke glyceollin accumulation and disease resistance. Compatible fungus races appear to possess surface carbohydrates with altered structures that are not recognized by the appropriate soybean resistance genotypes.

## ACKNOWLEDGMENTS

Our research was supported by grants from the U.S. National Science Foundation and the Japanese Ministry of Education.

## LITERATURE CITED

Akinrefon, O. A. 1968. Production of extracellular enzymes of *Phytophthora palmivora* (Butl.) Butl. J. Gen. Microbiol. 51:67-74.

Allen, E. H., and Thomas, C. A. 1971. Time course of safynol accumulation in resistant and susceptible safflower infected with *Phytophthora drechsleri*. Physiol. Plant Pathol. 1:235-240.

Amrhein, N., Deus, B., Gehrke, P., and Steinrücken, H. C. 1980a. The site of the inhibition of the shikimate pathway by glyphosate. II. Interference of glyphosate with chorismate formation in vivo and in vitro. Plant Physiol. 66:830-834.

Amrhein, N., Schab, J., and Steinrücken, H. C. 1980b. The mode of action of glyphosate. Naturwissenschaften 67:356-357.

Ballio, A., Gianani, L., Borrelli, R., Bottalico, A., and Graniti, A. 1972. Production of phytotoxins by *Phytophthora nicotianae* B. de Haan var. *parasitica* (Dast.) Waterh. Pages 431-432 in: Phytotoxins in Plant Diseases. R. K. S. Wood, A. Ballio, and A. Graniti, eds.

Academic Press, London.

Bateman, D. F., and Basham, H. G. 1976. Degradation of plant cell walls and membranes by microbial enzymes. Pages 316-355 in: Physiological Plant Pathology. Encyclopedia of Plant Physiology, Vol. 4. R. Heitefuss and P. H. Williams, eds. Springer-Verlag, Berlin.

Bills, D. D. 1977. Interactions between *Phytophthora infestans* and potato host. Am. Chem. Soc. Symp. Ser. 26:47-60.

Brenneman, J. A., and Black, L. L. 1979. Respiration and terminal oxidases in tomato leaves infected by *Phytophthora infestans.* Physiol. Plant Pathol. 14:281-290.

Bridge, M. A., and Klarman, W. L. 1973. Soybean phytoalexin, hydroxyphaseollin, induced by ultraviolet irradiation. Phytopathology 63:606-609.

Budde, A. D., and Helgeson, J. P. 1981. Phytoalexins in tobacco callus tissues challenged by zoospores of *Phytophthora parasitica* var. *nicotianae.* (Abstr.) Phytopathology 71:206.

Chamberlain, D. W. 1972. Heat-induced susceptibility to nonpathogens and cross-protection against *Phytophthora megasperma* var. *sojae* in soybean. Phytopathology 62:645-646.

Chamberlain, D. W., and Gerdemann, J. W. 1966. Heat-induced susceptibility of soybeans to *Phytophthora megasperma* var. *sojae, Phytophthora cactorum,* and *Helminthosporium sativum.* Phytopathology 56:70-73.

Chamberlain, D. W., and Paxton, J. D. 1968. Protection of soybean plants by phytoalexin. Phytopathology 58:1349-1350.

Chatterjee, A. K., and Starr, M. P. 1977. Donor strains of the soft-rot bacterium *Erwinia chrysanthemi* and conjugational transfer of the pectolytic capacity. J. Bacteriol. 132:862-869.

Clarke, D. D. 1966. Production of pectic enzymes by *Phytophthora infestans.* Nature 211:649.

Clarke, D. D. 1969. The accumulation of scopolin in potato tuber tissue after infection by *Phytophthora infestans* and its role in pathogenesis. Phytochemistry 8:7.

Daly, J. M. 1976. The carbon balance of diseased plants: Changes in respiration, photosynthesis and translocation. Pages 450-479 in: Physiological Plant Pathology. Encyclopedia of Plant Physiology, Vol. 4. R. Heitefuss and P. H. Williams, eds. Springer-Verlag, Berlin.

Doke, N., and Tomiyama, K. 1980a. Effect of hyphal wall components from *Phytophthora infestans* on protoplasts of potato tuber tissues. Physiol. Plant Pathol. 16:169-176.

Doke, N., and Tomiyama, K. 1980b. Suppression of the hypersensitive response of potato tuber protoplasts to hyphal wall components by water soluble glucans isolated from *Phytophthora infestans.* Physiol. Plant Pathol. 16:177-186.

Doke, N., Nakae, Y., and Tomiyama, K. 1976. Effect of blasticidin S on the production of rishitin in potato tuber tissue infected by an incompatible race of *Phytophthora infestans.* Phytopathol. Z. 87:337-344.

Doke, N., Garas, N. A., and Kuć, J. 1980. Effect on host hypersensitivity of suppressors released during the germination of *Phytophthora infestans* cystospores. Phytopathology 70:35-39.

Duniway, J. M. 1977. Changes in resistance to water transport in safflower during the development of Phytophthora root rot. Phytopathology 67:331-337.

Ellingboe, A. H. 1976. Genetics of host-parasite interactions. Pages 761-778 in: Physiological Plant Pathology. Encyclopedia of Plant Physiology, Vol. 4. R. Heitefuss and P. H. Williams, eds. Springer-Verlag, Berlin.

Farrell, G. M. 1971. Localization of photosynthetic products in potato leaves infected by *Phytophthora infestans.* Physiol. Plant Pathol. 1:457-467.

Farrell, G. M., Preece, T. F., and Wren, M. J. 1969. Effects of infection by *Phytophthora infestans* (Mont.) deBary on the stomata of potato leaves. Ann. Appl. Biol. 63:265-275.

Fehrmann, H., and Dimond, A. E. 1967a. Peroxidase activity and *Phytophthora* resistance in different organs of the potato plant. Phytopathology 57:69-72.

Fehrmann, H., and Dimond, A. E. 1967b. Studies on auxins in the *Phytophthora* disease of the potato tuber. I. Role of indole-acetic acid in pathogenesis. Phytopathol. Z. 59:83-100.

Fehrmann, H., and Dimond, A. E. 1967c. Studies on auxins in the *Phytophthora* disease of the potato tuber. II. Relation of indole-acetic acid to some physiological processes in pathogenesis. Phytopathol. Z. 59:105-121.

Friend, J., and Shimony, C. 1978. Hypersensitive cell death in the resistance of potato leaves to attack by *Phytophthora infestans.* Ann. Appl. Biol. 89:369-370.

Friend, J., Reynolds, S. B., and Aveyard, M. A. 1973. Phenylalanine ammonia lyase, chlorogenic acid and lignin in potato tuber tissue inoculated with *Phytophthora infestans.* Physiol. Plant Pathol. 3:495-507.

Garraway, M. O., and Pelletier, R. L. 1966. Distribution of $C^{14}$ in the potato plant in relation to leaf infection by *Phytophthora infestans.* Phytopathology 56:1184-1189.

Gross, D. 1979. Phytoalexine der Kartoffel. Biochem. Physiol. Pflanz. 174:327-344.

Hanchey, P., and Wheeler, H. 1971. Pathological changes in ultrastructure: Tobacco roots infected with *Phytophthora parasitica* var. *nicotianae.* Phytopathology 61:33-39.

Held, A., Kedar, N., and Birk, Y. 1965. Dehydrogenase activity of potato tuber tissue infected with *Phytophthora infestans.* Phytopathology 55:970-976.

Henfling, J. W. D. M., Bostock, R. M., and Kuć, J. 1980a. Cell walls of *Phytophthora infestans* contain an elicitor of terpene accumulation in potato tubers. Phytopathology 70:772-776.

Henfling, J. W. D. M., Bostock, R., and Kuć, J. 1980b. Effect of abscisic acid on rishitin and lubimin accumulation and resistance to *Phytophthora infestans* and *Cladosporium cucumerinum* in potato tuber tissue slices. Phytopathology 70:1074-1078.

Hohl, H. R., and Stössel, P. 1976. Host-parasite interfaces in a resistant and a susceptible cultivar of *Solanum tuberosum* inoculated with *Phytophthora infestans*: Tuber tissue. Can. J. Bot. 54:900-912.

Hohl, H. R., and Suter, E. 1976. Host-parasite interfaces in a resistant and a susceptible cultivar of *Solanum tuberosum* inoculated with *Phytophthora infestans*: Leaf tissue. Can. J. Bot. 54:1956-1970.

Howell, C. R. 1976. Use of enzyme-deficient mutants of *Verticillium dahliae* to assess the importance of pectolytic enzymes in symptom expression of Verticillium wilt in cotton. Physiol. Plant Pathol. 9:279-283.

Ingham, J. L., Keen, N. T., and Hymowitz, T. 1977. A new isoflavone from fungus-inoculated stems of *Glycine wightii.* Phytochemistry 16:1943-1946.

Ingham, J. L., Keen, N. T., Mulheirn, L. J., and Lyne, R. L. 1981. Inducibly-formed isoflavonoids from leaves of soybean. Phytochemistry 20:795-798.

Jones, D. R., Graham, W. G., and Ward, E. W. B. 1974. Ultrastructural changes in pepper cells in a compatible interaction with *Phytophthora capsici.* Phytopathology 64:1084-1090.

Jones, D. R., Graham, W. G., and Ward, E. W. B. 1975a. Ultrastructural changes in pepper cells in an incompatible interaction with *Phytophthora infestans.* Phytopathology 65:1274-1285.

Jones, D. R., Unwin, C. H., and Ward, E. W. B. 1975b. The significance of capsidiol induction in pepper fruit during an incompatible interaction with *Phytophthora infestans.* Phytopathology 65:1286-1288.

Keen, N. T. 1971. Hydroxyphaseollin production by soybeans resistant and susceptible to *Phytophthora megasperma* var. *sojae.* Physiol. Plant Pathol. 1:265-275.

Keen, N. T. 1982a. Mechanisms conferring specific recognition in gene-for-gene plant-parasite systems. Pages 67-84 in: Active Defence Mechanisms in Plants. R. K. S. Wood, ed. Plenum Press, London.

Keen, N. T. 1982b. Specific recognition in gene-for-gene host-parasite systems. Adv. Plant Pathol. 1:35-82.

Keen, N. T., and Bruegger, B. 1977. Phytoalexins and chemicals that elicit their production in plants. Am. Chem. Soc. Symp. Ser. 62:1-26.

Keen, N. T., and Legrand, M. 1980. Surface glycoproteins: Evidence that they may function as the race specific phytoalexin elicitors of *Phytophthora megasperma* f. sp. *glycinea.* Physiol. Plant Pathol. 17:175-192.

Keen, N. T., Wang, M. C., Bartnicki-Garcia, S., and Zentmyer, G. A. 1975. Phytotoxicity of mycolaminarans-$\beta$-1,3-glucans from *Phytophthora* spp. Physiol. Plant Pathol. 7:91-97.

Klarman, W. L., and Gerdemann, J. W. 1963. Induced susceptibility in soybean plants genetically resistant to *Phytophthora sojae*. Phytopathology 53:863-864.

Kuć, J., Henfling, J., Garas, N., and Doke, N. 1979. Control of terpenoid metabolism in the potato-*Phytophthora infestans* interaction. J. Food Prot. 42:508-511.

Lazarovits, G., Stössel, P., and Ward, E. W. B. 1981. Age-related changes in specificity and glyceollin production in the hypocotyl reaction of soybeans to *Phytophthora megasperma* var. *sojae*. Phytopathology 71:94-97.

Lisker, N., and Kuć, J. 1977. Elicitors of terpenoid accumulation in potato tuber slices. Phytopathology 67:1356-1359.

Lisker, N., and Kuć, J. 1978. Terpenoid accumulation and browning in potato sprouts inoculated with *Phytophthora infestans*. Phytopathology 68:1284-1287.

Lyne, R., Mulheirn, L. J., and Leworthy, D. P. 1976. New pterocarpinoid phytoalexins of soybean. J. Chem. Soc. Chem. Commun. pp. 497-498.

McIntyre, J. L., and Hankin, L. 1978. An examination of enzyme production by *Phytophthora* spp. on solid and liquid media. Can. J. Microbiol. 24:75-78.

Murch, R. S., and Paxton, J. D. 1977. Glyceollin concentrations in *Phytophthora* resistant soybean: Light influences. (Abstr.) Proc. Am. Phytopathol. Soc. 4:135-136.

Murch, R. S., and Paxton, J. D. 1979. Rhizosphere anaerobiosis and glyceollin accumulation in soybean. Phytopathol. Z. 96:91-94.

Murch, R. S., and Paxton, J. D. 1980. Rhizosphere salinity and phytoalexin accumulation in soybean. Plant Soil 54:163-167.

Mussell, H. W., and Staples, R. C. 1971. Phytoalexinlike compounds apparently involved in strawberry resistance to *Phytophthora fragariae*. Phytopathology 61:515-517.

Nakajima, T., Tomiyama, K., and Kinukawa, M. 1975. Distribution of rishitin and lubimin in potato-tuber tissue infected by an incompatible race of *Phytophthora infestans* and the site where rishitin is synthesized. Ann. Phytopathol. Soc. Jpn. 41:49-55.

Partridge, J. E., and Keen, N. T. 1976. Association of the phytoalexin kievitone with single-gene resistance of cowpeas to *Phytophthora vignae*. Phytopathology 66:426-429.

Paxton, J. D., and Chamberlain, D. W. 1967. Acquired local resistance of soybean plants to *Phytophthora* spp. Phytopathology 57:352-353.

Peters, B. M., Cribbs, D. H., and Stelzig, D. A. 1978. Agglutination of plant protoplasts by fungal cell wall glucans. Science 201:364-365.

Pitts, D. 1976. Nature and location of increased ribonuclease activity in potato leaves following infection with *Phytophthora infestans*. Trans. Br. Mycol. Soc. 66:239-248.

Ravisé, A., and Trique, B. 1972. Détermination des enzymes pectinolytiques deux souches *Phytophthora* De Bary variations d'activité dans des tissus des plantules de tomate en relation avec génomes de résistance. Agron. Trop. (Paris) 27:751-762.

Royle, D. J. 1976. Structural features of resistance to plant diseases. Pages 161-193 in: Biochemical Aspects of Plant-Parasite Relationships. J. Friend and D. R. Threlfall, eds. Academic Press, London.

Rubin, B. A., and Artsikhovskaya, E. V. 1964. Biochemistry of pathological darkening of plant tissues. Annu. Rev. Phytopathol. 2:157-178.

Sato, N., and Tomiyama, K. 1976. Relationship between inhibition of intracellular hyphal growth of *Phytophthora infestans* and rishitin concentration in infected potato cells. Ann. Phytopathol. Soc. Jpn. 43:598-600.

Sato, N., Kitazawa, K., and Tomiyama, K. 1971. The role of rishitin in localizing the invading hyphae of *Phytophthora infestans* in infection sites at the cut surfaces of potato tubers. Physiol. Plant Pathol. 1:289-295.

Scheffer, R. P. 1976. Host-specific toxins in relation to pathogenesis and disease resistance. Pages 247-269 in: Physiological Plant Pathology. Encyclopedia of Plant Physiology, Vol. 4. R. Heitefuss and P. H. Williams, eds. Springer-Verlag, Berlin.

Schwinn, F. J. 1965. Untersuchungen zum Wirt-Parasit-Verhältnis bei der Kragenfäule des Apfelbaumes (*Phytophthora cactorum*) und zu ihrer Bekämpfung. III. Die Eignung verschiedener Gründungungspflanzen zur Bekämpfung des Erregers im Boden. Phytopathol. Z. 54:258-274.

Seidel, H. 1961. Untersuchungen über den Nährstoffbedarf und die Toxinbildung des Pilzes *Phytophthora infestans* (Mont.) deBary in vollsynthetischen Nährlösungen. Phytopathol. Z. 41:1-26.

Siradhana, B. S., Schmitthenner, A. F., and Ellett, C. W. 1969. Formation of phytoalexin in *Peperomia* in relation to resistance to *Phytophthora nicotianae* var. *parasitica*. Phytopathology 59:405-410.

Stössel, P., Lazarovits, G., and Ward, E. W. B. 1980. Penetration and growth of compatible and incompatible races of *Phytophthora megasperma* var. *sojae* in soybean hypocotyl tissues differing in age. Can. J. Bot. 58:2594-2601.

Tegtmeier, K. J., and VanEtten, H. D. 1982. The role of pisatin tolerance and degradation in the virulence of *Nectria haematococca* on peas: A genetic analysis. Phytopathology 72:608-612.

Tippett, J., and Malajczuk, N. 1979. Interaction of *Phytophthora cinnamomi* and a resistant host, *Acacia pulchella*. Phytopathology 69:764-772.

Tomiyama, K. 1973. Effect of temperature on the process of the hypersensitive death of a potato plant cell infected by an incompatible race of *Phytophthora infestans*. Ann. Phytopathol. Soc. Jpn. 39:73-78.

Tomiyama, K., Takakuwa, M., Takase, N., and Sakai, R. 1959. Alteration of oxidative metabolism in a potato tuber cell invaded by *Phytophthora infestans* and in the neighbouring tissues. Phytopathol. Z. 37:113-144.

Tomiyama, K., Sakuma, T., Ishizaka, N., Sato, N., Katsui, N., Takasugi, M., and Masamune, T. 1968. A new antifungal substance isolated from resistant potato tuber tissue infected by pathogens. Phytopathology 58:115-116.

Tomiyama, K., Doke, N., and Lee, H. S. 1976. Mechanism of hypersensitive cell death in host-parasite interaction. Pages 136-142 in: Biochemistry and Cytology of Plant-Parasite Interaction. K. Tomiyama, J. M. Daly, I. Uritani, H. Oku, and S. Ouchi, eds. Kodansha Ltd., Tokyo.

Varns, J. L., Kuć, J., and Williams, E. B. 1971. Terpenoid accumulation as a biochemical response of the potato tuber to *Phytophthora infestans*. Phytopathology 61:174-177.

Vaziri, A., Keen, N. T., and Erwin, D. C. 1981. Correlation of medicarpin production with resistance to *Phytophthora megasperma* f. sp. *medicaginis* in alfalfa seedlings. Phytopathology 71:1235-1238.

Wang, M. C., and Bartnicki-Garcia, S. 1973. Novel phosphoglucans from the cytoplasm of *Phytophthora palmivora* and their selective occurrence in certain life cycle stages. J. Biol. Chem. 248:4112-4118.

Wang, M. C., and Bartnicki-Garcia, S. 1980. Distribution of mycolaminarans and cell wall β-glucans in the life cycle of *Phytophthora*. Exp. Mycol. 4:269-280.

Ward, E. W. B. 1976. Capsidiol production in pepper leaves in incompatible interactions with fungi. Phytopathology 66:175-176.

Ward, E. W. B., Lazarovits, G., Unwin, C. H., and Buzzell, R. I. 1979. Hypocotyl reactions and glyceollin in soybeans inoculated with zoospores of *Phytophthora megasperma* var. *sojae*. Phytopathology 69:951-955.

Ward, E. W. B., Lazarovits, G., Stössel, P., Barrie, S. D., and Unwin, C. H. 1980. Glyceollin production associated with control of *Phytophthora* rot of soybeans by the systemic fungicide, metalaxyl. Phytopathology 70:738-740.

Ward, E. W. B., Stössel, P., and Lazarovits, G. 1981. Similarities between age-related and race-specific resistance of soybean hypocotyls to *Phytophthora megasperma* var. *sojae*. Phytopathology 71:504-508.

Weste, G., Chaudhri, M. A., Haukka, M., and Vithanage, K. 1980. The influence of root infection by *Phytophthora cinnamomi* on mineral content of certain native species. Phytopathol. Z. 99:205-214.

Woodward, J. R., Keane, P. J., and Stone, B. A. 1980a. β-glucans and β-glucan hydrolases in plant pathogenesis with special reference to wilt-inducing toxins from *Phytophthora* species. Am. Chem. Soc. Symp. Ser. 126:113-141.

Woodward, J. R., Keane, P. J., and Stone, B. A. 1980b. Structures and properties of wilt-inducing polysaccharides from *Phytophthora* species. Physiol. Plant Pathol. 16:439-454.

Yoshikawa, M., Tsukadaira, T., Masago, H., and Minoura, S. 1977. A nonpectolytic protein from *Phytophthora capsici* that macerates plant tissue. Physiol. Plant Pathol. 11:61-70.

Yoshikawa, M., Yamauchi, K., and Masago, H. 1978a. Glyceollin: Its role in restricting fungal growth in resistant soybean hypocotyls infected with *Phytophthora megasperma* var. *sojae*. Physiol. Plant Pathol. 12:73-82.

Yoshikawa, M., Yamauchi, K., and Masago, H. 1978b. De novo messenger RNA and protein synthesis are required for phytoalexin-mediated disease resistance in soybean hypocotyls. Plant Physiol. 61:314-317.

Yoshikawa, M., Madama, M., and Masago, H. 1981. Release of a soluble phytoalexin elicitor from mycelial walls of *Phytophthora megasperma* var. *sojae* by soybean tissues. Plant Physiol. 67:1032-1035.

Zacharius, R. M., Osman, S. F., Heisler, E. G., and Kissinger, J. C. 1976. Effect of the R-3 gene in resistance of the Wauseon potato tuber to *Phytophthora infestans*. Phytopathology 66:964-966.

Zähringer, U., Ebel, J., and Grisebach, H. 1978. Induction of phytoalexin synthesis in soybean. Elicitor-induced increase in enzyme activities of flavonoid biosynthesis and incorporation of mevalonate into glyceollin. Arch. Biochem. Biophys. 188:450-455.

Zähringer, U., Ebel, J., Mulheirn, L. J., Lyne, R. L., and Grisebach, H. 1979. Induction of phytoalexin synthesis in soybean. Dimethylallylpyrophosphate: Trihydroxypterocarpan dimethylallyl transferase from elicitor-induced cotyledons. FEBS Lett. 101:90-92.

Zaki, A. I., Zentmyer, G. A., Pettus, J., Sims, J. J., Keen, N. T., and Sing, V. O. 1980. Borbonol from *Persea* spp.—Chemical properties and antifungal activity against *Phytophthora cinnamomi*. Physiol. Plant Pathol. 16:205-212.

# 22

# Histology and Cytology of Infection and Disease Caused by *Phytophthora*

M. D. Coffey and U. E. Wilson

CONTENTS



## INTRODUCTION

In this review of the histology and cytology of infection and disease caused by *Phytophthora,* we will concentrate on late blight of potato, caused by *P. infestans* (Mont.) de Bary, but will describe other disease situations where relevant. We will describe the modes of infection by *P. infestans* and the further development of the fungus, including the fine structure of the various types of haustoria.

Elizabeth Blackwell (1953) gave a detailed account of the fungal structures present in potato leaves and tubers infected with *P. infestans.* Haustoria were observed as simple, fingerlike organs with diameters much smaller than those of the intercellular hyphae. Short lateral branches of the intercellular hyphae were also observed and were thought to be sufficiently distinctive to be given the name "diverticula" (Blackwell, 1953). Pristou and Gallegly (1954) and Ferris (1955) established the mode of infection on leaves and also made useful cytologic studies of the nature of the hypersensitive reaction regulated by R-genes.

Tomiyama and co-workers added a further dimension to the subject by devising a method of infecting potato leaf petioles that allowed parallel studies to be made of the cytology (Kitazawa and Tomiyama, 1970) and biochemistry of the disease. Király et al (1972) introduced some controversy into the subject by contending that fungal death actually preceded host cell death in the hypersensitive response. Their claim, though obviously untenable in light of the critical work already done on the cytology of the disease, stimulated much discussion and some research on the nature of hypersensitivity and particularly its precise relationship to resistance (e.g., Maclean et al, 1974). Electron microscopy has been employed in a number of recent investigations of the cytology of late blight disease (Hohl and Stössel, 1976; Shimony and Friend, 1975).

We propose to draw on the more salient literature to describe both susceptible and resistant interactions. We will attempt to relate these cytologic and fine-structural observations to the available biochemical data and point out the gross deficiencies in our present knowledge.

## THE SUSCEPTIBLE STATE

We prefer to use the term "susceptible" rather than "compatible" when describing histologic events in diseases caused by *Phytophthora.* In many susceptible interactions, good evidence exists that individual host cells react in an apparently incompatible manner. Characteristic of this cell response is the formation of deposits of papillalike material (Aist, 1976) on host walls next to hyphae and around haustoria. These wall deposits have been seen in such diverse disease situations as susceptible tobacco roots infected with *P. parasitica* var. *nicotianae* (Hanchey and Wheeler, 1971) and soybean hypocotyls infected with *P. megasperma* f. sp. *glycinea* (Stössel et al, 1981). Similar deposits have been observed in susceptible tuber and leaf tissues infected with *P. infestans* (Blackwell, 1953; Wilson and Coffey, 1980).

The interaction between *P. infestans* and a susceptible host leaf begins when the fungal spore starts to germinate. The fungus produces both sporangia and zoospores, but only zoospores are believed to cause significant infection of the leaf (Pristou and Gallegly, 1954; Umaerus, 1969). Zoospores encyst on the leaf surface, and cysts germinate to form appressoria (Pristou and Gallegly, 1954). Penetration directly through the periclinal wall of an epidermal cell is most common (Figs. 2, 5, and 6). However, penetration via stomata has been reported as the usual mode of entry into the leaf in at least one instance (Hohl and Suter, 1976). Few quantitative data are available, but one study found that epidermal cells next to guard cells were penetrated preferentially (Fig. 1).

The electron microscope has revealed that infection hyphae may grow parallel to the leaf surface within the outer epidermal cell wall before penetration (Fig. 6; Hohl and Suter, 1976). Penetration of the host cell most likely involves enzymatic dissolution of the wall, because no evidence of mechanical deformation of the wall microfibrils has been found (Hohl and Suter, 1976). Once in the host cell, the fungus forms an infection vesicle (Shimony and Friend, 1975), from which one or more intracellular hyphae are produced (Figs. 2, 5, and 6). The mode of penetration is very similar to that encountered with the downy mildew *Bremia lactucae* (Sargent et al, 1973). By contrast, *P. megasperma* f. sp. *glycinea* usually enters between the anticlinal walls of the epidermal cells of soybean hypocotyls and only rarely penetrates them directly (Stössel et al, 1980).

The anatomy of the host tissue in part determines the invasion strategy of *P. infestans.* When the hypha penetrates the upper surface of a potato leaf, the first mesophyll cells it encounters are those of the tightly packed palisade layer. In this situation, the hyphae grow intracellularly, exiting directly from the epidermal cells into the palisade cells beneath (Figs. 5 and 6). At the point of entry into the palisade cell, the hypha has a restricted diameter, but otherwise no specialization occurs at the host-parasite interface, and little or no papillalike material is deposited (Figs. 5 and 6). The host cytoplasm in the palisade cells does accumulate around the invading hyphae, and the host nucleus is usually closely associated with the fungus (Fig. 5). Many small vacuoles may be present in the host cytoplasm (Fig. 5). Intracellular hyphae are not normally seen in any other part of the leaf tissue; rather, the hyphae grow intercellularly in the leaf spaces and produce intracellular haustoria that end in the host cells (Figs. 3 and 4). Rarely, hyphae are found growing from spongy mesophyll cell to mesophyll cell (Fig. 7), which implies "transcellular" growth of the fungus (Hohl and Stössel, 1976; Hohl and Suter, 1976). That the potential for transcellular growth exists is evident from the mode of penetration of the pathogen, but whether haustoria also possess this ability is not known.

The intercellular hyphae sometimes produce short lateral branches that some early workers misinterpreted as haustoria (see Blackwell, 1953). Blackwell termed these structures "diverticula." They are found in tuber (Fig. 8) and leaf tissue (Fig. 9). Diverticula are not particularly common, but their usual location close to vascular tissue (Fig. 9) suggests that their physiologic importance may be disproportionate to their frequency. In some instances, the fungus at these locations is seen to invade the xylem (Fig. 9).

Haustoria are characteristic not only of leaf and tuber infections caused by *P. infestans* (Hohl and Suter, 1976; Shimony and Friend, 1975) but also of *P. parasitica* (Hanchey and Wheeler, 1971), *P. capsici* (Jones et al, 1974), and *P. megasperma* f. sp. *glycinea* (Klarman and Corbett, 1974; Slusher et al, 1974; Stössel et al, 1980) in their respective hosts. In *P. infestans,* haustoria in both leaf and tuber infections (Hohl and Stössel, 1976; Hohl and Suter, 1976) range from simple, button-shaped structures (Fig. 11) to digitlike structures of increasing lengths (Figs. 12 and 13). All have in common a cell wall continuous with that of the intercellular hypha and cytoplasm containing mitochondria, storage vacuoles, dictyosomes, and many ribosomes (Figs. 11–13 and 15). They lack a nucleus, a characteristic they share with *Albugo candida* but not with the downy mildew fungi, which are also oomycetes (Ehrlich and Ehrlich, 1966). A nucleus is usually positioned in the intercellular hypha immediately adjacent to the neck of the haustorium (Figs. 11 and 13).

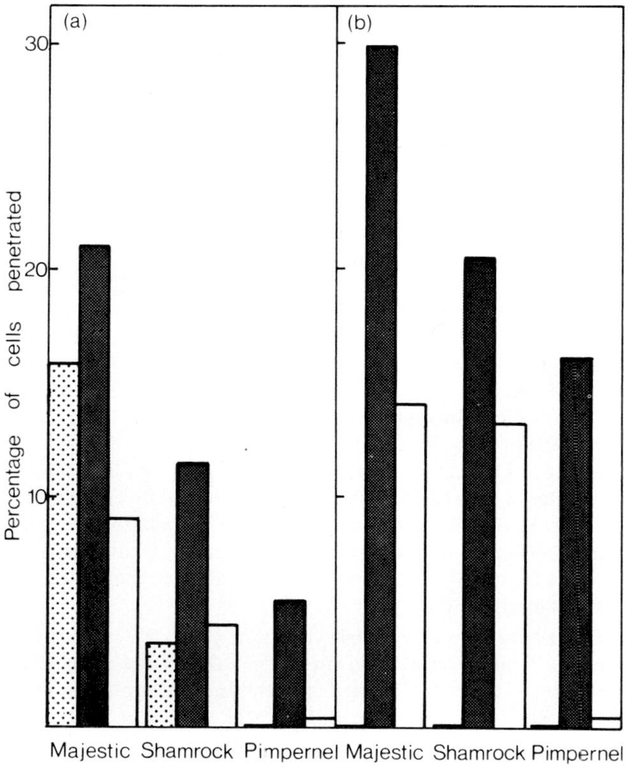


**Fig. 1.** The percentage of stomata (dotted bars), epidermal cells adjoining stomatal guard cells (shaded bars), and remaining epidermal cells (open bars) penetrated 24 hr after inoculation of the potato cultivars Majestic (susceptible), Pimpernel (resistant), and Shamrock (resistant) with *Phytophthora infestans* in two separate experiments (a and b), relative to their frequency in the upper epidermis.

Haustoria possess an electron-dense extrahaustorial matrix, which is particularly prominent around their point of entry into the host cytoplasm (Shimony and Friend, 1975). The matrix becomes thinner as it extends distally in intimate association with the haustorial cell wall (Figs. 13 and 15). In fact, it may be absent from the extreme distal portions of long, digitlike haustoria (Figs. 13 and 14).

Close examination of the relationship between the extrahaustorial matrix and the fungal and host cell walls reveals no clear interpretation of the possible origin of the matrix (Fig. 15). Silver proteinate staining (Thiéry, 1967) of such haustoria demonstrates that the matrix is in part carbohydrate in nature (Fig. 16). More revealing perhaps is that the haustorial wall stains much more intensely than the wall of the intercellular hypha (Fig. 16), indicating that its chemical composition probably differs from that of the hypha. Caution is necessary in assuming that because a structure is simple morphologically, it is necessarily simple in physiologic terms.

Haustoria are surrounded by an extrahaustorial membrane that is continuous with the host plasmalemma (Fig. 15) and is most clearly visualized when the electron-dense extrahaustorial matrix is absent (Figs. 13 and 14). The extrahaustorial membrane strongly adheres to the haustorium when the host cell is plasmolyzed with hypertonic sucrose solution (Nozue et al, 1979).

Information on how haustoria are formed is not easily obtained, presumably because haustorial formation is an extremely rapid developmental process. The earliest event appears to be dissolution of the host cell wall, accompanied by the formation of a moderately electron-dense penetration matrix (Fig. 10) that is bounded by the host plasmalemma and has an irregular, lobed appearance (Hohl and Suter, 1976). This structure is thought to become the extrahaustorial matrix (Hohl and Suter, 1976). Similar structures have been described in *P. capsici* (Jones et al, 1974) and *P. megasperma* f. sp. *glycinea* (Stössel et al, 1981). The mode of extension of the fingerlike haustorium is unknown but probably resembles hyphal tip growth (Bartnicki-Garcia, 1973). Concentrations of small vesicles are occasionally seen in the extreme distal portion of the haustorium (Fig. 12).

Host cell wall appositions are a usual feature of the early stages of penetration of leaf epidermal cells by *P. infestans* (Wilson and Coffey, 1980). Calloselike material accumulates around the site of entry into the epidermal cell and more particularly on the walls of adjacent host cells (Figs. 17 and 21). Similar material may also be deposited as a collar or even a complete encasement around haustoria (Hohl and Suter, 1976; Wilson and Coffey, 1980). These encasements give a positive reaction for callose (Fig. 18), but in the electron microscope they have a heterogeneous composition, containing much fine, fibrillar material in addition to electron-dense inclusions (Fig. 19). It is important to stress that in susceptible reactions involving *P. infestans,* there is good evidence that the formation of such material around haustoria is suppressed. Many haustoria do not possess even a small collar (Figs. 11–13 and 15). In fact, in one instance involving the infection of tubers, the extent to which haustoria are encased is believed to constitute an actual resistance

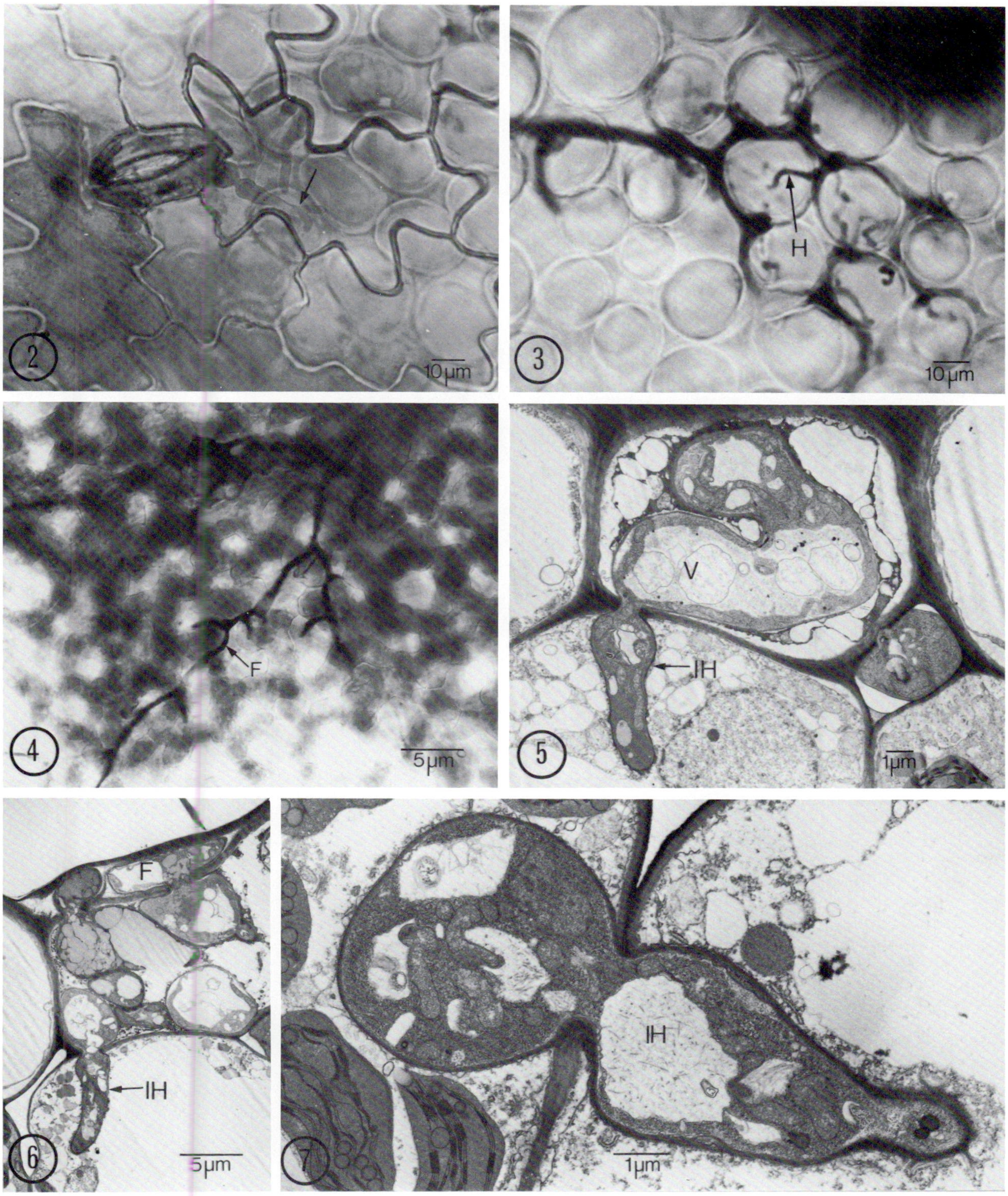


**Figs. 2–7. 2,** Penetration of *Phytophthora infestans* into epidermal cells next to stomata in foliage of the susceptible potato cultivar Majestic 24 hr after inoculation. Two intracellular hyphae have developed from an infection vesicle (arrow). **3,** Digitlike haustoria (H) extending into host mesophyll cells. **4,** Forty-eight hours after inoculation, branched intercellular hyphae (F) have spread through the mesophyll tissue, and some host cells have become necrotic. **5,** Infection vesicle (V) in a leaf epidermal cell of potato cultivar Majestic. An intracellular hypha (IH) has grown into the underlying host palisade cell. **6,** A multiple infection of a leaf epidermal cell of cultivar Majestic 24 hr after inoculation. Note the infection hyphae (F) growing within the outer wall of the epidermal cell and the intracellular hypha (IH) present in the palisade cell. **7,** An intracellular hypha (IH) growing from one leaf mesophyll cell of cultivar Majestic to the next 24 hr after inoculation.

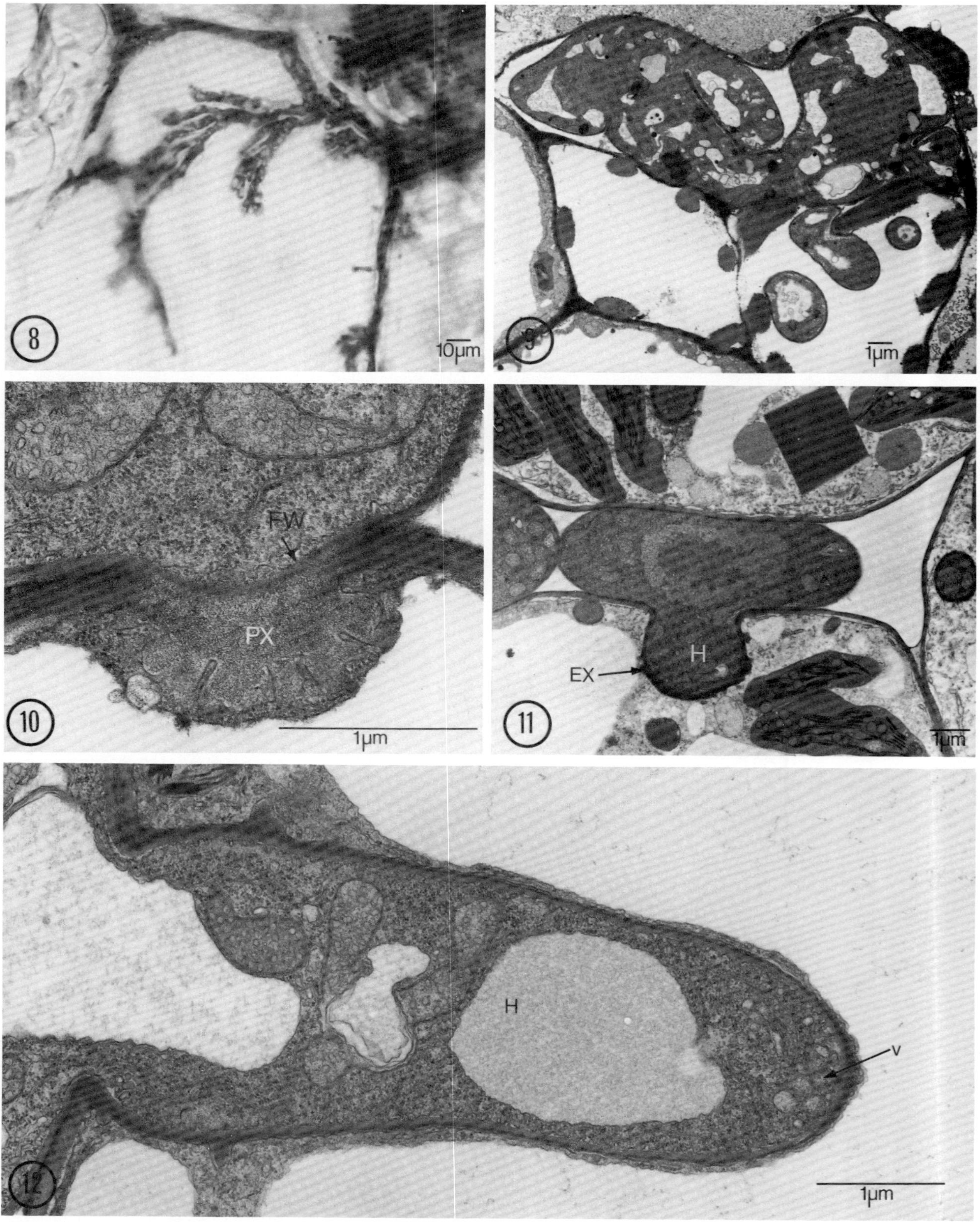


**Figs. 8–12. 8,** Branched intercellular hyphae (diverticula) in susceptible potato tuber tissue, cultivar Majestic. **9,** Intercellular hypha in leaf tissue. The fungus has also invaded the xylem. **10,** A penetration matrix (PX) with a characteristic multilobed outline. The host cell wall is absent at the point of formation of the matrix material, but a fungal cell wall is present (FW). **11,** A small, button-shaped haustorium (H) with a prominent extrahaustorial matrix (EX). **12,** A digit-shaped haustorium (H) with some vesicles (v) in the cytoplasm at its distal apex.

mechanism (Hohl and Stössel, 1976).

It is equally evident that susceptible host-parasite interactions involving other *Phytophthora* species are not characterized by any suppression of the formation of appositional material either on the host cell walls adjoining hyphae or around haustoria. Hanchey and Wheeler (1971) described the development of appositions on the host cell wall in susceptible tobacco roots infected by *P. parasitica.* They coined the term "wall lesion" to embrace all structures formed between the host plasmalemma and cell wall, whether immediately adjacent to haustoria or more distant. They concluded that such wall lesions represent a nonspecific

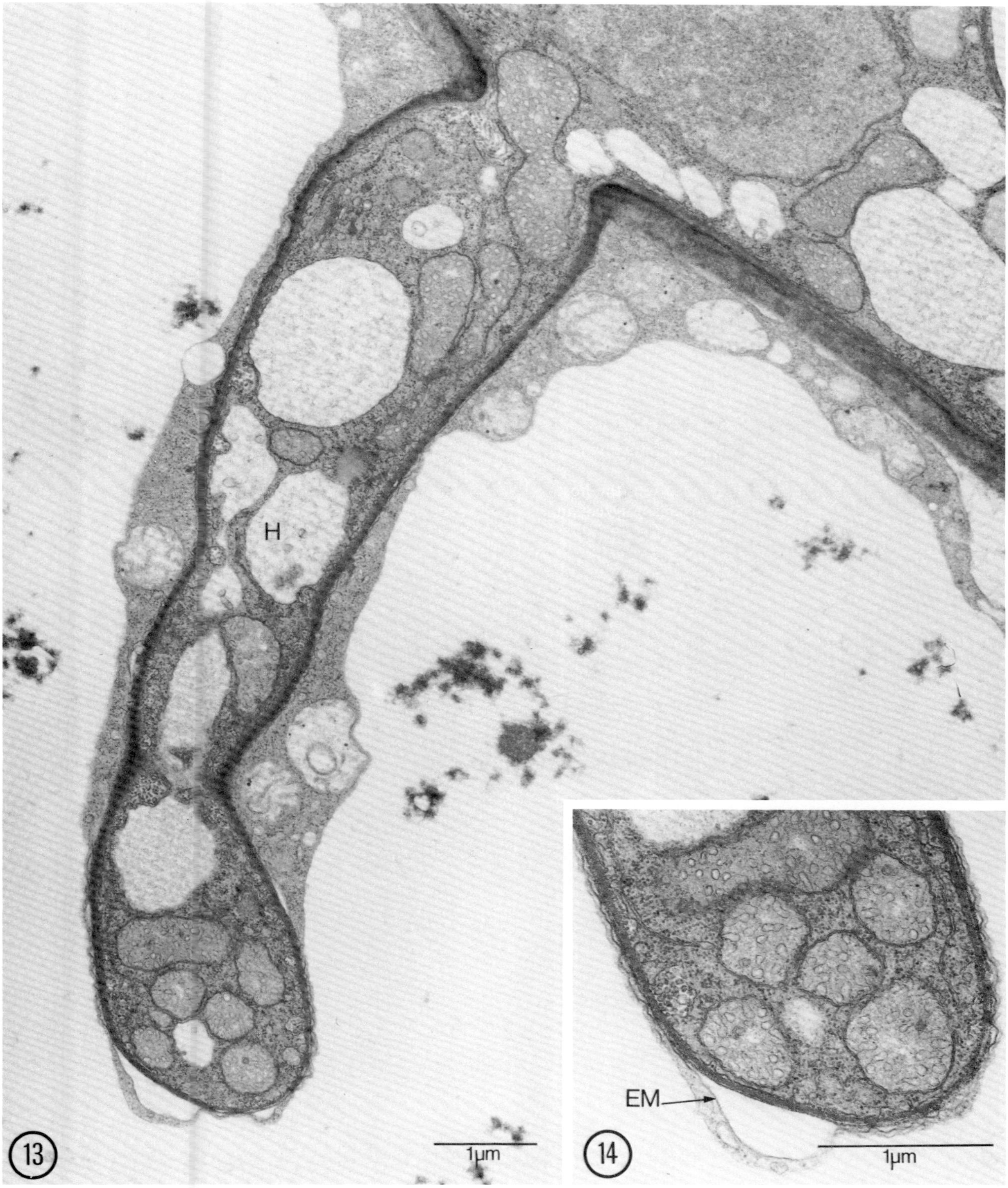


**Figs. 13 and 14. 13,** A long, digitlike haustorium (H) typical of infections caused by *Phytophthora infestans.* **14,** The distal apex of a long haustorium. Note the absence of the electron-dense extrahaustorial matrix. The extrahaustorial unit membrane (EM) is clearly visible.

response to cell injury, possibly functioning as part of a general repair mechanism delaying further injury to the host cell.

Wall appositions were frequently encountered in susceptible soybean hypocotyl tissues infected with *P. megasperma* f. sp. *glycinea* (Stössel et al, 1981). In fact, only a few fungal invasions of host cells showed neither necrosis nor formation of cell wall appositions. Stössel and co-workers concluded that in the soybean-*P. megasperma* f. sp. *glycinea* interaction, the difference between resistance and susceptibility was one of degree rather than an absolute difference as previously inferred (Keen and Bruegger, 1977).

To return to *P. infestans,* one final point concerning the susceptible state deserves attention: Are there any other obvious changes in the host cell cytoplasm in response to infection? The most striking feature of early stages of infection by *P. infestans* is the lack of host cell necrosis (Ferris, 1955; Kitazawa and Tomiyama, 1970; Pristou and Gallegly, 1954; Wilson and Coffey, 1980). The cytoplasm of infected host cells shows no detectable changes in either numbers or structure of organelles nor any difference in the activity of the endoplasmic reticulum or dictyosomes (Hohl and Stössel, 1976; Shimony and Friend, 1975). However, the cytoplasm does appear more abundant around some haustoria (Hohl and Suter, 1976) and around sites of penetration of intracellular hyphae (Fig. 5). A thin band of cytoplasm containing many ribosomes (Fig. 12) is always present, and organelles such as mitochondria (Fig. 13) often extend around haustoria. The overall impression is that in initial stages of invasion of a host cell, disturbance of the cytoplasm is minimal. There is no convincing evidence for increases in synthetic activity of the host involving either dictyosomes or endoplasmic reticulum.

Between about 24 and 48 hr after infection, the host cells become necrotic. The cells often appear plasmolyzed, and the structure of cell organelles becomes disorganized (Shimony and Friend, 1975). Rupture of the host tonoplast and degeneration of the cytoplasm have also been described (Hohl and Suter, 1976). In fact, our knowledge of the significance of the various ultrastructural changes observed at the onset of cell death is very poor and demands intensive cytochemical and biochemical study.

Cell necrosis, however, fails to restrict further development of the fungus, which continues to grow and eventually sporulates. Susceptible host genotypes vary in the speed with which their cells become necrotic (Ferris, 1955), although whether this variation has any direct bearing on the amount of sporulation produced subsequently is unclear.

## THE RESISTANT STATE

In race-specific resistance (Thurston, 1971) in potato, the cytologic evidence clearly points to confinement of the fungus to the site of initial penetration. In foliage of the cultivar Kennebec, which has an R1 genotype, the pathogen penetrated in a normal fashion but was then confined to that epidermal cell (Pristou and Gallegly, 1954). The epidermal cell and underlying mesophyll cells (which did not contain fungal hyphae) became necrotic.

Ferris (1955) obtained somewhat similar results with a range of R-genotypes. The potato foliage was inoculated on the lower surface; by 20 hr, many hyphae had penetrated beyond the epidermal cells into the intercellular spaces between spongy mesophyll cells. No further fungal growth took place. The resistant epidermal cells turned brown several hours before the susceptible cells did; the speed of the necrotic response depended on the genotype, being fastest in R1R4 and slowest in R1 (Ferris, 1955).

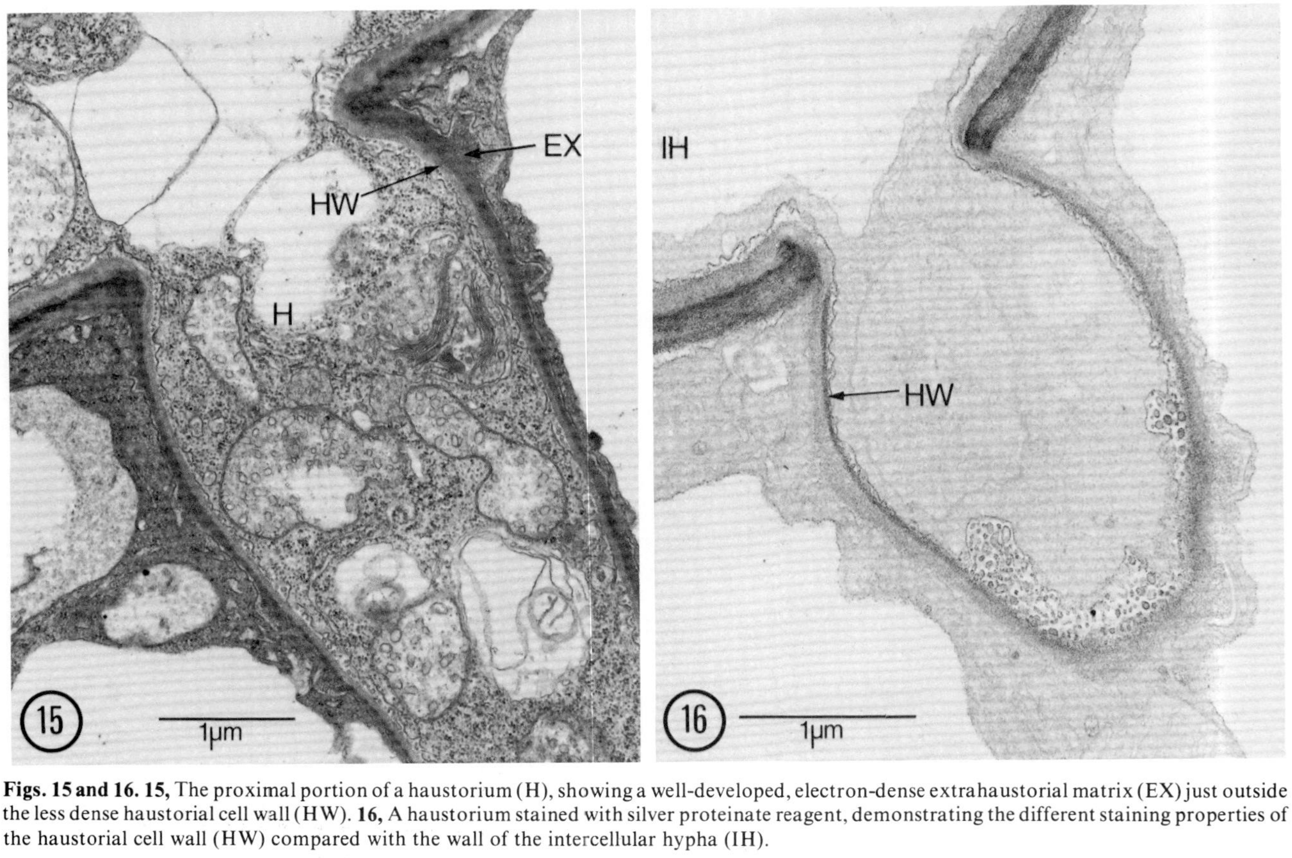


**Figs. 15 and 16. 15,** The proximal portion of a haustorium (H), showing a well-developed, electron-dense extrahaustorial matrix (EX) just outside the less dense haustorial cell wall (HW). **16,** A haustorium stained with silver proteinate reagent, demonstrating the different staining properties of the haustorial cell wall (HW) compared with the wall of the intercellular hypha (IH).

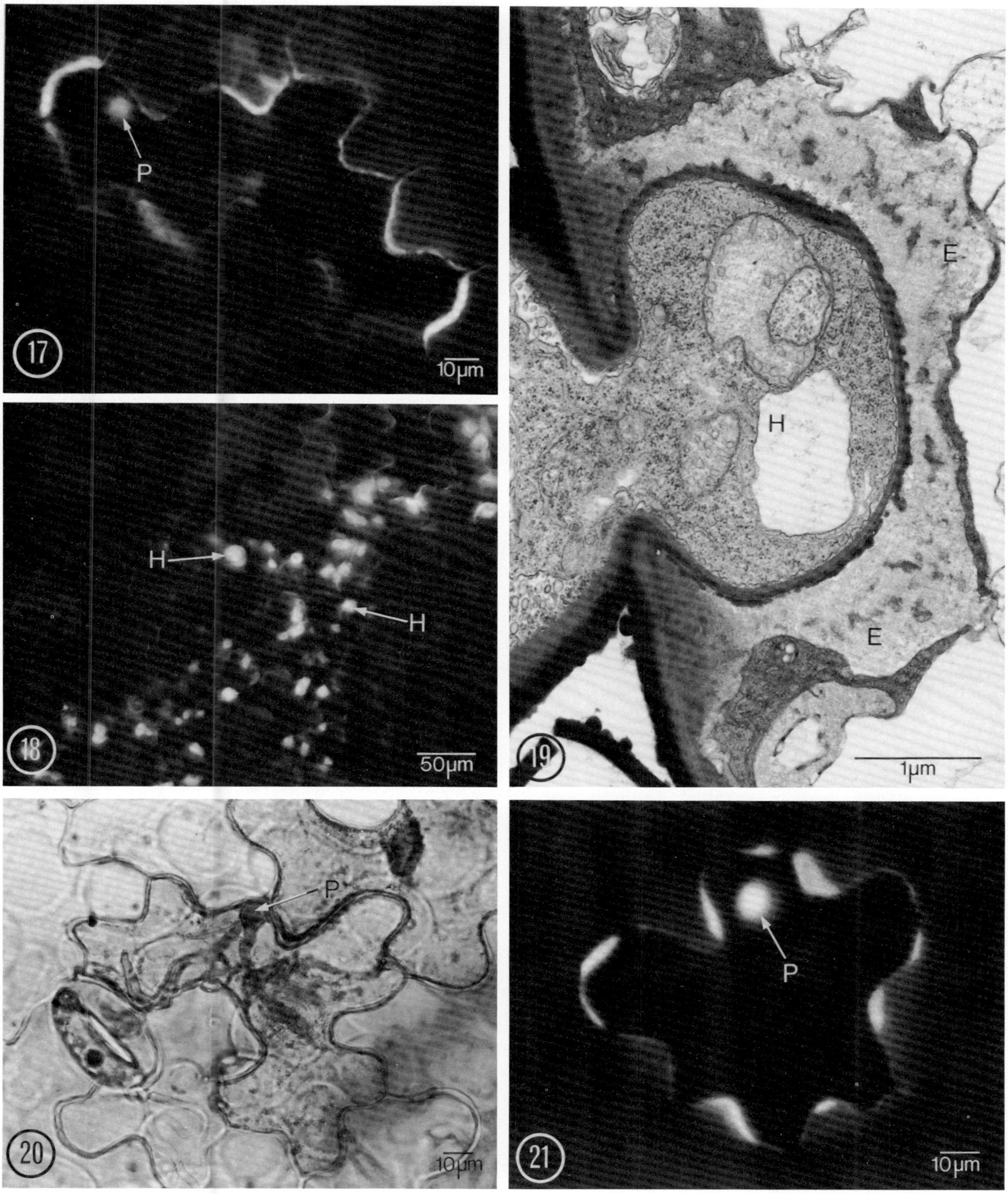


**Figs. 17–21. 17,** Cultivar Majestic 12 hr after inoculation, stained with aniline blue. Fluorescent material is present around the penetration site (P) and on the walls of adjacent epidermal cells. **18,** Cultivar Majestic at 72 hr, stained with aniline blue. Fluorescent material surrounds some haustoria (H). **19,** Cultivar Majestic at 48 hr. An encasement (E) surrounds the haustorium (H); notice its heterogeneous nature. **20,** Shamrock, a cultivar with high general resistance, 48 hr after inoculation. Penetration site (P) on leaf epidermal cell next to stomata. The penetrated cell contains brown granular material. **21,** Cultivar Shamrock 12 hr after inoculation, stained with aniline blue. Fluorescent material is present at a penetration site (P) and on the walls of epidermal cells next to the penetrated cell.

Kitazawa and Tomiyama (1970) used the cultivar Rishiri with the specific gene R1 and high general resistance (see Thurston, 1971). Cut surfaces of cortex tissue from potato leaf petioles were washed and held for about 15 hr, then inoculated with zoospores of either virulent (race 1) or avirulent (race 0) strains of *P. infestans*. The growth of the fungus was similar in both interactions for about 9 hr, when the avirulent strain ceased to develop. The host cells in the resistant interaction responded very rapidly; about half were dead after only 2 hr. At 9 hr, one to three host cells were brown and necrotic at each infection site, and the fungus was completely contained within the dead host cells. In the susceptible interaction, no browning of host cells was apparent for up to 2 days, and the fungus spread rapidly through the petiole tissue. Although this system has proved useful for biochemical work, it is not clear whether the events that occur are representative of what actually takes place in the intact plant. The principal question that should be asked is how much does cutting the petiole cortex tissue modify the physiologic expression of resistance?

Shimony and Friend (1975) conducted an ultrastructural investigation of resistance in the foliage of the potato cultivar Orion, which has the R1 gene. As early as 7 hr after inoculation, "only damaged epidermal cells could be seen," whereas epidermal cells of the susceptible cultivar Majestic first showed damage 12 hr after inoculation. Mesophyll cells of the resistant cultivar were also invaded and by 12 hr were obviously damaged; by 24 hr, all invaded host cells were dead. The pathogen was still alive 12 hr after inoculation, even though most penetrated epidermal cells were probably dead after 7 hr in the resistant reaction.

Hohl and Suter (1976) used race 4 of *P. infestans* to study resistance in foliage of the cultivar Eba, which has an R3 gene. The fungus entered the host mainly via stomata and was then largely confined to the substomatal cavity. Many hyphae were reported to be alive "after 5 and more days." Micrographs were presented of haustoria in the mesophyll cells 66 hr after inoculation. Unfortunately, the host cell reactions were not described, except that "small water-soaked spots" were evident on leaves of Eba after 2 days but not on the susceptible cultivar Bintje. The restriction of the fungus largely to the substomatal cavities and the fact that "hardly any" sporangia were produced on Eba (Hohl and Suter, 1976) imply a high level of resistance, which is typical of specific resistance involving R-genes. Surprisingly, Hohl and Suter (1976) took the viewpoint that "leaves of the two cultivars differ less in degree of resistance than corresponding tubers."

TABLE 1. The pattern of penetration of the upper epidermis of the susceptible cultivar Majestic and the resistant cultivars Shamrock and Pimpernel 12 and 24 hr after inoculation with *Phytophthora infestans*

| Cultivar | Penetrations[a] per mm² 12 hr | 24 hr | Cells penetrated (%) 12 hr | 24 hr | Penetrated cells with brown granular material (%) 12 hr | 24 hr | Zoospores that penetrated(%) 12 hr | 24 hr |
|---|---|---|---|---|---|---|---|---|
| Majestic | 21.8 | 65.0 | 7.7 | 12.6 | 1.6 | 28.3 | 3.2 | 9.4 |
| Shamrock | 21.0 | 20.5 | 7.5 | 5.5 | 43.8 | 48.1 | 3.0 | 3.0 |
| Pimpernel | 1.8 | 1.5 | 0.7 | 0.4 | 28.6 | 80.0 | 0.3 | 0.2 |

[a] After 12 hr, the number of penetrations per square millimeter in Majestic was not significantly different ($P > 0.05$) from the number in Shamrock but was significantly different ($P = 0.001$) from the number in Pimpernel. After 24 hr, all differences in the number of penetrations per square millimeter were significant ($P = 0.001$).

*Source:* Wilson and Coffey (1980).

Tuber resistance was studied in the cultivar Orion (R1) with small disks of tissue and race 4 of the pathogen (Shimony and Friend, 1976). At 24 hr, the most significant change in the host cytoplasm was the appearance of "large osmiophilic areas," particularly in association with haustoria and host mitochondria. The cytoplasm of the haustoria was "in good condition" at this time. After 48 hr, the infected host cells were necrotic and only senescent haustoria were seen. The host response in tubers was much slower than that in foliage. The osmiophilic areas were considered to be related to resistance because they were large in Orion, small in the susceptible cultivar Majestic, and absent in King Edward, which is extremely susceptible to blight (Shimony and Friend, 1976, 1977). Shimony and Friend speculated that perhaps these osmiophilic deposits were "related to cinnamic acid accumulation." Friend (1976) discussed the possibility that phenolic-carbohydrate esters situated in potato cell walls could act as inhibitors of cell wall-degrading enzymes produced by the pathogen. No experimental evidence has been forthcoming to support this idea.

Resistance in Eba with the R3 gene has also been studied with tuber slices 5 mm thick and race 4 of the pathogen (Hohl and Stössel, 1976). The evidence obtained from light microscopy indicated that the fungus was restricted to the top cell layer, and numerous brown cells were observed as early as 20 hr. By 35 hr, the browning had become very deep and the fungus had not penetrated further. In the susceptible cultivar Bintje, infection had penetrated to about 5 mm and only slight browning of the host tissue was observed by 45 hr.

A major difference in the ultrastructure of the susceptible and resistant reactions was the absence of wall appositions in the former, and the presence of well-developed encasements around the few small haustoria in host cells of the resistant cultivar (Hohl and Stössel, 1976). The host cytoplasm surrounding encased haustoria was found to be "electron dense and organelles [were] difficult to identify." In Eba, necrosis of the host cytoplasm and encasement of the haustoria appear to occur almost simultaneously. The cytoplasm of the encased haustorium also appeared necrotic at this time. The more rapid host cell death, as well as the formation of extensive host cell wall appositions around haustoria, is likely to be a component of the resistant reaction. In the susceptible reaction involving Bintje, the absence of significant host cell death and the lack of appreciable synthesis of host cell wall appositional material around haustoria favor rapid growth of the pathogen.

General resistance is characterized by the ability to function against all biotypes of the pathogen (Thurston, 1971). The three principal components that have been recognized are reduced penetration of the host, restricted growth of the fungus in the host, and reduced sporulation capacity (Umaerus, 1970). The only evidence of a cytologic nature is that of Wilson and Coffey (1980), who studied foliage resistance in two potato cultivars, Shamrock and Pimpernel, both possessing high general resistance. In Pimpernel, very few epidermal cells were penetrated, and even as early as 12 hr after inoculation, almost 30% of the few penetrated cells were necrotic (Table 1). In Shamrock, the number of epidermal cells penetrated per square millimeter after 12 hr was similar to the number in the susceptible cultivar Majestic; however, more than 40% of these penetrated cells were necrotic in Shamrock, compared with less than 2% in Majestic (Table 1). No further penetrations occurred with Shamrock, but during the next 12 hr, penetrations into Majestic tripled (Table 1), although the number of epidermal cells penetrated increased only 50%. More than 56% of epidermal cells with multiple penetrations were still apparently alive. Infected cells that had become necrotic had increased from less than 2% to more than 28% (Table 1; Wilson and Coffey, 1980).

A further interesting feature arose from this study (Wilson and Coffey, 1980): the epidermal cells immediately adjacent to guard cells were preferentially penetrated by the pathogen in all three cultivars (Fig. 1). However, Pimpernel had a very low stomatal frequency compared with both Shamrock and Majestic. In addition, epidermal cells other than those next to the guard cells were highly resistant to penetration in Pimpernel (Fig. 1). Thus, in that cultivar, the lack of fungal penetration was probably a major determinant of the resistant reaction, although the high percentage of penetrated host cells that turned brown must also be significant. In Shamrock, the penetration frequency was high, but subsequent fungal growth was largely restricted to the initially parasitized epidermal cells (Figs. 20 and 22).

Fluorescence illumination using aniline blue revealed the presence of calloselike material around the site of penetration and on the walls of adjacent epidermal cells (Fig. 21). A similar distribution of calloselike material was evident in the susceptible interaction involving Majestic (compare Figs. 17 and 21).

Ultrastructurally, the earliest phases of penetration of epidermal cells were characterized by a lack of host response. Cell appositional material did not accumulate, and the host cytoplasm that accumulated around the infection vesicle appeared normal (Fig. 23). This biotrophic phase was very short-lived; the host cytoplasm became necrotic, characterized by the accumulation of dense osmiophilic granules (Figs. 22, 24, and 25). Occasionally, the fungus spread to the underlying palisade layer (Figs. 24 and 25). The cytoplasm of these cells was initially viable (Figs. 24 and 25) but soon became necrotic (Fig. 26). There were exceptions to this pattern of growth when the fungus colonized the mesophyll tissue more extensively and the onset of host cell necrosis was delayed (Figs. 27 and 28). These infrequent but successful colonizations of the underlying mesophyll tissue probably account for the limited sporulation typical of this type of general resistance.

In a resistant reaction involving soybean hypocotyls and *P. megasperma* f. sp. *glycinea*, individual intercellular hyphae were similarly observed to spread deeply into the host tissue (Stössel et al, 1980). It is important to realize that the degree of resistance or susceptibility encountered in host tissue represents the summation of thousands of individual reactions at the cell level, each perhaps differing in their degree of compatibility.

## THE IMPLICATIONS

What does the cytology imply about the nature of the physiology of diseases caused by *Phytophthora?* At the outset, we must appreciate that each disease interaction is different, particularly at the species-species level. Compare the findings for the *P. megasperma*-soybean interaction (Stössel et al, 1980, 1981) with those for the *P. infestans*-potato interaction (Hohl and Suter, 1976; Shimony and Friend, 1975; Wilson and Coffey, 1980). In the *P. megasperma* interaction, the evidence suggests that the susceptible reaction of soybean hypocotyls involves many individual deaths of infected host cells at an early stage in pathogenesis (Stössel et al, 1981). Presumably, in this instance a high degree of compatibility (biotrophy?) is not required to facilitate successful colonization of soybean tissue by the *Phytophthora* parasite. Caution is necessary, however. Although there have been several other studies of this system, each used different host tissues—either soybean roots (Slusher et al, 1974) or hypocotyls with wounds into which fungal mycelium was placed (Klarman and Corbett, 1974). These studies cannot be compared, and more cytologic information on this system is urgently required.

Knowledge of the *P. infestans*-potato interaction is more comprehensive. The general feature that emerges is that in the susceptible interaction, both in foliage (Hohl and Suter, 1976; Wilson and Coffey, 1980) and in tubers (Hohl and Stössel, 1976; Shimony and Friend, 1976, 1977), there is a high degree of host cell compatibility. It is frequently assumed that this compatibility reflects an active biotrophic phase of fungal development, presumably critical for rapid colonization of the host. The only relevant data are probably those of Farrell et al (1969), who showed that in early phases of the disease, stomatal physiology is altered; instead of their usual diurnal rhythm, the stomata remain open, and carbon assimilation is enhanced at the infection sites. Current physiologic knowledge of the susceptible *P. infestans*-potato interaction favors the production of soluble glucans by the fungus that can suppress the resistance response of the host (Doke and Tomiyama, 1980). These soluble glucans are specific suppressors, characteristic of particular races of the pathogen. The insoluble glucans present in fungal cells elicit a nonspecific reaction in host tissues (Kuć et al, 1976).

The cytologic data certainly support the physiologic facts relating to susceptibility. But what of resistance? The current physiologic data indicate that resistance in potato tubers is a nonspecific event operating where the specific situation, susceptibility, is absent (Doke and Tomiyama, 1980).

Two recent cytologic studies of tuber resistance, one involving the R1-gene and the other the R3-gene, suggest that different resistance mechanisms function in each instance. With the R1-gene in the cultivar Orion, host cell death occurs after 48 hr and is the principal feature characterizing the resistance response (Shimony and Friend, 1976). With the R3-gene in the cultivar Eba, host cell necrosis was observed much earlier, at 20 hr, but the main feature thought to relate to resistance was the rapid development of encasements around haustoria (Hohl and Stössel, 1976). A single, detailed cytologic study of R-gene resistance, using several different cultivars for each resistance genotype, is urgently required.

In the case of general resistance to late blight, there are no physiologic data, and cytologic information is limited to one study. In foliage, different types of general resistance apparently exist (Wilson and Coffey, 1980). Data from laboratory and field assessments, using parameters such as the amount and rate of spread of necrosis and sporulation capacity, also support the concept of different resistance mechanisms (Umaerus, 1970).

Prevailing opinion among many physiologists supports the concept that phytoalexins can explain disease resistance both in soybean (Albersheim and Valent, 1978; Keen and Bruegger, 1977) and in potato (Kuć et al, 1976). The soybean system is controversial; one group provides evidence for specific elicitors of the resistant reaction (Keen and Legrand, 1980), while another group produces data that suggest that the resistant and susceptible states are a question of the degree of response (Ward et al, 1981). Despite some reservations about the actual role of the phytoalexin glyceollin in the resistant interaction (e.g., Lazarovits et al, 1981), glyceollin is nevertheless a consistent component of such interactions (Keen and Bruegger, 1977). A good cytochemical technique allowing identification of glyceollin isomers at sites within host cells is clearly needed.

The role of phytoalexins in the *P. infestans*-potato interaction seems less clear at present. Rishitin is not formed in resistant responses involving potato foliage (Kuć, 1972). In tuber tissue, the rapid formation of rishitin and several other related compounds is commonly reported in resistant

Studies have been conducted in Ecuador to see if the induction of early flowering and crop maturity of cacao could be used to help plants escape loss from pod diseases (Edwards, 1978). This research recognized that natural cacao flowering and pod development coincide with the rainy season and its attendant high humidity. The author suggested that the induction of early flowering could very significantly reduce the likelihood of black pod epidemics.

Many other aspects of crop management can contribute, both positively and negatively, to the development of epidemics. The discipline of plant pathology has been negligent in not asserting its role in the design of crop management systems. This point will become more evident in later sections of this chapter.

### Postharvest Considerations

When one considers the plant part that is attacked and the source of inoculum that can initiate an epidemic, some interesting comparisons begin to emerge. For instance, potato tubers infected with *P. infestans* not only rot in storage, but also, if used for seed, can initiate late blight epidemics. It is often stated that the use of certified seed for potato production may help restrict the amount of initial inoculum. However, most state seed certification programs have set tolerances for late blight infection far beyond levels necessary to create a late blight epidemic. Some potato growers have long recognized this fact and actively seek out seed growers who take extra precautions to limit foliar epidemics of late blight, hence lessening the likelihood of seed tuber infection.

We have not been able to reconcile epidemiologic principles with federal or state programs that favor monoculture by restricting acreage of tobacco. Small growers tend to use their "best" land for tobacco production, thereby risking the buildup of *P. parasitica* var. *nicotianae* and epidemics of black shank disease.

Moreover, the general disregard for pod cull piles in cacao plantations can only heighten the risks of black pod epidemics. We will look at this and other common practices in two later sections when we investigate sanitation ratios and efforts to reduce initial inoculum.

We investigated the epidemiology of diseases caused by *Phytophthora* in an attempt to determine the principles governing how likely growers are to use fungicides as a control method. The value of the crop and the availability of effective fungicides are unquestionably primary factors, but there are others. Many of the epidemiologic parameters discussed up to this point are considered important in the choice of when to use fungicides. However, what is even more interesting is our observation that until very recently, the decision to use fungicides was a decision *not* to use other control methods. In a later section, we will discuss the rejection of the single-solution approach to problem solving as one major contribution of modern epidemiology.

### Use of Plant Resistance

Quite often in the past, the alternative to fungicides was the use of plant disease resistance. There are many examples of success—and also many of failure.

Potato breeders once felt that they had solved the late blight problem through the use of R-genes from *Solanum demissum*. In the late 1930s, the potato variety Virgil was released in the United States with great fanfare as a "blightproof" potato cultivar. Experience has shown that such excitement was unfounded and short-lived. The resistance succumbed to new races of the pathogen. A pattern began to emerge, and single-gene resistance became suspect.

Information is insufficient to compare the four *Phytophthora* pathogens and draw conclusions about when plant disease resistance will or will not endure. It is generally believed that when the resistance is inherited as a single gene, the likelihood of its durability is greatly reduced. There are, however, exceptions to this observation. We expect that as more epidemiologic knowledge accumulates, the principles underlying host resistance and pathogen variation will lead to the proper deployment of plant disease resistance. This topic is given more detailed consideration by Umaerus et al (Chapter 24), but we would like to make the observation that until very recently the use of plant disease resistance was considered to be one choice from among several alternatives, rather than one of a group of complementary methods.

## ANALYTICAL EPIDEMIOLOGY

Vanderplank (1963) provided plant pathology with the methodology for beginning to understand the consequences of disease management strategies. We have looked at the available information on the four representative diseases to try to assess what is known about their epidemiology from an analytical perspective.

### Type of Epidemic

One major distinction among plant diseases is the type of epidemic progress curve associated with them. Vanderplank (1963) divided epidemics into simple interest (SI) and compound continuous interest (CCI) types. The distinction between these two models lies in whether or not the inoculum from current disease can result in subsequent infection in the relatively near future.

Epidemics of the SI type give disease progress curves like those in Figure 1. Data transformation allows linear regression determination of the descriptive parameters. We will deal later with the consequences of this model.

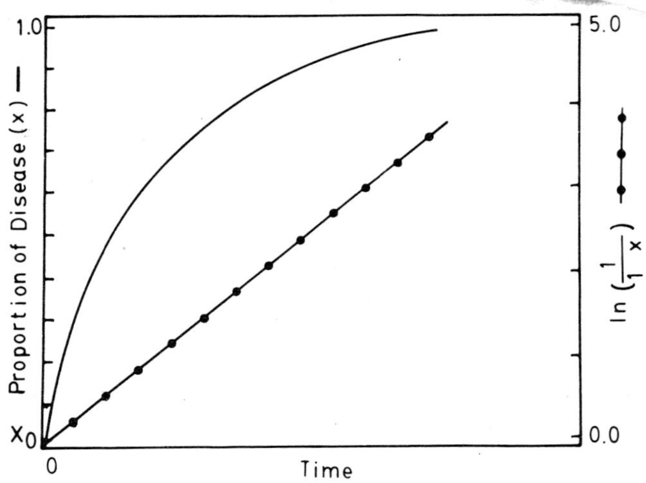


**Fig. 1.** Disease progress curve for a simple-interest type of epidemic plotted on an arithmetic (—) axis and a simple-interest-transformed (•—•) axis for proportion of disease over time. The expected result of the transformation is a straight line described as $\ln[1/(1-X)] = \ln[1/(1-X_0)] + rt$, where $X_0$ is the initial proportion of disease and $X$ is the proportion of disease after some time ($t$) at some simple rate of increase ($r$).

Epidemics of the CCI type show disease progress curves like those in Figure 2. Analysis of this model is simplified by logit transformation, which gives a linear model. CCI epidemics have the potential for exponential explosion, sometimes resulting in catastrophic disease.

A look at the available literature on potato late blight and black pod of cacao led us to the conclusion that these two diseases are easily accommodated by the CCI epidemic model. Campbell and Powell (1980) analyzed 19 tobacco black shank epidemics to determine whether they were better described by SI or CCI epidemic models. Vanderplank (1963) and Zadoks and Schein (1979) discuss the relationship between soilborne plant pathogens and hosts as SI. All too often, the common assumption is that most soilborne pathogens cause SI epidemics. Campbell and Powell (1980) found that 16 of the epidemics were better described by a CCI model. Only two epidemics could be better described as SI. (The remaining epidemic was described equally well by SI and CCI models.)

Avocado root rot is more difficult to classify. The literature available to us was not conclusive enough to determine analytically the epidemiologic nature of this disease. Although inconclusive, observations on the spread of the disease associated with root grafts and groundwater movement of the pathogen suggest a CCI type of epidemic.

The difference between SI and CCI types of epidemic models is more than academic, inasmuch as the underlying principles of plant disease management reflect the character of these epidemic progress curves.

### Apparent Infection Rates

One way to assess the epidemiologic potential of a disease is by analyzing disease progress curves, paying particular attention to apparent infection rates (*r* values) (Vanderplank, 1963). Apparent infection rates are a measure of how fast an epidemic has progressed over time when adjusted for multiple infections. The logistic transformation, which is now popular for the CCI type of epidemic, allows apparent infection rates to be estimated as linear regression coefficients on logit-transformed disease proportions. Many reports have been published on the estimation of apparent infection rates for late blight of potato. Some of these reports speak of *r* values approaching 0.6 units per unit per day (Fry, 1977; Latin et al, 1981), which makes potato late blight one of the most explosive plant diseases ever quantified.

Cultivars differ in their apparent infection rates—a trait sometimes referred to as horizontal resistance. Potato cultivars are known to differ greatly in this rate-limiting resistance. Some laboratories are working on strategies to complement this type of late blight resistance with fungicide protection (Fry, 1977; Latin et al, 1981; MacKenzie, 1981). The combination of rate-limiting resistance and fungicide provides an integrated approach to managing the disease. This strategy will be discussed in more detail in a later section.

Virtually no information is available in the literature on the quantitative aspects of apparent infection rates for avocado root rot epidemics. Consequently, we were unable to determine the nature of the avocado root rot model or to assess the relative epidemiologic potential of this disease. Its reputation as a destructive pathogen of avocado is, however, well known.

Analyzing epidemics also provides opportunities to evaluate the effects of fungicides on the development of disease. Kannwischer and Mitchell (1978) compared apparent infection rates for tobacco black shank epidemics as influenced by treatments with metalaxyl. This report demonstrates well how analytical epidemiology can be used to assess the impact of various tactics on epidemic development. Apparent infection rates are useful quantitative measures of treatment effects that reflect the change in disease with time.

Comparison of apparent infection rates is a useful procedure for epidemiologists. However, to compare SI and CCI rates directly is not possible. Other models are often appropriate for epidemic analysis, depending on the nature of the work and specific objectives. As information accumulates, individuals may choose to discard models. Authors of articles should be encouraged to include data bases for future analyses by others. As our knowledge evolves, reanalyses of past research may help speed solutions to new problems. Our efforts would certainly have been easier had earlier authors shared data sets so that we could have done more extensive or novel analyses.

Our analyses of black pod of cacao are a good example. We were able to locate some information from Africa that suggested that the disease has high explosive potential in that area (Hislop and Park, 1962b). Much of that information was not suitable for reanalysis by current procedures. In some other regions, such as southern Asia, epidemic progress of black pod is much slower (Gregory, 1974). We propose that for black pod, geographic differences that are attributable to the environment or microenvironment rather than to the pathogen or host could be the determining factors in this difference. If this conclusion is correct, black pod of cacao should also be classified as highly explosive.

The observation of geographic differences in black pod potential has led us to suggest that it is now possible to map regions that are high-risk areas for crops attacked by certain pathogens. Soil scientists have for years described the productivity of soils in broad areas. Plant pathogen maps are often limited to occurrence of the organism. Present and future knowledge concerning environmental and micro-environmental effects on epidemic development could be used to greatly expand our ability to recognize the limits to crop production areas. Much more effort needs to be devoted to quantifying epidemics so that analytical procedures can be

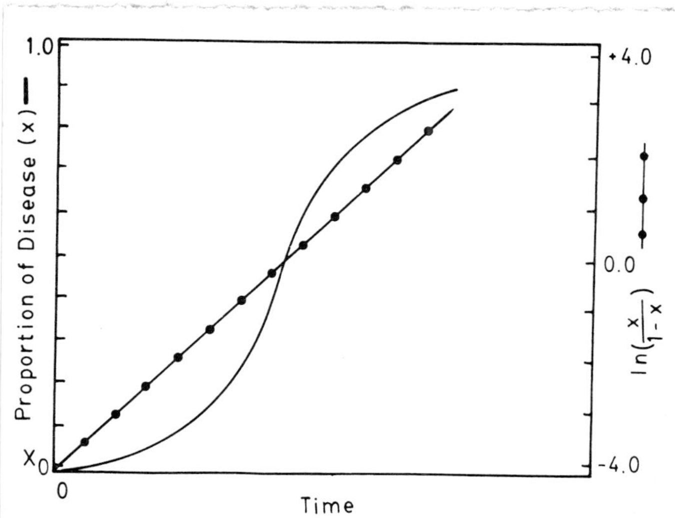


**Fig. 2.** Disease progress curve for a compound continuous interest type of epidemic plotted on an arithmetic (—) axis and a logit-transformed (•—•) axis for proportion of disease over time. The expected result of the logit transformation is a straight line described as $\ln[X/(1-X)]=[X_0/(1-X_0)]+rt$, where $X_0$ is the initial proportion of disease and $X$ is the proportion of disease after some time ($t$), at some compound continuous rate of increase ($r$), more commonly called the apparent infection rate. It is not possible to directly compare simple interest and compound continuous interest rates of increase.

used and new management strategies can be designed. We will return to this point in a later section where we discuss how disease management practices can be used to complement one another for attaining acceptable levels of disease.

### Rates of Disease Spread

Very little is known about either the short- or long-range spread of any of the representative diseases or the dispersal of their causal pathogens. Latin (1980) began with the task of quantifying the short-distance spread of late blight and discovered that it is a very focal disease, with most infection confined to a relatively small area. Significant differences were also attributed to cultivars, opening up the possibility of selecting for potato vines that would restrict the spread of late blight.

Black pod of cacao has also been studied for spread characteristics and, like late blight, is thought to spread only short distances.

The epidemiologic phenomenon of local epidemics turning into general epidemics is generally accepted but not well understood. The basic premise of late blight forecasting is to permit limited local (focal) buildup but prohibit a general epidemic. Much more information is needed on disease spread, because it seems that opportunities for disease management could be improved with more knowledge in this area.

### Sanitation Ratios

Crop rotation is a management practice that has been used for a considerable period of time with measurable success. The destruction of potato cull piles to delay epidemic onset has been an accepted practice in the northeastern United States for half a century. It is surprising, therefore, that analysis of the benefits of sanitation through cull pile destruction (reduction of initial inoculum) leads to the conclusion that the benefits are not worth the effort (Zadoks and Schein, 1979). We are reminded of the aeronautical engineer who "proved" that a bumblebee can't fly. Obviously, something must be wrong with the analysis.

We looked at sanitation ratios and suspected that the difference might lie in the epidemic models. We hypothesized that measurable benefits from sanitation might be seen with the SI type of epidemic that would not be expected for the CCI type of epidemic. We were wrong. Computer simulations of sanitation ratios for SI and CCI types of epidemics could not explain the apparent paradox.

We analyzed the four examples of *Phytophthora* diseases in an attempt to determine why sanitation would be practiced under some conditions but not others. We observed that *P. cinnamomi* could be classified as a soil inhabitant (Broadbent and Baker, 1974; Hickman, 1970; Hwang and Ko, 1978; Weste and Ruppin, 1975) and the other three pathogens as soil invaders. One would expect crop rotation to be more effective against soil invaders than against soil inhabitants, because the invaders might diminish more rapidly over time.

We also hypothesized that soilborne plant pathogens might be more easily controlled by crop rotation than would pathogens that cause epidemics on aerial plant parts. We reasoned that pathogens causing aerial epidemics would probably be dispersed over greater distances than would soilborne plant pathogens From these observations, we concluded that one would expect difficulties in using rotation to diminish avocado root rot problems, would realize some benefits from crop rotation for black shank of tobacco, and would experience some dissatisfaction with sanitation practices aimed at the pathogens of late blight of potato and black pod of cacao. These are, in fact, very much the experiences recorded in the literature.

The dissatisfaction with sanitation practices for the CCI type of epidemic (Zadoks and Schein, 1979) might be attributable to a basic misunderstanding of the complementary nature of disease management activities. Research designed to test the singular effects of sanitation practices fails to recognize complementary effects. As analytical epidemiology becomes more sophisticated, new opportunities are presented for plant disease management.

### Simulation

Several laboratories have begun developing simulations of epidemics of potato late blight for research purposes. Many of these attempts are greatly restricted by the lack of quantitative information necessary for simulation. We could not find any reference in the literature to computer simulations for the other three representative diseases.

Computer simulation has allowed the investigation of some complex systems that previously were too complicated to evaluate by any other means. Systems science is now part of the thinking of many members of the scientific community, and plant epidemiology has benefited from this new perspective. Systems science recognizes that the whole is more than the sum of the parts, and computer simulation provides opportunities to assess the whole system.

Often, some of the parts are not well understood. In such cases, simulation can be used to evaluate the known parts while holding the other parts constant. Through the rigorous mathematical description of each part of a complex system, knowledge about those parts and how they interact with one another can be explored for new understanding.

Sometimes, the value of computer simulations lies not so much in success as in failure. That is, if one cannot describe an epidemic well enough for computer simulations to give realistic results, it is apparent that more research is needed. Thus, simulation provides a convenient research tool for pointing out where research emphasis should be placed.

### Yield Loss

Very little is known about yield loss in general, let alone for this group of diseases caused by *Phytophthora*. This area has been very neglected and must receive immediate attention if integrated pest management strategies are to be developed.

The underlying foundation of all of integrated pest management is the critical action threshold, which is developed from the relationship between a pest and a consequent yield loss. When these relationships are well understood, pest management actions can be taken, because lack of action would result in economic injury to the crop. Heretofore, when the objective was absolute control of pests, the relationship of yield loss to other factors was of little concern, which no doubt accounts for the dearth of information on yield loss for the four representative diseases.

Research on yield loss for late blight of potatoes has received some attention. James et al (1972) found a complex relationship between the weekly change in late blight severity and potato yield loss. They established that early-season infection has a greater impact on yield than does late-season infection. These "impact coefficients" appear to decline linearly over the growing season, which allows a very complex multiple regression equation to be condensed mathematically into a more simplified form (MacKenzie and Petruzzo, 1975).

This relationship is probably not universal in its application. More likely, each disease-crop relationship will require different methods for interpreting the relationship between pests and yield loss. Much more work needs to be done in this area.

### Disease Assessment

Disease assessment is another area of analytical epidemiology that appears to be in a state of crisis. To deploy integrated pest management systems, one needs to understand the relationship between pests and crop loss, and to do this, one must be prepared to assess small amounts of disease with some measure of precision. We cannot now do that for most plant diseases.

It is generally recognized that management decisions for plant disease control are made early in the development of an epidemic. At these low levels of disease, precise estimates become extremely difficult. For the research epidemiologist trying to collect the vast amounts of data necessary for epidemiologic investigations, this difficulty becomes important. Moreover, until disease assessment methodologies are developed, very little progress can be made in deploying integrated disease management strategies.

## APPLIED EPIDEMIOLOGY

From comparative epidemiology we hope to learn the concepts and principles that allow the extension of knowledge between problems. From analytical epidemiology, we are able to comprehend complex biological systems with the intent of manipulating events for desired outcomes. Both of those aspects of epidemiology come together in the area we call applied epidemiology. This subdiscipline will provide the answers to disease management questions and will very likely become the raison d'être of plant pathology in the future. The following sections discuss possible applications of present knowledge.

### Disease Resistance

As mentioned earlier, cultivars of crops differ in their resistance to plant pathogens. Modern epidemiologic theory proposes that a plant can respond to a pathogen through either race-specific resistance or race-nonspecific resistance. The proposed usefulness of race-specific resistance is from the "genetic sanitation" gained by eliminating those races of the pathogen that are incompatible with that cultivar.

Race-nonspecific resistance is less well defined. Many now argue that its benefit is that it limits the rate at which an epidemic can progress (i.e., it lessens the apparent infection rate). Such rate-limiting resistance has been discovered to have measurable effects on the components of the disease cycle. Cultivars that possess rate-limiting resistance tend to have longer latent periods, reduced sporulation, shorter periods of sporulation, lessened disease efficiency, etc. When combined and magnified over the polycyclic epidemic, such small differences can have tremendous consequences on the progress of an epidemic. There is evidence in the literature that cultivars of the hosts of the four representative diseases do express such rate-limiting resistance (Asomaning, 1964; Gooding and Lucas, 1959; Wills and Crews, 1964; Zentmyer, 1978; Zentmyer and Mircetich, 1960; Zentmyer and Thorn, 1956; Zentmyer et al, 1977). These reports often express disappointment with this type of resistance, which some consider insufficient to control epidemics.

### Reducing Initial Inoculum

There are many ways to reduce the amount of inoculum that may otherwise initiate an epidemic. Exclusion of the pathogen, on-farm sanitation practices, and race-specific resistance (whereby some races of the pathogen population are eliminated) are just some examples.

As discussed earlier, analysis of sanitation ratios often fails to demonstrate benefits from sanitation practices that would justify the effort (Zadoks and Schein, 1979). These analyses fail to recognize one important aspect of pest management, namely, that single efforts, when combined with other pest management practices, complement one another to provide an acceptable final result that none of the individual efforts could provide.

In one study of black pod of cacao reported by Wood (1975), the removal of pods as a source of inoculum reduced the final incidence of infected pods from 90 to 60%. The author concluded that the benefits were not worth the effort. A more appropriate study would have tested the combined effects of sanitation practices complemented by host plant resistance and the intelligent use of fungicide scheduling.

### Disease Forecasting and Fungicide Scheduling

Potato late blight forecasting is perhaps the most advanced of all plant disease forecasts, reflecting 30 yr of effort by many people. This aspect of applied epidemiology deserves research attention for many other crops. The opportunities for forecasting the other diseases caused by *Phytophthora* are all too obvious.

Potato late blight forecasting began with the efforts of Russell Hyre and Jack Wallin who, working independently, discovered the dependence of the epidemic on key environmental factors (Hyre, 1954; Hyre and Bonde, 1955; Hyre et al, 1959; Wallin, 1962; Wallin and Waggoner, 1950). Raymond A. Krause led a research team at The Pennsylvania State University that adapted the Hyre-Wallin system for a large mainframe computer, with remote data entry and on-farm data collection. The system was known as BLITECAST (Krause et al, 1975).

As a grower service, BLITECAST has been very successful in recognizing blight-limiting weather conditions and recommending when the applications of fungicides may be skipped. Sometimes as many as six or more applications can be omitted in a single growing season. Recently, MacKenzie and Smilowitz (1982) reviewed the status of the BLITECAST system and its evolution into other information delivery systems. We will focus on items from that article to demonstrate how applied epidemiology can be useful in the management of plant disease.

Figure 3 was developed to demonstrate how growers of different potato cultivars can make use of fungicide scheduling systems like BLITECAST (MacKenzie, 1981). Potato cultivars with low apparent infection rates need less fungicide than do cultivars with high apparent infection rates (Fry, 1977). The late blight epidemic is assumed to begin with one lesion on one potato plant in a thousand. Epidemiologically speaking, this is logit −14. Obviously, if cull piles are not removed, or if highly infected potato seed are used, this assumption of BLITECAST will be invalidated, and attempts to skip fungicide applications could have tragic consequences. High-quality, certified seed and good on-farm sanitation are essential components of BLITECAST.

The first threshold of concern to potato growers is 0.1% disease. For the "average" Pennsylvania potato cultivar, this requires the occurrence of "blight weather," which is measured in severity values. Severity values are arbitrary units assigned to each day to describe the likelihood of infection by *P. infestans* as affected by weather. "Nonblight weather" is given a zero for the day, and intense blight weather is given a severity value of 4. By accumulating severity values, BLITECAST traces the progress of late blight epidemics through the growing season.

Severity values, therefore, may be thought of as measures of epidemiologic time. Rather than plotting the progress of an epidemic by the earth's position relative to the sun (i.e., days), severity values measure epidemic progress specific to that growing season. At severity values of 18–20, the "average" cultivar is likely to have reached 0.1% disease. At this point, BLITECAST begins a week-to-week game of roulette by gambling next week's fungicide protection on last week's fair weather. The gamble usually pays off, and most often the blight is held below the target of 1% disease severity.

Obviously, if the "average" Pennsylvania potato cultivar could be replaced with cultivars with a lower apparent infection rate, far less fungicide would be needed to protect the crop—even in years of intensely blight-favorable weather. This would seem to be a worthwhile topic for research.

By example, then, we are able to demonstrate how applied epidemiology merges with comparative and analytical epidemiology into management strategies for plant disease. We well recognize that the development of forecasting systems is much more advanced for potato late blight than for the other representative diseases. However, this does not rule out the possibility of developing such systems. For instance, what about the use of cacao cultivars with lower apparent infection rates? This virtue, complemented by sanitation and fungicide sprays scheduled according to need by environmental monitoring, could contribute very significantly to lessening the impact of a very destructive plant pathogen.

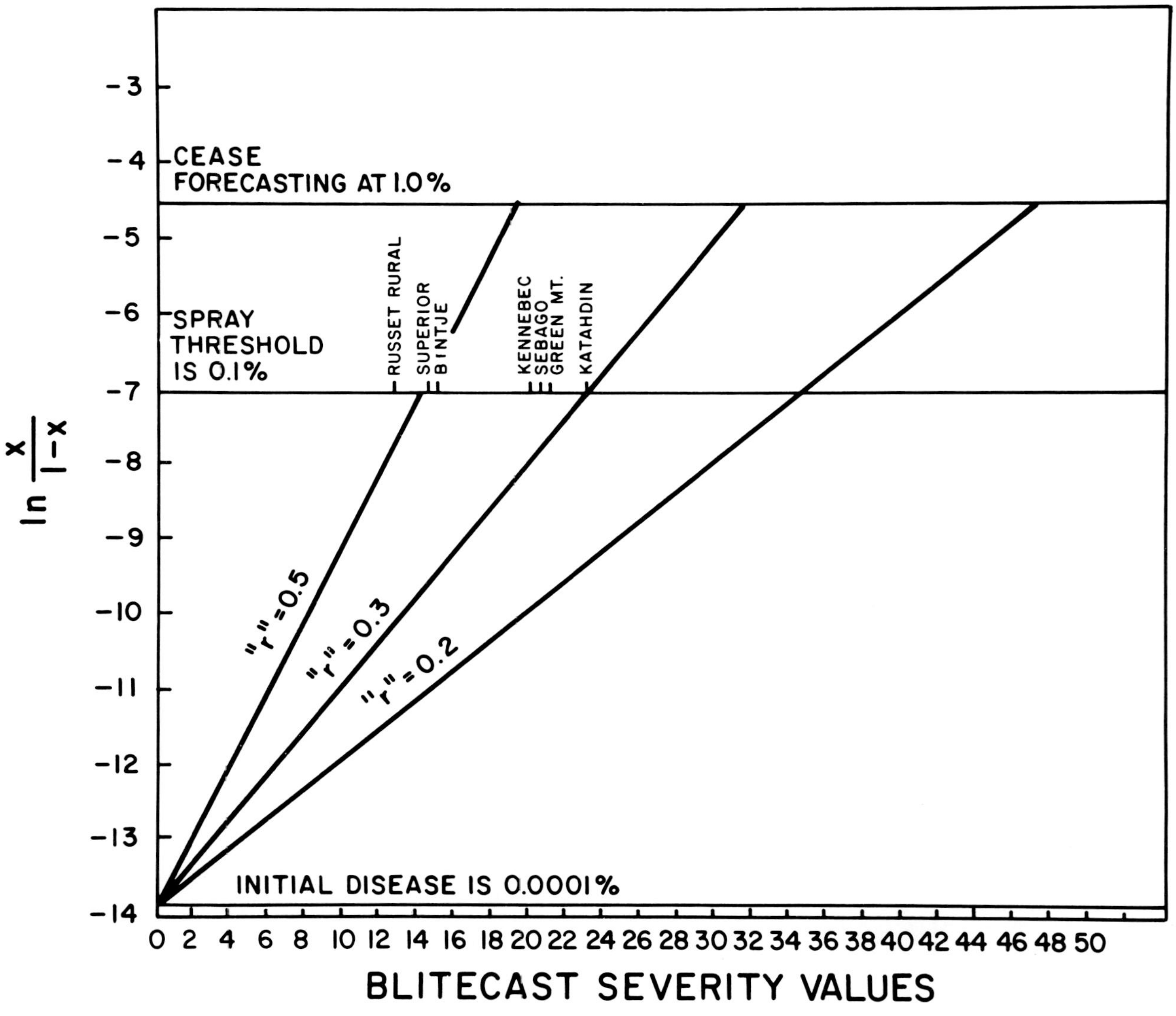


**Fig. 3.** Diagram showing the epidemiologic assumptions of the BLITECAST computer forecasting system for potato late blight: low initial disease (logit −14), differences in apparent infection rates (*r*) for selected cultivars as a function of BLITECAST severity values, and the two key BLITECAST thresholds at which critical activities should occur. Cultivar reactions were taken from Latin et al (1981). (Reprinted, by permission, from MacKenzie, 1981)..

We see an immense challenge in designing integrated strategies to take advantage of the knowledge that is now accumulating on these plant disease problems.

### Integrated Pest Management

The bringing together of strategies into a unified effort is the main objective of integrated pest management (IPM). IPM is often promoted to environmentalists as a method of reducing pesticide assaults on our environment. This benefit is, in reality, only secondary; IPM may call for more pesticide under some conditions. IPM does appear to be the direction of the future. Its benefits for crop production are derived from the use of complementary practices aimed at managing pests to keep them at acceptable levels.

The key difference between IPM and earlier pest control practices lies in the recognition by IPM of the economics of crop production. Previously, the choice of a single method to control a pest was based on gross economic factors, such as value per unit area of the crop and long-term experience with the impact of the disease. IPM, on the other hand, provides opportunities to make decisions that allow for differences in cultivars, fungicides, growing seasons, market value, etc. More important, IPM provides opportunities to mix strategies for the desired outcome. As we better understand the interrelationships of disease resistance, sanitation practices, fungicides, the effects of weather on disease development, and the many other factors that affect the course of epidemics, more sophisticated IPM strategies will unfold. The subdiscipline we have called applied epidemiology will, we hope, develop these possibilities into reality.

## PROJECTIONS

### Integrated Crop Management

As we look ahead to the future, it becomes apparent that the activities called IPM will become part of a total program named integrated crop management. Earlier we said that plant pathology has neglected aspects of crop production that have both direct and indirect effects on the development of an epidemic. Integrated crop management is based on a philosophy that takes a broad overview, recognizing the interrelationships among all aspects of crop culture, with the intent of developing optimal choices for the production of that crop. Today's agricultural universities are not structured to handle the interdisciplinary needs of integrated crop management. The splintering of agriculture by disciplines into agronomy, horticulture, entomology, plant pathology, etc. does not contribute to truly integrated research efforts. The international agriculture research centers, which have established themselves with such successes as the "Green Revolution," "Wonder Wheats," and "Miracle Rice," are novel institutions in many respects. Primarily, they provide opportunities for interdisciplinary research that focuses on a crop or cropping system. It is difficult to foresee how agricultural universities might alter their present structure but easy to see that without some form of interdisciplinary research, many complex problems of the future will not be addressed.

### Environment

Global concern for the preservation of the environment is growing. Many people express concern that the need for more food production in the future may contribute to even more pollution. These concerns are real and must continue to receive attention. Epidemiology may provide limited opportunities for reducing the use of pesticides, for more efficient use of land, or for more environmentally sound disease management practices. However, it would be wrong to hold out the promise of solving major pesticide pollution problems through epidemiology. As the needs for world food production increase, ever more pressure will be placed on the agroecosystem, with some very negative consequences. Epidemiology can provide a look at the options to be used in selecting the most desirable solution. The often-held belief that complete alternatives to fungicides can or will be found does not seem justifiable at this time. The task of epidemiology will be to find ways to use fungicides to the best advantage of the farmer and the consumer, with minimal impact on the environment.

### Regulations

Restrictions on the use of pesticides in agricultural production are increasing. Many of these restrictions are justified in preventing the abuse of, or unnecessary risk from, biologically active chemicals by unknowing people. Other restrictions appear to be excessive and contrary to what is known about the safety of these compounds compared with the alternatives.

However, it is not our point to argue for or against the regulation of pesticides but to speak of pesticides from the standpoint of epidemiology. Many decisions being made today are in direct conflict with what is known about the epidemiology of the pests that the chemicals are intended to control. The late blight of potato-Ridomil (Ciba-Geigy Ltd., Basle, Switzerland) scenario will exemplify our point.

Metalaxyl (the active ingredient of Ridomil) has been demonstrated to be very effective against species in the Pythiaceae, including the *Phytophthora* plant pathogens. During its early years of development, metalaxyl was hailed as a significant "breakthrough," because its activity was highly specific against this group of pathogens. It has shown both protective and eradicative properties.

Before long, problems of tolerance to metalaxyl were recognized (Davidse et al, 1981). The manufacturers began searching for alternatives. Their proposal at this point is to mix metalaxyl with mancozeb as a premix, in hopes of forestalling the selection for tolerance to metalaxyl.

Ridomil has not yet been licensed in the United States for potato late blight control, with the exception of some emergency labels first provided to potato growers in the state of Maine during August 1980. Final registration of Ridomil for potato late blight control is expected to come sometime soon. We will first offer some facts and then an epidemiologic perspective on what is happening with the registration of this compound to help explain how regulation of pesticides, in our view, has gotten off track.

Aroostook County, Maine, located on the Canadian border, has a major potato seed industry. Many of the growers participate in the BLITECAST system offered by the University of Maine. Officials must have recognized that, with the recent registration of Ridomil in Canada, Maine potato growers could easily transport quantities across the border for use in the United States. Late blight in Aroostook County was moderate during the 1980 growing season. We may never know why an emergency label was issued, but the action defies several epidemiologic principles.

Selection in *P. infestans* for tolerance to metalaxyl is expected to occur rapidly and, in fact, reports from Europe speak of problems soon after the introduction of metalaxyl (Davidse et al, 1981). We know that seed potatoes are perhaps

# 24

# Control of *Phytophthora* by Host Resistance: Problems and Progress

V. Umaerus, M. Umaerus, L. Erjefält, and B. A. Nilsson

CONTENTS



## INTRODUCTION

Conscious breeding for resistance to diseases caused by *Phytophthora* spp. was initiated soon after the serious epidemics of late blight caused by *P. infestans* on potatoes in Europe in 1845–1846 (Ross, 1958). A serious search for resistance in tomato did not begin until 100 years later (Gallegly, 1968). A survey of literature published during the last decade shows that research on or breeding for resistance to various species of *Phytophthora* is now done in at least 40 species of agricultural and horticultural crops.

Most of this paper will be concerned with the *P. infestans*-potato relationship, although considerable breeding efforts for resistance to *Phytophthora* are made in several crops besides potato. Some of these have a long history; for example, in breeding for resistance to *P. infestans* in tomato, surveyed by Gallegly (1968) and Turkensteen (1973), monogenic-dominant resistance was identified, but a compatible race of the fungus appeared even before the breeding program was completed (Conover and Walter, 1953). The first tobacco cultivar resistant to *P. parasitica* var. *nicotianae*, which may be classified as having general resistance, was released about 1930 (Tisdale, 1931). Specific resistance to *P. fragariae* in strawberry has been utilized in breeding since 1929, but general resistance is now regarded as desirable (Gooding, 1972). Breeding for resistance to other diseases caused by different species of *Phytophthora* that have epidemic potential is expected to follow the pattern demonstrated in potato and *P. infestans*, with increasing attention on general rather than specific resistance.

## TERMINOLOGY

Definitions and usage of terms relating to resistance are not uniform. The approximate meanings of many terms used are similar, but functional, genetic, and epidemiologic concepts have sometimes been included in the same term, which has led to confusion. "Field resistance" has been widely used in potato breeding as a practical concept. Field resistance was defined by Niederhauser (1961) as "the resistance exhibited by a plant towards all races capable of causing on it more than a hypersensitive reaction." The terms "horizontal" and "vertical" resistance introduced by Vanderplank (1968) have become widely accepted but are also criticized (Nelson, 1978; Parlevliet, 1979). For the plant breeder, the important thing is the end result—the cultivar produced should be less in danger of becoming diseased. Whether or not the components in the complex of characters leading to this effect can be properly labeled as one type of resistance or the other, or, indeed, as resistance, is less important. Theoretical considerations (in want of practical experience before the cultivar has been grown for a long period of time) should also justify with reasonable certainty the prediction that the danger of a plant's becoming diseased will not increase dramatically if the pathogen population changes.

As functional concepts, "general resistance" and "specific resistance," defined in accordance with Browning et al (1977), are used in this paper. General resistance is present "if there is no *known* differential interaction among genotypes of the host and genotypes of the pathogen" (Browning et al, 1977). The term is roughly synonymous with race-nonspecific, horizontal, partial, or field resistance, as those terms have been used for potato late blight. In an avoidance mechanism, host and parasite genes would be expected to act independently of each other. In the case of "true resistance," a gene-for-gene relationship may in reality exist (Parlevliet and Zadoks, 1977), but this is not distinguishable with normal experimental techniques.

Specific resistance, where "known genetic interactions exist between host genotypes and pathogen genotypes" (Browning et al, 1977), is used as a synonym for race-specific or vertical resistance. From a genetic point of view, i.e. to describe the mode of inheritance, oligogenic (monogenic), polygenic, or cytoplasmic resistance may exist. Specific resistance to a *Phytophthora* species is, as a rule, monogenically inherited.

Polygenic systems are assumed to be more stable than monogenic systems, because the pathogen population must change on various loci to adapt to the host. The more loci involved, the harder it is for the pathogen population to adapt. The term "durable resistance" has sometimes been used, but durability cannot be proven until after a long period of cultivation, and the term is not appropriate. The best hope of



TABLE 1. Examples of separation of resistance into components in some *Phytophthora*-host systems

| *Phytophthora* species | Host species | Component: Preinfection and infection | Colonization | Reproduction | Reference |
|---|---|---|---|---|---|
| *P. infestans* | Potato (foliage) | Infection efficiency | Lesion extension | Sporulation | Pietkiewicz, 1976; Umaerus and Lihnell, 1976; van der Zaag, 1959; Williams et al, 1975 |
| *P. infestans* | Potato (tuber) | Zoospore germination | Hyphal growth | ··· | Clarke and Kassim, 1977 |
| *P. infestans* | Tomato | Infection efficiency | Lesion extension | Sporulation | Turkensteen, 1973 |
| *P. erythroseptica* | Red raspberry | Percentage infection | Mean injury | ··· | Barritt et al, 1979 |
| *P. palmivora* | Cocoa | Infection efficiency | Lesion extension | ··· | Blaha and Lotodé, 1976; Lawrence, 1978 |

creating durable resistance is by the use of many genes. There is no definite evidence of the number of genes involved in general resistance to *Phytophthora.* Practical breeding efforts and many investigations on the *P. infestans*-potato system indicate that at least more genes are involved in general resistance than in specific resistance. Whether those genes all have equal effect and may be termed polygenes in the sense defined by Mather (1943) is uncertain. Whether cytoplasmic factors are important in resistance to *Phytophthora* is also uncertain.

## VARIABILITY IN *PHYTOPHTHORA*

### Mechanisms of Variation

The success or failure of efforts to breed for resistance to pathogens often depends on the capacity for variation in the

stages are diploid. These findings were made in a number of oomycetes, including some homothallic *Phytophthora* species. Before that, she had drawn similar conclusions from studies in other related species (Sansome, 1961, 1963a, 1963b). Marks (1965), however, in a study including the heterothallic species *P. infestans,* could not support Sansome's conclusions, and Gallegly (1968) speculated that homothallic species may be diploid and heterothallic ones haploid. This was, however, contradicted by Galindo-Alonzo and Zentmyer (1967), who found evidence for vegetative diploidy in another heterothallic species, *P. drechsleri,* and by Shaw and Khaki (1971) and Sansome and Brasier (1973) in separate cytologic studies of *P. infestans.* Additional convincing genetic as well as cytologic evidence for diploidy has since been reported in *P. infestans* and related species by Brasier and Sansome (1975), Sansome (1976), and Long and Keen (1977).

In view of present knowledge of the existence of somatic recombination and diploidy, the ability of *P. infestans* to accumulate and preserve genes for virulence that are



type and resistance level differ among the countries, the graphs are not directly comparable. Level of resistance is commonly assessed under field conditions in the country and is thus an expression of the combination of specific and general resistance in that particular situation with regard to races of *P. infestans.* Also, statistics on seed production in most cases refer to certified seed and therefore do not correspond to total seed production. Thus, interpretation of the graphs as a measure of the acceptance of late blight-resistant cultivars must be restricted to trends; however, these trends seem quite clear.

In the German list of cultivars (Bundessortenamt, 1980), specific resistance is indicated in terms of R-genes present. Of a total of 128 cultivars, 42 have R-genes. In 33, the R-genes are identified (R1, R3, R4, and R10). The cultivars with R-genes range from very susceptible to moderately resistant to late blight in field tests under German conditions, except those carrying the R3-gene, which have shown resistance under field conditions up to 1980. No clear difference in the expression of resistance is apparent under field conditions between the groups of cultivars with and without R-genes.

In the Netherlands list (Netherlands Potato Consultative Institute, 1980), 36 of 76 (47%) cultivars are stated to carry R-genes, but the genes are not identified. The distribution of cultivars over the range of resistance expressed in the field tests is similar to that of the German material.

In Germany and The Netherlands, several cultivars carry R-genes that are not effective against the races present. Sweden is the only country where so far there seems to be a relationship between presence of R-genes and high resistance in the field tests.

General resistance to late blight has often been associated with late maturity in the potato. It is apparent that the number of resistant cultivars is largest in the late group in Figure 1. It is possible that more breeding effort has been expended in producing resistant main crop and late cultivars, because early cultivars may avoid attack by being harvested early. An important reason is probably the simultaneous transfer of resistance and lateness from the most important source of resistance, *S. demissum.* With a prolonged period of breeding, this relationship may be broken by genetic separation, as indicated by the presence of resistance in all maturity groups in the Svalöv breeding program (Fig. 3).

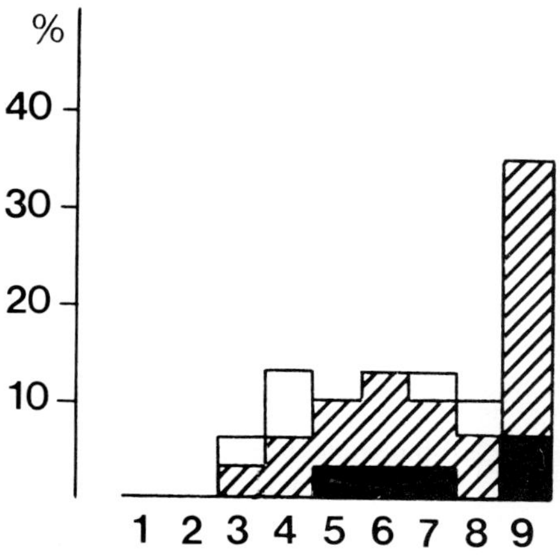


**Fig. 3.** Frequencies of early (unshaded), main crop (striped), and late (shaded) advanced breeding material from the breeding program at Svalöv, Sweden, classified on a susceptibility-resistance scale, presented in terms of number of clones. Resistance levels increase from susceptible on the left to resistant on the right of the horizontal axis.

The graphs showing the extent of seed production (Fig. 2) indicate generally that the market has not adopted the resistant cultivars. The potato market in most countries is still dominated by old or very old cultivars, most of them susceptible to late blight. For a new cultivar to become accepted, it must promise economic advantage to the grower and the seed trade, based on higher yield and/or lower production costs, and consumer acceptance better than or at least equal to established cultivars. The advantages must be obvious for a cultivar to be grown on a scale to even make it known to consumers. Seed production is expensive, cultivation costs are high, and management practices are well known for established cultivars but may cause some hazards with new and unknown material. Late blight resistance alone has not proved to be a great enough advantage.

Resistance may be specific, general, or a combination of both. As long as no races of the fungus that overcome the specific resistance of the host are present, the cultivar with specific resistance is not attacked, and the effect of general resistance is masked. In the absence of specific resistance or when compatible races of the fungus are present, the degree of general resistance will be the decisive factor in the field. Experience with R-gene resistant cultivars that have been heavily attacked after a limited number of years has shown that preventive spraying may be necessary in those cultivars. Late blight can be effectively controlled by fungicides, and growers have learned how to use them. The "breakdown" of resistance in cultivars that were claimed to be resistant when they were released has also caused a lack of confidence in general resistance, which is expected to be more durable.

Many of the old cultivars are satisfactory with respect to quality and consumer acceptance. Consumers are known to be conservative in their acceptance of new potato cultivars. In a recent study by Taylor and Hampson (1980), an old, well-established cultivar, King Edward, and a new cultivar, Drayton, were tested for their acceptability by consumers. The tubers were externally and internally similar, but when both cultivars were correctly named, consumers rated the acceptability of King Edward as superior to that of Drayton. However, when the names were interchanged, or when one was called King Edward and the other was given a code, the sample labeled "King Edward" was always rated as superior. When both samples were identified by codes, consumer panels could not separate them but gave both of them acceptability ratings lower than those of the real King Edward.

### North America

In the lists of cultivars from the United States and Canada, there is no consistent information on late blight resistance levels, and therefore the acceptance of resistant cultivars is difficult to estimate. A general impression is that the trend observed in Europe is even more pronounced in North America. Of the 10 most popular cultivars grown in North America, covering 90% of the seed acreage (Thornton and Sieczka, 1980), eight are characterized as susceptible to late blight and only two, Kennebec and Sebago, are considered "moderately resistant." Kennebec carries the gene R1, while the nature of resistance in Sebago is not indicated. Hodgson (1976) concludes from the results of 40 yr of research at Fredericton, Canada, that because of the low commercial acceptability of the resistant cultivars that have been released, the possibility of reducing the late blight problem by introducing one or more new resistant cultivars is probably remote.

### Mexico

In the Toluca Valley in Mexico, where the late blight problem is more severe than anywhere else in the world, the market is still dominated by old cultivars. In high-altitude mountain areas, old indigenous cultivars are grown by small-scale farmers, while predominantly Dutch cultivars are grown in the central valleys. The most important cultivar, Alpha, which in Europe is regarded as fairly resistant, requires intense spraying with fungicides to avoid heavy late blight attacks. However, an advanced breeding program is conducted in Mexico by the National Institute for Agricultural Research, and new cultivars with very good resistance have been released (M. Villarreal, *personal communication*).

### Developing Countries

Research on and development of late blight resistance have high priority in the program of the International Potato Center (CIP) in Peru. Selection for blight resistance has been conducted ever since the formation of the Center in 1972, and now every year tuber families are provided to various countries for further selection and breeding. Frequencies of seedlings selected for high levels of late blight resistance in CIP's breeding material were as high as 30% in 1979, and in the same year, almost 9,000 tubers from 108 families were provided to four different countries (International Potato Center, 1979).

Potato breeders all over the world are selecting, at varying degrees of intensity, for late blight resistance, and resistant cultivars are continuously being released. Several breeding programs have a long tradition in this field, and a high proportion of their advanced breeding material contains resistance to late blight, as exemplified by the program at Svalöv, Sweden (Fig. 3).

The acceptance of this material is still an open question. The decisive factor in advanced countries is the economy of the potato crop. With relatively cheap chemical control available, established cultivars are replaced only if new cultivars offer considerable advantages besides resistance to late blight. Growing concern about the effects of intensive use of chemicals, however, will intensify the need for other control measures, and here the use of resistant cultivars will be advantageous. Subsistence farmers in many developing countries, where potato is now becoming a new staple food, have for economic and practical reasons little interest in fungicidal control but a large interest in resistant cultivars.

## THE NEED FOR INTEGRATION

### In Breeding

At present, use of resistant cultivars to control late blight of potato is having a very modest impact in Europe and North America. Resistance to late blight is only one trait of many that must be included in a new potato cultivar, and one reason for the small impact of resistant cultivars on the market may be that some of the cultivars released have not met the high standards demanded for cooking quality, yield, and multiple resistance.

Several breeding programs emphasize the need to combine resistance to major pests and diseases, e.g. as described by Tarn (1976) with a combination of resistance to *Streptomyces scabies, P. infestans, Synchytrium endobioticum,* and potato virus yellows.

A systems analytical approach has been made by M. Umaerus (1980), who describes an integrated breeding program for yield, quality, and resistance. In all three sectors—yield, quality, and resistance—attempts are made to identify the components that are most important to the final economic yield, to find sources of genes for those components, and to create selection methods, preferably screening methods, that can be used at an early stage in the breeding program.

A potato cultivar is a highly heterozygous, tetraploid, vegetatively propagated clone. A large number of characters must be combined in a single genotype and selected for in a breeding program. The more traits that must be selected for, the more seedlings that must be produced and screened to find the superior clone. The laborious introduction of general resistance necessitates careful planning of the allotment of resources in a breeding program. In some cases, priorities placed on late blight resistance have been rather low, and breeding for resistance to late blight ceased when R-gene resistance was found to be unreliable, fungicides were effective, and the alternative was breeding for general resistance. The assumption is that the more genes that influence resistance, the higher the expectations for durable resistance. The introduction of as many genes for resistance as possible is a necessity when breeding for general resistance. This is the most apparent difficulty. The genes cannot be identified separately, and the expression of resistance is quantitative, which makes early screening difficult. Component analysis, where screening for resistance in some phases of the infection cycle can be made at an early stage in a breeding program, is a way to make the breeding work more efficient and makes it possible to integrate screening for components of resistance to late blight with screening for other traits in relation to the importance attached to them in the program.

The most important source of general resistance so far has been the Mexican species *S. demissum,* i.e. the same species that provided the R-genes for breeding for specific resistance. Complex races of *P. infestans* are used for screening to reveal the degree of general resistance even if R-genes are present. Many other *Solanum* species, especially those indigenous to Mexico, an epicenter of both *Solanum* and *P. infestans,* have been found to have high levels of general resistance. Tazelaar (1981) reported on the evaluation of 16 species from the German-Netherlands Potato Collection with regard to resistance as expressed in a screening test (Umaerus, 1969a) and in a field test. Several accessions of the species *S. bulbocastanum, S. demissum, S. pinnatisectum, S. polyadenium, S. polytrichon, S. stoloniferum,* and *S. verrucosum* from Central America had high levels of general resistance. Resistance was also found in species of South American origin, including *S. acaule, S. berthaultii, S. microdontum,* and *S. vernei.*

Some of the problems encountered in breeding for resistance are caused by the fact that many wild and primitive *Solanum* species used as sources of resistance also carry unwanted genes. The universal occurrence in the family Solanaceae of alkaloids, which may differ among species, is well known. Steroidal glycoalkaloids are known to exhibit in vitro toxicity to a wide variety of microorganisms and animals. This led early workers to postulate a role in vivo in inhibiting or reducing invasion by fungal pathogens. The role of steroidal glycoalkaloids is still a subject of debate. There are many conflicting reports relating to whether contents are higher in susceptible or resistant cultivars and whether levels increase or decrease after infection. It does not seem possible to find clear evidence of a correlation, even less a causal relationship, between resistance and levels of the major glycoalkaloids in the potato tuber, α-solanine and α-chaconine (Deahl et al, 1973; Frank et al, 1975; Maga, 1980). Even if glycoalkaloids are not involved in the resistance mechanism, they may easily be introduced as a side effect when other species of *Solanum* are used in breeding, and analysis of the glycoalkaloid content of new cultivars before their release is recommended.

The existence of a correlation between general resistance of the foliage to late blight and late maturity has also generated criticism of breeding for foliage resistance (Toxopeus, 1958). The correlation may not be causal in nature (Schick et al, 1958; Simmonds and Malcolmson, 1967; Umaerus, 1969a). Toxopeus (1964) found a weaker correlation in *S. demissum* than in *S. tuberosum.* Figure 3 also demonstrates the distribution of maturity types based on resistance levels in the present breeding material familiar to us, where high degrees of general resistance are also found in relatively early maturing material.

The possibility of breeding for resistance operative in both tubers and foliage is of great importance. Deahl and Sinden (1973) inoculated foliage and tuber tissues of cultivars carrying the R1-gene with race O of *P. infestans.* Hypersensitive flecks appeared on inoculated foliage, but tuber tissues were invaded to varying degrees. Observations made by Malcolmson in Scotland (Scottish Plant Breeding Station, 1976) indicate that general resistance in the haulm is not necessarily associated with resistance in the tubers. However, Piotrowski (1975) claims that tuber resistance is more frequently associated with leaf resistance than the reverse, and Chadaeva (1974) found a high degree of general resistance in both tuber and leaf in 85 of 366 cultivars tested. Chadaeva confirms, however, that R-gene resistance only occasionally operates in both tuber and leaf tissues.

The selection of the proper test methods is important to differentiate resistant clones from susceptible ones. Deahl et al (1974) could identify resistance to tuber rot in clones only if the cut tubers were given a 48-hr wound-healing treatment. Tuber resistance and foliar resistance were then found to be correlated ($r = 0.87$). The resistance factors affecting the colonization of potato tuber tissue by *P. infestans* through a wound surface were further investigated by Clarke and Kassim (1977). Differences in resistance affecting the duration of spore germination and the establishment of hyphal growth in the tissues were detected among the clones. Those differences were related to the general resistance in four of five clones tested. No active resistance mechanism seemed to operate during the phase of linear growth through the tissues.

Biochemical resistance mechanisms have attracted interest in many research programs. A clarification of the sequence of chemical events in the reactions in general resistance is expected to be of great importance for the development of screening methods to be used in breeding. Of particular interest is the observation that R-genes may be expressed in tissue cultures, as shown by Ingram (1967) in potato stem tissue culture infected by *P. infestans* and tobacco pith callus tissues infected by *P. parasitica.*

Many research programs at present are investigating the possibilities of using in vitro methods in potato breeding. Selection methods for factors in general resistance that could be used in vitro would be of considerable interest.

### In Control

In contrast to race-specific resistance, which produces hypersensitivity and arrests fungus invasion, general resistance provides a continuous variation in degree of attack from susceptibility to resistance. Although the ultimate goal is to obtain as high a level of resistance as possible, intermediate levels are also of value. Van der Zaag (1978) recommends a minimum degree of foliage resistance (six on an unspecified scale) for cultivars for Western Europe.

General resistance is also very much influenced by predisposing factors such as day length and light intensity (Bychenkova, 1976; Schumann and Thurston, 1977; Umaerus, 1970), mineral nutrition (Gallegly and Niederhauser, 1959; Umaerus, 1969b), and virus infection (Fernández de Cubillos and Thurston, 1975). Henriksen (1973) observed that resistance of the tubers was reduced when sodium chlorate was used to destroy potato haulms.

Under optimal environmental conditions for late blight, cultivars with a relatively high level of general resistance may be attacked. If this occurs late, at a time when haulm destruction is practiced anyway, no other treatments are needed. If the epidemic starts early, or if cultivars with an intermediate level of general resistance are used, additional protection by fungicide treatment may be needed, though at a lower rate than in susceptible cultivars. Fry (1978) has quantified general resistance of potato cultivars in terms of reduction of fungicidal treatment. Among 11 commercial cultivars and one breeding line, the difference in general resistance between the most resistant and the most susceptible clones was estimated to be equivalent to weekly applications of about 0.7 kg of mancozeb per hectare. Fungicide treatment can be adjusted to complement the general resistance of a cultivar based on a forecasting technique so that applications are made only when needed. Such an integrated control method for potato late blight has been described by Fry (1977) based on the forecasting technique BLITECAST, which combines a system based on relative humidity and temperature with a system based on rainfall and temperature.

## SUMMARY

A survey of literature published during the last decade shows that research on or breeding for resistance to various species of *Phytophthora* is now conducted in at least 40 species of agricultural and horticultural crops. The first attempts to select for resistance to late blight in potato apparently aimed at what we now recognize as general resistance. The identification of specific resistance controlled by dominant resistance genes was regarded as a breakthrough. Much effort was put into the transfer of R-genes from *Solanum demissum* to *S. tuberosum.* Virulent races of the fungus appeared, and now most potato breeders again concentrate their efforts on breeding for general resistance. This pattern in the breeding efforts for resistance to *P. infestans* in potato has been repeated in other crops attacked by *Phytophthora* spp. with epidemic potential.

Success or failure of efforts to breed for resistance to pathogens often depends on the capacity for variation in the pathogen population. Sexual and somatic recombination have been shown to cause exchange and rearrangement of nuclear information in *Phytophthora.* Anastomosis allows exchange of cytoplasm and nuclei, and consequently the exchange of cytoplasmically heritable factors and the formation of heterokaryons. The previously accepted view that all asexual stages of *P. infestans* are haploid has been questioned, and research during the last decade has provided convincing genetic as well as cytologic evidence that the fungus is diploid. In view of present knowledge of the existence of somatic recombination and diploidy, it is probable that the ability of *P. infestans* to accumulate and preserve genes for virulence that are temporarily unnecessary is greater than thought earlier.

Both general resistance and specific resistance to different species of *Phytophthora* have been reported in many crops.

Specific resistance behaves as a single-gene trait due to dominant genes. General resistance is governed by many genes. Polygenic systems are assumed to be more stable than monogenic systems. The more loci involved, the greater the problem of the pathogen population to adapt without losing fitness. General resistance to *P. infestans* in potato shows a pattern of quantitative inheritance. Additive gene effects have been reported as well as dominant or epistatic effects.

General resistance reduces the infection rate. Component analyses are now being done in an attempt to better understand general resistance through studies of defense reactions, epidemiology, and inheritance. The results have a strong influence on the planning of breeding programs, e.g. choice of parents and selection methods.

Although a fairly large number of potato cultivars have general resistance and many cultivars have specific resistance to one or several races of the pathogen, at present the impact of use of resistant potato cultivars to control late blight is very modest in Europe and North America, for several reasons. Control with chemicals is still relatively cheap and effective. The experience that R-gene resistant cultivars may become heavily attacked after a few years has made preventive chemical control necessary in cultivars claimed to be resistant and also has caused a lack of confidence in general resistance, which would be expected to be durable. Breeding for general resistance is more difficult and time-consuming than breeding for specific resistance. The cultivars released do not always meet the high standards demanded for yield, quality, and multiple resistance. Several breeding programs emphasize the need to combine resistance to major pests and diseases with high yield and quality. Risks associated with pesticide use have received considerable publicity, and awareness of these risks has stimulated new efforts to breed for resistance as well as efforts to increase the efficiency of pesticide used.

A potato cultivar is a highly heterozygous, tetraploid, vegetatively propagated clone. The more traits that must be selected for, the more seedlings that must be produced and screened. The laborious introduction of general resistance necessitates careful planning of the allotment of resources in a breeding program to weigh priorities to be placed on quality, yield, and resistance to a large number of diseases and pests. The assumption is that the more genes that govern resistance, the higher the expectations of durable resistance. There is thus no way to get around the necessity of introducing many genes when breeding for general resistance. Rationalization may be sought in breeding and screening techniques, as well as in the identification of those components of resistance that convey the largest increase in degree of resistance under the environmental conditions where a cultivar is to be grown. Greater knowledge of the biochemistry of the resistance mechanisms involved in general resistance is expected to aid in the development of screening techniques. In vitro techniques for haploid production combined with in vitro techniques for screening are at least theoretically promising for more efficient breeding in the future.

Integration of the use of resistance and fungicides is needed in control. Cultivars with intermediate resistance may need additional protection by fungicidal treatment but at a lower rate than with susceptible cultivars. Fungicide treatment can be adjusted to complement the general resistance of a cultivar, based on available forecasting techniques.

## LITERATURE CITED


Barritt, B. H., Crandall, P. C., and Bristow, P. R. 1979. Breeding for root rot resistance in red raspberry. J. Am. Soc. Hortic. Sci. 104:92-94.

Black, W. 1952. A genetical basis for the classification of strains of *Phytophthora infestans*. Proc. R. Soc. Edinburgh, Sect. B 65:36-51.

Black, W. 1971. Research on potatoes at the Scottish Plant Breeding Station. Pages 52-60 in: 50th Annu. Rep. 1970–71. Scottish Plant Breeding Station, Edinburgh, Scotland.

Black, W., and Gallegly, M. E. 1957. Screening of *Solanum* species for resistance to physiologic races of *Phytophthora infestans*. Am. Potato J. 34:273-281.

Black, W., Mastenbroek, C., Mills, W. R., and Peterson, L. C. 1953. A proposal for an international nomenclature of races of *Phytophthora infestans* of genes controlling immunity in *Solanum demissum* derivatives. Euphytica 2:173-179.

Blaha, G., and Lotodé, R. 1976. A major selection criterion for cacao in Cameroun: Resistance to black pod disease (*Phytophthora palmivora*). Café, Cacao, Thé 20:97-119.

Blaha, G., and Lotodé, R. 1977. Contribution to the knowledge of the mode of inheritance of resistance to black pod disease (*Phytophthora palmivora*) in coffee in Cameroun. Café, Cacao, Thé 21:179-196.

Bonde, R. 1932. A promising blight resistant potato. Am. Potato J. 9:49-54.

Brasier, C. M., and Sansome, E. 1975. Diploidy and gametangial meiosis in *Phytophthora cinnamomi, P. infestans* and *P. drechsleri*. Trans. Br. Mycol. Soc. 65:49-65.

Broili, J. 1921. Arbeiten mit Wildbaftarden von Solanum. Mitt. Biol. Reichsanst. Land Forstwirtsch. Berlin-Dahlem 21:154-158.

Browning, J. A., Simons, M. D., and Torres, E. 1977. Managing host genes: Epidemiologic and genetic concepts. Pages 191–211 in: Plant Disease. Vol. I. J. G. Horsfall and E. B. Cowling, eds. Academic Press, New York.

Bundessortenamt. 1980. Beschreibende Sortenliste 1980. Kartoffeln. Hannover, Germany.

Bychenkova, A. A. 1976. The role of environmental factors in changes in the resistance of potato to *Phytophthora*. Skh. Biol. 11:59-64.

Caten, C. E. 1974. Intra-racial variation in *Phytophthora infestans* and adaptation to field resistance for potato blight. Ann. Appl. Biol. 77:259-270.

Caten, C. E., and Jinks, J. L. 1968. Spontaneous variability of single isolates of *Phytophthora infestans*. I. Cultural variation. Can. J. Bot. 46:329-348.

Chadaeva, L. G. 1974. Blight resistance of potato varieties. Byull. Vses. Ord. Lenina Inst. Rastenievod. Imeni N.I. Vavilova 37:113-118. (In Russian)

Chumakova, A. I., and Loktina, G. I. 1978. Racial composition of *Phytophthora infestans* and evaluation of potato for resistance. Zashch. Ovoshchn. Bakhchev. Kult. Kartofelya Vred. Bolezn. Sornyakov, Moscow. pp. 113-122. (In Russian)

Chumakova, A., Kozlovskaia, I. N., and Kovalenko, E. D. 1979. Differentiation of *Phytophthora infestans* on pötato by morphological and cultural features. Skh. Biol. 14:744-779. (In Russian)

Clarke, D. D., and Kassim, M. Y. A. 1977. Resistance factors in potato tuber tissue to hyphal growth of *Phytophthora infestans*. Physiol. Plant Pathol. 11:267-277.

Conover, R. A., and Walter, J. M. 1953. The occurrence of a virulent race of *Phytophthora infestans* on late blight resistant tomato stocks. Phytopathology 43:344-345.

Davidson, W. D. 1928. The rejuvenation of the Champion potato. Econ. Proc. R. Dublin Soc. 2:319-330.

Deahl, K. L., and Sinden, S. L. 1973. A comparison of the expression of the $R_1$ gene for resistance to race O of *Phytophthora infestans* in potato tuber and foliar tissues. (Abstr.) Am. Potato J. 50:337.

Deahl, K. L., Young, R. J., and Sinden, S. L. 1973. A study of the relationship of late blight resistance to glycoalkaloid content in fifteen potato clones. Am. Potato J. 50:248-253.

Deahl, K. L., Gallegly, M. E., and Young, R. J. 1974. Laboratory testing of potato tubers for multigenic resistance to late blight. Am. Potato J. 51:324-329.

Denward, T. 1967. Differentiation in *Phytophthora infestans*. I. A comparative study of eight different biotypes. Hereditas 58:191-220.

Denward, T. 1970. Differentiation in *Phytophthora infestans*. II. Somatic recombination in vegetative mycelium. Hereditas 66:35-47.

Deshmukh, M. J., and Howard, H. W. 1956. Field resistance to potato blight (*Phytophthora infestans*). Nature 177:794-795.

Ellingboe, A. H. 1965. Somatic recombination in basidiomycetes.

Pages 36-48 in: Incompatibility in Fungi. Symp. 10th Int. Bot. Congr. K. Esser and J. R. Raper, eds. Springer-Verlag, Berlin.

Fernández, C. M. 1979. Changes in races of *Phytophthora infestans* (Mont.) de Bary observed in southern Chile from 1963 to 1967. Agric. Tec., Santiago 39:7-10.

Fernández de Cubillos, C., and Thurston, H. D. 1975. Effect of four potato viruses on the expression of general resistance of the potato (*Solanum tuberosum*) to *Phytophthora infestans.* Fitopatologia 10:91-95.

Flor, H. H. 1955. Host-parasite interaction in flax rust—Its genetics and other implications. Phytopathology 45:680-685.

Frank, J. A., Wilson, J. M., and Webb, R. E. 1975. The relationship between glycoalkaloids and disease resistance in potatoes. Phytopathology 65:1045-1049.

Fry, W. E. 1977. Integrated control of potato late blight—Effects of polygenic resistance and techniques of timing fungicide applications. Phytopathology 67:415-420.

Fry, W. E. 1978. Quantification of general resistance of potato cultivars and fungicide effects for integrated control of potato late blight. Phytopathology 68:1650-1655.

Galindo-Alonzo, J. 1964. Genetical and cytological studies on *Phytophthora capsici* in relation to variability. Ph.D. thesis. Univ. Calif., Riverside. 84 pp.

Galindo-Alonzo, J., and Zentmyer, G. A. 1967. Genetical and cytological studies of *Phytophthora* strains pathogenic to pepper plants. Phytopathology 57:1300-1304.

Gallegly, M. E. 1968. Genetics of pathogenicity of *Phytophthora infestans.* Annu. Rev. Phytopathol. 6:375-396.

Gallegly, M. E., and Galindo, J. 1958. Mating types and oospores of *Phytophthora infestans* in nature in Mexico. Phytopathology 48:274-277.

Gallegly, M. E., and Niederhauser, J. S. 1959. Genetic controls of host-parasite interactions in the *Phytophthora* late blight disease. Pages 168-182 in: Plant Pathology, Problems and Progress 1908–58. C. S. Holton, G. W. Fischer, R. W. Fulton, H. Hart, and S. E. A. McCallan, eds. Univ. Wis. Press, Madison.

Gooding, H. J. 1972. Studies on field resistance of strawberry varieties to *Phytophthora fragariae.* Euphytica 21:63-70.

Guzmán-N., J. 1964. Nature of partial resistance of certain clones of three *Solanum* species to *Phytophthora infestans.* Phytopathology 54:1398-1404.

Hahn, E., Henniger, H., and Oertel, H. 1973. Das Auftreten physiologischer Rassen von *P. infestans* (Mont.) de By. im Jahre 1970 auf dem Gebiet der DDR. Arch. Phytopathol. Pflanzenschutz 9:105-112.

Henriksen, J. B. 1973. The resistance of potato tubers to late blight (*Phytophthora infestans*) after top-killing. Tidsskr. Planteavl 82:227-229.

Hodgson, W. A. 1961. Laboratory testing of the potato for partial resistance to *Phytophthora infestans.* Am. Potato J. 38:259-264.

Hodgson, W. A. 1962. Studies on the nature of partial resistance in the potato to *Phytophthora infestans.* Am. Potato J. 39:8-13.

Hodgson, W. A. 1976. Breeding potatoes for late blight resistance. Proc. Can. Phytopathol. Soc. 43:37.

Ingram, D. S. 1967. The expression of R-gene resistance to *Phytophthora infestans* in tissue cultures of *Solanum tuberosum.* J. Gen. Microbiol. 49:99-108.

Institut National de la Recherche Agronomique. 1980. Bulletin des variétés, additif 1980, pommes de terre. Paris, France.

International Potato Center (CIP). 1979. Annual Report. pp 13-19. Lima, Peru.

Ito, S. 1918. A preliminary report on a late-blight resistant strain of potato. Ann. Phytopathol. Soc. Jpn. 1:5-8.

Jeffrey, S. I. B., Jinks, J. L., and Grindle, M. 1962. Intraracial variation in *Phytophthora infestans* and field resistance to potato blight. Genetica 32:323-338.

Jones, L. R., Giddings, N. J., and Lutman, B. F. 1912. Investigations of the potato fungus *Phytophthora infestans.* Vt. Agric. Exp. Stn. Bull. 168. 100 pp.

Killick, R. J., and Malcolmson, J. F. 1973. Inheritance in potatoes of field resistance to late blight (*Phytophthora infestans* (Mont.) de Bary). Physiol. Plant Pathol. 3:121-131.

Knutson, K. W. 1962. Studies on the nature of field resistance of the potato to late blight. Am. Potato J. 39:152-161.

Lamprecht, M. P., and van Wyk, R. J. 1971. Genetic variation in disease resistant Burley tobacco. Agroplantae 3:31-36. (In Afrikaans, English summary)

Lapwood, D. H. 1961. Laboratory assessments of the susceptibility of potato haulm to blight (*Phytophthora infestans*). Eur. Potato J. 4:117-128.

Lapwood, D. H. 1963. Potato haulm resistance to *Phytophthora infestans.* IV. Laboratory and field estimates compared, and further field analyses. Ann. Appl. Biol. 51:17-28.

Lapwood, D. H., and McKee, R. K. 1966. Dose-response relationships for infection of potato leaves by zoospores of *Phytophthora infestans.* Trans. Br. Mycol. Soc. 49:679-686.

Lawrence, J. S. 1978. Evaluation of methods for assessing resistance of cacao (*Theobroma cacao* L.) cultivars and hybrids to *Phytophthora palmivora* (Butl.). Bol. 62. Tec. Com. Exec. Plano Lavoura Cacaueira.

Leach, S. S., and Rich, A. E. 1969. The possible role of parasexuality and cytoplasmic variation in race differentiation in *Phytophthora infestans.* Phytopathology 59:1360-1365.

Lebedeva, N. A., and Lebedeva, V. A. 1976. Present state of the problem and prospects of breeding *Phytophthora*-resistant potato varieties. Usp. Sovrem. Genet. 6:193-224. (In Russian)

Lishak, I. O., and Nazar, S. G. 1977. Nature of the inheritance of resistance to *Phytophthora* in the hybrid progeny of potato. Kartoplyarstvo 8:9-11. (In Ukrainian)

Long, M., and Keen, N. T. 1977. Genetic evidence for diploidy in *Phytophthora megasperma* var. *sojae.* Phytopathology 67:675-677.

MacKey, J. 1981. Alternative strategies in fungal race-specific parasitism. Theor. Appl. Genet. 59:381-390.

Maga, J. A. 1980. Potato glycoalkaloids. Crit. Rev. Food Sci. Nutr. 12:371-405.

Malcolmson, J. F. 1970. Vegetative hybridity in *Phytophthora infestans.* Nature (London) 225:971-972.

Malcolmson, J. F., and Killick, R. J. 1980. The breeding values of potato parents for field resistance to late blight measured by whole seedlings. Euphytica 29:489-495.

Marks, G. E. 1965. The cytology of *Phytophthora infestans.* Chromosoma 16:681-692.

Mather, K. 1943. Polygenic inheritance and natural selection. Biol. Rev. 18:32-64.

Mosqueda-Vazquez, R. 1978. Resistance to root rot caused by *Phytophthora palmivora* Butl. in *Carica papaya* L.: Screening, heritability, and assessment of growth under nursery and field conditions. Diss. Abstr. Int. B 38:3978-3979.

Müller, K. O. 1925. Neue Wege und Ziele in der Kartoffelzüchtung. Beitr. Pflanzenzuecht. 8:45-72.

Müller, K. O. 1928. Uber die Züchtung krautfäuleresistenter Kartoffelsorten. (Vorläufige Mitteilung.) Z. Pflanzenzuecht. 8:143-156.

Müller, K. O. 1953. The nature of resistance of the potato plant to blight—*Phytophthora infestans.* J. Nat. Inst. Agric. Bot. 6:346-360.

Müller, K. O., and Haigh, J. C. 1953. Nature of field resistance of the potato to *Phytophthora infestans* de Bary. Nature (London) 171:781-783.

National Institute of Agricultural Botany. 1981a. Classified list of potato varieties, England and Wales, 1980/81. Cambridge, Great Britain.

National Institute of Agricultural Botany. 1981b. Recommended varieties of potatoes, 1980/81. Cambridge, Great Britain.

Nelson, R. R. 1975. Horizontal resistance in plants: Concepts, controversies and applications. Pages 1-20 in: Horizontal Resistance to the Blast Disease of Rice. S. Morris, ed. Centro Internacional de Agricultura Tropical (CIAT), Cali, Colombia.

Nelson, R. R. 1978. Genetics of horizontal resistance to plant diseases. Annu. Rev. Phytopathol. 16:359-378.

Netherlands Potato Consultative Institute. 1980. Catalogue of potato varieties in The Netherlands, 1979/80. The Netherlands.

Niederhauser, J. S. 1961. Genetic studies on *Phytophthora infestans* and *Solanum* species in relation to late-blight resistance in the potato. Recent Adv. Bot. 1:491-497.

Niederhauser, J. S. 1962. Evaluation of multigenic "field resistance" of the potato to *Phytophthora infestans* in 10 years of trials at Toluca, Mexico. (Abstr.) Phytopathology 52:746.

Nilsson, B.-A. 1981. Component analysis of general resistance to *Phytophthora infestans* in clones from the Colombian potato collection. Potato Res. 24:239-244.

O'Connor, C. 1933. Potato breeding and resistance to blight. Gard. Chron. 93:104-105.

Parlevliet, J. E. 1979. Components of resistance that reduce the rate of epidemic development. Annu. Rev. Phytopathol. 17:203-222.

Parlevliet, J. E., and Zadoks, J. C. 1977. The integrated concept of disease resistance; a new view including horizontal and vertical resistance in plants. Euphytica 26:5-21.

Pietkiewicz, J. B. 1976. Characteristic of horizontal resistance to blight (*Phytophthora infestans* (Mont.) de Bary) in the potato. Ziemniak 2:87-127.

Pietkiewicz, J. B. 1978. Pathogenic variation among field isolates of *Phytophthora infestans*. Phytopathol. Z. 93:289-294.

Piotrowski, W. 1975. Attempts of determination of the relationships between the types of resistance to *Phytophthora infestans* (Mont.) de Bary and defense reaction of potato plants. Ziemniak 1:67-102.

Pontecorvo, G., and Roper, J. A. 1953. Diploids and mitotic recombination. Adv. Genet. 5:218-233.

Reddick, D. 1928. Blight-resistant potatoes. Phytopathology 18:483-502.

Robinson, R. A. 1976. Plant Pathosystems. Springer-Verlag, Berlin. 184 pp.

Romero, S., and Erwin, D. C. 1969. Variation in pathogenicity among single oospore cultures of *Phytophthora infestans*. Phytopathology 59:1310-1317.

Ross, H. 1958. Kartoffel. Ausgangsmaterial für die Züchtung. Pages 43-58 in: Handbuch der Pflanzenzüchtung. Vol. 3, 2nd ed. H. Kappert and W. Rudorf, eds. Paul Parey, Berlin.

Salaman, R. N. 1929. Genetic studies in potatoes: Abnormal segregation in families arising from the cross *S. utile* × *S. tuberosum*. J. Genet. 20:311-343.

Salaman, R. N. 1931. Recent progress in the breeding of potato varieties resistant to blight (*Phytophthora infestans*.). Pages 436-437 in: Congr. Int. Patol. Comp., 2nd, Paris. Vol. 2, C. R. Commun.

Sansome, E. 1961. Meiosis in the oogonium and antheridium of *Pythium debaryanum* Hesse. Nature (London) 191:827-828.

Sansome, E. 1963a. Cytological observations on *Sclerospora sorghi* in relation to the life history of the Oomycetes. Proc. Sci. Assoc. Nigeria 6:48-52.

Sansome, E. 1963b. Meiosis in *Pythium debaryanum* Hesse and its significance in the life history of the biflagellatae. Trans. Br. Mycol. Soc. 46:63-72.

Sansome, E. 1965. Meiosis in diploid and polyploid sex organs of *Phytophthora* and *Achlya*. Cytologia 30:103-117.

Sansome, E. 1966. Meiosis in the sex organs of the Oomycetes. Pages 77-83 in: Chromosomes Today. Vol. 1. C. D. Darlington and K. R. Lewis, eds. Plenum Press, New York.

Sansome, E. 1976. Gametangial meiosis in *Phytophthora capsici*. Can. J. Bot. 54:1535-1545.

Sansome, E., and Brasier, C. M. 1973. Diploidy and chromosomal structural hybridity in *Phytophthora infestans*. Nature (London) 241:344-345.

Schaper, P. 1951. Die Bedeutung der Inkubationszeit für die Züchtung krautfäuleresistenter Kartoffelsorten. Z. Pflanzenzuecht. 30:292-299.

Schick, R. 1932. Über das Verhalten von *Solanum demissum, Solanum tuberosum* und ihren Bastarden gegenüber verschiedenen Herkünften von *Phytophthora infestans*. (Vorläufige Mitteilung zur Frage der biologischen Spezialisierung von *Phytophthora infestans*. Zuechter 4:233-237.

Schick, R., Möller, K. H., Haussdörfer, M., and Schick, E. 1958. Die Widerstandsfähigkeit von Kartoffelsorten gegenüber der durch *Phytophthora infestans* (Mont.) de Bary hervorgerufenen Krautfäule. Zuechter 28:99-105.

Schöber, B. 1975. Das Auftreten von Pathotypen von *Phytophthora infestans* (Mont.) de Bary in der Bundesrepublik Deutschland. Potato Res. 18:130-132.

Schumann, G., and Thurston, H. D. 1977. Light intensity as a factor in field evaluations of general resistance of potatoes to *Phytophthora infestans*. Phytopathology 67:1400-1402.

Scott, D. H., Draper, A. D., and Maas, J. L. 1976. Mass screening of young strawberry seedlings for resistance to *Phytophthora fragariae* Hickman. HortScience 11:257-258.

Scottish Plant Breeding Station. 1976. 55th Annual Report, 1975–76. Page 23. Edinburgh, Scotland.

Shattock, R. C., and Shaw, D. S. 1976. Novel phenotypes of *Phytophthora infestans* from mixed culture of antibiotic resistant mutants. Trans. Br. Mycol. Soc. 67:201-206.

Shaw, D. S., and Khaki, I. A. 1971. Genetical evidence for diploidy in *Phytophthora*. Genet. Res. 17:165-167.

Simmonds, N. W., and Malcolmson, J. F. 1967. Resistance to late blight in *Andigena* potatoes. Eur. Potato J. 10:161-166.

Sparnaaij, L. D., Garretsen, F., and Bekker, W. 1975. Additive inheritance of resistance to *Phytophthora cryptogea* (Pethybridge and Lafferty) in *Gerbera jamesonii* (Bolus). Euphytica 24:551-556.

Station Fédéral de Recherches Agronomiques. 1980. Liste officielle Suisse des variétés de pommes de terre 1980. Nyan, Switzerland.

Stevenson, F. J., Schultz, E. S., Clark, C. F., Cash, L., and Bonde, R. 1937. Breeding for resistance to late blight in the potato. Phytopathology 27:1059-1070.

Stuart, W. 1905. Disease-resistant potatoes. Pages 135-140 in: Vt. Agric. Exp. Stn. Bull. 115.

Sveriges Potatisodlares Riksförbund. 1980. Potatis 1980. Svensk sortlista för potatis. Stockholm, Sweden.

Świeżyński, K. M., Pietkiewicz, J. B., and Sieczka, M. T. 1974. Inheritance of hypersensitivity to *Phytophthora infestans* and that of resistance to viruses in potato. Genet. Pol. 15:295-304.

Swiszczewska, J., Perz, B., and Piotrowski, W. 1974. Rasy fizjologiczne *Phytophthora infestans* wystepujace na ziemniakach w 1973 r. Z Prac I. Ziem. 3:10-13.

Tai, G. C. C., and Hodgson, W. A. 1975. Estimating general combining ability of potato parents for field resistance to late blight. Euphytica 24:285-289.

Tarn, T. R. 1976. Breeding for resistance to diseases in potato. Proc. Can. Phytopathol. Soc. 43:39.

Taylor, L. M., and Hampson, C. P. 1980. A consumer acceptability survey: What's in a name? Eur. Assoc. Potato Res., Rep. Meeting Sect. Breeding and Varietal Assessment, 23–27 June 1980, Edinburgh. Potato Res. 24:199.

Tazelaar, M. F. 1981. The screening of *Solanum* species for horizontal resistance against late blight (*Phytophthora infestans*) and its use for breeding programmes. (Abstr.) Pages 34-35 in: Abstr. Conf. Pap., 8th Triennial Conf. Eur. Assoc. Potato Res., München.

Thornton, R. E., and Sieczka, J. B., eds. 1980. Commercial potato production in North America. Am. Potato J. Vol. 57, Suppl. 36 pp.

Thurston, H. D. 1971. Relationship of general resistance: Late blight of potato. Phytopathology 61:620-626.

Tisdale, W. B. 1931. Development of strains of cigar wrapper tobacco resistant to blackshank. Fla. Agric. Exp. Stn. Bull. 226.

Toxopeus, H. J. 1956. Reflections on the origin of new physiologic races in *Phytophthora infestans* and the breeding for resistance in potatoes. Euphytica 5:221-237.

Toxopeus, H. J. 1958. Some notes on the relations between field resistance to *Phytophthora infestans* in leaves and tubers and ripening time in *Solanum tuberosum* subsp. *tuberosum*. Euphytica 7:123-130.

Toxopeus, H. J. 1964. Treasure-digging for blight resistance in potatoes. Euphytica 13:206-222.

Turkensteen, L. J. 1973. Partial resistance of tomatoes to *Phytophthora infestans*, the late blight fungus. Agric. Res. Rep. 810. Centrum voor Landbouwpublikaties en Landbouwdocumentatie, Wageningen, The Netherlands. 91 pp.

Umaerus, M. 1980. An integrated breeding program for yield, quality and resistance in potato. Eur. Mediterr. Plant Prot. Org. Bull. 10:385-391.

Umaerus, V. 1960. Iakttagelser rörande fältresistens mot bladmögel (*Phytophthora infestans* (Mont.) de By) hos potatis. Sver. Utsaedesfoeren. Tidskr. (J. Swed. Seed Assoc.) 70:59-89. (English summary)

Umaerus, V. 1967. Rasinventering av potatisbladmögel 1965–66. Potatis 1967:41-43. Sver. Potatisodlares Riksförbund, Stockholm, Sweden.

Umaerus, V. 1969a. Studies on field resistance to *Phytophthora infestans*. 2. A method for screening young potato seedlings for field resistance to *P. infestans*. Z. Pflanzenzuecht. 61:167-194.

Umaerus, V. 1969b. Studies on field resistance to *Phytophthora infestans*. 4. Influence of mineral nutrition. Z. Pflanzenzuecht. 62:357-369.

Umaerus, V. 1970. Studies on field resistance to *Phytophthora infestans*. 5. Mechanisms of resistance and applications to potato breeding. Z. Pflanzenzuecht. 63:1-23.

Umaerus, V., and Lihnell, D. 1976. A laboratory method for measuring the degree of attack by *Phytophthora infestans*. Potato Res. 19:91-107.

van der Plank, J. E. 1963. Plant Diseases: Epidemics and Control. Academic Press, New York. 349 pp.

van der Plank, J. E. 1968. Disease Resistance in Plants. Academic Press, New York. 206 pp.

van der Plank, J. E. 1975. Principles of Plant Infection. Academic Press, New York. 216 pp.

van der Zaag, D. E. 1959. Some observations on breeding for resistance to *Phytophthora infestans*. Eur. Potato J. 2:278-286.

van der Zaag, D. E. 1978. Some remarks on the vertical and horizontal resistance of potatoes to *Phytophthora* and of wheat to yellow rust. Zaadbelangen 32:316-317. (In Dutch)

Vowinckel, O. 1926. Die Anfälligkeit deutscher Kartoffelsorten gegenüber *Phytophthora infestans* (Mont.) De By, unter besonderer Berücksichtigung der Untersuchungsmethoden. Arb. Biol. Reichsanst. Land Forstwirtsch. Berlin-Dahlem 14:588-641. (Rev. Appl. Mycol. 6:47-48)

Wilde, P. 1961. Ein Beitrag zur Kenntnis der Variabilität von *Phytophthora infestans* (Mont.) de Bary. Arch. Mikrobiol. 40:163-195.

Williams, R. L., Young, R. J., and Deahl, K. L. 1975. Role of selected components of the disease cycle of potato late blight in the expression of multigenic resistance to *Phytophthora infestans*. Proc. Am. Phytopathol. Soc. 2:129-130.

Yamamoto, M. 1974. Potato late blight, with special reference to the resistance of potatoes to the invasion of *Phytophthora infestans*. Rev. Plant Prot. Res. 7:45-56.

Young, R. J., Deahl, K. L., and Gallegly, M. E. 1972. Multigenic resistance to late blight in selected potato clones. (Abstr.) Am. Potato J. 49:365.

# 25

# New Developments in Chemical Control of *Phytophthora*

F. J. Schwinn

CONTENTS



## INTRODUCTION

*Phytophthora* species, along with other members of the order Peronosporales, are a major group of plant pathogens on a wide variety of crop plants. Some of the diseases they cause are extremely devastating, and the costs for the chemical control of these pathogens represent about one-third of the total fungicide market (Schwinn, 1979).

Basically, *Phytophthora* spp. cause three different types of diseases (Table 1). In the past, chemical control of the first type, foliar diseases, has depended on the protection of susceptible plant parts. This protection was achieved by applying residual fungicides such as copper compounds, dithiocarbamates, or phthalimides to plant surfaces where infection by these pathogens could be expected. Once the pathogens had penetrated the host tissue, they could no longer be reached by these chemicals, which lack the ability to penetrate plant tissue in fungitoxic concentrations (Staub and Hubele, 1980).

This shortcoming was even more dramatically apparent in the control of diseases of the second type—locally systemic stem, crown, and root diseases—and the third type—local diseases on roots, hypocotyl, or crown. Because these fungicides had only residual activity, they were not taken up by the plant and thus were not translocated within plant tissue or to unprotected plant parts attacked by the pathogens. In the case of soilborne diseases, this deficiency was particularly significant because it forced the farmer to treat large volumes of soil with rather high dosages of fungicides, a practice that was not only expensive but also gave only short-term disease control. For many agricultural crops, for which the application of soil fungicides was uneconomical because of low profitability per surface unit, no economic chemical control measure was apparently available. These farmers relied instead on a wide variety of cultural practices to minimize the risk of soilborne diseases. This chapter reviews the efficacy of several new types of fungicides, which show promise for marked improvements in chemical control.

## THE NEW GENERATION OF FUNGICIDES

Over the past 7 yr, after many years of intensive research by the agrochemical industry, several new chemicals with novel features for the control of diseases caused by *Phytophthora* have been launched into the market. They have been reviewed by Schwinn (1979) and Staub and Hubele (1980). The new chemicals belong to four different chemical classes (Table 2). Although all aim at the control of species within the Peronosporales, they vary in their spectrum of activity within this order (Table 3). In terms of the control of *Phytophthora* spp., Table 3 shows strong variation in activity among the four types of fungicides. The acylalanines have high activity and act on all species of this genus. Because of the superior performance of the acylalanines against *Phytophthora* spp., this review of the efficacy of the new fungicides against diseases caused by *Phytophthora* spp. emphasizes this class of chemicals. The problem of resistance of the target pathogens is also discussed.

### Carbamates

Figure 1 shows the structure of the two carbamates, prothiocarb HCl and propamocarb (Bastiaansen et al, 1974; Pieroh et al, 1978). Schering AG introduced these compounds under the trade names Previcur or Dynone for prothiocarb and Previcur N for propamocarb. They are mainly active as soil treatments against root and stem diseases caused by *Phytophthora* spp. and *Pythium* spp. in ornamentals (Englander et al, 1980; Favreau, 1978; Pieroh et al, 1975),

TABLE 1. Major types of diseases caused by *Phytophthora* spp.

| Type of disease | Site of symptoms | Type and site of infection | Site of survival of pathogen | Typical examples |
|---|---|---|---|---|
| 1 | Leaves | Local; leaves | Plant debris, infected tubers | *P. infestans* on potato |
| | Fruit | Local; fruit | Plant debris, infected pods | *P. palmivora* on cocoa |
| 2 | Shoot, leaves | Locally systemic;[a] roots, shoot, and crown | Plant debris, soil | *P. nicotianae* var. *nicotianae* on tobacco |
| 3 | Roots, hypocotyl | Local; roots, hypocotyl | Infected roots, soil | *P. cinnamomi* on avocado |
| | Crown, leaves | Local; crown | Infected roots, soil | *P. cactorum* on apple |

[a]Locally systemic is used to mean that the disease is not a vascular disease but does spread up the stem from the crown or roots.

TABLE 2. New chemicals for control of *Phytophthora*

| Chemical class | Common name | Trade names and formulations[a] | First report |
|---|---|---|---|
| Carbamate | prothiocarb | Previcur S70 (SCW) | Bastiaansen et al (1974) |
| | propamocarb | Previcur N (SCW) | Pieroh et al (1978) |
| Cyanoacetamide-oxime | cymoxanil[b] | Curzate (WP) | Serres and Carraro (1976) |
| Acylalanine | furalaxyl | Fongarid (WP, G) | Schwinn et al (1977a) |
| | metalaxyl | Ridomil (WP, G); Acylon (WP); Apron (SD) | Urech et al (1977) |
| | milfuram | Patafol, Caltan (WP) | Lukens et al (1978) |
| | benalaxyl | Galben (WP, G) | Bergamaschi et al (1981) |
| Ethyl phosphite | phosethyl Al[b] | Aliette (WP) | Bertrand et al (1977), Williams et al (1977) |

[a] SCW = water-soluble concentrate, WP = wettable powder, G = granule, SD = seed dressing.
[b] Proposed common name submitted for approval to the International Organization for Standardization (ISO).

certain vegetable crops, and tobacco (McIntyre and Lukens, 1977). They also have potential as a flower bulb seed treatment. Prothiocarb seems to be less effective than propamocarb for the control of foliar diseases (Ryan, 1977).

Prothiocarb is translocated apoplastically in plants (Ryan, 1977); no evidence has been found for symplastic transport (Iwan and Goller, 1975). Although acropetal transport has been established, the roots seem to present a substantial barrier to the uptake of prothiocarb (Kluge, 1978), which may explain the rather limited curative action of these carbamates. In laboratory experiments on artificial media, prothiocarb exhibits its full efficacy only at neutral pH of the substrate and, against some *Phytophthora* spp., only at temperatures around 10° C (Kerkenaar and Kaars Sijpesteijn, 1977). In the field, its activity lasts longest at slightly acid pH values (Anonymous, 1978). For fungi within the Peronosporales, the toxic entity of both prothiocarb and propamocarb appears to be the whole molecule, the action of which can be reversed by sterols (Kaars Sijpesteijn et al, 1974; Papavizas et al, 1978). Details of the biochemical mode of action are not known.

TABLE 3. Activity of the new fungicides against genera within the Oomycetes[a]

| Genus of the pathogen | Fungicide and activity[b] Prothiocarb, propamocarb | Cymoxanil | Phosethyl Al | Furalaxyl, metalaxyl, milfuram, benalaxyl |
|---|---|---|---|---|
| *Aphanomyces* | ± | ? | ? | − |
| *Pythium* | ± | ± | + | + |
| *Phytophthora* | | | | |
| On roots and stems | + | − | + | + |
| On foliage | − | + | − | + |
| *Peronospora* | | | | |
| On stems | + | + | ± | + |
| On foliage | ? | + | + | + |
| *Pseudo-peronospora* | + | + | + | + |
| *Plasmopara* | − | + | + | + |
| *Bremia* | + | + | + | + |
| *Sclerospora* | ? | ? | + | + |
| *Perono-sclerospora* | ? | ? | ± | + |
| *Sclerophthora* | ± | − | + | + |
| *Albugo* | ? | ? | ? | + |

[a] Derived from the references cited in this chapter.
[b] Activity: + = high; ± = varies from species to species; − = no useful activity; ? = no information.

## Cyanoacetamide-Oxime

Shortly after the introduction of the carbamates, cymoxanil was announced (Denis, 1976; Richards and Delp, 1976; Serres and Carraro, 1976). It has been commercially available (in Europe and certain other areas outside the United States) since 1979 (trade name Curzate, E. I. duPont de Nemours & Co., Inc.). The structure of cymoxanil is shown in Figure 1. This fungicide has selective activity against fungi in the Peronosporales. However, the sensitivity to cymoxanil of the genera in this order varies, as shown in Table 3. Differences in

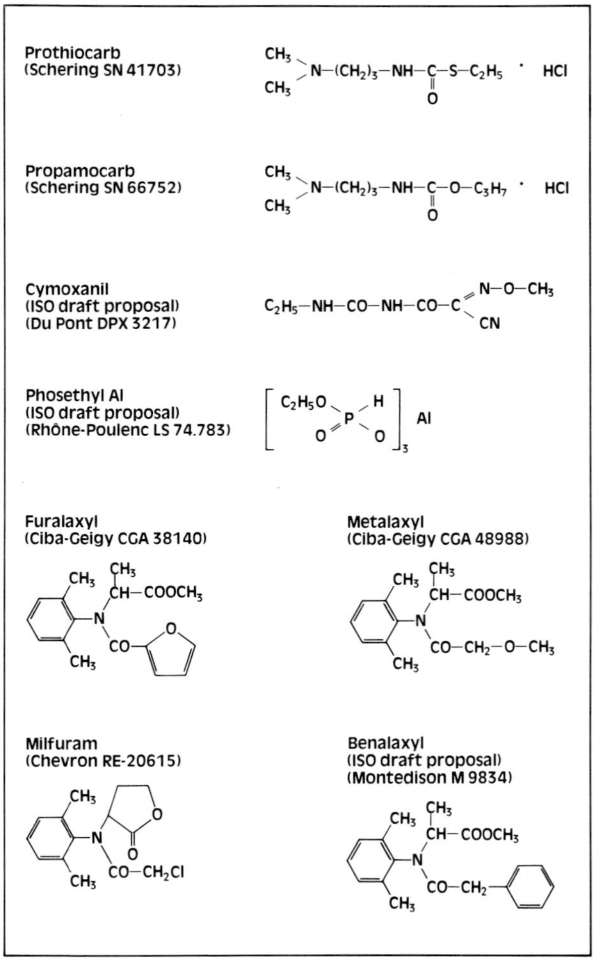


**Fig. 1.** Chemical structures of new fungicides for *Phytophthora* control.

sensitivity among species within a single genus can also be observed. For example, *P. infestans* is more than 100 times more sensitive in vitro than other *Phytophthora* spp. (Richardson, 1976). Therefore, late blight of potato and tomato is the only disease caused by a *Phytophthora* sp. against which cymoxanil is recommended (Douchet et al, 1977).

Cymoxanil has both contact and local systemic activity. Under practical conditions, it does not have sufficient persistence to be useful alone because its protective action is lost within a few days; however, it has a strong curative effect for 3–4 days after infection, depending on the length of the incubation period. It is used at low dosages (10–12 g a.i./100 L) in combination with reduced rates of conventional protective compounds, such as copper, folpet, or mancozeb. These admixtures have a synergistic effect and improve the level and duration of protective activity (Klopping and Delp, 1980; Serres and Carraro, 1976). The mode of action of cymoxanil is not yet known.

Cymoxanil provides improved control of foliar *Phytophthora* diseases, mainly late blight caused by *P. infestans*; however, it does not control either the locally systemic or the root-infecting *Phytophthora* diseases.

### Ethyl Phosphites

About 10 yr ago, ethyl phosphites were discovered by Rhône-Poulenc. This class of fungicides represented a new type of chemical with quite unusual biological properties. The most efficacious compound within this class is phosethyl Al (Fig. 1, Table 2), which was first described by Bertrand et al (1977) and Williams et al (1977).

Phosethyl Al has high fungitoxic activity on most species in the Peronosporales (Table 3); however, some species, such as *P. infestans,* are relatively insensitive. On the other hand, some pathogens, such as *Guignardia* spp., *Alternaria* spp., and *Cercospora* spp., belonging to taxonomic groups other than the Peronosporales, are also controlled (Bertrand et al, 1977; Chalazet et al, 1977). From these findings, one may conclude that the spectrum of activity may be determined, at least partially, by the host plant or by the interaction between the host plant and the chemical. Phosethyl Al has relatively weak in vitro activity; $ED_{50}$ values for the inhibition of mycelial growth are on the order of 1,000–3,000 ppm (Hai et al, 1979; Vegh et al, 1977).

Because phosethyl Al controls diseases under field conditions at dosages of 2.5–5 g a.i./L of water, more toxic transformation products might be formed in the host plant (Molot and Beyries, 1977). However, no evidence for such products has been found to date. Instead, chemical plant defense reactions stimulated by the chemical have been reported (Hai et al, 1979; Raynal et al, 1980), indicating an indirect mode of action on the pathogen. The biochemical mode of action of phosethyl Al is still unknown.

Phosethyl Al is the first commercial fungicide that is truly systemic. It is translocated by both acropetal and basipetal means after foliar application (Bertrand et al, 1977). This results in curative action and leads to temporary protection of new leaves formed after application. However, this effect appears to be obtained at the expense of the concentration in the sprayed leaves, which decreases the protective effect in these leaves. In view of this and other factors, phosethyl Al is recommended against foliar diseases in admixtures with residual fungicides, such as folpet or mancozeb. Such combinations show synergistic effects that allow an extension of the standard spray intervals (Chalazet et al, 1977; Lafon et al, 1977; Marais and van der Walt, 1978).

In a wide variety of crops, such as avocado, pineapple, and certain ornamentals, basipetal transport after foliar treatment is sufficient to fully protect the root system and the stem base against attack by soilborne *Phytophthora* spp. such as *P. cinnamomi* and *P. nicotianae* var. *parasitica* (Bertrand et al, 1977; Farih et al, 1981b; Hill and Piramarn, 1982; Williams et al, 1977; Zentmyer and Ohr, 1978). As an example, the efficacy of foliar application of phosethyl Al against stem canker of citrus caused by *P. citrophthora* is shown in Table 4. The versatility of phosethyl Al, derived from its full systemic nature and high curative activity, is demonstrated in the experiment described in Table 5, in which *Phytophthora* stem canker and root rot were controlled by foliar, drench, or paint treatments. In some crops, foliar application proved to be unsatisfactory in effectiveness and more risky with regard to crop tolerance (Bertrand et al, 1977; Vegh et al, 1977). Even though root uptake and upward translocation can easily be demonstrated, the process seems less efficient in some ways than downward transport (Ryan, 1977). This may explain the limited activity of phosethyl Al against the locally systemic stem and root disease caused by the soilborne *P. nicotianae* var. *nicotianae* on tobacco.

The introduction of phosethyl Al has markedly improved the arsenal of weapons against *Phytophthora* diseases of foliage, fruits, stems, and roots. This chemical is particularly valuable for the control of certain soilborne species in annual fruit crops, perennial tree crops, and both annual and perennial ornamental crops (Table 6) that were previously not practically amenable to chemical control (Beach, 1979; Chalandon et al, 1980; Frossard, 1978; Frossard et al, 1977; Hill and Piramarn, 1982; Laville, 1979).

### Acylalanines and Related Compounds

The acylalanines are a novel class of fungicides discovered by Ciba-Geigy. They have strong protective and curative activity; are taken up rapidly from foliage, shoots, and roots; and are translocated upward through the apoplastic system in the plant (Schwinn et al, 1977a; Staub et al, 1978). Basipetal

TABLE 4. Foliar application of phosethyl Al for control of stem canker (*Phytophthora citrophthora*) of citrus

| Treatment | Dosage (g a.i./L) | No. of treatments | Spray interval (days) | Tree mortality (%) |
|---|---|---|---|---|
| Phosethyl Al | 2.0 | 4 | 14 | 0 |
| | 2.0 | 2 | 28 | 0 |
| Untreated | 0 | 0 | ··· | 54 |

*Source:* Williams et al (1977).

TABLE 5. Curative control of stem canker and root rot (*Phytophthora palmivora*) in durian by application of phosethyl Al in Thailand, 1978–1979

| Application method | Rate (g a.i./L) | Disease recovery index[a] at time after application | | |
|---|---|---|---|---|
| | | 16 wk[b] | 12 wk[b] | 16 wk[b] |
| Trunk paint | 4.0 | 1.2 | 0.8 | 1.1 |
| Soil drench | 2.0 | 1.0 | 0.8 | ··· |
| Spray (foliar) | 2.0 | 2.7 | 1.3 | ··· |
| Spray (overall) | 2.0 | 1.5 | 1.7 | 2.8 |
| Untreated | ··· | 1.0 | 0.5 | ··· |

[a]Disease recovery index: 0 = no recovery; 4 = complete recovery.
[b]Trials 1, 2, and 3, respectively.

*Source*: Hill and Piramarn (1982).

phloem transport seems to be less pronounced (Staub et al, 1978), although in some crops—for example, in avocado—high dosages applied to foliage in pot tests indicated that downward translocation might be a significant factor in disease control (Zaki et al, 1981; Zentmyer and Ohr, 1978).

The acylalanines are highly effective both in vitro and in vivo against all fungal pathogens in the Peronosporales. Their spectrum of activity is so strongly correlated with this taxonomic unit that it might be regarded as further evidence for the validity of the unit. So far, the most effective molecule in this class is metalaxyl (Fig. 1, Table 2). Its general chemical, toxicological, and biological properties have been described (Bruck et al, 1980; Cohen et al, 1979; Farih et al, 1981a; Schwinn and Staub, 1982; Schwinn et al, 1977b; Urech et al, 1977). Another member of this class, furalaxyl (Fig. 1, Table 2), is particularly suited for use with ornamentals (Wiertsema and Wissink, 1977).

TABLE 6. Plants with root and crown rot diseases caused by *Phytophthora* spp. and previously not amenable to chemical control

| Pathogen | Hosts |
|---|---|
| *P. cactorum* | Apple, pear, strawberries |
| *P. capsici* | Pepper |
| *P. cinnamomi* | Avocado, eucalyptus, ornamentals, cinchona |
| *P. citrophthora* | Citrus |
| *P. fragariae* | Strawberries |
| *P. nicotianae* var. *parasitica* | Vegetables, pineapple, citrus |
| *P. nicotianae* var. *nicotianae* | Tobacco |

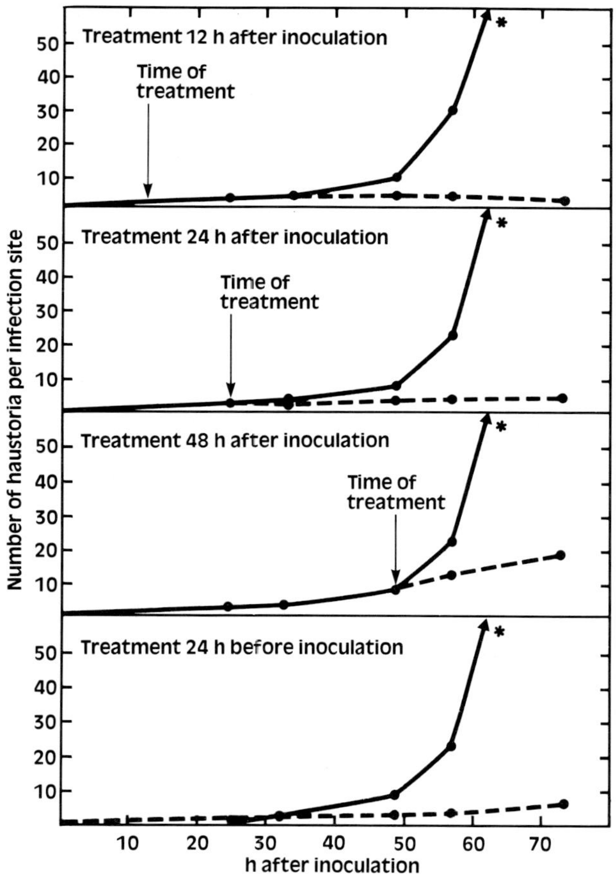


**Fig. 2.** Effect of preinfection and postinfection treatments with metalaxyl on haustoria formation of *Plasmopara viticola* on grape seedling leaves. Dotted lines represent treatment with 125 μg/ml metalaxyl. Solid lines represent the control (check); asterisks denote 158 haustoria 72 hr after inoculation. (Reprinted, by permission, from Staub et al, 1980)

Recently, new derivatives, namely milfuram—a butyrolactone (Kaspers and Reuff, 1979; Lukens et al, 1978)—and benalaxyl (Bergamaschi et al, 1981) have been developed by Chevron and by Farmoplant Montedison, respectively. Milfuram is being introduced into the market for control of foliar *Phytophthora* pathogens in admixtures with conventional fungicides. Because furalaxyl, milfuram, and benalaxyl have properties basically similar to those of metalaxyl but are less active in their overall fungicidal performance, they will not be treated further in detail.

Metalaxyl is the most active and versatile of the new fungicides for the control of *Phytophthora* diseases. It can be applied as a foliar spray at low rates (200–250 g a.i./ha) and in extended spray intervals, as a granule (2 g a.i./$m^2$, one to two applications per season), or as a seed dressing (175 g a.i./ha), depending on the nature of the specific disease problem (Margot, 1982; Schwinn, 1979; Urech et al, 1977). Applied as a spray, metalaxyl has remarkable gas-phase activity (Lafon et al, 1978). When used as a soil or a seed treatment, it controls locally systemic stem and root diseases, such as tobacco black shank, caused by *P. nicotianae* var. *nicotianae* (Kannwischer and Mitchell, 1978; Mitchell et al, 1977; Morton et al, 1978).

In contrast to conventional protective fungicides, which inhibit spore germination on the plant surface and prevent the release of zoospores from sporangia, the acylalanines inhibit the target fungi only after they have penetrated the leaf and have formed at least one haustorium. In studies with metalaxyl (Staub et al, 1980), the number of haustoria in treated leaves remained close to zero 72 hr after inoculation, whereas in the untreated check, 140 haustoria had developed per observation field. Curative foliar treatments applied up to 24 hr after inoculation had the same effect as protective treatments (Fig. 2).

The sites of action of metalaxyl in the disease cycle of a foliar pathogen like *P. infestans* can be summarized and contrasted with those of a residual fungicide like maneb as follows: metalaxyl inhibits formation of secondary haustoria and mycelial growth inside the leaf, formation of lesions, and sporulation, whereas maneb inhibits zoospore release from sporangia, zoospore germination, and initial penetration (Bruck et al, 1980; Staub et al, 1980).

When the sites of action are plotted on the time axis of the disease cycle, it becomes evident that metalaxyl interferes with fungal development during a much longer part of the disease cycle than maneb, because of its rapid penetration into the host tissue. Metalaxyl is effective against all the developmental stages of the disease within the host plant, whereas a residual surface fungicide interferes only with the initial phase, i.e., zoospore release from the sporangium or direct germination of the sporangia.

For crops such as potatoes, tomatoes, and vegetables on which other foliar diseases occur simultaneously with *Phytophthora* diseases, the acylalanines are more useful in mixtures with other fungicides, which broadens the spectrum of activity (Urech et al, 1977). Similarly, such mixtures improve the level of disease control on senescing foliage, on which the acylalanines are less effective, for unknown reasons, than on young tissue. In addition, mixtures are considered a useful tool to reduce the risk of the development of resistance in populations of *Phytophthora* spp.

Control by the acylalanines of diseases caused by soilborne *Phytophthora* spp. has been reported by several authors (Benson, 1979; Englander et al, 1980; Farih et al, 1981b). The effect of metalaxyl on *P. cinnamomi* and *P. nicotianae* var. *parasitica* as studied by Margot (1982) is a pertinent example. In vitro, metalaxyl inhibits mycelial growth and chlamydospore and sporangium formation (Table 7). The concentrations required for inhibition of sporangium formation are 25 times lower, and those for inhibition of chlamydospore formation are 100 times lower, than those required for inhibition of mycelial growth. From these results, we could predict that the life cycle of the pathogen is blocked in several stages, resulting in a suppression of chlamydospores and sporangia and, consequently, in reduction of zoospores. In bioassays, using a modified lupine baiting technique (Chee and Newhook, 1965), Margot (1982) monitored the recovery of these two spore types from soil under diseased citrus and avocado trees following application of metalaxyl granules. The results indicate that the period of inhibition of both chlamydospore production and zoospore release is proportional to the application rate and that the optimal rate is 2.0–2.5 g a.i./$m^2$. Activity was maintained for 20 wk. Wide use in practice has confirmed that root rot of avocado can now be effectively controlled by two granular applications of metalaxyl per year at 2.5 g a.i./$m^2$.

The curative action of metalaxyl even in a perennial crop with woody tissue is demonstrated in Table 8, which shows the results of a field test on citrus trees with established collar rot symptoms caused by *P. nicotianae* var. *parasitica*. After five granular treatments at 3-mo intervals, the size of the lesions was reduced and the trees had recovered. For protective treatment, two applications at a rate of 2.0–2.5 g a.i./$m^2$ spaced 20–25 wk apart gave full protection against zoospore infection.

For the black shank pathogen of tobacco, *P. nicotianae* var. *nicotianae*, the release of zoospores from sporangia and their motility and aggregation around tobacco root tips were not inhibited by metalaxyl (Staub and Young, 1980). Inhibitory effects were shown only after the pathogen penetrated the host tissue. Metalaxyl blocks the following steps in the disease cycle: sporangia formation from chlamydospores, growth of mycelium inside the roots, formation of sporangia on roots, and chlamydospores in infected roots.

Crute (1979) and Cohen et al (1979) reported necrotic reactions on metalaxyl-treated lettuce susceptible to *Bremia lactucae* and tomato plants susceptible to *P. infestans*. These reactions were indistinguishable from the reactions normally observed on untreated resistant cultivars. This effect may be direct, by stimulation of the resistance mechanisms in the host, or indirect, by slowing of the fungal growth enough to allow the operation of natural resistance mechanisms, even in susceptible hosts. In ultrastructural studies, Hickey and Coffey (1980) showed that metalaxyl stimulates the formation of encasements around fungal haustoria, which may be interpreted as an induced response of the host plant. Recent findings of Ward et al (1980), derived from studies of *P. megasperma* var. *sojae* on soybeans, indicate that metalaxyl strongly increases the phytoalexin content of compatible host cultivars. In studies on the biochemical mode of action, Fisher and Hayes (1979, 1982) showed that acylalanines had little effect on respiration, membrane permeability, or cell wall formation, whereas DNA, protein synthesis, and above all RNA were reduced. Kerkenaar (1981) and Davidse (1981b) concluded that interference with RNA synthesis might be a possible mode of action.

## THE RISK OF FUNGICIDE RESISTANCE

In view of the fact that some fungal pathogens have developed resistance to several of the newer fungicides such as the benzimidazoles and the pyrimidines, this aspect should be reviewed for the new fungicides active against *Phytophthora*. Up to now, no indications have been found of development of strains resistant to prothiocarb or propamocarb, cymoxanil, or phosethyl Al. However, resistance to the acylalanines has been noted in some species.

In preliminary studies with acylalanines and related compounds (Bruck et al, 1980; Lukens et al, 1978; Staub et al, 1979), in vitro resistance was easily found by selection or adaptation procedures, but in vivo resistance did not readily develop in the species of oomycetes studied, because all strains resistant in vitro had either lost their pathogenicity or maintained their sensitivity to the chemicals when exposed to them on the host plant. However, in investigations using mutagens (Bruin, 1980; Bruin and Edgington, 1980; Davidse, 1980, 1981a), strains of *P. capsici* and *P. megasperma* f. sp. *medicaginis* were found with both in vitro and in vivo resistance and with unchanged pathogenicity. Also, cross-resistance to other acylalanines and related structures such as milfuram was demonstrated (Bruin and Edgington, 1980).

Toward the end of 1979, the first occurrences of failure of metalaxyl under field conditions against cucumber downy mildew were reported from Israel and from the island of Crete. Analysis of isolates from these areas indicated that metalaxyl-resistant populations of *Pseudoperonospora cubensis* were present in the plots (Katan and Bashi, 1981; Malathrakis, 1980; Pappas, 1980; Reuveni et al, 1980). Resistance developed surprisingly quickly. Its first appearance under plastic tunnels used in Israel and Crete was probably caused by the high infection and selection pressure, which provided ideal conditions for the development of resistant strains.

During 1980, a more serious development occurred on potatoes in several Western European countries, especially The Netherlands (Davidse et al, 1981) and Ireland (M. D. Coffey, *personal communication*). *P. infestans* isolates from the problem fields showed that metalaxyl sensitivity in these populations decreased over 100-fold compared to sensitive reference isolates. Davidse et al (1981) suggested that

TABLE 7. In vitro activity of metalaxyl against two soilborne *Phytophthora* spp.

| Growth stage | Toxicity ($EC_{90}$ in $\mu$g/ml) | |
|---|---|---|
| | *P. nicotianae* var. *parasitica* | *P. cinnamomi* |
| Mycelial growth | 5.0 | 2.5 |
| Sporangium formation | 0.2 | 0.2 |
| Chlamydospore formation | 0.05 | 0.2 |

*Source*: Margot (1982).

TABLE 8. Curative effects of metalaxyl on collar lesions caused by *Phytophthora nicotianae* var. *parasitica* on citrus in South Africa

| Treatment | Rate[a] (g a.i./$m^2$) | Collar lesions[b] at Start of trial | 8 mo | 25 mo | 30 mo |
|---|---|---|---|---|---|
| Control | ··· | 3.3 | 3.5 | 4.5 | 5.0 |
| Metalaxyl 5G[c] | 1.25 | 3.3 | 3.0 | 2.7 | 2.2 |
| | 6.25 | 3.3 | 3.2 | 2.5 | 1.5 |
| LSD[d] | | | | 0.8 | 1.4 |

[a] Five applications were made at 3-mo intervals.
[b] 0 = no lesions; 1 = 10% ring barked; 2 = 11–25%; 3 = 26–50%; 4 = 51–75%; and 5 = 76–100% ring barked.
[c] 5G = 5% active ingredient in granules.
[d] Least significant difference at the 5% level.

*Source*: Margot (1982).

widespread use of metalaxyl (half of the fields in The Netherlands) and weather favorable to late blight were associated with the sudden appearance of metalaxyl-resistant strains of *P. infestans*.

The cases in which resistance to metalaxyl has been found have a number of features in common. First, all have occurred against foliar diseases under high disease pressure and on highly susceptible cultivars. Second, resistance has developed only where metalaxyl was used alone and continuously over a long period of time; it has not occurred when metalaxyl has been mixed with a conventional fungicide. Third, resistance has been associated with the application of metalaxyl mainly as a curative treatment, or spray intervals have extended beyond those recommended, or both. Fourth, outbreaks have appeared simultaneously in geographically separate places; however, the epidemiologic conditions and product use patterns have been similar.

All these findings clearly indicate a risk that resistance to acylalanines will occur more generally. The risk might be decreased or, even better, eliminated by the development of more suitable strategies for the use of the acylalanines and for the enforcement of these strategies in the fungicide market (Schwinn, 1979; Staub and Schwinn, 1980). For both theoretical and practical reasons, the companies that produced these fungicides have introduced prepackaged mixtures of acylalanines and conventional fungicides such as dithiocarbamates or phthalimides for use against foliar pathogens. Similar packages have not been developed to combat soilborne pathogens because the risk of resistance developing is considered to be low, partly because spore populations are smaller and the chance of resistant individuals in a fungal population being disseminated over a wide area in the soil is relatively slight. In spite of fairly widespread use of acylalanines against soilborne pathogens, so far resistance has not been reported in the field; however, its appearance remains a possibility.

The commercial companies must depend on the cooperation of registration officials and crop protection advisors to implement the use strategy of mixing acylalanines with other protective fungicides (Schwinn, 1979). The future availability of conventional fungicides such as the dithiocarbamates and the registration of fungicide mixtures as a tool to prevent resistance are key issues. Only with a joint effort will it be possible to successfully contain the threat of resistance and thus ensure the continued availability of the acylalanine fungicides as potent weapons against plant diseases caused by *Phytophthora*.

TABLE 9. Main uses of the new fungicides active for control of *Phytophthora*

| Fungicide | Main target crops | Major use |
|---|---|---|
| Prothiocarb | Ornamentals | As a soil treatment against diseases infecting roots and stems |
| Propamocarb | Ornamentals, vegetables | As a soil treatment against diseases infecting roots and stems |
| Cymoxanil | Potatoes, tomatoes | In spray applications against foliar diseases |
| Phosethyl Al | Vegetables, fruit crops, ornamentals | In foliar or soil applications against diseases infecting foliage, roots, and stems |
| Furalaxyl | Ornamentals | As a soil treatment against diseases infecting roots and stems |
| Metalaxyl | Field crops, vegetables, fruit crops, tobacco | In foliar, soil, or seed applications against diseases infecting foliage, roots, and stems |
| Milfuram, benalaxyl | Field crops | Against foliar diseases |

## CONCLUSION AND OUTLOOK

The main uses of the new fungicides effective against *Phytophthora* are summarized in Table 9. These new fungicides offer novel possibilities for disease control. Their advantages can be summarized as follows.

First, in the control of foliar diseases such as those caused by *P. infestans,* the new fungicides offer greater flexibility of use because of their high fungitoxic activity, quick uptake by leaves and green stems after foliar application or by roots after soil application, apoplastic and/or sometimes symplastic transport, and curative activity. These characteristics protect the fungicide from being washed off by rain, promote better distribution of the chemical in the host tissue, and provide internal protection of the new growth, all of which allow a reduction in the number of spray applications required per growing season. The stronger an epidemic of *P. infestans* that develops and the more applications of conventional fungicides that become necessary, the more the advantages of the new fungicides become evident. Under severe disease pressure, the number of applications can be reduced by up to 30%. Despite this reduction, the level of control of foliage disease is higher than with conventional fungicides and, above all, the control of tuber attack is better.

Second, the acylalanines and phosethyl Al open new routes for control of *Phytophthora* diseases that cause root, crown, and stem rots. Here, the high fungitoxic activity, duration of protection, favorable mobility pattern in soil and plants, quick uptake by roots (and, in the case of phosethyl Al, downward transport after foliar application), and ease of application (granules, drenches) make these products attractive to the user. Considering the broad distribution of this type of pathogen on annual and perennial crops in temperate and tropical zones, the severity of the diseases caused by them, and their former inaccessibility to chemical control, the possibilities for progress seem great. Similarly, the control of fruit rots caused by *P. palmivora* on cocoa and *P. cactorum* on apples, strawberries, and pears has been improved.

Third, the chemical control of systemic *Phytophthora* spp., such as *P. nicotianae* var. *nicotianae,* the causal organism of tobacco black shank, has traditionally been difficult and only marginally successful. Drench application of metalaxyl has been shown to be a simple and effective approach to the control of tobacco black shank.

The new fungicides, particularly the acylalanines, are in many respects well adapted for integrated control strategies. For example, they are attracting increasing attention from epidemiologists, who are developing forecasting systems for the integrated control of epidemics, such as of *P. infestans* on potatoes (Bruck et al, 1979). One major problem in the past was the lack of flexible and reliable fungicides for use in such systems. By their curative action, the new fungicides affect the development of the pathogen immediately after application, whereas conventional protectants are effective only after a time lag of at least one incubation period (Fry et al, 1979). Thus, the new chemicals permit more effective use of predictive systems for spray recommendations. On the other hand, full reliance on the curative potential of a fungicide is certainly critical, because it calls for very precise observation of the disease situation, a high level of understanding of the epidemiologic factors by the farmer, and suitable equipment for quick and accurate application of the fungicide. However, the need for mixing the acylalanines with conventional fungicides to counteract the development of resistant populations of *Phytophthora* spp. limits the full exploration of the potential of the acylalanines. Therefore, well-balanced, realistic recommendations that take all these aspects into account and offer enough safety to the farmer must be developed.

The systemic properties of the new products also offer additional flexibility in spray coverage. A continuous and uniform fungicide film over all plant parts is no longer necessary for protection, because these chemicals are redistributed rather predictably throughout the plant organs. This internal distribution is especially striking with the phloem-transported fungicides such as phosethyl Al.

The suitability of the new fungicides for use in connection with disease warning systems; i.e., more precise timing of their use in conjunction with low application rates and extended spray intervals, should stimulate the development of new disease control strategies with fewer hazards due to ecological and human exposure and with lower energy requirements.

In this context, the use of resistant cultivars and the use of fungicides should not be thought of as divergent and mutually exclusive methods of disease control but rather as complementary or synergistic approaches. The combined use of natural disease resistance and fungicides has recently been further developed by Fry (1977), and it merits renewed attention in connection with the new systemic fungicides described here. In my view the argument for an integrated approach to disease management is strengthened by the appearance of the new fungicides for controlling *Phytophthora;* however, greater involvement by more researchers in developing integrated control concepts is clearly needed.

## LITERATURE CITED


Anonymous. 1978. Previcur N. Fungizid. 1. Mitteilung Schering AG, Berlin. 29 pp.

Bastiaansen, M. G., Pieroh, E. A., and Aelbers, E. 1974. Prothiocarb, a new fungicide to control *Phytophthora fragariae* in strawberries and *Pythium ultimum* in flower bulbs. Meded. Fac. Landbouwwet. Rijksuniv. Gent 39:1019-1025.

Beach, B. G. W. 1979. The control of various *Phytophthora* diseases in tropical crops with aluminum tris (ethyl phosphonate). Br. Crop Prot. Conf., 10th 2:319-329.

Benson, D. M. 1979. Efficacy and in vitro activity of two systemic acylalanines and ethazole for control of *Phytophthora cinnamomi* root rot of azalea. Phytopathology 69:174-178.

Bergamaschi, P., Borsari, T., Garavaglia, C., and Mirenna, L. 1981. Methyl N-phenyl-acetyl-N-2,6-xylyl-dl-alaninate (M9834), a new systemic fungicide controlling downy mildew and other diseases caused by Peronosporales. Br. Crop Prot. Conf., 11th 1:11-18.

Bertrand, A., Ducret, J., Debourge, J.-C., and Horrière, D. 1977. Etude des propriétés d'une nouvelle famille de fongicides: Les monoethyl phosphites métalliques. Charactéristiques physicochimiques et propriétés biologiques. Phytiatr. Phytopharm. 26:3-18.

Bruck, R. I., Fry, W. E., and Mundt, C. C. 1979. Differential effects of protectant and systemic fungicides on potato late blight epidemics induced by *Phytophthora infestans.* (Abstr.) Int. Congr. Plant Prot., 9th. Washington, DC.

Bruck, R. I., Fry, W. E., and Apple, A. E. 1980. Effect of metalaxyl, an acylalanine fungicide, on developmental stages of *Phytophthora infestans.* Phytopathology 70:597-601.

Bruin, G. C. A. 1980. Resistance in Peronosporales to acylalanine-type fungicides. Ph.D. thesis. Univ. Guelph, Ontario, Canada. 110 pp.

Bruin, G. C., and Edgington, L. V. 1980. Induced resistance to Ridomil of some oomycetes. (Abstr.) Phytopathology 70:459-460.

Chalandon, A., Crisinel, P., and Paviot, J. 1980. Utilisation du Phosethyl Al (Aliette®) pour la lutte contre le *Phytophthora fragariae* Hickm. et le *Phytophthora cactorum* L. et C. du fraisier. Meded. Fac. Landbouwwet. Rijksuniv. Gent 45:207-219.

Chalazet, M., Crisinel, P., Horrière, D., and Thiollière, J. 1977. Résultats de l'expérimentation réalisée avec les ethyl phosphites contre différentes maladies de la vigne. Phytiatr. Phytopharm. 26:41-54.

Chee, K.-H., and Newhook, F. J. 1965. Improved methods for use in studies on *Phytophthora cinnamomi* Rands and other *Phytophthora* species. N.Z. J. Agric. Res. 8:88-95.

Cohen, Y., Reuveni, M., and Eyal, H. 1979. The systemic antifungal activity of Ridomil against *Phytophthora infestans* on tomato plants. Phytopathology 69:645-649.

Crute, I. R. 1979. Lettuce mildew—Destroyer of quality. Agric. Res. Counc. (U.K.) Res. Rev. 5:9-12.

Davidse, L. C. 1980. Resistance to Ridomil in *Phytophthora megasperma* f. sp. *medicaginis.* (Abstr.) Acta Bot. Neerl. 29:216.

Davidse, L. C. 1981a. Resistance to acylalanine fungicides in *Phytophthora megasperma* f. sp. *medicaginis.* Neth. J. Plant Pathol. 87:11-24.

Davidse, L. C. 1981b. Mechanism of action of metalaxyl in *Phytophthora megasperma* f. sp. *medicaginis.* Neth. J. Plant Pathol. 87:254-255.

Davidse, L. C., Looijen, D., Turkensteen, L. J., and van der Wal, D. 1981. Occurrence of metalaxyl-resistant strains of *Phytophthora infestans* in Dutch potato fields. Neth. J. Plant Pathol. 87:65-68.

Denis, S. J. 1976. DPX-3217 for control of grape downy mildew and potato/tomato late blight. (Abstr.) Proc. Am. Phytopathol. Soc. 3:324.

Douchet, J. P., Absi, M., Hay, S. J. B., Muntan, L., and Villani, A. 1977. European results with DPX 3217, a new fungicide for the control of grape downy mildew and potato late blight. Br. Crop Prot. Conf., 9th 2:535-540.

Englander, L., Merlino, J. A., and McGuire, J. J. 1980. Efficacy of two new systemic fungicides and ethazole for control of Phytophthora root rot of rhododendron, and spread of *Phytophthora cinnamomi* in propagation benches. Phytopathology 70:1175-1179.

Farih, A., Tsao, P. H., and Menge, J. A. 1981a. In vitro effects of metalaxyl on growth, sporulation, and germination of *Phytophthora parasitica* and *P. citrophthora.* Plant Dis. 65:651-654.

Farih, A., Menge, J. A., Tsao, P. H., and Ohr, H. D. 1981b. Metalaxyl and efosite aluminum for control of Phytophthora gummosis and root rot on citrus. Plant Dis. 65:654-657.

Favreau, J. 1978. Lutte contre *Phytophthora* et *Pythium* en culture ornamentale avec le Prévicur. L'hortic. Française 9:7-8.

Fisher, D. J., and Hayes, A. L. 1979. Fungicides, and mode of action. Phycomycete fungicides. Page 137 in: Annu. Res. Rep., Long Ashton Res. Stn., Bristol, England.

Fisher, D. J., and Hayes, A. L. 1982. Mode of action of the systemic fungicides furalaxyl, metalaxyl and ofurace. Pestic. Sci. 13:330-339.

Frossard, P. 1978. Lutte contre la pourriture du coeur à *Phytophthora* de l'ananas. Fongicides classiques et fongicides nouveaux. Fruits 33:183-191.

Frossard, P., Haury, A., and Laville, E. 1977. Résultats préliminaires concernant l'activité de l'éthyl phosphite d'aluminium (LS 74.783) sur les maladies à *Phytophthora* des agrumes, de l'avocatier et de l'ananas. Phytiatr. Phytopharm. 26:55-62.

Fry, W. E. 1977. Fungicides in perspective. Pages 19-50 in: Antifungal Compounds, Vol. 1. M. R. Siegel and H. D. Sisler, eds. Marcel Dekker, New York.

Fry, W. E., Bruck, R. I., and Mundt, C. C. 1979. Retardation of potato late blight epidemics by fungicides with eradicant and protectant properties. Plant Dis. Rep. 63:970-974.

Hai, V. T., Bompeix, G., and Ravisé, A. 1979. Rôle du tris-O-ethyl-phosphonate d'aluminium dans la stimulation des réactions de défense des tissus de tomate contre la *Phytophthora capsici.* C. R. Hebd. Seances Acad. Sci., Ser. D 228:1171-1174.

Hickey, E. L., and Coffey, M. D. 1980. The effects of Ridomil on *Peronospora pisi* parasitizing *Pisum sativum*: An ultrastructural investigation. Physiol. Plant Pathol. 17:199-204.

Hill, D. H., and Piramarn, P. 1982. Fosetyl aluminium—A versatile fungicide for the control and protection of tropical crops against *Phytophthora* diseases. Proc. S.E. Asian Symp. Plant Dis. Trop., 2nd. (In press)

Iwan, J., and Goller, D. 1975. Behaviour of SN 41703 (prothiocarb) in soils and plants. Environ. Qual. Saf. 175(Suppl. 3):319-321.

Kaars Sijpesteijn, A., Kerkenaar, A., and Overeem, J. C. 1974. Observations on selectivity and mode of action of prothiocarb (SN 41703). Meded. Fac. Landbouwwet. Rijksuniv. Gent 39:1027-1034.

Kannwischer, M. E., and Mitchell, D. J. 1978. The influence of a

because a control method for one species may favor an alternative pathogen. Halsall (1982a) has found, for example, that soil microorganisms and chemical compounds in a forest soil suppressed sporangial formation by *P. cinnamomi* but stimulated *P. cryptogea.*

## Saprophytism and Survival

Species of *Phytophthora* are generally considered to possess little competitive saprophytic ability in the sense defined by Garrett (1950). But some *Phytophthora* spp., e.g. *P. cinnamomi* and *P. palmivora*, can invade organic matter readily in competition with other soil microorganisms. The longevity of mycelium outside root pieces is debatable (Reeves, 1975; Turner, 1965; Zentmyer and Mircetich, 1966). Marks et al (1975) classified *P. cinnamomi* as a "saprophytic soil survivor" (in the sense defined by Griffin [1972]) based on field studies in eucalyptus forests in Victoria, Australia, implying that it is a parasite surviving in tissues that are colonized when the host is alive.

Although many *Phytophthora* spp. can survive in soil as chlamydospores or oospores (Zentmyer and Erwin, 1970), zoospores (MacDonald and Duniway, 1979), sporangia (Sneh and McIntosh, 1974), and mycelium (Shea et al, 1978), alternative hosts (Cother and Griffin, 1973) and live host tissue may aid survival. The capacity of *P. cinnamomi* to invade large roots and the lower stem of *Banksia grandis* Willd. in the jarrah forest permits it to survive through the prolonged dry summer (Shea, 1979b).

Prolonged drying of soil (Zentmyer and Erwin, 1970) and high soil temperatures (Hine et al, 1964) can reduce the degree of survival of *Phytophthora* spp. It is possible to destroy the inoculum of *P. parasitica* var. *nicotianae* by flooding fields and growing swamp rice in a tobacco and rice rotation system (van Schreven, 1948).

Nitrogen, and particularly the form of nitrogen, has been implicated as an important factor operating in a number of instances of biological control of Phytophthora root rot. Gilpatrick (1969a) reported that soils amended with 5% alfalfa meal accumulated 60–197 $\mu$g of ammonia per gram within 7 days, and this could be toxic to mycelium and zoospores of *P. cinnamomi.* Tsao and Oster (1981) and Tsao and Zentmyer (1979) implicated $NH_3$ and $HNO_2$ in suppression of *Phytophthora* in soils amended with chicken manure and urea.

Antagonism to *Phytophthora* propagules by a variety of soil microorganisms has been demonstrated in a number of studies (Ahmed and Ahmed, 1963; Broadbent and Baker, 1974a, 1974b; Broadbent et al, 1971a; Honour, 1973; Malajczuk and McComb, 1977; Malajczuk et al, 1977; Marx, 1973; Old, 1979; Pratt, 1971; Shea and Malajczuk, 1977; Sneh et al, 1977). This subject is reviewed in more detail by Malajczuk (Chapter 19).

Promotion of an antagonistic soil microbiota is the classic approach to biological control, but monitoring soil antagonists and assessing their significance are difficult. Techniques currently available for measuring microbial populations involve direct counting, staining of microorganisms (Babiuk and Paul, 1970; Söderström, 1977), or soil extractions in which enzymes (Skujins, 1967) or components (e.g., ATP) characteristic to living cells (Ahmed et al, 1982) are extracted and quantitatively analyzed. Other methods include the disintegration of soil followed by separation and collection of cells (Balkwill et al, 1975; Faegri et al, 1977) or respiration studies in which substrates are added to quantify the bacterial and fungal populations (Anderson and Domsch, 1973). In another method, the carbon bound in the microbial biomass is released by microbial mineralization, and its weight can be calculated (Anderson and Domsch, 1978). However, these methods do not permit identification of the specific groups of microorganisms that are inhibitory to *Phytophthora* pathogens. Where control of *Phytophthora* is achieved by cover cropping and mulching to attain a "balanced" state with improved chemical, physical, and biological properties of the soil, a nonspecific assemblage of organisms appears to be involved, so microbial biomass measurements may be applicable.

Few attempts have been made to relate the activity of antagonistic microflora and/or fauna to the activity of the pathogen in the environment where the disease occurs. The importance of the survival phase of *Phytophthora* in soil varies with the susceptibility of the host and soil conditions. For example, on freely drained sites in jarrah forests, *P. cinnamomi* is a transient soil inhabitant, whereas on lowland sites, it can be recovered consistently from the soil throughout the year (Shea et al, 1980c). Where the pathogen is a transient soil inhabitant, control procedures based on biological antagonism to propagules in the soil are less likely to be successful; to be effective, the antagonists must operate during the periods of the year when the pathogen is present in the soil. Even when the soil-inhabiting phase is prolonged, control by eradication of the fungus is unlikely to be achieved solely through biological control. Indeed, there is some evidence with other diseases (e.g., take-all caused by *Gaeumannomyces graminis*) that antagonism occurs only in response to live inoculum of the pathogen (Gerlagh, 1968).

## Sporangial Formation and Zoospore Release and Germination

The ability of *Phytophthora* spp. to increase in inoculum density rapidly by asexual reproduction is probably the major factor contributing to their success as plant pathogens. Yet the sequence of events leading to the formation of sporangia and release and transmission of zoospores is complex and sensitive to small changes in the physical and microbiological environment. This sensitivity could be exploited by the use of cultural techniques to modify the soil environment.

Sporangial formation for many species occurs only in relatively narrow ranges of soil moisture potentials; zoospores will be released only when the soil is near saturation, and zoospore transmission over significant distances depends on overland flow (Duniway, 1979). Similarly, fluctuations in soil temperature of a few degrees can affect sporangial formation (Chee and Newhook, 1964). The nature of the temperature regime can also affect sporulation significantly (Hyre and Ettinger, 1967).

A variety of chemical factors in the soil have been shown to affect sporangial formation and zoospore release and germination of *Phytophthora* spp., including major soil nutrients (Halsall and Forrester, 1977; Tsao and Zentmyer, 1979), trace elements (Halsall, 1977), sterols (Hendrix, 1964, 1965), chemicals originating from decomposing plant roots (Whitfield et al, 1981), and organic matter (Gilpatrick, 1969b; Zentmyer and Thompson, 1967). One of the most important chemical factors affecting sporangial formation and germination is the form and concentration of nitrogen in soil (Broadbent, 1977; Gilpatrick, 1969a; Tsao and Oster, 1981; Tsao and Zentmyer, 1979).

Soil microorganisms can affect zoospore production by causing plasmolysis or rupture of sporangia or by affecting sporangial formation (Broadbent and Baker, 1974b). Some *Phytophthora* species, such as *P. citrophthora* (Sm. & Sm.) Leon., produce sporangia readily in axenic culture, but their number is increased in unsterile soil leachate. Others (e.g., *P.*

*cinnamomi*) require the presence of living soil microorganisms (Mehrlich, 1935) for maximum production of sporangia. Before 1969, when Chen and Zentmyer (1970) produced sporangia of *P. cinnamomi* under axenic conditions by washing mycelium in a salt solution, sporangial production was negligible on mycelia placed in sterile soil leachate. Living soil microorganisms, particularly *Pseudomonas* spp. (Ayers, 1971; Ayers and Zentmyer, 1971; Chee and Newhook, 1966; Manning and Crossan, 1966; Marx and Haasis, 1965) and *Chromobacterium violaceum* (Zentmyer, 1965), have been implicated in the production of sporangia by the fungus in unsterile extracts.

The microbial factor in soils responsible for sporangial production appears to be ubiquitous and possibly nonspecific; the microbiota that affect stimulation are temperature-sensitive and are destroyed in soil by aerated steam at 49° C for 30 min (Broadbent and Baker, 1974a). Variation in the formation of sporangia may be caused by differences in the numbers of stimulatory microorganisms or in the numbers of other microorganisms inhibitory to them or destructive to their stimulatory metabolites, or by differential sorption of bacteria or their metabolites by soil colloids.

Control of some *Phytophthora* species might possibly be achieved by manipulation of the microbial population. Broadbent (*unpublished*) has examined more than 300 soils for suppression of Phytophthora root rot. Although sporangial production in soil extracts varied, only four soils (Broadbent and Baker, 1974a) inhibited sporangial formation enough to have epidemiologic significance in terms of biological control.

Halsall (1982b) has shown consistent suppression of sporangial formation by *P. cinnamomi* in soil from a wet sclerophyll forest in New South Wales, Australia. Actinomycetes in this soil were antagonistic to a number of soil bacteria that had the capacity to stimulate sporangium formation, and in addition a heat-stable component of the soluble fraction of the soil contributed to suppression of sporulation.

Shea and Malajczuk (1977) showed suppression of the stimulation factor in soil extracts from pots in which *Acacia pulchella* R. Br. had been grown. Saponin extracts (Alexander et al, 1978), root volatiles (Whitfield et al, 1981), and root extracts (S. R. Shea, *unpublished*) from this species suppressed sporangial formation. However, soils collected at various times of the year from beneath *A. pulchella* stands growing in the field have not consistently suppressed *P. cinnamomi* sporulation (S. R. Shea, *unpublished*).

Even if it is not possible to suppress the microorganisms that stimulate sporangium production, their role in disease epidemiology must be elucidated. For example, seasonal variations in soil stimulatory capacity (Chee and Newhook, 1966; Otrosina and Marx, 1975) probably have an important effect on temporal variations in propagule density of the pathogen in soil. Do the nature and density of the stimulatory microorganisms affect the relationship between soil temperature and sporangium production?

The importance of the sporangium formation and germination phase of the life cycle of *Phytophthora* species and its demonstrated sensitivity to soil microorganisms suggest that manipulation of the soil microbiological environment to disfavor sporangial formation and/or germination is one of the most promising approaches to control. However, assessments of the effect of soil microorganisms on sporangial formation and germination must be based on realistic experimental techniques. It is unrealistic to expect that the soil microflora present in a soil extract that has been incubated for several days will be representative of the field situation from which the samples were collected. For example, there was no seasonal variation in sporangial stimulation in jarrah forest soil when a soil extract method was used, but marked seasonal variation was observed for the same soil when stimulatory capacity was tested in intact soil cores (L. McGann and S. R. Shea, *unpublished*). Blowes (1980) demonstrated that sporangial production on *P. cinnamomi* mycelial mats immersed in a sand suspension amended with an inorganic source of nitrogen was not comparable in successive experiments because of variation in the sporangia-producing capacity of the *P. cinnamomi* cultures used.

The development of control methods based on manipulation of the soil microbiological, physical, and chemical environment to disfavor survival and reproduction of *Phytophthora* pathogens is handicapped by the lack of data on spatial and temporal variations in pathogen propagule density, on the environment where the disease does and does not occur, and on the response of these pathogens to environmental regimes that are relevant to the field environment. To be effective, changes in the soil environment aimed at reducing survival and/or reproduction must operate during periods when the pathogens are present in the soil and/or are operating their asexual cycle. But in almost no situations have the factors influencing the soil-inhabiting phase of *Phytophthora* pathogens been comprehensively documented. Assessing the potential of control methods based on manipulation of the soil environment is difficult without detailed documentation of the environment where the disease occurs in relation to the temporal and spatial variation in pathogen propagule density. For example, in a disease situation where the soil moisture and temperature regime is suitable for sporangial production only for limited periods of the year, relatively small changes in the physical environment of the soil can have a significant effect on pathogen propagule density. But when the pathogen is in an environment with prolonged periods when soil moisture and temperature are suitable for survival and reproduction, major changes in the soil physical environment would be required to achieve a significant effect on the pathogen.

Many of the data on the response of *Phytophthora* pathogens to changes in the environment have little relevance to the actual environment where the disease is occurring. For example, temperature, moisture, soil microbial activity, nitrogen form and concentration, and other such conditions vary markedly over time in the field. But most of the data on the effects of these factors on survival and reproduction have been obtained under constant conditions. Even if these data could be extrapolated to the field, very little research has been undertaken to determine if the concentrations or levels of soil factors that have been shown to suppress *Phytophthora* survival and reproduction could be maintained in the field situation.

## ENVIRONMENTAL FACTORS AFFECTING THE HOSTS

### Host Predisposition

The effect of nongenetic host factors on susceptibility to *Phytophthora* spp. has received little attention, possibly because the effects of stresses induced by the environment on the host-pathogen interaction are often difficult to separate from effects on the host itself (Schoeneweiss, 1975; Yarwood, 1959). But changes in susceptibility to *Phytophthora* spp. related to age and stage of development of the host have been

or trees has not simply brought about a favorable physiologic state in these plants that may mask symptoms of feeder root disease. Comparisons of mycorrhizal with nonmycorrhizal plants under field conditions are questionable. Conditions

protection and the species of ectomycorrhizal fungi that prevent pathogenic infection and to determine if these fungi could be maintained as the dominant symbionts on the root system. Is the percentage of mycorrhizal roots correlated with commonly observed. Many species of plants are susceptible to *Phytophthora* spp. at the seedling stage but resistant at later stages of development. *Pinus* spp. are susceptible to *P.*

But in other hosts, such as macadamia (*Macadamia integrifolia* Maiden and Betche), expansion of stem cankers caused by this pathogen is variable (Pegg, 1973), which



the degree of control of root infection, or can the production of antibiotics in mycorrhizal roots afford protection to adjacent nonmycorrhizal roots? Can a mycorrhizal fungus, highly efficient in water and nutrient uptake and present in a small portion of a root system, overcome the effects of root rot in the remainder of the root system? The association of ectomycorrhizae with other microorganisms also needs more study. Do the beneficial effects of mycorrhizal roots on host vigor contribute to resistance by reducing the capacity of pathogens to invade the major root system and collar of susceptible species?

### Endomycorrhizae

Evidence that endomycorrhizal plants are more resistant to *Phytophthora* diseases is sparse. No protection against *P. cinnamomi* was conferred on avocados infected with *Glomus fasciculatus* (Davis et al, 1978; Mataré and Hattingh, 1978). Mycorrhizal roots of a nonresistant cultivar of soybean were more susceptible to *P. megasperma* f. sp. *glycinea* than were nonmycorrhizal roots (Ross, 1972).

Sweet orange seedlings infected by *G. fasciculatus* were susceptible to root rot by *P. parasitica*, but size, root health, and phosphorus uptake were greater than in nonmycorrhizal plants (Davis and Menge, 1980). This tolerance may have been due to the ability of mycorrhizal roots to absorb more phosphorus and possibly other minerals than nonmycorrhizal roots can (Davis and Menge, 1980). But vesicular-arbuscular mycorrhizae on citrus can also affect seedling stomatal conductance, photosynthesis, and proline accumulation, which suggests that the effect is on stomatal regulation rather than on root resistance (Levy and Krikun, 1980).

Most biological or cultural approaches to control of *Phytophthora* diseases have been directed at suppressing the pathogen. But comparatively recent research suggests that the environment may have a much greater effect on host susceptibility than previously assumed. However, the host factors that affect infection and mycelial growth within host tissue by *Phytophthora* pathogens must be identified before it will be possible to determine if cultural procedures can be used to manipulate host susceptibility and lessen disease.

Some data suggest that the physiologic status of the host affects the composition and quantity of host exudates from roots and subsequently the number of zoospores attracted to the roots. But almost no information is available on the effect of root physiology and anatomy on the probability that a zoospore located at the root surface will penetrate host tissue. Seasonal variation in *Phytophthora* lesion extension within host tissue suggests that mycelial growth within host tissue is affected by host status, but the processes responsible for this have not been documented. Can hosts "compartmentalize" *Phytophthora* infections as has been demonstrated for wood decay fungi (Shigo and Marx, 1975), and is the capacity to contain *Phytophthora* pathogens affected by host physiology?

There is circumstantial evidence that host factors such as growth rate, growth periodicity of roots and shoots, flowering and fruit formation, carbohydrate status of host tissue, and root-shoot ratios all affect host susceptibility. Most of these host characteristics can be modified by cultural procedures. Thus, considerable potential exists to lessen *Phytophthora* diseases by host manipulation, but further research to determine the host factors that affect infection and mycelial extension is required before this potential can be realized.

## SUPPRESSIVE SOILS

The occurrence of situations where susceptible vegetation is not diseased in the presence of pathogens, or where the probability of pathogen introduction is high, under climatic conditions that are normally suitable for disease development, provides the best evidence of the potential of biological and/or cultural control techniques. The inhospitability of certain soils to some plant pathogens is such that either the pathogens cannot establish themselves, or they become established but fail to produce disease, or they become established and initiate disease but diminish in severity with continued culture of the crop (Baker and Cook, 1982). Such soils are referred to as "suppressive" soils, as opposed to "conducive" soils where, given favorable climatic conditions and a susceptible host, disease progresses unchecked. Suppressive soils have also been called resistant, immune, intolerant, and antagonistic.

In a number of areas in Australia, both in forests and avocado groves, *P. cinnamomi* is present but has not caused disease in susceptible vegetation and under climatic conditions normally suitable for disease development.

### Forest Soils

#### Western Australia

The jarrah (*E. marginata* Sm.) forest is a dry, sclerophyllous evergreen forest in southwestern Australia. The climate is Mediterranean, with the highest rainfall in the winter months and little or no rainfall in the hot, dry summer. The soils are predominantly lateritic and are extremely infertile compared with the virgin rain forests and avocado groves of southeastern Queensland; they are low in organic matter, nitrogen, and exchangeable cations (Malajczuk, 1979a). Severe root rot, caused by *P. cinnamomi*, is invariably associated with these sandy and gravelly soils. The most susceptible sites occur along water courses or in poorly drained areas, but root rot has been found in most jarrah forest communities, soil types, and topographic situations, including freely drained sites (Podger, 1968; Shea, 1975), except in the vegetation on the more fertile, red-brown loams of the steeper valleys that dissect the jarrah forest (Podger, 1968; Shea, 1975). Severe disease occurs on lateritic soils immediately upslope of these apparently resistant soils.

Malajczuk (1975) compared the pathogenicity of *P. cinnamomi* to *E. marginata* and *E. calophylla* seedlings grown in lateritic and loam soils. Seedlings were grown in pots of sterile and unsterile soil and inoculated with *P. cinnamomi*. In sterile loam and lateritic soil, both eucalypt species were equally susceptible to infection. In unsterile soils, *E. marginata* was susceptible only in the lateritic soils, and *E. calophylla* was unaffected. This implies that the suppression of root rot in the loam soil involves a microbial factor and that resistance of the tolerant *E. calophylla* is not entirely due to innate resistance but also involves a microbial component in the rhizosphere.

#### Southeastern Australia

Serious disease caused by *P. cinnamomi* has been recorded for duplex soils, such as the shallow lateritic sands over impervious clays in the Brisbane Ranges (Weste and Taylor, 1971), deep sandy soils at Wilson's Promontory National Park (Weste and Law, 1973), and infertile shallow soils with an impervious clay pan on the coastal plains of East Gippsland (Marks et al, 1972) in Victoria. A krasnozem from the foothills of Mount Dardenong, Victoria, was found by Marks and Smith (1981) to strongly suppress root disease caused by *P. cinnamomi* in two *Eucalyptus* spp. However, although *P. cinnamomi* is widely distributed geographically in New South Wales (Pratt and Heather, 1973), disease caused by this fungus is restricted to moist gullies. Even though the species that are highly susceptible in Victoria are present, there have been no major disease outbreaks (Gerrettson-Cornell and Dowden,

1978). *P. cinnamomi* has been associated with the death of native plants such as *Banksia aemula* R. Br. following heavy rains in the low sclerophyll shrub woodland ("wallum") of the coastal lowlands of southeastern Queensland (Pegg and Alcorn, 1972).

The relative freedom from severe *Phytophthora* epidemics in southeastern Australia may in part be the result of microbial suppression of *P. cinnamomi*. Halsall (1982b) described a soil from Tallaganda, New South Wales, which despite its low organic matter and nutrient levels was suppressive to *P. cinnamomi*. *P. cryptogea*, by contrast, was not inhibited.

Blowes (1980) attributed the greater disease severity in forests of southwestern compared with southeastern Australia to the colonization of roots of a highly susceptible understory species (e.g., *Banksia grandis*) in Western Australian jarrah forests on which a higher inoculum potential was produced and to summer droughts that exacerbated the damage. Blowes (1980) also suggested that the more evenly distributed rainfall in southern New South Wales (southeastern Australia) and the smaller food base in the absence of a susceptible understory might result in "subclinical" damage to root systems. As described by Pratt and Heather (1973) and Broadbent and Baker (1975), *P. cinnamomi* can be readily isolated from areas with poor drainage. In adjacent areas, the situation is intermittent or varying: plants susceptible to root rot become recolonized but are eliminated in wet seasons. *E. sieberi*, which is very susceptible to root rot, is restricted to the deeper, well-drained soils of the ridges.

**Rain Forest**

There have been few reports of *P. cinnamomi* on susceptible rain forest vegetation in Queensland, despite adequate opportunities for *P. cinnamomi* to spread in drainage water from nearby infected plants like avocado and pineapple (Newhook and Podger, 1972). "Patch death" in virgin rain forest in northern Queensland, which Brown (1976) attributed to *P. cinnamomi*, was associated with pig wallows that, because of heavy clay subsoil close to the surface, retained water even when the soil under the apparently healthy rain forest was relatively dry.

Greenhouse pathogenicity tests were conducted on 100 soils collected from various rain forest associations in New South Wales and southeastern Queensland. The soils were derived from shales and basalt, and the texture varied from sand to clay (P. Broadbent, *unpublished*; Broadbent and Baker, 1974a). These tests were performed under a favorable temperature-moisture regime with a susceptible host (*E. sieberi* or *Jacaranda acutifolia*) growing in small trays of soil infected with *P. cinnamomi*. The soils had been treated with aerated steam at 49 or 60°C for 30 min or with flowing steam at 100°C for 30 min or were left untreated in an attempt to appraise the role of the microflora in the suppression. In soils conducive to root rot, seedlings developed root rot in all four treatments. In soils suppressive to root rot, seedlings were healthy and grew well in infected untreated soil but developed root rot in the steamed (100°C) soil. This suppression was removed by steam-air at 49 or 60°C.

Sporangium production of *P. citrophthora* and *P. cinnamomi* is sometimes much reduced in the suppressive soil or soil leachate. For example, Coulter counts of zoospores of *P. citrophthora* from leachates of two conducive soils were over 4,000/ml each, whereas only 290–774 zoospores per milliliter were counted in leachates of four suppressive soils. The reduction of sporangium formation and zoospores may be sufficient to prevent effective invasion of the host without eliminating the pathogen from the soil under most environmental conditions. Failure of *P. cinnamomi* to form many sporangia in the suppressive soil did not result in greater chlamydospore production (Broadbent and Baker, 1975).

Red basaltic soils suppressive to root rot have a high content of organic matter, sometimes with a humus layer 9 in. deep; high levels of exchangeable calcium (20–27 meq), with the calcium tied up in the organic cycle; a pH of 5.5–7.0; high total nitrogen levels, apparently locked in humic residues; and high biological activity. No association could be found between root rot suppression and the clay mineral composition of the soil, but the only soils found to be suppressive were derived from red basalts (Broadbent and Baker, 1974a). "Patch death" of virgin rain forest (Brown, 1976) was restricted to soils derived from granite-diorite-granodiorite parent material.

**Avocado Soils**

Before they were cleared, the krasnozems of the north coast of New South Wales and southeastern Queensland were vegetated with a subtropical rain forest. For a luxuriant rain forest to flourish on soils poorly supplied with certain plant nutrients, there must be a rapid turnover of nutrients achieved by rapid decomposition of fallen leaf litter and softwood trees by saprophytic fungi and bacteria. Higher contents of organic carbon, nitrogen, and exchangeable cations in upper horizons of the profile suggest a closed cycle in which the cation requirements of the virgin vegetation are supplied by leaf litter. The replacement of this vegetation with crops such as avocados (*Persea americana* Mill.) appears to have interrupted the cycle and replaced it with an exploitive system (Colwell, 1958; Nicholls and Tucker, 1956).

Surveys of avocado groves in northern New South Wales and southeastern Queensland, coupled with soil tests to appraise the antagonistic potential to root rot, have shown that avocado soils suppressive to root rot do exist and have physical, chemical, and microbiological properties similar to those of nearby suppressive rain forest soils. These properties have been maintained principally by the use of organic amendments. Before planting and while the avocado trees are young, extensive cover cropping is practiced, using maize (*Zea mays*) and *Dolichos lablab* in summer and New Zealand blue lupine (*Lupinus angustifolius*) and oats (*Avena sativa* L.) in winter, plus fowl manure, dolomite, and synthetic fertilizers. When the tree canopies touch, mulches with a high ratio of carbon to nitrogen, e.g. barley straw, hay, Rhodes grass (*Chloris gayana*), or mixed native grass, are added. The timing of the application of mulches may be important: mulches should be partly decomposed before the summer rains, otherwise they may be colonized by *P. cinnamomi* or retain too much soil moisture. Without the extensive application of organic amendments, organic matter, cation exchange capacity, and nitrogen levels fall, and the soils become conducive to root rot.

Red basaltic soils conducive to root rot can be rendered suppressive in time (upwards of 2 yr) by the use of organic amendments. Pegg (1977a) demonstrated the effectiveness of these cultural practices with two examples. In the first example, many trees that declined after the 1972 deluge (1,750 mm of rain in 3 days), when soils lost their suppressiveness because of prolonged waterlogging, have since been restored to health by the use of surface organic mulches and fowl manure. In the second example, *P. cinnamomi* could no longer be recovered from a site at Mount Tamborine where declining avocado trees were removed in 1973 before the soil was restored by intensive cover cropping and frequent light dressings of dolomite and fowl manure and was replanted with clean nursery stock. This demonstrates that the suppressive and conducive states of soil are relative and that the balance can be readily tipped.

Maintenance of a deep humus layer in the red basaltic soils seems to contribute to the suppression of root rot. The type of mulch or cover crop used is partly irrelevant (although the rate of breakdown, the frequency of addition, and the necessity of additional nitrogenous amendments vary with the mulch). "Single shot" applications of fowl manure, sawdust, or alfalfa, especially after a tree is showing visible signs of root rot, are rarely effective. A continuing program of mulching, cover cropping, and addition of nitrogenous and calcareous amendments is essential to maintain the humus layer, cation exchange capacity, and root rot suppression. Suppression of root rot is rarely, if ever, absolute; the avocado tree can tolerate a degree of root rot with no obvious effects on "aboveground" tree health. After a threshold is reached where further loss of roots may result in irreparable damage to the tree, no amount of organic amendments will cause a significant response in tree health.

## CULTURAL PROCEDURES

Few examples in the literature show that cultural procedures have been used to achieve complete control of *Phytophthora* diseases. However, a number of techniques can be used to reduce the severity of disease, and there are some promising approaches to control.

### Exclusion and Sanitation

In many areas, *Phytophthora* species are not endemic, and the simplest approach to control is to prevent their introduction. The recognition that quarantine and/or hygiene may ultimately be inadequate to prevent introduction of *Phytophthora* spp., and the capacity of *Phytophthora* spp. to rapidly increase in propagule density once introduced, have caused this approach to control to be underemphasized. Disputes about the endemic or introduced status of some pathogens have also impeded the introduction of hygiene and/or quarantine (Shea, 1979a). MacKenzie et al (Chapter 23) also warn that the value of sanitation has been underestimated.

In some disease situations, particularly in forests, the only workable long-term control strategy is based on manipulation of the existing vegetation. This option is not feasible if the pathogen has been introduced, has become widely dispersed, and has severely debilitated the forest. Hence, even if hygiene and/or quarantine only delays disease development, it is justified. Even after the pathogen has been introduced, hygiene procedures may still be justified, because they limit the widespread distribution of soilborne *Phytophthora* spp. (other than by water flow) after periods of natural quiescence. For example, natural extension of *P. cinnamomi* infections in the jarrah forest upslope is less than 1 m/yr, but uncontrolled roadmaking can transmit this pathogen over several kilometers in a few days (Shea et al, 1980a).

The importance of nursery hygiene in preventing the spread and intensification of *Phytophthora* diseases is well documented. Even when nursery stock is destined for areas where pathogens are already present, disease control is greater if the plant is free of the pathogen at the time of establishment.

Transfer of inoculum to uninfected sites cannot be avoided if the locations of the diseased areas that are the sources of inoculum are not accurately defined. Large-scale aerial color photography has been used to identify and locate affected areas based on symptoms on plants in the shrub and understory layer of the jarrah forest in Western Australia. The objective of this program is to provide accurate maps of disease occurrence in over 500,000 ha of forest. Identification of potential sources of inoculum in the forest makes it possible to minimize the transfer of inoculum into healthy forest areas by activities such as logging and mining (Bradshaw, 1974; Bradshaw and Chandler, 1978).

Temporal as well as spatial variation in inoculum distribution can be used in hygiene management. For example, during the summer months, the potential for spread of *P. cinnamomi* by vehicles in the jarrah forest on upland, freely drained sites is minimal because inoculum levels in the soil are low (Shea et al, 1980a).

Spatial and temporal variations in the receptivity of sites to inoculum should also be used in designing hygiene procedures. For example, the probability that transferred *P. cinnamomi* inoculum will remain viable on moisture-gaining sites in the jarrah forest is high throughout the year, but inoculum survival on dry, upland, freely drained sites is poor during the summer months (Shea, 1975).

Seasonal and year-to-year variations in inoculum levels and variation in host susceptibility with season and/or stage of development can be exploited to achieve control. For example, Ko (1971) found that resistance of papaya roots to *P. palmivora* was directly correlated with the age of the plant. Awareness of this host factor led to a unique method of biological control. The planting holes were filled with pathogen-free soil, so that seedlings escaped infection during their susceptible stage. By the time the roots reached the infested soil, they were resistant. Grainger (1968) proposed that the timing of potato planting in Scotland could be exploited to control late blight. Potatoes planted early in the season escape disease because environmental conditions are unfavorable for *P. infestans*, and at later stages of crop development, when the fungus is active, the host is "unreceptive." Based on this, an index of host susceptibility, the Cp/Rs ratio (weight of total carbohydrate in plant/residual dry weight), was proposed (Grainger, 1968). *P. megasperma* can cause major damage to early sown chickpeas in New South Wales, particularly on irrigated clay soils. Late-March sowings were almost destroyed, whereas plantings delayed until temperatures were cooler in May escaped the disease (G. E. Stovold, *personal communication*). This effect was attributed to temperature and illustrates how cultural practices can be modified to exploit differences in the timing of pathogen activity and host susceptibility to achieve a form of control by exclusion.

Where susceptibility differs among trunk roots, major roots, and feeder roots on such horticultural crops as citrus (Broadbent et al, 1971b) and peaches (Taylor, 1980), disease may be controlled by management practices such as shallow planting, keeping the trunks dry (Taylor, 1980), and preventing wounding of trunks and stems.

Sanitation involves reducing the inoculum level by removing infected host tissue. Turner (1965) recommended the removal of diseased cacao pods infected by *P. palmivora* and disposal of the cacao crop residue to reduce inoculum potential. To avoid tuber rot by *P. infestans*, potatoes should not be dug until the tops are either killed by frost or are treated with defoliant chemicals to prevent inoculum such as sporangia from spreading from the tops to the tubers (Zentmyer and Bald, 1977). Reduction of the density of the *Banksia grandis* understory, which is highly susceptible to *P. cinnamomi*, has been proposed as one method of reducing inoculum potential in the jarrah forest (Shea, 1975).

### Manipulation of the Physical Environment

A number of cultural techniques can be used to change

physical environmental factors in the soil (e.g., temperature, moisture, aeration, and light) to disfavor *Phytophthora* spp. Where the physical environment is only marginally favorable for the pathogen(s), relatively small changes could have a major effect on disease severity.

Careful site selection is important. In some sites, the inherent structure of the soil profile or its position in the landscape may result in a soil moisture regime so favorable for the survival and reproduction of *Phytophthora* that no cultural technique is likely to bring about control. Poorly drained soils are hazardous for avocados, especially when *P. cinnamomi* is present, because root rot damage is closely correlated with soil type (Goodall, 1955). Soil drainage characteristics have been used to determine if specific soil series are suitable for growing avocados in California (Burns et al, 1960; Zentmyer et al, 1967) and in South Africa (Wolstenholme and Le Roux, 1974).

One of the most important factors of the physical environment affecting the development of *Phytophthora* diseases is the soil moisture regime. Although root rot may result from impeded drainage and/or excessive rainfall, losses are accentuated by *Phytophthora* spp. In general, any cultural practice that minimizes the period when the soil is at a soil moisture potential below 100 mb and that prevents overland flow of water will reduce disease. Duniway (Chapter 14) covers this area in greater detail.

A number of management procedures can create a potential disease hazard by providing a favorable soil moisture regime. Evaluation of the effects of such practices as road making, drain construction, logging, and track formation on soil drainage and subsequent disease can assist in the development of simple procedures to minimize disease. For example, the restriction of roads to lower topographic positions in the landscape will minimize the area predisposed to root rot. In New South Wales, headland drains are used to prevent runoff water from entering avocado orchards, surface drains prevent excessive ponding and water-soaking of the soil, and subsurface drains allow free water to percolate from the root zone (Chalker, 1974). Other helpful practices include ripping (Borst, 1975) and mounding of the beds on which the crop is planted (Baker, 1938).

Disease development can also be modified by regulating the irrigation regime (Mircetich, 1978). For example, the dependence of many *Phytophthora* spp. on overland flow of water for widespread propagule dispersal emphasizes the advantage of drip irrigation and fine undertree sprinklers as opposed to furrow irrigation.

Although inefficient vertical drainage is a critical factor that favors *Phytophthora* root disease, the water table is more than 10 m deep in large areas in the severely diseased jarrah forest in Australia, and impedance of water flow is minimal in the upper 50 cm of the soil profile (Shea, 1975). Even where susceptible vegetation is growing on freely drained sites with coarse-textured soils, high or frequent rainfall may at times favor sporangium formation and release of zoospores. In these situations, the soil moisture regime can be modified only by cultural treatments that affect the density and composition of the vegetation and the surface litter (Newhook, 1970; Shea, 1975).

Temperature also has a marked effect on survival and reproduction of *Phytophthora* pathogens. Zentmyer (1980) compiled (from 24 reports) the cardinal temperatures for growth of *P. cinnamomi* (see his Table 4) and also reviewed the effect of temperature on sporulation and spore germination. Soil temperature regimes can be modified by changing the canopy and the litter cover. Coincidence of relatively high temperatures and low matric potentials is required for sporulation by *P. cinnamomi*. Roth and Kuhlman (1963) postulated that *P. cinnamomi* was unlikely to be a threat to Douglas fir forests in Oregon because of the absence of periods when high soil moisture coincided with temperature regimes suitable for sporangium formation. Similarly, an unfavorably low soil temperature regime was cited as the principal factor preventing survival in, and colonization of, mountain soils in eastern Victoria by *P. cinnamomi* (Kassaby et al, 1977). Shea (1975) suggested that promotion of a dense canopy to lower soil temperatures on freely drained sites in the jarrah forest would decrease the periods during which sporangia could form.

Identification of the upper temperature limits for survival and reproduction of *P. cinnamomi* permitted the identification of low-hazard sites for Phytophthora root rot of pineapple (*Ananas comosus* (L.) Merr.) (Hine et al, 1964). High-temperature treatments of nursery stock can be used to eradicate *P. cinnamomi* (Baker, 1962; Benson, 1978).

## Fire

Fire is a natural part of the environment in many regions of the world. Prescribed fire has been extensively used in some forests for removing litter and brush, which are fire hazards, and for tree regeneration. The effect of fire on a forest depends on its frequency and intensity and on the season (Shea et al, 1980b). By varying these factors, the structure and composition of forests and the nature of the litter layer can be manipulated. This, together with the fact that fire can be used over large areas at a relatively low cost, makes prescribed fire potentially one of the most practical tools available to the forest manager to modify the forest environment.

Changes in the fire regime, however, may have both beneficial and detrimental effects. Periodic burning of the litter layer could significantly increase the susceptibility of a forest to Phytophthora root rot by allowing soil temperatures to increase and by decreasing the interception of rainfall, thereby prolonging the period when the soil moisture and temperature regimes are suitable for sporangium production by *P. cinnamomi* (Shea, 1975).

The microbiological effects of periodic litter removal by fire are not well understood (Warcup, 1981). Soil organic matter levels did not differ significantly in regularly burned and unburned *E. marginata* forest sites (Hatch, 1959). *P. cinnamomi* causes disease in unburned areas (Podger, 1968), and a high level of sporangium production was observed in forest soils bearing a 7-yr accumulation of litter (Shea et al, 1978). Nesbitt et al (1979) found that soils into which jarrah forest litter had recently been incorporated caused lysis of *P. cinnamomi* mycelium and abortion of sporangia, but it is unlikely that these treatments were relevant to the field situation.

Malajczuk (1979a) stated that "the development of ectomycorrhizae of jarrah appears to be strongly associated with an increased organic matter or litter component beneath trees." Because ectomycorrhizal roots are resistant to infection, removal by fire of the proportion of ectomycorrhizal roots in the litter layer would, by inference, increase the proportion of the fine root system that is susceptible to infection. The significance of this to disease development depends on the relative importance of the ectomycorrhizal roots removed by fire to total nutrient uptake.

The intensity, frequency, and season of burn can markedly affect the structure and composition of the understory. For example, periodic (5- to 9-yr cycle) low-intensity fire to reduce fuel accumulation and increase protection from fire in the jarrah forest favors the maintenance of an understory dominated by *Banksia grandis* Willd., which is highly susceptible to *P. cinnamomi* (Shea, 1975, 1979a). Moderate to high-intensity fire causes regeneration of leguminous species and reduces the density of the susceptible *B. grandis*. *Acacia pulchella*, a common legume species, is resistant to root invasion by *P. cinnamomi* (Shea, 1975; Shea and Malajczuk, 1977; Shea et al, 1976; Tippett and Malajczuk, 1979). Some evidence indicates that the microbiological and physical environment in soil under legume stands is less suitable for sporangium production and survival of *P. cinnamomi* (Shea and Malajczuk, 1977; Shea et al, 1978) and that chemical exudates from *A. pulchella* depress sporangium production and germination and mycelial growth (Alexander et al, 1978; Whitfield et al, 1981).

Fire can stimulate growth of *Eucalyptus* species and affect flowering and seeding cycles (Kimber, 1978), and this in turn may affect host susceptibility. Fire is also an important method by which regeneration of eucalypt species is achieved. The potential for increasing the diversity of mixed eucalypt forests by direct seeding of selected tree species following regeneration burns is considerable. For example, mixtures of tree species tolerant and susceptible to *P. cinnamomi* have been successfully established in sites infested with *P. cinnamomi* in coastal forest areas in southeastern Victoria (Shea, 1979a).

Fire can affect the host, the pathogen, and the environment, but little progress will be made in elucidating its effect on disease development and its potential for control until the processes by which it affects each of these factors are understood and evaluated epidemiologically.

## Organic Amendments

High levels of organic matter in soil have been associated with some soils that are suppressive to *P. cinnamomi* (Broadbent and Baker, 1974a, 1974b). Manipulation of organic matter levels in forest soils has been proposed as a potential method of control of this pathogen (Malajczuk, 1979a; Nesbitt et al, 1979). Organic matter can affect the vigor of the host by improving the structure and moisture- and nutrient-holding capacity of soil, but its principal effect is to provide a more complex and antagonistic soil microflora and fauna (Broadbent and Baker, 1975).

When high-value agricultural and horticultural crops are grown in short rotations, soil organic matter levels can be manipulated by direct additions of organic matter and/or cover cropping. Even in forests, it may be possible to regulate organic matter directly by protection from fire (Malajczuk, 1979a), or fire may be used to change the species composition of the understory to favor species of plants from which soil organic matter levels are increased (Shea, 1979a).

A good example of Phytophthora root rot control by organic amendments is found in Australia, where some avocado growers add copious amounts of plant residues and chicken manure each year to their orchard soils to build up the organic matter levels or to maintain them at a level near that of the surrounding undisturbed rain forest. Orchards managed in this way have little root rot caused by *P. cinnamomi*, whereas orchards where tillage has reduced the organic matter levels have severe root rot (Broadbent and Baker, 1975).

The potential for controlling *Phytophthora* pathogens by the use of organic amendments has its most likely application in nurseries where plants are grown in containers. Hardwood bark composts have been used to eliminate *Phytophthora* pathogens during the composting phase, thus eliminating the necessity for steam sterilization (Hoitink, 1980; Hoitink et al, 1977). In addition, these composts suppress the activity of *Phytophthora* pathogens after they are introduced. Hoitink's studies demonstrated that the type of organic matter and the control of the composting process were critical. Control was not achieved if the hardwood bark component was excessively diluted.

A number of other studies have shown that *Phytophthora* diseases can be controlled in pots by regulation of the quantity and composition of the organic matter (Gerrettson-Cornell et al, 1976; Gilpatrick, 1969b; Tsao et al, 1975; Zentmyer, 1963).

Although *Phytophthora* diseases can be controlled by regulation of the media used to grow container-grown plants, and although higher levels of organic matter are correlated with lower levels of disease in some field situations, it cannot be assumed that organic matter amendments per se will always contribute to a reduction of the incidence or severity of disease. Some types of organic amendments may lead to an increase in diseases caused by *Phytophthora* spp. The rate at which *Pinus echinata* Mill. was affected by little leaf and the rate of decline thereafter were increased by the addition of forest litter, and compost and horse manure had no effect on disease development (Roth et al, 1948).

At what stage in the breakdown of organic matter is *Phytophthora* suppressed? Nesbitt et al (1979) added varying quantities of fresh jarrah litter to a lateritic soil and noted increased lysis of hyphae and sporangia of *P. cinnamomi*, which they attributed to increased concentrations of nutrients and microbes. However, the organic matter treatments stimulated greater sporangium production. Undecomposed leaf litter piled around a citrus tree can act as a substrate for *Phytophthora* and may retain moisture, so that collar rot losses may be increased (Fawcett, 1923). Broadbent (*unpublished*) has isolated *P. cinnamomi* from partly decomposed sorghum mulch and mixed native grass hay around avocado trees. Leachates from these mulches may stimulate the formation of sporangia.

## Fertilizers

The application of fertilizer is one of the most commonly used cultural techniques in the management of agricultural and forest crops. But our knowledge of the interactions among fertilizer application, plant growth, and *Phytophthora* diseases is inadequate and often contradictory. Part of this confusion can be attributed to the difficulty of separating the effects of fertilizers on the host from their effects on the pathogen.

The addition of large quantities of nitrate or ammoniacal fertilizers was generally effective in reducing the incidence of little leaf among healthy pine trees and improving many trees that were already diseased (Roth et al, 1948). However, addition of a large number of other elements (Na, P, K, S, Ca, Zn, Mg, Mn, Cu, Mo, Co, Ni, B, Fe) had no beneficial effect. Broadbent and Baker (1974a) and Pegg (1977a) have implicated nitrogen and cation exchange capacity as factors that influence the suppression of root rot of avocados in red basaltic soils.

Several studies have shown that the form of nitrogen has an important effect on survival and reproduction of *Phytophthora* pathogens (Broadbent, 1977; Gilpatrick, 1969a; Tsao and Oster, 1981). Soil pH can affect the form of nitrogen, the pathogen directly, and the antagonistic microflora, as has been demonstrated by Smiley (1978) for take-all disease of wheat. Form and concentration of nitrogen may also affect host susceptibility (Hornby and Goring, 1972). In a model system, Brown (1975) found that the amount of take-all of wheat depended on a complicated interaction among host nutrition, form of nitrogen on which the pathogen was grown, and activity of the microbial flora of the roots. Such an interaction may also occur for Phytophthora root rot.

Newhook (1970) applied superphosphate to a forest of *Pinus radiata* in New Zealand to correct a phosphate deficiency and improved the crown density, root growth, and mycorrhizal development. The increased tree growth in fertilized plots led to a greater removal of water from the soil, which contributed to a reduction in Phytophthora root rot.

Soil applications of elemental sulfur in replant pineapple culture have been effective in controlling pineapple root and heart rot caused by *P. cinnamomi* (Lewcock, 1935; Pegg, 1977b). This result was attributed to a lowering of the soil pH below 3.8, which markedly reduced sporangium formation (Chee and Newhook, 1964; Pegg, 1977a). Fortunately, the pH values that were limiting to the sporulation of *P. cinnamomi* were favorable for the growth of the pineapple. The population of *Trichoderma* spp. explodes in sulfured soils (Pegg, 1977a), but their role in reducing disease caused by *Phytophthora* has yet to be determined.

Fertilizer application in some instances may increase disease. In field trials, fertilization of *Eucalyptus* spp., primarily with N, P, and K (applied as a mixture), has increased disease or at least had no significant effect (Marks et al, 1973).

The effect of fertilizing plants with minor elements on disease caused by *Phytophthora* has received comparatively little attention. However, several studies have suggested that calcium may have a positive effect on disease control. Avocado soils suppressive to Phytophthora root rot in eastern Australia are high in exchangeable cations, particularly calcium and magnesium (Broadbent and Baker, 1975). The maintenance of high calcium levels in soil and the addition of calcareous amendments may affect root rot by improving soil structure but in addition there appears to be a complex relationship between soil drainage, aeration, calcium supply, and resistance to fungal attack (Borst, 1970). Chapman (1965) found in nutrient solution experiments with citrus that root rot in areas of low oxygen supply is much reduced in the presence of a high concentration of calcium. Zentmyer and Lewis (1975) reduced root rot of avocado and *Persea indica* seedlings by drenching with calcium salts. Lee (1979) found that seedlings of *Persea indica* grown in Hoagland's solution no. 1 containing 160 ppm $Ca^{2+}$ were more resistant to infection or tissue colonization by *P. cinnamomi* than those grown at lower $Ca^{2+}$ levels. Bellamy et al (1971) found that calcium chloride reduced the rate of spread of *P. cinnamomi* in the roots of New Zealand blue lupine. Calcium carbonate markedly increased the proportion of healthy fine roots and markedly suppressed infection of jarrah roots by *P. cinnamomi*. This was attributed to an increase in the development of ectomycorrhizal fungi on roots (Boughton et al, 1978). However, Halsall (1980) found that calcium neither reduced host mortality nor reduced the growth suppression effect caused by infection of susceptible eucalypt seedlings by *P. cinnamomi*.

The addition of gypsum or lime at varying rates to avocado seedlings in pots of conducive soil infested with *P. cinnamomi* reduced root rot very slightly; the response was more marked in the presence of $NH_4^+$ (P. Broadbent, *unpublished*). Gypsum applied in addition to cover cropping (*Lablab purpureus* L. and *Lupinus angustifolius* L.) and application of fowl manure to avocado trees in a soil management trial on red basaltic soil at Alstonville, New South Wales, increased the availability of phosphate and improved vigor and yields of trees (Trochoulias and Broadbent, 1978). Root rot severity was less in treatments receiving calcareous amendments (Broadbent, *personal communication*).

Fertilization, like fire, is a cultural procedure with great potential for the control of *Phytophthora* diseases because it can be controlled by the forest manager or farmer and because fertilizers have been shown to affect pathogens and disease development. But fertilizers, like fire, also affect the host and the environment. Further research is needed to elucidate these interactions and the processes involved before it will be possible to determine if cultural procedures based on fertilizer applications can be used to reduce *Phytophthora* diseases.

## CONCLUSIONS

Grainger (1979) observed that "It is easy . . . for plant pathologists . . . to concentrate on [their] desire for scientific understanding and to neglect [their] duty to be useful to farmers." Although much progress has been made in understanding the biology of *Phytophthora* pathogens and the hosts they attack, this knowledge has rarely been translated into practical control procedures. We suggest that this is primarily a consequence of the failure of plant pathologists to integrate their research findings and apply the principles of integrated control to the field.

Although the dynamic nature of the interactions among host, *Phytophthora* pathogens, and the environment is recognized, most research has examined only the effect of single factors on the host and on the pathogen. Almost no research has been done on the relationship between temporal variations in pathogen activity and host susceptibility, induced by changes in the environment, and disease development. We suggest that many of the puzzling variations in *Phytophthora* disease intensity in time and space could be resolved by contemporaneous analysis of pathogen activity and host receptivity together with the principal factors of the environment that affect pathogen and host.

Although the data base is small in situations where we know something about the basic epidemiology of *Phytophthora* diseases, disease outbreaks appear to occur during periods when a number of environmental factors favor the pathogen and when the host has minimal resistance. Biological or cultural control techniques directed at manipulating one or even a few of these factors, without consideration of the temporal variations in pathogen activity and host resistance, are unlikely to significantly lessen disease. For example, the potential for successful biological or cultural control is negligible if there are unavoidable periods during which the host is susceptible to very low levels of inoculum.

In the relatively few instances where *Phytophthora* diseases have been controlled by biological and/or cultural methods and where *Phytophthora* diseases are suppressed in natural ecosystems, a number of environmental factors that disfavor the pathogens and increase the resistance of the hosts appear to be operating. We suggest that biological and/or cultural control is most likely to succeed when the various techniques that can be used to manipulate pathogens and hosts are integrated to reduce pathogen activity and increase host resistance during the critical periods when disease outbreaks occur.

An understanding of the basic epidemiology of *Phytophthora* diseases, however, is a prerequisite for the development of integrated biological and/or cultural control procedures. Effective control is likely to be achieved by these methods only if more of our research effort is directed toward studies that elucidate the interactions between host and pathogen over time in the environment where the disease is occurring.

Our failure to undertake this type of research in the past can be attributed in part to the difficulty of making a meaningful analysis of the data that are derived. But as MacKenzie et al (Chapter 23) point out, computer simulation has the potential to remove this constraint. A more serious problem is our failure to develop integrated research programs. Elucidation of the basic epidemiology of *Phytophthora* diseases requires specialized knowledge in a number of disciplines, and the logistic problems are considerable. But most plant pathology research programs are structured around a single research scientist with limited technical support staff. The premium placed on the quantity of published papers produced by plant pathologists also is a disincentive. In view of the limited success in developing methods to control *Phytophthora* diseases, we suggest that it may be appropriate to examine the institutional constraints on the development of integrated research programs.

## SUMMARY

The ability of *Phytophthora* spp. to increase propagule density rapidly is the basis for their success as plant pathogens. Yet the sequence of events leading to the release and transmission of zoospores is complex and sensitive to small changes in the physical and biological environment. Similarly, the susceptibility of many hosts varies with changes in the environment. Thus the potential exists to use cultural techniques to modify the environment to disfavor the pathogens and increase the resistance of hosts.

Several cultural procedures, for example sanitation and drainage control, have been used to lessen disease, but in few instances has control been achieved by biological or cultural methods. However, the absence of disease in some situations in nature (and in a few instances in crops) in the presence of the pathogen under environmental conditions normally suited for disease development indicates that control by ecological methods is possible.

Considerable progress has been made in understanding the biology of *Phytophthora* pathogens, but very little research has been done on the interaction between temporal variations in pathogen activity and host susceptibility in the environment where the disease is occurring. Elucidating these interactions is the major prerequisite for the development of practical biological and/or cultural control methods.

## LITERATURE CITED

Ahmed, M., Oades, J. M., and Ladd, J. N. 1982. Determination of ATP in soils: Effect of soil treatments. Soil Biol. Biochem. 14:273-279.

Ahmed, N., and Ahmed, Q. A. 1963. Selection of antagonistic soil microbes against a few fungal pathogens of *Piper betle* L. Mycopathol. Mycol. Appl. 21:334-340.

Alexander, R., Croft, K. D., Kagi, R. I., and Shea, S. R. 1978. Saponins of *Acacia pulchella*. Aust. J. Chem. 31:2741-2744.

Anderson, J. P. E., and Domsch, K. H. 1973. Quantification of bacterial and fungal contributions to soil respiration. Arch. Mikrobiol. 93:113-127.

Anderson, J. P. E., and Domsch, K. H. 1978. A physiological method for the quantitative measurement of microbial biomass in soil. Soil Biol. Biochem. 10:215-221.

Ayers, W. A. 1971. Induction of sporangia in *Phytophthora cinnamomi*. Phytopathology 61:1188-1193.

Ayers, W. A., and Zentmyer, G. A. 1971. Effect of soil solution and two pseudomonads on sporangium production by *Phytophthora cinnamomi*. Phytopathology 61:1188-1193.

Babiuk, L. A., and Paul, E. A. 1970. The use of fluorescein isothiocyanate in the determination of the bacterial biomass of grassland soil. Can. J. Microbiol. 16:57-62.

Baker, K. F. 1938. Effect of raised beds in the reduction of heart rot and root rot in H.P. 4718. Pineapple Res. Inst. Rep., Pathol. Project 14-f. Pineapple Res. Inst., Hawaii. 1 p.

Baker, K. F. 1962. Principles of heat treatment of soil and planting material. J. Aust. Inst. Agric. Sci. 28:118-126.

Baker, K. F., and Cook, R. J. 1974. Biological Control of Plant Pathogens. W. H. Freeman and Company, San Francisco. 433 pp.

Balkwill, D. L., Labeda, D. P., and Casida, L. E., Jr. 1975. Simplified procedures for releasing and concentrating microorganisms from soil for transmission electron microscopy viewing as thin-section and frozen-etched preparations. Can. J. Microbiol. 21:252-262.

Barham, R. O. 1972. Effect of nematodes and *Phytophthora cinnamomi* on mycorrhizae of shortleaf pine seedlings. (Abstr.) Phytopathology 62:801.

Barham, R. O., Marx, C. H., and Ruehle, J. L. 1974. Infection of ectomycorrhizal and nonmycorrhizal roots of shortleaf pine by nematodes and *Phytophthora cinnamomi*. Phytopathology 64:1260-1264.

Bartle, J. R., and Shea, S. R. 1979. Development of the ecosystem after mining. Pages 1-16 in: Environmental Workshop, Vol. 1. Aust. Mining Ind. Council, Bunbury, Western Australia.

Batini, F., and Podger, F. D. 1968. Shelterbelt mortalities on the Swan Coastal Plain. Aust. For. Res. 3:39-45.

Bellamy, G., Heather, W. A., and Pratt, B. H. 1971. The effects of calcium chloride treatment in reducing the rate of spread of *Phytophthora cinnamomi* within roots of *Lupinus angustifolius*. Pages 3-15 in: Proc. Aust. Plant Pathol. Conf., Hobart, Tasmania, 1971, Vol. 3(e). Australian Plant Pathology Society, Hobart.

Benson, D. M. 1978. Thermal inactivation of *Phytophthora cinnamomi* for control of Fraser fir root rot. Phytopathology 68:1373-1376.

Blaker, N. S., and MacDonald, J. D. 1981. Predisposing effects of soil moisture extremes on the susceptibility of rhododendron to Phytophthora root and crown rot. Phytopathology 71:831-834.

Blowes, W. M. 1980. A comparison of the occurrence, sporulation and survival of *Phytophthora cinnamomi*. Ph.D. thesis, Aust. Nat. Univ., Canberra. 195 pp.

Borst, G. 1970. Selection, management of avocado soils. Citrograph 55:263-265.

Borst, G. 1975. Amelioration of root rot hazards in granitic soils. Calif. Avocado Soc. Yearb. 58:33-35.

Boughton, T. J., Malajczuk, N., and Robson, A. D. 1978. Suppression of the infection of jarrah roots by *Phytophthora cinnamomi* with application of calcium carbonate. Aust. J. Bot. 26:611-615.

Bowen, G. D. 1973. Mineral nutrition of ectomycorrhizae. Pages 151-206 in: Ectomycorrhizae—Their Ecology and Physiology. G. C. Marks and T. T. Kozlowski, eds. Academic Press, New York. 444 pp.

Bradshaw, F. J. 1974. Large scale 70-mm colour photography as an aid to detecting jarrah dieback. Pages 439-448 in: Commission VII Proceedings. Symposium on Remote Sensing and Photo Interpretation, Banff, Alberta, Canada. Int. Soc. Photogrammetry.

Bradshaw, F. J., and Chandler, R. J. 1978. Full coverage at large scale. Symposium on Remote Sensing for Vegetation Damage Assessment, Seattle, WA.

Broadbent, P. 1977. Organic amendments for avocadoes. Pages 3-5 in: Seminar Papers. Annu. Conf. Aust. Nurserymen's Assoc. Ltd., Hobart.

Broadbent, P., and Baker, K. F. 1974a. Behaviour of *Phytophthora cinnamomi* in soils suppressive and conducive to root rot. Aust. J. Agric. Res. 25:121-137.

Broadbent, P., and Baker, K. F. 1974b. Association of bacteria with sporangium formation and breakdown of sporangia in *Phytophthora* spp. Aust. J. Agric. Res. 25:139-145.

Broadbent, P., and Baker, K. F. 1975. Soils suppressive to Phytophthora root rot in eastern Australia. Pages 152-157 in: Biology and Control of Soil-Borne Plant Pathogens. G. W. Bruehl, ed. Am. Phytopathol. Soc., St. Paul, MN. 216 pp.

Broadbent, P., Baker, K. F., and Waterworth, Y. 1971a. Bacteria and actinomycetes antagonistic to fungal root pathogens in Australian soils. Aust. J. Biol. Sci. 24:925-944.

Broadbent, P., Fraser, L. R., and Waterworth, Y. 1971b. The reaction of seedlings of *Citrus* spp. and related genera to *Phytophthora citrophthora*. Proc. Linn. Soc. N.S.W. 96:119-127.

Brown, B. N. 1976. *Phytophthora cinnamomi* associated with patch death in tropical rain forests in Queensland. Aust. Plant Pathol. Soc. Newsl. 5:1-4.

Brown, M. E. 1975. Effect of host and fungal nutrition on expression of disease. Pages 250-251 in: Rep. Rothamsted Exp. Stn. for 1974, Part 1. Harpenden, England.

Burns, R. M., Miner, J. H., Gustafson, C. D., Zentmyer, G. A., and Thorn, W. A. 1960. Correlation of soil series and avocado root rot damage in the Fallbrook area. Calif. Avocado Soc. Yearb. 44:110-113.

Campbell, W. A., and Copeland, O. L., Jr. 1954. Littleleaf disease of shortleaf and loblolly pines. U.S. Dep. Agric. Circ. 940.

Chalker, F. C. 1974. Drainage of avocado plantings on the North Coast. Agric. Gaz. N.S.W. 85(1):43-44.

Chapman, H. D. 1965. Chemical factors of the soil as they affect microorganisms. Pages 120-142 in: Ecology of Soil-Borne Plant Pathogens. K. F. Baker and W. C. Snyder, eds. Univ. Calif. Press, Berkeley. 571 pp.

Chee, K.-H., and Newhook, F. J. 1964. Variability in *Phytophthora cinnamomi* Rands. N.Z. J. Agric. Res. 8:96-103.

Chee, K.-H., and Newhook, F. J. 1966. Relationship of microorganisms to sporulation of *Phytophthora cinnamomi* Rands. N.Z. J. Agric. Res. 9:32-43.

Chen, D.-W., and Zentmyer, G. A. 1970. Production of sporangia by *Phytophthora cinnamomi* in axenic culture. Mycologia 62:397-402.

Colwell, J. D. 1958. Observations on the pedology and fertility of some krasnozems in northern New South Wales. J. Soil Sci. 9:46-57.

Cother, E. J., and Griffin, D. M. 1973. The role of alternative hosts in survival of *Phytophthora drechsleri*. Aust. J. Biol. Sci. 26:1109-1113.

Cruickshank, I. A. M., and Perrin, D. R. 1967. Studies on phytoalexins. X. Effect of oxygen tension on the biosynthesis of pisatin and phaseollin. Phytopathol. Z. 60:335-342.

Dakwa, J. T., and Sewell, G. W. F. 1981. Influence of rootstock type and time of inoculation on the resistance of five apple scion cultivars to collar rot caused by *Phytophthora cactorum*. J. Hortic. Sci. 56:357-362.

Davis, R. M., and Menge, J. A. 1980. Influence of *Glomus fasciculatus* and soil phosphorus on Phytophthora root rot of citrus. Phytopathology 70:447-452.

Davis, R. M., Menge, J. A., and Zentmyer, G. A. 1978. Influence of vesicular-arbuscular mycorrhizae on Phytophthora root rot of three crop plants. Phytopathology 68:1614-1617.

Drew, M. C., and Lynch, J. M. 1980. Soil anaerobiosis, microorganisms and root function. Annu. Rev. Phytopathol. 18:37-66.

Duniway, J. M. 1977. Predisposing effect of water stress on the severity of Phytophthora root rot in safflower. Phytopathology 67:884-889.

Duniway, J. M. 1979. Water relations of water molds. Annu. Rev. Phytopathol. 17:431-460.

Faegri, A. Torsvik, V. L., and Goksöyr, J. 1977. Bacterial and fungal activities in soil: Separation of bacteria and fungi by a rapid fractionated centrifugation technique. Soil Biol. Biochem. 9:105-112.

Fawcett, H. S. 1923. Gum diseases of citrus trees in California. Pages 370-423 in: Calif. Agric. Exp. Stn. Bull. 360.

Garrett, S. D. 1950. Ecology of the root inhabiting fungi. Biol. Rev. 25:220-254.

Gates, J. E., and Millikan, D. F. 1972. Seasonal fluctuations in susceptibility of the inner bark tissues of apple to colonization by the collar rot fungus, *Phytophthora cactorum*. Phytoprotection 53:76-81.

Gerlagh, M. 1968. Introduction of *Ophiobolus graminis* into new polders and its decline. Neth. J. Plant Pathol. 74(Suppl. 2):1-97.

Gerrettson-Cornell, L., and Dowden, H. 1978. Forestry and *Phytophthora* in New South Wales. Forest Timber 14(1):13-15.

Gerrettson-Cornell, L., Humphreys, F. R., and Townsend, S. R. 1976. Results of a preliminary investigation on the use of *Pinus radiata* bark against *Phytophthora cinnamomi* Rands. Phyton 34(1):3-6.

Gilpatrick, J. D. 1969a. Role of ammonia in the control of avocado root rot with alfalfa meal soil amendment. Phytopathology 59:973-978.

Gilpatrick, J. D. 1969b. Effect of soil amendments upon inoculum survival and function in Phytophthora root rot of avocado. Phytopathology 59:979-985.

Goodall, G. E. 1955. Avocado root disease in Santa Barbara County. Calif. Avocado Soc. Yearb. 39:149-153.

Grainger, J. 1968. $C_p/R_s$ and the disease potential of plants. Hortic. Res. 8:1-40.

Grainger, J. 1979. Scientific proportion and economic decisions for farmers. Annu. Rev. Phytopathol. 17:223-252.

Greenhalgh, F. C. 1978. Evaluation of techniques for quantitative detection of *Phytophthora cinnamomi*. Soil Biol. Biochem. 10:257-259.

Griffin, D. M. 1972. Ecology of Soil Fungi. Syracuse Univ. Press, Syracuse, NY. 193 pp.

Halsall, D. M. 1977. Effects of certain cations on the formation and infectivity of *Phytophthora* zoospores. 2. Effects of copper, boron, cobalt, manganese, molybdenum, and zinc ions. Can. J. Microbiol. 23:1002-1010.

Halsall, D. M. 1980. Calcium nutrition and the infection of eucalypt seedlings by *Phytophthora cinnamomi*. Aust. J. Bot. 28:19-25.

Halsall, D. M. 1982a. A forest soil suppressive to *Phytophthora cinnamomi* and conducive to *Phytophthora cryptogea*. 1. Survival, germination and infectivity of mycelium and chlamydospores. Aust. J. Bot. 30:11-25.

Halsall, D. M. 1982b. A forest soil suppressive to *Phytophthora cinnamomi* and conducive to *Phytophthora cryptogea*. II. Suppression of sporulation. Aust. J. Bot. 30:27-37.

Halsall, D. M., and Forrester, R. I. 1977. Effect of certain cations on the formation and infectivity of *Phytophthora* zoospores. 1. Effects of calcium, magnesium, potassium and iron ions. Can. J. Microbiol. 23:994-1001.

Hatch, A. B. 1959. The effect of frequent burning on the jarrah (*Eucalyptus marginata*) forest soils of Western Australia. J. R. Soc. West. Aust. 42:97-100.

Hendrix, J. W. 1964. Sterol induction of reproduction and stimulation of growth of *Pythium* and *Phytophthora*. Science 144:1028-1029.

Hendrix, J. W. 1965. Influence of sterols on growth and reproduction of *Pythium* and *Phytophthora* spp. Phytopathology 55:790-797.

Hickman, C. J. 1958. *Phytophthora*—plant destroyer. Trans. Br. Mycol. Soc. 41:1-13.

Hilgeman, R. H., Dunlap, J. A., and Sharples, G. C. 1967. Effect of time of harvest of Valencia oranges on leaf carbohydrate content and subsequent set of fruit. Proc. Am. Soc. Hortic. Sci. 90:110-116.

Hine, R. B., Alaban, C., and Klemmer, H. 1964. Influence of soil temperature on root and heart rot of pineapple caused by *Phytophthora cinnamomi* and *Phytophthora parasitica*. Phytopathology 54:1287-1289.

Hoitink, H. A. J. 1980. Composted bark, a lightweight growth medium with fungicidal properties. Plant Dis. 64:142-147.

Hoitink, H. A. J., Van Doren, D. M., Jr., and Schmitthenner, A. F. 1977. Suppression of *Phytophthora cinnamomi* in a composted hardwood bark potting medium. Phytopathology 67:561-565.

Honour, R. C. 1973. Lysis of *Phytophthora parasitica* mycelium and oospores in soil. Ph.D. dissertation, Univ. Calif., Riverside. 204 pp. Diss. Abstr. Int. B 34(10):4775.

Hornby, D., and Goring, C. A. I. 1972. Effects of ammonium and nitrate nutrition on take-all disease of wheat in pots. Ann. Appl. Biol. 70:225-231.

Hyre, R. A., and Ettinger, G. 1967. Sporulation of *Phytophthora phaseoli*. Plant Dis. Rep. 51:253-256.

Kassaby, F. Y., Fagg, P. C., and Marks, G. C. 1977. Survival in and colonization of mountain forest soil by *Phytophthora cinnamomi* Rands in eastern Australia. Aust. For. Res. 7:173-183.

Khan, A. G. 1974. The occurrence of mycorrhizas in halophytes, hydrophytes and xerophytes, and of *Endogone* spores in adjacent soils. J. Gen. Microbiol. 81:7-14.

Kimber, P. C. 1978. Increased girth increment associated with crown scorch in jarrah. For. Dep. West. Aust. Res. Pap. 37.

Kliejunas, J. T., and Nagata, J. T. 1979. *Phytophthora cinnamomi* in Hawaiian forest soils: Seasonal variations in population levels. Phytopathology 69:1268-1272.

Klotz, L. J., De Wolfe, T. A., and Wong, P.-P. 1958. Influence of two varieties of citrus scions on the pathogenicity of three isolates of *Phytophthora parasitica* to sweet orange rootstocks. Phytopathology 48:520-521.

Klotz, L. J., Bitters, W. P., and De Wolfe, T. A. 1966. Resistance to Phytophthora root rot. Calif. Citrogr. 51:257-258.

Ko, W. H. 1971. Biological control of root rot of papaya seedling caused by *Phytophthora palmivora*. (Abstr.) Phytopathology

61:1322.

Kreidemann, P. E. 1969. $^{14}$C translocation in orange plants. Aust. J. Agric. Res. 20:291-300.

Kuan, T. L., and Erwin, D. C. 1980. Predisposition effect of water saturation of soil on Phytophthora root rot of alfalfa. Phytopathology 70:981-986.

Lee, B.-S. 1979. The influence of calcium and nitrogen on diseases caused by *Phytophthora cinnamomi* and two morphological forms of *Phytophthora palmivora*. Ph.D. thesis. Univ. Calif., Riverside.

Levy, Y., and Krikun, J. 1980. Effect of vesicular-arbuscular mycorrhizae on *Citrus jambhiri* water relations. New Phytol. 85:25-31.

Lewcock, H. K. 1935. Pineapple wilt disease and its control. Queensl. Agric. J. 43:9-17.

MacDonald, J. D. 1982. Effect of salinity stress on development of Phytophthora root rot of chrysanthemum. Phytopathology 72:214-219.

MacDonald, J. D., and Duniway, J. M. 1979. Use of fluorescent antibodies to study the survival of *Phytophthora megasperma* and *P. cinnamomi* zoospores in soil. Phytopathology 69:436-441.

Malajczuk, N. 1975. Interactions between *Phytophthora cinnamomi* Rands and the roots of *Eucalyptus calophylla* R. Br. and *Eucalyptus marginata* Donn. ex Sm. Ph.D. thesis, Univ. West. Aust., Perth. 371 pp.

Malajczuk, N. 1979a. Biocontrol of *Phytophthora cinnamomi* in eucalypts and avocados in Australia. Pages 635-652 in: Soil-Borne Plant Pathogens. B. Schippers and W. Gams, eds. Academic Press, London. 686 pp.

Malajczuk, N. 1979b. The microflora of unsuberized roots of *Eucalyptus calophylla* R. Br. and *Eucalyptus marginata* Donn. ex Sm. seedlings grown in soils suppressive and conducive to *Phytophthora cinnamomi* Rands. II. Mycorrhizal roots and associated microflora. Aust. J. Bot. 27:255-272.

Malajczuk, N., and McComb, A. J. 1977. Root exudates from *Eucalyptus calophylla* R. Br. and *Eucalyptus marginata* Donn. ex Sm. seedlings and their effect on *Phytophthora cinnamomi* Rands. Aust. J. Bot. 25:501-514.

Malajczuk, N., Nesbitt, H. J., and Glenn, A. R. 1977. A light and electron microscope study of the interaction of soil bacteria with *Phytophthora cinnamomi* Rands. Can. J. Microbiol. 23:1518-1525.

Malajczuk, N., Bowen, G. D., and Greenhalgh, F. C. 1978. A combined fluorescent antibody and soil sieving technique to count chlamydospores of *Phytophthora cinnamomi* in soil. Soil Biol. Biochem. 10:437-438.

Manning, W. J., and Crossan, D. F. 1966. Effects of a particular soil bacterium on sporangial production in *Phytophthora cinnamomi* in liquid culture. Phytopathology 56:235-237.

Marks, G. C., and Smith, I. W. 1981. Factors influencing suppression of root rot of *Eucalyptus sieberi* and *E. obliqua* caused by *Phytophthora cinnamomi*. Aust. J. Bot. 29:483-495.

Marks, G. C., Kassaby, F. Y., and Reynolds, S. T. 1972. Die-back in the mixed hardwood forests of eastern Victoria: A preliminary report. Aust. J. Bot. 20:141-154.

Marks, G. C., Kassaby, F. Y., and Fagg, P. C. 1973. Dieback tolerance in eucalypt species in relation to fertilization and soil populations of *Phytophthora cinnamomi*. Aust. J. Bot. 21:53-65.

Marks, G. C., Kassaby, F. Y., and Fagg, P. C. 1975. Variation in population levels of *Phytophthora cinnamomi* in *Eucalyptus* forest soils of eastern Victoria. Aust. J. Bot. 23:435-449.

Marks, G. C., Smith, I. W., and Kassaby, F. Y. 1982. Trunk infection of *Eucalyptus* species by *Phytophthora cinnamomi* Rands—A preliminary report. Aust. For. Res. (In press)

Marx, D. H. 1969a. The influence of ectotrophic mycorrhizal fungi on the resistance of pine roots to pathogenic infections. I. Antagonism of mycorrhizal fungi to root pathogenic fungi and soil bacteria. Phytopathology 59:153-163.

Marx, D. H. 1969b. The influence of ectotrophic mycorrhizal fungi on the resistance of pine roots to pathogenic infections. II. Production, identification, and biological activity of antibiotics produced by *Leucopaxillus cerealis* var. *piceina*. Phytopathology 59:411-417.

Marx, D. H. 1972. Ectomycorrhizae as biological deterrents to pathogenic root infections. Annu. Rev. Phytopathol. 10:429-454.

Marx, D. H. 1973. Mycorrhizae and feeder root diseases. Pages 351-382 in: Ectomycorrhizae—Their Ecology and Physiology. G. C. Marks and T. T. Kozlowski, eds. Academic Press, New York. 444 pp.

Marx, D. H. 1975. Tailoring tree seedlings in nurseries with specific mycorrhizal fungi. (Abstr.) Proc. Am. Phytopathol. Soc. 2:47.

Marx, D. H., and Haasis, F. A. 1965. Induction of aseptic sporangial formation in *Phytophthora cinnamomi* by metabolic diffusates of soil micro-organisms. Nature (London) 206:673-674.

Masago, H., Yoshikawa, M., Fukada, M., and Nakanishi, N. 1977. Selective inhibition of *Pythium* spp. on a medium for direct isolation of *Phytophthora* spp. from soils and plants. Phytopathology 67:425-428.

Mataré, R., and Hattingh, M. J. 1978. Effect of mycorrhizal status of avocado seedlings on root rot caused by *Phytophthora cinnamomi*. Plant Soil 49:433-435.

Mehrlich, F. P. 1935. Nonsterile soil leachate stimulating to zoosporangia production by *Phytophthora* sp. Phytopathology 25:432-435.

Mircetich, S. M. 1978. Influence of different rootstocks and soil water management on Phytophthora root and crown rot of cherry trees. (Abstr.) Page 193 in: Proc. Int. Congr. Plant Pathol., 3rd, Munich.

Mircetich, S. M., and Matheron, M. E. 1970. Relative resistance of northern California black paradox and English walnut seedlings to *Phytophthora* spp. (Abstr.) Proc. 11th Annu. Walnut Res. Conf., Davis, Calif.

Mircetich, S. M., and Matheron, M. E. 1976. Phytophthora root and crown rot of cherry trees. Phytopathology 66:549-558.

Mircetich, S. M., and Matheron, M. E. 1978. Phytophthora root and crown rot of walnut trees in California. (Abstr.) Phytopathol. News 12:146.

Mircetich, S. M., Matheron, M. E., Ramos, D. E., Moller, W. J., Olson, W. H., Aldrich, T. M., Osgood, J. W., and Schreader, W. R. 1979. *Phytophthora* spp. associated with walnuts and relative resistance of northern black and paradox rootstocks to four *Phytophthora* spp. (Abstr.) Page 4 in: Proc. 10th Annu. West Coast Walnut Res. Conf., Stockton, Calif., January 24-25, 1978.

Napier, D. J. 1969. The occurrence and seasonal variation of the fine roots and mycorrhizae of loblolly pine (*Pinus taeda* L.) in the West Bay area of Allen Parish, Louisiana. M.S. thesis, Northwestern State College of Louisiana, Natchitoches. (Cited in Marx, 1973)

Nesbitt, H. J., Malajczuk, N., and Glenn, A. R. 1979. Effect of organic matter on the survival of *Phytophthora cinnamomi* Rands in soil. Soil Biol. Biochem. 11:133-136.

Newhook, F. J. 1959. The association of *Phytophthora* spp. with mortality of *Pinus radiata* and other conifers. I. Symptoms and epidemiology in shelterbelts. N.Z. J. Agric. Res. 2:808-843.

Newhook, F. J. 1970. *Phytophthora cinnamomi* in New Zealand. Pages 173-176 in: Root Diseases and Soil-Borne Pathogens. T. A. Toussoun, R. V. Bega, and P. E. Nelson, eds. Univ. Calif. Press, Berkeley. 252 pp.

Newhook, F. J., and Podger, F. D. 1972. The role of *Phytophthora cinnamomi* in Australian and New Zealand forests. Annu. Rev. Phytopathol. 10:299-326.

Nicholls, K. D., and Tucker, B. M. 1956. Pedology and chemistry of the basaltic soils of the Lismore district, New South Wales. Page 153 in: CSIRO Soil Publ. 7. CSIRO, Melbourne.

Old, K. M. 1979. Effects of soil microorganisms on *Phytophthora* spp. Pages 52-63 in: *Phytophthora* and Forest Management in Australia. K. M. Old, ed. CSIRO, Melbourne. 114 pp.

Otrosina, W. J., and Marx, D. H. 1975. Populations of *Phytophthora cinnamomi* and *Pythium* spp. under shortleaf and loblolly pines in littleleaf disease sites. Phytopathology 65:1224-1229.

Palzer, C. 1976. Zoospore inoculum potential of *Phytophthora cinnamomi*. Ph.D. thesis, Univ. West. Aust., Perth. 222 pp.

Pegg, K. G. 1973. Macadamia trunk canker disease. Queensl. Agric. J. 99:595-596.

Pegg, K. G. 1977a. Biological control of *Phytophthora cinnamomi* root rot of avocado and pineapple in Queensland. Pages 7-12 in: Seminar Papers, Annu. Conf. Aust. Nurserymen's Assoc. Ltd., Hobart. 38 pp.

Pegg, K. G. 1977b. Soil application of elemental sulphur as a control of *Phytophthora cinnamomi* root and heart rot of pineapple. Aust. J. Exp. Agric. Anim. Husb. 17:859-865.

Pegg, K. G., and Alcorn, J. L. 1972. *Phytophthora cinnamomi* in indigenous flora in southern Queensland. Search 3:257.

Podger, F. D. 1968. Aetiology of jarrah dieback. M.S. thesis, Univ. Melbourne. 214 pp.

Podger, F. D. 1972. *Phytophthora cinnamomi*, a cause of lethal disease in indigenous plant communities in Western Australia. Phytopathology 62:972-981.

Pratt, B. H. 1971. Isolation of basidiomycetes from Australian eucalypt forest and assessment of their antagonism to *Phytophthora cinnamomi*. Trans. Br. Mycol. Soc. 56:243-250.

Pratt, B. H., and Heather, W. A. 1973. The origin and distribution of *Phytophthora cinnamomi* Rands in Australian native plant communities and the significance of its association with particular plant species. Aust. J. Biol. Sci. 26:559-573.

Rands, R. D. 1922. Streepkanker van Kaneel, Veroorzaakt door *Phytophthora cinnamomi* n. sp. (Stripe canker of cinnamon caused by *Phytophthora cinnamomi* n. sp.) Meded. Inst. Plantenziekten 56:41.

Reeves, R. J. 1975. Behaviour of *Phytophthora cinnamomi* Rands in different soils and water regimes. Soil Biol. Biochem. 7:19-24.

Ross, J. P. 1972. Influence of *Endogone* mycorrhiza on Phytophthora rot of soybean. Phytopathology 62:896-897.

Rossetti, V. 1969. Studies on *Phytophthora* species on citrus in Brazil. Proc. Int. Citrus Symp. 1st 3:1211-1216.

Rossetti, V., and Bitancourt, A. A. 1951. Studies of *Phytophthora* gummosis of citrus. II. Influence of the state of vegetation of the host on the size of the experimental lesions. Arq. Inst. Biol. Sao Paulo 20:7[illegible]-94.

Roth, E. L., and Kuhlman, E. G. 1963. Field tests of the capacity of Phytophthora root rot to damage Douglas fir. Forestry 61:3, 199-205.

Roth, E. R., Toole, E. R., and Hepting, G. H. 1948. Nutritional aspects of the littleleaf disease of pine. J. For. 46:578-587.

Schoeneweiss, D. F. 1975. Predisposition, stress and plant disease. Annu. Rev. Phytopathol. 13:193-211.

Sewell, G. W. F., and Wilson, J. F. 1973. Phytophthora collar rot of apple: Seasonal effects on infection and disease development. Ann. Appl. Biol. 74:149-158.

Sewell, G. W. F., and Wilson, J. F. 1974. Seasonal factors and Phytophthora collar rot. Pages 183-185 in: Rep. East Malling Res. Stn. for 1973. Maidstone, England.

Shea, S. R. 1975. Environmental factors of the northern jarrah forest in relation to pathogenicity and survival of *Phytophthora cinnamomi*. West. Aust. For. Dep. Bull. 85. 83 pp.

Shea, S. R. 1979a. Forest management and *Phytophthora cinnamomi* in Australia. Pages 73-100 in: *Phytophthora* and Forest Management in Australia. K. M. Old, ed. CSIRO, Melbourne. 114 pp.

Shea, S. R. 1979b. *Phytophthora cinnamomi* (Rands)—A collar rot pathogen of *Banksia grandis* Willd. Aust. Plant Pathol. 8:3-34.

Shea, S. R., and Dell, B. 1981. Structure of the surface root system of *Eucalyptus marginata* Sm. and its infection by *Phytophthora cinnamomi* Rands. Aust. J. Bot. 29:49-58.

Shea, S. R., and Malajczuk, N. 1977. Potential for control of eucalypt dieback in Western Australia. Pages 13-19 in: Seminar Papers, Annu. Conf. Aust. Nurserymen's Assoc. Ltd., Hobart. 38 pp.

Shea, S. R., Malajczuk, N., and Kitt, R. J. 1976. Promotion of understorey native legumes—A possible method of control of *Phytophthora cinnamomi* in the northern jarrah forests of W. A. Aust. Plant Pathol. Soc. Newsl. 5 (Suppl.): Abstr. 5.

Shea, S. R., Gillen, K. J., and Kitt, R. J. 1978. Variation in sporangial production of *Phytophthora cinnamomi* Rands on jarrah *E. marginata* Sm. forest sites with different understorey compositions. Aust. For. Res. 8:219-226.

Shea, S. R., Gillen, K. J., and Leppard, W. I. 1980a. Seasonal variation in population levels of *Phytophthora cinnamomi* Rands in soil in diseased, freely-drained *Eucalyptus marginata* Sm. sites in the northern jarrah forest of south-western Australia. Prot. Ecol. 2:135-156.

Shea, S. R., Peet, G. B., and Cheney, N. P. 1980b. The role of fire in forest management. Pages 443-470 in: Fire and the Australian Biota. A. M. Gill, R. H. Groves, and I. R. Noble, eds. Aust. Acad. Sci., Canberra.

Shea, S. R., Shearer, B. L., and Titze, J. 1980c. Seasonal development of *Phytophthora cinnamomi* Rands in the *Eucalyptus marginata* forest in relation to site and environment. (Abstr.) 4th Nat. Aust. Plant Pathol. Soc. Conf., Perth.

Shearer, B. L., Shea, S. R., and Fairman, R. G. 1981. Infection of the stem and large roots of *Eucalyptus marginata* by *Phytophthora cinnamomi*. Aust. Plant Pathol. Soc. Newsl. 10:2-3.

Shigo, A. L., and Marx, H. G. 1975. Compartmentalization of decay associated with *Fomes annosus* in trunks of *Pinus resinosa*. Phytopathology 65:1038-1039.

Skujins, J. J. 1967. Enzymes in soil. Pages 371-414 in: Soil Biochemistry, Vol. 1. A. D. McLaren and G. H. Peterson, eds. Marcel Dekker, New York. 509 pp.

Slankis, V. 1974. Soil factors influencing formation of mycorrhizae. Annu. Rev. Phytopathol. 12:437-457.

Smiley, R. W. 1978. Colonization of wheat roots by *Gaeumannomyces graminis* inhibited by specific soils, microorganisms and ammonium-nitrogen. Soil Biol. Biochem. 10:175-179.

Sneh, B., and McIntosh, D. L. 1974. Studies on the behavior and survival of *Phytophthora cactorum* in soil. Can. J. Bot. 52:795-802.

Sneh, B., Humble, S. J., and Lockwood, J. L. 1977. Parasitism of oospores of *Phytophthora megasperma* var. *sojae, P. cactorum, Pythium* sp., and *Aphanomyces euteiches* in soil by oomycetes, chytridiomycetes, hyphomycetes, actinomycetes, and bacteria. Phytopathology 67:622-628.

Söderström, B. E. 1977. Vital staining of fungi in pure cultures and in soil with fluorescein diacetate. Soil Biol. Biochem. 9:59-63.

Taylor, P. 1980. Relative susceptibility of peach roots and trunks to Phytophthora rot. (Abstr.) 4th Nat. Aust. Plant Pathol. Soc. Conf., Perth.

Tippett, J., and Malajczuk, N. 1979. Interaction of *Phytophthora cinnamomi* and a resistant host, *Acacia pulchella*. Phytopathology 69:764-772.

Trochoulias, T., and Broadbent, P. 1978. Comparison of soil management regimes for avocadoes. (Abstr.) 20th Int. Hortic. Congr., Sydney.

Tsao, P. H. 1970. Applications of the vital fluorescent labeling technique with brighteners to studies of saprophytic behavior of *Phytophthora* in soil. Soil Biol. Biochem. 2:247-256.

Tsao, P. H., and Guy, S. O. 1977. Inhibition of *Mortierella* and *Pythium* in a *Phytophthora*-isolation medium containing hymexazol. Phytopathology 67:796-801.

Tsao, P. H., and Oster, J. J. 1981. Relation of ammonia and nitrous acid to suppression of *Phytophthora* in soils amended with nitrogenous organic substances. Phytopathology 71:53-59.

Tsao, P. H., and Zentmyer, G. A. 1979. Suppression of *Phytophthora cinnamomi* and *Ph. parasitica* in urea-amended soils. Pages 191-199 in: Soil-Borne Plant Pathogens. B. Schippers and W. Gams, eds. Academic Press, London. 686 pp.

Tsao, P. H., Bhalla, H. S., and Zentmyer, G. A. 1975. Effects of certain organic amendments on survival and activity of *Phytophthora cinnamomi* in soil. (Abstr.) Proc. Am. Phytopathol. Soc. 2:40.

Turner, P. D. 1965. Behavior of *Phytophthora palmivora* in soil. Plant Dis. Rep. 49:135-137.

van Schreven, D. A. 1948. Investigations on certain pests and diseases of Vorstenlanden tobacco. Tijdschr. Plantenziekten 54:149-174.

Warcup, J. H. 1981. Effect of fire on the soil microflora and other nonvascular plants. Pages 203-214 in: Fire and the Australian Biota. A. M. Gill, R. H. Groves, and I. R. Noble, eds. Aust. Acad. Sci., Canberra.

Weste, G., and Law, C. 1973. The invasion of native forest by *Phytophthora cinnamomi*. III. Threat to the National Park, Wilson's Promontory, Victoria. Aust. J. Bot. 21:31-51.

Weste, G., and Taylor, P. 1971. The invasion of native forest by *Phytophthora cinnamomi*. I. Brisbane Ranges, Victoria. Aust. J. Bot. 19:281-294.

Whitfield, F. B., Shea, S. R., Gillen, K. J., and Shaw, K. J. 1981. Volatile components from the roots of *Acacia pulchella* R. Br. and their effect on *Phytophthora cinnamomi* Rands. Aust. J. Bot. 29:195-208.

Wilcox, H. E. 1968. Morphological studies of the roots of red pine, *Pinus resinosa*. II. Fungal colonization of roots and the development of mycorrhizae. Am. J. Bot. 55:688-700.

Wolstenholme, B. N., and Le Roux, J. 1974. Tentative evaluation of soil types for avocado growing in South Africa. Pages 7-10 in: Citrus Sub-Trop. Fruit J., No. 482 (Feb.).

Yarwood, C. E. 1959. Predisposition. Pages 521-562 in: Plant Pathology, Vol. 1. J. G. Horsfall and A. E. Dimond, eds. Academic Press, New York. 674 pp.

Young, B. R., Newhook, F. J., and Allen, R. N. 1977. Ethanol in the rhizosphere of seedlings of *Lupinus angustifolius* L. N.Z. J. Bot.

15:189-191.

Zak, B. 1961. Aeration and other soil factors affecting southern pines as related to littleleaf disease. U.S. Dep. Agric. For. Serv. Tech. Bull. 1248. 30 pp.

Zak, B. 1964. Role of mycorrhizae in root disease. Annu. Rev. Phytopathol. 2:377-392.

Zentmyer, G. A. 1963. Biological control of Phytophthora root rot of avocado with alfalfa meal. Phytopathology 53:1383-1387.

Zentmyer, G. A. 1965. Bacterial stimulation of sporangium production in *Phytophthora cinnamomi*. Science 150:1178-1179.

Zentmyer, G. A. 1980. *Phytophthora cinnamomi* and the Diseases It Causes. Monograph 10. Am. Phytopathol. Soc., St. Paul, MN. 96 pp.

Zentmyer, G. A., and Bald, J. G. 1977. Management of the environment. Pages 121-144 in: Plant Disease—An Advanced Treatise, Vol. 1. J. G. Horsfall and E. B. Cowling, eds. Academic Press, New York. 465 pp.

Zentmyer, G. A., and Erwin, D. C. 1970. Development and reproduction of *Phytophthora*. Phytopathology 60:1120-1127.

Zentmyer, G. A., and Lewis, L. L. 1975. Avocado root rot research program. Calif. Avocado Soc. Yearb. 58:89-92.

Zentmyer, G. A., and Mircetich, S. M. 1966. Saprophytism and persistence in soil by *Phytophthora cinnamomi*. Phytopathology 56:710-712.

Zentmyer, G. A., and Thompson, C. R. 1967. The effect of saponins from alfalfa on *Phytophthora cinnamomi* in relation to control of root rot of avocado. Phytopathology 57:1278-1279.

Zentmyer, G. A., Paulus, A. O., and Burns, R. M. 1967. Avocado root rot. Calif. Agric. Exp. Stn. Ext. Serv. Circ. 511 (revised). 16 pp.

# 27

# Problems and Prospects in *Phytophthora* Research

C. M. Brasier

CONTENTS



## INTRODUCTION

A veritable galaxy of papers reviewing almost every aspect of *Phytophthora* research has been presented at this Symposium. In the face of this impressive array of information, I have been given the daunting task of drawing some threads together and providing some form of prognosis. However, like all of the speakers at this Symposium, I am a specialist in only one or two fields, and I don't possess any means of foretelling the future. Many of the things I planned to say have already been more ably put by earlier authors, and since I cannot give all aspects of *Phytophthora* research their due consideration, I shall have to be selective. I am well aware that this will result in the omission of some important aspects of *Phytophthora* research, but this does not mean that they are any less significant.

I shall concentrate on topics fundamental to an understanding of how *Phytophthora* exists and behaves in populations. I shall also concentrate on conceptual problems requiring an experimental approach. In so doing, I will allow myself to speculate freely, and if I am provocative, it will only be with the intention of viewing a problem from a different angle. At times I shall be unashamedly teleological, because in some circumstances thinking teleologically—or even "thinking like a fungus"—can stimulate more ideas in reader and writer alike than a more conventional approach, and ideas are surely the currency of science. If a phytophthora could think, maybe it would also possess the rudiments of a personality and hence of individuality. My first topic will be the question of individuality in *Phytophthora*.

## WHAT IS AN INDIVIDUAL IN *PHYTOPHTHORA*?

Most of us have a laboratory-derived concept of an "individual" of *Phytophthora*: a single-isolate culture, largely genetically homogeneous and derived from a single zoospore, chlamydospore, or oospore or from a mycelial subculture. But can a more genetically complex mycelium exist? What is the "norm" for an individual in nature? Let us look at two theoretical situations where a root tip is attacked by zoospores (Fig. 1). The root tip could equally well be a cocoa pod, a potato leaf, or a rubber stem.

In situation A, zoospores of the same genotype arrive at the root surface and put out germ tubes. But what happens next? Can the germ tubes fuse to form a collaborative homokaryotic unit mycelium, or do they remain as individual mycelia? In situation B, zoospores of different genotypes arrive at the root surface. Do these germ tubes normally fuse to form a heterokaryotic unit mycelium with certain theoretical selective advantages, or do they remain discrete, and if they remain discrete, do they achieve their individuality through some form of "vegetative incompatibility system" and mutual antagonism? That is, can and do individuals of *Phytophthora* recognize and respond to each other (apart from sexual interaction)?

The answers to these questions are important for a number of reasons. First, they relate to the problem of host-parasite interactions. Are some genotypes eliminated at the earliest possible infection phase as a result of some form of competition with each other and not just because of the host's resistance mechanisms?

Second, these questions are important for an understanding of the early stages of the infection process, about which little is known. To what extent is infection a cooperative venture between zoospores, and to what extent is it a competitive race between individuals, leading perhaps to the "survival of the fittest" (Fig. 1B)?

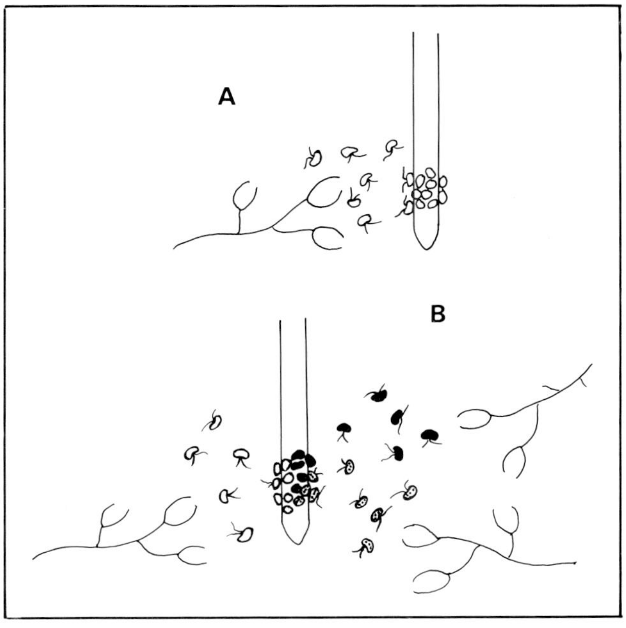


**Fig. 1.** A root tip attacked by zoospores of the same genotype **(A)** and by zoospores of three different genotypes **(B)**. Does fusion occur between zoospore germ tubes in situation A or B? Do zoospores of different genotypes compete or collaborate?

Third, to discover whether or not vegetative incompatibility systems occur in *Phytophthora* is important for an understanding of evolution in the genus, since they would probably be potent isolating mechanisms.

Observation of the presence or absence of hyphal fusions, such as the hyphal anastomoses reported by Wilde (1961) and Stephenson et al (1974), will not be enough. What is needed is supporting physiologic and genetic evidence. The current state of knowledge regarding heterokaryon formation in *Phytophthora* is largely confined to *P. infestans* and *P. megasperma*, in which new races or new auxotrophic or drug resistance combinations have been obtained when different genotypes of mycelium or zoospores have been brought together (Dyakov and Kuzovnikova, 1974; Leach and Rich, 1969; Long and Keen, 1977; Shattock and Shaw, 1976). These authors have invoked heterokaryosis and parasexuality as possible explanations. Their evidence, though only in two species, indicates that fusion may occur between different genotypes.

However, this evidence conflicts with conclusions drawn by some researchers working on ascomycetes and basidiomycetes that although heterokaryons may be forced in the laboratory, in nature such fusion is often prevented by vegetative incompatibility systems. Broadly speaking, they find that only genetically similar mycelia fuse (Fig. 1A), whereas the majority of genetically different mycelia (Fig. 1B) actually antagonize each other (e.g., Caten and Jinks, 1966; Esser and Blaich, 1973; Todd and Rayner, 1980). Work is now required to see whether similar vegetative incompatibility systems occur in *Phytophthora*; in this way we may better define the theoretical level of the individual.

While there is an important place in such work for formal genetic analysis with auxotrophic or other genetic markers, we should also supplement such work by studying interactions between isolates from wild populations, looking for signs of aversion zones or barrages and for antagonism and cytoplasmic breakdown between paired isolates of homothallic species or between isolates of the same mating type of heterothallic species (to avoid the problem of mating type interaction). If vegetative incompatibility exists in *Phytophthora*, we will have the interesting challenge of determining its genetic basis. Its ecological implications will be far-reaching.

## Amphigyny

If there is a vegetative incompatibility system in *Phytophthora*, with antagonism commonly occurring between genetically different individuals, how does sexual fusion take place between different mycelia among the outbreeding heterothallic species? Possibly the answer might be found in the curious phenomenon of amphigyny (Fig. 2). All heterothallic species of *Phytophthora* are amphigynous, an indication perhaps that amphigyny has a function in heterothallism. In contrast, all truly paragynous species are homothallic, and in homothallic species the problem of such antagonism probably does not arise at the sexual level.

With the operation of a vegetative incompatibility system, premature fusion between the hyphae of two mating types belonging to different vegetative incompatibility groups would probably lead to some form of cytoplasmic breakdown. Amphigyny might be a mechanism maintaining continued physical contact between two potentially antagonistic individuals, delaying the mingling of cytoplasm until the point when the oogonium had formed, meiosis was complete, and control over the fertilization process could be exercised by the relevant nuclei. This would also form part of an argument for a system of heterothallism and amphigyny having evolved after a system of vegetative incompatibility.

## TAXONOMY

When referring to an isolate of, say, "*P. megasperma*," we need to be confident that we are referring to the same entity. In one sense, therefore, taxonomy is a vital line of communication among *Phytophthora* workers. In another sense, that of defining and identifying homogeneous genetic units in a very wide spectrum of variation, *Phytophthora* taxonomy is an aspect of the study of populations. On the

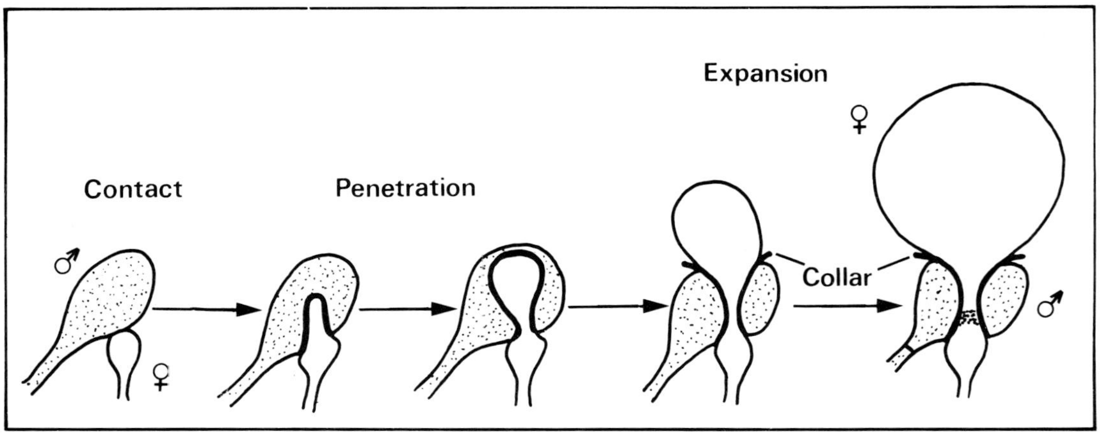


**Fig. 2.** Diagram of amphigyny (after Hemmes and Bartnicki-Garcia, 1975). Contact (left) is established between the oogonial (♀) and antheridial (♂) initials. The antheridial initial is penetrated (center) by the oogonial initial, which passes right through it. On reemergence (right), the oogonial initial expands into the oogonium, leaving the oogonial stalk surrounded by the antheridium, the remnants of the antheridial wall and the oogonial stalk wall forming a "collar." The morphological process involved in amphigyny has itself been a source of debate. Recent electron microscopic evidence (Hemmes and Bartnicki-Garcia, 1975; Ho, 1979) has supported Pethybridge's (1913) classical concept of penetration of the antheridial initial by the oogonial initial, rather than the encirclement concept proposed by Stephenson and Erwin (1972). But what is the *function* of amphigyny?

communicative aspect, we are greatly indebted to two outstanding *Phytophthora* workers, C. M. Tucker and G. M. Waterhouse, whose careful methodology has provided *Phytophthora* taxonomy with a solid and workable foundation. The population approach, first advocated by Leonian (1925), has been largely overlooked, although it was used successfully by Shepherd and Pratt (1973) to separate ecotypes of *P. drechsleri.*

Few would deny that applying the current system of *Phytophthora* taxonomy is difficult (Leonian, 1934) or that the system is frequently misapplied, so that taxonomic confusion occurs all too often in the literature. It is probably partly in response to these difficulties that the search for new diagnostic techniques has intensified during the past 15 yr. Perhaps an inevitable consequence of these explorations has been an increasing flux and uncertainty about what was previously considered firmer taxonomic ground. Indeed, *Phytophthora* taxonomy may be undergoing a slow metamorphosis, which might be usefully accelerated.

The present taxonomic uncertainties may have a number of more specifically identifiable causes. First, are we too constrained by the "rules of the game" as we have to play them today? In particular, should we continue to be accommodated within the the International Code of Botanical Nomenclature at a time when many modern systematists (e.g., Whittaker, 1969) consider that fungi are not plants and assign them to a kingdom of their own? The broad answer to this question must lie with the satisfaction of mycologists at large with the status quo, but I shall return to this point later with regard to *Phytophthora.*

Second, we are faced with the inevitable conflict of interest between the communication and population aspects of taxonomy; that is, the conflict between attempting to define the nomenclatural or artificial species on the one hand and the evolutionary or natural species on the other. Enshrined within the nomenclatural species through the Botanical Code is the concept of the "type" isolate, in which "the specific epithet is indissolubly tied to the type. A type, be it the original specimen, culture or description with figures, or all three, must be the basis for classification" (Waterhouse, 1970). Leonian (1934) argued forcefully that the type isolate concept causes serious taxonomic problems, and I agree. In reality, *Phytophthora* consists not of type isolates but of changing populations of individuals that come and go. Indeed we have not yet even defined the relationship between an isolate and an "individual" in *Phytophthora*, as discussed above; yet we attempt to define a whole species on the basis of a static type isolate.

Furthermore, many taxonomic characters such as sporangial size or form show continuous rather than discrete variation, and within each species they generally have a frequency distribution approaching a normal distribution. If such characters show a broad frequency distribution, as they often do in *Phytophthora*, a large proportion of isolates, perhaps including the type, will deviate considerably from the "worldwide" mean. Our emphasis should therefore surely be shifted away from a fortuitously selected "type" toward investigation of the whole range of variation in a taxon, in which large numbers of isolates (i.e., population samples) are studied.

A third and related cause of difficulty is that the morphological approach to taxonomy is by itself no longer adequate. In this I am in agreement with Waterhouse (1970). Indeed, with the limited number of morphological characters available, it might even be possible to have morphologically similar but physiologically distinct species within the genus. We surely cannot afford to ignore the increasingly valuable contribution of chromosome differences, protein patterns, and other biochemical distinctions, and we should soon reach the point where we can agree to use selected physiologic techniques for naming purposes, as bacteriologists commonly do through the International Code of Nomenclature of Bacteria and Viruses. To achieve this we might either continue to adjust the Code of Botanical Nomenclature or, more positively, construct an "International Code of Mycological Nomenclature." This would seem a reasonable topic for discussion at a future International Mycological Congress.

As an example, Table 1 shows the advancement in our knowledge of chromosome numbers in *Phytophthora* spp. No chromosome numbers were known before 1960. The chromosome numbers of three species were known by 1970 and 17 were known by 1980. The recent taxonomic revision of "*P. palmivora*" on cocoa was spearheaded by the discovery of chromosome differences in certain isolates (Sansome et al, 1975). These chromosome differences were the cause of further detailed morphological investigations and led to the recognition of new taxa within a species that for decades had been treated as a morphological dustbin (Brasier and Griffin, 1979; Kaosiri and Zentmyer, 1980).

A tendency to use rather few characters in making taxonomic distinctions may also be contributing to

TABLE 1. Chromosome numbers in *Phytophthora*

| Taxa | Basic chromosome number ($n$)[a] | References |
|---|---|---|
| Homothallic | | |
| *P. cactorum* | 9–10 | Sansome (1965) |
| *P. erythroseptica* | 9–10 | Sansome (1965) |
| *P. heveae* | 9–10 | C. M. Brasier and E. R. Sansome (*unpublished*) |
| *P. megasperma* var. *sojae* | 10–15[b] | Sansome and Brasier (1974) |
| *P. megasperma* var. *megasperma* | 22–27 (–30)[b] | Sansome and Brasier (1974) |
| Heterothallic | | |
| *P. botryosa* | 9–11 | Sansome (1980) |
| *P. cambivora* | 10–12 | Sansome (1977, 1980) |
| *P. capsici* | 9–10 | Sansome (1976) |
| *P. cinnamomi* | 9–10 | Brasier and Sansome (1975), Sansome (1980) |
| *P. cryptogea* | 9–11 | Sansome (1977, 1980) |
| *P. drechsleri* | 9–12 | Brasier and Sansome (1975), Galindo and Zentmyer (1967), Sansome (1980) |
| *P. infestans* | 9–10 | Brasier and Sansome (1975), Sansome (1980), Sansome and Brasier (1973) |
| *P. meadii* | 8–10 | E. R. Sansome (*unpublished*) |
| *P. megakarya* | 5[c] | Sansome (1980), Sansome et al (1975, 1979) |
| *P. nicotianae* var. *parasitica* | 8–9 | Sansome (1977, 1980) |
| *P. palmivora* | 9–10[c] | Sansome (1980), Sansome et al (1975, 1979) |
| "MF 4" (*P. capsici*?) | 9–12 | Brasier and Medeiros (1978) |

[a] A chromosome range (e.g., $n = 9–10$) is normally given, because with the smallest chromosomes it is difficult to distinguish between bivalents and univalents at metaphase/anaphase I (Brasier and Sansome, 1975; Sansome and Brasier, 1974).

[b] Chromosome differences that supported a taxonomic separation.

[c] Chromosome differences that resulted in a taxonomic separation.

difficulties. Thus, a recent report of the identification for the first time of *P. megakarya*—i.e., morphological form (MF) 3—outside West Africa (in New Guinea and Malaysia) cites only one character, sporangial pedicel type, as evidence. Several other *Phytophthora* species with sporangial pedicels similar to *P. megakarya* (e.g. *P. botryosa*) have already been identified in Southeast Asia. Common sense clearly demands the initial consideration of as wide a range of characters as possible in any identification or description, even though ultimately a single character may achieve the discrimination required. The relative merits of the characters available to *Phytophthora* taxonomy are reviewed by Waterhouse et al (Chapter 10).

The use of such terms as "atypical" (regarding an isolate) and "group" and "complex" (regarding a group of taxa) should be avoided. These vague terms often act as a smoke screen to cover important taxonomic problems and hence delay their solution. For example, certain tropical phytophthoras were for a long time referred to in the literature as being "atypical *P. palmivora*" or as belonging to the "*P. palmivora* group" or "complex." Following recent taxonomic investigations, they have been reassigned to another described species, *P. capsici* (Kaosiri, 1978), and to the new species *P. megakarya* mentioned above (Brasier and Griffin, 1979).

Formae speciales (f. sp.), taxonomic units below the species level based on host plant specialization, are also single-criterion taxonomic distinctions. Is their recently increased use in *Phytophthora* clarifying or compounding our current taxonomic difficulties? In this context, I would like to consider the current taxonomic status of *P. megasperma*.

Waterhouse (1963) separated *P. megasperma* into variety (var.) *sojae* and variety *megasperma* on the grounds that the latter has larger oogonia. Kuan and Erwin (1980) recently suggested discarding the varietal separation on the grounds that a continuum of oogonial size exists within the species (Fig. 3A). They also proposed two formae speciales, *P. megasperma* f. sp. *medicaginis* and f. sp. *glycinea*, based on pathogenic specificity to alfalfa and soybean respectively, a distinction that is of undoubted pathological value.

However, Kuan and Erwin's data also show a striking relationship between oogonial size and host type (Fig. 3B). Their isolates with larger oogonia come mainly from woody hosts (cherry, raspberry, etc.), and those with smaller oogonia come from alfalfa or soybean and one from sugarcane. Sansome and Brasier (1974) showed that isolates of the large-oogonial type *P. megasperma* (var. *megasperma*) had a different chromosome number from those of the small-oogonial type from soybean (var. *sojae*) (Fig. 3B), yet this character was not considered by Kuan and Erwin (1980). Moreover, Hamm and Hansen (1981), using an inoculation technique similar to that of Kuan and Erwin (1980), have recently shown that certain isolates of *P. megasperma* highly pathogenic to Douglas-fir were also highly pathogenic to soybean (Table 2). Whether this would justify the inclusion of these isolates in *P. megasperma* f. sp. *glycinea* is open to question. (See also Erwin, Chapter 11.)

Do we therefore have in *P. megasperma* two varieties, var. *sojae* and var. *megasperma*, differing broadly in host type, oogonial size, and chromosome number—and, as I shall suggest later, perhaps with different genetic systems governing pathogenicity—or do we have one taxon, *P. megasperma*? And while this problem is still being resolved, should we also superimpose formae speciales on these taxa? One anxiety, therefore, with formae speciales is that we may erect them without taking full account of broader morphological and physiologic differences, as pointed out by Hamm and Hansen (1981).

Another concern is that formae speciales may be erected without full knowledge of the genetic basis of the host specificity to which they refer. In some instances, the specificity might be confined to only one or two loci. Yet

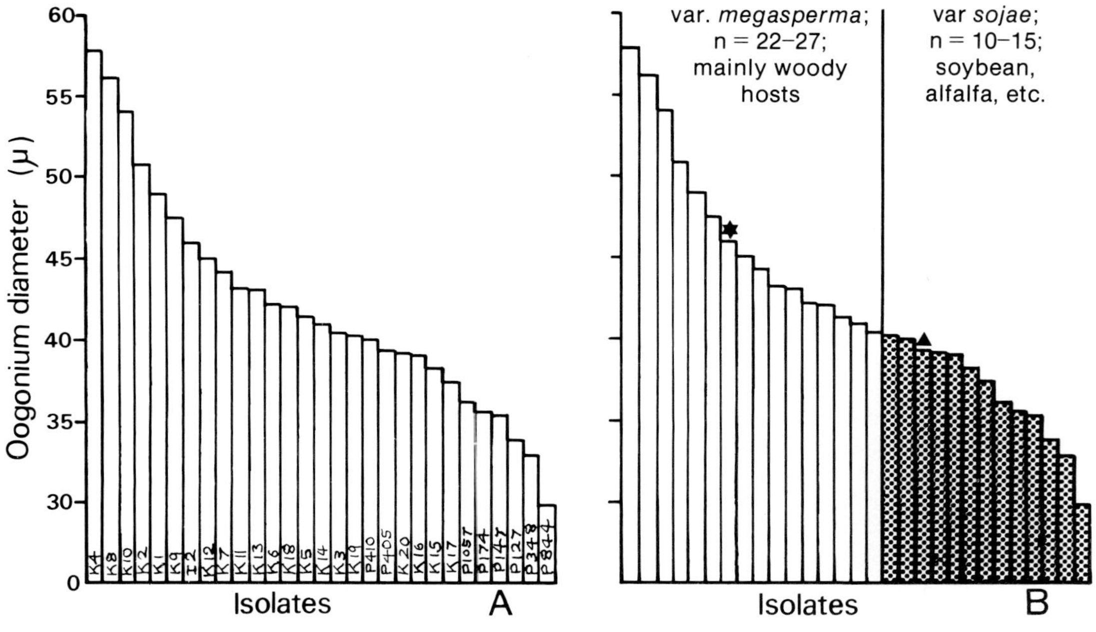


**Fig. 3. A,** Data of Kuan and Erwin (1980), demonstrating a continuum of oogonial sizes in 29 isolates of *Phytophthora megasperma*. **B,** A reinterpretation of the data in A, suggesting the possible separation of the isolates into two taxa, var. *megasperma* (unshaded bars) and var. *sojae* (shaded bars) (after Waterhouse [1963]), having chromosome numbers of about 22–27 and 10–15, respectively (after Sansome and Brasier [1974]), and being pathogenic mainly to woody hosts (var. *megasperma*) and to soybean, alfalfa, and similar hosts (var. *sojae*). Kuan and Erwin's (1980) isolate I2 (★) from *Brassica* sp. was described by Sansome and Brasier (1974) as having about 22–27 chromosomes, and isolate P405 (▲) from soybean was described as having about 10–15 chromosomes.

another concern is that we may be committing ourselves to recognizing an indefinite and rather unwieldy number of such formae speciales in those taxa possessing wide host ranges, particularly those on crops on which major gene virulence systems may commonly operate.

Thus *P. megasperma* illustrates quite well some of the current conflicts in *Phytophthora* taxonomy. How are we to see a way forward? In an ideal world, we could retain our existing morphological species concepts for traditional communication purposes; embark on an intense period of experimental taxonomy pursuing the genetic, physiologic, and morphological limits of our species; and then compare our more precisely defined species with the "old" morphological order (Fig. 4). I am sure that much of the old order would be confirmed but that some radical reappraisals would also result. Equally importantly, we would have defined the limits of the morphological criteria much more accurately and added some valuable new criteria. A useful outcome of such work could be the completion of chromosome and isoenzyme atlases of the species. Then would be the time to indulge in numerical taxonomy and study evolutionary relationships, a task made particularly difficult today because of the many inadequate descriptions in the literature.

In the real world we cannot expect rapid progress. Nevertheless, the application of experimental methods to taxonomy is increasing, and a deeper taxonomic understanding, including a greater consensus on what constitutes a species, must emerge in consequence. If this process can be accelerated by the willing efforts of individuals to tackle particularly difficult species such as *P. megasperma*, so much the better.

## SEXUALITY

Sexuality is one of the most complex areas of *Phytophthora* biology. Both the terminology and the concepts associated with the reproductive processes are at present ill defined. Part of the problem lies in a lack of clarity simply in what we mean by homothallism and heterothallism and particularly in how we describe selfing in heterothallic species. To say that an isolate of a heterothallic species behaves "homothallically," as we often do, is surely mixing metaphors. However, the division of the genus into homothallic and heterothallic *species* is very clear-cut, and I suggest that we should confine the use of the word homothallic to describing species.

I offer the following definitions of sexuality in an effort to resolve some of this confusion. Given in brackets are the equivalent terms of Burnett (1976).

*Homothallic species* (*homomixis*)—species that produce oospores promptly and abundantly in normal single-isolate culture without special external stimuli.

*Heterothallic species* (*dimixis*)—species in which single isolates exist either as an A1 or A2 compatibility type and in which the pairing of A1 and A2 types in culture results in prompt induction of oospore formation. Single isolates of either compatibility type may be induced to *self* in response to certain external stimuli.

*Secondarily homothallic isolates* (*homodimixis*)—single isolates of heterothallic species that are self-compatible without special external stimuli as a result of genetic rearrangement of the incompatibility system and that characteristically exhibit breakdown or regression to the A1 or A2 compatibility type.

*Secondarily homothallic species* (*homomixis*) (yet to be confirmed in *Phytophthora*)—species exhibiting true homothallism but derived from a permanent rearrangement of the heterothallic incompatibility system (e.g., permanent "A1A2" polyploids).

TABLE 2. Pathogenicity[a] of *Phytophthora megasperma* isolates

| Host origin of isolate | Douglas-fir | | Harosoy soybean stem | | | Alfalfa root rot |
|---|---|---|---|---|---|---|
| | Stem lesions | Root rot | Wilt | Mortality | | |
| | | | 1 day | 2 days | 9 days | |
| Douglas-fir group 1 | | | | | | |
| 1 | + | + | + | − | + | (+) |
| 2 | + | − | + | − | + | (+) |
| 3 | (+) | + | + | − | + | (+) |
| 4 | + | + | + | − | + | (+) |
| 5 | + | + | + | − | + | (+) |
| 6 | + | + | + | − | + | + |
| Douglas-fir group 2 | | | | | | |
| 1 | − | − | − | − | − | (+) |
| 2 | − | (+) | − | − | − | (+) |
| 3 | − | (+) | − | − | − | (+) |
| 4 | − | − | − | − | − | (+) |
| 5 | − | − | − | − | − | (+) |
| 6 | − | − | − | − | − | (+) |
| Soybean | | | | | | |
| 1 | − | − | − | − | + | (+) |
| 2 | − | (+) | − | (+) | + | (+) |
| 3 | − | − | − | (+) | + | + |
| 4 | − | − | (+) | (+) | + | (+) |
| 5 | − | − | − | (+) | + | (+) |
| Alfalfa | | | | | | |
| 1 | − | − | − | − | − | + |
| 2 | − | (+) | − | − | − | + |
| 3 | − | − | − | − | − | + |
| 4 | − | − | − | − | − | + |

[a]+ = Considerable disease, (+) = moderate disease, − = little or no disease. Data reinterpreted from Hamm and Hansen (1981).

In the definition of heterothallic species, I have stressed the word "self" because here the alternative use of the word homothallic would be confusing. In all heterothallic fungi from the Basidiomycetes downward, the incompatibility system is often subject to breakdown, resulting in some form of selfing. Single monokaryons of tetrapolar basidiomycetes can be induced to produce fruit bodies and basidiospores. Supposed single mating type isolates of ascomycetes will produce perithecia, asci, and ascospores. Each form of selfing has a specific genetic or physiologic cause that relates to and must be defined in terms of the nature of the incompatibility system.

In *Phytophthora* we must distinguish between at least two forms of selfing behavior: 1) Selfing such as that of an A2 isolate of *P. cinnamomi* in response to another *Phytophthora* isolate, *Trichoderma*, or other agents. This is covered under the definition of heterothallic species, because in *Phytophthora* the mechanism of heterothallism does not preclude the possibility of selfing (discussed below). 2) Selfing of derived heterothallic isolates that have a genetically determined ability to self, such as an A1A2 progeny produced as a result of trisomy from a cross between an A1 and an A2 (Mortimer et al, 1977). These are secondarily homothallic isolates of heterothallic species, common products of sexual incompatibility systems in fungi; hence I have offered a separate definition to cover them.

### Mechanism of Heterothallism

The mechanism of heterothallism is an area in which important progress is currently being made. Figure 5 shows a model for the sexual mechanism in heterothallic species put forward a decade ago (Brasier, 1972) and based mainly on the work of Ashby (1929), Stamps (1953), Galindo and Gallegly (1960), Savage et al (1968), Sansome (1970), and Brasier (1967, 1971, 1972). The model encompasses the following main points: 1) that there are only two compatibility groups, A1 and A2; 2) that isolates of each group are bisexual; 3) that chemical stimulation between the A1 and A2 compatibility types leads to induction of sexual differentiation; and 4) that once induction has occurred, both hybrid and selfed gametangial combinations can be produced.

Because of more recent developments, a number of conceptual problems within this model urgently require consideration. Galindo and Gallegly (1960) demonstrated convincingly that the A1/A2 incompatibility system overrides a ♂/♀ system of sexual morphogenesis and relative sexuality similar to that found in *Achlya* and *Dictyuchus*. They drew careful attention to the distinction between sex (maleness and femaleness) and the A1/A2 incompatibility system. They did not look for evidence of chemical stimulation, nor did they consider the possibility of selfing. However, my observations (Brasier, 1972) showed that the homothallic species *P. heveae* would chemically stimulate both A1 and A2 compatibility types of "*P. palmivora*" to self when separated from them by cellophane membranes (Fig. 6, top). Because I was also convinced by Galindo and Gallegly's (1960) conclusion that incompatibility and sex were separate systems, I put forward the following as a simple chemical model for heterothallism (Fig. 5): 1) that the A1 and A2 types produce *compatibility substances*, one substance from each type, that serve to initiate sexual differentiation in the other type; 2) that the compatibility substances act as inducers or derepressors of operons controlling structural genes responsible for sexual

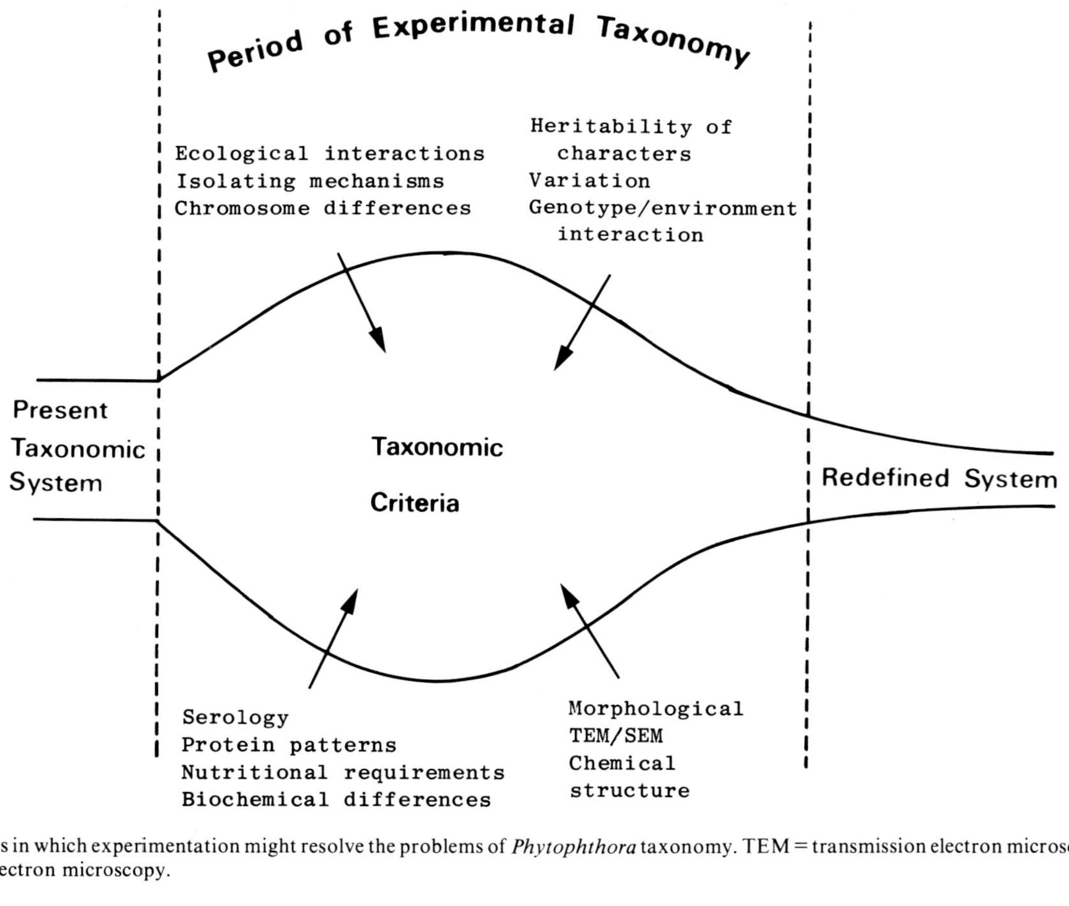


**Fig. 4.** Ways in which experimentation might resolve the problems of *Phytophthora* taxonomy. TEM = transmission electron microscopy; SEM = scanning electron microscopy.

differentiation, i.e., the genes controlling the ♂×♀ interaction and relative sexuality; and 3) that the gene products are enzymes controlling the production of "sex hormones."

Undisputed evidence for chemical stimulation could not be obtained with cellophane membranes because the membranes were occasionally penetrated by the fungus. This problem was overcome by Ko (1978), using polycarbonate membranes. Ko's (1978) work unequivocally demonstrates direct chemical stimulation of selfing between A1 and A2 isolates of several species (Fig. 6, bottom). Although Ko's (1978) results are consistent with the earlier theory of two compatibility substances, he postulates "sex hormones" $\alpha 1$ and $\alpha 2$ without distinguishing whether they are *compatibility* substances or *sex* substances. Another difficulty is that the term "sex hormone" implies analogy to the hormone systems in *Achlya, Dictyuchus,* and *Sapromyces,* which do not have an incompatibility system like that in *Phytophthora.* To maintain clarity, I suggest that the incompatibility and sex systems should be kept separate unless they are shown to be otherwise and that for the time being we should retain the following separate terminology for them:

*Compatibility substances*—substances involved in the breakdown of self-incompatibility between A1 and A2 compatibility types, leading to sexual differentiation. Compatibility substances are denoted CS1 and CS2 (CS1b, CS2b, etc.). CS1 is produced by A1 types and breaks down self-incompatibility in A2 types; CS2 is produced by A2 types and breaks down self-incompatibility in A1 types (Fig. 5).

*Sex hormones* (postulated)—substances produced during sexual morphogenesis by ♂ and ♀ gametangial initials (compare Raper, 1951). Hormones are denoted H1, H2, H3, etc. (Fig. 5).

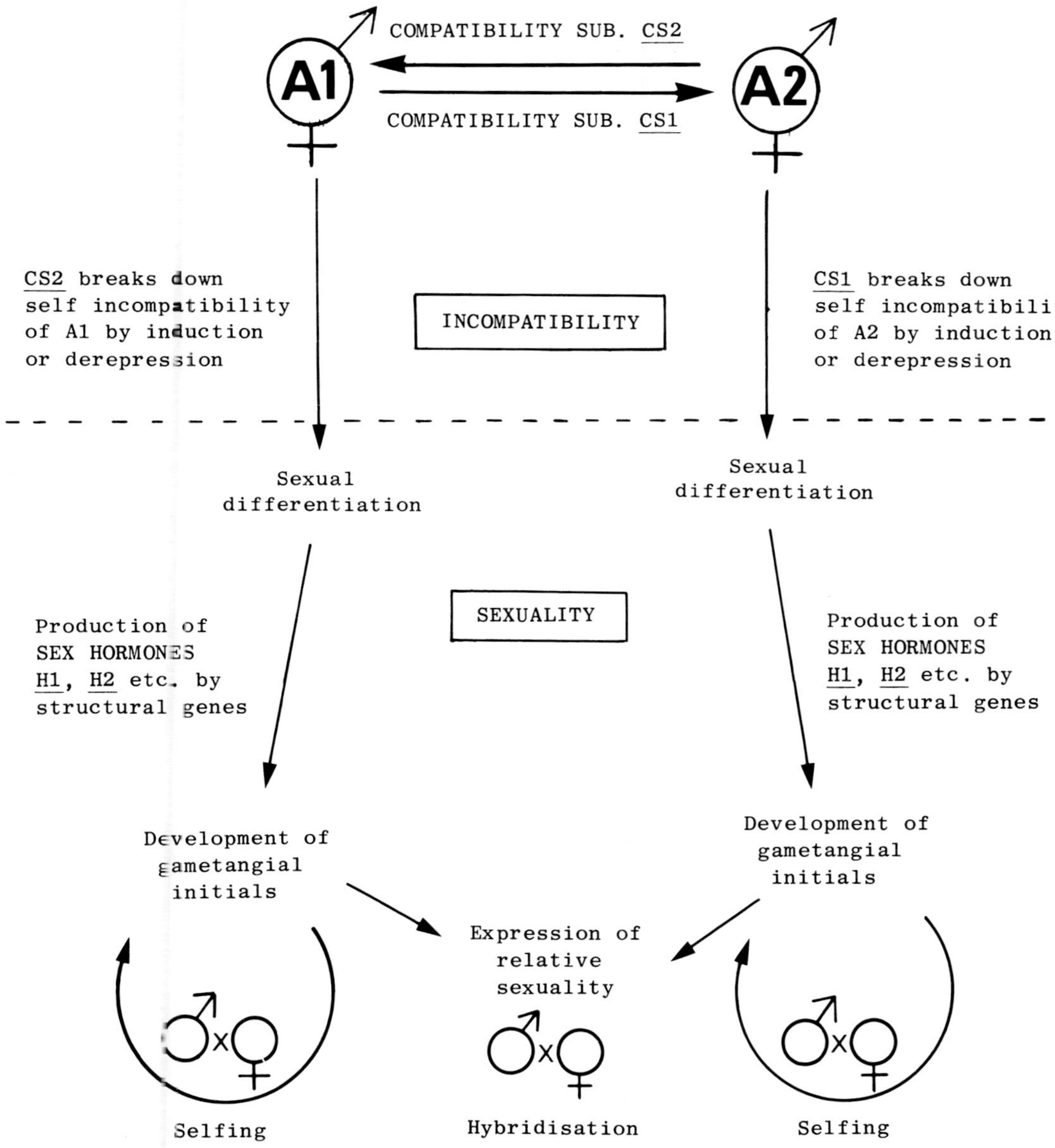


**Fig. 5.** A model of the sexual mechanism in heterothallic species of *Phytophthora*, incorporating a separation of a system of incompatibility (above) and a system of sexuality (below), with associated production of compatibility substances (CS1 and CS2) (above) and sex hormones (H1, H2, etc.) (below). (After Brasier, 1967, 1972)

In further articles, Ko (1980) and Ann and Ko (1980) propose 16 types of chemically regulated sexuality among *Phytophthora* spp. They substitute the terms "cross inducing" for heterothallic and "self inducing" for homothallic species and imply that true hybridization does not occur in A1 × A2 pairings and that sexual exchange is confined to selfing. However, a great deal of evidence exists to show that hybrid oospores are formed in A1 × A2 pairings, the evidence coming from hyphal tracing (Galindo and Gallegly, 1960; Stamps, 1953), chemical staining (Brasier and Sansome, 1975; Huguenin, 1973; Sansome et al, 1979), and analysis of genetic recombination (Khaki and Shaw, 1974). Moreover, outcrossing has certain important evolutionary advantages and therefore has evolved in many fungi. If true hybridization does occur, then Ko's (1980) scheme, though attractively simple, does not take adequate account of the conclusions established by earlier work.

The number of postulated compatibility substances and their chemical structure and mode of operation (e.g., via specific receptor sites) all require elucidation that can come about only through precise experimentation. The possibility that they are steroidal is discussed by Elliott (Chapter 5). But surely the greatest challenge is the possibility of demonstrating a genuine sex hormone system in *Phytophthora* equivalent to that demonstrated in *Achlya* (Raper, 1951), where steroids are already implicated (McMorris, 1978).

Evidence for a sex hormone system might be obtained from chemical labeling of steroids and from the study of induced morphogenetic mutations. We might also study those interspecific pairings that produce only partial development of gametangia and that could result from recognition failures caused by minor structural variation in the hormones concerned. In addition, the use of selfed single isolates might greatly accelerate such work.

We must also determine the genetic basis of the incompatibility and sex systems. In a major paper, Sansome (1980) recently proposed a convincing chromosomal model for the maintenance of heterothallism in *Phytophthora* based

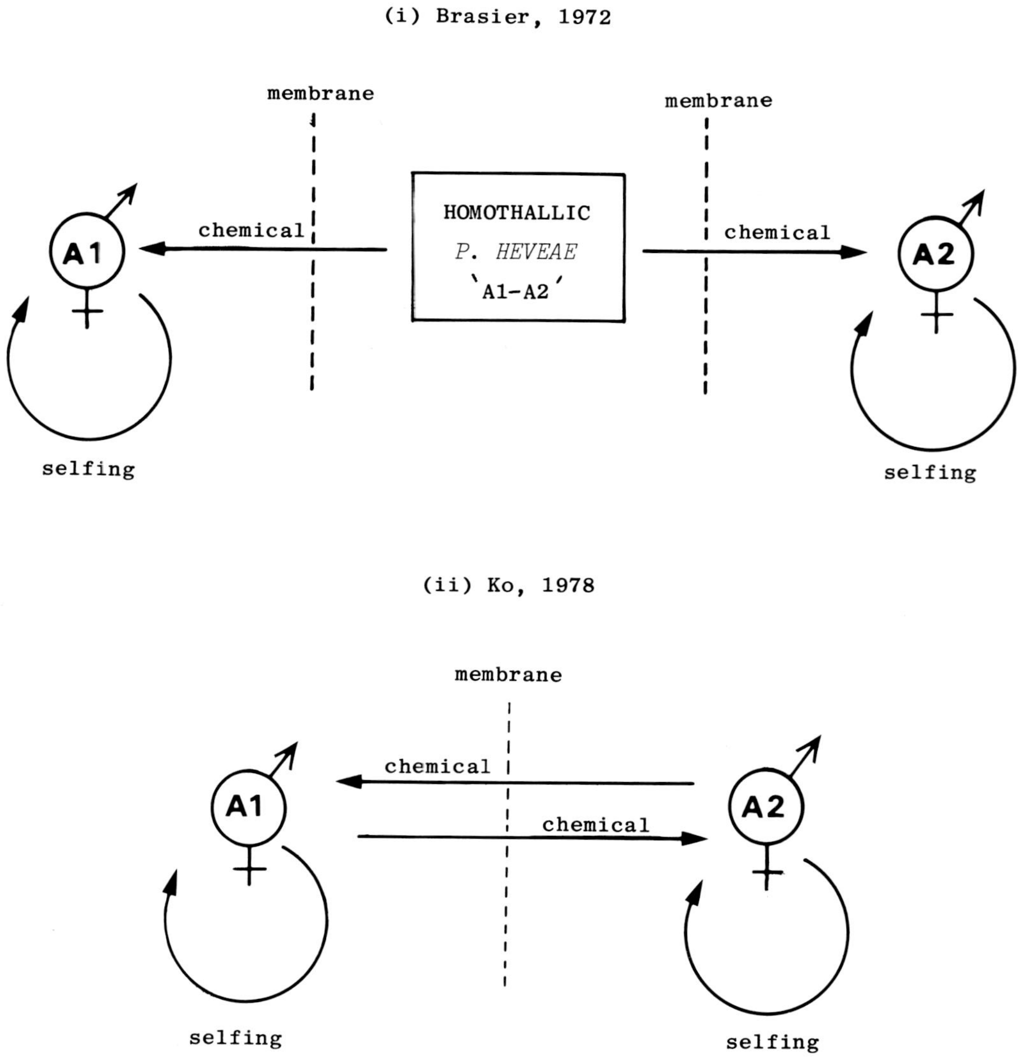


**Fig. 6.** The demonstration of chemical stimulation of selfing in A1 and A2 isolates of *Phytophthora* using membrane systems. System of Brasier (1972) (above), using cellophane membranes, demonstrates direct induction of selfing in both A1 and A2 isolates by the homothallic *P. heveae*. *P. heveae* was interpreted as possessing both A1 and A2 potentialities within its genome ("A1-A2"). System of Ko (1978) (below), using polycarbonate membranes, demonstrates direct induction of selfing between A1 and A2 isolates. The chemicals crossing the membranes were interpreted as two compatibility substances (1 and 2) by Brasier (1972) and as "sex hormones" $\alpha 1$ and $\alpha 2$ by Ko (1978).

on reciprocal translocation heterozygosity (see Shaw, Chapter 6). However, we are probably still a long way from mapping the compatibility loci and even further from identifying the loci involved in sex. Nevertheless, the diploid genetic system of *Phytophthora* (Sansome, 1965) should readily allow for complex expression of sexuality, such as the relative degrees of maleness and femaleness identified in *P. infestans* by Galindo and Gallegly (1960), and it would be interesting to attempt an analysis of the sex system by crossing, for example, a strong male A1 with a strong female A2 and examining the sex of $F_1$ relative to the parents. This might also provide a test of whether the compatibility loci segregate independently of the sex loci.

### Other Selfing Phenomena

A number of external agents have been reported that can induce selfing in normally self-sterile A2 isolates of heterothallic species, including *Trichoderma* volatiles, chloroneb, and avocado root extract (Brasier, 1971, 1975a, 1975b; Noon and Hickman, 1974; Zentmyer, 1979). Quite plentiful oospore formation can also occur in aged cultures of both A1 and A2 isolates (Brasier, 1972; Savage et al, 1968; Tsao et al, 1980), and sporadic selfing sometimes occurs in A2 cultures of certain species (in particular *P. capsici, P. meadii*, and *P. parasitica*). All of these phenomena should probably be viewed in terms of either 1) the chemical triggering, or 2) the genetic breakdown of the self-incompatibility mechanism, leading to sexual differentiation along the lines shown in Figure 5.

A chemical trigger mechanism has already been suggested for the *Trichoderma* effect, in which it is proposed that the *Trichoderma* volatiles act by mimicking the action of the compatibility substance produced by an A1 isolate (Brasier, 1975a). A chemical effect has also been suggested to explain the aging phenomenon (Brasier, 1972; Tsao et al, 1980). However, a more convincing explanation of the aging phenomenon can be provided in terms of genetic breakdown, based on Sansome's (1980) theory that the A2 is the mating type heterozygote and that the segregation of the A1 homozygote from the A2 by mitotic recombination is normally prevented by a chromosomal reciprocal translocation. This explanation requires that aging in some way allows mitotic recombination to occur freely (or the products of mitotic recombination to survive), resulting in the segregation of A1 homozygote nuclei within an originally A2 mycelium, and the development of local self-fertile sectors. The evidence of Kamjaipai and of Ko that A1 and A1/A2 nuclei are present in the self-fertile regions of an aged colony (see Shaw, Chapter 6) strongly supports this explanation. Moreover, according to Sansome's (1980) theory, such mitotic recombinants are likely to be unbalanced. They would therefore tend to be selected out in a vigorously growing mycelium, which would account for the commonly observed reversion of aged mycelia to their original compatibility type (Brasier, 1972; Tsao et al, 1980). The above explanation can also account for the action of chloroneb (see Chapter 6) and could explain the sporadic selfing that occurs in some A2 cultures. A drawback is that it does not as yet explain selfing in aged mycelia of A1 isolates.

It now remains to be discovered whether the *Trichoderma* effect (and that of avocado extract) may also involve a genetic breakdown phenomenon in which, for example, the *Trichoderma* volatiles act by enabling mitotic segregation of A1 types from A2 nuclei to occur. An analysis of the compatibility types of zoospores from regions of *Trichoderma*-induced selfing similar to that conducted by Kamjaipai and by Ko with aged cultures should discriminate between a chemical trigger mechanism and genetic breakdown.

## GENETICS OF PATHOGENICITY

We have a long way to go in understanding the genetic basis of pathogenicity in *Phytophthora* despite its importance for disease control and for understanding population behavior. So far the emphasis has been on resistance mechanisms in the host, and the time is surely overdue for an onslaught on the pathogenicity systems in the fungus.

*Phytophthora* species attack an extraordinary range of host types varying from herbaceous, nonwoody tissues such as leaves and fruits to woody stems and roots. Major gene race-specific virulence systems are already well known, for example in *P. infestans* on potato and *P. megasperma* var. *sojae* on soybean (Black, 1960; Schmitthenner, 1972), but virtually nothing is known of the pathogenicity mechanisms involved in hundreds of other diseases ranging from black pod disease of cocoa and black stripe of rubber to root and collar rot of apple or *Eucalyptus*.

In light of this gap in our knowledge, I venture the following broad hypothesis as a basis for discussion: First, that in systems where more herbaceous tissues of the host are attacked, we shall sometimes—but not always—find involvement of major gene virulence loci of the sort elegantly demonstrated in the *P. infestans*-potato and *P. megasperma*-soybean systems. The gene products will often be chemical elicitors that, in incompatible combinations, invoke specific resistance gene expression such as the production of phytoalexins and hypersensitive reactions in the host (see Keen and Yoshikawa, Chapter 21). I shall refer to this as system A. Second, that in the massive invasion of woody tissues such as stems and roots, we may find genetic systems (referred to here as system B) operating in which the gene products will often be specialized batteries of enzymes, and perhaps also toxins, attacking particular complexes of relatively nonspecific resistance mechanisms, both chemical and mechanical. These systems may tend to be under polygenic control but may sometimes be rather complex in their inheritance, not always showing degrees of additivity.

System A is, of course, the classical "gene-for-gene" or "vertical" system, about which a great deal is known in other fungal pathogens. In *Phytophthora*, however, we still do not know whether virulence is recessive, as in the majority of other fungal pathogens, or dominant. Yet the way is surely open for a more critical analysis of the problem now that the diploid cytogenetic systems of phytophthoras are better understood and methods of selfing heterothallic isolates have been developed. For example, a genetic analysis of the r-gene loci in the *P. infestans*-potato system could be made through germinating selfed $F_1$ oospores. The occurrence of polyploidy in some isolates of this species (Sansome, 1977) should add interest to the analysis!

Evidence of the nature of pathogenicity in system B (which is not necessarily "horizontal") is even more urgently needed. This may be the type of system that enables *P. cinnamomi* to attack such a broad range of woody and nonwoody hosts. The genetic basis of pathogenicity in *P. cinnamomi* is certainly worthy of investigation. A number of other host-pathogen systems are ideal for analysis; for example, the cocoa- and rubber-*P. palmivora* systems, in which disease is caused on both pods or fruits and woody stems and on which, theoretically, both genetic systems A and B might operate. Analysis of the pathogenicity of $F_1$ progeny of A1 × A2 crosses to cocoa pods, roots, and stems could be very rewarding. Important pioneer work in this area has been carried out by workers in Congo Brazzaville and Ivory Coast (Office de la Recherche Scientifique et Technique Outre-Mer [ORSTOM], 1973).

Another system that is ideal for investigation is that of *P. megasperma* on its various hosts. The recent excellent data of Hamm and Hansen (1981), interpreted by them largely in terms of physiologic host specificity, possibly provide a good example of the operation of two different genetic systems of pathogenicity. I would like to draw attention to the following points from their data, which are summarized in Table 2: First, Douglas-fir group 1 is the only group of isolates able to cause significant stem lesions on Douglas-fir. This group perhaps has a complex pathogenicity system equivalent to that of system B. Second, Douglas-fir group 2 isolates lack this system and cause only some fine root damage to Douglas-fir, similar to that caused by the soybean and alfalfa isolates. Third, of the Douglas-fir group 2, soybean, and alfalfa isolates, only the soybean isolates can attack soybean, presumably because they have the necessary virulence genes to overcome specific resistance loci in Harosoy soybean; i.e., they have the type A system. However, Douglas-fir group 1 isolates also destroy Harosoy soybean, but apparently by a completely different mechanism than that used by the soybean isolates. This is perhaps the same mechanism, system B, as that used in their successful attack on the Douglas-fir stems. Fourth, the mechanism that enables all the isolates to attack alfalfa roots to some degree may be a very basic and generalized type of system B, common to many phytophthoras.

Here again, much will be learned from analyzing, via oospore progenies, the genetic basis of the different types of pathogenic systems indicated in these data. Thus it would be of particular interest to analyze the basis of the pathogenic differences between the Douglas-fir group 1 and Douglas-fir group 2 isolates and the structure of the virulence loci in the soybean isolates.

The possibility remains that Douglas-fir group 1 may even be a different taxon, *P. megasperma* var. *megasperma*, with about 22–27 chromosomes, larger oogonia, and a tendency to specialize in woody hosts, and that Douglas-fir group 2, together with the soybean and alfalfa isolates, may be *P. megasperma* var. *sojae*, with about 10–15 chromosomes and smaller oogonia and specializing in attacking legumes and other herbaceous hosts (Fig. 3).

### Chemotaxis

In considering the genetics of pathogenicity, we should also remember that in the chemotactic responses of zoospores, *Phytophthora* has a specialized guidance system that enables it to reach the host surface. Although the physiology of chemotaxis is being carefully investigated (see Carlile, Chapter 7), the genetic basis of chemotaxis also merits examination. For example, with *P. cinnamomi* zoospores in a mixture of grass and conifer roots, does genetically controlled selective homing occur toward the conifer roots, and if so, what form of genetic control is involved?

## POPULATIONS AND EVOLUTION

In *Phytophthora*, the fields of evolution and population genetics are largely unexplored, although experimentation in these areas might contribute immeasurably to our understanding of the genus. Thus, a basic framework of knowledge of the population genetics of *Phytophthora* is important for understanding the causes of epidemics, the development of host-parasite systems, and much of disease etiology. Although counting propagules or diseased plants as the seasons change or as an epidemic develops is valuable, in future we might also take more interest in assessing any subtle qualitative changes underlying the quantitative ones.

For a start, we might compare the genetic structure of soil populations of our organisms—the resting, latent, or saprophytic populations—with that of the disease-causing or parasitic populations. For example, several years ago I received from F. Arentz a small collection of isolates of "*P. palmivora*" from Papua, New Guinea, for morphological examination. One isolate was from a cocoa pod and was morphologically and culturally typical of cocoa isolates worldwide. The remainder of the isolates came from soil beneath cocoa or rain forest, and all were extremely variable and quite different from anything I had ever seen before. This led me to wonder whether a system of selection (Fig. 7) was operating between the soil and the parasitic populations and between different crops or hosts as follows: 1) Various hosts in the ecosystem are attacked, together with different parts of those hosts, such as buds, fruits, stems, or roots. 2) Infected plant detritus accumulating on the soil surface and in the soil is associated with the production of resting stages of *Phytophthora* (oospores and chlamydospores). The soil acts as a reservoir of variation from many sources, including that resulting from sexual recombination. 3) Reinfection occurs from the soil phase. Selection occurs for forms most fitted to attack a particular host and a particular part of a host (e.g., roots, stems, or fruits). 4) In a crop ecosystem, the asexual infection cycle maintains adaptation to a particular host or part of a host. 5) Selection and adaptation result in specialized subgroups of the pathogen; e.g., the different forms of *P. palmivora* on cocoa pods, coconut buds, or rubber stems (Brasier and Griffin, 1979).

Selection such as this between the resting and parasitic phases might at times be very intense, and its measurement might tell much about population equilibria and the evolution of host-pathogen systems. It would probably be most intense in monoculture crop situations, where it could be a powerful driving force toward speciation. Thus, the increase in intensity of monoculture cropping over the past century may well have accelerated the rate of speciation in the genus and could have resulted in some of our present taxonomic difficulties. Perhaps both the problematic *P. megasperma* varieties and the cocoa and rubber forms of *P. palmivora* are prime examples of evolution in action. In epidemics, selection might sometimes be intensified and evolution very rapid.

An understanding of the evolutionary process is clearly important for deciding what are "natural" rather than "artificial" taxonomic relationships within *Phytophthora*. We still have everything to learn about natural relationships in the genus, and thus I feel free to embark upon a final bit of wild speculation. Suppose we consider as an approximation to an evolutionary sequence the series from Saprolegniales to Peronosporales, and from *Pythium* to *Phytophthora* as proposed, for example, by Gäumann (Gäumann and Wynd, 1952). Then we might also consider *Pythium*-like characters in *Phytophthora* as primitive and, by comparison, certain non-*Pythium*-like characters rather advanced. Table 3 lists proposed primitive and advanced characters in *Phytophthora* made on this basis.

Figure 8 shows a speculative phylogenetic scheme based on these principles and on Waterhouse's (1963) groups I–VI. The groups themselves, of course, may not represent natural associations. Considerable problems also arise in placing certain species such as *P. cactorum* (group I), which possesses a tantalizing mixture of "primitive" and "advanced" characteristics. Nevertheless, some possible evolutionary trends are indicated. *P. primulae* (group III) would be a very primitive species, being homothallic and having several laterally attached paragynous antheridia; intercalary, noncaducous, often rather irregular (indeed rather *Pythium*-like) sporangia; and a limited host range. *P. palmivora* (group II) and *P. cinnamomi* (group VI) would represent advanced species, being heterothallic and having a single amphigynous antheridium, caducous papillate sporangia or noncaducous proliferating sporangia, chlamydospores, and a wide host range. The weighting of the morphological characters in such a series is of course highly subjective. What is required now is an input of more accurately measurable physiologic characteristics such as chromosome differences, isoenzymes, biochemistry, nutrition, fine structure, and so on (Fig. 4), which will allow numerical analysis of possible relationships.

A number of interesting problems are raised immediately by even such a tentative effort at a phylogenetic scheme, and that

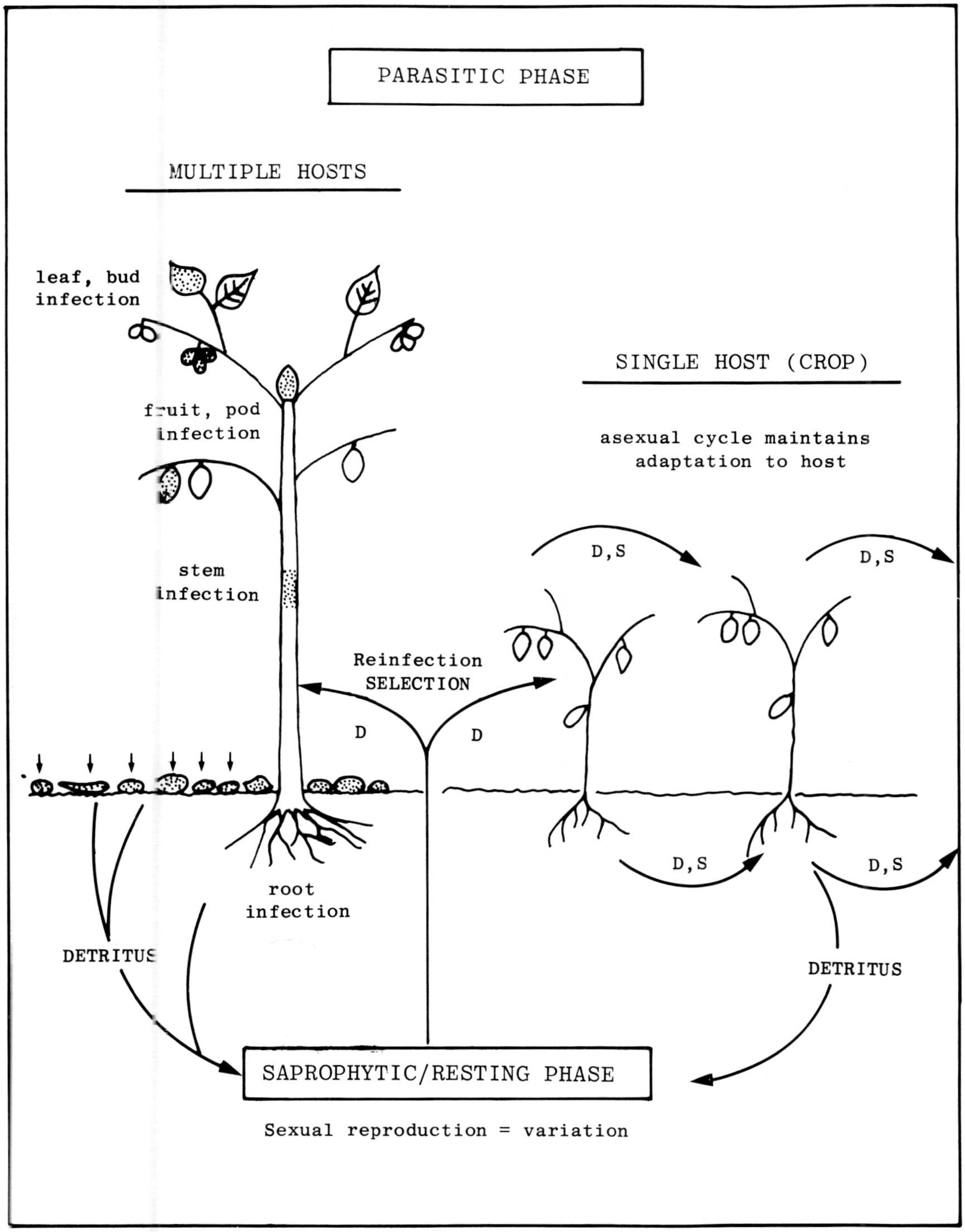


**Fig. 7.** A proposed model for selection operating between the saprophytic or resting phase and the parasitic phase of the *Phytophthora* disease cycle. D = directional selection, S = stabilizing selection.

such questions can be asked may at least justify its existence. Two clear evolutionary lines are apparent. One is toward the formation of caducous papillate sporangia, perhaps adapted to an aerial mode of dispersal and represented by *P. palmivora*, *P. heveae*, *P. infestans*, etc. (groups II and IV); the other is toward the formation of noncaducous, proliferating, wide-pored sporangia, perhaps adapted to a mode of dispersal in soil and represented by *P. megasperma*, *P. cinnamomi*, etc. (groups V and VI).

A major problem of even this simple conclusion is that not

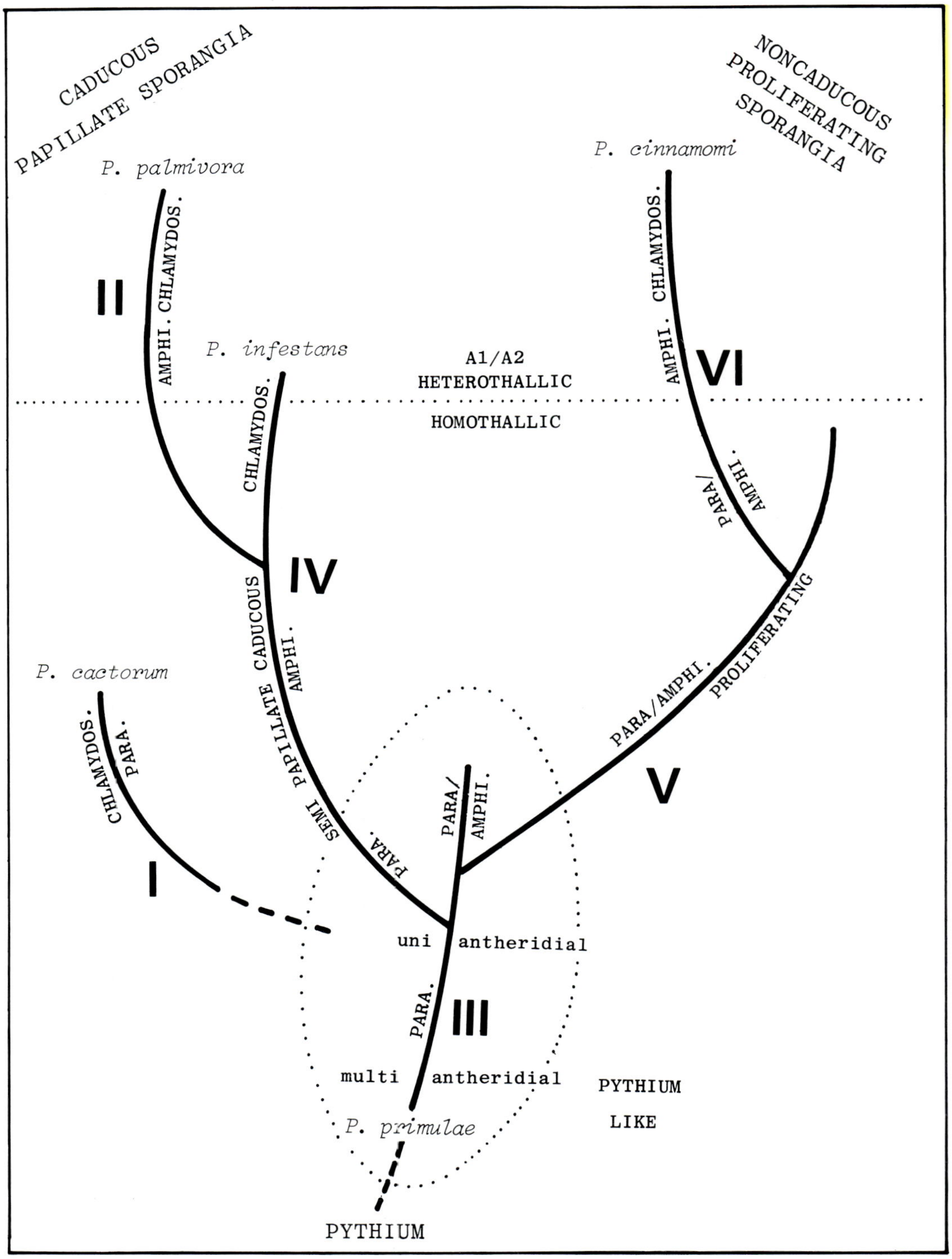


**Fig. 8.** A phylogenetic scheme for *Phytophthora* based on the "primitive" and "advanced" characters in Table 3 and on groups I–VI of Waterhouse (1963). Group III is "primitive," groups IV and V are "intermediate," and groups II and VI are "advanced." For example, in this scheme *P. primulae* is a very "primitive," *Pythium*-like species, and *P. palmivora* and *P. cinnamomi* are very "advanced" species. Group I represents only one species, *P. cactorum*. PARA = paragynous antheridia, AMPHI = amphigynous antheridia, PARA/AMPHI = mixture of paragynous and amphigynous antheridia, CHLAMYDOS = chlamydospores. Semipapillate caducous (group IV) and proliferating (group V) refer to types of sporangia.

only do both lines show evidence of evolution from homothallism to heterothallism, but, curiously, evolution to the very same type of A1/A2 heterothallism, such that an A1 isolate of one line can respond to an A2 isolate of the other. Either parallel evolution of heterothallism has occurred in the two lines from a similar homothallic ancestor, or they have a common heterothallic ancestor and heterothallism evolved before sporangial type. I have pointed out a rather similar problem in relation to the evolution of response to *Trichoderma*, in which species attacking roots of woody hosts, whether they have caducous papillate sporangia or noncaducous proliferating sporangia, appear to be more likely to respond to *Trichoderma* than species attacking aerial parts of plants (Brasier, 1978).

Other pertinent questions include the nature and function of amphigyny, which emerges about halfway up the postulated scheme. Is it associated with the evolution of heterothallism? Why are some species both amphigynous and paragynous, and do they represent an intermediate step in the evolutionary sequence? Are some homothallic species secondarily homothallic and therefore relatively advanced?

Finally, what may have been the prime causes of the evolutionary divergence in terms of environmental and internal genetic factors? In addition to classical geographic isolation, selection linked to host plant specialization must be a key environmental influence on divergence, as I have already discussed (Fig. 7). This may include host specialization not just in terms of species but in terms of the type of host (woody or herbaceous) and the parts of a host (roots or aerial) attacked. Strong environmental selection is also likely to be imposed by soil type, vegetation type, other microorganisms, and climate. Indeed, the climatic specialization of *Phytophthora* species is a fairly evident feature of the genus. Thus, most species can probably be categorized broadly as cool temperate, hot temperate-subtropical, or tropical in distribution.

The consequences of environmental selection are likely to be promoted or reinforced by genetic isolating mechanisms. In the inbreeding homothallic species, these mechanisms are likely to be rather limited, although in some taxa such as *P. megasperma*, speciation may have been achieved through the development of polyploidy. A more flexible system is available to the outbreeding heterothallic species, which could regulate interbreeding between divergent forms or species by means of various reproductive isolating mechanisms operated over and above the A1 × A2 incompatibility system. These mechanisms could include small qualitative differences in sex hormones, leading to failure of gametangial copulation; lack of meiotic synchrony, leading to gametic abortion; gametic nonrecognition (e.g., as a result of an incompatibility system operating on the nuclear envelope), leading to failure of gametic fusion; and structural changes in chromosomes and the development of balanced lethal systems, both of which could lead to zygotic abortion or germ tube death. Indeed, it is quite possible that free hybridization between heterothallic species is already prevented by one or more such mechanisms. Functionally, they would constitute additional incompatibility systems and could account for the commonly observed abortion and failure of germination of hybrid oospores conceived in interspecific pairings and the contrasting viability of selfed oospores formed in the same pairings (Boccas, 1981; Boccas and Zentmyer, 1976; Erselius and Shaw, 1982).

The selective selfing of one mating type in response to *Trichoderma* volatiles or a similar external agent could also lead to evolutionary divergence, as has already been suggested for *P. cinnamomi* (Brasier, 1975a, 1978; Zentmyer, 1979). Systems of vegetative incompatibility could also be involved in speciation in *Phytophthora*, although these have yet to be demonstrated.

TABLE 3. "Primitive" and "advanced" characteristics in *Phytophthora*[a]

| "Primitive" | "Advanced" |
|---|---|
| Characters | |
| Homothallism | Heterothallism |
| Paragynous antheridia | Amphigynous antheridia |
| More than one antheridium | Single antheridium |
| Laterally attached antheridium | Basal antheridium |
| Semipapillate noncaducous sporangia | Papillate caducous sporangia (aerial forms); nonpapillate, proliferating, persistent sporangia (soil forms) |
| Intercalary sporangia | Terminal sporangia |
| Chlamydospores absent | Chlamydospores present |
| Trends | |
| Limited host range | Wide host range, or narrow host range tending to biotrophy |
| Unspecialized, "soft rotting" | Specialized virulence systems |
| | Sophisticated woody pathogens |
| Low optimal temperatures | |
| | Degeneration of sexual reproduction |
| | Increased emphasis on asexual reproduction |

[a]Based on the assumption that the series from Saprolegniales to Peronosporales (and from *Pythium* to *Phytophthora* [Gaümann and Wynd, 1952]) approximates an evolutionary sequence.

## CONCLUSION

The field of reproductive isolation and evolutionary divergence in *Phytophthora* is on the verge of being opened up. The problems presented should provide a considerable challenge for many an enterprising research student, and the answers may provide important clues to how speciation occurs. It is in this general area, that of speciation, that all the topics I have considered—the nature of the individual, taxonomy, sexual reproduction, host-parasite systems, population behavior, and selection mechanisms—come together to form a whole. Phytophthoras exist as individuals, individuals form populations, populations are subject to change, and change leads to speciation.

Like evolution, research is also a dynamic process. It has been some 30 years since our distinguished colleague Professor George Zentmyer, whose retirement and considerable contribution to *Phytophthora* research we have honored at this Symposium, published his first paper on *Phytophthora*. If the next 30 years of research bring about as much progress as the last, then we have a great deal to look forward to. At the end of it, not only will our view of *Phytophthora* have changed, but some of the phytophthoras we think we know today may also be quite different.

## LITERATURE CITED


Ann, P. J., and Ko, W. H. 1980. *Phytophthora insolata*, a new species from Taiwan. Mycologia 72:1180-1185.

Ashby, S. F. 1929. Strains and taxonomy of *Phytophthora palmivora* Butler (*P. faberi* Maubl.). Trans. Br. Mycol. Soc. 14:18-38.

Black, W. 1960. Races of *Phytophthora infestans* and resistance problems in potatoes. Pages 29-38 in: Rep. Scott. Breed. Stn., Pentlandfield.

Boccas, B. R. 1981. Interspecific crosses between closely related heterothallic *Phytophthora* species. Phytopathology 71:60-65.

Boccas, B. R., and Zentmyer, G. A. 1976. Genetical studies with interspecific crosses between *Phytophthora cinnamomi* and *P. parasitica*. Phytopathology 66:477-484.

Brasier, C. M. 1967. Physiology of reproduction in *Phytophthora*. Ph.D. thesis, University of Hull, England. 220 pp.

Brasier, C. M. 1971. Induction of sexual reproduction in single $A^2$ isolates of *Phytophthora* species by *Trichoderma viride*. Nature (London) New Biol. 231:283.

Brasier, C. M. 1972. Observations on the sexual mechanism in *Phytophthora palmivora* and related species. Trans. Br. Mycol. Soc. 58:237-251.

Brasier, C. M. 1975a. Stimulation of sex organ formation in *Phytophthora* by antagonistic species of *Trichoderma*. I. The effect *in vitro*. New Phytol. 74:183-194.

Brasier, C. M. 1975b. Stimulation of sex organ formation in *Phytophthora* by antagonistic species of *Trichoderma*. II. Ecological implications. New Phytol. 74:195-198.

Brasier, C. M. 1978. Stimulation of oospore formation in *Phytophthora* by antagonistic species of *Trichoderma* and its ecological implications. Ann. Appl. Biol. 89:135-139.

Brasier, C. M., and Griffin, M. J. 1979. Taxonomy of '*Phytophthora palmivora*' on cocoa. Trans. Br. Mycol. Soc. 72:111-143.

Brasier, C. M., and Medeiros, A. G. 1978. The karyotype of *Phytophthora palmivora* morphological form 4. Trans. Br. Mycol. Soc. 70:295-297.

Brasier, C. M., and Sansome, E. R. 1975. Diploidy and gametangial meiosis in *Phytophthora cinnamomi, P. infestans* and *P. drechsleri*. Trans. Br. Mycol. Soc. 65:49-65.

Burnett, J. H. 1976. Fundamentals of Mycology. Edward Arnold Ltd., London. 673 pp.

Caten, C. E., and Jinks, J. L. 1966. Heterokaryosis: Its significance in wild homothallic Ascomycetes and Fungi Imperfecti. Trans. Br. Mycol. Soc. 49:81-93.

Dyakov, Y. T., and Kuzovnikova, T. A. 1974. Heterokaryosis in *Phytophthora infestans* (Mont.) de Bary. II. Genetical investigations. Mikol. Fitopatol. 8:81-89. (In Russian)

Erselius, L. J., and Shaw, D. S. 1982. Protein and enzyme differences between *Phytophthora palmivora* and *Phytophthora megakarya*: Evidence for self-fertilization in pairings of the two species. Trans. Br. Mycol. Soc. 78:227-238.

Esser, K., and Blaich, R. 1973. Heterogenic incompatibility in plants and animals. Adv. Genet. 17:107-152.

Galindo, A. J., and Gallegly, M. E. 1960. The nature of sexuality in *Phytophthora infestans*. Phytopathology 50:123-128.

Galindo, A. J., and Zentmyer, G. A. 1967. Genetical and cytological studies of *Phytophthora* strains pathogenic to pepper plants. Phytopathology 57:1300-1304.

Gaümann, E. A., and Wynd, F. L. 1952. The Fungi. Hafner, New York.

Hamm, P. B., and Hansen, E. M. 1981. Host specificity of *Phytophthora megasperma* from Douglas fir, soybean, and alfalfa. Phytopathology 71:65-68.

Hemmes, D. E., and Bartnicki-Garcia, S. 1975. Electron microscopy of gametangial interaction and oospore development in *Phytophthora capsici*. Arch. Microbiol. 103:91-112.

Ho, H. H. 1979. The nature of amphigyny in *Phytophthora*. Mycologia 71:1057-1064.

Huguenin, B. 1973. Application de la technique du marquage fluorescent à la détermination du comportement sexuel du *Phytophthora palmivora*. Cahiers ORSTOM, Ser. Biol. 20:59-61.

Kaosiri, T. 1978. Morphological, taxonomic and cytological studies of *Phytophthora palmivora*. Ph.D. thesis. Univ. Calif., Riverside. 148 pp.

Kaosiri, T., and Zentmyer, G. A. 1980. Protein, esterase, and peroxidase patterns in the *Phytophthora palmivora* complex from cacao. Mycologia 72:988-1000.

Khaki, I. A., and Shaw, D. S. 1974. The inheritance of drug resistance and compatibility type in *Phytophthora drechsleri*. Genet. Res. 23:75-86.

Ko, W. H. 1978. Heterothallic *Phytophthora*: Evidence for hormonal regulation of sexual reproduction. J. Gen. Microbiol. 107:15-18.

Ko, W. H. 1980. Hormonal regulation of sexual reproduction in *Phytophthora*. J. Gen. Microbiol. 116:459-463.

Kuan, T.-L., and Erwin, D. C. 1980. Formae speciales differentiation of *Phytophthora megasperma* isolates from soybean and alfalfa. Phytopathology 70:333-338.

Leach, S. S., and Rich, A. E. 1969. The possible role of parasexuality and cytoplasmic variation in race differentiation in *Phytophthora infestans*. Phytopathology 59:1360-1365.

Leonian, L. H. 1925. Physiological studies on the genus *Phytophthora*. Am. J. Bot. 12:444-498.

Leonian, L. H. 1934. Identification of *Phytophthora* species. Va. Univ. Agric. Exp. Stn. Bull. 262:2-36.

Long, M., and Keen, N. T. 1977. Evidence for heterokaryosis in *Phytophthora megasperma* var. *sojae*. Phytopathology 67:670-674.

McMorris, T. C. 1978. Sex hormones of the aquatic fungus *Achlya*. Lipids 13:716-722.

Mortimer, A., Shaw, D. S., and Sansome, E. R. 1977. Genetical studies of secondary homothallism in *Phytophthora drechsleri*. Arch. Microbiol. 111:255-259.

Noon, J. P., and Hickman, C. J. 1974. Oospore production by a single isolate of *Phytophthora capsici* in the presence of chloroneb. Can. J. Bot. 52:1591-1595.

Office de la Recherche Scientifique et Technique Outre-Mer (ORSTOM). 1973. Cahiers ORSTOM, Ser. Biol. 20. ORSTOM, Paris.

Pethybridge, G. H. 1913. On the rotting of potato tubers by a new species of *Phytophthora* having a method of sexual reproduction hitherto undescribed. Sci. Proc. R. Dublin Soc. 13:529-564.

Raper, J. R. 1951. Sexual hormones in *Achlya*. Am. Sci. 39:110-121.

Sansome, E. R. 1965. Meiosis in diploid and polyploid sex organs of *Phytophthora* and *Achlya*. Cytologia 30:103-117.

Sansome, E. R. 1970. Selfing as a possible cause of disturbed ratios in *Phytophthora* crosses. Trans. Br. Mycol. Soc. 54:101-107.

Sansome, E. 1976. Gametangial meiosis in *Phytophthora capsici*. Can. J. Bot. 54:1535-1545.

Sansome, E. 1977. Polyploidy and induced gametangial formation in British isolates of *Phytophthora infestans*. J. Gen. Microbiol. 99:311-316.

Sansome, E. 1980. Reciprocal translocation heterozygosity in heterothallic species of *Phytophthora* and its significance. Trans. Br. Mycol. Soc. 74:175-185.

Sansome, E., and Brasier, C. M. 1973. Diploidy and chromosomal structural hybridity in *Phytophthora infestans*. Nature (London) 241:344-345.

Sansome, E., and Brasier, C. M. 1974. Polyploidy associated with varietal differentiation in the *megasperma* complex of *Phytophthora*. Trans. Br. Mycol. Soc. 63:461-467.

Sansome, E., Brasier, C. M., and Griffin, M. J. 1975. Chromosome size differences in *Phytophthora palmivora*, a pathogen of cocoa. Nature (London) 255:704-705.

Sansome, E., Brasier, C. M., and Sansome, F. W. 1979. Further cytological studies on the 'L' and 'S' types of *Phytophthora* from cocoa. Trans. Br. Mycol. Soc. 73:293-302.

Savage, E. J., Clayton, C. W., Hunter, J. H., Brenneman, J. A., Laviola, C., and Gallegly, M. E. 1968. Homothallism, heterothallism and interspecific hybridization in the genus *Phytophthora*. Phytopathology 58:1004-1021.

Schmitthenner, A. F. 1972. Evidence for a new race of *Phytophthora megasperma* var. *sojae* pathogenic to soybean. Plant Dis. Rep. 56:536-539.

Shattock, R. C., and Shaw, D. S. 1976. Novel phenotypes of *Phytophthora infestans* from mixed culture of antibiotic resistant mutants. Trans. Br. Mycol. Soc. 67:201-206.

Shepherd, C. J., and Pratt, B. H. 1973. Separation of two ecotypes of *Phytophthora drechsleri* Tucker occurring in Australian native forests. Aust. J. Biol. Sci. 26:1095-1107.

Stamps, D. J. 1953. Oospore production in paired cultures of *Phytophthora* species. Trans. Br. Mycol. Soc. 36:255-259.

Stephenson, L. W., and Erwin, D. C. 1972. Encirclement of the oogonial stalk by the amphigynous antheridium in *Phytophthora capsici*. Can. J. Bot. 50:2439-2441.

Stephenson, L. W., Erwin, D. C., and Leary, J. V. 1974. Hyphal anastomosis in *Phytophthora capsici*. Phytopathology 64:149-150.

Todd, N. K., and Rayner, A. D. M. 1980. Fungal individualism. Sci. Prog. (Oxford) 66:331-354.

Tsao, P. H., Ugale, R., Hobbs, S., and Farih, A. 1980. Control of homothallic oospore formation in *Phytophthora parasitica* by culture manipulations. Trans. Br. Mycol. Soc. 75:153-156.

Waterhouse, G. M. 1963. Key to the species of *Phytophthora* de Bary. Mycol. Pap. 92. Commonw. Mycol. Inst., Kew, Surrey, England. 22 pp.

Waterhouse, G. M. 1970. Taxonomy in *Phytophthora*. Phytopathology 60:1141-1143.

Whittaker, R. H. 1969. New concepts of kingdoms of organisms. Science 163:150-160.

Wilde, P. 1961. Ein Beitrag zur Kenntnis der Variabilität von *Phytophthora infestans* (Mont.) de Bary. Arch. Mikrobiol. 40:163-195.

Zentmyer, G. A. 1979. Stimulation of sexual reproduction in the $A^2$ mating type of *Phytophthora cinnamomi* by a substance in avocado roots. Phytopathology 69:1129-1131.

# Author Index

Numbers in italics refer to complete references in Literature Cited sections; numbers not in italics refer to citations in text. A text citation such as "Jones et al (1980)" is indexed here only under Jones; however, the accompanying reference in the Literature Cited section is indexed under the names of all the authors.

# Subject Index